Advances in Intelligent Systems and Computing

Volume 1079

The series "Advances in Intelligent Systems and Computing" contains publications on theory, applications, and design methods of Intelligent Systems and Intelligent Computing. Virtually all disciplines such as engineering, natural sciences, computer and information science, ICT, economics, business, e-commerce, environment, healthcare, life science are covered. The list of topics spans all the areas of modern intelligent systems and computing such as: computational intelligence, soft computing including neural networks, fuzzy systems, evolutionary computing and the fusion of these paradigms, social intelligence, ambient intelligence, computational neuroscience, artificial life, virtual worlds and society, cognitive science and systems, Perception and Vision, DNA and immune based systems, self-organizing and adaptive systems, e-Learning and teaching, human-centered and human-centric computing, recommender systems, intelligent control, robotics and mechatronics including human-machine teaming, knowledge-based paradigms, learning paradigms, machine ethics, intelligent data analysis, knowledge management, intelligent agents, intelligent decision making and support, intelligent network security, trust management, interactive entertainment, Web intelligence and multimedia.

The publications within "Advances in Intelligent Systems and Computing" are primarily proceedings of important conferences, symposia and congresses. They cover significant recent developments in the field, both of a foundational and applicable character. An important characteristic feature of the series is the short publication time and world-wide distribution. This permits a rapid and broad dissemination of research results.

** Indexing: The books of this series are submitted to ISI Proceedings, EI-Compendex, DBLP, SCOPUS, Google Scholar and Springerlink **

More information about this series at http://www.springer.com/series/11156

K. Srujan Raju · Roman Senkerik ·
Satya Prasad Lanka · V. Rajagopal
Editors

Data Engineering and Communication Technology

Proceedings of 3rd ICDECT-2K19

 Springer

Editors
K. Srujan Raju
Professor & Head
Department of Computer Science &
Engineering
CMR Technical Campus
Hyderabad, India

Satya Prasad Lanka
Stanley College of Engineering
and Technology
Hyderabad, Telangana, India

Roman Senkerik
Faculty of Applied Informatics
Tomas Bata University in Zlín
Zlín, Czech Republic

V. Rajagopal
Department of EEE
Stanley College of Engineering
and Technology
Hyderabad, Telangana, India

ISSN 2194-5357   ISSN 2194-5365 (electronic)
Advances in Intelligent Systems and Computing
ISBN 978-981-15-1096-0   ISBN 978-981-15-1097-7 (eBook)
https://doi.org/10.1007/978-981-15-1097-7

This Springer imprint is published by the registered company Springer Nature Singapore Pte Ltd.
The registered company address is: 152 Beach Road, #21-01/04 Gateway East, Singapore 189721, Singapore

Stanley College of Engineering and Technology for Women, Osmania University, Hyderabad, India

Stanley College of Engineering and Technology for Women—a temple of learning—was established in the year 2008 on a sprawling 6-acre campus of historic Stanley College campus at Abids, Hyderabad. The college provides a serene and tranquil environment to the students, boosting their mental potential and preparing them in all aspects to face the cut-throat global competition with a smile on the face and emerge victoriously. Stanley College of Engineering and Technology for Women has been established with the support of Methodist Church of India that has been gracious and instrumental in making the vision of an engineering college on this campus a reality.

The college is affiliated to the prestigious Osmania University, Hyderabad. It has been approved by AICTE, New Delhi, and permitted by the Government of Telangana. Today, it is the reputed college among the campus colleges of Osmania University. The Decennial Celebrations have been done in 2018. The college was, accredited by NBA in 2018 and NAAC in 2019 with Grade A. The college offers four-year engineering degree courses leading to the award of Bachelor of Engineering (B.E.) in computer science and engineering, electronics and communication engineering, electrical and electronics engineering, and information technology. The college also offers postgraduate programmes in M.E/M.Tech and MBA. As of today, there is a yearly intake of 420 undergraduate students (full-time) and 108 postgraduate students.

We at Stanley aim to serve the state, the nation and the world by creating knowledge, preparing students for lives of impact, and addressing critical societal needs through the transfer and application of knowledge.

All UG Courses offered by Stanley are accredited by NBA, New Delhi. Stanley College of Engineering and Technology for Women is accredited by NAAC with "A" Grade.

ICDECT Organizing Committee

Chief Patrons

Prof. S. Ramachandram, Vice Chancellor, Osmania University
Bishop M. A. Daniel, Resident and Presiding Bishop, Hyderabad Episcopal Area

Patrons

Mr. K. Krishna Rao, Secretary and Correspondent, Stanley College of Engineering and Technology for Women (SCETW)
Dr. Satya Prasad Lanka, Principal, Stanley College of Engineering and Technology for Women (SCETW)

Conference Chair

Dr. A. Vinaya Babu, Director, Stanley College of Engineering and Technology for Women (SCETW)

Honorary Chair

Dr. Suresh Chandra Satapathy, Professor, School of Computer Engineering, Kalinga Institute of Industrial Technology (KIIT), Deemed to be University, Bhubaneswar

Conveners

Dr. B. Srinivasu, Professor in CSE, Stanley College of Engineering and Technology for Women (SCETW)
Dr. M. Kezia Joseph, Professor in ECE, Stanley College of Engineering and Technology for Women (SCETW)

Co-conveners

Dr. Kedar Nath Sahu, Professor in ECE, Stanley College of Engineering and Technology for Women (SCETW)
Dr. K. Vaidehi, Associate Professor in CSE, Stanley College of Engineering and Technology for Women (SCETW)

Programme Chair

Dr. K. Chennakesava Reddy, Professor and Head in ECE, Stanley College of Engineering and Technology for Women (SCETW)

Organizing Committee

Dr. A. Gopala Sharma, HOD, ECE
Dr. B. V. Ramana Murthy, HOD, CSE
Dr. A. Kanakadurga, HOD, IT
Dr. V. Rajagopal, HOD, EEE
Dr. M. Kasireddy, HOD, MBA
Dr. R. Manivannan, Coord. Infra
Dr. G. V. S. Raju, TPO

National Advisory Committee

Dr. A. Venugopal Reddy, VC, JNTUH
Dr. A. Govardhan, Rector, JNTUH
Dr. V. Bhikshma, OU
Dr. P. Ramesh Babu, OU
Dr. P. Sateesh, IIT Roorkee

Dr. D. V. L. N. Somayajulu, NITW
Dr. E. Selvarajan, AU, TN
Dr. Kiran Chandra, FSMI
Dr. N. Sreenath, UoP
Dr. K. Narayana Murthy, UoH
Dr. C. R. Rao, UoH
Dr. R. V. Ravi, IDRBT
Dr. Kumar Molugaram, OU
Dr. LakshmiNarayana, OU
Dr. P. Premchand, OU
Dr. S. Sameen Fatima, OU
Dr. P. Chandhrashekar, OU
Dr. B. Rajendra Naik, OU
Dr. K. Shymala, OU
Dr. P. V. Sudha, OU
Dr. A. Q. Ansari, JMO
Dr. B. Padmaja Rani, JNTUH
Dr. B. Vishnuvardhan, JNTUH
Dr. S. Vishwanath Raju, JNTUH
Dr. L. Pratap Reddy, JNTUH
Dr. T. S. Subashini, AU, TN
Dr. Ch. Satyanarayana, JNTUK
Dr. A. Ananda Rao, JNTUA

Web and Publicity Committee

Prof. D. V. RamSharma, Professor in EEE, SCETW
Dr. D. Shravani, Associate Professor in CSE, SCETW
Mrs. Kezia Rani, Associate Professor in CSE, SCETW

Foreword

The aim of this 3rd International Conference on Data Engineering and Communication Technology (ICDECT) is to present a unified platform for advanced and multidisciplinary research towards the design of smart computing, information systems and electronic systems. The theme focuses on various innovation paradigms in system knowledge, intelligence and sustainability that may be applied to provide a realistic solution to variegated problems in society, environment and industries. The scope is also extended towards the deployment of emerging computational and knowledge transfer approaches, optimizing solutions in a variety of disciplines of computer science and electronics engineering. The conference was held on 15 and 16 March 2019 at Stanley College of Engineering and Technology for Women, Hyderabad, Telangana, India.

After having a thorough review of each submitted article, only quality articles are published in this volume. Eminent academicians and top industrialists are delivering lectures on contemporary thrust areas. The resource pool is drawn from IITs, NITs, IIITs, IDRBT and universities along with software companies like TCS, ThoughtWorks, GSPANN, Variance IT, FSMI, etc.

A galaxy of nearly 40 eminent personalities are chairing and acting conference as jury. The papers are classified into 7 tracks which will be delivered in 2 days in spacious technically state-of-the-art air-conditioned rooms.

On 14 March 2019, Conference Tutorial on DATA SCIENCE, Conference Workshop on Technology Trends, Conference Workshop on Python and Conference on IoT and Cloud Computing are scheduled.

Hyderabad, India
April 2019

Dr. A. Vinaya Babu
Director SCETW and Conference Chair

Preface

This book constitutes the thoroughly refereed post-conference proceedings of the 3rd International Conference on Data Engineering and Communication Technology (ICDECT) held at Stanley College of Engineering and Technology for Women, Hyderabad, Telangana, India, on 15–16 March 2019. The aim of this conference is to enhance the information exchange of theoretical research and practical advancements at national and international levels in the fields of computer science, electrical, electronics and communication engineering. This encourages and promotes professional interaction among students, scholars, researchers, educators, professionals from industries and other groups to share the latest findings in their respective fields towards sustainable developments.

The refereed conference proceedings of the ICDECT-2K19 are published in a single volume. Out of 286 paper submissions from all over India, only 81 papers are being published after reviewing thoroughly; this Volume 1 under the theme "Advances in Intelligent Systems and Computing—3rd International Conference on Data Engineering and Communication Technology (ICDECT-2K19)" comprises the comprehensive state-of-the-art technical contributions in the areas computer science engineering streams. Major topics of these research papers include the latest findings in the respective fields towards sustainable developments include Internet of things, cryptography and network security, image processing, natural language processing, data mining, machine learning, etc.

Hyderabad, India   Dr. K. Srujan Raju

Acknowledgements

We thank all the authors for their contributions and timely response. We also thank all the reviewers who read the papers and made valuable suggestions for improving the quality of the papers.

We are indeed thankful to keynote speakers for delivering lectures which create curiosity on research and session chairs for their fullest support and cooperation.

We would like to thank the array of distinguished Vice Chancellors Prof. S. Ramachandram, Osmania University, Prof. V. Venugopal Reddy, JNTUH, Prof. A. K. Pujari, Central University of Rajasthan, Prof. Amiya Bhaumik, Lincoln University College, Malaysia, for delivering inaugural and valedictory speech of ICDECT-2K19.

We express our sincere thanks to Sri K. Krishna Rao, Correspondent, Stanley College of Engineering and Technology for Women (SCETW), for accepting and organizing ICDECT-2K19 conference with exponentially higher success rate.

We would like to extend our sincere thanks to eminent Profs. A. Vinaya Babu, Director and Satya Prasad Lanka, Principal and family members of SCETW, the institute which is empowering girl students.

We would like to thank Dr. K. Vaidehi, Dr .YVSS Pragathi, Dr. D. Shravani, Mrs. Kezia Rani, Dr. Kezia Joseph and Dr. K. N. Sahu, the coordinators of this conference, for their combined efforts, and they put together as a team to make this event a huge success.

A dream does not become reality through magic. It takes sweat, determination and hard work. Let us extend our thanks to all the teaching, technical and administrative staff, and student volunteers of SCETW, who had worked tirelessly to accomplish this goal.

Finally, we sincerely thank the team comprising Prof. Suresh Chandra Satapathy, Dr. M. Ramakrishna Murthy and editors of this conference Prof. K. Srujan Raju, Prof. Satya Prasad Lanka, Prof. V. Rajagopal, Prof. Roman Senkerik for guiding and helping us throughout.

Hyderabad, India
May 2019

Contents

About the Editors

Dr. K. Srujan Raju is the Professor and Head, Department of CSE, CMR Technical Campus, Hyderabad, India. Prof. Raju earned his PhD in the field of network security and his current research includes computer networks, information security, data mining, image processing, intrusion detection and cognitive radio networks. He has published several papers in refereed international conferences and peer reviewed journals and also he was in the editorial board of CSI 2014 Springer AISC series; 337 and 338 volumes, IC3T 2014, IC3T 2015, IC3T 2016, ICCII 2016 & ICCII 2017 conferences. In addition to this, he has served as reviewer for many indexed National and International journals. Prof. Raju is also awarded with Significant Contributor, Active Young Member Awards by Computer Society of India (CSI). Prof. Raju also authored 4 Text Books and filed 7 Patents so far.

Dr. Roman Senkerik is a Researcher/Lecturer of Applied Informatics and Lectures on the fundamentals of theoretical informatics, theory of algorithms and cryptology. Seminars and laboratories in courses of Applied Informatics, Methods of Artificial Intelligence, Cryptology, Mathematical Informatics, Theory of Algorithms, Fundamentals of Computer Science. Supervision of Bachelors and Masters theses and consultations for Ph.D. students. His research in the field of artificial intelligence, optimization, chaos theory, soft computing and evolutionary techniques. Presently working as professor, Department of Applied Informatics, Tomas Bata University, Czech Republic.

Dr. Satya Prasad Lanka is currently working as Principal in Stanley College of Engineering and Technology for Women, Hyderabad, India and he is serving as a professor in Electronics and Communication Engineering department. He received his Ph.D degree in Image Processing from JNTUK, Kakinada and Master degree in Electronics and Communication Engineering from IT, BHU, Varanasi. He obtained his Bachelor degree from Bangalore University, Karnataka. He has more than 31 years of teaching and research experience. He has published several papers in International conferences and journals. He has guided more than 35 projects. His area of interest are

Digital Signal Processing, VLSI design, Digital Image Processing, Pattern Recognition and Communications. He is a member of ISTE and Fellow of Institution of Engineers and Institution of Electronics and Telecommunication Engineers.

Dr. V. Rajagopal received his bachelor's degree in Electrical Engineering in the year 1999 from The Institution of Engineers (India). He completed his M.Tech Degree in Power Electronics and Drives from Uttar Pradesh Technical University in the year 2004. He pursued his Ph.D degree in Power Electronics from I.I.T. Delhi in the year 2012. He worked in various educational institutions and also worked as a NBA, NAAC and Autonomous coordinator. He is trained on Jaguar Aircraft, Jaguar Simulator and worked for up-gradation of Jaguar Simulator with French Delegation of Thales while serving in Indian Air Force. His has rich teaching, research and industry experience of 27 years. He has published over 12 International Journals, 06 National Journals, 27 International and 6 National Conferences, and 01 Indian Patent filed. He coordinated Symposium and workshops. He is a reviewer of various International Journals like IEEE, IET and Taylor and Francis. He is a life member of ISTE and Fellow of Institution of Engineers (India). Currently he is guiding 04 Ph.D. Research Scholars and given 20 lectures in various institutions all over India. Currently he is working as a Professor and Head of EEE Department at Stanley College of Engineering and Technology for Women in Hyderabad.

Automatic Water Controller Switch and pH Determination for Water in Overhead Tank Using IoT

Farha Nausheen and Amtul Sana Amreen

Abstract Water is a valuable natural resource which is used very thoughtlessly. Ground water is pumped up to overhead tanks through electric motors for daily usage. To overcome the alarming water crisis, the unnecessary wastage of water due to overflow in overhead tanks needs to be controlled. Automatic water level controller switch is designed to overcome this concern by automating the manual switch used to control water fill-up in the overhead tank. It is implemented using water level detector and Raspberry Pi module. The water level detector made using probes detects "empty-tank" condition in overhead tank and triggers fill-up process. The switch can be controlled through mobile phone app facilitating the user to regulate the overflow anywhere outside the home also. We also propose to determine the pH of the water stored in the overhead tank enabling us to identify acidic content of the water supplied for daily use and suitably initiate necessary cleaning action.

Keywords Water level detector · Overhead tanks · Raspberry PI · pH determination

1 Introduction

According to the World Health Organization Fact sheets [1], it is reported that by 2025 half of the world's population will live in the water-stressed areas. Water scarcity [2] is results due to the inadequacy of natural water resources and poor

F. Nausheen (✉)
IT Department, Muffakham Jah College of Engineering & Technology,
Hyderabad, India
e-mail: farha@mjcollege.ac.in

A. S. Amreen
CSE Department, Stanley College of Engineering & Technology for Women,
Hyderabad, India
e-mail: amsana@stanley.edu.in

© Springer Nature Singapore Pte Ltd. 2020   1
K. S. Raju et al. (eds.), *Data Engineering and Communication Technology*,
Advances in Intelligent Systems and Computing 1079,
https://doi.org/10.1007/978-981-15-1097-7_1

management of available resources. Many homes and other public places use ground water for their daily usage which is pumped up to overhead tanks using water pumps which are controlled by electric motors. The water motor switch is manually turned ON and it is monitored for the tank to be filled-up for about 30 to 90 min. Controlling the pumps has become important to avoid wastage of water.

In this paper, we propose to develop an automatic water level controller switch, the switch is used to switch ON the motor when the water level in the overhead tank falls below pre-defined low level and switch OFF the motor when the water level rises up to pre-determined high level.

We also aim to determine the pH of the water being filled-up. This will help us to find the presence of hydrogen ions in the water which thereby determines whether the water is acidic or alkaline. Internet of things (IoT) allows to connect water level controller and pH sensor to be controlled from a handheld device embedded with an app or a computer.

2 Literature Survey

In [3], based on availability of water level in tank, the water pump is adjusted and is implemented using sequential logic circuits. In [4], ultrasonic sensor is used to measure water level in non-contact approach, while existing automated systems involve contact-based water level sensors. In [5], Zigbee technology is used to monitor the overhead tank water level employing three-tank simulation model. [6] discusses how GSM technology is used to monitor water levels and notify users on low state.

3 Proposed Methodology

The overall goal of this paper is to overcome the unnecessary wastage of water that occurs due to overflow in overhead tanks thereby preventing the alarming water crisis. Automatic water controller switch is designed to implement water level detector in the overhead tank to identify the water level. The overhead tank is indicated with three levels basically, level 3 depicting tank fill condition, level 2 indicating half fill condition, and level 1 indicating empty-tank condition. Initially, GPIO pins on Raspberry PI 2 module are configured to be connected to water level detector and pump motor. The water level sensing is performed through the probes in contact with the water in overhead tank. This may lead to two possibilities. If the water level reading is 1, it represents an "empty-tank" condition. The relay controls the submersible pump motor by switching it ON and initiate water fill process in the overhead tank from the water sump storage. If the water level reading is 3, it represents a "filled tank" condition and the relay controls the switching OFF of the submersible water pump.

Fig. 1 Proposed
methodology

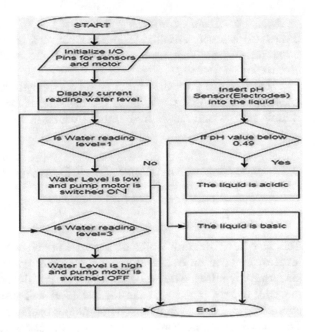

Additionally, we propose to determine the pH of the water available in the overhead tank helps in identifying if the water supplied for daily use is acidic or not. Since the usage of acidic water may lead to health concerns, necessary cleaning action may be initiated for pure water. Figure 1 illustrates the proposed methodology.

4 Design and Development

4.1 Hardware Interface

Raspberry Pi: The Raspberry Pi [7] is a credit card-sized electronic board which offers full complement of features and interfacing capabilities. It comes with a quad-core ARM Cortex-A53 processor and USB Micro power supply. Raspberry PI has a row general purpose IO pins (GPIO PINS) which interfaces the Pi from the outside world.

Pump Motor: A pump is a tool that moves about fluids by mechanical action. It is immersed in the fluid to be pumped.

Relay: A relay is an electrically operated switch used to control a circuit by a separate low-power signal, or where single signal controls several circuits.

Power supply section: Power supply unit, PSU, forms an important part of different components of electronics equipment. It primarily contains the following components: transformer, rectifier circuit, filter, and regulator circuits.

Analog-to-digital converter (LM 317): An analog-to-digital converter (ADC) [8] is a very essential feature that converts an analog voltage on a pin to a numerical value. This enables to interface the analog components with the electronics.

pH sensor: pH gives the concentration of free hydrogen and hydroxyl ions in the water. The pH [9] of a solution is the measure of the acidity. The pH is a logarithmic scale which ranges from 0 to 14 with a neutral point being 7.

4.2 Development of Water Level Detector and pH Sensor

Water Level Detector

The water level detector works on a simple principle to identify and represent the level of water in an overhead tank or any other water container. The sensing is performed by utilizing three probes which are positioned at three different levels in the tank (where probe 3 at the highest level and probe 1 at the lowest level), common probe (i.e. a supply carrying probe) is positioned at the bottom of the tank. The level 3 marks the "tank full" condition whereas level 1 marks the "tank empty" condition. When the water level falls below the minimum detectable level (MDL), indication is made that the tank is empty (level 1), if the water reaches level 2 (which is above level 1 and below level 3), indication of half-full level is made.

pH Sensor

A pH measurement loop consists of three components, namely the pH sensor, that comprises a measuring electrode, a reference electrode, and a temperature sensor; a preamplifier; and an analyzer. This loop is basically a battery where in the positive end is the measuring electrode and the negative end is the reference electrode. The measuring electrode being sensitive to the hydrogen ion concentration tends to build up a voltage difference (potential) directly associated to the concentration of hydrogen ion in the solution. The reference electrode supplies a stable potential to compare with the measuring electrode and it does not vary with the changing concentration of hydrogen ions. A solution in the reference electrode is in the contact with the sample solution and the measuring electrode via a junction. The preamplifier performs signal-conditioning and converts high-impedance electrode signal into low-impedance signal.

4.3 Circuit Diagram

The complete circuit diagram is shown in Fig. 2. The circuit comprises of water level detector, power supply section. LM 317 facilitates the purpose of analog-to-digital converter. In the power supply section, transformer is used to step

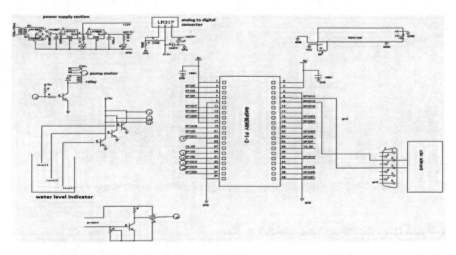

Fig. 2 Complete circuit diagram showing different components like power supply section, analog-to-digital converter, Raspberry PI, water level detector, relay, and sensor network

up or step down the input voltage line to the desired level and also couples this voltage to the rectifier section. The rectifier converts the AC signal to pulsating DC voltage which is further converted to filtered DC voltage using filter. Regulator is used to maintain the power supply section output at a constant level irrespective of large change that occur in load current or in input line voltage. Through the Raspberry PI, conductivity input pin and another pin to relay to pump motor output connections are made. The monitor is used to observe the water level readings and status of overhead tank fill-up. In our case, the values are observed as a message through Cayenne [10] app on mobile phone or in e-mail.

4.4 Circuit Operation

The power supply section consists of transformer, rectifier, filter, and regulator. The 230 V power supply is passed through the step-down transformer and submersible pump motor. A set of four bridge rectifiers are used to convert AC to DC power. Regulators are used to control the power and supply only 12 V to entire circuit. The relay controls the switching of pump motor. The water level is indicated through three probes placed at 3 levels in the tank through conductivity. When the water level is low it gets detected through conductivity and passes the input to relay. The relay passes logic 1 and pump motor is switched ON. When the water level is full, it gets detected and the relay passes logic 0 and pump motor is switched OFF. The pH sensor is used to record pH of given liquid by means of electrodes and these values are passed to analog-to-digital converter and finally observed on Cayenne app.

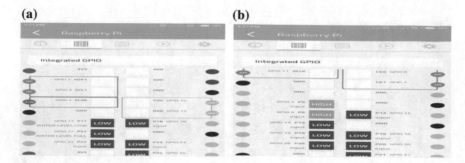

Fig. 3 a, b Raspberry PI connections being configured through the Cayenne app

4.5 Programming Raspberry PI

A C++ code is written to program Raspberry PI 2.0 using Arduino [11] IDE. WiringPI access library is used to program and setup GPIO pins. QThread class is used to manage threads and suitably define timer functions. A handler is created to process and respond incoming requests. The code is downloaded to configure Raspberry PI connections. Figure 3 illustrates the Raspberry PI connections through Cayenne app.

5 Results and Discussion

5.1 Specifications

The specifications of the hardware, software and apps involved in the development of automatic water level controller switch are listed in Table 1.

5.2 Interfacing Raspberry PI with the Water Level Detector

The water level is detected by three probes placed at three levels in the tank through conductivity. Pins 11 and 13 on Raspberry PI are connected to the conductivity pins

Table 1 Specifications for automatic water level controller switch

Hardware		Software	Apps
Component	Rating		
Power supply	12 V 1 A	Arduino IDE	Cayenne
ADC	12 V 1 A		Yahoo Mail
Raspberry PI 2	5 V 2 A		

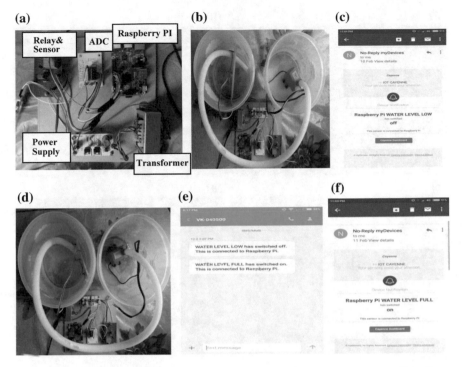

Fig. 4 **a** Labeled circuit setup. **b** Setup showing empty overhead tank scenario **c** E-mail sent indicating water level low condition. **d** Setup showing water fill-up process. **e** Message sent on mobile phone and e-mail through Cayenne app on tank full

coming from water level detector. Pin 22 is connected from Raspberry PI to relay for pump motor output. When the low level is detected, logic 1 is passed to relay switch which automatically switches on the submersible pump motor which initiates water fill-up into the overhead tank. When the full level is detected, logic 0 is passed to relay from raspberry pi and the pump motor is switched off by relay. Figure 4a shows water controller using labeled components. Figure 4b and c depict the water level low condition and e-mail notified to the user. Figure 4d, e, and f depict tank fill condition and all notifications made to user's phone and e-mail for the condition.

5.3 pH Determination

The pH value is calculated using electrode. Totally, three pins are connected to pH sensor. Out of the three pins, two are connected to Vdd and Gnd of power section. One pin is connected to analog-to-digital converter to show values in digital format. The pH value is represented on the scale of 0–1.

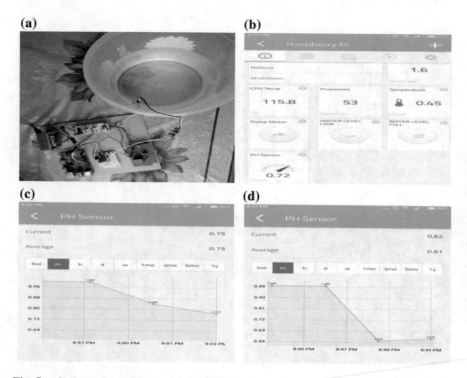

Fig. 5 **a** It shows how pH sensor is inserted into tumbler to determine pH value. **b** pH value observed through pH sensor can be seen on Cayenne app. **c** Graph shows observed pH value for drinking water; X-axis represents time domain and Y-axis represents observed pH values. **d** Graph shows observed pH value for ground water

i. 0.7–1.0 shows green color (safe level)
ii. 0.4–0.69 shows yellow color (moderate level)
iii. 0–0.39 shows red color that means impurity level (needs to be changed).

The pH value of pure water is 0.49. Value above 0.49 is basic. Value below 0.49 is acidic. The ph range for ground water system is normally between 0.42 and 0.69. The figure below shows the setup of how pH sensor is inserted into the water for determination of pH. From the graphs in Fig. 5b and c, it is observed that pH for bore well water is 0.62 and that of drinking water is 0.75. Figure 5 shows how the circuit is setup by inserting pH sensor in a water tumbler to determine pH. Figure 5c and d show pH values determined for drinking and ground water. The x-axis represents time and Y-axis represents observed pH values. The pH values obtained are observed in a period of minutes. However, these values can be observed on hourly, daily or yearly basis also.

6 Conclusion

Ground water is considered one of the World's most extracted raw materials catering to a range of human requirements. There is a stringent need to develop mechanisms to sustain ground water levels and avoid worsening water crisis. Our proposal aimed to design automatic water level controller switch that helps in preventing wastage of water due to overflow in overhead tanks that occurs due to negligence in switching off the motor supply to the tank. IoT-based model developed automates the controlling switch and also facilitates user to access the switch from elsewhere too. Additionally, pH of the water stored in overhead tanks was determined for the presence of acidic content in the water. This will help in imposing sanitation techniques for the stored water as its usage may lead to health hazards. The developed model can be further enhanced to detect the leakage in the overhead tank and necessitating appropriate action on such occurrence.

References

1. http://www.who.int/mediacentre/factsheets/fs391/en/
2. Chinnasamy, Pennan, Agoramoorthy, Govindasamy: Groundwater storage and depletion trends in Tamil Nadu State, India. Water Resour. Manage. **29**(7), 2139–2152 (2015)
3 Getu, B.N., Attia, H.A.: Automatic water level sensor and controller system. In: 2016 5th International Conference on Electronic Devices, Systems and Applications (ICEDSA). IEEE (2016)
4. Shrenika, R.M., et al. Non-contact water level monitoring system implemented using LabVIEW and Arduino. In: 2017 International Conference on Recent Advances in Electronics and Communication Technology (ICRAECT). IEEE (2017)
5. Reddy, B.P., Eswaran, P.: Metro overhead tanks monitoring system using ZigBee based WSN. In: 2013 IEEE conference on information & communication technologies (ICT). IEEE (2013)
6. Johari, A., et al.: Tank water level monitoring system using GSM network. Int. J. Comput. Sci. Inf. Technol. **2** (3), 1114–1115 (2011)
7. Maksimović, M., et al.: Raspberry Pi as internet of things hardware: performances and constraints. Des. Issues **3** (2014)
8. STMicroelectronics, L.M., LM317, L.C.: 1.2 V to 37 V adjustable voltage regulator (2005)
9. Prasad, A.N., Mamun, K.A., Islam, F.R., Haqva, H.: Smart water quality monitoring system. Proc. IEEE (2015)
10. Cayenne, 'Gartner Says 8.4 Billion Connected "Things" Will Be in Use in 2017, Up 31 Percent From 2016' [Online]. Available: http://www.gartner.com/newsroom/id/3598917. Accessed: 27 Apr 2017 (2017)
11. Gupta, Y., et al.: IOT based energy management system with load sharing and source management features. In: 2017 4th IEEE Uttar Pradesh Section International Conference on Electrical, Computer and Electronics (UPCON). IEEE (2017)

A Dynamic ACO-Based Elastic Load Balancer for Cloud Computing (D-ACOELB)

K. Jairam Naik

Abstract Cloud computing is for the delivery of computational service like servers, software, databases, storage, networks, intelligence, analytics, and more via the Internet ("the cloud") for tendering faster innovation, economies of scale, and flexible resources. The *workload* in cloud computing is defined as the amount of computational work at a given time that the computer has been given to do. This computational workload comprises of some number of users connected to and interacting with the computer's applications additionally to the computer's application programs running in the system. This workload may alter dynamically based on the available resources and the end-users. Hence, the key challenging duty in the cloud computing is balancing the workload among the Virtual Machines (VM's) of the systems. So, there is a need for introducing an elastic dynamic load balancer for distributing the workloads efficiently in the cloud. A load balancer can distribute the loads among the VM's of multiple servers or compute resources. The utilization of such elastic and dynamic load balancer can enhance the fault tolerance capability of applications and the availability of cloud resources. As your needs change, adding or removing of computing resources without disrupting the overall flow of requests is made possible utilizing this elastic and dynamic load balancer. Generally, an elastic load balancing task chains three types of balancers, they are Application Load Balancers, Classic Load Balancers, and Network Load Balancers. The Application Load Balancer can be chosen based on the need of applications. In this chapter, it proposed a dynamic and elasticity approach (D-ACOELB) for workload balancing in cloud data centers based on Ant Colony Optimization (ACO).

Keywords Elastic · Workload · Makespan time · Scheduling · Load balancer

K. Jairam Naik (✉)
Department of Computer Science and Engineering, National Institute of Technology (NIT), Raipur, Chattisgarh 492010, India
e-mail: Jnaik.cse@nitrr.ac.in

© Springer Nature Singapore Pte Ltd. 2020
K. S. Raju et al. (eds.), *Data Engineering and Communication Technology*, Advances in Intelligent Systems and Computing 1079, https://doi.org/10.1007/978-981-15-1097-7_2

1 Introduction

In the era of computational technology, a very big evolution is the pay per usage services of cloud computing. Cloud computing presents a variety of resources, services, and IT infrastructures as per the user needs in a secured and efficient way. The different types of pay-as-you-go service offered by this technology are Software as a Service (SaaS), Platform as a Service (PaaS), and Infrastructure as a Service (IaaS) [1]. Storing and accessing the files from the data centers located at diverse places is one of the important services of the cloud computing. This service of the cloud computing can be traced from the basic service model called IaaS. When the users of the cloud are increasing hastily, there should be a proportional increase in the number of resources readily available for serving. Also, there is a need of efficient resource allocation scheme and continuous load monitoring mechanism. The increased workloads from the users may affect the performance of the cloud if there is no proper resource and load management mechanisms. Workload balancing equally plays a vital responsibility in preserving the networks of the cloud [2]. Hence, there is a need of efficient and elastic load balancing mechanisms that can deal with the workloads among the diverse machines in the datacenter.

The workload can be dispersed by applying numerous virtualization methods across the nodes of physical machines of the data center nodes. A different set of virtualization technologies has been invented for spreading the workload among the nodes in the data centers. Workload in the cloud network will be scattered among the systems nodes, so that no nodes should be underutilized or overutilized in the networking system.

2 Related Works

The various algorithms for maintaining the system's workload at a normally loaded state are categorized into two types. They are dynamic load balancing algorithms and static load balancing algorithms.

Utilizing the static load balancing algorithms, the available workload is shared among the nodes by considering the preexisting knowledge of the user jobs and system resources which was known prior. This category of algorithms does not consider the dynamically varying status of workload and the resource characteristics [3]. Hence, static algorithms are suitable for the cloud environments where the workload from the user is less, and the provider's resources are fixed. Mostly, this method is preferred by the cloud service providers when the workload is less. Whereas, the present workloads are given to the cloud nodes by the users and also the previous history of the jobs executed using that node was considered by the dynamic load balancing algorithms. Again, these dynamic algorithms are further categorized as Centralized, Hierarchical, and Distributed load balancing [4]. Later on, the solutions for this problem could be evolved by naturally stimulated intelligent systems like artificial bee colony (ABC), Ant Colony System & Optimization (ACS & ACO), Particles Swarm Optimizations (PSO), and Genetic algorithms (GA) [5].

The dynamic and elasticity ACO algorithm for load balancing in Cloud Computing (DEACOLB) [6] was not clear with its load balancing policies, and it also reduces the throughput of the system.

The optimization technique called Artificial Bee Colony (ABC) is utilized and simulated by the foraging behavior of honey bees. The various practical problems [7] also been successfully implemented by this ABC algorithm. An Ant Colony Optimization (ACS) [4, 8] introduces a best method for finding qualitative resources available in more quantity by considering the chemical substances widely called pheromone. If the value of the pheromone is more, that path contains qualitative resources in more quantity. The ACS was inspired by the ACS Optimization algorithm (ACO) to improve the accuracy of ACS results. The proposed D-ACOELB was further inspired by the ACO. The PSO (Particle Swarm Optimization) is another computational evolution in the era of cloud computing. This method optimizes the research problems by solving iteratively for improving the candidate solution according to the given quality of measure [9]. To produce high-quality outcomes for the optimization of searching problems, a very commonly used approach is Genetic algorithms (GA). The GA's will be relying on the bio-inspired operators like crossovers, selections, and mutations [10].

Ant Colony Optimization (ACO): The algorithm ACO was initiated by M Dorigo et al. in the beginning of 1990's. In the natural environment, the ants flow the best path for finding the food resources. The best path is the one which can have quality of resources in more quality. Hence, many ants can travel on this path by depositing chemical substances on the path travelled. The path is called thick because the substance (Bandwidth) on this path is more. The ant in this general example represents user jobs (or) applications, and the food resources represent the VM's in the proposed approach. For the simulation study, the foraging performance of ant colonies can be related and instigate a family of ant algorithms called as *"Ant Colony Optimization (ACO)."* Ant colony, for finding the shortest route between the source and destination, a self-organized group of ants follows a very intelligent approach by pheromone techniques.

The ACO is a standard and nature-inspired Meta heuristic approach [11] for finding the hard solutions of the problems like combinatorial optimization (CO). It also helps to find the solutions in a reasonable amount of computation time for hard combinatorial optimization problems. In order to find solutions for the CO problems, this feature of real ant colonies was subjugated in artificial ant colonies. Artificial ants are with some unobserved features, and additional behaviors are used to solve the problems, rather than employing the literal behaviors of biological ants.

2.1 The ACO Algorithm

Initialize pheromone values: Pheromone is used for determining the path, for illustration determining the paths to food from the nest of the ants (sources). When the ant starts movements, it will check for the value of pheromone. The pheromone with highest value is picked up for traveling. It means, the path having greater food resources (quality & quantity) will be selected by the ant for traveling. Initially, the pheromone value can be initialized with very little measure of time at ($t = 0$). The pheromone value $\mathcal{P}_{ij}^k(t)$ at time t can be determined by the following equation (Eq. 1).

Allocation of VM's for subsequent task: The probability for which the VM is allocated for the next task is determined as below:

$$
\mathcal{P}_{ij}^k(t) = \begin{cases} \frac{\left[\tau_{ij}(t)^{\alpha+}e^{i\theta}*[\eta_{ij}]^{\beta}\right]}{\sum_{s\in \text{Allowed}_k}[\tau_{is}(t)^{\alpha}*[\eta_{is}]^{\beta}]} + \sum_{i=1}^{n}J\partial e^{i\theta}, & if\, j\,\epsilon\,\text{allowed}_k \\ 0, & \text{otherwise} \end{cases} \tag{1}
$$

where,

$\mathcal{P}_{ij}^k(t)$—Is the probability that VM$_j$ is being selected by ant k (1...m)/task for executing the task i.

m—Num. of ants/tasks in the system.

n—Num. of VM's.

$\tau_{ij}(t)$—Pheromone Indicator at time t for the path lies between and VM$_j$ and task i.

α & β—Constant values of threshold

η_{ij}—$(1/d_{ij})$ calculated visibility with heuristic algorithm for the t moment, and d_{ij} represents the sum of expected execution time for task i on VM$_j$ and the transfer time from i to VM$_j$

Pheromone Updating: Updating of pheromone will take place when the visit is done by the ant k at iteration t as given in (Eq. 2) below:

$$
\gamma\tau_{ij}{}^k(t) = \begin{cases} \frac{Q}{L^k(t)}, & if\,(i,j) \in T^k(t), \\ 0, Otherwise \end{cases} \tag{2}
$$

Here, (i, j)—An *edge from task to the resource*

$L^k(t)$—*Makespan time* (expected) of the visit

Q—*Parameter adapted* (depends on the previous task scheduling)

$T^k(t)$—*Traveling performed* by the ant k at iteration t.

2.2 Ant Colony Optimization (ACO) Procedure (Pseudo Code)

1. Initialize pheromone values
   Iteration value = 1; Solution = null;
2. Repeat the following steps (2.a–2.f) until all the ants (tasks) complete its tour

   a. Determine the distance from sources to destination
   b. Generate pheromone values
   c. Select the highest pheromone for scheduling the ant (task) and assign VM to it
   d. Update pheromone once the task i is allocated (local pheromone update)
   e. Again compute the distance from sources to destination
   f. Update global pheromone once the task finishes execution on the assigned VM
   g. Increment the iteration value by 1 for the next task (ant)

3. As a current solution, display the output
4. Choose the jobs for allocating to VM's and insert the values of VM into the table
5. Adjust the Aging factors from few tasks always being selected.

The proposed D-ACOELB (12.5 ms) reduces the makespan time of the tasks to 16% compared to the makespan time of the DEACOLB (28.57 ms), when the number of tasks submitted is 1500. Also, the proposed D-ACOELB (1436 tasks) improves the throughput of the system to 13.73% compared to the throughput of the existing DEACOLB (1230 tasks), when the number of tasks submitted is 1500. The other existing methods considered in this article for comparing the results with the proposed method even perform poorer than the DEACOLB. The results of the experimentations are shown in Sect. 4.

3 The Proposed D-ACOELB for Load Balancing

The architecture of the proposed D-ACOELB is as shown in Fig. 1. It mainly consists of four parts. They are user, data center broker, cloud controller, and the VM's in the hosts. In D-ACOELB, the user submits their tasks to the data center broker with their QOS needs. The broker acts as a dispatcher between data center and user and also helps in allotting the tasks on to the virtual machines.

The data center consists of number of hosts, and each host contains number of virtual machines. The tasks are scheduled on to these VM's for execution as per the scheduling polices of the data center broker. Each user may have their own data center broker for interacting with the cloud services. The broker communicates with the cloud controller and handovers the tasks received from the user. Further, the

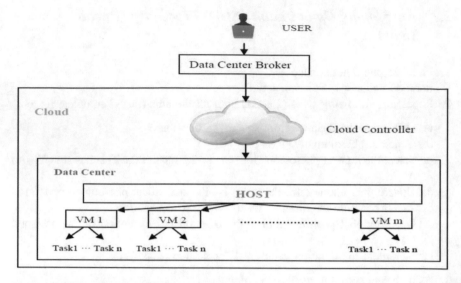

Fig. 1 Architecture of D-ACOELB

cloud controller allots the normally loaded VM for executing the task. In this section, it is proposed a dynamic and elasticity approach for load balancing based on ACO called D-ACOELB. The proposed D-ACOELB applies the following adaptation to the ACO algorithm for updating the pheromone and selection of the next task for execution.

3.1 Pheromone Updating

The pheromone value is updated using the equation given below (Eq. 3) once if the iteration is completed:

$$\tau_{ij}(t) = \tau_{ij}(t)(1 - v) + Y\tau_{ij}(t) \tag{3}$$

Here, v is the decay of the trail, $1 < v < 0$ and $Y\tau_{ij}(t)$ is determined by the averaging the entire ant's values from 1 to n.

$$v_1 = \frac{\mathcal{D}^k(t) - \mathcal{D}^+}{\mathcal{D}^k(t) + \mathcal{D}^+} + 0.1 \tag{4}$$

The total distance between source and destination can be calculated using the following (Eq. 5).

$$\mathcal{D}^k(t) = ArgMax_{j \in J} \{ Sum_{I \in IJ}(d_{ij}) \} \tag{5}$$

Once the visit is completed by all the ants, pheromone on the edges belong to the finest tour established is received from a superior ant supports (determined by M/ D^+). When every ant completes the visit, a discriminatory ant emphasizes pheromone on the edge by a Mass M/D$^+$. Where D$^+$ meant for representing the distance of the best tour (T$^+$). This support is known as the global pheromone update and calculated using Eq. (6) as follows:

$$\tau_{ij}(t) = \tau_{ij}(t) + \frac{M}{D^+}, \text{if } (i,j) \in T^+ \tag{6}$$

3.2 Choosing of VM for the Next Task

The (Eq. 1) with an optimization factor called Success Rate (S_r) is used for improvement of the cloud performance. Optimal pheromone factor is written as follows in (Eq. 7).

$$P_{ij}^k(t) = \left\{ \frac{\left[\tau_{ij}(t)^{\alpha +} e^{i\theta} * [\eta_{ij}]^\beta \right]}{\sum_{S \in Allowed_k} \left[\tau_{iS}(t)^\alpha * [\eta_{iS}]^\beta \right]} * S_r + \sum_{i=1}^n J\partial e^{i\theta} \right. \tag{7}$$

The (Eq. 7) is used for choosing the VM for the next task to be executed using the proposed method D-ACOELB. Where, θ is the balancing factor's degree, and it is determined to commence the earlier tour. The attribute e is the control parameter. S_r is the success rate of a machine and is defined as $R_r = \frac{N(s)}{N(s)+N(f)}$. Here, $N(s)$ and $N(f)$ indicates the number of time the VM succeeded or failed to execute the tasks in time assigned for it. This will helps to improve the performance of the VM's in the cloud.

$$J = \frac{\eta}{P_i * P_{max}} \tag{8}$$

Here, J is the pheromone smoothing function, and it can be determined using the (Eq. 8) as given above.

4   Experimental Results

The Space shared architecture of the CloudSim [12, 13], a Java-based simulation toolkit is used for the implementation of the proposed D-ACOELB. It is considered the isolated and non-preemptive tasks for simulation. For simulating with the proposed algorithm, it is considered the environment as shown in Table 1. For studying the performance of the proposed method for comparison, it is considered the existing ACO, MACO, and DEACOLB Algorithms.

Parameter	Value
Number of data centers	2
Number of hosts	4
Number of VM's	12
RAM size of each VM	128–256 MB
Number of tasks	300–1500
Task size	32–128 MB

Throughput: The first parameter which was considered for space-shared approach is throughput. Throughput of a system is the number of tasks successfully executed per the unit time. Figure 2 shows experimental results of all approaches in which x-axis designates the number of cloudlets (or) tasks submitted and y-axis designates the throughput. The simulation result shows that the performance of the proposed method D-ACOELB is better than the existing ACO, MACO, and DEACOLB Algorithms. The D-ACOELB performed 283 cloudlets when 300 cloudlets are submitted, 478 cloudlets when 500 cloudlets are submitted, 953 cloudlets when 1000 cloudlets are submitted and 1436 cloudlets when 1500 cloudlets are submitted. But, the other method performed less cloudlet.

Makespan Time: The proposed D-ACOELB is examined for diverse workload situations by changing the parameters. The D-ACOELB can reduce the average Make Span Time of the cloudlets while comparing with the standard available

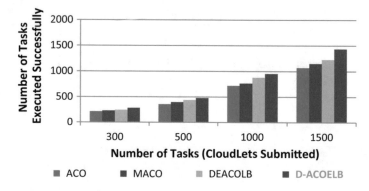

Fig. 2 Throughput comparison

methods such as ACO and MACO. The makespan time of D-ACOELB is 36 µs when 300 cloudlets are submitted, 53 µs when 500 cloudlets are submitted, 120 µs when 1000 cloudlets are submitted and 175 µs when 1500 cloudlets are submitted. But, the other existing methods ACO, MACO, and DEACOLB consumed more makespan time. Hence, the Proposed D-ACOELB decreases Average Make span with the increasing in the number of cloudlets submitted (Fig. 3).

Standard Deviation (SD): The standard deviation (SD) is a measurement used for quantifying the measure of workload deviation or dispersal on the VM's of the cloud. Smaller the standard deviation indicates that the lower the workload difference among the VM's, while a high standard deviation signifies that the bigger the workload difference among the VM's. Since the standard deviation will never become zero, the proposed D-ACOELB seeks for the workload balancing solution which was differing from the best-so-far solutions. The simulation results state that the average standard deviation of D-ACOELB is 2.1 when 300 cloudlets are submitted, 3.8 when 500 cloudlets are submitted, 4.3 when 1000 cloudlets are submitted and 4.9 when 1500 cloudlets are submitted (Fig. 4).

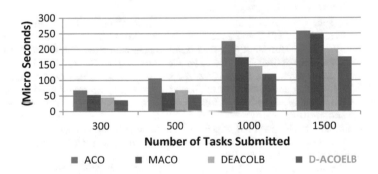

Fig. 3 Makespan time comparison (micro sec)

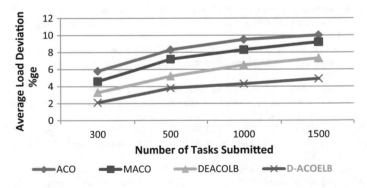

Fig. 4 Comparison of average standard load deviation

5 Conclusion

An innovative algorithm for workload balancing is anticipated in this article, which was stimulated from the Ant Colony System. The proposed algorithm is implemented and evaluated using open-source simulation tool kit called CloudSim. An experimental result of this article shows that the proposed method can considerably reduce the makespan time and average standard load deviation percentage. Also, this method improves the throughput of the system under various workload conditions. Hence, the proposed D-ACOELB is optimal in scheduling and cloud load balancing.

References

1. Dorigo, M., Blum, C.: Ant colony optimization theory: A survey. J. Theor. Comput. Sci. **344**, 243–278 (Elsevier) (2005)
2. Soni, G., Kalra, M.: A Novel Approach for Load Balancing in Cloud Data Center. 978–1-4799-2572-8/14/$31.00 c_2014. IEEE
3. Sim, K.M., Sun, W.H.: Ant Colony Optimization for Routing and Load-Balancing: Survey and New Directions. IEEE Trans. Syst. Man Cybern. Part A: Syst. Hum. **33** (5) (2003)
4. Ramezani, F., Lu, J., Hussain, F.K.: Task-based system load balancing in cloud computing using particle swarm optimization. Int. J. Parallel Prog. **42**(5), 739–754 (2013)
5. Calheiros, R.N., Ranjan, R., Beloglazov, A., De Rose, C.A.F., Buyya, R.: CloudSim: A Toolkit for Modeling and Simulation of Cloud Computing Environments and Evaluation of Resource Provisioning Algorithms. Published online 24 August 2010 in Wiley Online Library (wileyonlinelibrary.com)
6. Padmavathi, M., Basha, S.M.: Dynamic and elasticity ACO load balancing algorithm for cloud computing. In: International Conference on Intelligent Computing and Control Systems ICICCS 2017, © 2017. IEEE
7. Kushwah, P.: A survey on load balancing techniques using ACO algorithm. Int. J. Comput. Sci. Inf. Technol. (IJCSIT) **5** (5), 6310–6314 (2014)
8. Ruekaew, B.K., Kimpan, W.: Virtual machine scheduling management on cloud computing using artificial bee colony. In: Proceedings of the International Multi Conference of Engineers and Computer Scientists, vol. 1 (2014)
9. Khan, S., Sharma, N.,: Effective Scheduling algorithm for load balancing (SALB) using ant colony optimization in cloud computing. Int. J. Adv. Res. Comput. Sci. Softw. Eng. **4** (2) (2014)
10. Soni, G., Kalra, M.: A novel approach for load balancing in cloud data center. In: IEEE international advance computing conference (IACC-2014), 978-1-4799-2572-8/14/$31.00 c_2014. IEEE
11. Farrag, A.A.S., Mahmoud, S.A.: Intelligent cloud algorithms for load balancing problems: a survey. In: IEEE Seventh International Conference on Intelligent Computing and Information Systems (ICiCIS 'J 5) (2015)
12. Dasgupta, K., Mandai, B., Dutta, P., Mandai, J.K., Dam, S.: A genetic algorithm (GA) based load balancing strategy for cloud computing. Int. Conf. Comput. Intell. Model. Techn. Appl. **10**, 340–347 (2013)
13. Calheiros, R.N., Ranjan, R., De Rose, C.A.F., Buyya, R.: CloudSim: A novel framework for modeling and simulation of cloud computing infrastructures and services. Comput. Sci. Distribu. Parallel, Cluster Comput. (2009)

Tamil Stopword Removal Based on Term Frequency

N. Rajkumar, T. S. Subashini, K. Rajan and V. Ramalingam

Abstract As text data in digital form is increasing exponentially nowadays, managing and retrieving these documents becomes difficult. A number of natural language processing (NLP) processes, viz. archival, retrieval, query response, information summarization, etc., highly rely on automatic classification of text documents. This has induced researchers to apply machine learning logic to automatically categorize documents based on languages and within documents belonging to the same language to devise methods to segregate them according to its contents. More than at present, 70% of the total text classification process involves 'Preprocessing of text', alone [1]. This indicates its importance of preprocessing and the efficiency based on text classification logic is solely dependent on an efficient preprocessing step. This article deals with corpus creation for Tamil documents and Tamil language stopword removal. Dictionary-based and frequency-based stopword removal methods have been proposed in this work.

Keywords NLP · Feature extraction · Pre-processing · Text classification

N. Rajkumar (✉) · T. S. Subashini · V. Ramalingam
Department of Computer Science and Engineering, Annamalai University, Chennai, India
e-mail: raju.prg@gmail.com

T. S. Subashini
e-mail: rtramsuba@gmail.com

V. Ramalingam
e-mail: aucsevr@gmail.com

K. Rajan
Department of Computer Engineering, Muthiah Polytechnic College,
Annamalai Nagar, Chidambaram, India
e-mail: rajankperumal@yahoo.co.in

© Springer Nature Singapore Pte Ltd. 2020   21
K. S. Raju et al. (eds.), *Data Engineering and Communication Technology*,
Advances in Intelligent Systems and Computing 1079,
https://doi.org/10.1007/978-981-15-1097-7_3

1 Introduction

Nowadays, digital text data is increasing drastically; the need to extract only the required information from this huge content of data has resulted in a new field namely text mining which analyzes text to mine valuable information. Compared to numerical data, mining of text data is difficult as it is language-dependent, unstructured and ambiguous.

Text mining includes summarization, classification, and clustering of text. These works deal with text classification. Text classification tries to assign a predefined class label to a word depending on the category to which the words belong. Text classification tries to categorize words taken from social media sites for sentiment analysis, from emails to detect fraudulent mails and spam, from commercial Web sites which helps in mining data for predicting user intent from search queries and for suggesting products of their liking. Other applications of text classification include analyzing customer feedback for better customer satisfaction, archiving newspaper article based on predefined categories, etc.

As an initial step, tokenization, stopword removal, and stemming are carried out before feature extraction, and feature selection is done in Tamil text classification process.

Prepositions, articles, and conjunctions, which occur frequently in a language and have no meaning when it stands alone, are called stop words. These words have no significance in NLP applications such as text classification and content-based retrieval. Therefore, these types of poor words lead to the degradation inaccuracies. To improve performance before processing text documents for clustering or classification, the stop words need to be cutoff or removed.

This paper concentrates on identification and elimination of Tamil language stopword. The dictionary-based and frequency-based stopword removal logic has been proposed which serves as a preprocessing logic for Tamil document classification.

This work has been categorized into four sections: Sect. 1 explains the need for preprocessing of text, stopwords and stopword removal methods. The related research work on stopword removal is given in Sects. 2 and 3 describes the methodology adopted along with the results, and Sect. 4 finally concludes the paper.

1.1 Stopwords

Documents are mainly in different forms such as list of individual words, sentences, multiple paragraphs, and also seems special characters (like tweets for example).

Preprocessing step transforms and cleans the input text so that the classification algorithms will be able to perform better and with ease. The preprocessing phase tries to segregate the document text into individual unit of words. Then, it proceeds to get rid of words of very meager relevance or stopwords with regard to classi-fication and then represents the remaining words of high discriminative power in the form of a feature vector.

The general characteristics of stopword are they possess very low discriminative information. The probability of verbs and nouns to be a candidate stopword is less compared to articles, conjunctions, etc. The stopwords have no meaning when it stands alone; it just serves only a syntactic function. Stopwords frequently occur in document; they unnecessarily take too much processing time which can in turn affect the classification or retrieval performance drastically. Further, stopwords minimize the keywords impact of frequency differences between comparatively less common words thereby affecting the weighting process.

Next, stemming which reduces the words to its stem by removing suffixes and prefixes if any. Finally, the output of stemming is fed to the morphological analyser, which produces root word by analyzing whether the given stem is inflected or not.

Tokenization divides the documents into tokens that excludes white spaces and punctuation. Stopwords have low discrimination value and have no relevance to a specific class of documents and as such does not aid in classification. They are also called as noise words or negative dictionary that appear frequently in documents and does not carry useful information to aid learning tasks.

The preprocessing phase includes various sub-phases namely tokenization, stopword removal, stemming and morphological analysis as shown in Fig. 1.

Among the various preprocessing steps, this paper focuses mainly on stopword removal. The elaborate account on the various stopword removal methods employed by researchers for preprocessing text for document classification is portrayed in Sect. 1, and a detailed survey of the methods seen in the literature is discussed in Sect. 2. Implementation details and the results are dealt with in Sect. 3, and Sect. 4 concludes the paper.

1.2 Stopword Removal Methods

A good stopword removal method aims to reduce the feature dimension by reducing words which have no or very little discriminating power and which is effective in reducing data sparseness. Also, the selection of features should result in good classification performance. The widely used stopword removal methods seen in the literature are as follows:

Zipfs-Method:

High-frequency terms, terms occurring only once and terms with minimal inverse document frequency are removed in Z-method.

The Mutual Information Method (MI):

This supervised method computes the mutual information between a given term and the document category. This mutual information is a measure as to how much

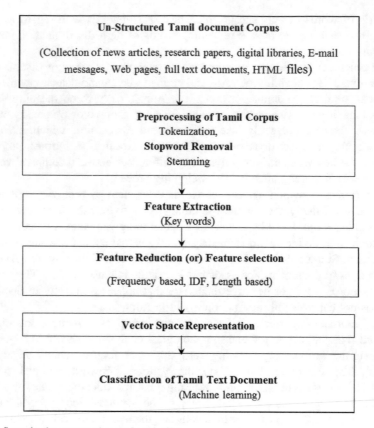

Fig. 1 Steps in the proposed text classification

relevance that particular term has with respect to the category under consideration. A lower value (negative) suggests that the term has low relevance to the particular category and could be considered to be a candidate stopword that could be eliminated compared to that of a term having higher mutual information value (positive).

Term-based Random Sampling (TBRS):

Lo et al. proposed a method to detect stopwords manually from Web documents using Kullback–Leibler divergence measure, given in the following equation,

$$dx(t) = Px(t) . \log 2 Px(t)/P(t)$$

where

$Px(t)$ is the normalized term frequency of a term t within a chunk of data x, and $P(t)$ is the normalized term frequency of term t in whole document. Those terms which have the lowest informative data are considered as stopwords.

In this work, term-based random sampling (TBRS) method has been employed to remove stopwords in Tamil text documents.

2 Literature Review

This section presents a survey of the work done so far under stop word removal for English text classification and Tamil text classification.

Though some survey papers could be seen in the literature for English stopword removal, today's automatic Tamil text classification is needed for a wide variety of government sector, research, and business needs. Literature related to stopword removal alone is considered within the scope of this paper.

The study in [2] several retrieval techniques and their potential in improving Arabic retrieval effectiveness were explored. A comparative study has been carried out in [1] where the impact of term weighting and stopwords on Arabic retrieval were compared. In [3], an algorithm for removing stopwords has been proposed which is based on a finite state machine in Arabic language.

The authors in [4] have discussed the various phases of preprocessing for text summarization for Gujarathi language, and they created a list of Gujarati language stopwords based on term frequency.

In [5], the authors proposed and implemented a stopword removal algorithm for Sanskrit language using dictionary-based approach. Here, the authors compared the predefined list of stopwords with the input document and removed those words that were present in the dictionary.

In [6], a stopword removal algorithm for Hindi documents is proposed which is based on DFA implementation. For DFA implementation, it uses JSON objects which take care of current state, accepting state, start state, final state, transitions and input characters. A total of 200 Hindi documents were taken as input which is movie review data set collected from the net and returns the documents with stopwords removed from them.

A method to automatically and dynamically identify a comprehensive list of Gujarati stopwords using a set of eleven rules is presented in [7]. The authors experimented on around 500 documents, and the results indicate that the proposed rule-based method was successful in identifying the stopwords with very good precision. The authors are confident that their method could be effectively employed as a preprocessing step for stop word removal for NLP tasks involving Guajarati documents.

Term-based random sampling is based on the Kullback–Leibler divergence measure, which determines how informative a term is employed by the authors in [8] to identify the stopword list automatically.

Within class and between class similarities are considered by the authors in [9] to find words candidate stopwords.

The authors in [10] created a list of stop words using an aggregated method for Persian language based on part of speech tagging. This method tries to improve the accuracy of retrieval and at the same time minimize the potential side effects of removing informative terms.

The proposed approach was effective with an enhanced average precision, decreased index storage size and improved overall response time.

The authors in [11] proposed a method for stopword removal. Here, a classifier is used to identify those words that have a minimum threshold based on its information content and discriminative capacity. Words whose threshold value is above a minimum threshold are considered to be candidate stopwords.

A statistical and information model-based method for stopword extraction from Chinese language is proposed by the authors in [12]. The statistical model extracts stopword according to the probability and distribution. The information model measures the significance of a word by using information theory. The results from these two models are aggregated to generate the Chinese stopword list.

The authors in [13–17] proposed a high-term frequency method to remove stopwords for the Tamil text classification process. In [13], low-term frequency, inverse document frequency, and length-based methods have been proposed to segregate stopwords for Tamil text classification. In [18], frequency-based stopword removal methodology has been implemented for sentiment classification of Tamil movie tweets, and dictionary-based approach is proposed in [5].

The authors employed an ASCII-based preprocessing method to eliminate the stopwords in [19].

Table 1 Corpus for Tamil text classification

Name of a category of Tamil	Total no of document per category	Total no of words per category
Agriculture	110	17,500
Astrology	103	18,642
Business	105	18,742
Science	106	19,675
Cinema	108	17,564
Sports	102	21,786
Spiritual	107	20,437
Health	109	19,567
Politics	104	17,453
Literature	109	20,213

3   Implementation and Results

As the first step, the Tamil corpus was created under ten different domains such as agriculture and astrology as shown in Table 1. In total, 1153 documents were collected from the Internet.

3.1   Dictionary-Based Approach

A generic stopword list containing 350 stopwords based on the ten different categories of Tamil documents was created, and the sample of the same is given in Table 2.

Table 2 Dictionary of stopwords created

Dictionary of stop words in Tamil language				
இந்த	என	அங்கே	வரை	கொண்டு
ஒரு	இது	என்பது	இருந்து	சில
வேண்டும்	அந்த	பல	ஏற்படும்	இதுவே
என்று	உள்ள	அதுவும்	இதை	போன்ற
தற்போது	தற்போது	செய்து	உன்	அந்த
இருக்கும்	ஆனால்	உள்ளது	தான்	எங்களை
என்ற	மிகவும்	கிடைக்கும்	அல்லது	இல்லை

Table 3 Statistics of dictionary-based approach for stopword removal

Name of a category of Tamil	Total no of document per category	Total no of words per category	No. of Stopwords removed per category	% removal of stop words per category
Agriculture	110	17,500	1,222	6.9
Astrology	103	18,642	2,031	10.8
Business	105	18,742	2,120	11.3
Science	106	19,675	2,009	10.2
Cinema	108	17,564	1,882	10.7
Sports	102	21,786	3,221	14.7
Spiritual	107	20,437	3,011	14.7
Health	109	19,567	1,994	10.2
Politics	104	17,453	1,342	7.6
Literature	109	20,213	3,003	14.9

In this approach, first the document is converted into tokens, and these tokens are stored in an array. Stopword is read from stopword list one by one and compared to target text in the form of array using sequential search technique. If it matches, the word in array is removed, and the comparison is continued till length of array. After removal of stopword completely, another stopword is read from stopword list and iterated again until all the stopwords are compared. Stopword removed final text is displayed. Table 3 shows statistics such as no. of stopwords removed from target text, total count of words in target text, and % removal of stopwords in each category.

On an average, 11.2% of words have been identified as stopwords for the Tamil corpus considered in this work.

3.2 Term Frequency-Based Approach

The strategy used for determining a stopword list is to sort all the tokens in accordance with its number of occurrences, that is, the total number of times each and every term appears in the document collection and then using a threshold, the high-frequency and low-frequency terms are eliminated.

This strategy is applied to all the categories to identify high- and low-frequency terms present in the respective document categories. The words which occur repeatedly and those which appear rarely have no significance or relevance in classifying the documents and as such, those words can be considered as stopwords and could be eliminated. Further, the threshold is not fixed for all the document categories, and it varies depending upon the document domain and is decided empirically as shown in Table 4.

Table 4 High-frequency and low-frequency threshold chosen for different document categories

Category of Tamil document	High-frequency terms (Threshold)	Low-frequency terms (Threshold)
Agriculture	100	5
Astrology	70	10
Business	120	5
Science	110	10
Cinema	70	5
Sports	90	10
Spiritual	80	4
Health	100	10
Politics	100	10
Literature	120	5

Table 5 Statistics of term frequency-based approach for stopword removal

Category of Tamil document	No of words after subtracting stop words eliminated using dictionary based approach	High frequency terms identified as stop words	Low frequency terms identified as stop words	Total No of stop words removed	% of stopwords removed
Agriculture	16,278	2,500	1,100	4,822	29.6
Astrology	16,611	2,750	1,073	5,854	31.2
Business	16,622	3,612	1,143	6,935	37.0
Science	17,666	2,712	2,100	6,821	34.6
Cinema	15,682	2,630	2,773	7,285	41.4
Sports	18,565	2,560	1,850	7,631	35.0
Spiritual	17,426	3,125	1,991	8,127	39.7
Health	17,573	2,712	1,763	6,469	33.0
Politics	16,111	2,539	1,842	5,715	32.7
Literature	17,210	2,803	1,672	5,806	28.7

Table 5 also shows statistics such as no. of words in each category considered, after subtracting those words that were eliminated using the dictionary-based approach, and total number of high-frequency stopwords and low-frequency stopwords. The table also gives the total stopwords removed and the % removal of stopwords in each category. On an average, 34.2% of words have been identified as stopwords by this approach for the Tamil corpus considered in this work.

4 Conclusion

Though Tamil is considered as an age-old language and has been widely acknowledged world wide as a language of great heritage, only a few NLP-based works could be seen in the literature. Still a lot of work is required to explore the potential of this language to open vistas in computational linguistics domain. Identification of stopwords in Tamil language may help researchers for various text preprocessing activities such as information retrieval, text summarization, spelling normalization, stemming, and lemmatization. In this work, dictionary-based and term frequency-based approaches were employed to identify stopwords in ten different document classes considered in this study. First, a corpus for Tamil documents was created, and dictionary-based and frequency-based stopword removal methods have been implemented, and the results indicate that both the methods were able to eliminate stopwords successfully. Tamil being a morpho-logically rich language segmentation methods could also be employed to identify other stopwords and to make the better feature selection for Tamil document classification.

References

1. Katharina, M., Martin, S.: The mining mart approach to knowledge discovery in databases. Intell. Technol. Inf. Anal. 47–65 (Springer) (2004)
2. El-Khair, A.: Effects of stop words elimination for Arabic information retrieval: a comparative study. Int. J. Comput. Inf. Sci. **4**(3), 119–133 (2006)
3. Al-Shalabi, R., Kanaan, G., Jaam, J.M., Hasnah, A., Hilat, E.: Stop-word removal algorithm for arabic language. In: Proceedings of 1st International Conference on Information & Communication Technologies: from Theory to Applications CTTA '04, pp. 545–550 (2004)
4. Ashish, T., Kothari, M., Pinkesh, P.: Pre-processing phase of text summarization based on Gujarati language. Int. J. Innovative Res. Comput. Sci. Technol. (IJIRCST) **2** (4) (2014)
5. Raulji, J.K., Saini, J.R.: Stop-word removal algorithm and its implementation for Sanskrit language. Int. J. Comput. Appl. (0975–8887) **150** (2) (2016)
6. Jha, V., Manjunath, N., Shenoy, P.D., Venugopal, K.R.: HSRA: Hindi stopword removal algorithm. In: 2016 International Conference on Microelectronics, Computing and Communications (MicroCom), Durgapur, pp. 1–5 (2016)
7. Rakholia, R.M., Saini, J.R.: A rule-based approach to identify stop words for Gujarati language. In: Proceedings of the 5th International Conference on Frontiers in Intelligent Computing: Theory and Applications. Advances in Intelligent Systems and Computing, vol. 515. Springer, Singapore (2017)
8. Lo, R.T.W., He, B., Ounis, I.: Automatically building a stopword list for an information retrieval system. JDIM **3**, 3–8 (2005)
9. Popova, S., Krivosheeva, T., Korenevsky, M.: Automatic stop list generation for clustering recognition results of call center recordings. Speech and Computer. SPECOM 2014. Lecture Notes in Computer Science, vol. 8773. Springer, Cham (2014)
10. Yaghoub-Zadeh-Fard, M., Minaei-Bidgoli, B., Rahmani, S., Shahrivari, S.: PSWG: An automatic stop-word list generator for Persian information retrieval systems based on similarity function & POS information. In: 2015 2nd International Conference on Knowledge-Based Engineering and Innovation (KBEI), Tehran, pp. 111–117 (2015)
11. Makrehchi, M., Kamel, M.S.: Automatic extraction of domain-specific stopwords from labeled documents. In: Proceedings of the IR Research, 30th European Conference on Advances in Information Retrieval (ECIR'08). Springer-Verlag, Berlin, Heidelberg, pp. 222–233 (2008)
12. Zou, F. et al.: Automatic construction of Chinese stop word list. In: Proceedings of the 5th WSEAS International Conference on Applied Computer Science, pp. 1009–1014 (2006)
13. Rajan, K., Ramalingam, V., Ganesan, M., Palanivel, S., Palaniappan, B.: Automatic classification of Tamil documents using vector space model and artificial neural network. Expert Syst. Appl. **36**, 10914–10918 (2009)
14. Sanjanasri, J., Anand Kumar, M.: A computational framework for Tamil document classification using random kitchen sink. In: International Conference on Advances in Computing, Communications and Informatics (ICACCI) (2015)
15. Hanumanthappa, M., Swamy, M.N.: Indian language text documents categorization and keyword extraction. IJCTA, **9** (3), 1473–1481 (2016)
16. Swamy, M.N., Hanumanthappa, M.: Indian language text representation and categorization using supervised learning algorithm. Int. J. Data Min. Techn. Appl. **02**, 251–257 (2013)
17. Kanimozhi, S.: Web based classification of Tamil documents using ABPA. Int. J. Sci. Eng. Res. **3** (5), ISSN 2229-5518 (2012)
18. Ravishankar, N., Raghunathan, S.: Corpus based sentiment classification of tamil movie tweets using syntactic patterns. IIOABJ, **8** (2017)
19. Ramakrishna Murty, M.V., Murthy, J.V.R., Prasad Reddy, P.V.G.D.: Text document classification based on a least square support vector machines with singular value decomposition. Int. J. Comput. Appl. (IJCA) [impact factor 0.821, 2012], **27** (7), 21–26 (2011)

Query-Based Word Spotting in Handwritten Documents Using HMM

V. C. Bharathi, K. Veningston and P. V. Venkateswara Rao

Abstract The retrieval of keywords from handwritten documents is tasked between one or more query word in a database. The works involves segment the individual words from the document images and formation of an index to all words, it uses search mechanism to access the query word from the scanned documents. However, unconstrained handwritten document realization remains a challenging problem with inadequate work to providing robust research experience in handwritten document. The proposed work characterizes to focus on keyword retrieval. The input word images are 2×2 block, each partition region again sub-block into 4×4, 5×5 and 6×6. In each sub-block, calculate average intensity of pixels and find the maximum average intensity value in horizontal and vertical direction. Thereby 32, 40 and 48 dimensional features are extracted from different sub-block and extracted MAIV features are fed to HMM to construct the models and validation of handwritten keywords. The query words are recognized using the Euclidean distance of the keyword and search keyword word acquire from the index position to retrieve the appropriate words from the document. The performance measure such as precision, recall and F-measure is calculated for keywords in different sub-block from different own handwritten dataset.

Keywords Handwritten word retrieval · Word spotting · Segmentation · Maximum average intensity vector (MAIV) · Hidden Markov models

V. C. Bharathi (✉) · K. Veningston · P. V. Venkateswara Rao
Department of Computer Science and Engineering, Madanapalle Institute of Technology and Science, Chittoor, AP, India
e-mail: bharathivc@gmail.com

K. Veningston
e-mail: veningstonk@gmail.com

P. V. Venkateswara Rao
e-mail: raovenkat21@gmail.com

© Springer Nature Singapore Pte Ltd. 2020
K. S. Raju et al. (eds.), *Data Engineering and Communication Technology*,
Advances in Intelligent Systems and Computing 1079,
https://doi.org/10.1007/978-981-15-1097-7_4

31

1 Introduction

In this modern world, it is easy to access the information from a database or storage media based on index-based searching and retrieving mechanisms facilities are easily available from the Internet. Word spotting in handwritten documents is the pattern classification task which consists of sensing keywords in input documents. However, keyword searches made on a digitized handwritten document such as scanned input images are continuing to adopt challenges to provide search keyword to the document retrieval. Word spotting involves based on indexing and retrieving the query information from a large collection of significant data. In document analysis, the research area is mainly focused on word spotting in unconstrained handwritten documents. Handwritten document can be very helpful in every document image analysis, especially keyword identification.

1.1 Related Work

Word spotting by using CNNs which are able to find the multiple attributes at the same time. Sudholt et al. [1] design 2 CNNs for the word spotting are able to find binary as well as real-valued attribute representation. Stauffer et al. [2] proposed a stable approach for spotting keywords in historical handwritten documents using template matching. Giotis et al. [3] describes the query image is firstly aligned with the test image based on similarity of the images descriptors are matched through a deformable non-rigid descriptors algorithm. Cao et al. [4] proposed term frequency evaluation technique integrated in word segmentation and TF/IDF similarity stores the estimated values for retrieval. Fischer et al. [5] proposed learning-based word spotting systems were introduced to adopt to multi writers with promising results using HMM. Frinken et al. [6] proposed bidirectional long short-term neural networks and token passing algorithm to spot the words. Rath [7] proposed word spotting and indexing which involve groups word images into clusters of similar words by using template matching to find similarity using dynamic time wrapping are compared.

1.2 Outline of the Work

The proposed paper deals with query-based word spotting in unconstrained handwritten document and handwritten words gathered from different writer with various writing styles.

This paper is organized as follows: Sect. 2 describes the proposed query-based word spotting. In this approach, preprocessing and line segmentation task involved. Maximum average intensity vector is described in Sect. 3. Section 4 deals with HMM classification. Section 4 describes index, and Sect. 5 deals with HMM classification. Section 6 describes the experimental results, and Sect. 7 concluded the work.

2 Proposed Query-Based Word Spotting

The given input document is preprocessed with binarization and Gaussian filter to eliminate the noise from unwanted effects of the original image. Horizontal and vertical line segmentation is applied to segment the word for feature extraction. Segmented word image are subblock into 2×2 block and each block partition into different sizes of subblock. In each subblock to calculate average intensity of pixels and find maximum average intensity value in horizontal and vertical direction to extract feature and index can be assigned for all words. Figure 1 shows a block diagram of the query-based word spotting in handwritten documents proposed in this paper. This block diagram consists of the following steps:

2.1 Input Document

This dataset is prepared for my Ph.D research work in Annamalai University, and doctoral commitees approved the datasets. Handwritten datasets created by different writers, to ensure various writing styles across different age group and different genders. The input images are captured using a scanner with 300 dpi resolution and saved in JPEG/JPG format as shown in Fig. 2. Handwritten words are collected from 150 different writers with different writing styles and size as shown in Table 1.

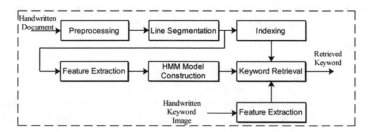

Fig. 1 Query-based word spotting

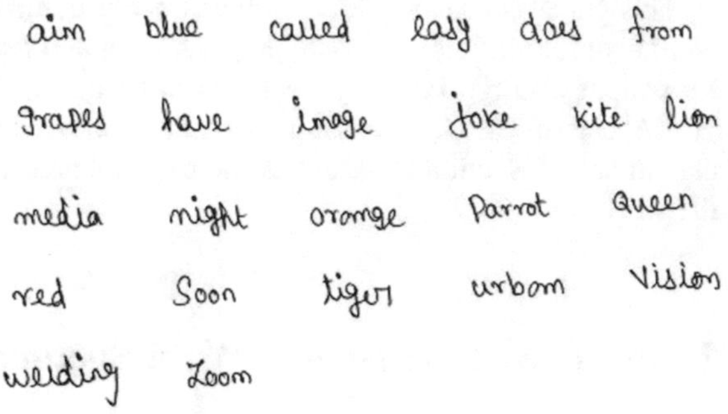

Fig. 2 Input document

Table 1 Handwritten dataset

S. No.	Genders	Age	No. of samples
1	Male	25–40	50
2	Female	18–26	50
3	Children	5–8	50

2.2 Preprocessing

The role of the preprocessing is required when the handwritten document image is captured from scanners. The noise is introduced in the image while the acquisition of images. Preprocessing is to reduce noise and to improve the quality of the handwritten input document for robust recognition and ensure the text is in a suitable form. Binarization and Gaussian filtering involved in preprocessing stage.

Binarization

The input RGB image is transformed into greyscale, and grey images are inverted as white foreground and black background for word segmentation.

Gaussian Filtering

Input document images are taken in various environment might contain noises and shadows. After segmenting the document opportune, preprocessing is needed to eliminate the noise by using Gaussian lowpass filter.

Segmentation

Segmentation is the process of splitting of groups of lines into multiple words. In holistic word segmentation, the profile-based segmentation [8–10] is commonly used for horizontal and vertical line. In horizontal line segmentation, the sum of the pixel values along the horizontal direction and find the white space between the text lines in the horizontal direction. If found, fragments the text lines in binary image and the text lines applied in the vertical direction for vertical line segmentation. Finally, all the segmented words are stored in the normalized size of the 120 width and 120 height of all holistic words.

Fig. 3 **a** and **b** represent MAIV in horizontal and vertical directions respectively

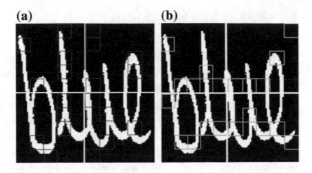

(a)   **(b)**

Table 2 Feature dimension in different size of sub-block

S. No.	Different sub-block	Feature dimension
1	4 × 4	32
2	5 × 5	40
3	6 × 6	48

3 Maximum Intensity Vector

The segmented words are partitioned into 2 × 2 sub-block, and each of size 60 rows and 60 columns block size is fixed. In each 60 rows and 60 columns, region is subdivided into 4 × 4, 5 × 5 and 6 × 6 sub-block, and each of different size of rows and columns is fixed. In each sub-block, calculate average intensity of pixels and find the maximum average intensity value in horizontal and vertical direction as shown in Fig. 3a and b of each 5 × 5 sub-block [11]. Thereby different dimension features are extracted from different size of sub-block as shown in Table 2.

4 Index Creation

Handwritten documents are commonly stored as list of words assign index value. In that particular index location store the corresponding word the list of documents that the word appears. After vertical line segmentation, the segmented words assign index for retrieval fastly based on user query. Query-based word spotting can be used for linear indexing algorithm is used to immense amount of available hand-written document images, in order to make them responsive to searching the keywords.

5 Hidden Markov Models

HMM is a powerful statistical tool for modelling generative sequences that can be characterized by an underlying process generating an observation sequence. HMMs are widely used in the field of machine learning and pattern recognition for sequence modelling of handwritten word recognition and word spotting [12] and [13]. HMMs have become very popular in word spotting in handwritten documents. When using HMMs for a classification problem, an individual HMM is designed for each pattern class. For each observation, sequence of feature vectors and the likelihood that this sequence was produced by an HMM of a class can be calculated. The class whose HMM achieved the highest likelihood is considered as the class that produced the actual sequence of observations. Each handwritten word is represented by a sequence of feature vectors known as observations O, defined as O = $o_1, o_2, \ldots, o_t$, where o_t is a feature vector observed at time t.

The evaluation of the HMM orients to determine the probability of the observation sequence generated by a given model and is accomplished using the forward algorithm. The decoding procedure targets to find the most probable sequence of hidden states for a given sequence of visible states V^T. The Viterbi algorithm finds at each time step t, the state that has the highest probability, thus resulting is to determine the model parameters from the training samples and the forward–backward algorithm or Baum–Welch algorithm.

6 Experimental Results

The experimental results are implemented in Windows 8 operating system. For the appraisal of the performance of the proposed query-based word spotting in a handwritten document in English language needs the formation of an immense amount of sample handwritten document images for determination of training and validation of samples. For the experimental work, handwritten documents are collected from different genders written various writing styles and size. From 150 writers, 100 writer samples are used training, and 50 writer samples are used testing with different ages. A list of keywords picked from different document is shown in Fig. 4. The keywords are chosen to be model samples of different writing styles and size.

A HMM with four states and two Gaussian mixtures per state is formed for each models. The models are first initialized, and the arguments of the HMM are estimated using MAIV features extracted from the word images in the training, and these models are re-estimated using Baum–Welch re-estimation procedure to give up the models for all 24 classes.

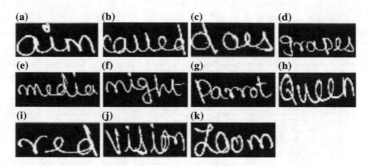

Fig. 4 List of keywords is represented from (**a**) to (**k**)

6.1 Performance Measures

The proposed approaches are used in MATLAB and estimate the eleven keywords using the test samples. The given query keyword image to extract the features using maximum intensity vector feature dimension is fed to HMM validation through the most likelihood for all the words and searching the best matching of the query word image. Query word feature dimension is used to match all query keyword and Euclidean distance to do the nearest matching of the word samples. The spotted words are located in relevant document over the indexed details and display the corresponding words. The retrieval measure is defined as the

$$\text{Precision}(P) = \frac{\text{correctly retrieved words}}{\text{total retrieved words}} \tag{1}$$

$$\text{Recall}(R) = \frac{\text{correctly retrieved words}}{\text{total existed words}} \tag{2}$$

$$F = \frac{2PR}{P+R} \tag{3}$$

The performance of the proposed query word spotting method and different size of sub-block is used to retrieval measure like precision(P), recall(R) and F-measure (F) of the different size of sub-block as shown in Table 3.

The 11 keywords average performance of precision, recall and F-measure in different size of sub-block is shown in Fig. 5a. In different size of sub-block, 5 × 5 sub-block given maximum spotting accuracy and measure the individual performance of the different genders testing sample are 20 male, 20 female and 10 children samples as shown in Fig. 5b.

Table 3 Precision, recall and F-measure in different size of sub-block

Keyword	4 × 4			5 × 5			6 × 6		
	P	R	F	P	R	F	P	R	F
1	80	84.21	82.05	80	76.19	78.04	70	82.35	75.67
2	75	88.23	81.07	85	89.47	87.17	70	87.5	77.77
3	80	84.21	82.05	80	80	80	80	84.21	82.05
4	80	72.72	76.18	85	85	85	80	66.6	72.68
5	85	89.47	87.17	95	86.36	90.47	85	89.47	87.17
6	90	90	90	90	94.73	92.30	80	80	80
7	85	89.47	87.17	85	85	85	85	89.47	87.17
8	85	77.27	80.95	90	85.71	87.80	85	73.91	79.06
9	90	78.26	83.72	90	85.71	87.79	90	78.26	83.72
10	85	85	85	85	89.74	87.17	85	85	85
11	80	80	80	85	94.44	89.47	80	80	80
Average	83.18	83.53	83.21	86.36	86.55	86.38	80.90	81.52	80.93

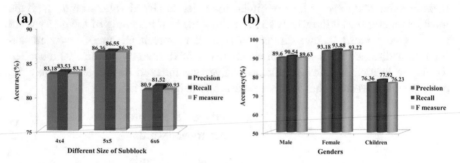

Fig. 5 a Average precision, recall and F-measure in different size of sub-block **b** performance of different gender in 5 × 5 sub-block

7 Conclusion

In this paper, proposed a novel approach for query-based word spotting in unconstrained handwritten documents. The method has been to recognize the keywords and find them in the handwritten documents with an index to retrieve the keywords. The methods have been implemented the following steps that include line segmentation, maximum intensity vector in horizontal and vertical direction, creation of an index, construction of HMM models for each keyword, validation of testing samples and retrieve the keywords from an index. The 11 keywords are picked from different document and chosen to be model samples of different writing styles. The performance of the proposed query word spotting measured and obtained maximum average F-measure in 5×5 of 40 dimension feature given 86.38% for the 11 keywords. In future work, to improve the accuracy and measure the performance in different classifier to enable the robust recognition accuracy.

References

1. Sudholt, S., Fink, G.A.: Attribute CNNs for Word spotting in Hand-Written Documents. **21** (3), 199–218 (2018)
2. Stauffer, M., Fischer, A., Riesen, K.: Keyword spotting in his-torical handwritten documents based on graph matching. Pattern Recognit. **81**, 240–253 (2018)
3. Giotis, A.P., Sfikas, G., Nikou, C., Gatos, B.: Shape based word spotting in handwritten document images. Doc. Anal. Recognit. 561–565 (2015)
4. Cao, H., Govindaraju, V., Bhardwaj, A.: Unconstrained handwrit-ten document retrieval. Int. J. Doc. Anal. Recognit. (IJDAR) **14** (2), 145–157 (2011)
5. Fischer, A., Keller, A., Frinken, V., Bunke, H.: Lexicon-free handwritten word spotting using character HMMs. Pattern Recogn. Lett. **33**(7), 934–942 (2012)
6. Frinken, V., Fischer, A., Manmatha, R., Bunke, H.: A novel word spotting method based on recurrent neural networks. IEEE Trans. Pattern Anal. Mach. Intell. **34**(2), 211–224 (2012)
7. Sulem, L.-L., Zahour, A., Taconet, B.: Text line segmentation of historical documents: a survey. Int. J. Doc. Anal. Recognit. 1–25 (2006)
8. Mamatha, H.-R., Srikantamurthy, K.: Morphological operations and projection profile based segmentation of handwritten Kannada document. Int. J. Appl. Inf. Syst. **4**(5), 13–19 (2012)
9. Bharathi, V.C., Geetha, M.K.: Hierarchical character grouping and recognition of character using character intensity code. Artif. Intell. Evol. Algorithms Eng. Syst. **9**(20), 789–797 (2015)
10. Bharathi, V.C., Geetha, M.K.: Holistic handwritten word recognition using block contour vector. Int. J. Appl. Eng. Res. (IJAER) **9** (20), 4767–4772 (2015)
11. Bharathi, V.C.: Unconstrained handwritten document retrieval based on user query interaction. Int. J. Eng. Res. Comput. Sci. Eng. **5**(2), 215–220 (2018)
12. Wshah, S., Kumar, G., Govindaraju, V.: Statistical script independent word spot-ting in offline handwritten documents. Pattern Recognit **47**(3), 1039–1050 (2014)
13. Cao, H., Bhardwaj, A., Govindaraju, V.: A probablistic method for keyword retrieval in handwritten document images. Patter Recognit. **42**(12), 3374–3382 (2009)

Biometric Passport Security by Applying Encrypted Biometric Data Embedded in the QR Code

Ziaul Haque Choudhury and M. Munir Ahamed Rabbani

Abstract In this paper, a noble biometric encryption technique is proposed for national security without the electronic chip by applying encrypted biometric data embedded in the Quick Response code (QR code) to secure the biometric data and personal information. The proposed biometric encryption method is achieved by applying the Advanced Encryption Standard (AES) algorithm and SHA-256 algorithm embedded into the QR code. This technique will enhance the safety and security for a biometric passport from unauthorized access without knowing the passport holder. The proposed method will provide more security for the border crossing and illegal immigrants.

Keywords Biometric · Encrypted data and biometric in the QR code · E-Passport

1 Introduction

A biometric passport is an electronic travel archive that comprises biometric data of the passport bearer that can be naturally perused and handled by a PC to validate the citizenship of a person. It is a half-breed report that joins organizes with electronic abilities. E-passport makes utilization of two advances: radio frequency identification (RFID) and biometrics. Today RFID innovation is progressively being utilized as an anti-fogging device in different application territories, for example, stock chains, individual recognizable proof, access control, financially related accreditations, and sensor systems. Around 20 years back, an authority of the US national

Z. H. Choudhury (✉)
Department of Information Technology, B.S. Abdur Rahman Crescent Institute of Science and Technology (Deemed to be University), Chennai, India
e-mail: ziaulms@gmail.com

M. Munir Ahamed Rabbani
School of Computer, Information and Mathematical Sciences, B.S. Abdur Rahman Crescent Institute of Science and Technology (Deemed to be University), Chennai, India
e-mail: rabbani3012@gmail.com

© Springer Nature Singapore Pte Ltd. 2020
K. S. Raju et al. (eds.), *Data Engineering and Communication Technology*,
Advances in Intelligent Systems and Computing 1079,
https://doi.org/10.1007/978-981-15-1097-7_5

government had assessed that among the then 3 million US passport applications got each day in any event 30,000–60,000 identifications are fake and just 1,000 deceitful travel papers have been recognized. As indicated by the reports, 80% illicit drug dealers and an extensive number of militants are supported with phony passports and visas and travel unreservedly everywhere throughout the world [1]. So as to enhance the respectability of the passport, the different nations have been issuing identifications including an RFID chip that contains the passport holder's personal data. Malaysia was the primary nation on the planet to issue e-passport in 1998, and to date, it has been executed in excess of 45 nations.

These Machine Readable Travel Documents (MRTDs) were presented by the International Civil Aviation Organization (ICAO) in its Document 9303 [2] which gives a lot of tenets and measures for a biometric passport. A biometric passport is the international identification travel document which contains biometric data to replace the conventional passport. This technique contains hybrid documents (i.e., paper, antenna, and embedded chip) that allow wireless communication with the special reader in the border control to authenticate a traveler. The Extended Access Control (EAC) mechanism [3] illustrated the authentication procedure between the e-passport and an Inspection Terminal (IS).

Since the e-passport or biometric passports are incorporated RFID technology, an access control mechanism is requiring privacy protection. In the existing work found various kinds of security issues on RFID technology. There is a number of threat scenarios found in the literature which is relevant for the issuance of travel documents in [4, 5]. The relevant issues for the issuing of electronic passports are listed—data leakage threats, identity theft, tracking, and host listing.

To prevent attacks such as skimming and eavesdropping, it needs to provide a more secure security system. Hence, this paper proposed a biometric encryption method based on Advanced Encryption Standard (AES) with SHA-256 (Secure Hash Algorithms) and embedded into the QR code. Figures 2 and 3 images are collected from the FEI database [6]. This technique will enhance the security from various attacks because it is not an active element.

The main novelty and contributions of this paper are:

- AES and SHA-256 algorithm are applied for biometric encryption, and it will enhance the security.
- The encrypted biometric data are embedded in the QR code.

The rest of the paper described as follows: Sect. 2 presents related work. Section 3 elaborated on face recognition and template generation. Section 4 discussed the biometric encryption. Section 5 proposed biometric encryption using the AES, SHA-256 algorithm and elaborated on encrypted biometric data embedded in the QR code. Section 6 concludes this paper.

2 Related Work

A lot of research has just been directed on Machine Readable Travel Documents (MRTD). One of the principal security examinations on the biometric passport was exhibited by Juels et al. [7] in 2005. A few downsides in the ICAO standard were distinguished by them which incorporate clandestine scanning and tacking, biometric information spillage, eavesdropping, skimming, and cloning. They likewise discovered certain deformities in the cryptographic structure of the ICAO standard. Hoepman et al. [8] in 2006 examined the latent assaults against Basic Access Control (BAC) and biometrics. The symmetric key between the peruser and the chip has entropy under 80 bits and can without much of a stretch is speculated.

Lehtonen et al. [9] recommended in 2006 a conceivable arrangement against the clandestine reading and eavesdropping attacks by incorporating the RFID empowered MRTD with the optical memory gadget. The correspondence channel between the peruser and the optical memory gadget is secure since they require an observable pathway association for perusing. The principal disadvantage of this proposed thought is that an equipment change must be done on biometric passports.

Kumar et al. [10] proposed the online secure biometric passport convention which provides common verification between the inspection system and the biometric passport. This convention requires just the best dimension authentication issued by Country Verifying Certificate Authorities (CVCA) to be stored in a biometric passport, which would diminish the memory necessities and keep a denial-of-service attack by a pernicious reader on a biometric passport. In spite of the fact that the present design of biometric passport provides great security of biometric information of an individual, distributed computing could be utilized sooner rather than later to play out a brute force attack. A solution for this is to join the cancellable biometrics with cryptographic conventions where the biometric information of a person can be secured by cryptographic keys that are traded by the Password-Authenticated Connection Establishment (PACE) convention [11].

Kundra et al. [12] exhibited a paper in 2014 which essentially focus around enhancing the security of e-travel papers utilizing biometrics. It exhibits a cryptographic security investigation of different advances utilized in biometric passport configuration uniquely utilizing face recognition, palm print, and iris biometric. These strategies intend to give enhanced security in ensuring biometric data of the e-visa holder.

There are some menace situations seen in the literature which is important for the issuance of travel documents [4]. The pertinent issues for the issuing of electronic travel papers, for example, identity theft, data leakage threats, tracking, and host listing are examined [5]. Adrian et al. [13] examined two stages of attacks on network and transport layers based on RFID. The attacks on the labels are cloning, spoofing, impersonation, eavesdropping, application layer, unapproved label perusing, and label change. In [14] discloses four techniques to attack used by adversaries to comprise the security of a framework that utilizes ISO/IEC 14443 RFID, for example, eavesdropping, skimming cloning, and relaying.

Regardless of the way that PC chips and RFID labels have been now used for storing biometric information, they are not suitable for low-cost arrangements. Likewise, The RFID chip is legible information leakage threats normally without such mindfulness [15]. The existing showed that ICAO standard biometric passport based on RFID has countless issues and security infringement issues [16, 17]. Hence, to overcome these challenges, this paper proposed biometric encryption using the AES with SHA-256 algorithm and embedded into the QR code. The QR code has the ability to store the biometric information, and it cannot be used as an active component; they are significantly more affordable and do not require specific equipment for recuperating data. To be sure, QR codes are efficient, latent read-just segments whose information cannot be altered and can be decoded by the explicit gadgets.

3 Face Recognition

Viola–Jones algorithm presented in [18] has been used for face recognition from a face image. The Viola–Jones detector algorithm is a quick and authentic technique for focalizing a face in a picture. In face, the arrangement must be performed so as to look after explicit the position of the facial features in a face picture. We proposed a system which distinguishes facial features, for example, eyes, nose, and mouth. Along these lines, a face picture is adjusted and trimmed as indicated by the situation of the identified facial highlights.

3.1 Local Binary Pattern (LBP) Feature Extraction

Face pictures are constituted by features that describe a facial visual aspect. We have used Local Binary Pattern (LBP) features for the portrayal of a face. LBP features have been presented in [19]. LBP features appear to be suitable for the depiction of face pictures caught in an uncontrolled situation and for utilization in close constant face acknowledgment frameworks. The LBP method captivates and improves edges of facial features. The picture is in this manner partitioned into non-covering regions. A histogram of reactions to the LBP technique is registered independently in each region.

3.2 Nearest Neighbor (NN) Classification

The NN method is utilized for categorization of an obscure item, portrayed by features, into $n + 1$ classes, where n is the quantity of classes decided by templates; there is one additional class for items that do not have a place with any template. In the utilization of NN classifier in face acknowledgment, an element vector of an obscure individual is contrasted with all face templates. The obscure element vector is doled out to that template, whose common separation uniqueness, as per Eq. (1), is insignificant. We have used a separation edge for individual templates, which speaks to maximally permitted uniqueness between the element vector and the template. The element vector may surpass the separation edge everything being equal. In such a case, it is not allocated to any template. Such a component vector represents a unique without a face template—an impostor.

The contrast of templates and obscure feature vectors depends on a distance measure. In this experiment, we used the chi method [19] for LBP feature, and the equation is given below,

$$\chi^2(T, S) = \sum_i \frac{(T_i - S_i)^2}{T_i} \tag{1}$$

where T = template feature, S = feature vector, and i = dimensionality of T and S.

3.3 Face Template Creation

This area depicts strategies for face template creation. There are distinctive face template creation techniques accessible [20, 21]. Nonetheless, formats made by various techniques have diverse properties and effect on order execution. Therefore, a suitable technique must be structured and used to keep up a high correct characterization rate (CCR). We have chosen the most similar face detection method [20] to create the template. This technique acquires one face template for each individual by contrasting all component vectors that depict one individual's face pictures. Feature vector which has negligible amassed separation as indicated by Eq. (1) to all other feature vectors of one individual is chosen as a template. This template is encrypted using AES algorithm and encoded into the QR code.

4 Biometric Encryption

Biometric encryption employs the physical characteristics of a person as a way to code or decode and provide or deny access to a computer system. Biometric encoding techniques address directly on the authorize predilection and proposals of the privacy and information security authorities for applying biometrics to assert the identity instead of identification function alone. The biometric data in itself cannot assist as a cryptographic key because of its variableness. Nevertheless, the number of data comprised in a biometric data is prominent. It is the procedure that firmly bonds a cryptographic key to biometric data so that the key or biometric cannot be recovered from the stored template. It is only possible to recreate the key only if the right live biometric sample is demonstrated on verification.

The digital key is haphazardly created during enrollment; hence, the user is unaware of the key. The biometric key is entirely autonomous, and it can generally be changed. Later on, a biometric sample is developed and the biometric encryption algorithm consistently and securely holds the key to develop a biometric template, and it is also known as a private template. In principle, the biometric key encrypted with the biometric and the biometric encryption template provides splendid privacy protection. The generated biometric encryption can be stored in a database, smart card, token, etc.

For the verification part, the user gives his/her recent biometric data, it is verified across the decriminalize biometric encryption template, which permits the biometric encryption to recall the similar key and the biometric assists as a decryption key. Toward the finish of the check, the biometric test disposed of indeed. In another way, the attacker whose biometric data is unlike enough will not be retrieving the key. Hence, these kinds of encryption and decryption are fuzzy since the biometric sample is dissimilar each time; the different encryption key is conventional cryptography. Afterward, the digital key is recovered, and it can be applied as the basis for any physical or logical application. Hence, biometric encryption is a proficient, secure, and protection agreeable tools for biometric key administration.

In this paper proposed an image encryption technique by applying the AES algorithm and embedded into the QR code. This technique will enhance to secure the biometric data for a biometric passport. Figure 1 shows the framework of biometric encryption.

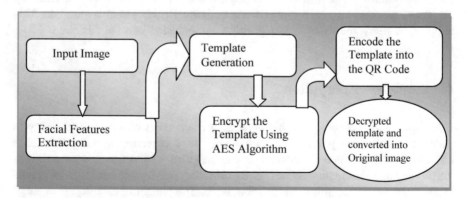

Fig. 1 Biometric encryption and decryption framework

5 Proposed Biometric Encryption Using AES Algorithm

Encrypted biometric information is encoded in the QR code, which is achieved by applying the Advanced Encryption Standard (AES) calculation [22, 23]. AES was created by Joan Daemen and Vincent Rijmen, known as Rijndael cipher block. It is generally called a symmetric key algorithm which is the fact that the comparative key is used for encoding and decoding the data. AES is a form of cipher with different block sizes and key sizes. In AES, 128 piece of square size and 3 kinds of key sizes are accessible, for example, 128 pieces, 192 pieces, and 256 pieces. Plain text and key size are chosen independently and the size of plaintext, key size chooses the number of rounds to be executed. In light of the plaintext and keys, there is a negligible of 10 rounds for 128-piece key and 14 rounds are the greatest for a 256-piece key. We have scrambled the e-identification biometric data which contains biometric data of an individual and encoded into the QR code. This encoded biometric data to keep mystery from the interloper. The biometric data has been collected from FEI dataset [6], and it is used for facial encryption. A single-face image has encrypted for encoding into QR code. The encryption of the picture is adjusted by applying the AES calculation [24]. The biometric data is the figure finished by applying the biometric encryption method which relies upon the keys. The AES encryption is connected to create a way to encode the biometric.

5.1 Encryption Phase

Two input is required; consider the input image I_1 [6] which is encrypted, and k_1 is the secret key. The k_1 is changed to SHk_1 by applying an SHA-256 algorithm. Therefore, I_1 is encoded into base 64 string B_1. Hence, B_1 and SHk_1 are enriched into the AES-256 encryption algorithm to create ciphertext C_t.

ALGORITHM (with $N = 2$):

1. The input biometric data to be read and encode it by applying the base 64 standard.
2. The key file to be read and start the encryption using AES-256-bit key by utilizing the hash algorithm, here we considered SHA-256 of the key file.
3. The biometric is encrypted by utilizing the base 64-encoded content and hash produced in step 1 and step 2 severally.
4. Produce another Image I_2 of size (S_1, S_2) with pixel information P_i, where

   a. S_1—character endorse for the key document (Default: 255).
   b. S_2—Number of characters in the key document
   c. P_i—Pixel information to be occupied (Default: 0)

5. Individual row I_R in the height of image repeat:

   a. Get J_A to be ASCII code of the I_Rth character in the key document.

   b. Satisfy the first J_A pixels of the image in the I_Rth row with black color. I_R. I_2 [I_R] [J_A] = 0 for each I_R, J_A in S_1, S_2 with the end goal that $J_A <$ ASCII (key[I_R])

6. Make N (= 2) Images (S, T) of a similar size (S_1, S_2) and pixel information to such an extent that

   a. For the 1st image S, pixel information is created arbitrarily. It tends to be either 0 (black) or 1 (white).
    I_R. S [I_R] [J_A] = random(0, 1)

   b. 2nd image pixel information T [I_R] [J_A] is characterized such that I_R. T [I_R] [J_A] = S[I_R] [J_A] xor
    I_2 [I_R] [J_A] for each I_R, J_A in (S_1, S_2)

7. The C_t is the encrypted encoding output, the biometric images S and T, respectively.

5.2 Decryption Phase

Two inputs are given for decryption phase; C_t is the ciphertext to decrypt and the array of shares k_2, the secret key. Therefore, the original key K_3 image is constructed. The K_2 is then decoded to the k_1 by applying ASCII. The k_1 is changed to SHk_1 by applying an SHA-256 algorithm. The C_t, SHk_1 are fed into the AES-256 decryption algorithm to produce the base 64 encoding of the B_1 image. The B_1 is converted to I_1 output.

ALGORITHM (with N = 2):

1. Compute the input ciphertext which from the image C_t.
2. Load the images K_2, K_3 from the input K_2.
3. Produce another Image I_2K_1 of size (S_1, S_2) same as K_2, K_3 with the end goal that

   a. I_2K_1 [I_R] [J_A] = K_2 [I_R] [J_A] xor K_3 [I_R] [J_A] for each I_R, J_A in (S_1, S_2)

4. Introduce key K_1 as an array of characters of sizes same as the height of image I_2K_1 (S_2).
5. For each row I_R^{th} in the height of image I_2K_1 repeats

   a. Let count = 0

   b. For each pixel J_A of the image in the I_Rth row with black color I_R. Increase count by 1

    c. Find the character K_1I_R by utilizing the ASCII code of the count generated after b.

       i.e., K_1I_R = char(count)

    d. Set $K_1[I_R]$ = K_1I_R

6. With the Key, K_1 initializes the AES-256 algorithm with a hash (K_1) (SHA-256)
7. Decrypt the ciphertext C_t and save the decrypted base 64 encoding as an Image I_1
8. The output I_1 is the decrypted image.

5.3 Sample Input and Output for Image Encryption

See Fig. 2.

5.4 QR Code Generation

Japanese Corporation Denso Wave built up a two-dimensional barcode known as Quick Response (QR) code [25]. In this, encoding of data is performed on both horizontal and vertical headings and accordingly holding a few times larger number of information than a conventional standardized bar code. These codes have quickly picked up prominence worldwide and have been embraced by numerous frameworks particularly in Japan because of its capacity to encode Kanji symbols as a matter of course which makes it particularly reasonable. QR codes are utilized for storing URLs, addresses, signs, business cards, open transport vehicles, and so forth. QR codes comprise various territories that are held for explicit purposes. QR codes are utilized for their quicker intelligibility and more prominent ability to store data. The littlest codes are of size 21×21 modules; these are called form 1 QR code. The measure of QR code increases by four modules for each next rendition of the QR Code. The biggest QR code is of size 177×177 known as Version 40 QR

Input image [22]　　**Encrypted Data**　　**Decrypted image**

Fig. 2 Biometric encryption and decryption

| Input biometric [22] | Encrypted | Embedded into QR Code | Decrypted Image |

Fig. 3 Encrypted biometric embedded into the QR code and decrypted the image

code. Figure 3 shows encrypted biometric data encoded into the QR code. The encrypted biometric data has encoded by applying online QR code generator software [26].

6 Conclusions

This paper proposed novel biometric encryption and encode the QR code for a secure biometric passport. The biometric data has encrypted by applying the AES with SHA-256 algorithm to secure the data from the legible, data leakage threats. The proposed method will enhance the security in the biometric passport because the encrypted biometric data embedded in the QR code cannot be utilized as an active element. They are substantially less expensive and do not require particular hardware for recovering information. Indeed, QR codes are economical, passive read-only components whose data cannot be modified and can be decoded by the specific devices. We are focusing on secure RFID and enhancement of its security.

Acknowledgements Authors would like to acknowledge the support of the Artificial Intelligence Laboratory of FEI in São Bernardo do Campo, São Paulo, Brazil, for providing the database for the research work. This work has been done in the BSACIST Research Laboratory, Chennai, India. We thank our research department who provided expertise that greatly aided in the research.

References

1. King, B., Zhang, X.: RFID: An anti-counterfeiting tool. RFID security: Techniques, protocols, and system-on-chip design. Springer Science+Business Media, LLC (2008)
2. ICAO, Doc 9303. Machine Readable Travel Documents, Seventh Edition, Part 3: Specifications common to all MRTDs (2015)
3. Bundesamt fur Sicherheit in der Informationstechnik (BSI), Germany: Advanced Security Mechanisms for Machine Readable Travel Documents, Extended Access Control (EAC), version 1.0, TR-03110 (2006)

4. Bolle, R.M., Connell, J.H., Pankanti, S., Ratha, N.K., Senior, A.W.: Guide to Biometrics. Springer Verlag (2004)
5. Shivani, K., Aman, D., Riya, B.: The study of recent technologies used in E-passport system. In: 2014 IEEE Global Humanitarian Technology Conference—South Asia Satellite (GHTC-SAS), September 26–27, Trivandrum (2014)
6. de Oliveira, Jr. L.L., Thomaz, C.E.: Captura e Alinhamento de Imagens: Um Banco de Faces Brasileiro. Undergraduate technical report, Department of Electrical Engineering, FEI, São Bernardo do Campo, São Paulo, Brazil (in Portuguese) (2006)
7. Juels, A., Molnar, D., Wagner, D.: Security and privacy issues in E-passports. In: Proceedings of the First International Conference on Security and Privacy for Emerging Areas in Communications Networks (SecureComm'05), Washington, DC, USA, IEEE (2005)
8. Hoepman, J.H., Hubbers, E., Jacobs, B., Oostdijk, M., Wichers Schreur, R.: Crossing borders: Security and privacy issues of the European e-passport. In: Advances in Information and Computer Security, First International Workshop on Security (IWSEC'06), Kyoto, Japan. Lecture Notes in Computer Science 4266, pp. 152–167. Springer-Verlag (2006)
9. Lehtonen, M., Staake, T., Michahelles, F., Fleisch, E.: Strengthening the security of machine readable documents by combining RFID and optical memory devices. In: Presented at Developing Ambient Intelligence: Proceedings of the First International Conference on Ambient Intelligence Development (Amid'06), 2006. Sophia Antipolis, France, Springer to appear in International Journal of Information Security (IJIS) (2006)
10. Kumar, V.K.N., Srinivasan, B.: Design and development of e-passports using biometric access control system. Int. J. Adv. Smart Sens. Netw. Syst. (IJASSN), 2 (3) (2012)
11. Belguechi, R., Lacharme, P., Rosenberger, C.: Enhancing the privacy of electronic passports. Int. J. Inf. Technol. Manage. (IJITM) Special Issue on: Advances and Trends in Biometrics 11 (1/2), 122–137 (2012)
12. Kundra, S., Dureja, A., Bhatnagar, R.: The study of recent technologies used in E-passport system. In: IEEE Global Humanitarian Technology Conference—South Asia Satellite (GHTC-SAS), September 26–27 (2014)
13. Adrian, A., Marius, l.: Biometric passports (e-passports). 978-1-4244-6363/10/$26.00 © IEEE (2010)
14. Tiko, H.: Using NFC Enabled Android Devices to Attack RFID Systems (2018) www.cs.ru.nl/bachelors-theses/2018/Tiko_Huizinga_4460898_Using_NFC_enabled_Android_devices_to_attack_RFID_systems.pdf
15. The Ministry of the Interior and Kingdom Relations, 2b or not to 2b, The Netherlands (2005). Available from: http://www.minbzk.nl/contents/pages/43760/evaluatierapport1.pdf/
16. ICAO. Biometrics Deployment of Machine Readable Travel Documents, version 2.0, ICAO MRTD (2004)
17. Juels, A., Molnar, D., Wagner, D.: Security and privacy issues in E-passports. In: Proceeding of First International Conference on Security and Privacy for Emerging Areas in Communications Networks (SECURECOMM'05), Sept (2005)
18. Viola, P., Jones, M.: Robust real-time object detection. Int. J. Comput. Vision (2001)
19. Ahonen, A., Hadid, A., Pietikäinen, M.: Face description with local binary patterns: application to face recognition. In IEEE Transaction on Pattern Analysis and Machine Vision, vol. 28, no. 12, pp. 2037–2041
20. Malach, T., Prinosil, J.: Face templates creation surveillance face recognition system. In: Proceedings of the 3rd International Conference on Pattern Recognition Applications and Methods. Portugal: SCITEPRESS, pp. 724–729. ISBN 978-989-758-018-5 (2014)
21. Zhao, W., Chellappa, R.: Face Processing: Advanced Modelling and Methods. Academic Press (2006)
22. Advanced Encryption Standard (AES). FIPS. Nov 23 (2001). https://csrc.nist.gov/csrc/media/publications/fips/197/final/documents/fips-197.pdf
23. Selent, D.: Advanced encryption standard. Rivier Acad. J. 6 (2) (2010)

24. Kalubandi, V.K.P. et al.: A novel image encryption algorithm using AES and visual cryptography. In: 2nd International Conference on Next Generation Computing Technologies (NGCT) (2016)
25. QR codes (1994). https://en.wikipedia.org/wiki/QR_code
26. QR code generator software: https://www.qr-code-generator.com/how-to-create-a-qr-code/

An Extensive Review on Cloud Computing

Mahendra Kumar Gourisaria, Abhishek Samanta, Aheli Saha, Sudhansu Shekhar Patra and Pabitra Mohan Khilar

Abstract Cloud computing is one of the most sought after fields in the world of computing and IT. This paper is an extensive survey intended to highlighting the importance of cloud computing in the business world and how it has benefited major companies as well as the potential ones. The different service models and key features along with database services play a major role. Virtualization, its features, and its role in making cloud computing possible are extensively discussed. However, like any other technology, even the cloud technology faces a vast array of issues that need to be attained. Some of the major issues like resource scheduling, security, and interoperability pertaining to cloud are reviewed here, along with possible ways of tackling them. A comprehensive study on the different fields to be explored in the study of cloud computing is presented here. We have done a systematic literature review of 62 selected papers and believe that this paper will help the new researchers in cloud computing area.

Keywords Cloud computing · Service models · Infrastructure as a service (IaaS) · Platform as a service (PaaS) · Software as a service (SaaS) · Virtualization · Migration · Scheduling

M. K. Gourisaria (✉) · A. Samanta · A. Saha
School of Computer Engineering, KIIT Deemed to be University Bhubaneswar, Odisha
751024, India
e-mail: mkgourisaria2010@gmail.com

A. Samanta
e-mail: abhisheksamanta60@gmail.com

A. Saha
e-mail: ahelis340@gmail.com

S. S. Patra
School of Computer Application, KIIT Deemed to Be University Bhubaneswar, Odisha
751024, India
e-mail: sudhanshupatra@gmail.com

P. M. Khilar
Department of Computer Science and Engineering, National Institure of Technology
Rourkela, Odhsha 769008, India
e-mail: pmkhilar@nitrkl.ac.in

© Springer Nature Singapore Pte Ltd. 2020
K. S. Raju et al. (eds.), *Data Engineering and Communication Technology*,
Advances in Intelligent Systems and Computing 1079,
https://doi.org/10.1007/978-981-15-1097-7_6

1 Introduction

Cloud computing has evolved considerably to an entity that drives all businesses today from what it was a decade back. In the year 1961, John McCarthy, the recipient of the prestigious Turing Award, mentioned that computing can be traded as a utility as we do with electricity or water. In the year 1999, Salesforce.com began providing applications to clients by the use of a website. Soon enough, Amazon joined and provided unparalleled and unseen abilities to its customers. Google and Microsoft joined in 2009.

Today, the technological and IT world is unimaginable without cloud computing a field that has developed with incredible speed in the last decade, transforming the business and commercial world of computing forever. It is the trigger that catalyzed the availability of computing as a utility [1]. Cloud computing encompasses an elite range of web-based services that provide the functionalities of computing such as software, storage, and virtual hardware as services without high investments in acquiring the hardware or involvement in the technical complexities [2]. A wide variety of deployment and cloud delivery models have been made procurable keeping in mind the customer requirement, thereby giving them access to "infinite resources" [3, 4]. They have a diverse range of security and accessibility levels, thus tending to all groups of customers starting from individual software developers to business and IT infrastructures [5]. The components of cloud comprise of data, semantics, information, application, ontologism, schema, data catalogue, data dictionary, metadata, and information model [6].

The cloud is a developed and modified adaptation of the pre-existing distributed computing concepts of cluster and grid. Cloud computing is a result of advances made in the fields of distributed computing, Internet technologies like the service-oriented architecture (SOA), hardware concepts of virtualization, and systems technology autonomic computing and data handling [7]. It gives its users perception of unlimited resources, which are conveniently and feasibly obtainable anytime and anywhere. It is mobile and collaborative due to the abstraction of infrastructure.

This paper is intended to highlight and discuss all the indispensable sectors of cloud computing. Section 1 compares cloud computing with the pre-existing models of grid and mobile computing, which will give an understanding of why it took a greater leap towards success than any of them. The extensive roles of cloud computing in some of the world's leading companies are also described in the latter part of Sect. 1. Section 2 goes through the key features of cloud which contributed to its rapid and steady success in the industry followed by Sect. 3 which discusses the underlying techniques. There is a wide range of deployment and service models available in cloud, which are extensively discussed and compared in Sects. 4 and 5 of this paper, respectively. Section 6 presents an overview of the cloud database. Virtualization is imperative to the study of cloud computing. A detailed study of the various techniques and types of virtualization is presented in Sect. 7. Section 8 deals with the processes and techniques of migration of data from on-site machines to the cloud environment. Such procedures and algorithms are discussed in Sect. 9. Section 10 provides an extensive study on the different existing threats and issues that may arise in the cloud environment, and also the existing and proposed solutions to manage and overcome them. The study concluded in Sect. 11.

1.1 A Comparative Study—Grid and Mobile Computing

In grid computing, the resources are shared by participants in a grid, with some given restrictions. In this case, the users are able to access resources depending on their needs. The technology that revolves around grid computing is parallel processing. The machines might be physically wired together or interconnected with the help of the Internet. Some of the differences between cloud and grid computing are mentioned as following in Table 1.

Mobile computing is vastly relied on cloud to make its features work by storing the user data in a public or private cloud. Mobile Cloud Computing (MCC) brings mobile and cloud computing together with wireless networks to give rise to powerful resources in computing. MCC promises execution of mobile software and applications with a well-devised user experience. However, it is not without the security concerns of data monopoly and privacy. In certain cases, distribution of a particular work is not even and it has to be offloaded to a resourceful platform. Migration of the executive unit needs to be done in such a case [8].

1.2 Involvement of Major IT Sectors

Cloud computing, once an IT jargon is now a common business strategy and almost all companies of varying sizes are signing in for the massive cloud revolution that has altered the business scenario of the world. While the major companies like Google, Amazon, Microsoft, Aneka, Salesforce.com have become the largest distributors and retailers of cloud services and incurred massive gains, small organizations and new start-ups have had their initial start-up investment substantially lowered [1, 9]. Some of the examples of companies that have left prominent marks in the cloud vending industry till date are discussed here.

Table 1 Comparison between cloud, grid computing

Basis	Grid	Cloud
Management	Decentralized	Centralized
Programming model	Unicore, Globus	Open Nebula, Eucalyptus
Flexibility	Low	High
Architecture	Distributed	Client–server

Amazon—The Amazon Web Services (AWS) were the trailblazers of Iaas in 2004 and have since then provided an elegant and elaborate range of services, the most pronounced ones being Elastic Compute Cloud (EC2) for access to virtual servers, Simple Storage Service (S3). Some other examples are Simple Queuing Service (SQS), Relational Database (RDS), CloudFront and Elastic MapReduce.

Google—The Google Cloud Platform (GCP) provides a wide domain of public cloud computing services. The core products include Google App Engine for access to software development kits, Google Compute Engine for accessing virtual machines, Google Container Engine and Google Cloud Storage. Apart from this, it is also specialized services for fields like data analytics and data processing (Google BigQuery, Google Cloud Dataproc and Google Cloud Dataflow), artificial intelligence (Cloud Machine Language Engine) and IoT (Google Cloud IoT Core) [10].

Microsoft provides platform services like Windows Azure, Sql Server and Visual Studio. Under software as services, we have services like Office365, Microsoft Azure Sql database, dynamic CRM [10]. The IBM SmartCloud provides infrastructure, software and platform as services offered under three delivery models namely SmartCloud Foundation, SmartCloud Services and SmartCloud Solutions.

2 Key Features

2.1 On-Demand Service

On-demand service is mentioned by the National Institute of Standards and technology as *a* consumer will be able to provide computing capabilities, such as network storage and server time, without any sort of human intervention with a particular service provider [11]. To put it in simple words, on-demand service provides virtual services such as storage and server time as per the requirements of the end user [5].

2.2 Omnipresent Network

Flexible methods to connect and use cloud services promote its omnipresent nature. The basic requirement for access today is a mobile phone. For example, Google photos on a mobile phone allow seamless interaction between the user and the cloud, with just the necessity of being an internet connection.

2.3 Location Independent Resource

All the resources offered by the company are at the disposal of the end user which can be used by them as per their necessity [5].

2.4 Rapid Elasticity and Scalability

Cloud technologies are scalable, as they possess the potential to flex with as per the interest of resources [11]. The work of elasticity is to enable the creation of virtual machines to cope up with "an expansion in the workload" (the growth of demand) [4].

2.5 Utility Model (Pay Per Use)

The user pays month as per his use. Unnecessary wastage of computing resources is avoided and the model paves way for a more economic method. Microsoft, IBM and HP are one of the first companies to embrace the utility model.

2.6 Energy Efficiency

With the advent of cloud computing, data centres and business industries rely on minimal infrastructure and minimizing energy consumption. The effect on both the environment and commerce is well witnessed due to this change [12].

3 The Underlying Technique

The technology that fires cloud is that of servers. An interconnection of servers is what drives the cloud. Modern companies and IT giants frame their cloud structure depending on the amount of data their servers are programmed to handle [12]. Companies design their system based on a statistical study of the number of users logging in at a particular time. Excessive overload may lead to crashing. The components needed for cloud computing are clients, data centres and distributed servers.

4 Deployment Models

4.1 Public Cloud

A cloud is public when software and applications are being accessible publicly with anyone availing its utilities [10]. Schemes can be various, including pay per use or purchasing. Google App Engine and Amazon Elastic Compute Cloud (EC2) are important examples of public cloud [13]. The advantages of public cloud are inexpensive setup, scalability, seamless uptime and no wastage of resources. The principal concerns pertaining to public cloud are security threats.

4.2 Private Cloud

Private cloud is a protected cloud base providing access to specific users. Information is stored in an organization's privatized servers. It is mostly targeted towards keeping control over an organization's characteristics and information along with keeping the necessary security [11]. Microsoft's Azure is a well established private. By the usage of Azure, users create models and services to ascend .NET. Users can collaborate with information from such private clouds with their data.

4.3 Hybrid Cloud

Hybrid cloud uses the integrated facilities present in both public and private clouds [10, 14]. It is valued in places of fluctuating workloads. The scalability that hybrid cloud brings to the company workforce poses a significant reduction in its infrastructural and managemental expenditure [15].

4.4 Community Cloud

Community cloud provides cloud services to specific individuals or groups. The "communities" have similar goals [10, 11]. Businesses that work on similar projects fall back to community cloud. With the expenditure distributed over a less number of users than public cloud, community cloud provides a high level of security and privacy. An example of community cloud is Google's Gov .Cloud [16].

5 Service Models

5.1 SaaS (Software as a Service)

It is a common software distribution scheme in which a third-party provider hosts applications and makes it available as a web-based service. Some of the widely used Saas applications are billing and invoicing system, customer relationship management, applications and help desk applications. SaaS provides a wide range of advantages, besides eliminating the compulsion of organizations to handle installation, setup and management, and hardware maintenance. Most of these services are based on a utility model or may be subscription based and are effortlessly available on demand. It thus provides an economic alternative to paying for licensed applications. The software and hardware upgradations are also dealt with the service providers, further reducing workload on the end-user side. The main issues involved are:

- **Lack of Control**—The control in this scheme resides with a third-party vendor which provides a lower degree of control than in-house software applications.
- **Security**—The data being handled by third-party host, privacy of sensitive and important data pose a serious threat.
- **Network Dependence**—Being web-based service, accessibility to the software and data solely depends on continuous availability of network, in absence of which one may lose access to both.
- **Switching**—Switching between Saas vendors can prove to be a demanding task as the user interface and business execution can be specific to the provider.
- **Multitenancy**—In multitenancy, multiple users share a single version of software. It has economic impacts on an unprecedented scale. Vendors often fall back on multitenancy due to advantages of data aggregation. Information about different customers is stored in one scheme of a database, facilitating mining of data and running queries. However, it is not without its security threats.

5.2 IaaS (Infrastructure as a Service)

This particular scheme presents the main resources of hardware required for computation as a service. The most common facilities employed include virtual machines, virtual storage, servers and routers on demand to clients as a web-based service. On the one hand where traditional hosting services rent out the physical servers or a part of it on basis of a subscription model, IaaS providers rent virtual machines based on a utility model.

Due to its scalability, this scheme is highly flexible and efficient. It offers a high level of responsibility and control to end consumer as the particular user can issue administrative control over the VM, thereby controlling when it runs, the installation of applications and software, or even the management and supervision of a custom operating system. The most trending and popular example of IaaS is the Amazon Elastic Compute Cloud (EC2) which forms a major part of Amazon Web Services (AWS) [17]. Other established examples include Microsoft Azure, Google Compute Engine (GCE), Netmagic Solutions and Rackspace. In spite of its numerous advantages, even this is not without its demerits.

- **Network Dependency**—This too, being a web-based service is non-functional without the availability of continuous network.
- **Prompt Updation of VM**—The VM can become out of date as the control of the VM lies with the user in running, suspended and off states, and automatic updation is not provided by the service provider.
- **Legacy Security Vulnerabilities**—Security issues pertaining to the usage of legacy software in the provider's infrastructure may arise.

5.3 PaaS (Platform as a Service)

PaaS provides development tools, programming models, and runtime environment for applications. It provides a computing platform, encompassing the operating system, database, software frameworks, servers, middleware and computing environments suitable for certain needs as a service [18]. Developers will no longer have to painstaking acquire the main hardware and get associated with the complexities of the infrastructure setup, configuration and management. PaaS offers facilities for application development, application design, testing and deployment, web services, and database integration, security, persistence, storage, scalability, state management, application versioning and application instrumentation.

The most common PaaS providers are Force.com, Heroku, Google App Engine [19] and Windows Azure among many others. They support a wide array of programming languages for user appropriacy and utility. For instance, Google App Engine provides a Java- and Python-based environment thereby giving the developer a choice. Microsoft Azure supports a .NET based environment along with Java and Ruby SDK. Heroku is a platform that implements Ruby on Rails web application. Force.com, on the other hand, has formed two languages of its own—a Java-like language named Apex and a query language bearing resemblance to Excel, named VisualForce. Risks are involved in this service just like the others. The major challenges involved in Saas are:

- **Security Issues**—The entire data is managed by third-party vendors and their infrastructures. Also, legal and legislative issues are major factors. The laws of data management and cryptography encryption vary from country to country, and since the physical servers in which ones data is stored can be anywhere in the world, this can pose security threats to sensitive data.

- **Data Lock-in**—Most service providers do not offer the facilities to transfer data and workloads from one vendor to another at a later time. The users thus have their workload and data locked into a particular service provider, which may cause inconvenience at the user end.
- **Efficient Resource Utilization**—The physical resources like the central processing unit cores, processing power, network bandwidth and storage space have to configure and allocated dynamically so as to ensure the least consumption of resources. This can problematic as the same physical resources can divide among users with a variegated range of workloads, and the most optimum distribution is to be mapped. If this is not attended to, the vendor can incur huge economic losses. It also possesses a threat to the environment due to the large amounts of CO_2 released from cooling systems [12].

5.4 DBaaS (Database as a Service)

Database as a service is provided by certain cloud platforms. In this case, the virtual machine is not physically operated. The database service provider takes charge of the database while clients rent the services. Database services are being provided by Amazon as a part of their cloud services. It is one of the key components of anything as a service. DBaaS contains a database manager. This exerts control over elemental instances of database by an API. A management console causes the API to be available for the user. This management console is commonly a web application.

5.5 XaaS (Anything as a Service)

Xaas is known as "anything as a service", Xaas encompasses anything and everything that can be provided and accessed as services through cloud computing, over the internet. Along with the already discussed IaaS, SaaS and PaaS, it also includes storage as a service, desktop as a service (DaaS), network as a service (NaaS), disaster recovery as a service (DRaaS), database as a service (DbaaS) and even emerging services like backup and restore as a service, voice as a service, marketing as a service and healthcare as a service, among many more. It reflects the promising capabilities and power of on-demand cloud services.

6 Cloud Database

A database is a collection of sorted data. The cloud database comprises of multiple components interacting with each other. The basic architecture comprises of a front and a back end. The front end consists of the user's application and networking system present in his computer that is used to access the cloud. The back end comprises of servers that run the cloud along with infrastructure pertaining to data storage. Advantages of using cloud database are economic, improved performance and efficiency, software updates, improved document compatibility, improved group collaboration and high storage.

7 Virtualization

Virtualization leads to the sharing of a unit instance among multiple users and customers, thereby, causing a marked reduction in infrastructural costs and profiting industries and companies [20, 21]. A number of applications and operating systems can run on a single server like never before [22]. It solved the issue of under-utilization of resources, which prevailed earlier, to a great extent. The working of the virtual machine monitor (VMM) or the hypervisor is the key to the functioning of virtualization technologies in the cloud. It is responsible for resolving conflicts between VMs and allocation of resources to VMs [22]. The concept of virtualization was brought forth by IBM, with the development of **SIMMON** and the **CP/CMS** operating system. Over the years, it had benefited the $\times 86$ computers immensely.

7.1 The Hypervisor

The hypervisor is a firmware of low-level program which manages the sharing of a unit physical entity of cloud resources among multiple tenants [23]. Unlike other applications that need the operating system to function, the hypervisor bypasses the connection between the operating system and the hardware [21]. It creates a virtual environment for the guest OS and leads it to believe that it has access to the real physical resources [24]. It has a share of the host computer's CPU cycles, RAM, disc storage and bandwidth. Depending on the host on which the hypervisor operates, they can be classified as Native Hypervisors and Hosted Hypervisors.

7.2 ×86 Hardware Virtualization

The ×86 architecture provides a four-level model namely Ring 0, 1, 2, and 3 to operating systems and routines to operate and access the hardware resources. The OS has direct access to the hardware and storage and operates on Ring 0, whereas software applications run by the user typically operate on Ring 3 [25]. This is the structural model that exists in absence of any virtualization. Virtualizing the X86 architecture involves inserting a layer of virtualization beneath the operating system that manages and provides share resources. However, some sensitive instructions cannot easily be virtualized as they behave differently when they are not operated on Ring 0 [25]. To solve this issue, the following techniques were developed for handling such non-virtualizable information.

7.2.1 Full Virtualization Using Binary Translation

In this type of virtualization, the OS operates on top of a virtual layer and is completely severalized from the underlying hardware [26]. It operates by trapping translating the sensitive OS instructions into a form that produces the required result on the virtual layer [27]. This translation is performed by the hypervisor, now operating on Ring 0, during run-time. The OS requires no modification, as the virtual environment for it is so created that is it unaware of its absence from Ring 0 [28, 29]. This integrated system of binary translation and direct execution is known as full virtualization. The efficiency of a full-virtualized system is 80–97% of the host. KVM is a full-virtualization solution for Linux, comprising of virtualization extensions of AMD-V and Intel VT [22]. Some examples include VMware ESXi and Microsoft Virtual Server. Some cons are

- The binary translation during execution can cause implementation overheads due to additional time consumption [30].
- All resources have to be emulated for accessing them [30].

7.2.2 Para-Virtualization or OS-Assisted Virtualization

The guest operating system interacts directly with the hypervisor through hyper calls to the hypervisor. The OS, in this case, requires a recompilation for the replacement of sensitive and non-virtualizable OS instructions with hyper calls, which are specific to a VMM [26, 27]. The non-virtualizable instructions are handled at compile time. Para-virtualization provides a software interface to the virtual layer which is not identical to the underlying hardware, but similar. Hyper-call interfaces for core critical operations like timekeeping, management of memory and handling of interrupts are provisioned by the hypervisor [29]. Some cons are:

- The advantage in usage of para-virtualization over full virtualization is dependent on the workload and can vary widely [29].
- It is not easily portable since the modifications to the OS kernel are hypervisor-specific. The modified OS compatibility issues on other hypervisors or hardware [30].
- The requirement of deep OS kernel modifications leads to issues in compatibility and maintenance [30].

Xen, an open-source project, is an example of para-virtualization. Xen hypervisors schedule tasks, manage memory and delegate I/O operations to the privileged drive [22]. VMware supports both full and para-virtualization, the choice based on relative performance [31].

7.2.3 Hardware-Assisted Virtualization

In this type of virtualization, a supplementary CPU execution routine is introduced which provides the VMM a new root mode beneath Ring 0. The VMM operates in the privileged root mode (**Ring 0P**) and the OS operated in de-privileged root mode (**Ring 0D**) [25, 29]. The requirement for binary translation and par virtualization is thus eliminated [30]. Intel and AMD were the first to create processor extensions namely VT-x and AMD-V. Pentium 4, the first generation x86 Intel processors to support this technology (VT-x), was released in late 2005. Some cons are

- The usage of an inelastic programming model leads to inflexibility in management of frequency and cost of VMMs to guest changeovers.
- Inflated guest to hypervisor transition expenses.

7.3 Other Types of Virtualization

7.3.1 Desktop Virtualization

This type of virtualization separates an individual's desktop environment from the physical server they are operating on. Virtualized desktops are not hosted on the hard drive of a PC, but alternatively on a remote central server. The OS is isolated from its client. It reduces the requirement and cost of configuring a large number of desktops in small- and medium-sized businesses. Using hardware-based GPU sharing one can access high-performance applications from any device over a secure network.

7.3.2 Server Virtualization

A server with sufficient capability allocates resources to a single client in the pool of multiple users in the same server. A single physical server can host multiple virtual servers, each operating in an isolated manner. Restart, upgradation or crash of a virtual server do not affect the other ones. Infrastructural costs are significantly reduced as server virtualization paves way to eliminate multiple servers. This way companies become more efficient and economic in their dealings with clients and customers [32].

7.3.3 Application Virtualization

In this type of virtualization, an application operates on a network workstation or a thin client, in lieu of a local host. Only a limited number of programs it accesses reside in the host machine, while the others reside on a connected server. The application is itself unaware of this virtualized and hence requires no modification. The operation of the thin client is in an environment that is isolated from the hosting OS, thereby solving compatibility issues. Multiple conflicting applications can be run simultaneously on the local machine. Also, encapsulation of the application from host provides better security solutions. This technique has huge applications in stock trades, banking, marketing and e-commerce.

7.3.4 Storage Virtualization

The client applications act upon the virtual storage management software which in turn operates on the physical storage resources. Direct-attached storage (DAS), storage area network (SAN) and network-attached storage (NAS) are used in virtualization. However, scalability is an important factor to be looked after in virtual storage solutions and a wide range of functionalities have to be offered by them to surpass the great variety offered by normal storage arrays [32].

7.3.5 Network Virtualization

Guest applications act on a virtual network managed by a virtual network client (VPN clients), which gives it a semblance of the original or a different network, while itself operating on the real physical network present at the node. It is especially advantageous when the perception of a different physical network is essential for the access to certain resources. It eliminates the requirement of network routing and the latency due to it. **VLAN** provides additional levels of security by restricting the transferring data to the network itself and secluding it from other networks.

7.4 Virtual Machine Provisioning

Virtual provisioning allocates memory space to devices as per demand. It is a technique used for the effective management of memory in a storage area, providing an illusion of increased storage capacity than the hardware actually provides. The actual storage capacity of the hardware is only provided when the data is actually written. In this way, economic interests are well served by reducing the idle storage present in the array. Therefore, it can be safely called one of the successful torch-bearers of computing. VM provisioning can be achieved in the following ways:

- Procuring the machine in a public cloud service such as Amazon EC2.
- Usage of private cloud management or a VM management present in the local data centre in order to procure the organization's virtual machine.

The VM life cycle:

- A request is generated and sent to the IT department mentioning the need for creating a new server.
- The request is thoroughly processed, by looking at the server's swarm of resources.
- Starting the groundwork of the requested VM.
- Once started, the virtual machine is prepared deliver the recommended service as per a service level agreement (**SLA**) [33, 34].

Process:

Server selection along with its operating system is the primary step in the process of virtual machine provisioning. Once the hardware is ensured, convenient software is loaded by using a configured template. The next step is customizing the machine to a valid network and resources for storage.

8 Migration Services

Cloud migration refers to the movement of data from the hardware of a client to the cloud data storage system. It also refers to the retrieval of this data from the cloud. It involves four principal phases which are definition, design, migration and management. There are a number of reasons why we should migrate into a cloud, chiefly due to economic and business purposes. The migration of applications might occur in multiple ways. The application might be independent or some of its programs must be altered. There might be conditions where the design is migrated to the cloud service environment. The migration of an enterprise is demonstrated as follows:

$$P - > P'_c + P'_l - > P'_{OFC} + P'_l \tag{1}$$

In the above, Prefers to the application preceding migration. P'_c refers to the application succeeding migration. P'_l is the portion of the application operating in the local data centre. P'_{OFC} is the portion of application efficiently optimized for the cloud [12].

8.1 Hot Migration

Hot migration refers to the movement of applications and the operating system from virtual machines to the hardware of a computer without hampering the operating system or application performance. It is useful as it frees as a particular server without providing any disadvantage to users, thereby providing smooth performance. It also prevents inefficient servers from causing discontinue in operations. This feat is achieved by hot migration by the distribution of workloads among different servers so that the functioning stays at an optimal level. The stages of hot migration are pre-migration, reservation, stop and copy, commitment and activation.

8.2 Cold Migration

Cold migration refers to the migration of non-functional virtual machines, which has been turned off. It is disadvantageous as it suspends applications and operating systems during the period of their transfer from the virtual machine to the hardware of a computer. With cold migration, we have the choice of shifting associated discs in between two data stores. It is also easier to implement. The purpose of using cold migration is to reduce complexity and allow the transfer to be more effective. However, the downtime might affect the performance of business and thereby, cold migration is a challenge faced by companies.

8.3 Migration Risks

Risks associated with migration can be broadly classified into the following:

- **General risks**—In general risks, we look into tuning and monitoring of performance.
- **Security risks**—There are a number of legal consents that migration has to abide by, which includes accessing execution logs and preserving the rights to review trails. The gravity of data leakage in the world of information technology in a cloud environment is recognized.

9   Load Balancing and Task Scheduling

Cloud utilizes distributed computing resources to manage its functions in a decentralized manner, making the proper handling and distribution of the enormous traffic an undeniable necessity. Scheduling of tasks in an effective manner aims at minimizing transmission delays and computational costs and maximizing resource utilization at the same time [35]. A number of algorithms referred to as scheduling algorithms are used for the attainment of this goal.

9.1   Scheduling of Virtual Machines

In a distributed system, it is always beneficial and profitable to have total workload divided between several nodes. Task scheduling deals with choosing the most suitable resource for the execution of a particular task. Non-uniform distribution of load to the VMs may lead to over-utilization or under-utilization of the resources, both of which lead to high energy consumption. But also, if the workload is dispensed uniformly, it may be that the number of host machines in use at particular instant is higher than that required to efficiently manage the tasks at hand, thus elevating the energy utilization again. Barring the idle and fully loaded conditions, the CPU alternates between these two conditions, at any instant. Average CPU utilization is given by the average of utilization in the idle and loaded conditions [36]. Thus, the VMs have to be allocated to the physical nodes in an optimized manner [37].

Each VM can exist in either of the two states—active or idle [38]. The energy consumption in the idle state of a VM is nearly 60% compared to its active state. Also, the physical server utilization in the active state is about 7–12%. As a result, multiple VMs may be mapped to a single server [39]. Makespan (MS) and energy consumption (EC) are two of the most important factors in assessing the performance of a VM. Makespan of a system is defined as the maximum time consumed by a VM to carry out all the tasks in the queue [40, 41]. Minimizing the makespan optimizes the performance of a load balancer. The response time (RT) of a VM also plays an important role. It is the total time taken to respond to a task, and hence, lower the response time of a VM, better its performance. A low RT is essential for an optimized MS [42].

9.1.1   Dynamic Scheduling

This scheduling technique refers to the distribution of dynamic workload across multiple servers and processors, to achieve efficient allocation of tasks. Unlike static scheduling, dynamic scheduling depends on the current state of the system and VMs along with the task queue [43, 44]. It reduces the traffic on a single server or processor resulting in a decrease in delays in receiving sending of data [45]. Resource utilization is maximized and this in turn brings down resource consumption. Throughput is maximized for a fixed response time by this technique.

Dynamic scheduling can further be broken down into centralized approach and distributed or decentralized approach [46].

9.1.2 Load Imbalance Factor

The summation of all the loads of all VMs is stated as:

$$L = \sum_{i=1}^{k} l_i \tag{2}$$

where i stand for the total VMs in a data centre. The load/unit capacity is defined as:

$$LPC = \frac{L}{\sum_{i=1}^{m} c_i} \tag{3}$$

Threshold

$$T_i = LPC * c_i \tag{4}$$

where c_i is the capacity of the node

The load imbalance factor of a particular VM is given by:

$$\begin{aligned}
&\text{If } V_m < T_i - \sum_{v=1}^{k} L_v, \quad \text{Underloaded} \\
&\quad\quad > T_i - \sum_{v=1}^{k} L_v, \quad \text{Overloaded} \\
&\quad\quad = T_i - \sum_{v=1}^{k} L_v, \quad \text{Balanced}
\end{aligned} \tag{5}$$

The transfer of load from the overloaded VM is carried out until its load is less than the threshold. The under-loaded VM can accept load only up to its threshold, thus avoiding it being overloaded [37].

A conventional load balancer algorithm combines these prototypical policies:

- **Information Policy**—It specifies the quantity of data to be made available to the decision making nodes. It stipulates how and when this information is collected [47].
- **Selection or Transfer Policy**—This takes into consideration the pre-existing load on the node as well as size of the workload which is currently to be assigned and determines if it is to be transferred to a least loaded or idle VM.
- **Resource type Policy**—It establishes the resources accessible during load balancing [46].

- **Placement or Location Policy**—It resolves the node to which the workload is to be transferred, utilizing the information from resource type policies [46, 47].

Metrics of efficient load balancing [46, 48]:

- High throughput
- Fault tolerant
- High scalability
- Minimal time to migrate workloads from one node to another
- Minimal time of response of the algorithm to the task.

Proper estimation of load, comparison between loads at different nodes, interaction between them and the size of the task in question are essential to the development of an efficient load balancing algorithms. There are plenty of algorithms that are suitable for the cloud environment. In the cloud environment, scheduling algorithms can be classified under two groups—Batch-Mode Heuristic Algorithms (BMHA) and Online-Mode Heuristic Algorithms (OMHA). Round Robin(**RR**), First Come First Served (FCFS), Min–Min and Max–Min algorithms fall under BMHA, where tasks are queued and assembled in a set and the scheduling algorithm begins operating on them after a fixed time interval (determined by the triggering policy in such cases.

9.2 Scheduling Algorithms

Some of the majorly used algorithms in this field are as follows-

9.2.1 Round Robin

This algorithm is based on the equal distribution of workload among the resources [49]. A task is allocated to a VM in a cyclic manner, without considering the priority of the task, task requirements, present workload on the VM or its handling capabilities [35]. This may lead to unbalancing as high priority or time-consuming tasks may end up with long response times. The weighted round-robin algorithm considers the VM capabilities and assigns tasks accordingly, but the task length and priority are still not considered.

9.2.2 Min–Min

It begins with a set of assigned tasks and computes the minimum time for all the given tasks on all the available resources. The one with the lowest execution time is chosen and assigned to the VM where it produced the lowest time [35, 49]. Correspondingly, the accessibility time period for that particular resource is updated for the remaining tasks in queue. This procedure is repeated until all the tasks are allocated. It is better for the execution of smaller tasks [50]. The Load Balance Improved Min–Min (LBIMM) algorithm divides the task into two groups based on high priority and low priority. The tasks on the high priority queue are scheduled first, followed by those on the low priority list [51].

9.2.3 Max–Min

Following the computation of the minimum execution time of each task, it chooses the one with maximum time (unlike Min–Min) which uses the task with lowest minimum time) and allocates it to the VM where it takes minimum time for execution. The scheduled task is then removed from the queue and the procedure is carried on for the remaining tasks [49, 51]. Contrary to Min–Min, it is more suitable for execution of lager tasks. It is, however, a better option when the number of small tasks is much higher than large ones [50].

9.2.4 MET (Minimum Execution Time)

It is a heuristic algorithm in which the scheduler maps a task to a VM based on the expected computation time and assigning the task with the least execution time to the VM [40, 41]. However, it does not consider the machine ready time which makes it unreliable and prone to production of non-optimized solutions [42].

9.2.5 MCT (Minimum Completion Time)

This heuristic algorithm takes into account both the machine ready time and the expected computation time. The task is allocated to the VM with least time of completion [42].

Some other widely used algorithms in task scheduling are First Come First Serve (FCFS), Opportunistic Load Balancing (OLB), Particle Swarm Optimization (PSO), Honey Bee Behaviour Based Load Balancing (HBB-LB), and Simulated Annealing (SA), Active Clustering and Shortest Response Time First.

10   Issues

Downsides are a chief characteristic of any technology, present or in theory. Some of the most common issues pertaining to cloud computing technology are as follows:

10.1 Security Issues

Security issues are chiefly categorized into two broad categories, the ones faced by the service providers and the other by the clients who fall back on cloud computing for their business. In order to prevent leakage of information belonging to a client, service providers have fallen back on systems and careful monitoring to protect the interests of the clients [52].

10.1.1 Data Integrity

Is one of the most sought after domains in cloud computing, and rightfully so, as the dependability of clients depends on it [53]. Data is stored in two groups—Paas or Saas environment and Iaas environment. In Iaas environment, the data is stored in cloud after undergoing encryption [54].

Data might be replicated among cloud's data centres, thereby making any changes in data visible elsewhere. Data integrity provides consistency, validity and regularity of data [54]. Denial of service (DoS) is another security issue faced by cloud computing. It affects the services of clients by sending a very large number of packets of data [55]. This might cause the hypervisor to disrupt its service. An efficient decentralized firewall might help prevent such an occurrence.

10.1.2 Malware Injections

Malware injections are snippets of code inserted into cloud services and work as software as a service. Access to user data can be obtained by this process [56, 57].

10.1.3 Cookie Poisoning

In the case of cookie poisoning, unauthorized access takes place by altering the contents of the cookie. Encrypting the data or clearing the cookie is a valid solution.

10.1.4 Malicious Insiders

This may come in the form of a business partner, a former or a current employee. These insiders misuse data in a cloud.

10.1.5 Traffic Hijacking

In the case of traffic hijacking, there is a misdirection of internet traffic towards a third party. The HTTPS protocol is used commonly to prevent traffic hijacking.

10.2 Data Handling

10.2.1 Dependability

As of the present scenario, there are numerous companies and growing businesses that rely on cloud to keep their efficiency at its zenith, along with minimizing expenditure. Cloud technologies have made itself reliable and dependable in the long run. Little knowledge is harboured regarding the fact where the data is actually stored in the servers.

10.2.2 Data Loss

Loss of useful data might prove to be quite a difficult situation for clients and business relying on cloud storage. A data wipe or malware might cause irreversible data loss [58]. In 2011, Amazon witnessed one of the major incidents of data loss.

10.2.3 Data Breaches

A data breach is an unwanted occasion in which confidential data is disclosed to the third party. It may or may not result in a data loss. However, a breach of data is always an unwanted incident by the client. It might lead to hamper of business by leakage of sensitive content [59].

10.2.4 Data Lock-In

Services to migrate from one service provider to another are not common and thus users may find themselves confined to a single provider. Data losses may be incurred on migration to another vendor.

10.3 Traditional Issues

Network and computer attacks constitute traditional issues. Some of the concerns constitute of:

- **Authentication**—Authentication frameworks do not extend inside a cloud naturally. This is an existing issue.
- **Phishing**—Phishing is a potential risk in a cloud environment.
- **VM attacks**—Potential weaknesses in the hypervisor paves the way for VM attacks.
- **Abuse of services in cloud**—Cloud services are mostly abused due to the free trial periods offered by numerous companies providing cloud services and also the unlimited cloud storage plans, as once hosted by Amazon.

10.4 Existing and Proposed Solutions

10.4.1 Data Privacy

Data privacy refers to the act of data abstraction, which means isolating information and display them selectively to the parties concerned. Oblivious RAM or **ORAM** is used extensively for the protection of data inside a cloud platform. ORAM keeps on shuffling memory continuously while being accessed by the user. It technically behaves as the interface between the physical RAM and the program.

10.4.2 Encryption

Due to the potential risk of data breach, it is unwise to store information on the cloud without any sort of security. Encryption thus becomes a necessity [60].

Homomorphic Encryption—It is a particular type of encryption where cipher-text computation is allowed [61].

Distributive Storage—A promising way to value data integrity is to store the data in multiple cloud storage.

Deletion Confirmation—Deletion confirmation refers to the non-availability of data after a user has deleted his data after the deletion confirmation. However, data is stored in multiple copies in a server or multiple servers in the cloud. Deletion does not erase all the copies at one go, giving rise to the challenge of illegal data recovery without valid permission of the client.

10.4.3 Debugging

Developers use the option of debugging in order to implement changes at a future stage. Debugging contributes a back door for developers. Some of the possible solutions to the other issues are Information Oriented Security, Encrypted Business Intelligence and Transparency.

10.4.4 NIST Standards

NIST plays a vital role in defining standards and conventions in cloud computing. The principal focus of NIST is to implement a synopsis on public cloud along with the involved security issues.

The privacy and security issues diagnosed by NIST are as following:

- Compliance
- Trust
- Governance
- Data protection
- Software and hardware construction
- Software isolation
- Access and identity management
- Incident response
- Availability.

10.4.5 Cloud Security Alliance

The CSA is an organization focusing on areas of security guidance in cloud computing [62]. The CSA guide deals with the following domains in cloud computing—cloud construction framework, legal aspects, business risk management, audit and compliance, portability, information cycle handling, electronic discover, traditional security, disaster, recovery and business continuity, incident response, remediation and notification, data centre operations, access management and identity, key management and encryption, application security, virtualization and storage.

11 Conclusion

This paper, as promised, has done an extensive literature review on cloud computing. The techniques pertaining to cloud and the different service models serving it are noteworthy. The cloud migration services have been mentioned along with the idea of scheduling algorithms. The potential of Xaas (anything as a service) is remarkable and a prospect of the next generation. The dependence of the IT industry on cloud computing obliges the latter to take a stride forward. Our study concludes that despite the security risks and other factors, cloud computing has immense potential and indispensable for the generations to come. In a world with decreasing resource per individual along with increasing demand, cloud solves the hardware problem pertaining to storage. Security and privacy challenges are real-time problem. We propose the introduction of blockchain technology to secure data by using the concept of sub-block rather than storing all data at central database. Keeping in mind the security challenges and issues, cloud technology and its use must be developed to provide clients with a safe and dependable working platform.

References

1. Marston, Sean, Li, Zhi, Bandyopadhyay, Subhajyoti, Zhang, Juheng, Ghalsasi, Anand: Cloud computing—the business perspective, Decis. Support Syst. **51**(1), 176–189 (2011)
2. Gourisaria, M.K., Patra, S.S., Khilar, P.M.: Energy saving task consolidation technique in cloud centers with resource utilization threshold. Adv Intell Syst Comput (2018)
3. Gourisaria, M.K., Patra, S.S., Khilar, P.M.: Minimizing energy consumption by task consolidation in cloud centers with optimized resource utilization. Int. J. Electr. Comput. Eng. (IJECE) **6**(6), 3283–3292 (2016)
4. Galante, G., de Bona, L.C.E.: A survey on cloud computing elasticity. In: 2012 IEEE/ACM Fifth International Conference on Utility and Cloud Computing, pp. 263–270
5. Srivastava, A.: A detailed literature review on cloud computing. Asian J. Technol. Manag. Res. **4**(2) (2014)
6. Gai, K., Li, S.: Towards cloud computing: a literature review on cloud computing and its development trends. In: Proceedings of the Fourth International Conference on Multimedia Information Networking and Security, pp. 142–146 (2012)
7. Gong, C., Liu, J., Zhang, Q., Chen, H., Gong, Z.: The characteristics of cloud computing. In: International Conference on Parallel Processing Workshops (2010)
8. Gaurav, N.K., Kumar, J.: A literature survey on mobile cloud computing: open issues and future directions. Int. J. Eng. Comput. Sci. **3**(5) (May 2014)
9. Gupta, P., Seetharaman, A., Raj, J.R.: The usage and adoption of cloud computing by small and medium businesses. Int. J. Inf. Manag. **33**(5), 861–874 (2013)
10. Sharma, R., Trivedi, R.K.: Literature review: cloud computing—security issues, solution and technologies. Int. J. Eng. Res. **3**(4), 221–225 (2014)
11. Mell, P., Grance, T.: The NIST definition of cloud computing. In: National Institute of Standards Technology, Information Technology Laboratory, Technical Report Version 15 (2009)
12. Buyya, R., Broberg, J., Goscinski, A.: Cloud computing principles and paradigms. Wiley
13. Goyal, S.: Public versus private versus hybrid vs community—cloud computing: a critical review. Int. J. Comput. Netw. Inf. Secur. (2014)
14. Padhy, R.P., Patra, M.R., Satapati, S.C.: Cloud computing: security issues and research challenges. Int. J. Comput. Sci. Inf. Technol. Secur. (IRASCT) **1**(2) (2011)
15. Mishra, S., Pandey, M.: Impact of security risk on cloud computing adoption. Int. J. Eng. Comput. Sci. **5**(10) (October 2016)
16. Nagendra Babu, P., Chaitanya Kumari, M., Venkat Mohan, S.: A literature survey on cloud computing. Int. J. Eng. Trends Technol. (IJETT) **21**(6) (March 2015)
17. Mishra, N., Siddiqui, S., Tripathi, J.P.: A compendium over cloud computing cryptographic algorithms and security issues. BIJIT-BVICAM's Int. J. Inf. Technol. (November 2014)
18. Menzel, M., Warschofsky, R., Thomas, I., Willems, C., Meinel, C.: The service security lab: a model-driven platform to compose and explore service security in the cloud. In: IEEE 2010 6th World Congress on Services (July 2010)
19. Yang H., Tate, M.: A descriptive literature review and classification of cloud computing research. In: Communications of the Association for Information Systems (July 2012)
20. Sivathanu, S., Liu, L., Yiduo, M., Pu, X.: Storage management in virtualized cloud environment. In: 2010 IEEE 3rd International Conference on Cloud, Computing (2010)
21. Swathi, T., Srikanth, K., Raghunath Reddy, S.: Virtualization in cloud computing. Int. J. Comput. Sci. Mobile Comput. **3**(5), 540–546 (2014)
22. Li, Q., Hao, Q., Xiao, L., Li, Z.: Adaptive management of virtualized resources in cloud computing using feedback control. In: IEEE 2009 1st International Conference on Information Science and Engineering (ICISE) (December 2009)
23. Lee, J.H., Park, M.W., Eom, J.H., Chung, T.M.: Multi-level intrusion detection system and log management in cloud computing. In: International Conference on Advanced Communication Technology (ICACT)

24. Loganayagi, B., Sujatha, S.: Creating virtual platform for cloud computing. In: IEEE International Conference on Computational Intelligence and Computing Research (ICCIC) (December 2010)
25. Uhlig, R., Neiger, G., Rodgers, D., Santoni, A.L., Martins, F.C.M., Andersons, A.V., Bennett, S.M., Kagi, A., Leung, F.H., Smith, L.: Intel virtualization technology. Computer 38(5), 48–56 (2005)
26. Fayyad-Kazan, H., Perneel, L., Timmerman, M.: Full and para virtualization with Xen: a performance comparison. J. Emerg. Trends Comput. Inf. Sci. 4(9), 719 (2013)
27. Barham, P., Dragovic, B., Fraser, K., Hand, S., Harris, T., Ho, A., Neugebauer, R.: Ian Pratt and Andrew warfield, Xen and the art of virtualization. ACM SIGOPS Oper. Syst. Rev. SOSP'03 37(5), 164–177 (2003)
28. Abels, T., Dhawam, P., Chandrasekaran, B.: An overview of Xen virtualization [Online]. Available on http://www.dell.com/downloads/global/power/ps3q05-20050191-abels.pdf
29. VMWare: Understanding full virtualization, Para virtualization and hardware assist [Online]. http://www.vmware.com/files/pdf/VMware_paravirtualization.pdf (2007)
30. Rodriguez-Haro, F., Freitag, F., Navarro, L., Hernandez-Sanchez, E., Farias-Mendoza, N., Guerrero-Ibanez, J.N., Gonzalez-Potes, A.: A summary of virtualization techniques. In: The 2012 Iberoamerican Conference on Electronics Engineering and Computer Science, Procedia Technology 3, pp. 267–272 (December 2012)
31. Kumar, R., Charu, S.: An importance of using virtualization technology in cloud computing. Global J. Comput. Technol. 1(2) (2015)
32. Durairaj, M., Kannan, P.: A study on virtualization techniques and challenges in cloud computing. Int J. Sci. Technol. Res. 3(11) (2014)
33. El-Refaey, M.: Virtual machines provisioning and migration services. Available on http://research.iaun.ac.ir/pd/faramarz_safiold/pdfs/HomeWork_6541.pdf
34. Kandukuri, B.R., Ramakrishna Paturi, V., Rakshit, A.: Cloud security issues. In: International Conference on Services, Computing, pp. 517–520 (2009)
35. Khan, D.H., Kapgate, D.: Efficient virtual machine scheduling in cloud computing. Int. J. Comput. Sci. Mobile Comput. (IJCSMC) 3(5), 444–453 (2014)
36. Liu, H·A measurement study of server utilization in public clouds. In: IEEE Ninth Conference on Dependable, Autonomic and Secure Computing, pp. 435–442, 2011
37. Chitra Devi, D., Rhymend Uthariaraj, V.: Load balancing in cloud computing environment using improved weighted round robin algorithm for non-preemptive dependent tasks. Sci. World J. (2016)
38. Gondhi, N.K., Sharma, A.: Local search based ant colony optimization for scheduling in cloud computing. In: IEEE Second International Conference on Advances in Computing and Communication Engineering (2015)
39. Zhong, Z., Chen, K., Zhai, X., Zhou, S.: Virtual machine-based task scheduling algorithm in a cloud computing environment. Tsinghua Sci. Technol. 21(6) (December 2016)
40. Maheswaran, M., Ali, S., Jay Siegel, H., Hensgen, D., Freund, R.F.: Dynamic mapping of a class of independent tasks onto heterogeneous computing systems. J. Parallel Distrib. Comput. 59, 107–113 (1999)
41. Braun, T.D., Jay Siegel, H., Beck, N., Boloni, L.L., Maheswaran, M., Reuther, A.I., Robertson, J.P., Theys, M.D., Yao, B.: A comparison of eleven static heuristics for mapping a class of independent tasks onto heterogeneous distributed computing systems. J. Parall. Distrib. Comput. 61, 810–837 (2001)
42. Mishra, S.K., Sahoo, B., Parida, P.P.: Load balancing in cloud computing: a big picture. J. King Saud Univ. Comput. Inf. Sci. (2018)
43. Singh, R.M., Paul, S., Kumar, A.: Task scheduling in cloud computing: review. Int. J. Comput. Sci. Inf. Technol. 6(6), 7940–7944 (2014)
44. Al Nuaimi, K., Mohamed, N., Al Nuaimi, M., Al-Jaroodi, J.: A survey of load balancing in cloud computing: challenges and algorithms. In: 2012 Second Symp. Netw. Cloud Comput. Appl. (December 2012)

45. Lepakshi, V.A., Prashant, C.S.R.: A study on task scheduling algorithms in cloud computing. Int. J. Eng. Innov. Technol. **2**(11), 119 (2013)
46. Kaur, R., Luthra, P.: Load balancing in cloud computing. In: Proceeding of the International Conference on Recent Trends in Information, Telecommunication and Computing, ITC (2014)
47. Acharya, J., Mehta, M., Saini, B.: Partical swarm optimization based load balancing in cloud computing. In: 2016 International Conference on Communication and Electronics Systems (ICCES) (October 2016)
48. Dam, S., Mandal, G., Dasgupta, K., Dutta, P.: An ant-colony-based meta-heuristic approach for load balancing in cloud computing. In: Chapter—9, Applied Computational Intelligence and Soft Computing in Engineering, IGI Global, pp. 204–232 (2018)
49. Suresh Kumar, D., George, E., Raj, D.P.: A literature review on load balancing mechanisms in cloud computing. Int. J. Adv. Res. Comput. Sci. **9**(1), 1 (2018)
50. Elzeki, O.M., Reshad, M.Z., Elsoud, M.A.: Improved max-min algorithm in cloud computing. Int. J. Comput. Appl. **50**(12) (July 2012)
51. Sindhu, S.: Task scheduling in cloud computing. Int. J. Adv. Res. Comput. Eng. Technol. (IJARCET), **4**(6) (2015)
52. Abouelmehdi, K., Dali, L., Abdelmajid, E., Elsayed, H., Fatiha, E., Abderahim, B.: Classification of Attaks over cloud environment. World Acad. Sci. Eng. Technol. Int. J. Human Soc. Sci. **9**(6) (2015)
53. Kumar, S.R., Saxena, A.: Data integrity proofs in cloud storage. In: International Conference on Communication Systems and Networks, pp. 1–4 (COMSNETS)
54. Chaudhary, J., Mishra, A.: Literature review: cloud computing—security issues and data encryption schemes. MIT Int. J. Comput. Sci. Inf. Technol. **6**(1) (2016)
55. Backe, A., Lindén, H.: Cloud computing security: a systematic literature review. Uppsala University, Department of Informatics and Media [Online]. Available on https://www.diva-portal.org/smash/get/diva2:825307/FULLTEXT01.pdf
56. Khalil, Issa M., Khreishah, Abdallah, Azeem, Muhammad: Cloud computing security: a survey. Computers **3**(1), 1–35 (2014)
57. Rahman, M., Cheung, W.M.: Analysis of cloud computing vulnerabilities. Int. J. Innov. Sci. Res. **2**(2) (2014)
58. Reddy, S.R., Mohan, Y.R., Naik, J.S.: An overview of cloud computing and security issues. Int. J. Sci. Eng. Appl. Sci. **1**(5) 2015
59. Kanthe, R.R., Patel, R.C.: Data security and privacy protection issues in cloud computing. Int. J. Comput. Sci. Inf. Technol. Res. **3**(2) (2015)
60. Li, J., Zhao, G., Chen, X., Xie, D., Rong, C., Li, W., Tang, L., Tang, Y.: Fine-grained data access control systems with user accountability in cloud computing. In: IEEE International Conference and Workshops on Cloud Computing Technology and Science (CLOUDCOM) (2010)
61. Sun, Y., Zhang, J., Xiong, Y., Zhu, G.: Data security and privacy in cloud computing. Int. J. Distrib. Sensor Netw. (2014)
62. El-Gazzar, R.F.: A literature review on cloud computing adoption issues in enterprises. In: Bergvall-Kareborn, B., Nielson, P.A.: (eds.) Creating value for All through IT. TDIT 2014. IFIP Advances in Information and Communication Technology, vol. 429. Springer, Berlin, Heidelberg (2014)

Performance Comparison of Filter-Based Approaches for Display of High Dynamic Range Hyperspectral Images

R. Ragupathy and N. Aswini

Abstract This paper presents performance comparison of nonlinear edge-preserving filters namely bilateral filter and weighted least square filter on improved visualization for display of high dynamic range hyperspectral image on low dynamic range media. This filter is used to decompose image into components of base and detail. The contrast of base image is reduced and details are preserved in detail component. Image construction is made by combining these two components. Performance analysis revealed that the bilateral filter-based approach performs better than the conventional average method and the weighted least square filter in terms of normalized average gradient and root mean square contrast.

Keywords Bilateral filter · Contrast reduction · High dynamic range images · Hyperspectral imaging · Low dynamic range media · Nonlinear filters · Visualization · Weight least square filter

1 Introduction

Hyperspectral remote sensing is the combination of spectral imaging and digital imaging, and it aims to obtain the spectrum for each and every pixel of the image. This provides enormous information about the objects and also gives spectral information about each pixel of the image. A common method for a quick overview of hyperspectral data is an RGB color representation; selecting three image bands from a hyperspectral cube randomly considers it to RGB color space. But in this

R. Ragupathy (✉)
Department of Computer Science and Engineering, Faculty of Engineering and Technology, Annamalai University, Annamalainagar, Tamil Nadu 608002, India
e-mail: cse_ragu@yahoo.com

N. Aswini
Division of Computer and Information Science, Faculty of Science, Annamalai University, Annamalainagar, Tamil Nadu 608002, India
e-mail: aswini.it03@gmail.com

© Springer Nature Singapore Pte Ltd. 2020
K. S. Raju et al. (eds.), *Data Engineering and Communication Technology*,
Advances in Intelligent Systems and Computing 1079,
https://doi.org/10.1007/978-981-15-1097-7_7

method, huge information loss occurs. Many research works are carried out in recent days to display high dynamic range (HDR) images, having high contrast, on low dynamic range (LDR) display since the cost of HDR display is very high. If the contrast of HDR image is reduced and directly used for display, then the information loss will be occurred in brighter and darker areas. So that, a tone mapping method creates an HDR depiction of the image on the LDR screens. We know that our human vision is sensitive to local contrast so the local tone mapping using nonlinear filter gives pleasant visualization of data without information loss.

The foremost method used to display the hyperspectral image is to calculate the principal component analysis basis functions for the hyperspectral dataset, and the result is mapped to RGB color space. In that method, there are several issues such as pseudo colors and it has high complexity [1]. It is the same for the independent component analysis method [2]. Another important problem in hyperspectral image is dimensionality reduction. Using constant-Luma border basis dimensionality reduction technique and signal-to-noise ratio adaptation technique in [3] also has the same problem, and they are linear methods. The spectral fusion using weight obtained through bilateral filtering method was proposed in [4]. In this method, the weight of bands is determined according to its amount of detail included in it. Some of the band selection methods such as one-bit transform, normalized information, minimal abundance covariance, and linear prediction are also used to select the suitable bands for visualization in spite of information loss.

Recently, researchers work with the edge-preserving filters to display the HDR images. Among them, anisotropic diffusion is one of the important methods. Anisotropic diffusion is inspired by an interpretation of gaussian blur as a least conduction partial differential equation. Perona and Malik introduced an edge-preserving filtering technique [5]; its discrete diffusion nature makes it a slow process. Even more recently, Xudong Kang et al. used the edge-preserving filters for improving the accuracy of hyperspectral image classification [6]. As an alternative to anisotropic diffusion, bilateral filtering was developed by Tomasi and Manduchi [7], which motivated us to carry out this work. We use nonlinear edge-preserving filters for decomposing the image into the base and the detail components as in [8]. This paper examines a simple and effective method for the visualization of hyperspectral images which is motivated by the techniques of HDR image processing presented in [7, 9].

The rest of the paper is organized as follows: Sect. 2 discusses the basics of both nonlinear edge-preserving filters, namely bilateral filter and WLS. The detailed framework for displaying HDR images in LDR media is presented in Sect. 3. Experimental results are discussed in Sect. 4. Finally, the conclusion is made in Sect. 5.

2　Nonlinear Filters

2.1　Bilateral Filter

Tomasi and Manduchi introduced a bilateral filter called nonlinear edge-preserving filter which calculates the output as a weighted sum of input [7]. The weights depend on the intensity of the pixels, and it is also an edge stopping function. The bilateral filter is formulated as in Eq. (1).

$$J(x) = \frac{1}{k(x)} \sum_{\xi} f(x, \xi) g(I(\xi) - I(x)) I(\xi) \tag{1}$$

where the pixel within the region of support is denoted as ξ. Gaussian operating in the spatial domain is represented as f. Gaussian operating in the intensity domain is denoted as g. In terms of spatial distance, f ensures that pixels closer get higher weights, whereas lower weights are assigned to the pixels which are further away. Similarly, in terms of intensity distance, g ensures that pixels closer get higher weights while pixels with large distance get lower weights. Here, $k(x)$ is normalization factor expressed as in Eq. (2).

$$k(x) = \sum_{\xi} f(x, \xi) g(I(\xi) - I(x)) \tag{2}$$

2.2　Weighted Least Square Filter

The concept of weighted least square (WLS) filter was introduced by Zeev et al. From an input image g, we get new image u, which is as close as g, and everywhere, apart from significant gradients in g, as even as possible. Here, u is obtained by applying a nonlinear operator F_λ on g as in Eq. (3).

$$u = F_\lambda(g) = (I + \lambda L_g)^{-1} g \tag{3}$$

The operator F_λ varies spatially so its frequency response is difficult to analyze [10]. As smoothness weights are roughly equal, Eq. (3) is changed to Eq. (4).

$$F_\lambda(g) \approx (I + \lambda L_g)^{-1} g \tag{4}$$

where $L = D_x^T D_x + D_y^T D_y$ is an ordinary Laplacian matrix. Eq. (5) gives the frequency response of F_λ as in [11].

$$\mathcal{F}_\lambda(\omega) = 1/(1 + a\lambda\omega^2) \tag{5}$$

Thus, in the frequency domain, scaling by a factor of c is equivalent to multiplying a factor of c^2 by λ as in Eq. (6).

$$\mathcal{F}_\lambda(c\omega) = 1/\left(1 + ac^2\lambda\omega^2\right) = \mathcal{F}_{c^2\lambda}(\omega) \tag{6}$$

3 Framework

Initially, bands of R, G, and B channels are extracted from the HDR hyperspectral imagery and corresponding summation of R, G, and B channels is computed. These values are used to calculate HDR intensity channel which is then converted into log domain. Then, the nonlinear edge-preserving filter is applied in the log domain to decompose the image into the base and detail components, respectively. Next, the contrast reduction is applied only in the base component. Afterward, the contrast reduced base component and retained detail component are combined to forming the image in the log domain. Finally, image construction is carried out after the conversion from the log to the intensity domain of the formed image. The block diagram of the framework of filter-based approach is shown in Fig. 1.

3.1 Summation of RGB Channels

The hyperspectral imagery has several numbers of bands which are combined together for display purpose. The HDR hyperspectral image bands of R, G, and B channels are constructed as the sum of bands falling on the electromagnetic spectrum's of red, green, and blue wavelength shown in Eq. (7).

$$R_{\text{HDR}} = \sum_{t \in R} \text{HSI}(t); \quad G_{\text{HDR}} = \sum_{t \in G} \text{HSI}(t); \quad B_{\text{HDR}} = \sum_{t \in B} \text{HSI}(t) \tag{7}$$

where t is the band index of the hyperspectral imagery. $R_{\text{HDR}}, G_{\text{HDR}}, B_{\text{HDR}}$ are the HDR color channels are constructed as the sum of bands that fall within the hyperspectral imagery's similar wavelength range.

3.2 RGB to Intensity Domain

The HDR intensity channel is calculated as the RGB color channel's average sum as shown in Eq. (8).

$$I_{\text{HDR}} = (R_{\text{HDR}} + G_{\text{HDR}} + B_{\text{HDR}})/3 \tag{8}$$

Fig. 1 Framework of
filter-based approach

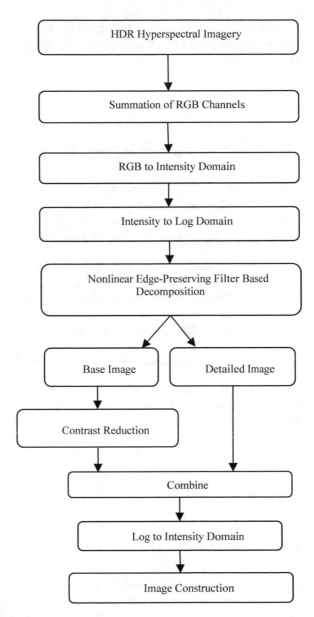

Fig. 1 Framework of
filter-based approach

3.3   *Intensity to Log Domain Conversion*

In the intensity domain of color images, contrast reduction is performed to preserve color balance. Contrast reduction in intensity of pixels decomposes color after contrast reduction. If HDR image processing with a low-pass nonlinear edge-preserving filter is done in intensity domain, the filter will remove high frequencies while preserving high contrast edge. Usually, contrast reduction is performed in logarithmic domain as per Eq. (9) since contrast is a multiplicative effect.

$$I_{\log} = \log 10(I) \tag{9}$$

3.4 Nonlinear Edge-Preserving Filter-Based Decomposition

Nonlinear edge-preserving filter like bilateral and WLS filter used to break down the HDR image into a component of base and detail. The base component is obtained through nonlinear edge-preserving filtering of the original image. Thus, it consists of preserved edges and smooth parts. This operation is denoted in Eq. (10).

$$I_{\text{base}} = \text{NLEPF}(I) \tag{10}$$

Here, NLEPF() represents nonlinear edge-preserving filter.

The difference between the original image and the base component is named as detail component which is denoted in the form of Eq. (11).

$$I_{\text{detail}} = I - I_{\text{base}} = 1 - \text{NLEPF}(I) \tag{11}$$

But here, the nonlinear edge-preserving filter is used to acquire the base and detail component from the log intensity domain of image in the form of Eqs. (12) and (13).

$$I_{\log,\text{base}} = \text{NLEPF}(I_{\log}) \tag{12}$$

$$I_{\log,\text{detail}} = I_{\log} - I_{\log,\text{base}} = 1 - \text{NLEPF}(I_{\log}) \tag{13}$$

3.5 Contrast Reduction

The contrast must be reduced to display HDR images on conventional equipment. If this is accomplished directly on the original image, due to the process of contrast reduction, the image detail is lost. The contrast reduction is applied only for the base component as defined in Eq. (14). By retaining the detail component as same, the detail loss is reduced.

$$\text{RI}_{\log,\text{base}} = \text{cf} * I_{\log,\text{base}} \tag{14}$$

Here, cf denotes the compression factor, which is used to reduce the contrast of the image.

3.6 Combine

The base component contrast is lowered. The detail component's contrast is maintained. Therefore, an enhanced contrast reduction with retained detail is achieved with preserved detail as in Eq. (15).

$$I_{\text{log,new}} = \text{RI}_{\text{log,base}} + I_{\text{log,detail}} \tag{15}$$

3.7 Log to Intensity Domain Conversion

For high dynamic range images, the companding is assumed to be applied in the log domain; i.e., the original image has gone through a log transformation before they processed for final image construction. When the sub-bands are summed, they produce a distorted signal. So the ratio of blur images is more effective when computed in the log intensity domain. The artifacts are small errors in local contrast within sub-bands, and these are not visually disturbing. So that, the artifacts do not take the form of visible contours. Even though, the visible difference between the original and the companded image is notable. Hence, the image is converted from log domain to linear domain as given in Eq. (16).

$$I_{\text{new}} = 10^{I_{\text{log,new}}} \tag{16}$$

3.8 Image Construction

The multi-scale techniques sometimes have the reputation of being difficult to use without introducing halo artifacts. The I_{HDR} intensity channel is used to obtain the new intensity channel with reduced contrast and preserved detail. I_{LDR} is denoted as I_{new} in Eq. (16). The final contrast reduced hyperspectral image is obtained by combining the new intensity channel with the color components, which is used for display. It is possible to obtain a pixel-wise ratio factor as in Eq. (17).

$$\text{Ratio} = I_{\text{LDR}}/I_{\text{HDR}} \tag{17}$$

The final low dynamic range color channels are obtained using the ratio factor as given in Eq. (18).

$$R_{\text{LDR}} = \text{Ratio} * R_{\text{HDR}}; \quad G_{\text{LDR}} = \text{Ratio} * G_{\text{HDR}}; \quad B_{\text{LDR}} = \text{Ratio} * B_{\text{HDR}}; \tag{18}$$

4 Experimental Results and Discussion

Datasets from various hyperspectral sensors namely airborne visible/infrared imaging spectrometer (AVIRIS) [12], reflective optics system imaging spectrometer (ROSIS) [13], and hyperion [14] are taken for experimentations. These dataset specifications are given in Table 1.

For all the datasets, normalized average gradient (NAG) and root mean square (RMS) are computed and tabulated in Table 2 for all three methods. For average display method, to obtain the color channels, the corresponding RGB wave bands are simply averaged using equal weights. It contains more contrast to the brighter and the darkest areas so that the information loss occurs. The edges will be lost if reduction in contrast is applied globally. Hence, the image contrast is reduced

Table 1 Dataset specifications

Hyperspectral sensor	Dataset name	Size (rows × columns × bands)	Wavelength ranges/ band numbers
AVIRIS	Kennedy Space Centre	512 × 614 × 176	Red: 610 nm to 700 nm/22–31 Green: 500 nm to 570 nm/11–18 Blue: 450 nm to 500 nm/6–11
	Indian Pines	145 × 145 × 220	
	Salinas	512 × 217 × 224	
	Salinas-A	83 × 86 × 224	
ROSIS	Pavia Centre	1096 × 715 × 102	Red: 614 nm to 718 nm/46–72 Green: 506 nm to 574 nm/19–36 Blue: 454 nm to 502 nm/6–18
	Pavia University	610 × 340 × 103	
Hyperion	Botswana	1476 × 256 × 145	Red: 610 nm to 700 nm/22–31 Green: 500 nm to 570 nm/11–18 Blue: 450 nm to 500 nm/6–11

Table 2 Performance evaluation

Method	Metrics	Kennedy Space Center	Indian Pines	Salinas	Salinas-A	Pavia Center	Pavia University	Botswana
Average	NAG	0.066	0.2095	0.0459	0.0138	0.1221	0.1092	0.1619
	RMS	0.2940	0.3716	0.3716	0.1076	0.7346	0.5627	0.5753
Bilateral	NAG	0.0706	0.2443	0.0519	0.0788	0.1299	0.1299	0.1830
	RMS	0.2027	0.2916	0.2118	0.2125	0.4038	0.4038	0.3390
WLS	NAG	0.0668	0.2157	0.0497	0.0784	0.1223	0.1090	0.1723
	RMS	0.2812	0.2729	0.2089	0.2442	0.3871	0.2495	0.3262

locally so that the output image is better suited for the visualization of images without loss of edges. Hence, high contrast edges are preserved using the nonlinear filters like bilateral and WLS. For sample, visualization results of all three methods for Salinas and Botswana are presented in Figs. 2 and 3, respectively. The results of average method, bilateral filter method, and WLS filter method are compared using normalized average gradient which measures the clarity of the image [15] and the root mean square contrast which measures the dynamic range of the image. The

Fig. 2 Visualization results of Salinas

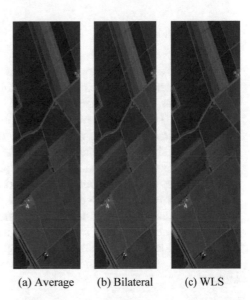

(a) Average   (b) Bilateral   (c) WLS

Fig. 3 Visualization results of Botswana

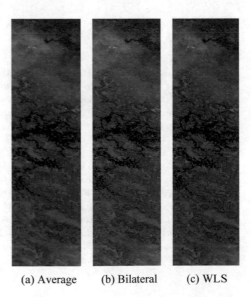

(a) Average   (b) Bilateral   (c) WLS

bilateral filter method provides more visualization than WLS method. The constructed images are close to the images acquired using a color camera for ROSIS sensor data because the bands cover 80% of the visible spectrum data. Hence, it is more suitable for visual interpretation. The result of AVIRIS and hyperion sensors datasets does not correspond to the normal colors; it may produce pseudo colors because both sensors cover visible and short-wave infrared and medium-wave infrared wavelength.

5 Conclusion

This paper examines and compares the results of three methods, namely average method, bilateral filter method, and weight least square method. It has been addressed that the issues such as loss of edges, computational complexity, and visualization. Unlike low dynamic range images, high dynamic range images can contain detailed information in the deepest shadows and the brightest light sources at the same time. Also, they capture the full range of luminance present in real-world scenes. Low dynamic range depictions of high dynamic range images avoid the cost of loss of contrast information by using nonlinear edge-preserving filter. The normalized average gradient and root mean square contrast values show that the bilateral method performs better than weighted least square filter method and average method. As a concluding remark, it can be stated that the bilateral method provides more visualization of high dynamic range hyperspectral images compared to weighted least square filter which is better than average display method.

References

1. Tyo, J.S., Konsolakis, A., Diersen, D.I., Olsen, R.C.: Principal-components-based display strategy for spectral imagery. IEEE Trans. Geosci. Rem. Sens. **41**(3), 708–718 (2003)
2. Du, H., Hairong, Q., Xiaoling, W., Rajeev, R., Wesley, E.S.: Band selection using independent component analysis for hyperspectral image processing. In: 32nd Applied Imagery Pattern Recognition Workshop, pp. 93–98 (2003)
3. Jacobson, N.P., Gupta, M.R.: SNR-adaptive linear fusion of hyperspectral images for color display. IEEE Int. Conf. Image Process. **3**, 477–480 (2007)
4. Kotwal, K., Chaudhuri, S.: Visualization of hyperspectral images using bilateral filtering. IEEE Trans. Geosci. Remote Sens. **48**(5), 2308–2316 (2010)
5. Perona, P., Malik, J.: Super-scale and edge detection using anisotropic diffusion. IEEE Trans. Geosci. Remote Sens. **12**(7), 629–639 (1990)
6. Kang, X., Li, S., Benediktsson, J.A.: Spectral-spatial hyperspectral image classification with edge preserving filtering. IEEE Trans. Geosci. Rem. Sens. **52**(5), 2666–2677 (2014)
7. Tomasi, C., Manduchi, R.: Bilateral filtering for gray and color images. In: IEEE International Conference on Compute Vision, pp. 839–846 (1998)

8. He, K., Sun, J., Tang, X.: Guided image filtering. IEEE Trans. Patt. Anal. Mach. Intell. **35**(6), 1397–1409 (2013)
9. Farbman, Z., Fattal, R., Lischinski, D., Szeliski, R.: Edge-preserving decompositions for multi-scale tone and detail manipulation. ACM Trans. Grap. (TOG) **27**(3), 67 (2008)
10. Fattal, R., Agrawala, M., Rusinkiewicz, S.: Multiscale shape and detail enhancement from multi-light image collections. ACM Trans. Grap. (TOG) **26**(3), 51 (2007)
11. Oppenheim, A.V., Schafer R.W.: Discrete-Time Signal Processing. Pearson Education India (1999)
12. Green, R.O., Eastwood, M.L., Sarture, C.M., Chrien, T.G., Aronsson, M., Chippendale, B.J., Faust, J.A., Pavri, B.E., Chovit, C.J., Solis, M., Olah, M.R.: Imaging spectroscopy and the airborne visible/infrared imaging spectrometer (AVIRIS). Rem. Sens. Environ. **65**(3), 227–248 (1998)
13. Bach, E.: Optoelectronic imaging spectrometers: german concepts for remote sensing. Int. Arch. Photogramm. Rem. Sens. **29**, 138–144 (1993)
14. Ham, J., Chen, Y., Crawford, M.M., Ghosh, J.: Investigation of the random forest framework for classification of hyperspectral data. IEEE Trans. Geosci. Rem. Sens. **43**(3), 492–501 (2005)
15. Ertürk, S., Süer, S., Koç, H.: A high-dynamic-range-based approach for the display of hyperspectral images. IEEE Geosci. Remote Sens. Lett. **11**(11), 2001–2004 (2014)

Survey on Ontology-Based Sentiment Analysis of Customer Reviews for Products and Services

Sumalatha Bandari and Vishnu Vardhan Bulusu

Abstract In recent years, the use of Internet and World Wide Web increases the research on sentiment analysis. Few years ago, whenever customer wanted to buy any product or to use any services, he/she is used to take the review from those who are already using that products or services. In recent years, various e-commerce sites are available from where one can buy the products and get the customer reviews of various products and services. Ontology plays an important role to analyze the product reviews based on the semantics of the reviews. The people living around the world are not familiar with the English language. Various e-commerce sites are available on the Internet in different languages. The customers write the reviews of the products or services they purchased/used in their own local languages in the e-commerce sites. The main aim of this paper is to conduct the survey on the sentiment analysis of customer reviews present in different languages using ontology. Also mention the problems which exist in those languages to analyze the products and services' reviews. An experiment is performed on the sentiment analysis of Hindi online customer reviews of the mobile phones using ontology which tries to minimize the challenges occurred in the earlier systems.

Keywords Ontology · Product reviews · Sentiment analysis · Hindi language

1 Introduction

Nowadays, the opinions of the various products are observed in the blogs, social networking sites, e-commerce sites, and online forums. Whenever a customer wants [1] to take the decision for purchasing of any product, the opinion plays an

S. Bandari (✉)
Research Scholar, Computer Science and Engineering, JNTU Hyderabad,
Hyderabad, India
e-mail: sumaabandari@gmail.com

V. V. Bulusu
Professor, Department of CSE, JNTUH College of Engineering, Manthani, India
e-mail: mailvishnu@jntuh.ac.in

© Springer Nature Singapore Pte Ltd. 2020
K. S. Raju et al. (eds.), *Data Engineering and Communication Technology*,
Advances in Intelligent Systems and Computing 1079,
https://doi.org/10.1007/978-981-15-1097-7_8

important role. Before purchasing any new product, the customer has to refer the sentiments or opinions of the people who are already using that product. The reviews of the products given by the customers are very much helpful for the companies to know the positive and negative points of their products in order to improve the quality of the products. The customers who purchased [2] the products will share their experience about that product online in the form of textual reviews. Reviews are often useful for the buyers for purchasing [3] products. The emotions present in the reviews regarding the product features provide in detail in understanding the reviews and also help in the analysis of the feature sentiments of the product. It is difficult for the users to extract [4] the meaningful information from the opinion reviews.

The customers are expressing their opinions of the products with the help of text in their natural language. The mining is required to analyze the opinions of the customer. The research work on English language opinion mining is conducted in the very fast rate, but the other language opinion mining research is performed in the moderate speed.

The domain knowledge in the form of hierarchy of concepts is represented by the 'Ontology.' Ontology model contains [5] objects, classes, properties of an object and the relation between the objects. Web Ontology Language (OWL) or Resource Description Language (RDF) formats are used to store the ontology model. AspectMention, SentimentMention, and SentimentValue are the three main classes of ontology [6]. The Web documents contain lack of semantic information which is understandable by humans but not the machines. OWL/RDF is the semantic Web technologies which were used to handle the challenges in key word-based search. Document is annotated [7] based on the semantic information from the domain ontology was the purpose of semantic Web. The natural language processing (NLP) techniques are not suitable for unstructured languages because morphology and part-of-speech (POS) tagging is the challenging task in the NLP. Each customer review is separately analyzed [8] using ontology with the extended keyword search.

Opinion mining is also called as sentiment analysis. The process of determining [9] the polarity of the reviews like positive, negative, or neutral is called opinion mining or sentiment analysis. Ontologies are used in different fields like scientific and e-commerce. The knowledge extraction was performed [10] at four levels in the opinion mining, which are word level, sentence level, feature level, and document level. Machine learning and lexicon-based approaches were used in the feature-level opinion mining. Decision tree and support vector machine (SVM) classifiers were used in the machine learning approach, where as dictionary of words and its sentiments were created [11] in the lexicon-based approach. Recent aspect-based sentiment analysis techniques divide the sentiment and assign sentiment scores to different aspects of the product or service mentioned [12] in the review.

2 Literature Review

This literature review is categorized into two parts, i.e., foreign languages' sentiment analysis and Indian languages sentiment analysis of customer reviews for products and services.

2.1 Foreign Languages Sentiment Analysis

Laila Abd-Elhamid et al. experimented [13] 'Feature-Based Sentiment Analysis in Online Arabic Reviews' in 2016. This system proposed five rules to extract and analyze the feature sentiment pairs. Lexicon-based and corpus-based sentiment analyses are the two main approaches which are used to handle the feature-based sentiment analysis. Lexicon-based classification is to be performed to assess the performance of each rule. The positive, negative, and neutral weights are assigned to every word according to their word existence in the review. Based on the dictionaries, the ontology makes the synonyms group of every word for the gathered sentiments and the features.

The dataset was gathered from different online social media Web sites. To convert the colloquia Arabic language to standard Arabic language, Arabic Tool Kit Service (ATKS) tool 'colloquial to Arabic converter' was used by the authors. To perform the experiments and to evaluate the results, only 200 reviews were used. The consequence of this system shows automatic extraction and reorganization of the polarity of the feature sentiment words with which the high precision is achieved. The proper sentiment assignments of the single adjective words were 94.73%. The noun words were extracted for feature extraction. The various accuracies achieved by applying proposed five rules are 90%, insufficient reduction to accurateness, 98%, reduce the accuracy and some features polarity was altered. High accurateness is achieved for pair's extraction, but after the third rule, a little bit accuracy is decreased.

Alaa El-Halees and Ahmed Al-Asmar experimented [14] on 'Ontology-Based Arabic Opinion Mining' in 2017. This system focuses on the problems that occur in the Arabic language, which is based on opinion mining technique. The opinion mining technique in this system was based on the ontology which performs the feature-level classification of Arabic reviews. The public dataset used in this system was hotels and books. The ontology building was the first step in this system. ConceptNet is used to build the domain-specific ontology trees for the given reviews. To parse the reviews and to generate the POS tag for words, Stanford Parser Tool was used. In the second step, the opinion word related to the specific feature was obtained. That opinion word may be noun, verb, and adjective. In the last step to determine the polarity of opinion words, Arabic Sentiment Lexicon (ArSenL) was used.

The dataset contains 2000 reviews of hotel and books' domain. For better matching, all the words in the review document were stemmed to root. The approaches used in this system are machine learning, lexicon based, ontology baseline, and ontology with important features. For subject evaluation, the authors selected 100 reviews of the each product feature for each domain. To get the better opinion mining performance, ontology information is used in this system. The drawbacks with this system are only one ontology, and two datasets are used. The less number of data mining methods is used; due to this, the performance is degraded. The effectiveness of this method gives an average F score of 84.62 and 83.31% for hotels and books datasets according to the subject evaluation.

Haiyun Pen et al. performed [15] the review on 'A Review of Sentiment Analysis Research in Chinese Language' in 2017. In this system, the survey on monolingual sentiment classification was conducted in Chinese language using three classification frameworks. Monolingual and bilingual approaches were used in the Chinese sentiment analysis. The survey on Chinese monolingual approach conducted initially. At last, the authors refer the research on English language and then use the multilingual approach that uses the resources of English language into the Chinese language. The corpus contains the words, sentences, phrases, and paragraphs in the emotional expressions. Eight emotional classes were used in the Chinese to investigate emotional expressions. The first step in the sentiment classification is the preparation of corpus. For machine learning techniques, the corpus-based method is utilized.

The three methods used for the construction of sentiment lexicon in the Chinese language are the manually constructed sentiment lexicons, dictionary-based method, and the semantic lexicon which was constructed based on the corpus. HowNet dictionary was used in this method. The issue with this dictionary is the context of the words which were not considered by the list. The problem with the Chinese language is the lack of inter-word spacing. The string in the Chinese text formed with the equally spaced graphemes (characters). For sentiment classification, the language lexicon is required. The problems with the sentiment lexicon in Chinese language were the Chinese words which are syntactically and semantically ambiguous, insufficient resources for construction of Chinese lexicon.

2.2 Indian Languages' Sentiment Analysis

Sumitra Pundlik et al. experimented [16] on 'Multiclass Classification and Class-based Sentiment Analysis for Hindi Language' in 2016. This system was used to classify the Hindi speech document into various classes like agriculture, society, education, and festivals and then extract the sentiments for the identified classes. In this system, ontology was used to classify the Hindi speech document into different classes. To find the polarity of the classes, HindiSentiWordNet (HSWN) was used. To improve the polarity extraction accuracy, HSWN and language model (LM) classifier methods were combined by the authors. The Hindi speeches delivered by the user are the dataset of this system.

Two approaches were used for sentiment classification. In the first approach, the polarity of the word (i.e., positive, negative, and neutral) was found using HSWN where as in the second approach, HSWN and LMClassifier were combined. Positive, negative, and neutral scores of the words were present in HSWN file, and this was used for sentiment analysis. Less number of words was present in the HSWN which is the problem with the HSWN, and to overcome this problem, LMClassifier was used in this system. The limitation of this system is that it can not test the document more than 1000 words and less than 500 words. Only the static ontology was created in this system. The future scope of this system includes implementation of self-learning ontology for better accuracy and analysis of both multimodal and multiclass sentiment analysis.

Anil Kumar et al. proposed [17] 'Analysis of Users Sentiments from Kannada Web Documents' in 2014. Kannada language is morphologically rich language. For Indian languages, the tags in the Penn tags are inadequate then in that case, additional tags such as VM (main verb) and VAUX (auxiliary verb) were used for verbs. The sentence-based approach [18] was used to check the sentiment scores of extracted opinion expressions using SentiWordNet dictionary.

Using Google translator, Kannada words are translated into English. To implement the adjective analysis, Turney's algorithm and the Kannada part-of-speech tagger were used. The polarity was computed based on the variation between positive and negative counts. The value is greater than zero, the comment is positive, less than zero, the comment is negative, otherwise neutral. The polarity of the whole review is considered based on the sentence which is having more number of keywords. The text corpus of 182 positive and 105 negative reviews is taken. The average precision of semantic methods is 68.36%, and the machine learning methods is 75.58%. The precision of machine learning methods is better as compared to semantic methods due to the usage of training dataset instead of explicit patterns.

Quratulain Rajput experimented [19] on 'Ontology-Based Semantic Annotation of Urdu Language Web Documents' in 2014. This system used the semantic annotation to annotate the Urdu language written documents. Semantic annotation uses the context keywords and domain ontologies. The corpus was collected from the online Urdu newspapers, and it contains the online classified ads. To perform the experiment in this system, car ads are considered which are available in the Jang news Web site. This system was used the domain ontologies to annotate the information which is present in the online Web documents.

To extract the information, the extraction algorithm was used. This algorithm searches for the context keyword; if it is found, then the required value is present beside to the keyword and extracted using suitable rule. For every property in the ontology, this extraction is continuous. After completing the extraction process, it stored in OWL database; in this database, structure query was applied. From the Jang Web site, 850 ads were selected to conduct the experiment. The sentiment analysis of this system was mostly dependent on the context keywords. The context words are not properly defined, and then, the performance of this system was degraded. There was an extra overhead incurred for the annotation process to find the context keywords. The biggest challenge of Urdu language was one word which is represented with two tokens.

3 Ontology in Sentiment Analysis

Ontology is the set of concepts and properties in the subject area or the domain which shows the relation between the concepts. Sentiment analysis or opinion mining is used to analyze the opinions, reviews, and feedback given by the customer about products, movies, and any services. The customers want to buy any product online, before buying the product he/she has to see online reviews of that product. The customers all over the world who buy that product posted their reviews in the e-commerce Web site from where they purchase that product online. In those Web sites, all reviews are analyzed and give the rating like 0–5 range scale. But this rating is based on the syntax of the reviews. This type of rating is not giving the accurate result.

Instead of using the syntax of reviews for sentiment analysis, use the semantic knowledge for review analysis, and then, it gives better result. The ontology incorporating [20] method is not less sensitive to the size of the training data than the method that does not incorporate an ontology. To analyze the reviews based on the semantic knowledge, ontology is used. This paper listed the challenges faced by the various languages to analyze the sentiments of customer reviews using ontology and also try to overcome those problems.

4 Sentiment Analysis of Customer Reviews in Hindi Language

Hindi is the national language of India. The most of the people who lives in India are able to speak and also write the Hindi language. The people not familiar with the English language, they are using Hindi language to write the online reviews of the products. Therefore, it is important to perform sentiment analysis of product reviews which are given in the Hindi language using ontology. To perform the Hindi language customer reviews' sentiment analysis, the mobile phone reviews' dataset is taken. The mobile phones are having many features like camera, touch, storage, price, and model.

This experiment concentrates on the feature-based sentiment analysis of the mobile phone reviews. The English language reviews are collected from the www. amazon.in Web site. Google translator is used to convert the English language reviews to Hindi language reviews. In the next step, the Hindi reviews are preprocessed using tokenization and stop word elimination. In the tokenization, the review data is divided into tokens which are the smallest part of the review. The stop word elimination is applied to remove the stop words in the review data. Stop words are the type of words which does not give the opinion of the product. After that, POS tagging of the words has been assigned using Centre for Development of Advanced Computing (CDAC) Hindi POS tagger. The adjectives in the POS tagged data represent the product opinion, and the nouns represent the product features.

5 Results and Discussions

Opinion mining and sentiment analysis have application in different domains like business intelligence, recommender system, health, finance, social networks, and media [21]. In this survey on ontology-based sentiment analysis, the foreign and Indian language customer reviews are considered for the various products and services from social networking and e-commerce Web sites. Comparison of sentiment analysis performance measures in different languages is shown in Table 1.

To perform the Hindi language customer reviews sentiment analysis 100 English language reviews of each 10 different mobile phones total 1000 reviews are taken as dataset. This dataset contains both positive and negative reviews. The list of the mobile phone names whose reviews are considered for an experiment is shown in Table 2.

Following are the steps for sentiment analysis of Hindi language customer reviews:

Step 1: Mobile phones' English language reviews are collected from www.amazon. in Web site.
Step 2: English language reviews are converted into Hindi language using Google translator.

Table 1 Comparison of various languages customer reviews' sentiment analysis performance

Language	Collected reviews	Domain	Accuracy
Arabic	200	Facebook, YouTube	94.73%
Arabic	2000	Hotels	84.62%
		Books	83.31%
Chinese	1021	House, movies and education	–
Kannada	182 positive, 105 negative	–	Semantic method: 68.36%
			Machine learning method: 75.58%
Urdu	850	Car ads	–

Table 2 List of the mobile phones from where customer reviews taken

S. No.	List of the mobile phones
1	BlackBerry Passport
2	iPhone SE
3	Lenovo K8
4	Micromax Canvas Infinity
5	Moto G5 Plus
6	Nokia 6
7	Oppo F3
8	Redmi Note 4
9	Samsung Galaxy On7 Prime
10	Vivo V7

Sample English language review:

"As a blackberry fan, I always trust the quality and features for all BB products."

After translation of English to Hindi language review:
ब्लेकबेरी प्रशंसक के रूप में, मैं हमेशा बीबी उत्पादों के लिए गुणवत्ता और सुविधाओं पर भरोसा करता हूं |

Step 3: Breaking up the input review text into small pieces is called the tokenization. The use of tokenization is to assign the POS tag for each token in the review text.

After applying tokenization in the preprocessed stage, the review output is:
'ब्लैकबेरी,' 'प्रशंसक,' 'के,' 'रूप,' 'में,' 'मैं,' 'हमेशा,' 'बीबी,' 'उत्पादों,' 'के,' 'लिए,' 'गुणवत्ता,' 'और,' 'सुविधाओं,' 'पर,' 'भरोसा,' 'करता,' 'हूं|'

Step 4: Stop words are the list of words which are commonly used. These are specific to each language. Every language contains many stop words.

After applying the stop word elimination, the review output is:
ब्लैकबेरी प्रशंसक रूप में, हमेशा बीबी उत्पादों गुणवत्ता सुविधाओं भरोसा |

Step 5: The work of POS tagger is to assign the part of speech to each token of the review.

POS tagged reviews after preprocessing:
|QTC ब्लैकबेरी|NN प्रशंसक|JJ रूप|NN में,|NN हमेशा|NN बीबी|NN उत्पादों|NN गुणवत्ता|NN सुविधाओं|NN भरोसा|NN

To know the features of the products, the nouns in the review text are used, which are represented by NN (noun) tag. To perform the sentiment analysis of the product reviews, the adjectives in the review text are used. JJ (adjective) tag represents the adjectives. Based on these adjectives, the opinions of the customers are identified. The features and opinion words of the mobile phones are shown in Table 3.

Table 3 Identified features and opinion words of the mobile phones

Identified features	Opinion words
'आकार,' 'फोन,' 'बैटरी,' 'रूप,' 'उत्पादों,' 'गुणवत्ता,' 'आवाज़,' 'कुंजी,' 'कैमरे,' 'वीडियो,' वॉइस, कॉल, 'टच,' 'सुक्रीन,' 'सुविधाओं,' 'भरोसा,' 'अपेक्षति,' 'उत्पाद'	'प्रशंसक,' 'गरम,' 'थोड़ी,' 'सी,' 'आसानी,' 'उत्पादकता,' 'असली,' 'दुर्लभ,' 'अद्भुत,' 'पूर्ण,' 'शानदार,' 'बैकअप,' 'बढ़िया,' 'अच्छा,' 'बैकअप,' 'चिकनी,' 'वशिष,' 'अलग,' 'भौतिकिवादी'

6 Evaluation Measurements

Precision, recall, and F measure are the performance evaluation measurements for the datasets [22].

Precision (*P*): Precision is the fraction of the documents retrieved that are related to the information need for the user.

$$\text{Precision}(P) = \frac{\text{Correct Answers}}{\text{Answers Produced}} \tag{1}$$

Recall (*R*): Recall is the fraction of the documents that are related to the query that are successfully retrieved.

$$\text{Recall}(R) = \frac{\text{Correct Answers}}{\text{Total Possible Correct Answers}} \tag{2}$$

***F* measure**: The weighted harmonic mean of precision and recall is called *F* measure. The traditional *F* measure or balanced *F* score is

$$F \text{ measure} = \frac{(\beta^2 + 1)\text{PR}}{\beta^2 R + P} \tag{3}$$

β is the weighting between precision and recall. Typically, the value of $\beta = 1$. There are tradeoffs between precision and recall in the performance metric.

7 Conclusion

The increasing usage on the multilingual languages over the Web gives the emphasis of sentiment analysis in other than English language. This paper elaborates the challenges faced in the various languages to perform the sentiment analysis using ontology. Based on the domain knowledge, the sentiments are analyzed in ontology. The challenges found in the Chinese languages sentiment extraction are ambiguity in the syntax and the semantics of the words in the reviews. Insufficient resources availability and older dictionaries were used. Challenges in the Hindi language sentiment analysis are limitation of the words in the document; only the static ontologies were created not the self-learning ontologies. The accuracy of the result is not good in some of the semantic and machine learning methods which are the challenge with Kannada sentiment analysis. The challenges faced in the Urdu sentiment analysis are in Urdu; one word is represented by two tokens; and the sentiment analysis mostly depends on context keywords. Extra overhead is incurred to find the context keywords. The Hindi language sentiment analysis using ontology overcomes the some of the challenges exist in the other languages. There is no limit of the words in the review text, and there is no ambiguity in syntax and semantics of the words in reviews. This experiment identified the opinion words in the review text. Ontology construction is the future work of this experiment. By constructing the ontology, automatic identification and extraction of the user opinions from the reviews are possible.

References

1. Alfrjani, R.: A new approach to ontology-based semantic modelling for opinion mining. In: 2016 UKSim-AMSS 18th International Conference on Computer Modelling and Simulation
2. Nithish, R., Sabarish, S., Abirami, A.M., Askarunisa, A., Navaneeth Kishen, M.: An ontology based sentiment analysis for mobile products using tweets. In: 2013 Fifth International Conference on Advanced Computing (ICoAC)
3. Zehra, S., Wasi, S., Jami, I., Nazir, A., Khan, A., Waheed, N.: Ontology-based sentiment analysis model for recommendation systems. In: Proceedings of the 9th International Joint Conference on Knowledge Discovery, Knowledge Engineering and Knowledge Management (KEOD 2017), pp. 155–160. https://doi.org/10.5220/0006491101550160, ISBN: 978-989-758-272-1
4. Ali, F., Kwak, D., Khan, P., Riazul Islam, S.M., Kim, K.H., Kwak, K.S.: Fuzzy ontology-based sentiment analysis of transportation and city feature reviews for safe traveling. JO—Transp. Res. Part C: Emerg. Technol. https://doi.org/10.1016/j.trc.2017.01.014
5. Thakor, P., Sasi, S.: Ontology-based sentiment analysis process for social media content. In: 2015 INNS Conference on Big Data
6. Schouten, K., Frasincar, F.: Ontology-driven sentiment analysis of product and service aspects. In: The Semantic Web: 15th International Conference, ESWC 2018, Heraklion, Crete. https://doi.org/10.1007/978-3-319-93417-4_39
7. Lee, T.B.: The semantic web: feature article: Scientific American, May 2001
8. Mannar Mannan, J., Jayavel, J.: An adaptive sentimental analysis using ontology for retail market. Int. J. Eng. Technol. 7(1.3), 176–180 (2018)
9. Kontopoulos, Efstratios, Berberidis, Christos, Dergiades, Theologos, Bassiliades, Nick: Ontology-based sentiment analysis of Twitter posts. Expert Syst. Appl. 40, 4065–4074 (2013)
10. Farra, N., Challita, V., Assi, R.A., Hajj, H.: Sentence-level and document-level sentiment mining for Arabic texts. In 2010 IEEE International Conference on Data Mining Workshops (ICDMW), pp. 1114–1119 (2010)
11. Lazhar, F., Yamina, T.G.: Identification of opinions in Arabic texts using ontologies. J. Inf. Technol. Softw. Eng. 2012(2), 2 (2012). https://doi.org/10.4172/2165-7866.1000108
12. de Kok, Sophie, Punt, Linda, van den Puttelaar, Rosita, Ranta, Karoliina, Schouten, Kim, Frasincar, Flavius: Review-aggregated aspect-based sentiment analysis with ontology features. Progr. Artif. Intell. 7, 295–306 (2018)

13. Abd-Elhamid, L., Elzanfaly, D., Eldin, A.S.: Feature-based sentiment analysis in online Arabic reviews. IEEE 2016
14. El-Halees, A., Al-Asmar, A.: Ontology based Arabic opinion mining. J. Inf. Knowl. Manag. **16**(3), 1750028 (2017). https://doi.org/10.1142/s0219649217500289
15. Pen, Haiyun, Cambri, Erik, Hussain, Amir: A review of sentiment analysis research in Chinese language. Cogn Comput **9**, 423–435 (2017). https://doi.org/10.1007/s12559-017-9470-8
16. Pundlik, S., Dasare, P., Kasbekar, P.: Multiclass classification and class based sentiment analysis for Hindi language. In: International Conference on Advances in Computing, Communications and Informatics (ICACCI) (2016)
17. Anil Kumar, K.M., Rajasimha, N., Reddy, M., Rajanarayana, A., Nadgir, K.: Analysis of users' sentiments from kannada web documents. In: Eleventh International Multi-Conference on Information Processing-2015 (IMCIP-2015)
18. Khan, A., Baharudin, B.: Sentiment classification by sentence level semantic orientation using SentiWordNet from online reviews and blogs. Int. J. Comput. Sci. Emerg. Tech **2**(4) (2011)
19. Rajput, Q.: Ontology based semantic annotation of Urdu language web documents. In: 18th International Conference on Knowledge-Based and Intelligent Information and Engineering Systems—KES2014
20. Schouten, K., Frasincar, F., de Jong, F.: Ontology-enhanced aspect based sentiment analysis. Proc. 17th Int. Conf. Web Eng. **10360**, 302–3320 (2017)
21. Piryani, R., Madhavi, D., Singhc, V.K.: Analytical mapping of opinion mining and sentiment analysis research during 2000–2015. Inf. Process. Manag. **53**, 122–150 (2017)
22. del Salas-Zárate, M.P., Medina-Moreira, J., Lagos-Ortiz, K.: Sentiment analysis on tweets about diabetes: an aspect-level approach. Hindawi Comput. Mathem. Methods Med. **2017**, 9 (2017). (Article ID: 5140631)

Spam Detection in Link Shortening Web Services Through Social Network Data Analysis

Sankar Padmanabhan, Prema Maramreddy and Marykutty Cyriac

Abstract Twitter is one of the most popular social networking and micro-blogging Web sites in the world. TinyURL is a URL or link shortening web service used by Twitter. Recently, it is being exploited by spammers as a podium to transmit malicious information. TinyURL spam detection in Twitter is a challenging task. In this paper, an efficient scheme is proposed to detect spam in the TinyURL with particular focus on Twitter tweets. A set of features are first extracted from the tweets. This feature set is analyzed to select a reduced set of features. The reduced feature set is fed as input to train three classifiers, namely simple logistic regression, decision tree, and SVM. The classification results show that the SVM classifier has the highest accuracy in detecting spam in TinyURLs.

Keywords Data analysis · Feature extraction · Feature selection · Social network services · Twitter · Web data mining

1 Introduction

Twitter is a micro-blogging social networking platform that facilitates the users to send and receive tweets. It utilizes its own URL with the prefix "t.co" to post the URLs, which are defined by TinyURL. The spammers have utilized the Twitter as a platform to propagate malware phishing and other illicit activities. A survey on online social networks [1] showed that nearly eleven percent of the Twitter posts were spam and also found that the primary source of spam in Twitter was the TinyURL itself.

S. Padmanabhan · P. Maramreddy (✉)
Sri Venkateswara Engineering College for Women, JNTUA, Tirupati, Andhra Pradesh, India
e-mail: premaareddi@gmail.com

S. Padmanabhan
e-mail: sankarp1@gmail.com

M. Cyriac
Jerusalem College of Engineering, Anna University, Chennai, TamilNadu, India
e-mail: marycyriac123@gmail.com

© Springer Nature Singapore Pte Ltd. 2020
K. S. Raju et al. (eds.), *Data Engineering and Communication Technology*,
Advances in Intelligent Systems and Computing 1079,
https://doi.org/10.1007/978-981-15-1097-7_9

103

In general, URL spam detection schemes fall into two categories: static and dynamic [2]. Static methods analyze the Web site codes, whereas dynamic methods exploit the system resources by visiting the concerned web pages. The analysis of web page by dynamic approach is time-consuming, since these methods need to download large number of web pages with enormous content of codes [3, 4]. The spam detection is generally performed on social networks by creating large set of honey profiles [5]. Frequently used methods for spam detection in Twitter are based on account [6–8], relation [9], and message [10]. Account-based approaches generally use the date of account creation, which can be easily created by spammers. But this approach fails for compromised accounts.

Relation-based schemes are robust graphical approaches, and they consume significant amount of computing power. Message-based methods consider the lexical features, such as similarity between the spam messages, for spam detection, and these can be fabricated without much difficulty. The ease of use and the simplicity in creating the tweet content have also resulted in the proliferation of spam, especially phishing. It opens the way for sending malware to unsuspecting clients ranging from simple URL redirects to Dropper worms [11, 12]. A large percentage of spammers abstain from posting URLs directly to the suspicious sites. Instead, they use the same TinyURL method to conceal their URLs by posting directly on the Web site as normal tweets [13–20]. A few social network platforms utilize collaborative filtering, where users report certain logs as spam [21]. Accuracy of detection in this method depends heavily on the technical capability of users and their personal preferences. In a recent research, spam detection is treated as anomaly detection problem for streaming data [22]. A recent attempt on spam detection is a static spam filtering mechanism named as Prophiler [23, 24]. This approach uses 77 features that are derived from various characteristics of a web page such as HTML content, JavaScript code, host-related features, and other URLs. Extraction and analysis of such large number of features make the system complex and time-consuming.

In this paper, the number of features is significantly reduced, without compromising on the accuracy of detection. Instead of 77 features used in the previous work, the proposed system uses only eight features. Out of these eight features, six features are related to host, URL, and GeoIP, which were already used by the previous researchers. In our analysis, we are introducing a page ranking feature, which was found to be significant in the detection of malicious TinyURLs. Reduction in the number of features results in reduced processing time, which is crucial while handling big data like tweets in a Twitter database. After extracting the features, they are given as input to various supervised learning algorithms to determine the accuracy of classification.

The remainder of this paper is organized into various sections to provide additional details as follows: Sect. 2 discusses the feature extraction methodology used in this work, Sect. 3 describes the feature extraction and analysis performed on the extracted data, Sect. 3.2 details the proposed system for spam detection, Sect. 3.3 discusses the classification results, and finally, Sect. 4 concludes the paper.

2 Types of Features

The features used in this work are derived by observing various characteristics of the domain and the web page. These features are related to host, URL, GeoIP, top-level domain, and information on page ranking.

2.1 Host-Related Features

The motivation for using host-related feature is based on the observation that malicious servers have hostnames different from the trusted sites. Therefore, by observing the hostname of a particular site, it is easy to distinguish a trusted site from a suspicious one. For instance, brand names like Twitter and Instagram are trusted while cricket-earn-money.org is not. According to the observations made by McGrath and Gupta [25], the distribution of hostname of phishing sites is very much different from that of the trusted ones. Most trusted hostnames are often short, and the malicious hostnames are long. As the hostname length increases, the chance of it being a non-trusted site also increases.

2.2 URL-Based Features

Some of the characteristics of the URLs are more useful in predicting the phishing activities than others. One such characteristic feature is the number of slashes in a URL. Most spammers have limited resources and likely to use the same URL repeatedly to spam through multiple TinyURLs. Another feature of interest is the number of hops. A large number of hops indicate deep content to be obtained from the Web site. For legitimate sites, most of the web content is readily available at a distance of about two hops away from the user's screen. However, phishing URLs usually hide the content deep inside the benign URLs with more hops, so that unsuspecting users may be drawn into them. Therefore, it is logical to consider the number of hops as a proper feature to detect the suspicious URLs.

2.3 GeoIP-Based Features

These features describe the country that hosts the web server and the owner of the site. Sites were classified based on the amount of spam in each country as presented in literature [26]. They observed that some of the countries produce more spam than the rest, which makes the country of origin as one of the features. The country of origin of a particular Web site can be obtained by looking at the registered physical address for the IP address of the server.

2.4   Top-Level Domain (TLD)

The TLD is a list of internationally recognized domain name server of the Internet. Countries have their own TLDs and almost every site is a sub-domain of the TLD. The Web site www.example.com has the TLD as.com, and it is easy to classify certain TLDs as spam when they appear out of order like www.maliciousexample. combiz. It is also very common for the malicious Web sites to have unusual domain names.

2.5   Google Page Ranking

This is a feature newly introduced in this work. Page ranking is viable information in detecting the malicious sites. It was initially applied to full-text search engines like Google [27]. Google search engine uses the page ranking algorithm to rank Web sites based on search results. The algorithm counts the number and the quality of links to the page to estimate the importance and popularity of a Web site.

The Google page ranking value is calculated using the following equation.

$$ \mathrm{PR}(p) = (1 - \gamma) + \gamma \sum_{\{d \in \mathrm{in}(p)\}}^{\infty} \left(\frac{\mathrm{PR}(d)}{|\mathrm{out}(d)|} \right) \tag{1} $$

where "p" is the page being scored, and "in(p)" is the set of pages pointing to p "out (d)" is the set of links out of d. The symbol γ is a damping factor, usually set to 0.85, which represents the probability that the random user requests another random page.

In modified version of Google page rank API, every page is ranked on the Internet from 1 to 10 with 1 being the least popular and 10 being a trusted site similar to Google or Twitter itself. Malicious pages tend to have low page ranks.

2.6   Alexa Page Ranking

Another popular page ranking method is Alexa page ranking. Alexa provides traffic data, global rankings, and other information on 30 million Web sites. A low value of the Alexa ranking means that the site is popular and credible. The Twitter has an Alexa rank of 11 while a site like maliciousurl.com has an Alexa rank of over 2000. Alexa does ranking of pages up to one million, and clearly, the rank below a few hundred ensures that the site is not malignant at all. There are a few low ranked sites, which have an Alexa ranking of 499,692, verifying the claim that there is no real upper bound on the ranking. However, most of these sites have been newly created and may not be visited by many people. The most suspicious Web sites have a rank greater than 100,000 and above.

The site score value using Alexa page ranking is calculated as,

$$\text{Site Score} = \left| \frac{A * \text{views} + B * \text{reach} + C * \text{time}}{k} \right| \tag{2}$$

where A, B, C, and k are constants maintained by the Alexa.

3 Feature Extraction and Analysis

3.1 Data Collection

The data collection module comprises of Twitter4j application programming interface using a web crawler which was configured to extract the minimum number of publicly available tweets. Total number of URLs considered in this study is 17,843 URLs, of which 9868 are benign URLs and 7975 are Spam URLs. The tweets collected from these URLs are used to create a dataset consisting of 25,821 tweets.

3.2 Feature Analysis

In order to validate feature selection, all eight features of interest are analyzed to pinpoint which feature is associated with what spam at a given time. For instance, during the Cricket World Cup, the majority of spammers targeted the trending topics which are related to cricket. But in the normal scenario, it is difficult to determine which topic acquires the spam and whether the trending topic contains spam or shows false positives. It is possible to trace the spamming communities by determining whether the same list of URLs is being spammed from various topics.

3.2.1 Number of Hops

Although features like hostname and hostname length can be easily determined by parsing the URL directly, extraction of other features requires advanced processing techniques. In order to find the number of hops for a web page, a remote HTTP URL connection is used as a normal URL in the Internet. Then, the URL is checked for HTTP status code for redirection such as 301 (permanent redirect), 302 (temporary redirect), and 200 (success). If the URL is redirected, the location of the next URL in the chain can be determined by inspecting the "Location" parameter in the header field of the URL. The process is continued until the URL does not get redirected any further. The number of hops is thus obtained from the final URL information.

Figure 1 shows the number of hops used by various benign and malicious URLs to reach the required page. It is clearly seen that the number of hops ranges from 1 to a maximum of 6 for benign URLs. On the other hand, majority of the malicious URLs have about 4–9 hops. Majority of the URLs having 4 redirects are classified as benign, while the number of URLs in both classes is equal for hop values of 5 and 6. When the number of hops is beyond six, the URLs are all malicious, thus validating the selection of this attribute.

3.2.2 Domain Name and Its Length

The domain name or hostname and length of the domain name features are obtained from the full URL, and they are compared with the name and length of the trusted URLs. Since the sample domain used in this study is cricket, hostnames such as Twitter, Facebook, Instagram, ICC Cricket, and Cricinfo were chosen as benign. All other URLs are treated as spam and labeled as others.

Figure 2 shows the split-up of the URLs between trusted and other categories. There are 7975 spam URLs in the other categories whereas nearly 1600 for each benign category. Another point of interest in the domain name-related feature is the domain name length in the URL. Figure 3 shows the classification of URLs based on this feature. For most trusted URLs, the hostnames seem to have a length of at least 6 and not more than 14 characters. However, majority of the spam URLs have

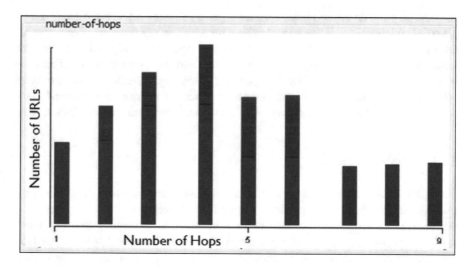

Fig. 1 Number of hops (*red—benign, blue—trusted*)

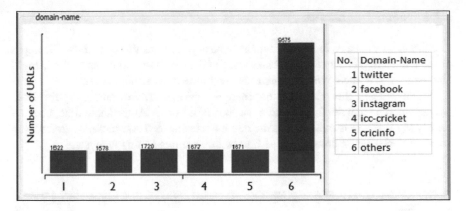

Fig. 2 Domain name

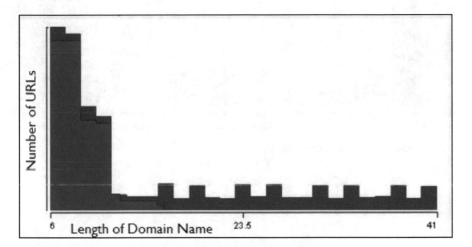

Fig. 3 Distribution of trusted/benign URLs with domain name length

names with lengths that extend beyond 6. It is observed that one such URL persistently turning up in the tweets is "www.winnit-bunnies-playingcricket.co.uk."

3.2.3 Top-Level Domain Information

Another feature considered in this paper for spam URL analysis is the top-level domain information. In general, the top-level domains have sub-domains, and they are belonging to a well-established infrastructure. Figure 4 shows the URLs for different top-level domain categorization used in the analysis.

3.2.4 Number of Slashes

The number of slashes in a URL indicates the depth of the Web site link. The depth of most trusted Web sites has 2–3 slashes. But the depth of malicious URLs is high. One example of such Web site name is "/path/to/some/malicious/url."

A drawback of this feature is that there is a certain level of ambiguity for those URLs which have a depth of 3 and 4, as shown in Fig. 5. At the same time, one can clearly say that site is malicious if the depth of slashes is 5 and more. Therefore, the number of slashes is considered as one of the features in this analysis.

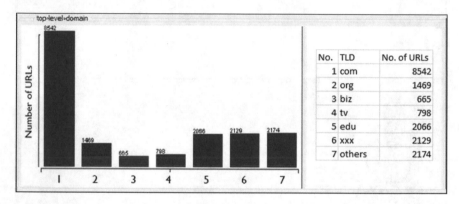

Fig. 4 Top-level domain characterization

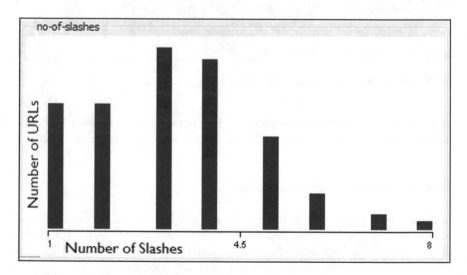

Fig. 5 Distributions of URLs with number of slashes

3.2.5 Page Ranking

The page ranking information deals with the real value of the URL that leads to the page. The distribution of URLs for Google and Alexa page ranks are shown in Figs. 6 and 7, respectively. It is seen from these figures that a high page rank

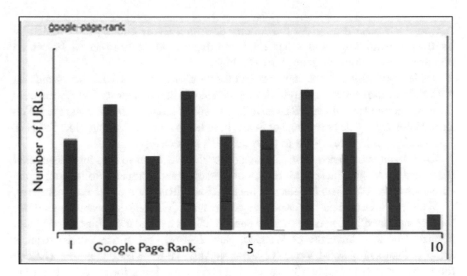

Fig. 6 Google page rank for URLs

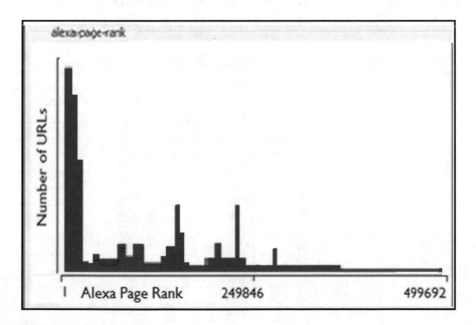

Fig. 7 Alexa page rank of URLs

indicates a benign URL, but a low page rank does not necessarily indicate that the page is spam. This is the case with lesser-known blogs which are not widely used.

3.3 GeoIP

A GeoIP-related feature gives the country of origin. Every country is governed by its own firewall laws and other Internet privacy and security laws. In certain countries, the number of spam URLs is high.

Table 1 provides a list of countries and the number of spam URLs for a search on cricket. Since the search for TinyURLs was conducted in India, most of the tweets were from Indian users and those URLs have lead to sports blogs or Indian cricket fan Web sites. However, a large percentage of tweets have led to reputed Web sites like ESPN or ICC Cricket, which are hosted in USA and UK, respectively.

Eight features extracted from URLs and are given as input to J48, logic regression (LR), and SVM classifiers. The results of classifiers are compared to determine the efficiency of the classifiers in detecting the malicious URLs for the given set of features.

Waikato Environment for Knowledge Analysis (WEKA) software is used for training and classification of the feature data. WEKA is a popular machine learning tool published by the University of Waikato, New Zealand [28]. Since WEKA requires either a Comma-Separated Value (CSV) or an Attribute-Relation File Format (ARFF) file to be given as the input, the feature data is rearranged in a format suitable for the software. Once the eight features are extracted, they are stored in a file such that each row corresponds to a TinyURL and each column represents the corresponding features. Table 2 lists the sample feature data formatted in the WEKA CSV format.

Table 1 List of spam URLs country-wise classification results

S. No.	Country of origin	Number of spam URLs
1	South Korea	804
2	China	1380
3	Taiwan	703
4	India	2358
5	Romania	503
6	Ukraine	321
7	Poland	276
8	Hongkong	1402
9	Singapore	1086
10	UK	1597
11	Malaysia	743
12	Australia	1002
13	Russia	1002
14	USA	2320
15	Others	2345

Table 2 Sample features used for classification

Number-of-hops	Domain name	Domain-length	Top level-domain	Google page rank	Alexa page rank	No-of-slashes	Country of origin	Class
4	Instagram	9	com	9	21	1	India	'Not spam'
1	Cricinfo	8	com	9	543	3	Australia	'Not spam'
3	Instagram	9	com	7	21	1	UK	'Not spam'
3	ICC Cricket	11	com	6	67	1	USA	'Not spam'
3	Cricinfo	8	com	6	362	2	India	'Not spam'
3	Twitter	7	com	6	11	1	USA	'Not spam'
2	Facebook	8	com	7	9	3	Others	'Not spam'
1	Twitter	7	com	5	11	2	India	'Not spam'
6	Others	37	org	3	334,741	8	Taiwan	'Spam'
5	Others	19	com	2	11,023	4	Hong Kong	'Spam'
5	Others	15	Others	5	426,693	4	Ukraine	'Spam'
6	Others	34	edu	1	6732	5	Poland	'Spam'
6	Others	39	biz	3	88,023	4	China	'Spam'
5	Others	29	xxx	3	882,337	6	Others	'Spam'
4	Others	24	com	3	4442	8	USA	'Spam'
4	Others	20	com	5	193,635	7	Taiwan	'Spam'
5	Others	41	tv	5	822,970	8	Romania	'Spam'
3	Twitter	7	com	8	11	1	India	'Not spam'
4	Instagram	9	com	7	21	2	Others	'Not spam'
4	Twitter	7	com	8	11	3	Australia	'Not spam'
1	Cricinfo	8	com	8	110	2	India	'Not spam'
3	ICC Cricket	11	com	7	67	2	Australia	'Not spam'
4	ICC Cricket	11	com	6	67	3	Others	'Not spam'

(continued)

Table 2 (continued)

Number-of-hops	Domain name	Domain-length	Top level-domain	Google page rank	Alexa page rank	No-of-slashes	Country of origin	Class
1	ICC Cricket	11	com	8	67	3	Others	'Not spam'
2	Cricinfo	8	com	9	110	2	Australia	'Not spam'
4	Others	25	Others	5	7823	4	Singapore	'Spam'
5	Others	28	org	4	884,869	6	Russia	'Spam'
4	Others	25	com	2	52,473	7	China	'Spam'
2	Facebook	8	com	7	9	3	Others	'Not spam'
1	Twitter	7	com	5	11	2	India	'Not spam'
6	Others	37	org	3	334,741	8	Taiwan	'Spam'
5	Others	19	com	2	11,023	4	Hong Kong	'Spam'
5	Others	15	Others	5	426,693	4	Ukraine	'Spam'
6	Others	34	edu	1	6732	5	Poland	'Spam'
6	Others	39	biz	3	88,023	4	China	'Spam'
5	Others	29	xxx	3	882,337	6	Others	'Spam'
4	Others	24	com	3	4442	8	USA	'Spam'
4	Others	20	com	5	193,635	7	Taiwan	'Spam'
5	Others	41	tv	5	822,970	8	Romania	'Spam'
3	Twitter	7	com	8	11	1	India	'Not spam'
4	Instagram	9	com	7	21	2	Others	'Not spam'
4	Twitter	7	com	8	11	3	Australia	'Not spam'
1	Cricinfo	8	com	8	110	2	India	'Not spam'

The classification methodology used is a statistical supervised learning technique. By using the supervised learning, URLs can be subjected to a binary classification of either benign or spam. Based on the eight features discussed above, three supervised learning classifiers, namely J48, SVM with RBF kernel, and logistic regression (LR), are used to classify the URLs as spam or benign. The testing dataset consists of 2730 TinyURLs, out of which 1505 are benign and 1225 are spam URLs. If the benign URL is identified as benign, then it is true positive (TP). Otherwise, it is false negative (FN). When the spam URL is correctly recognized as spam, it is true negative (TN), or else it is false positive (FP). The confusion matrix obtained for the testing data is shown in Table 3.

The outputs of three classifiers are evaluated based on various parameters such as true-positive rate (TPR), false-positive rate (FPR), true-negative rate (TNR), false-negative rate (FNR), F-score, and accuracy. These parameters obtained for various classifiers are shown in Table 4. It is evident from Table 4 that an accuracy of 96.99% is achieved by using the chosen features with the SVM classifier. F-score is another important parameter that is used to evaluate the proposed system. A large value for F-score (Table 4) reflects the ability of the system to differentiate between malicious and benign sites effectively by using the selected features.

The results obtained from the proposed approach are compared with those published in the previous researches [19, 24]. These researchers have used 14 and 77 features, respectively. An accuracy level of 91.87% was obtained when the number of features was 14. The results of our work could not be compared with the Google system as it uses proprietary browsing API. Besides, it is not possible to collect the FP and FN ratios from the Google system. Besides, the FNR in our system is 0.99, which is much less compared to the previous results. Since a

Table 3 Confusion matrix for URL

Classifier	True positives (TP)	False negative (FN)	True negative (TN)	False positive (FP)
J48	1466	39	1103	132
LR	1478	27	1141	84
SVM	1490	15	1158	67

Table 4 Evaluation results for the classifiers

Evaluation parameter (All in %)	J48	LR	SVM	Ref [19] wb	Ref [22] Canali
True positive rate	97.41	98.20	99.01	98.77	NA
True negative rate	90.05	93.14	94.54	92.76	NA
False positive rate	9.95	6.85	5.46	1.23	5.46
False negative rate	2.59	1.79	0.99	7.24	4.13
F-score	94.61	96.38	97.32	96.28	NA
Accuracy	94.10	95.93	96.99	91.87	NA

reduced value for FNR indicates higher detection performance, it shows that our system is better than other methods. Similarly, FPR value of our system is better than the previous work, which used 77 features.

One drawback in the Wepawet software-based study in the previous approach [24] is the reduced volume of data used in the analysis. It uses only 1% of the available web pages for analysis and classification. Therefore, the value of false negative calculated with Wepawet cannot be fully relied upon. The percentage of malicious to benign web pages utilized in earlier works is also insufficient. For example [24], used a total of 51,958 web pages for testing. However, out of 51,958 pages, only 787 pages used by them are initially malicious, which accounts to 1.51%. In our implementation, a number of benign and spam URLs are used to train the dataset which is almost equal. Utilization of equal number of benign and spam URLs makes the training dataset more robust and effective than the previous works.

4 Conclusion

In this paper, an optimized feature extraction scheme suitable for the detection of malicious TinyURLs in Twitter is proposed and implemented. First, TinyURLs are extracted from the tweets related to specified topics. After analyzing the relevance of features, six key features are chosen from TinyURL characteristics and domain-related information. A new feature called page ranking is introduced in this work. Three supervised statistical classifiers, namely J48 decision tree, logistic regression, and the SVM, are used for classifying the URLs. Out of the three classifiers used, accuracy of SVM classifier was more than the other classifiers. Therefore, it is concluded that the SVM is the most appropriate classifier for the given set of features. Since the approach used is static, feature analysis takes less time when compared to its dynamic counterparts. By using a compact feature set, a fast and efficient classification of TinyURLs is achieved.

References

1. Jin, L., Chen, Y., Wang, T.G., Hui, P., Vasilakos, A.V.: Understanding user behavior in online social networks: A survey. IEEE Commun. Mag. **51**(9), 144–150 (2013)
2. Kim, B.I., Im, C.T., Jung, H.C.: Suspicious malicious web site detection with strength analysis of a javascript obfuscation. Int. J. Adv. Sci. Technol. **26**, 19–32 (2011)
3. Ma, J., Saul, K.L., Savage, S., Geoffrey, M.V.: Beyond blacklists: learning to detect malicious web sites from suspicious URLs. In: 15th ACM SIGKDD International Conference on Knowledge Discovery and Data Mining, pp. 1245–1254. https://doi.org/10.1145/1557019. 1557153 (2009)
4. Irani, D., Webb, S., Pu, C.: Study of static classification of social spam profiles in myspace. In: Fourth International AAAI Conference on Weblogs and Social Media, May 23–26, pp. 82–89. George Washington University, DC (2010)

5. Lee, K., Caverlee, J., Webb, S.: Uncovering social spammers: social honey pots þ machine learning. In: 33rd Int'l ACM SIGIR Conference Research and Development in Information Retrieval, 978-1-60558-896-4/10/07 (2010)
6. Wang, A.H., Don't Follow me: spam detecting in Twitter. In: International Conference on Security and Cryptography, Athens, Greece, pp. 1–10 (2010)
7. Benevenuto, F., Magno, G., Rodrigues, T., Almeida, V.: Detecting spammers on Twitter. In: Seventh Annual Collaboration, Electronic Messaging, Anti-Abuse and Spam Conference (CEAS) July, Redmond, Washington, USA (2010)
8. Strinhini, G., Kruegel, C., Vigna, G.: Detecting spammers on social networks, In: 26th Annual Computer Security Applications Conference (ACSAC), pp. 1–9 (2015)
9. Song, J., Lee, S., Kim, J.: Spam filtering in Twitter using sender-receiver relationship. In: 14th International Conference on Recent Advances in Intrusion Detection (RAID), pp. 301–317. Menlo Park, CA, 20–21 Sept 2011
10. Gao, H., Chen, H., Lee, K., Palsetia, D., Choudhary, A.: Towards online spam filtering in social networks. In: 19th Annual Network and Distributed System Security Symposium. Hilton, San Diego, USA, 5–8 Feb 2012
11. Klein, F., Strohmaier, M.: Short links under attack: geographical analysis of spam in a URL shortener network. In: 23rd ACM Conference on Hypertext and Social Media. Milwaukee, WI, USA, 25–28 June 2012
12. Ameya, S., Joyce, J., Upadhyaya, S.: The early bird spreads the worm: an assessment of Twitter for malware propagation. Proced. Comput. Sci. **10**(3), 705–7121 (2012)
13. Moshchuk, A., Bragin, T., Gribble, S., Levy, H.: A crawler-based study of spyware in the web. In: Proceedings of the Symposium on Network and Distributed System Security (NDSS) (2006)
14. Provos, N., Mavrommatis, P., Rajab, M.A., Monrose, F.: All your iFrames point to us. In: 17th Conference on Security Symposium, San Jose, CA, July 28–Aug 01, pp. 1–15 (2008)
15. Ikinci, A., Holz, T., Freiling, F.: Monkey-spider: detecting malicious websites with low-interaction honey clients. In: Proceedings of Sicherheit, Schutz und Zuverlässigkeit, 2–4 Apr 2008
16. Eckersley, P.: How unique is your web browser? In: 10th International Conference in Privacy Enhancing Technologies (PET), pp. 1–18. Berlin, Germany, 21–23 July 2010
17. Kapravelos, A., Cova, M., Kruegel, C., Vigna, G.: Escape from monkey Island: evading high-interaction honey clients. In: Holz, T., Bos, H. (eds.) Detection of Intrusions and Malware, and Vulnerability Assessment. DIMVA, Lecture Notes in Computer Science, vol. 6739. Springer, Berlin, Heidelberg (2011)
18. Zhang, J., Seifert, C., Stokes, J.W., Lee, W.: ARROW: generating signatures to detect drive-by downloads. In: 20th International World Wide Web Conference, pp. 187–196. New York, USA, ACM (2011)
19. Lee, S., Kim, J.: Warning bird, a near real time system for a suspicious Twitter stream. IEEE Trans. Depend. Sec. Comput. **10**(3), 157–165 (2013)
20. Vangapandu, K.B., Brewer, D., Li, K.: A study of URL redirection indicating spam. In: CEAS Sixth Conference on Email and Anti Spam, Mountain View, California, USA, 16–17 July 2009
21. Kamoru, B.A., Jaafar, A., Jabar, M.A.B., Murad, M.A.A.: A mapping study to investigate spam detection on social networks. Int. J. Appl. Inf. Syst. **11**(11), 16–34 (2017)
22. Miller, Z., Dickinson, B., Deitrick, W., Hu, W., Wang, A.H.: Twitter spammer detection using data stream clustering. Inf. Sci. **260**, 64–73 (2013)
23. Feinstein, B., Peck, D.: Caffeine monkey: automated collection, detection and analysis of malicious javascript. In: Proceedings of the Black Hat Security Conference. Las Vegas, NV, USA (2007)
24. Canali, D., Cova, M., Vigna, G., Kruegel, C.: Prophiler: a fast filter for the large-scale detection of malicious web pages. In: 20th International World Wide Web Conference, pp. 197–206. Hyderabad, India, March 28–April 1 (2011)

25. McGrath, D.K., Gupta, M.: Behind phishing: an examination of Phisher Modi Operandi. In: Of the USENIX Workshop on Large-Scale Exploits and Emergent Threats (LEET), San Francisco, CA (2008)
26. Moura, G.C.M., Sadre, R., Pras, A.: Taking on internet bad neighborhoods. IEEE Symp. Netw. Oper. Manag. (NOMS) 5–9(May), 1–7 (2014)
27. Page, L., Brin, S., Motwami, R., Winograd, T.: The page rank citation ranking: bringing order to the web, technical report. Computer Science Department, Stanford University (1999)
28. Hall, M., Frank, E., Holmes, G., Pfahringer, B., Reutemann, P., Witten, I.H.: The WEKA data mining software: an update. SIGKDD Explor. 11(1), 10–18 (2009)

Definitional Question Answering Using Text Triplets

Chandan Kumar, Ch. Ram Anirudh and Kavi Narayana Murthy

Abstract Definitional question answering deals with answering questions of the type "Who is X" and "What is X." The techniques used in the literature extract long sentences that may not only give irrelevant facts, but also pose difficulty in evaluating the performance of the system. In this paper, we propose a technique that uses text triplets. We further choose relevant triplets based on a manually built list of terms that are found in definitions in general. The selected triplets give simple, short, and precise definitions of the target. We also show that evaluation becomes easy.

Keywords Question answering · Definitional question answering · Triplets

1 Introduction

Finding answers to arbitrary questions is a highly complex task. Most question answering (QA) systems restrict themselves to searching for sentences in a given corpus and possibly selecting parts of these sentences, which are intended to contain the answers sought by the users.

The text retrieval conference (TREC) series introduced a number of text processing tasks in a formal way and participating groups worked on these specified tasks in a competitive spirit. The QA task was first introduced in 1999 [1]. The task of QA is to find answers in a collection of unstructured text to questions posed in natural language. Initially, only factoid and list questions were considered. An answer to a factoid question is generally a short span of text, i.e., a word, a phrase

---

C. Kumar (✉) · Ch. R. Anirudh · K. N. Murthy
School of Computer and Information Sciences, University of Hyderabad, Hyderabad, India
e-mail: chandanmcmi20@gmail.com

Ch. R. Anirudh
e-mail: ramanirudh28@gmail.com

K. N. Murthy
e-mail: knmuh@yahoo.com

© Springer Nature Singapore Pte Ltd. 2020
K. S. Raju et al. (eds.), *Data Engineering and Communication Technology*,
Advances in Intelligent Systems and Computing 1079,
https://doi.org/10.1007/978-981-15-1097-7_10

(e.g., How many calories are there in a Big Mac? What is the capital city of India?). List Questions ask for list of items (e.g., List all states in India, List all Universities in India). TREC first introduced and defined definitional question answering in its 2003 edition. Definitional questions are questions such as *What is X? Who is X?*, henceforth termed as type 1 and type 2 question, respectively. An answer to a definitional question should be a collection of facts that define the term being questioned as precisely as possible. Types of facts that define the questioned terms vary from term to term. The challenge is to find suitable facts for a given term. Note that no attempt is made to define a term precisely, that is an extremely hard problem, if not impossible. The goal is only to find parts of a given corpus which can help someone in getting some idea of what or who X is.

These Definitional QA systems suffer from a lack of standard representation of the answer. Either whole sentences are given out or parts of sentences selected somewhat arbitrarily are given out as answers. Length of answers varies quite a bit. All these make the task of evaluating the precision of answers very difficult [2]. These are the main issues addressed in this paper.

2 Related Work

Many groups participated in TREC 2003. All the groups went through the same pipeline of processes: question processing, information retrieval (IR) to find relevant documents from the document collection, candidate sentence selection, sentence ranking, and redundancy removal [1]. Most of the groups used their own IR engines and retrieved relevant documents from the AQUAINT [3] corpus, taking it as the source of the answer. TREC also provided a collection of relevant documents to 50 selected definitional questions. The participating groups applied a variety of heuristics to select candidate sentences. Word overlapping measure and text summarization techniques were used by participating groups for redundancy removal.

BBN (Bolt, Beranck, and Newman) [4] defined kernel facts as phrases extracted from a candidate sentence in a specific way. They defined four types of kernel facts: appositives and copula, propositions, structured pattern, and relation. They extracted these kernel facts using linguistic processing with the help of an information extraction (IE) tool. They used full candidate sentences if kernel facts of the above types could not be extracted. They ranked the extracted kernel facts by calculating the similarity measure to the profile of the question [4]. BBN got the highest F measure score of 0.55.

Qualifier [5] is a question answering system developed at the National University of Singapore. It applied co-reference resolution to relevant documents returned by an IR tool. It put all sentences which contain any part of the question target in the positive set and other sentences in the negative set. This system ranked the candidate sentences (positive set) two times. First, it ranked the sentences statistically. A sentence is scored by using a combination of scores of each word present in the sentence, and the score of a word is being computed from its frequency. Candidate sentences were then ranked again using a repository of definitional patterns. Finally, the qualifier system applied an MMR text summarization technique to eliminate redundancy and produced the final answer. Qualifier got the second highest F measure score of 0.47.

TextMap [6], an NLP group at the University of Southern California, differs from other works only in the ranking techniques used. This group used four resources to rank the candidate sentence: a collection of biographies, a collection of descriptors of proper people, wordnet, and semantic relationship patterns. This group generated variable-length answers and got the third highest F measure score of 0.46.

The MIT group [7] developed three modules: database lookup, dictionary lookup, and document lookup. Each module generated an answer to a definitional question, and a final answer was generated by merging the answers from all the modules. In the database lookup module, a database was built by applying 13 surface patterns to the Acquaint corpus offline and answer of a question was found by querying the question target in this database. In the dictionary lookup module, answer projection technique was applied to map answer to a question from the dictionary to the corpus. In the document lookup module, sentences containing question targets from the relevant documents returned by IR were returned as the answer. This module comes into action if the first two modules fail. MIT also generated variable-length answers. It got an F measure score of 0.30. An extension of the work by the same team presented a component-level evaluation of each module [8].

Han et al. [9] used a probabilistic model consisting of three parameters: a topic model, definition model, and sentence (language) model. The goal is to find the probability that a sentence is a definition given a topic (target) ($P(D, S|T)$). Cui et al. [10, 11] explored probabilistic lexico-syntactic pattern matching, also known as soft pattern matching models, for answer extraction. Chen et al. [12] used N-gram language models for re-ranking the answers extracted. Paşca et al. [13] extracted answers from text snippets extracted from web that are anchored with time information. The idea is that these texts inform about an event associated with the target.

TREC 2003 used only the second pass of evaluation technique of "The Definitional Pilot," a series of the pilot evaluations as part of the AQUAINT program [2]. This aspect is discussed in detail in a later section.

3   Key Observations

Current systems either throw out entire sentences or some parts of selected sentences. Answers are thus variable-length text strings, even the syntactic structure may vary significantly from item to item. When lengthy sentences are given out as answers, parts may be irrelevant or distractive. For example, for 'Sunderbans,' the retrieved answers could be:

The Sunderbans in West Bengal and the Gahirmatha coast in Orissa are what could be called the stars of mainland India's coastal and marine ecosystems.

In the mangrove forests of Sunderbans, West Bengal, Ramakrishna Mission Lokasiksha Parishad (RKMLSP) and Sri Ramakrishna Ashram Nimpith (SRAN) are two organizations that help local communities understand and overcome problems arising from their unique surroundings.

It is easy to see that the second sentence has many parts which are not relevant or useful for defining what 'Sunderbans' are.

Existing systems also use varied resources to rank candidate sentences: word vectors, dictionaries, collection of biographies, wordnet glosses, semantic relationship patterns, etc. [14]. It would be good if we can minimize and standardize the use of external resources for system development as also for evaluation.

3.1 Problems of Evaluation

Drawing from [2], consider the question "Who is Christopher Reeve?" List of concepts extracted by human experts for the purposes of evaluation may be

1. Actor
2. Accident
3. Treatment/Therapy
4. Spinal cord injury
5. Activist
6. Written an autobiography
7. Human embryo research activist

   Let us say the response from system is

1. Actor
2. The actor who was paralyzed when he fell off his horse
3. The name attraction
4. Stars on Sunday in ABCs remake of rear window
5. Was injured in a show jumping accident and has become a spokesman for the cause

How do we now count the matching concepts and calculate precision and recall? String matching will not work. Earlier researchers have manually marked the concepts and tried to match. For example, *paralyzed when he fell off his horse* is taken as a concept and is manually equated to the concept *accident*. Thus, counting the total number of facts in answer string becomes subjective and hard if lengthy and verbose answers are generated. It would be good if answers always conform to a specified structure. For example, going back to the question target "Sunderbans", if the system could generate

```
The Sunderbans in West Bengal and the Gahirmatha coast in
Orissa
    mainland India's coastal and marine ecosystems
    mangrove forests of Sunderbans
```

that would be so much better both as an answer to the given question and for evaluation by comparing with reference answers.

The 'definitional pilot' was the first of the series of pilot evaluations for question answering where the objective of each pilot was to come with effective evaluation technique for certain type of questions [2]. The definitional pilot was completed in two rounds by two human assessors. In each round, both the assessors evaluated and ranked eight runs submitted by different participating groups. In the first round, each run was evaluated by two scores, one for the content of answer and one for the order of the answer. Two rankings of the eight runs by two different human assessors in the first round varied a lot due to the order score. In the second round of evaluation, the system runs were evaluated by only the content score. This time the two rankings of the system runs were more similar to each other. Thus, more stable evaluation technique was found. In the second round, the assessor made a list of nuggets by reading the system responses and classified nugget as vital or non-vital. An information nugget was defined as a fact for which the assessor could make a binary decision as to whether a response contained the nugget. Vital facts are essential to make a definition good, whereas non-vital facts act as *do not care* (they should not be awarded or penalized). The content score was computed by *F*-measure, a combination of *R* (Recall) and *P* (Precision).

$$\text{Recall} = \frac{\text{number\_retrieved\_vital}}{\text{total\_number\_vital\_on\_list}} \tag{1}$$

$$\text{Precision} = \frac{\text{number\_retrieved\_relevant}}{\text{total\_number\_retrieved}} \tag{2}$$

$$F_\beta = \frac{\left(\beta^2 + 1\right) \times \text{Precision} \times \text{Recall}}{\beta^2 \times \text{Precision} + \text{Recall}} \tag{3}$$

Calculating recall was straightforward, but this was not the case with precision. It is not easy to say how many facts are present in a sentence because strings can be substrings of other strings. A trial evaluation before the pilot showed that assessors found enumerating all concepts represented in a response to be so difficult as to be unworkable. For example, how many concepts are contained in "stars on Sunday in ABC's remake of Rear Window"? Very long answers need to be penalized in some way. A crude approximation of precision was made by giving an allowance of 100 characters for each fact. The precision was downgraded proportionately for answers longer than this allowance [2].

4   The Proposed System

We represent all text in the form of triplets. A text triplet is a three-tuple: (subject; relation; object). Here, subject and object are some entities and relation is a relationship between these two entities. This is similar to the RDF framework. It must be noted that the terms subject and object do not necessarily conform to the linguistic notions of subject and object.

We use the Stanford Open Information Extraction tool to generate text triplets for a given sentence. Text triplet representation of the answers increases the precision of the answers—we can hopefully retain only the relevant parts of the candidate sentences. This also makes evaluation easier—it is easier to check if specific triplets are found in the reference answers or not.

4.1   System Architecture

The proposed system architecture is shown in Fig. 1. We could simply do a regular expression-based search in the text corpus for sentences containing the target word and process only these sentences further. However, we see that the Stanford Open IE tool, which we use later to extract triplets, is capable of co-reference resolution. Thus, sentences, not containing the target word but containing pronouns that refer to the target word, can also be included. This requires that entire documents are selected at this stage, not just the sentences containing the target words. The following example illustrates how the Stanford Open IE tool handles co-references:

Fig. 1 Proposed system architecture

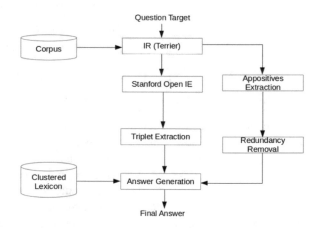

John is a nice boy. He played chess well.
John is nice
John is boy
John is nice boy
John played chess
John played well chess

It may be observed that the pronoun 'he' has been replaced with 'John.'

An information retrieval (IR) tool is used to extract documents that are relevant for the question target, which is treated as the query string for IR. Terrier [15], an open-source search engine, readily deployable on large-scale collections of documents, is used here. The IR tool is first trained on a large collection of text documents before using it. The IR tool returns a list of documents which are hopefully relevant for the given query string, along with a score. Only the top 30 documents are retained for further processing.

Then, the documents returned by the above module are given to Stanford Open IE tool as input, and text triplets are generated in this module. All the text triplets generated by the Stanford Open IE tool are not equally useful. We select only the triplets which contain question target as the subject.

Stanford Open IE tool Stanford Open IE [16] is a tool developed at Stanford University for the task of open domain information extraction. Open domain information extraction deals with identifying all the entities and relations among them present in text and extracting them. The tool first breaks a sentence into some entailed clauses. These entailed clauses are maximally shortened further, and text triplets are generated. Text triplet is a three tuple <*subject*; *relation*; *object*> where subject and object are entities, and relation is binary relation between these two entities.

Triplets generated by Stanford Open IE tool for the sentence "Abdul Kalam better known as A. P. J. Abdul Kalam (October 15, 1931 to July 27, 2015) was the 11th President of India from 2002 to 2007" are shown in Table 1. It may be observed that the text triplets 7, 8, 9, 10, and 11 are redundant, given the text triplet 6.

Answer Extraction: A good definition should include all the relevant points, must avoid or at least minimize redundancy. Therefore, simply giving out whatever triplets match the target word is not a good idea. As we have seen above, this can lead to a lot of redundancy. Also, we need to handle synonyms and equivalent terms in order to increase the recall. In this paper, we propose a method to address these issues. We develop and use what we call 'clustered lexicons.' Clustered lexicons are collections of synonyms, morphological variants, and equivalent expressions in general. For example, *originated, located, situated, lies* may all denote location of an organization. As an example, the Stanford Open IE tool may generate a triplet containing 'located' as a relation, and we can expand this to include triplets containing 'situated,' etc. More importantly, if we are looking for answer to a question of the type 'who is X?,' we know what kinds of information

Table 1 Triplets generated by StanfordOpenIE for sentence "Abdul Kalam better known as A. P. J. Abdul Kalam (15 October 1931 27 July 2015) was the 11th President of India from 2002 to 2007."

No	Subject	Relation	Object
1	Abdul Kalam	known as	15 October 1931
2	Abdul Kalam	known as	A. P. J. Abdul Kalam
3	Abdul Kalam	was President from	2002–2007
4	Abdul Kalam	was	President
5	Abdul Kalam	better known as	A. P. J. Abdul Kalam
6	Abdul Kalam	was	11th President of India from 2002 to 2007
7	Abdul Kalam	was 11th President of	India
8	Abdul Kalam	was President of	India
9	Abdul Kalam	was 11th President from	2002 to 2007
10	Abdul Kalam	was	11th President
11	Abdul Kalam	was	President of India from 2002 to 2007
12	Abdul Kalam	better known as	15 October 1931

we want in the answers. We may want to know his place and date of birth, his affiliation and position or status, his achievements, his contributions, etc. So, instead of directly working with all the large number of triplets we may have obtained, we should instead start from what kinds of information we wish to get and try and locate such triplets. We can easily avoid redundancy, and we can include only one triplet which talks about the place of birth, for example. This way, we can seek and include whatever information we need in the final answer, avoiding redundancy and including equivalent expressions as needed to enhance recall.

The clustered lexicons were developed using the same Stanford Open IE tool. We first collect certain categories of terms and put them in different clusters. For example, we put *Economics, Business, Finance, Banking, Import, and Export* in a separate cluster for type 1 questions. Similarly, we put *Scientist, Inventor, Engineer, and Subject Expert* in a separate cluster for type 2 questions. We choose these terms manually look up the definitions of the terms in a dictionary, and these definitions are given to the Stanford Open IE tool for triplet extraction. The triplets so generated are used to further enhance these clustered lexicons. The Stanford Open IE tool may give relations such as *known as, also known as, better known as*. We cut down on redundancy and retain only the term 'known' in the clustered lexicon. The clustered lexicons include nouns, verbs, and adverbs.

Appositives are noun phrases that define their adjacent nouns and occur frequently in news articles. They are useful in definitional question answering. For example, in the sentence *"The trees, some of which grow only in the Sundarbans, the world's largest mangrove swamp, were felled to pave the way for fisheries,"* appositive is, *"Sundarbans, the world's largest mangrove swamp."* Stanford Open IE tool is not able to recognize the appositives. We use a natural language Parser to extract appositives. The PCFG parser [17], which is one of the six parsers in the Stanford Lexicalized Parser v3.9.1, is used here. The parser tags appositives with the 'appos' tag. Some appositives may occur in multiple documents. We retain only the unique set of appositives.

The final list of triplets is obtained by expanding the triplets given by the Stanford Open IE tool using the clustered lexicons, and adding the appositives extracted separately. There is a need for some normalization of the text triplets. Triplets are sorted on the basis of the number of words they contain. If all words of the smaller triplet occur in a bigger triplet, then the smaller triplet is considered duplicate and eliminated. Triplets that contain the same subject and the same relation, differing only in objects, are merged.

4.2 Improvements to Evaluation

The problem with calculating precision is counting the total number of facts in the answer string. Text triplet representation of the answers makes this simple. The total number of facts present in the answer string can be taken as the total number of objects in the text triplets. For example, consider three text triplets (Gandhi; was born; on October 2, 1869) (Gandhi; was born; at Porbandar) (Gandhi; was born; as Mohan Das). Here, the total number of facts = the total number of the objects = 3 (namely, on October 2, 1869, at Porbandar and as Mohan Das). Now, recall and precision are calculated as

$$\text{Recall} = \frac{\text{number\_retrieved\_vital\_objects}}{\text{total\_number\_vital\_objects\_on\_list}} \qquad (4)$$

$$\text{Precision} = \frac{\text{number\_retrieved\_relevant\_objects}}{\text{total\_number\_retrieved\_objects}} \qquad (5)$$

5 Experiments and Results

A collection of news articles from *The Hindu* daily newspaper published during the period 2006–2010 is used here in our experiments.

Clustered lexicons were developed as described above. We choose 70 terms for type 1 and 20 terms for type 2 questions, and we collected definitions of these terms manually. These definitions were given to the Stanford open IE tool as input, and clustered lexicons were built by manual inspection and careful selection. Clustered lexicons for type 1 and type 2 questions contained 267 and 76 words, respectively.

We prepared 50 test questions selecting the terms manually, 30 for type 1 and 20 for type 2. Making a list of all the nuggets from a large collection of the documents is tedious. To make this task easy, we used another source of answers, namely Wikipedia. Now, we could do two different experiments. We collected Wikipedia

Table 2 Test Results

	What			Who			Overall		
	R	P	F	R	P	F	R	P	F
Wikipedia	0.65	0.85	0.65	0.52	0.89	0.52	0.58	0.87	0.58
The Hindu	0.44	0.80	0.44	0.29	0.83	0.29	0.39	0.81	0.39

pages for 50 question targets and made a final list of facts by reading each Wikipedia page for each question target. Further, we classified nuggets as vital or non-vital. We evaluated the system responses generated by our system from Wikipedia pages. The content score is measured by F measure. For the second experiment, we first run our system on the collection of the Hindu news articles and collect system responses. We made a list of nuggets by reading system responses. We calculated the average number of facts from nuggets list of Wikipedia. Later, we used this number as the number of retrieved nuggets (denominator in the *precision*), if the number of nuggets retrieved in the response was less than the average number of the Wikipedia nuggets.

For the key word 'Sundarbans,' the Terrier tool retrieved 134 documents. Top 30 of these were considered for further processing by the Stanford Open IE Tool for triplet extraction and Stanford Parser for extracting appositives. Stanford Open IE generated 1309 triplets, of which 179 had the key term 'Sundarbans.' Stanford parser generated 50 appositives, of which 7 had 'Sundarbans.'

From Table 2, it can be seen that when tested on news articles and Wikipedia, our system could produce a precision of 0.81 and 0.87, respectively. Wikipedia generally contains more of definitional facts than news articles. Also, short answers such as these would be generally more preferable to lengthy sentences, parts of which may be completely unrelated. See Tables 3, 4, and 5 for answers retrieved by the system.

Table 3 Results for 'Sundarbans'—The Hindu

Cluster No.	Text triplets
1	Sundarbans; are popularly referred to; to last refuge of tiger
3	Sundarbans; are; popularly referred
3	Sundarbans; is in; South 24 Parganas district of West Bengal
3	Sundarbans; is home to; estimated 425 species of wildlife including 300 species of birds
8	Sundarbans; has; has long declared, has battered by rains caused by deep depression in last two days

Table 4 Results (appositives) for 'Sundarbans'—The Hindu

Appositives
appos(Sundarbans-11, swamp-18)
appos(Aila-4, Sundarbans-9)
appos(Aila-4, Sundarbans-9)
appos(Sundarbans-7, landscape-11)
appos(Sundarbans-25, forest-33)
appos(Sundarbans-13, delta-18)
appos(Sundarbans-25, forest-33)

Table 5 Results for 'Sundarbans'—Wikipedia

Cl. No.	Text triplets
3	Sundarbans; is network of; marine streams
3	Sundarbans; is vast forest in; coastal region of Bay of Bengal
8	Sundarbans; contain; world's largest coastal mangrove forest with area

6 Conclusions

In this paper, we have presented our preliminary work on question answering for 'Who is X' and 'What is X' kinds of questions. We believe working with short and structured pieces of texts, such as the triplets we have described, would be better than working with lengthy sentences. We have shown how we can possibly extract short, crisp, and more relevant and precise answers to definitional questions. Evaluation also becomes easier.

Working with triplets may be better not only for the present task but for many other tasks in Natural Language Processing. More rigorous studies are needed to firmly establish this.

References

1. Voorhees, E.M., Dang H.T.: Overview of the TREC 2003 question answering track. In: TREC 2003, pp. 54–68 (2003, November)
2. Voorhees, E.M.: Evaluating answers to definition questions. In: 2003 Conference of the North American Chapter of the Association for Computational Linguistics on Human Language Technology: companion volume of the Proceedings of HLT-NAACL 2003-short papers, vol. 2, pp. 109–111. (2003, May)
3. David, G.: The AQUAINT corpus of english news text LDC2002T31. Linguistic Data Consortium, Web Download. Philadelphia (2002)
4. Xu, J., Licuanan, A., Weischedel, R.M.: TREC 2003 QA at BBN: answering definitional questions. In: TREC 2003, pp. 98–106 (2003, November)

5. Yang, H., Cui, H., Maslennikov, M., Qiu, L., Kan, M.Y., Chua, T.S.: Qualifier in TREC-12 QA main task. In: TREC 2003, pp. 480–488 (2003, November)
6. Echihabi, A., Hermjakob, U., Hovy, E.H., Marcu, D., Melz, E., Ravichandran, D.: Multiple-engine question answering in TextMap. In: TREC 2003, pp. 772–781 (2003, November)
7. Katz, B., Lin, J.J., Loreto, D., Hildebrandt, W., Bilotti, M.W., Felshin, S., Fernandes, A., Marton, G., Mora, F.: Integrating web-based and corpus-based techniques for question answering. In: TREC 2003, pp. 426–435 (2003, November)
8. Wesley, H., Katz, B., Lin, J.: Answering definition questions with multiple knowledge sources. In: Human Language Technology Conference of the North American Chapter of the Association for Computational Linguistics, Boston, Massachusetts, HLT-NAACL (2004)
9. Han, K.S., Song, Y.I., Rim, H.C.: Probabilistic model for definitional question answering. In: 29th Annual International ACM SIGIR Conference on Research and Development in Information Retrieval, pp. 212–219. ACM, New York, USA (2006)
10. Cui, H., Kan, M.Y., Chua, T.S.: Generic soft pattern models for definitional question answering. In: 28th Annual International ACM SIGIR Conference on Research and Development in Information Retrieval (SIGIR'05). pp. 384–391, ACM, New York (2005) http://dx.doi.org/10.1145/1076034.1076101
11. Cui, H., Kan, M.Y., Chua, T.S.: Soft pattern matching models for definitional question answering. ACM Trans. Inf. Syst. **25**(2), 8 (2007). https://doi.org/10.1145/1229179.1229182
12. Chen, Y., Zhou, M., Wang, S.: Reranking answers for definitional QA using language modeling. In: 21st International Conference on Computational Linguistics and the 44th Annual Meeting of the Association for Computational Linguistics, pp. 1081–1088, Sidney (2006)
13. Pasca, M.: Answering definition questions via temporally-anchored text snip-pets. Third Int. Joint Conf. Nat. Lang. Process. **1**, 411–417 (2008)
14. Lita L.V., Hunt W.A., Nyberg E.: Resource analysis for question answering. In: ACL 2004 Interactive Poster and Demonstration Sessions. Association for Computational Linguistics (2004)
15. Macdonald, C., McCreadie, R., Santos, R.L, Ounis, I.: From puppy to maturity: experiences in developing terrier. In: OSIR at SIGIR, pp. 60–63 (2002)
16. Angeli, G., Premkumar, M.J., Manning, C.D.: Leveraging linguistic structure for open domain information extraction. In: 53rd Annual Meeting of the Association for Computational Linguistics (2015, July)
17. Klein, D., Manning, C.D.: Accurate unlexicalized parsing. In: 41st Annual Meeting of the Association for Computational Linguistics (2003)

Minimum Cost Fingerprint Matching on Fused Features Through Deep Learning Techniques

Pavuluri Vidyasree and Somalaraju ViswanadhaRaju

Abstract Biometrics is a scientific order that includes techniques for recognizing the individuals with their physical or behavioral attributes. The most widely recognized physical and behavioral traits of a man utilized for authentication are as per the following: fingerprint palm print, iris, retina, DNA, ear, signature, speech, keystroke elements, motion and hand-geometry. Among them, fingerprint is recognized and accepted as a universal trait. Fingerprint has a unique pattern that can easily recognize the individual. The key motivation behind this paper is to enhance the accuracy rate of human recognition by adhering to the factors like revocability, speed and with minimum cost. The high-level recognition rate is achieved through fusion of multi features like ridge endings and ridge bifurcations of fingerprint. Autoencoder (AE) is an unsupervised deep learning technique, exercised on the fused multi-feature representation template to address the various spoofing attacks and also achieve revocability. Minimum cost matcher (MCM) is applied to maximize the efficiency of the multi-representation system. The experimental results explain the high-level accuracy of the proposed system in human identification.

Keywords Autoencoder · Fingerprint recognition system · Minimum cost matcher · Multi-Representation system · Ridge endings · Ridge bifurcations

1 Introduction

The present world is tightly combined with the biometric advancements that play a great role in various areas like providing the security for the sensitive data, authenticating the person and also in network security. In this era, fingerprint

P. Vidyasree (✉)
Research Scholar of JNTU Hyderabad, Asst. Professor of Stanley
Women's Engineering College, Hyderabad 500001, India
e-mail: vidyasreepavuluri@gmail.com

S. ViswanadhaRaju
Professor, JNTUH CEJ, Hyderabad 500085, India
e-mail: svraju.jntu@gmail.com

© Springer Nature Singapore Pte Ltd. 2020
K. S. Raju et al. (eds.), *Data Engineering and Communication Technology*,
Advances in Intelligent Systems and Computing 1079,
https://doi.org/10.1007/978-981-15-1097-7_11

accessibility gained a prominent place by not only providing the authentication and also securing the sensitive information. Fingerprint recognition alludes to the robotized technique for checking a match between two human fingerprints. However, fingerprints innately have fundamentally higher data content, and the programmed acknowledgment frameworks cannot utilize all the accessible unfair data because of confinements, for example, poor picture quality and mistakes in highlight extraction and coordinating stages. A unique mark surface has diverse nearby features like edges, valleys, circles are named as details which vary from individual to individual. A single bended portion on the finger called as an edge, and the area between the two edges is named as ridge bifurcation or valley. Edge decides different kinds of minor focuses among them edge endings and bifurcations are generally critical one. Many of the researchers have done the experiments on the fingerprint recognition system. Dale et al. proposed a fingerprint recognition application using more data other than minutiae is much useful. In this paper, discrete cosine transform (DCT) based element vector for unique fingerprint representation and matching [1]. The major thing in this proposed model was carried out without preprocessing. Amornraksa and Tachaphetpiboon used DCT for feature extraction on discrete image, and performance was calculated by the k-nearest neighbor (k-NN) classifier [2]. Choosang and Vasupongayya used fingerprint recognition scheme for identifying the person in case of accidents [3]. Sugandhi et al. utilized a finger vein recognition technique and PIC16F877A microcontroller to prepare a high authentication system [4]. Ma et al. demonstrated a diffusion tensor imaging [DTI] unique finger impression development to analyze quantitatively white issue tissue by using the tract-based spatial estimations in stroke course of action [5]. Rodionov et al. proposed an estimation in perspective of a fingerprinting method composed with data blend and conjecture calculations used for patient tracking [6]. Ashwin et al. proposed another model of a unique mark-based allowing framework for driving [7]. Lin et al. showed a unique mark-based remote client confirmation plot that makes home human services clients remember them through a cell phone [8]. Rajeswari et al. shown that the multi unique mark framework acquires precision than a solitary unique finger impression and furthermore gives the security to the information in the cloud. Extraction of the one of a kind highlights is regularly required for human acknowledgment [9]. Kumari and Suma presented an experimental study of dimensionality reduction in multimodal biometrics using principal component analysis (PCA) based on feature level fusion [10]. Therefore, in this paper, we proposed the feature level fusion of multi features like ridge endings and ridge bifurcations of fingerprint in order to maximize the accuracy of human recognition rate using autoencoder and minimum cost matcher [11]. The rest of the paper organized as follows. Section 2 deals with the proposed methodology. Section 3 discusses the experiment conducted on the fingerprint dataset. Finally, Sect. 4 concludes the paper.

2 Methodology

Fingerprint recognition is one of most prominent and exactness of biometric advances. These days, it is utilized in some real applications. In any case, perceiving fingerprints in low quality pictures is as yet an extremely perplexing issue. The proposed methodology as shown in Fig. 1 enhances the recognition rate of an individual through unsupervised deep learning technique.

Step 1: Image Acquisition

The fingerprints acquired from the scanned are converted into the grayscale images.

Step 2: Preprocessing

Preprocessing aims to remove the noise or dirt on the fingerprint and then performs the binarization and thinning process. Binarization is done in order to convert the gray scale image to the black and white image. Thinning is the process is done on the binarized fingerprint image using the block filter, in order to address the disturbances caused by binarization and also enhances the thickness of the ridge lines of one-pixel width that assures the true representation of the features or minutiae points.

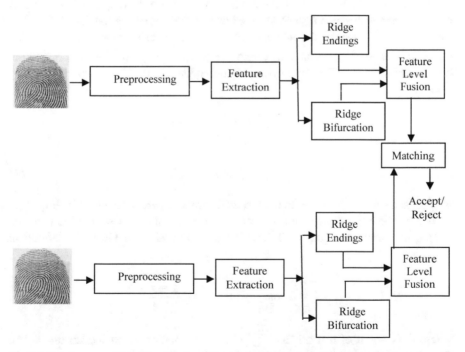

Fig. 1 Human recognition using the fused features of fingerprint recognition system

Step 3: Feature Extraction

Ridge endings and ridge bifurcations are the important features of the fingerprint recognition. The crossing number (CN) idea is the most consistently embraced technique for details extraction in which the edge stream configuration is eight-associated. The bifurcation details highlights are recovered by checking the neighborhood of each edge pixel in the picture using a 3×3 window. The CN esteem is then determined which is described as a large portion of the entirety of the contrasts between sets of bordering pixels in the eight-neighborhood. In the event that the focal pixel or center pixel is 1 and has precisely 3 one-esteem neighbors, at that point, the focal pixel is a ridge bifurcation, and focal pixel or center pixel is 1 and has just 1 one-esteem neighbor, at that point, the focal pixel is an edge finishing as appeared. The true information of the ridge endings is extracted using the Hough transformation which is insensitive to noise.

Step 4: Feature Level Fusion

The extracted ridge ending and ridge bifurcation features from heterogeneous algorithms are concatenated using the horizontal fusion method. Autoencoder is applied on the fused template for rich and robust feature extraction and also reduces the reconstruction error. It helps to mitigate the dimensionality of the fingerprint input image into the smaller representations at the hidden layer. Decoding is performed on the encoded image at the hidden layers results the output image which is as similar as input image $I_i = O_i$ On the basis of the approximation function, it sets the input value to the target value. Hidden unit "I" gets activated to "q_i." On specific input "h," external representation cannot be performed, $q_i^{(2)}(h)$ helps in activation of the hidden unit. The activation of the hidden unit "i" is expressed using "Eq. (1)."

$$I_i^{\wedge} = \frac{1}{x} \sum_{j=1}^{x} \left[q_i^{(2)}(h(j)) \right] \tag{1}$$

$$I_i^{\wedge} = I \tag{2}$$

where I is a small value close to zero termed as sparse parameter given in "Eq. (2)."

If $I = 0.05$ means activation of each neuron $h = 0.05$, then condition on the hidden unit must be equal to 0. The penalty term added deviates $I_i^{\wedge}$ defined in "Eq. (3)."

$$\sum_{i=2}^{I_2} I \log \frac{I}{I_i} + (1 - I) \log \frac{1 - I}{1 - I_i} \tag{3}$$

where I_2 is count of neurons and "$\sum i$" is a summation of all hidden units. The penalty term is obtained from Kullback–Leibler (KL) divergence represented in "Eq. (4)."

$$\sum_{i=0}^{I_2} \text{KL}(I||I_i^\wedge) \tag{4}$$

KL divergence has the rule that either $\text{KL}(I||I_i^\wedge) = 0$, or it maximizes monotonically and mitigates the penalty term, $I_i^\wedge$ becomes equal to I.

This helps to extract the enhanced enriched features from the fused template aims to increase the accuracy of individual authentication by filling the gaps or missed features.

Step 5: Matching

Minimum cost matcher (MCM) helps to obtain the similarity scores between the query and enrolled templates. This process is carried out through column or row-wise minimization. Column-wise minimization is considered in this proposed methodology to enhance the accuracy of recognition rate. Consider the minimum value of each column and assign the minimum area of either of the enrolled or query areas. That assigned column is eliminated, and the process is continued for all the columns in the matrix to obtain the matching scores. Match scores are obtained by the Monge technique communicated in "Eq. 5."

$$b_{pq} = |a_p - e_q| \tag{5}$$

where $b_{p,q}$ indicates the distance between query and enrolled templates, and a_p and, e_q are the query and enrolled feature fused vectors, respectively. The threshold value is obtained at which false acceptance rate (FAR) and false rejection rate (FRR) become equal, i.e., equal error rate (EER) to recognize the individual within minimum repetitions. "Equations (6) and (7)" give the recognition rate of proposed system.

$$\text{accuracy} = 100 - \text{EER} \tag{6}$$

$$\text{EER} = \frac{\text{FAR} + \text{FRR}}{2} \tag{7}$$

3 Experiment Result and Discussion

The proposed finger recognition system performance is measured using JNTUK UCEV real fingerprint dataset taken from JNTUK CEV students of age 20–25. The dataset consists of 200 images of 40 persons in JPEG format with the size of 260 * 300 pixels. The proposed system is demonstrated from "Figs. 2, 3, 4, 5, 6, 7, 8, 9." The experiment starts by considering the region of interest (ROI) of 180 * 180 pixels. The acquired images of enrolled and query images shown in "Fig. 2" are sent to the preprocessing stage for thinning, and binarization is

Fig. 2 Enrolled and query images of use

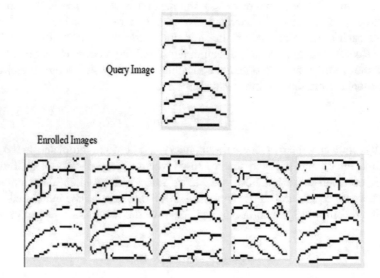

Fig. 3 Preprocessed images and enrolled users

depicted in the "Fig. 3." "Figures 4 and 5" illustrate extracting the ridge ending and ridge bifurcation features using the Hough transformation and cross number techniques. "Figure 6" shows the feature level fused images of enrolled and query fingerprints of user. "Figure 7" illustrates holistic features extraction of the user fingerprint using the autoencoder technique. Minimum cost matcher (MCM) is used to obtain the similarity scores between query and enrolled users fused bimodal feature templates which ameliorate the accuracy and efficiency of proposed system. "Figure 8" depicts when DCT and PCA techniques are employed on the fused fingerprint template for feature extraction in the place of autoencoder. The efficiency of proposed system is calculated on 250 fingerprint sample images.

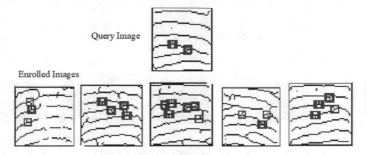

Fig. 4 Ridge bifurcation extraction for enrolled and query users

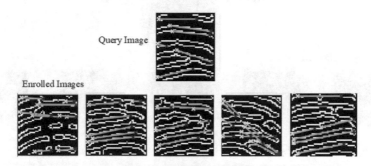

Fig. 5 Ridge ending extraction for enrolled and query users

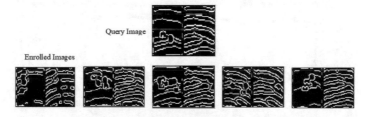

Fig. 6 Feature level fusion of ridge endings and ridge bifurcations for enrolled and query users'

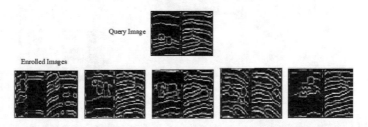

Fig. 7 Rich feature extraction using autoencoder for enrolled and query users

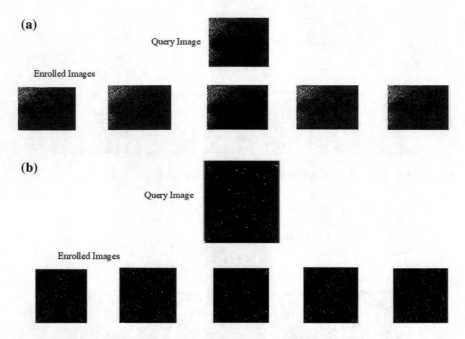

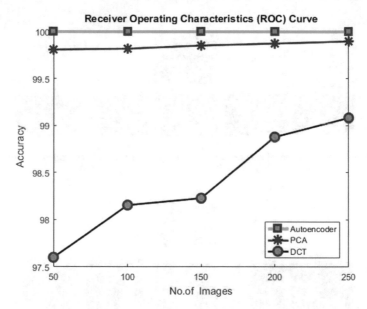

Fig. 8 a, b Feature extraction using DCT and PCA for enrolled and query users

Fig. 9 ROC curve depicting the high performance of proposed system compared to PCA and DCT

"Figure 9" proves that MCM matcher enhances the accuracy of proposed multi representation of fingerprint recognition system over the traditional techniques PCA and DCT.

4 Conclusion

Fingerprint recognition alludes to the robotized technique for checking a match between two human fingerprints. An unsupervised deep learning technique is employed on the feature level fused fingerprint template for the extraction of the holistic features that enhances the recognition of the user. Ridge endings and ridge bifurcations are the robust features of the fingerprint are extracted using the cross number technique, and rich information of the ridge endings are highlighted using the Hough transformation. Minimum cost matcher maximizes the accuracy of individual recognition and performance of the proposed system with minimum number of iterations. Convolution neural network is applied in the proposed system for deeper representation and dimensionality reduction that utilize less database space.

References

1. Dale, M.P., Joshi, M.A., Sahu, M.K.: DCT feature based fingerprint recognition. In: Intelligent and advanced systems. International conference on ICIAS 2007, IEEE, pp. 611–615. Nov 2007
2. Amornraksa, T., Tachaphetpiboon, S.: Fingerprint recognition using DCT features. Electron. Lett. **42**(9), 522–523 (2006)
3. Choosang, P., Vasupongayya, S.: Using fingerprints to identify personal health record users in an emergency situation. In: Computer science and engineering conference (ICSEC), 2015 international, IEEE, pp. 1–6. Nov 2015
4. Sugandhi, N., Mathankumar, M., Priya, V.: Real time authentication system using advanced finger vein recognition technique. In: Communications and signal processing (ICCSP), 2014 international conference on, IEEE, pp. 1183–1187. Apr 2014
5. Ma, H.T., Ye, C., Wu, J., Yang, P., Chen, X., Yang, Z., Ma, J.: A preliminary study of DTI Fingerprinting on stroke analysis. In: Engineering in medicine and biology society (EMBC), 2014 36th Annual international conference of the IEEE, pp. 2380–2383, Aug 2014
6. Rodionov, D., Kolev, G., Bushminkin, K.: A hybrid localization technique for patient tracking. In: Engineering in medicine and biology society (EMBC), 2013 35th annual international conference of the IEEE, pp. 6728–6731. July 2013
7. Ashwin, S., Loganathan, S., Kumar, S. S., Sivakumar, P.: Prototype of a fingerprint based licensing system for driving. In: Information communication and embedded systems (ICICES), 2013 international conference on IEEE, pp. 974–987. Feb 2013
8. Jin, Q., Jeon, W., Lee, C., Choi, Y., Won, D.: Fingerprint-based user authentication scheme for home healthcare system. In: Ubiquitous and future networks (ICUFN), 2013 fifth international conference on IEEE, pp. 178–183. July 2013

9. Rajeswari, P., Raju, S. V., Ashour, A. S., Dey, N.: Multi-fingerprint unimodel-based biometric authentication supporting cloud computing. In: Intelligent techniques in signal processing for multimedia security, pp. 469–485. Springer International Publishing (2017)
10. Kumari, P.A., Suma, G.J.: An experimental study of feature reduction using PCA in multi-biometric systems based on feature level fusion. In: Advances in electrical, electronic and systems engineering (ICAEES), international conference on IEEE, pp. 109–114. Nov 2016
11. ViswanadhaRaju, S., Vidyasree, P., Gudavalli, M., Chandra Sekhar, B.: Minimum cost fused feature representation and reconstruction with autoencoder in bimodal recognition system. In: Proceedings of the international conference on big data and advanced wireless technologies BDAW '16'. https://doi.org/10.1145/3010089.3010102. Nov 2016

Enhanced Edge Smoothing for SAR Data Using Image Filter Technique

G. Siva Krishna and N. Prakash

Abstract The SAR is usually corrupted by some surplus speckle formed. These speckles have multiplicative noise, which appears like a grainy pattern in the SAR image. This performs an accurate interpretation of SAR images. The aim of this work was to remove the noise and to accurately classify the LULC facts with quality evolution. The SAR images play an important key role in earth observation applications using high resolution for all-weather conditions and all times. The SAR images an effect of coherent handing out of a mixture of regions and uses a variety of applications like as crop estimation, Land Use Land Cover (LULC), one of the military application is that the target detection, etc. The SAR images are high-resolution LULC facts, and still, it includes the noise. The LULC images continuously for collecting. The traditional techniques (Lee Filter, Gamma Filter) which are in use are not effective to identify the LULC facts and features of the SAR images. These never remove all the noises in SAR images especially the noise like "salt and pepper." Therefore, in this paper, the researcher proposes a new technique "Enhanced Discontinue Image Filter" which is window size based and effectively visualizes the SAR images with 89.1% accuracy to determine the ground truth value.

Keywords SAR · LULC · Filter technique · Lee filter · Enhanced gamma filter

1 Introduction

The main aim of the SAR image is collecting information from the earth's surface. The SAR images have the facility to employ any time. These information having own speckle, a state of art phenomenon that outcome from the coherent summation

G. Siva Krishna (✉) · N. Prakash
Department of IT, BSACIST, Chennai, India
e-mail: sivakrishna_it_phd_2016@crescent.education

N. Prakash
e-mail: praksah@crescent.education

© Springer Nature Singapore Pte Ltd. 2020   141
K. S. Raju et al. (eds.), *Data Engineering and Communication Technology*,
Advances in Intelligent Systems and Computing 1079,
https://doi.org/10.1007/978-981-15-1097-7_12

of the backscattered signal. The SAR images have a top-notch deal of speckle noise because of backscattered radar echoes. The state of art phenomenon does not best deteriorate the picture; however, it makes goal detection, discrimination and terrain classification tough. The speckle of SAR images collected by different polarization. The SAR polarizations are of two types [1]. The first one, the HH polarization used horizontal transmission to horizontal receiving, and second, the VV polarization used vertical transmission to vertical receiving. The other combinations also (like as VH, HV) VH image is radar transmits vertically polarized signals then receives the horizontally polarized return signal. HV image radar transmits is horizontally polarized signals but receives vertically polarized signals. These polarizations are considered for calculating LULC facts. However, the sigma_db values measure the image with different polarization as shown in Fig. 1.

The SAR radar is collecting several images with different frequencies, which are C, X, Ku, etc. as shown in below Fig. 2.

There are many existing techniques used to remove the noise in the SAR images. The important techniques are Lee Sigma Filter and Enhanced Gamma Filter techniques [1]. The Lee Sigma Filter technique is introduced in 1983. It is primarily

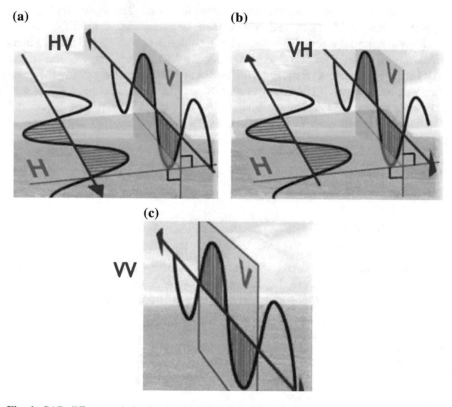

(a) **(b)**

(c)

Fig. 1 SAR different polarization **a** HV and VH, **b** HH and **c** VV

Fig. 2 Different bands frequency for SAR

based on the easy idea of sigma prospect, and it will become fairly effective in speckle filtering because it makes use of superior synthetic-aperture radar (SAR) knowledge with high-measurement data of considerable dimensions and green speckle filtering algorithms. There are many limitations in the method. It calculates biased estimation too in SAR images. When analyzing scatter plot and histogram, it was realized that a set of the SAR image removal methods, that are very useful in remote sensing systems, are visual representations of histograms, the different existing filters, different image corrections, unsupervised classification, several environmental indices computation, among others.

The main objective of this work was the creation of the Enhanced Discontinue Image Filter inspired on edge detection techniques including a new set of filter operations. The EDIF plug-in can be spat into following group of operations: pre-processing, processing and classification with quality evolution. The proposed method, Bayesian nonlocal (BNL) mean filter also removes the speckle noise Yougwei et al. [2, 3]. The limitations of this method are as follows: It is not capable to identify the sigma range selection and strong reflective scatter. Jiao et al. [4] focuses on deficiencies of the Lee Sigma Filter. Limitations of the approach are as follows: It could calculate simplest sigma tiers from one-to-four search for the facts series, and additionally, it needs many resolutions for statistics calculation. Mohanan et al. [5] proposed frequency and spatial domain techniques for removing noise. The wiener filter which is performed in the spatial and adaptive threshold was used in wavelet transform for speckle reducing. Jones et al. [6] proposed wavelet soft threshold and curvelet tender threshold method to get residual image, then make use of the denoised residual picture to enhance the curvelet denoised picture. Bhateja et al. [7–10] proposed contourlet technique is sigmoid function for enhanced image quality. The contourlet reworks to correctly constitute the directional facts and intrinsic geometrical systems of the SAR image. Chen et al. [11–16] proposed particle filter sample texton, the particle filter is introduced to trace the key points, and therefore, the half-tracked density function $q(x)$ is adjusted perpetually by observant the connexion between the pixels and therefore the key points within the neighborhood of a particular size.

Weiping et al. [17–19] proposed wavelet tender threshold and curvelet gentle threshold approach to get the residual picture, after which uses the denoised residual picture to enhance the curvelet denoised picture. Proposed algorithm solves this trouble to a positive quantity. Benediktsson et al. [10, 20–23] proposed strategies for choosing the parameters in the adaptive sigmoid thresholding the use of SURE. The SURE estimate changed into best carried out at the high-excessive (HH) subband of the wavelet converted SAR photo and that minimization changed into used to select the proper value of the coefficient for the sigmoid thresholding.

Diego et al. [15, 16, 24–26] proposed that the novel gentle class with the clean combination of contributes. The writer estimate of homogeneity, and improve through nonlocal and multi-resolution processing steps. The hindrance is for huge annotated database need to educate and test. Sun et al. [8] proposed two-steps sparse decomposition method. In this first sparse, decomposition is classical spares representation over complete dictionary and second is measurement from sparse coefficients as the criterion to identify a principal sub-dictionary with linear combination. The limitation is in subspace decomposition is to identify the important atoms form the similar blocks only. Fang et al. [27] proposed nonsubsampled contourlet transform (NSCT) to calculate geometric information of SAR images. It depends on neighborhoods coefficients.

Uslu et al. [28–30] proposed two types of curvelet-based totally feature extraction techniques. They are CBIR based totally on Gaussian distribution parameter estimation for each curvelet subband and histograms analysis. The limitation in this proposed method is original date applied on data partitioning scheme is in limited bin. Moisan et al. [31–35] proposed the multiplicative speckle elimination, in this method the logarithmically converted picture, and then to use it in a variational version constructed for noise elimination. Limitation of this proposed technique is dictionary-based totally strategies to get better textures.

This paper proposes "Enhance Discriminating Image Filter" technique which finds the scatter resolution distribution very effective than existing methods. It calculates different resolution sigma values through scattering coefficients (sigma naught). This paper is designed as follows: the proposed work, explanation, experimental result and discussion. Finally, the paper is ended with conclusion and destiny work. The major contribution of this research work is stated as follows:

- Here, the effective noise removal over the SAR images is performed by the soft threshold for this denoising technique and is identified from the image filtering technique named as Enhanced Discontinuities Image Filter. There are two different noises considered in this proposed methodology such as speckle noise and Gaussian noise.
- The three features such as texture, color and shape are extracted by Lee, Gamma MAP and EDIF, respectively.
- Furthermore, the specific retrieval process from the SAR images is made by image classification. And then the correlation among the label images to unlabeled images is performed to narrow down the respected images.

2   Materials and Methods

The SAR image data collected from RADARSAT-2, Nellore in Andhra Pradesh, India, as geographically this image has many LULC facts. The RADARSAT-2 product had (RS2-SLC-FQ28-DEC-24-AUG-2011) 8 m resolution, 56 mm wavelength and 5.4 GHz frequencies. However, the SAR radar is collecting several images with different frequencies, which are C, X, Ku, etc. Figure 2 shows different bands for SAR and dataset description is shown in Table 1.

The proposed method has two steps: (i) The first step is to collect the SAR image and make a subset of the image. This subset image has noise, so now remove the noise, and before going to this need, find the calibrated value for the subset image. Then reproject the subset image for ortho-rectification. Now, calculate the statistical values as shown in Table 2. (ii) In the second step, filter technique is applied, however, to find edges of the image. The edge-detect values are calculated for each region, and the values of the region are related to LULC facts. The SAR subset image calibrates values which are equal to LULC facts. By considering the LULC facts, the researcher can identify the LULC facts easily in the subset image. The flow of the block diagram for the proposed method is shown in Fig. 3.

The problem statement in the image is analyzed in this section and also it states how our proposed methodology gives the solution for this problem. The problem of image retrieval is detailed below:

- Selection of feature descriptors.
- Works only at appropriate noises.

To consider the SAR image for noise removing for edge smoothing technique, the proposed method is using for despeckling the SAR image. The SAR image has noise; it can be removed by filter techniques like Lee, Gamma MAP with window size. The window sizes considered have 7 * 7 and 9 * 9, apply both window sizes on the SAR image. However, the SAR images noise removed and the statistic values measured by the normalized radar cross-section.

$$\sigma(\text{DB}) = 10 \log_{10}(\text{energy ratio})$$

Table 1 SAR dataset description

RADARSAT-2	
Acquisition type	Fine quad polarization
Product type	SLC
Date	24-Aug-2011
Pixel spacing	4.733 m
Swath	25 km
Approximate resolution	Range: 4 m Azimuth: 4 m
Incidence angle	20–41°

Table 2 Statistical values

pixel_x	pixel_y	Sigma0_HH_dB_mean	Sigma0_HH_dB_sigma
2461	181	−2.4251	0.5112656
2460	182	−2.77664	1.0606371
2459	183	−3.6347	1.1628313
2459	184	−4.04568	1.6535006
2458	185	−5.40165	0.95462483
2457	186	−6.01107	0.860686
2456	187	−6.84015	1.3764964
2456	188	−6.7555	1.3702612
2455	189	−6.09886	2.5088332
2454	190	−5.84233	1.9841009
2453	191	−6.41975	0.9454716
2452	192	−6.06551	0.6582491
2452	193	−5.61332	0.86838305
2451	194	−5.34032	1.1551288
2450	195	−4.12271	0
2449	196	−4.0227	0.15000607
2449	197	−4.0227	0.15000607
2448	198	−3.96091	0.22986937
2447	199	−3.7278	0.6240947
2446	200	−4.37137	1.3444822

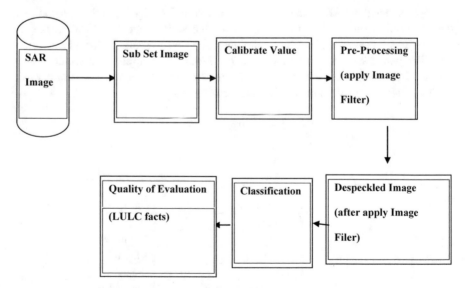

Fig. 3 Proposed methodology for block diagram

where

$$\text{energy ratio} = \frac{\text{recevied energy by sensor}}{\text{energy reflected anisotropic way}}$$

The statistical values taken from the histogram and scatter plot are shown in Fig. 4. From these statistical values, we can understand the SAR image has not well detected the edges, and for this use, Enhanced Discontinuities technique was applied. Then, the images are compared with speckle-removed SAR image. However, the result visualizes more clearly.

3 Experimental Environments and Result Discussion

The proposed methodology is simulated by using MATLAB simulator software version 2018a. The entire SAR image retrieval is made by using I3 system with 3 GB RAM. Besides, the better performance of the proposed method is evaluated by using any one of the despeckling techniques. The proposed methodology is analyzed in terms of LULC facts.

Dataset description

The proposed methodology uses RADARSAT-2 dataset to evaluate the image retrieval process. There are five different categories that are used from the RADARSAT-2 dataset such as farmland, ocean, medium density residential, mixed forest and water bodies. The proposed methodology applied Enhanced Discontinues Image Filter which is window size based only. The Lee Filter and Gamma MAP Filter scatter plots are compared with proposed method, and then the observation of those scatter plots is very carefully done. The scatter plot identifies the objects with different polarization; those are VH to HH, VH to HV and VH to VV. These scatter plot values are located in various sigma naught, and the sigma naught values give object identification based on LULC facts. However, the proposed method scatter plot values are clearly located with a sigma naught value, and then the scatter plot value is shown in Fig. 4.

From the scatter plots with different polarization, we observed that the scatter plots clearly visualize the SAR image of LULC facts. This comparison can also be done with proposed work as shown in below diagrams. The proposed work scatter plot very effectively identifies the objects form the SAR images. Histogram deed could be a projected methodology that produces associate output image with a standardized distribution of pixel intensity and means the bar chart of the output image is planate and extended consistently. A histogram equalization would improve the image performance.

The SAR image was scaled from 16 bits to 8 bits, and the entire filter was clipped to district extension. The scale was performed in order to reduce the processing time. Therefore, the application was created incorporating the scale

Lee Scatter plot:

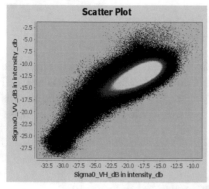
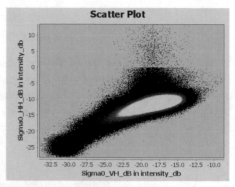

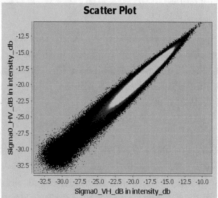

Gamma Map Scatter plot:

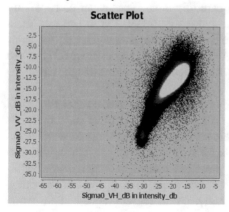
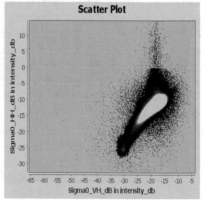

Fig. 4 Scatter plot and histogram Lee, gamma MAP and EDIF

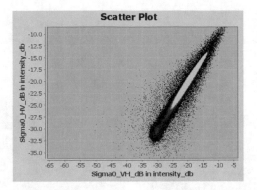

Enhanced Discontinuities Image Filter (EDIF) Scatter plot:

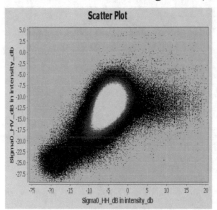
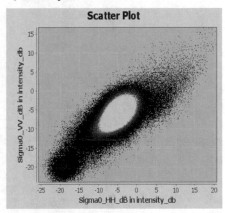

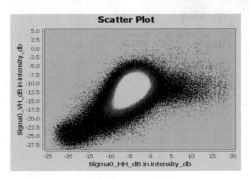

Fig. 4 (continued)

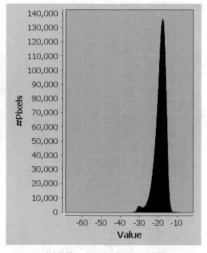
(a) Histogram for Lee Filter

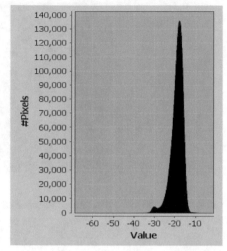
(b) Histogram for Gamma Map

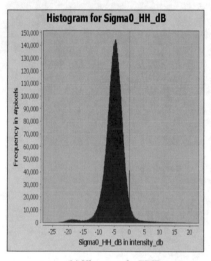
(c) Histogram for EDIF

Fig. 4 (continued)

operation in histogram functionality. Figure 4-histogram a, b and c show the combination of histogram VV, HH and VH created and saved as an image. From Fig. 4-histogram a, b and c, it was concluded that the image has very low contrast and brightness. Then histogram equalization would improve the SAR image performance. Now, the researcher has to classify the SAR image using image analysis; however, the statistical analysis of SAR image is classified with 16 and 14 classes.

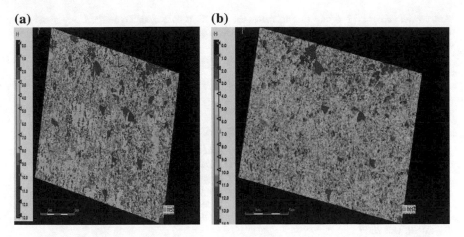

Fig. 5 Classification image with 14 classes and 16 classes

For these classified purpose applied KNN clustering, the SAR image was classified up to 89.1% of ground truth value and is shown in Fig. 5.

The classification functionality was tested, and the result was compared with a supervised classification obtained with the KNN from image analysis. The training areas were based in Nellore in Andhra Pradesh Land Cover 2011 v.2 classes 14 and 16. Four training areas were defined: water, vegetation, buildings and agriculture. Through the proposed filter technique, an unsupervised classification with the same number of classes was performed. Figure 5 presents the unsupervised classifications. The accuracy assessment generated from the unsupervised classification showed an overall accuracy of 89.1% and Kappa statistic of 0.88, which indicates a good agreement between the maps generated from image and the reference data.

4 Conclusion

The SAR image has noise; by removing the noise, it is easy to identify the image. The main objective of this work was the creation of the Enhanced Discontinue Image Filter inspired on edge detection techniques including a new set of filter operations. Therefore, "Enhanced Discontinuities edge detection technique" is applied to classify the SAR image and get 89.1% ground truth value. This research compares image smoothing with all existing techniques and reveals that Enhanced Discontinuities edge technique is more efficient than Lee Sigma Filter and Enhanced Gamma Filter techniques for finding features of SAR image. The researcher gives the suggestion to the research community that they can apply edge detection techniques based on the window size of SAR image for finding more features with different locations or areas.

Acknowledgements I would like to thank the NRSC Bala Nagar, Hyderabad, my supervisor, teaching staff, non-teaching staff and Head Institution of B. S. Abdur Rahman Crescent Institute of Science and Technology.

References

1. https://www.nest.org
2. Lee, J., Wen, J.H., Ainsworth, T.L., Chen, K.-S., Che, A.J.: Improved sigma filter for speckling filtering of SAR image. IEEE **47**(1), 202–213 (2009)
3. Scarpa, G., Verdoliva, L.: SAR despeckling based on soft classification. In: Proceedings of IEEE international geoscience and remote sensing symposium 2015, pp. 2378–2381
4. Li, Y.W., Jiao, L.C.: Bayesian nonlocal mean filter for SAR image despeckling. In: Proceedings of Asian-pacific conference Synthetic aperture Radar, Xian, China, pp 1096–1099. Oct 2009
5. Mohanan, P., Rajesh, M.R., Mridula, D.: Speckle noise reduction in images using wiener filtering and adaptive wavelet thresholding. IEEE, 2860–2863. Feb 2016
6. Finn, S., Glavin, M., Jones, E.: Echocardiographic speckle reduction comparison. IEEE **58**(1), 82–101 (2011)
7. Bhateja, V., Gupta, A., Tripathi, A.: Despeckling of SAR Images in contourlet domain using a new adaptive thresholding. IEEE, pp. 1257–1261. Feb 2013
8. Sun, H., Sang, C.-W.: Two-step sparse decomposition for SAR image despeckling. IEEE GRSL **14**(8). Aug 2017
9. Zhong, H., Li, Y., Jiao, L.C.: SAR image despeckling using bayesian non-local mean filter with sigma preselection. IEEE **8**(4), 804–813 (2011)
10. Baronti, S., Alparone, L., Garzelli, A.: A hybrid sigma filter for unbiased and edge-preserving speckle reduction. In: Proceedings of IGARSS, Florence, Italy, pp. 1409–1411. July 1995
11. Chen, D., He, C., Zhuo, T., Zhao, S., Yin, S.: Particle filter sample texton feature for SAR image classification. IEEE GRSL **12**(5), 1141–1145 (2015)
12. Siva Krishna, G., Prakash, N.: Enhanced noise removal technique based on window size for SAR data. IJPAM **114**(7), 227–235 (2017)
13. Liu, F., Zhang, W., Jiao, L.C., Hou, B., Wang, S., Shang, R.: SAR image despeckling using edge detection and feature clustering in bandlet domain. IEEE Geosci. Remote Sens. Lett. **7**(1), 131–135 (2010)
14. Cozzolino, D., Parrilli, S., Scarpa, G., Poggi, G., Verdoliva, L.: Fast adaptive nonlocal SAR despeckling. IEEE Geosci. Remote Sens. Lett. **11**(2), 524–528 (2014)
15. Verdoliva, L., Gaetano, R., Ruello, G., Poggi, G.: Optical-driven nonlocal SAR despeckling. IEEE Geosci. Remote Sens. Lett. **12**(2), 314–318 (2015)
16. Parrilli, S., Poderico, M., Angelino, C.V., Verdoliva, L.: A nonlocal SAR image denoising algorithm based on LLMMSE wavelet shrinkage. IEEE Trans. Geosci. Remote Sens. **50**(2), 606–616 (2012)
17. Junzheng, W., Weidong, Y., Hui, B., Weiping, N.: A despeckling algorithm combining curvelet and wavelet transforms of high resolution SAR images. IEEE **6**, 302–305 (2010)
18. Cheng, J., Wang, N., Tellambura, C.: Probability density function of logarithmic ratio of arithmetic mean to geometric mean for Nakagamim fading power. In: Proceedings of 25th Biennial symposium communications, pp. 348–351. May 2010
19. Hao, Y., Feng, X., Xu, J.: Multiplicative noise removal via sparse and redundant representations over learned dictionaries and total variation. Signal Process. **92**(6), 1536–1549 (2012)
20. Song, J., Xu, B., Cui, Y., Li, Z., Zuo, B., Yang, J.: Patch ordering-based SAR image despeckling via transform-domain filtering. IEEE JAEORS **8**(4), 1682–1695 (2015)

21. Lee, J.S.: A simple speckle smoothing algorithm for synthetic aperture radar images. IEEE Trans. Syst. Man. Cybern. **SMC-13**(1), 85–89 (1983)
22. Lee, J.S.: Digital image enhancement and noise filtering by use of local statistics. IEEE Trans. Pattern. Anal. Mach. Intell. **PAMI-2**(2), 165–168 (1980)
23. Milne, K., Dong, Y., Forster, B.C.: Toward edge sharpening: a SAR speckle filtering algorithm. IEEE Trans. Geosci. Remote Sens. **39**(4), 851–863 (2001)
24. Poggi, G., Scarpa, G., Gragnaniello, D., Verdoliva, L.: SAR image despeckling by soft classification. IEEE JAEORS **9**(6), 2110–2130 (2016)
25. Gomez, L., et al.: Supervised constrained optimization of Bayesian nonlocal means filter with sigma preselection for despeckling SAR images. IEEE Trans. Geosci. Remote Sens. **51**(8), 4563–4575 (2013)
26. Zhong, H., Li, Y., Jiao, L.: SAR image despeckling using Bayesian nonlocal means filter with sigma preselection. IEEE Geosci. Remote Sens. Lett. **8**(4), 809–813 (2011)
27. Fang, L., Xia, C., Licheng, J., Yuhen, S.: SAR image despeckling using scale mixtures of gaussians in the nonsubsampled contourlet domain. CJE **24**(1), 205–211 (2015)
28. Uslu, E., Albayrak, S.: Curvelet-based synthetic aperture radar image classification. IEEE Geosci. Remote Sens. Lett. **11**(6), 1071–1075 (2014)
29. Coll, B., Buades, A., Morell, J.M.: A review of image denoising algorithms, with a new one. SIAM Interdisc. J. Multiscal Model. Simul. **04**(02), 490–530 (2005)
30. Kervrann, C., Boulanger, J., Coupe, P.: Bayesian non-local means filter, image redundancy and adaptive dictionaries for noise removal. In: Proceedings of international conference scale space methods variational methods computer vision, pp. 520–532 (2007)
31. Huang, Y., Moisan, L., Ng, M.K., Zeng, T.: Multiplicative noise removal via a learned dictionary. IEEE Trans. Image Process. **21**(11), 4534–4543 (2012)
32. Li, Y., Zhong, H., Jiao, L.: SAR image despeckling using bayesian nonlocal means filter with sigma preselection. IEEE GRSL **8**(4), 809–813 (2011)
33. Deledalle, C.A., Denis, L., Tupin, F.: Iterative weighted maximum likelihood denoising with probabilistic patch-based weights. IEEE Trans. Image Process. **18**(12), 2661–2672 (2009)
34. Zhong, H., Li, Y.W., Jiao, L.C.: Bayesian nonlocal means filter for SAR image despeckling. In: Proceedings of Asia-Pacific Conference synthetic aperture Radar, Xian, China, pp. 1096–1099. Oct 2009
35. Buades, A., Coll, B., Morel, J.M.: A review of image denoising algorithms, with a new one. SIAM Interdisc. J. Multiscale Model. Simul. **4**(2), 490–530 (2005)

Implementation of Home Automation Using Voice Commands

M. Karthikeyan, T. S. Subashini and M. S. Prashanth

Abstract Home automation is the mechanism of controlling household appliances automatically using various control system techniques. Nowadays, humans have become much dependent on electronic appliances and technology related to it. This paper gives a proposal to control home appliances using voice commands. We plan to implement an IOT-based system to automate light and fan operations using speech. A model that comes up with smart automation using Google Assistant serves this purpose. The user issues the voice command to the Google Assistant which gets interpreted and based on that the proposed system instructs the relay to switch on/off the home appliance. Adafruit and IFTT application recognize voice commands and instruct the power relay for the appliances accordingly through ESP8266 Wi-Fi. The communication between the appliances and microcontroller is established via Wi-Fi and Adafruit library; whereas, IFTT is used to control and monitor the status of devices.

Keywords Arduino · ESP8266 · Adafruit · IFTT

M. Karthikeyan (✉)
Department of Computer Science and Engineering, SRM IST, Kattankulathur, India
e-mail: karthickrock125@gmail.com

T. S. Subashini
Department of Computer Science and Engineering, Annamalai University,
Chidambaram, India
e-mail: rtramsuba@gmail.com

M. S. Prashanth
L&T Technology Services, Chennai, India
e-mail: chintu04144@gmail.com

© Springer Nature Singapore Pte Ltd. 2020
K. S. Raju et al. (eds.), *Data Engineering and Communication Technology*,
Advances in Intelligent Systems and Computing 1079,
https://doi.org/10.1007/978-981-15-1097-7_13

1 Introduction

Home automation system consists of sensors, Arduino, and network devices that work accordingly to make ones' home comfortable, customized, efficient, and secure. It enables to "speak" with the home appliances and interact with a remote control or smart device such a mobile phone. Some controls allow to turn a device on or off at a specific time, so that other devices may be triggered by some external events. Internet of Things (IOT) plays a significant role for automation by connecting the sensors and actuators and can be controlled anywhere in the world. Home automation is necessary to help the physically abled and aged to control the home appliances. In this work, we proposed a voice-based IOT system to switch on and off appliances like light or fan using Google Assistant. Anyone can just switch on their appliances without physically going and switching them on. The most popular smart home applications such as connected sensors, appliances, and gadgets can monitor and control many household purposes like heating, lighting, security, and so on to enhance safety and comfort. In addition to that, creation of the human–speech interface helps the users to communicate and command their household appliances through voice. In smart home application, voice-based interaction with the devices plays a major role to control and monitor different devices in home through personalization. For example, to switch the power on light and fan can be called "switch on light" or "switch on fan."

2 Literature Survey

A web UI-based home surveillance and automation system was implemented in [1] using Raspberry Pi which uses signals from emanating from installed camera and motion sensors. In embedded micro server-based IP connectivity, the authors in [2] implemented low-cost house monitoring and controlling system within Android application. The authors in [3] have designed IOT-based architecture for smart homes by harnessing the features of mobile IPV6. Employing the Siri technology of Apple Inc., the authors in [4] attempted to control the household appliances using Siri's built-in voice commands. Speech recognition or text to speech uses acoustic and language models to decode the speech signal into the sequence of words. Google Assistant API is used to control the appliances in home [5]. Speech recognition software is [6] used to convert voice commands to text which is then used to control the home appliances in wireless mode. Visual Basic has been utilized as checking and control segment in [7] to implement a low-cost smart home automation system which employs the Microsoft speech recognition software. Voice Controlled Wireless Intelligent Home Automation System (IHSA) based on Zigbee was proposed by authors in [8]. Home automation framework which consists of sensor node collects the room data and sends to the sink node which in turn transfers it to the web server. The web server sends the decision based on the received data to the actuators of the sink node [5]. Dynamic Hierarchical Models (DHLM) are used to classify phonemes into a standard term that contributes to semantic tagging in order to train the action to be performed in response to the word [9].

3 System Architecture

The proposed system is based on home automation using voice commands that is implemented with Ardunio and IFTT. Google voice Assistant is used to get the voice commands. Google Assistant is a fast-emerging technology. Currently, most of the smartphone market is leading in Android which comes with Google Assistant technology. Using sketch programming, Arduino is programmed to check the status of the power relay state as either s "On" or "Off." Google voice API recognizes the text and interprets with the stored text in Adafruit library cloud server. Semantic tagging method is used to train the voice commands. For "Turn on light" command, the action to be performed is stored in IFTT cloud service. Then, the speech recognition technique parses the commands, sends and sets power "On" to the relay1 which is connected through ESP8266. It enables the electric bulb to glow. Figure 1 gives the architecture diagram of the proposed home automation system, and Fig. 2 shows the block diagram for speech recognition.

Fig. 1 Architecture diagram of the proposed home automation system

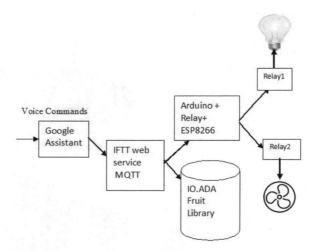

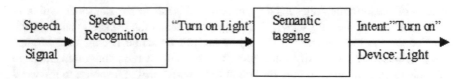

Fig. 2 Block diagram for speech recognition

4 System Design

Table 1 gives the list of hardware and software components needed to implement the proposed system to automate the home appliances, namely light and fan.

The major components are shown in Fig. 3.

Table 1 Design components

Hardware components	Software components
• Arduino UNO	• ARDUINO
• NODEMCU ESP8266	• ADAFRUIT LIBRARY
• Jump wires	• IO.ADAFRUIT
• Bulb	• IFTTT
• Relay	
• Lights	
• Holder	
• Fans	
• USB connector	
• 9 W Battery	

Fig. 3 Major hardware components

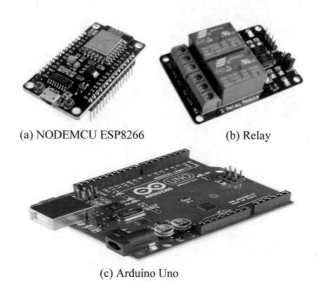

(a) NODEMCU ESP8266   (b) Relay

(c) Arduino Uno

4.1 Hardware Components Used

Arduino Uno is a microcontroller developed by Arduino Corporation based on Atmega328 Family. Electronic gadgets are becoming smaller, better adaptable and more modest capable of doing more and comparable to processor output. Arduino Uno introduced a revolution in the electronics which consists of 14 digital I / O pins, six analog pins, a USB interface and a microcontroller called Atmega328. Serial communication is possible in Arduino using Tx and Rx pins.

NodeMCU is an open source, and it contains firmware that executes on the ESP8266 Wi-Fi SoC systems and hardware which is based on the ESP-12 module. The term "NodeMCU" also refers to the firmware instead of development kit.

Relay is an electrical device, consists of an electromagnet, that enables when the current is passed through or signal in one circuit to open or close another circuit.

4.2 Software Components Used

MQTT

Message Queue Telemetry Transport (MQTT) is an application layer protocol and transfers messages from IOT devices to the cloud server for monitoring and processing data. It involves in M2M communication designed in such a way that client and server can communicate with each other.

SKETCH

A Sketch is the name used in Arduino programming that contains setup(), loop() functions are used to initate the Arduino board, for example, Pin 13 is used to blink a led in Arduino board [10].

ADAFRUIT LIBRARY

Arduino libraries are used to connect the devices drivers or commonly used utility functions. Library manager contains predefined code that makes to connect the sensor, Arduino, module, etc. For example, the default library for liquid crystal module makes it easy to communicate with character LCD displays. There are several number of additional libraries available on the Internet for download [5].

IO.ADAFRUIT

Adafruit IO is a system that utilized the data properly. Our focus is to connect the data to the end devices with the help of small programming. IO includes client libraries that supports REST and MQTT APIs [11]. Adafruit.io is a *cloud service* that just means we run it for you and you do not have to manage it. You can connect to it over the Internet. A dashboard is a feature integrated into Adafruit IO which allows you to chart, graph, gauge, log, and display your data. You can view your dashboards from anywhere in the world.

IFTTT

IFTTT provides web service that supports Google Assistant to recognize the speech and its connected to Adafruit IO [6].

5 Implementation and Results

The circuit designed for the proposed system to automate the operations of light and fan is given in Fig. 4.

Fig. 4 Circuit diagram for the proposed system

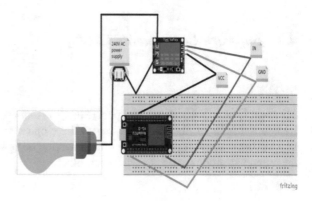

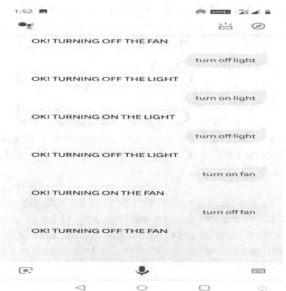

Fig. 5 Screenshot of Google Assistant voice commands

Arduino ESP8266 is connected with the power relay through jump wires and gets powered externally. Then, Arduino is flashed with the speech recognition module. Then, we created a dashboard in Adafruit.io for the power relay as relay1 for light and relay2 for fan using sketch programming speech signals data are recognized and transform to 1 or 0 to identify power status for the relay and get updated in the cloud. Using Google Assistant, "Turn on Light" is the keyword used to switch on the light which is connected at Arduino ESP8266. IFTTT is a web service with the help of the MQTT protocol, and the data is transferred to the Arduino ESP8266 Wi-Fi hotspot [12]. The programmed Arduino turns on the power relay enables the bulb to glow. Similarly, "Turn on Fan," voice command is used to turn the fan on. Figure 5 shows the screenshot of Google Assistant voice commands, and Fig. 6 shows the experimental results.

Fig. 6 Shows the experimental results

Power Off status

Turn on Light

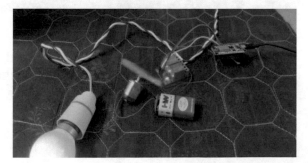

Turn on Light & Turn on Fan

6 Conclusion and Future Work

This paper has proposed the idea of home automation with Google Assistant that can support an IOT of home automation systems. It includes a connection between wireless communication, monitoring, and tracking. They are huge systems that include multiple technologies and applications that can provide control and security for the electronic devices easily. Future of home automation is huge. Everyone is making a shift to automatic technology which makes life easier. The future of this home automation using Google Assistant model deals with the voice recognition part of it. This model switches appliances on and off when anyone uses the voice commands. But to provide greater security to it, it is much better and advisable to work with voice-specific automation. If the house belongs to person A, then only A can access the commands. This can be achieved with voice recognition. We need to train the voice of the particular person and then give the voice commands.

References

1. Kandala, H.B., Patchava, V., Babu, P.R.: A smart home automation technique with raspberry Pi using IoT. In: 2015 International conference on smart sensors and systems (IC-SSS) (2015)
2. Lee, S.R., Piyare, R.: Smart home-control and monitoring system using smart phone. In: 1st International conference on convergence and its application (ICCA), vol. 24
3. Park, B., Caytiles, R.D.: MobileIP-based architecture for smart homes. Int. J. Smart Home **6**(1) (2012)
4. Celebre, A.M.D., Medina, I.B.A., Dubouzet, A.Z.D., Surposa, A.N.M., Gustilo, E.R.C.: Home automation using Raspberry Pi through Siri enabled mobile devices. In: 8th IEEE international conference humanoid, nanotechnology, information technology communication and control, environment and management (HNICEM), international journal of pure and applied mathematics special issue
5. Srinath, M.S., Kishore, M.N.: Interactive home automation system with google assistant. Int. J. Pure Appl. Math. **119**(12), 14083–14086 (2018)
6. Amrutha, S., Aravind, S., Ansu Mathew, S.S, Rajasree, R., Priyalakshmi, S.: Speech recognition based wireless automation of home loads-E home. Int. J. Eng. Sci. Innov. Technol. (IJESIT) **4**(1) (2015). ISSN: 2319-5967
7. Kamarudin, M.R., Aiman, M., Yusof, F.M.: Low cost smart home automation via microsoft speech recognition. Int. J. Eng. Comput. Sci. UECS-UENS **13**(03), 6–11 (2013)
8. Usha Devi, Y.: Wireless home automation system using ZigBee. Int. J. Sci. Eng. Res. **3**(8) (2012). ISSN: 2229-5518
9. Lavanya, I.K.: Intelligent home automation system using bit voicer. Int. Conf. Intell. Syst. Control (ISCO)
10. https://www.arduino.cc/en/tutorial/sketch
11. https://learn.adafruit.com/adafruit-all-about-arduino-libraries-install-use/arduino-libraries
12. https://ifttt.com/

Configure and Management of Internet of Things

M. Varaprasad Rao, K. Srujan Raju, G. Vishnu Murthy
and B. Kavitha Rani

Abstract This document is in the required format. Internet of things (IoT) became buzzword everywhere in the market. It is designed with embedded things and Internet. These systems are used to interact with every other devices or services or human being on a large platform in the world. This type of architecture creates the scope in ease of using the improved applications and makes increases reliability and effectiveness of sustainable devices. Nowadays, the applications of IoT in everywhere in the society such as home, office, environment, and power grid automations, which replaces human being. Because of this, it is very much required to build and maintainable, low powered, less cost things. The authors have proposed a process of configure and management of IoT using MongoDB cloud database.

Keywords IoT · Sensors · MongoDB · NoSQL · Configuration management · Reliability · Efficiency

1 Introduction

Nowadays the end users want to monitor and control their electronic gadgets automatically. There are many such devices available in the market are called sensors. So device management of such sensor devices from remote area is becoming tedious job. The managing of users' data is not only sufficient but also

M. Varaprasad Rao (✉) · K. Srujan Raju · B. Kavitha Rani
CMR Technical Campus, Hyderabad, Telangana, India
e-mail: vpr_m@yahoo.com; varam78@gmail.com

K. Srujan Raju
e-mail: ksrujanraju@gmail.com

B. Kavitha Rani
e-mail: phdknr1@gmail.com

G. Vishnu Murthy
Anurag Group of Insittutions, Hyderabad, Telngana, India
e-mail: gvm189@gmail.com

© Springer Nature Singapore Pte Ltd. 2020
K. S. Raju et al. (eds.), *Data Engineering and Communication Technology*,
Advances in Intelligent Systems and Computing 1079,
https://doi.org/10.1007/978-981-15-1097-7_14

configuration of services provided by them is necessary in Internet of things (IoT) [1]. The use of embedded devices or sensors has been increasing day-by-day and with this there arose the need for managing devices data. In order to monitor device, we must have a proper managing system to look after it. This paper provides better configuration management of sensors that will resultant to easy understand and update the data by user.

Connecting different sensor devices under one module: When devices are interconnected we can use this feature for automations. For example, there are two sensors connected the temperature and air conditioner (AC). We can use data of temperature to AC be switched on or off. By configuring these devices the user can control all services within a single interface.

Managing sensors data and storing to remote server: It is assumed that data send by devices are as sensor's name and value. This is converted by field-user and they provide same to user interface as SensorName = value. Data send by the sensors are to be stored in database in proper order so that it can be retrieved as well.

Challenges in storing data: Will become complex when user increases number of sensors or additional features to devices. For example, If a user had temperature sensor, so data will be saved into database with two parameters as UserName = User and Temperature = value. Further, if user wants to add one feature as status of a machine is on or off, then the database should be able to incorporate these changes and store schema as UserName = User, Temperature = value, AC_Status = Boolean.

Feature of monitoring and controlling devices at run time: User should be able to get report of all sensors at run time, and also should have feature to control devices from a remote place.

2 Literature Review

Client–server model: In a client–server model, client requests service from server response service, where a services may be available for communication within a computer or across network. There are many benefits to think about our application as two separate parts as client and server. Through this server API, it is ease to access all the data with appropriate API, this resultants to design a scalable application [2, 3, 4].

IoT applications such as home appliances, smart homes [5], and automobiles using human–computer interactive intelligence environments.

MEAN Stack: is the open source and easier to adapt. It is feasible to proceed further by separating the main components into the technical fields [6, 7].

Mongo DB: Mongo DB [8, 9] is an open source, document-oriented database designed with both scalability and developers agility. Instead of storing data in tables and rows as in relational database, in MongoDB, we store JSON-like documents with dynamic schemas. It is a schema-less database and its features are MongoDB database is a collection of documents. Every collection of document consists of various documents. Structure of a single object is clear. No complex joins. MongoDB supports dynamic queries on documents using a document-based query language that is nearly as powerful as SQL. MongoDB is easy to scale. Conversion/mapping of application objects to database objects not needed.

AngularJS: AngularJS [10] is a structural framework for dynamic Web apps. This enables the user to interpret the application's components using skeleton of HTML language syntax.

ExpressJS: ExpressJS [11] is a NodeJS Web application server framework, designed for building single-page, multi-page, and hybrid Web applications. It is the standard server framework for NodeJS. ExpressJS is a minimal and flexible NodeJS Web application framework that provides a robust set of features to develop Web and mobile applications. It enables a platform to development of Web applications very fast. Core features of express framework are, setting up HTTP request and responses, maintaining of routing tables, and passing arguments [12, 13, 3].

NodeJS: NodeJS [14] is a server-side platform built on Google Chrome's JavaScript Engine (V8-Engine). NodeJS is used in developing applications by JavaScript easily and can be scaled NodeJS uses an event-driven, non-blocking I/O model that makes it lightweight and efficient, perfect for data-intensive real-time applications that run across distributed devices. NodeJS is used to develop server-side script and network applications. It is an open ware, multiplatform environment. NodeJS applications are written in JavaScript, and can be run within the NodeJS run time on OS X, Microsoft Windows, and Linux. NodeJS also provides a rich library of various JavaScript modules which simplifies the development of Web applications using NodeJS to a great extent.

JavaScript Object Notation (JSON): The MEAN stack [6, 7] enables a perfect harmony of (JSON) [4, 15] format for development. MongoDB stores data in a JSON format. Express and NodeJS facilitate easy JSON query creation. AngularJS enables the client hassle free to send and receive JSON documents.

Node Package Manager: Package npm [16] provides the following two main functionalities.

- Online repositories for node.js packages/modules which are searchable on search.NodeJS.org.
- Command-line utility to install Node.js packages, version management and dependency management of Node.js packages.npm comes bundled with Node.js installable after v0.6.3 version. Express and NodeJS requires npm for its usage.

3   System Design

Configurable device management in Internet of things uses MVC [17, 2] design pattern. It has four models, nine views and four controllers. These controller takes care of the process of enabling user to interact as well as for the sensor data to generate report.

Model view controller (MVC): MVC was one of the oldest GUI application development tool. This was used to design and implement the application based on user's responsibilities. Model view controller is called as a software design pattern for developing Web application. A model view controller pattern is made up three parts they are

1. Model—The bottom level of the MVC pattern which is responsible for maintaining the data.
2. View—This enables the user to view partial or entire data of the user.
3. Controller—It is the software code that controls the interactions between the model and view.

MVC separates the logic information from the application layer. The controller receives all requests for the application and then works with the model to get any data needed by the view. The view then uses the data given by the controller to generate a presentable response. It does the support for all user's and application logic concerns (Figs. 1 and 2).

Our MVC model interacts with server using controller. It sends GET/POST method to the REST API present in the server and act accordingly.

Proposed Process

Device sends its data using GET/POST method to server which does processing of the data to refine and store the chunk data into proper format (Fig. 3).

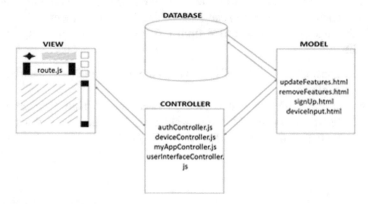

Fig. 1 Model view controller

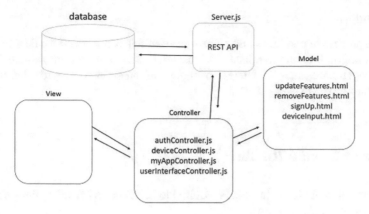

Fig. 2 Client–server model

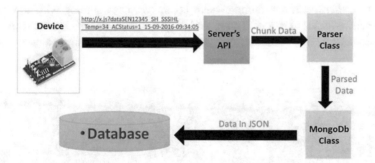

Fig. 3 Device data flow diagram

The advantages

- Creates a master script which can receive any number of parameter from device and maintains database for it.
- Creates REST API which is used for organizing and controlling device data.
- Allows user to update feature or downgrade feature on its own.
- User is provided with a module which can send request from user interface to field-user.
- Field-user validates the request and without any modification of data fields in database sensor data is dynamically stored.
- It creates automation for data storage in database.
- It also attempts to provide better configuration of the devices.

Limitations

- The process of configure and management of IoT is not tested for data migration of MongoDB to another database. This does not attempt for creating new device management system of Internet of things but attempts to provide this dynamic feature to the existing software.

4 Experimental Results

When user fills the form, its data is required to be stored in Metadata table shown in the following figure (Fig. 4).

Now the metadata table should be used for creating user's form Data as a part of registration (Figs 5 and 6).

In the experiment, the authors have taken two case studies with the different parameters as case1. Data of Temp, AC_Status, doorOpen of 3 parameters and in case2. Data of Temp, Pressure, AC_Status, doorOpen of 4 parameters. It is able to

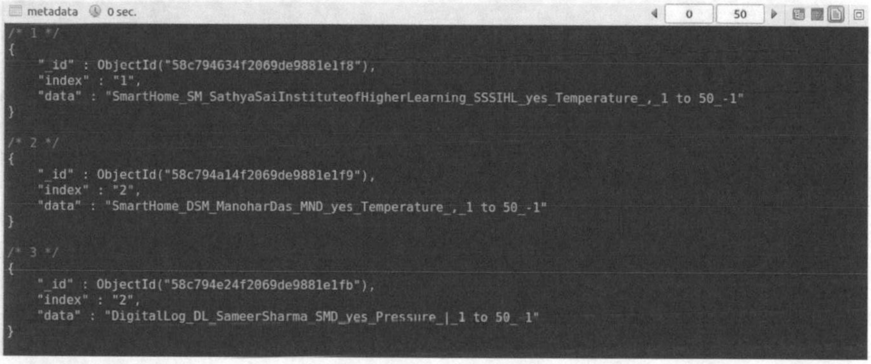

Fig. 4 Meta data table

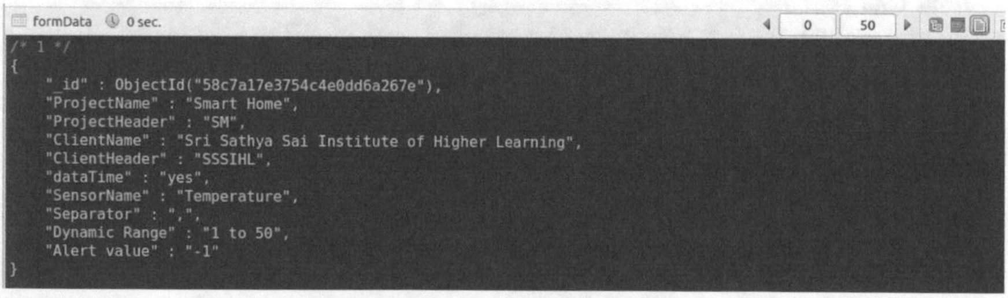

Fig. 5 Form data

_id	ProjectName	ProjectHeade	ClientName	ClientHeader	dataTime	SensorNam	Separator	Dynamic Ran(	Alert value
1 ☐ Objectl...	⊞ Smart Home	⊞ SM	⊞ Sri Sathya Sai Inst...	⊞ SSSIHL	⊞ yes	⊞ Temp...	⊞ ,	⊞ 1 to 50	⊞ -1
2 ☐ Objectl...	⊞ DigitalLog	⊞ DSM	⊞ ManoharDas	⊞ MND	⊞ yes	⊞ Temp	⊞ ,	⊞ 1 to 50	⊞ -1

Fig. 6 Form data in the table

```
/* 2 */
{
    "_id" : ObjectId("58c7a3833754c4e0dd6a2681"),
    "header" : "SSSIHL",
    "Temp" : "23.5",
    "AC_Status" : "ON",
    "doorOpen" : "Yes",
    "time" : "2017:03:03:10:52:21"
}

/* 3 */
{
    "_id" : ObjectId("58c7a39e3754c4e0dd6a2682"),
    "header" : "SSSIHL",
    "Temp" : "23.5",
    "Pressure" : "76",
    "AC_Status" : "ON",
    "doorOpen" : "Yes",
    "time" : "2017:03:03:10:52:21"
}
```

Fig. 7 Case1 and case2 list of parameters

```
/* 5 */
{
    "_id" : ObjectId("58c7a3b63754c4e0dd6a2684"),
    "header" : "SSSIHL",
    "Temp" : "23.5",
    "Pressure" : "76",
    "AC_Status" : "ON",
    "doorOpen" : "Yes",
    "time" : "2017:03:03:10:54:21"
}

/* 6 */
{
    "_id" : ObjectId("58c7a3c03754c4e0dd6a2685"),
    "header" : "SSSIHL",
    "Temp" : "23.5",
    "Pressure" : "76",
    "time" : "2017:03:03:10:54:21"
}
```

Fig. 8 Removes features of doorOpen and AC_Status

store same in database. The following figure shows the variations that occurred in the database along with changes (Fig. 7).

In the following figure depicts about removes features of doorOpen and AC_Status parameters (Fig. 8).

The following figure shows variation of data in database (Fig. 9).

The following figures describe about generating reports of temperature in weekly and monthly. It also generates an analysis report of temperature for the entire week/month (Figs 10, 11 and 12).

	_id	header	Temp	AC_Status	doorOpen	time	Pressure
1	Objecti...	SSSIHL	23.5	ON	Yes	2017:03...	
2	Objecti...	SSSIHL	23.5	ON	Yes	2017:03...	
3	Objecti...	SSSIHL	23.5	ON	Yes	2017:03...	76
4	Objecti...	SSSIHL	23.5	ON		2017:03...	76
5	Objecti...	SSSIHL	23.5	ON	Yes	2017:03...	76
6	Objecti...	SSSIHL	23.5			2017:03...	76

Fig. 9 The data changes in database

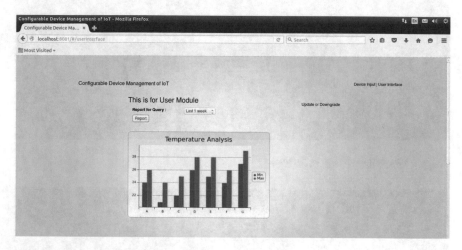

Fig. 10 Week-wise temperature report

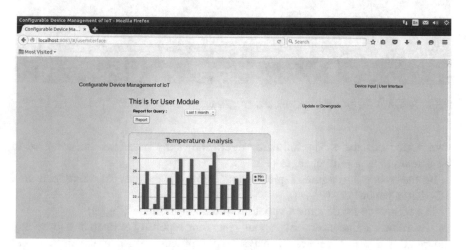

Fig. 11 Month-wise temperature report

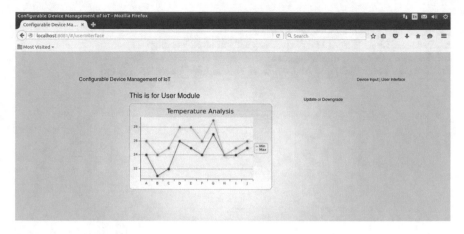

Fig. 12 Temperature analysis report

5 Conclusion

The applications of IoT in everywhere in the society such as home, office, environment, power grid automations, which replaces human being. Because of this, it is very much required to build and maintainable, low powered, less cost things. The proposed a process of configure and management of IoT using MongoDB cloud database is more ease of use and managed.

References

1. Mukhopadhyay, S.C. (ed.): Internet of things, smart sensors, measurement and instrumentation, vol. 9. Springer International Publishing (2014)
2. Stack Overflow Questions and Answers Forum. http://www.stackoverflow.com
3. Quora Questions and Answers Forum. http://www.quora.com
4. Wikipedia for Introduction to Topics. https://en.wikipedia.org
5. Ravinder, B., Srujn Raju, K.: An application of Intrnet of things for smart home. IJARES (2016). ISSN: 2349–6495
6. Sevilleja, C., Lloyd, H.: MEAN machine: a beginner's practical guide to the JavaScript stack-Lean pub (2015)
7. Mean Stack. http://meanstack.com
8. Dasadia, C., Nayak, A.: MongoDB Cookbook. Packt Publishing (2016)
9. MongoDB-NoSQL Database. http://mongodb.com
10. Angular Java Script. https://angularjs.org
11. Express Java Script. http://expressjs.org
12. Workflow with Light Weight Tools. http://github.com

13. W3 Schools. https://www.w3.org
14. Node Java Script. http://NodeJS.org
15. Json—A Light Weight Data Interchange Tool. http://json.org
16. Package Manager, http://npmjs.com
17. Tutorials Point Questions And Answers Forum. http://tutorialspoint.com

Automatic Race Estimation from Facial Images Using Shape and Color Features

B. Abirami and T. S. Subashini

Abstract In this work, an attempt has been made to estimate the human race from facial images. This work is done for three major ethnicities: African, American, and Asian. The performances of the classifiers were tested with the faces images of African, American, and Asian population belonging to different age groups and genders. For this, first the facial part such namely the nose region is detected using the well-known Viola–Jones object detection technique. From the detected nose, 9 Zernike moments, 81 HoG features, and 2 color features are extracted to estimate the race using classifiers namely ANN and SVM. Four hundred and fifty sample images taken from the FERET database were considered for this study, out of which 330 images were used for training and for 120 testing. The accuracy of the model obtained through artificial neural network is 91.06%, whereas the accuracy obtained by applying SVM is 95.1%. From the results obtained, it is evident that SVM outperforms ANN in identifying the race of a person from his/her facial image and could be effectively employed in automatic race estimation systems based on facial images.

Keywords Face detection · Zernike moments · HoG · ANN · SVM · Race estimation

1 Introduction

A wide range of social information can be obtained by analyzing the face image, and some of them include the facial expression such as happy, anger, sad, the person's identity, the person's age, gender, and race. As such, face analysis has

B. Abirami (✉) · T. S. Subashini
Department of Computer Science and Engineering, Annamalai University,
Annamalainagar, Tamil Nadu 608002, India
e-mail: abi.vsb@gmail.com

T. S. Subashini
e-mail: rtramsuba@gmail.com

© Springer Nature Singapore Pte Ltd. 2020
K. S. Raju et al. (eds.), *Data Engineering and Communication Technology*,
Advances in Intelligent Systems and Computing 1079,
https://doi.org/10.1007/978-981-15-1097-7_15

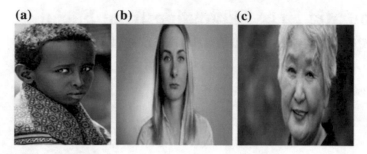

Fig. 1 Images of samples from **a** African, **b** American, and **c** Asian (Image Source: Internet)

gained attention from researchers belonging to various disciplines such as psychology, computer science, medicine etc. The computer vision and machine learning techniques help in automatically identifying the human race from face images which has gained popularity in recent times as it has a widely used in surveillance applications. The aim of this research work is to identify descriptors that can represent and highlight race, distinct information, which is used to build the model for testing and recognition purpose. In this work, three ethnic categories have been considered: the African, the American, and the Asian.

Figure 1 shows samples' face images of African, American, and Asian population. As within race population have similar features compared to between race populations, we exploited this idea in our work to automatically detect the human race from their distinguishing facial features.

This work is organized into five sections: Earlier works done on race identification is dealt with Sect. 2. Section 3 explains the proposed work, the findings are discussed in Sect. 4, and finally, Sect. 5 concludes the paper.

2 Literature Review

The emergence of face detection algorithms with very good performance has fueled the research on face component-based ethnic classification systems and during the last decade. Gabor wavelets' transformation and retina sampling were applied in [1] to extract features which were modeled using SVM to classify Asian, European, and African face images. In [2], local binary pattern as features was explored and the classification accuracy obtained is 99%.

Principal component analysis (PCA) and independent component analysis ICA combined with SVM was employed by the authors in [3], and linear discriminant analysis (LDA) was used in [4] to classify face image into Asian and non-Asian. Eye and the mouth positions were employed by the authors to normalize the face in [5]. An error rate of 0.026% using SVM was reported in correctly classifying the race. Local binary patterns and Haar wavelets as features were used in [6] to categorize Asian and Non-Asian and classifier used were AdaBoost, and the accuracy reported was 97%.

To classify face images into three minority Chinese groups, the authors in [7] employed algebraic features and geometric features, and to assess the efficacy of the features in classifying the three Chinese minority groups considered for the study, K-nearest neighbors (k-NN) and C5.0 classifiers were used. The study reported that elastic model-based geometric features performed better with both the classifiers.

The authors in [8] experimented with GAIT energy image in assessing the ethnicity and obtained 84% accuracy. A hybrid architecture comprising radial basis function network and inductive decision trees was employed by the authors in [9] for ethnicity classification of human faces. To classify Asian, European, and African face images, the work in [10] used biologically inspired features based on Gabor filters and support vector machines. The work also attempted to identify the gender as well as the age based on the face image and MORPH-II database was used for the study.

From the review of literature, it is learnt that most of the methods make use of the entire face area for ethnic classification. Here, features from the nose region alone are considered. From the nose region, different features representing shape and color are extracted and applied to ANN and SVM classifiers to demonstrate the performance of the proposed model.

3 Methodology

The proposed methodology consists of three steps. As a first step, the nose part is detected using the well-known Viola–Jones object detection method [11].

As the second step, various features namely Zernike moments, HoG features, and color features are extracted. As the final step, the extracted features are fed to classifiers namely ANN and SVM to automatically estimate the race from the given facial image.

3.1 Nose Part Detection

Viola–Jones algorithm is capable of detecting various face parts such as the full face area, left eye, right eye, nose, and mouth. To identify the different regions of the face, the algorithm employs the following steps:

1. The Haar-like features for the feature extraction;
2. AdaBoost machine learning method for detecting the face;
3. Cascade classifier to combine many of the features efficiently.

Fig. 2 **a** Property 1 of nose
and **b** Property 2 of nose

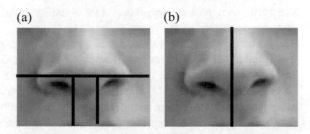

Fig. 3 Face and face parts'
detection using Viola–Jones
algorithm (Image Source:
Internet)

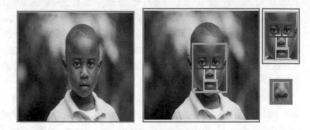

Most of the earlier methods make use of the whole face for feature extraction. In this work, a method that extracts shape and color features from the nose region alone is considered since the nose region contains skin and the nose shape is a very good discriminating feature compared to eyes, mouth, or whole face. Further, on subjectively analysing various faces belonging to different ethnic races, it could be seen that there wide variation with respect to nose shape and as such it was decided to harness this aspect to classify human race based on nose region alone.

Viola–Jones algorithm exploits the two major properties of nose to detect it easily. (a) The two regions of holes on the nose represent the dark pixels, and the center region represents the white pixels. (b) Region of black areas on both the left and right sides of the nose which is very same. Figure 2 illustrates these two properties of the nose region. Figure 3 shows the various face parts detected by Viola–Jones algorithm.

3.2 Feature Extraction

To estimate the race, in this work 9 Zernike moment features, 81 HoG features, and 2 color features are extracted. In total, a feature vector of dimension 92 is created which will be subsequently given as input to the ANN and SVM classifiers. The performance of the two classifiers is analyzed to test the efficiency of the proposed system in correctly identifying the race from facial images.

Zernike Moment Features

The mathematical properties of Zernike moments have been harnessed by researchers to use them as shape descriptors in many image processing applications. Hu was the first to apply moments to describe images [12].

In the 1930s, Zernike polynomials were used to describe optical aberrations by Zernike [13]. Today, Zernike polynomials are being extensively used as an effective shape descriptor in many object recognition systems. The orthogonal properties of ZMs allow one to evaluate up to any order to calculate these moments to get a good descriptor for a given image.

The list of 9 Zernike polynomials up to third order used in this work is given in Table 1.

Histogram of Oriented Gradients (HOG) Descriptor

HOG gives information on local appearance as well as shapes of objects [14] and is implemented as follows:

1. First, the image is divided into cells.
2. Histogram of gradient directions for each cell is computed using the cell pixels.
3. Based on the gradient direction, each cell is discretized into angular bins.
4. Weighted gradient of each angular bin is computed.
5. The pixel of each cell contributes to its corresponding angular bin.
6. Adjacent cells are grouped together as blocks to form a block histogram.
7. The set of these block histograms represents the descriptor.

Color Features

Different ethnicities have varying skin colors. Generally, YCbCr color space which is invariant to illumination variation is used for skin color detection. The input RGB image is converted to YCbCr color space using Eqs. 1 and 2.

$$Cb = -0.148r - 0.291g + 0.439b \qquad (1)$$

$$Cr = 0.439r - 0.368g - 0.071b \qquad (2)$$

Table 1 Zernike polynomials up to third order

n	Zernike polynomials
1	$2\rho \sin \theta$
1	$2\rho \cos \theta$
2	$6\rho^2 \sin 2\theta$
2	$3(2\rho^2 - 1)$
2	$\sqrt{6\rho^2} \cos 2\theta$
3	$\sqrt{8\rho^3} \sin 3\theta$
3	$\sqrt{8(3\rho^3 - 2\rho)} \sin \theta$
3	$\sqrt{8(3\rho^3 - 2\rho)} \cos \theta$
3	$\sqrt{8\rho^3} \cos \theta$

The skin color classification is a good technique to classify ethnicities, but it sometimes overlaps for American and Asian.

3.3 Classification

Classifiers namely support vector machine and artificial neural networks are employed in this work.

SVM

Support vector machine (SVM) is widely used supervised classifier that is well suited for two-class problems. The principle of SVM is that it constructs a hyperplane in a high-dimensional feature space. The feature points which lie on the hyperplane is considered as support vectors which are used for classification and regression tasks [15, 16].

ANN

ANN is a widely used pattern classification model which is a fully connected network in which weights are associated with each connection [17]. It consists of three layers namely input layer, hidden layer, and output layer.

4 Experimental Results

The performances of the classifiers were tested with the face images of African, American, and Asian population belonging to different age groups and genders. Totally, 450 images were collected which comprises 150 face images under each of the three categories and the resolution of each image is 240X360. Of the total 450 samples taken, 330 images were used in training the classification model. For training, the face samples were taken almost in equal amount in each of the three categories. The remaining images were employed for testing purpose.

4.1 Performance of ANN Classifier

Multi-layered perceptron was implemented for the purpose of training. The confusion matrix obtained for different classes namely African, American, and Asian for a network structure of 92 input neurons, 10 hidden neurons, and 3 output neurons is tabulated in Table 2.

The best result is obtained with parameters, viz. 500 epochs, learning rate of 0.17, and Lavenberg–Marquardt training algorithm. The overall accuracy obtained is 91.06%. The ANN classifier was able to identify positive samples with an overall sensitivity of 93.3% and false samples with an overall specificity of 89.9% as shown in Table 3.

Table 2 Confusion matrix of ANN

	TP	TN	FP	FN
African	27	59	1	3
American	29	52	8	1
Asian	28	51	9	2

Table 3 Performance of ANN in (%)

Attributes/class	African	American	Asian	Overall
Accuracy	95.6	90.0	87.7	91.06
Sensitivity	90.0	96.6	93.3	93.3
Specificity	98.3	86.6	85.0	89.9

4.2 Performance of SVM Classifier

SVM with polynomial kernel is used for classification of face image into any of the three races namely African, American, or Asian. Three hundred and thirty images were used for training the network 30 for validation, and 90 images were used for testing the trained network. The confusion matrix obtained for different classes, namely African, American, and Asian, is given in Table 4.

The overall accuracy obtained is 95.1%. The SVM classifier was able to identify positive samples with an overall sensitivity of 96.6% and false samples with an overall specificity of 94.4% as shown in Table 5.

The graph in Fig. 4 compares the results obtained through SVM and ANN, and it can be concluded that the performance of the SVM is far superior to ANN in identifying the race of a person from his/her facial image.

Table 4 Confusion matrix of SVM

	TP	TN	FP	FN
African	30	59	1	0
American	29	55	5	1
Asian	28	56	4	2

Table 5 Performance of SVM in (%)

Attributes/class	African	American	Asian	Overall
Accuracy	98.8	93.3	93.3	95.1
Sensitivity	100	96.6	93.3	96.6
Specificity	98.3	91.6	93.3	94.4

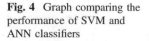

Fig. 4 Graph comparing the performance of SVM and ANN classifiers

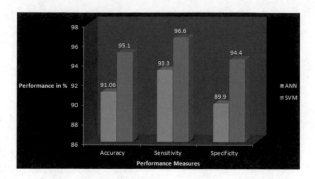

5 Conclusion

To automatically identify race from facial images, three continental races namely African, American, and Asian were considered for this study. From the face image, a method that extracts shape and color features from the nose region alone is considered since the nose region contains skin and the nose shape is a very good discriminating feature compared to eyes, mouth, or whole face. Nose region is detected using the well-known Viola–Jones object detection technique, and from the detected nose region, 92 shape and color features were extracted and the efficacy of these features to model the three different races was tested with classifiers namely ANN and SVM. ANN and SVM performed well in correctly identifying however, SVM gave a very good performance of 95.1% accuracy compared to ANN. Also, the sensitivity and specificity of SVM are higher compared to ANN. Also, the sensitivity and specificity of SVM are better for ANN. Thus, it is inferred that our race identification system based on SVM and the proposed features can play a significant role in real-world applications such as defense, security, and biometric-based authentication.

References

1. Hosoi, S., Takikawa, E., Kawad, M.: Ethnicity estimation with facial images. In: Proceedings of the Sixth IEEE International Conference on Automatic Face and Gesture Recognition, pp. 195–200 (2004)
2. Zhang, G., Wang, Y.: Multimodal 2d and 3d facial ethnicity classification. In: Fifth International Conference on Image and Graphics (ICIG'09), pp. 928–932 (2009)
3. Ou, Y., Wu, X., Qian, H., Xu, Y.: A real time race classification system. In: Proceedings of the 2005 IEEE International Conference on Information Acquisition, pp. 378–383 (2005)
4. Lu, X., Jain, A.K.: Ethnicity identification from face images. In: Proceeding of SPIE International Symposium on Defense and security: Biometric Technology for Human Identification, pp. 114–123 (2004)
5. Manesh, F.S., Ghahramani, M., Tan, Y.P.: Facial part displacement effect on template based gender and ethnicity classification. In: Proceedings of the 11th International Conference on Control Automation Robotocs and Vision (ICARCV), pp. 1644–1649 (2010)

6. Yang, Z., Ai, H.: Demographic classification with local binary patterns. In: Proceedings of the 2007 International Conference on Advances in Biometrics, ICB'07, pp. 464–473. Springer, Berlin (2007)
7. Duan, X.D., Wang, C.R., Li, Z.J., Wu, J., Zhang, H.L.: Ethnic feature extraction and recognition of human faces. In: Proceedings of the 2nd International Conference on Advanced Computer Control (ICACC), pp. 125–130 (2010)
8. Zhang, D., Wang, Y., Bhanu, B.: Ethnicity classification based on gait using multi-view fusion. In: Proceedings of the IEEE Conference on Computer Vision and Pattern Recognition, pp. 108–115 (2010)
9. Gutta, S., Wechsler, H.: Gender and ethnic classification of face images. In: Proceeding of the Third IEEE International Conference on Automatic Face and Gesture Recognition, pp. 194–199 (1998)
10. Guo, G., Mu, G.: A study of large scale ethnicity estimation with gender and age variations. In: Proceedings of the IEEE Computer Society Conference on Computer Vision and Pattern Recognition Workshops (CVPRW), pp. 79–86 (2010)
11. Damanik, R.R., et al.: An application of Viola Jones method for face recognition for absence process efficiency. IOP Conf. Ser.: J. Phys.: Conf. Ser. **1007**, 012013 (2018)
12. Hu, M.K.: Visual pattern recognition by moment invariants. IRE Trans. Inf. Theory **8**, 179–187 (1962)
13. Khotanzad, A., Hong, YH.: Invariant image recognition by Zernike moments. IEEE Trans. Pattern Anal. Mach. Intell., 489–498 (1990)
14. Dadi, II.S., Pillutla, G.K.M.: Improved face recognition rate using HOG features and SVM classifier. IOSR J. Electron. Commun. Eng. (IOSR-JECE) **11**(4), 34–44 (2016). e-ISSN: 2278-2834, p-ISSN: 2278-8735
15. Vapnik, V.: Statistical Learning Theory. Wiley, New York (1998)
16. Boser, B.E., Guyon, I., Vapnik, V.: A training algorithm for optimal margin classifiers. In: Proceedings of the 5th ACM Workshop on Computational Learning Theory, pp. 144–152. ACM Press, New York (1992)
17. Cheng, T., Wen, P., Li, Y.: Research status of artificial neural network and its application assumption in aviation. In: 2016 12th International Conference on Computational Intelligence and Security (CIS), Wuxi, pp. 407–410 (2016)

A Security Model to Make Communication Secure in Cluster-Based MANETs

Deepa Nehra, Kanwalvir Singh Dhindsa and Bharat Bhushan

Abstract The MANETs are more prone to different types of attacks in comparison to wired networks. In cluster-based MANETs, the problem becomes more difficult as member needs to communicate with the cluster head. The infrastructure-less and scattered behavior of MANETs puts huge challenge for the research community. Further, cluster heads also need to communicate with neighboring cluster heads. To secure the communication among these nodes, an appropriate cryptography technique will be needed. The security model presented here uses AES-based encryption technique to generate secret keys that can be used to encrypt or decrypt the messages transferred between member nodes. This security model ensures the secure communication among member nodes and cluster heads in the cluster-based MANETs. Simulation results show the packet loss percentages, and delay will decrease with increase in number of cluster.

Keywords Cluster · MANETs · Security · Communications

1 Introduction

MANET contains a set of mobile devices which communicates to each other wirelessly and can also change their positions freely [1]. MANET is more susceptible to attacks than wired network due to multiple reasons. Multicasting is an

D. Nehra (✉)
Department of RIC, I.K. Gujral Punjab Technical University, Kapurthala, Punjab, India
e-mail: deepa.nehra@gmail.com

K. S. Dhindsa
Department of CSE, Baba Banda Singh Bahadur Engineering College, Fatehgarh Sahib, Punjab, India
e-mail: kdhindsa@gmail.com

B. Bhushan
Department of Computer Application, Guru Nanak Khalsa College, Yamuna Nagar, Haryana, India
e-mail: bharat_dhiman@gmail.com

© Springer Nature Singapore Pte Ltd. 2020
K. S. Raju et al. (eds.), *Data Engineering and Communication Technology*,
Advances in Intelligent Systems and Computing 1079,
https://doi.org/10.1007/978-981-15-1097-7_16

essential feature required in group communications, e.g., VoD (video on demand), videoconferencing, frequent stock updates, discussion forums, and advertising. The blend of MANETs and multicast services [2–4] presents more difficulties to the infrastructure security.

The security in MANETs is the major concern of this technology, and researchers need to pay more attention to this aspect. Communication between the nodes in MANET should be carried out in secure manner. The communication among the mobile users in MANETs needs to be more secure. Due to lack of any stringent security policy and centralized management, it opens doors for the attackers and they can simply exploit attack on nodes and the services provided by them. Owing to open medium of nodes, dynamic topology, and limited battery power, MANET is subject to several types of attacks, such as traffic monitoring, DoS attacks on routing, packet dropping, partitioning of network, and rushing attack. MANETs cannot be used in real life if applications are not secure. Because of the above reasons, security in MANET is a challenging one. The communication among mobile node in MANETs can be secured by applying some cryptographic techniques such as public key encryption. This technique is based on distributed or centralized key management in which public key certificate is provided by trusted certificate authorities (CAs) so that mobile nodes can be authenticated easily.

The security model proposed here is based on cluster-based MANETs. In cluster-based MANETs, each cluster contains some cluster members and group leader called cluster head (CH). CH creates and maintains the cluster member keys. It also updates the cluster member key whenever a member joins or leaves the cluster. A cluster member key should be shared among all members (nodes) in the cluster in order to multicast information. The information encrypted by a cluster member key by the authorized users can only be decrypted by the same key by other member nodes. But according to MANET characteristic, members of a group may be changed. On joining of new members, a new cluster member key must be generated and distributed to other member nodes along with the new member. This process prevents the new member to access the former information exchanged within the group that is known as forward security. The same process is followed if a member left the group as it has no rights to access the information anymore that is known as backward security.

2 Related Work

A numerous number of methods that secure the communication between different nodes of the cluster through encryption and keys are proposed in the literature. Some of them along with their features are highlighted below.

Wang and Fang [5] suggested a unique hierarchical key management-based method to secure communication in a group. In order to enhance the security, a packet will be encrypted for twice. The work also contains the detail about the group maintenance whenever any changes in the topology will be detected.

Performance analysis is also carried out by the authors, in which they compared their work against the conventional key management methods. The results of the performance depict that the hierarchical key management-based scheme performs well in securing group communications.

Hadjichristofi et al. [6] suggested a new key management system that offers robust Security Association (SA) establishment among MANET nodes. A hierarchy-based public key infrastructure is used, in which mobile node can perform management roles. It delivers high availability to for members of the MANETs through various schemes. Moreover, non-repudiation is also applied that enables new nodes to securely communicate with certificate authorities (CAs). Further, the nodes can secure their communication with other nodes by generating cryptographic keys. Node behavior and node authentication are additional parameters that strengthen the security levels of the key management scheme.

Rahman and Rahman [7] proposed a new security technique for securing wireless communications of MANETs. They proposed a group key distribution technique based on Diffie–Hellman. The idea behind this method is to safely generate a secret session key and distribute it between users wants to establish secure communication with others. The method works by constituting a spamming tree from all the nodes which found to be valid. The password "p" helps in the verification of nodes and avoids the happening of further attacks. Further, this method does not require multicast/broadcast capability.

Bechler et al. [8] discovered cluster-based security architecture for MANETs. The idea behind this method is based on a distributed certification capability. A cluster-based approach is used in which MANET is divided into clusters and every cluster needs to have a cluster head. Cluster head needs to perform administrative tasks and share a secret key between other cluster nodes. Certification of nodes can be done using a secret key. Decentralization is attained through a secret key and threshold cryptography. The framework tackles the problems of access control, authorization and fine-tunes the abilities of mobile nodes.

Umar et al. [9] proposed a CH-based routing protocol that protects ad hoc networks from various kinds of attacks using asymmetric encryption. Cluster heads use private keys to decrypt the data which is encrypted by member nodes through their public keys. Cluster heads help the member nodes to save their battery power because they decrypt the data for member nodes.

3 Security Requirements

The following requirements should be fulfilled in order to secure the communication among member nodes and cluster heads in cluster-based MANETs [10].

1. *Forward secrecy*: This is the case when a node leaves the cluster it should not be allowed to access any secret key in future. It safeguards from the fact that decryption of data after leaving a cluster is not possible.

2. *Backward secrecy*: If a node joins a new cluster, then it will be prevented from the access of its old secret key. It certifies that the data decryption is not possible until the node joins this new cluster.

3. *Non-cluster confidentiality*: The nodes which are never the part of a cluster are restricted from the access of any secret key that helps in the decryption or multicasting the data sent to the cluster.

4. *Key independence*: It guarantees that any kind of cluster secret keys should not be discovered by any other cluster keys.

5. *Trust relationship*: In MANETs, due to lack of a trusted central authority, all the nodes have equal rights to participate in the computation of common secret key.

4   Security Model

The security model presented in this section can be used to protect communication and member nodes from various kinds of attack. It provides more strength to our proposed cluster-based DDoS defense mechanism and keeps cluster heads and member nodes safe from attackers. Public keys can be used to encrypt the communication medium between different nodes of a cluster. The concept of key management proposed by Renuka and Shet [11] is used in this security model for the generation of security keys. Figure 1 shows the working of the security model with the help of flowchart.

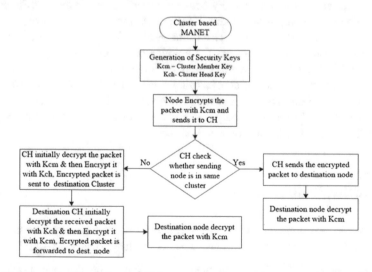

Fig. 1 Flowchart showing defense model

The different process (i.e., formation of cluster-based MANETs, generation of cluster member and CH keys, node-to-node communication and various mobility issues) involved in the security model are discussed below.

4.1 Clustering

The network is portioned in the form of clusters. In each cluster, a cluster head will be elected. Weight-based algorithm is used for the election of CH. The device with lowermost weight would be selected as CH. Cluster head acts as a local manager monitors the overall working of the cluster. It is also responsible for the authentication and secure communication among the member nodes of the cluster. Cluster head generates and distributes a cluster member key to each node of the cluster. Whenever any member node joins or lefts the cluster, it again regenerates and distributes the key.

4.2 Assumptions

The following assumptions are taken during the selection of an appropriate security scheme.

1. Every member node in the cluster is assigned with a unique ID, public/private key value.
2. Each cluster uses two keys. First is cluster member key (Kcm), which can be used by nodes during the encryption and decryption of data exchanged with other members through cluster head. The second is cluster heads key (Kch) to be used by the cluster heads of the neighboring cluster during their communication.

4.3 Security Keys

The node-to-node security of cluster nodes is done with the help of self-generated public/private keys. Public keys are distributed in the clusters by the certificates issued by a certificate authority (CA). Cluster head performs an important role in the generation, management, and distribution of keys among its member nodes. The following keys are used for ensuring the secure communication in cluster-based MANETs.

- **Cluster members key (Kcm)**—An Advanced Encryption Standard (AES)-based symmetric key is created by cluster head called cluster members key. It is safely distributed to other member nodes by encrypting it with the corresponding public key of each member. It can be used to encrypt the messages before sending it to other member nodes of the cluster.
- **Cluster heads key (Kch)**—Cluster heads key is generated Diffie–Hellman algorithm and is used by the cluster heads during the inter-cluster communication. The key needs to be refreshed each time when a cluster head leaves or joins a cluster.

4.4 Node-to-Node Communication

A. **Intra-cluster communication**: The cluster member key (Kcm) is used by the member nodes to communicate within a cluster. Suppose node A2 wants to communicate with node A5 in cluster A (as shown in Fig. 2), the process of communication is described below.

1. Node A5 encrypts (EN) the packets with cluster member key Kcm (cluster member key of cluster A) and sends it to cluster head CHa.

$$A2 : pkt \rightarrow [EN_{Kcm}(pkt)]$$
$$A2 : [EN_{Kcm}(pkt)] \rightarrow CHa$$

2. Cluster head CHa receives the encrypted packet from node A2 and forwards it to node A5; later node A5 decrypts it with cluster member key.

Fig. 2 Communication between cluster nodes

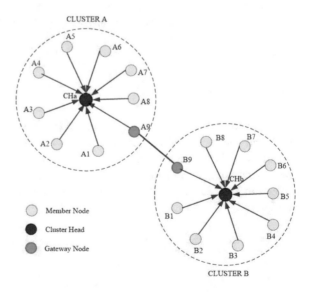

$$\text{CHa} : [\text{EN}_{\text{Kcm}}(\text{pkt})] \rightarrow \text{A5}$$
$$\text{A5} : \text{DC}_{\text{Kcm}}[\text{EN}_{\text{Kcm}}(\text{pkt})] \rightarrow \text{pkt}$$

B. **Inter-cluster communication**: Suppose member A2 of cluster A wants to communicate with member B5 of cluster B; initially, it sends a packet to cluster head CHa. Cha initially decrypts the packet with cluster member key (Kcm) and later encrypts it with cluster heads key (Kch) and then forwards it to cluster head CHb. CHb then decrypts it with its cluster member key and then encrypts it with its cluster member key. The encrypted packet is then forwarded to member node B5. The whole process of encryption and decryption performed by the sender, intermediate, and receiver nodes is illustrated below.

$$\text{A2} : \text{pkt} \rightarrow \left[\text{EN}_{\text{Kcm(a)}}(\text{pkt})\right]$$
$$\text{A2} : \left[\text{EN}_{\text{Kcm(a)}}(\text{pkt})\right] \rightarrow \text{CHa}$$
$$\text{CHa} : \text{DC}_{\text{Kcm(a)}}\left[\text{EN}_{\text{Kcm(a)}}(\text{pkt})\right] \rightarrow \text{pkt1}$$
$$\text{CHa} : \text{EN}_{\text{Kch(a)}}(\text{pkt1}) \rightarrow \text{CHb}$$
$$\text{CHb} : \text{DC}_{\text{Kch(b)}}\left[\text{EN}_{\text{Kch(a)}}(\text{pkt1})\right]$$
$$\text{CHb} : \text{EN}_{\text{Kcm(b)}}(\text{pkt2}) \rightarrow \text{B5}$$
$$\text{B5} : \text{DC}_{\text{Kcm(b)}}\left[\text{EN}_{\text{Kcm(b)}}(\text{pkt2})\right] \rightarrow \text{pkt}$$

where EN—Encryption, DC—Decryption, CHa—Cluster head of cluster A, CHb—Cluster head of cluster B, Kcm(a)—Cluster member key of cluster A, Kcm(b)—Cluster member key of cluster B, Kch(a)—Cluster head key of A, Kch(b)—Cluster head key of B.

4.5 Mobility Issues

The number of nodes in any cluster can be fixed as they can join or leave any cluster at anytime. Even cluster head can also change its position. These issues are discussed here.

- *Cluster member node joins*: During the joining of a new member, the member sends joining request to the CH. CH head asks the node to send its public key. When cluster head authenticates its public key, then it can become the member of the cluster. For the new nodes that are unknown to the networks are required to obtain warranty certificates from their neighboring nodes so that they can be authenticated easily. Cluster head also re-computes the cluster member key and broadcast it old as well as a new member.
- *Cluster member leaves*: An old member node leaves the cluster with left message and its ID. Cluster head calculates a new cluster member key by eliminating the public key of leaving nodes and distributes it to each node of the cluster.

Table 1 Simulation parameters

Parameter	Value
Simulation area (M^2)	1000 m × 1000 m
Number of nodes	180, 100,60
Mobility model	Random way point
Routing protocol	AODV
Packet size	64 bytes
Traffic type	Constant bit rate (CBR)
Queue	Drop tail (50)
Transmission range	50–250 m
Time window	5 s
Simulation time	100 s

5 Experimentation

The performance of the security model can be done by conducting a simulation using the OMNeT++ simulation framework [12]. The simulation is performed by taking 1000 × 1000 m^2 areas with a varying number of nodes. Random way point mobility model along with AODV protocol is used to conduct the simulation. The simulation for each number of nodes runs for 100 s with a fixed windows size of 5 s. Table 1 displays the different parameters to be used during the simulation setup.

The experiments are performed on the MANET size with 180, 100, and 60 numbers of nodes. Initially, 180 nodes are divided into 1, 2, 4, and 6 clusters having 180, 90, 45, and 30 nodes, respectively. The same process is repeated for 100 and 60 numbers of nodes. The results of the simulation are collected in form of scalar (.sca) and vector (.vec) values which can later be analyzed with the data of interest. Figure 3 shows the snapshot of the running simulation in which mobile nodes are

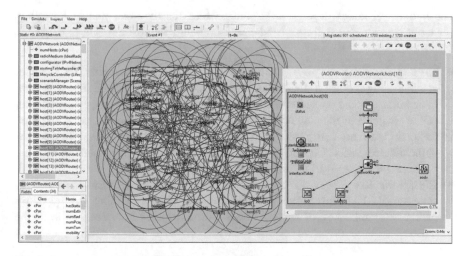

Fig. 3 Screenshot of AODV-based network topology

configured with AODV protocol. AODV-based hosts send UDP traffic to the node host (87), which is configured as sink node in the simulated topology.

6 Performance Analysis

The simulation results are analyzed to evaluate the efficiency of security model on the performance of cluster-based MANETs. The performance of cluster-based MANETs against some important network performance metrics as explained below [13].

6.1 Packet Loss Percentage

The percentage of packet lost during the communication in a network can be calculated by dividing the no. of packet lost to the total packet sent from sender. An efficient network must have the value of packet loss percentage as low as possible.

Figure 4 shows the percentage of packet losses with different number of nodes and cluster sizes. The results depict that the packet loss percentage decreases with the increase in number of clusters. Similarly, the packet loss rate will also decrease as we decrease the number of nodes in each cluster. Hence, the cluster-based MANETs perform better than non-clustered MANETs.

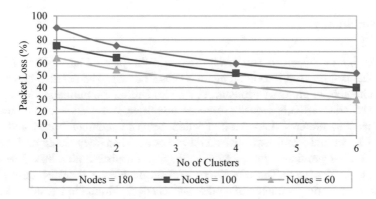

Fig. 4 Packet loss (%) with varying no. of clusters and nodes

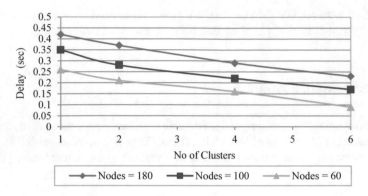

Fig. 5 Variations in end-to-end delay with varying cluster sizes

6.2 End-to-End Delay

Time consumed by packets during transmission form the sender to receiver is called end-to-end delay. For ideal networks, the value of end-to-end delay should be minimum.

Figure 5 shows the variation in delay value for different number clusters having a varying of nodes. The value of delay shows the decreasing behavior with an increase in the number of clusters. Hence, clustering approach can achieve better throughput.

7 Conclusion

The security model presented in this paper is the part of cluster-based defense model that can be used to defend against DDoS attacks. The defense model needs the requirement to protect various mobile nodes and their communication against a variety of attacks. The security process identified uses AES-based encryption technique that protects the communication between member nodes and cluster heads. Cluster head authenticates member nodes when they enter or leaves the cluster. It also provides cluster member key to member nodes that they use to encrypt their messages. The effectiveness of the technique is tested through simulations on a variety number of nodes and clusters. Simulation results show the packet loss percentages, and delay will decrease with increase in number of cluster. So the scheme works well in the cluster-based MANET approach.

References

1. Corson, S., Macker, J.: Mobile Ad hoc Networking (MANET): Routing Protocol Performance Issues and Evaluation Considerations. RFC 2501, 1999. https://www.ietf.org/rfc/rfc2501.txt
2. Chiang, T., Huang, Y.: Group keys and the multicast security in ad hoc networks. In: Proceedings of IEEE International Conference on Parallel Processing, pp. 385–390. IEEE Press, Los Alamitos (2003)
3. Kaya, T., Lin, G., Noubir, G., Yilmaz, A.: Secure multicast groups on ad hoc networks. In: Proceedings of 1st ACM Workshop on Security of Ad Hoc and Sensor Networks, pp. 94–102. ACM Press (2003)
4. Lazos, L., Poovendram, R.: Energy-aware secure multicast communication in ad hoc networks using geographical location information. In: Proceedings of IEEE International Conference on Acoustics Speech and Signal Processing, pp. 201–204. IEEE (2003)
5. Wang, N.C., Fang, S.Z.: A hierarchical key management scheme for secure group communications in mobile ad hoc networks. J. Syst. Softw. **80**(10), 1667–1677 (2007)
6. Hadjichristofi, G.C., Adams, W.J., Davis, N.J.: A framework for key management in mobile ad hoc networks. Int. J. Inf. Technol. **11**(2), 31–61 (2006)
7. Rahman, R.H., Rahman, L.: A new group key management protocol for wireless ad-hoc networks. Int. J. Comput. Inf. Sci. Eng. **2**(2), 74–79 (2008)
8. Bechler, M., Hof, H.J., Kraft, D., Pählke, F., Wolf, L.: A Cluster-based security architecture for ad hoc networks. In: Proceedings of Twenty-third Annual Joint Conference of the IEEE Computer and Communications Societies, INFOCOM, vol. 4, pp. 2393–2403. IEEE (2004)
9. Umar, M.M., Mehmood, A., Song, H.: SeCRoP: secure cluster head centered multi-hop routing protocol for mobile ad hoc networks. Secur. Commun. Netw. **9**(16), 3378–3387 (2016)
10. Gomathi, K., Parvathavarthini, B.: An efficient cluster based key management scheme for MANET with authentication. In: Proceedings of Trends in Information Sciences & Computing (TISC2010), Chennai, India (2010)
11. Renuka, A., Shet, K.S.: Cluster based group key management in mobile ad hoc networks. Int. J. Comput. Sci. Netw. Secur. **9**(4), 42–49 (2009)
12. Varga, A.: The OMNeT++ discrete event simulation system. In: Proceedings of European Simulation Multiconference (ESM), Prague, Czech Republic, pp. 354–371. Addison-Wesley, Springer, Berlin (1996)
13. Kumar, D., Gupta, S.C.: Transmission range, density & speed based performance analysis of ad hoc networks. Afr. J. Comput. ICT **8**(1), 173–178 (2015)

Performance Improvement of Naïve Bayes Classifier for Sentiment Estimation in Ambiguous Tweets of US Airlines

Jitendra Soni, Kirti Mathur and Yuvraj S. Patsariya

Abstract Airline deals with wide scale of customers, including family members, business class, business deals, men, women, and youngsters. Every kind of customer is involved in this domain. So their feedbacks are very important; direct feedback of customers is always positive but analyzing their tweets is important, how the Tweets are? Whenever we move toward personal tweets, analysis becomes so much hectic because of customers in very high volume. Most of the times tweets are ambiguous; it depends on thinker. If the thinker is positive, then he always gives positive tweets and negative thinker gives negative tweets. So ultimately, our work is to find sentiments of descriptive tweets and words and address their expression in quantities format whether they are happy or not. The novel factor inside this work is to examine ambiguous tweets and neutralize them according to proposed algorithm. The complete work is based on Twitter dataset for US Airlines and performs different level of mining and processing for most accurate results. Here, improved sentiment analysis model has been proposed based on Naïve Bayes classifier to classify tweets based on sentiments and neutralized tweets from ambiguous to positive or negative. A Java tool has been developed to implement the proposed solution that is tested through manual testing and performance parameters. This work considers recall/accuracy, f-score, and computation time as performance parameter for proposed solution. The complete solution is compared with existing work and concluded that it performs better than conventional algorithms.

Keywords Sentiment analysis · Ambiguous words · Airline dataset · Naïve Bayes classifier

J. Soni (✉) · Y. S. Patsariya
Institute of Engineering & Technology, Devi Ahilya Vishwavidyalaya,
Indore, M.P., India
e-mail: jitendrasoni2017@gmail.com

K. Mathur · Y. S. Patsariya
International Institute of Professional Studies, Devi Ahilya Vishwavidyalaya,
Indore, M.P., India

© Springer Nature Singapore Pte Ltd. 2020
K. S. Raju et al. (eds.), *Data Engineering and Communication Technology*,
Advances in Intelligent Systems and Computing 1079,
https://doi.org/10.1007/978-981-15-1097-7_17

195

1 Introduction

Twitter data or social networking sites are used to know about behavior and opinion of users. It may help to track the interest and connectivity of user with respective of their viewpoint. Let us consider the example of feedback system, where opinion and different viewpoint of users can observe from their feedback. Feedback system lies on the concept of marking which leads to calculate the relationship on basis of various points, despite descriptive feedback can also play important role to understand best way of opinion. Descriptive or subjective feedback can help to understand the demand and need. It can be written as subjective form of feedback to express thoughts or user view. Thus, it may be considered as the great source for analysis.

This work examined different existing solutions with observing data classification algorithm; these approaches are the major approach for feature extraction deriving sentiments and opinions of different users. Here, Naïve Bayes classifier is considered to classify user opinion [tweets].

On the basis of contextual ambiguity, issue arises of polarity calculation for sentimental analysis. Different polarity is calculated for different context using opinion keywords; it is a big challenge for researchers in the area of sentimental analysis. This issue of polarity is resolved ineffectively from term-level features.

Earlier, opinion mining effectively deals in progressive way with different scenario by developing document-level analysis [1–3]. In sentimental analysis, many issues arise with subjective detection and sentiment classification.

In general manner, sentimental analysis focuses on determining writer's attitude with respect to complete context of document with calculating polarity. This attitude of speaker and writer evaluates user's judgment in a state of emotions of writer when writing with emotional communication. Important task of sentimental analysis is the classification of polarity of text in a document, and also feature level is expressed through opinion of document either it is positive or negative or neutral. Sentiment classification for polarity in an emotional state can be as sad, happy, or angry.

2 Literature Study

Citizen's behavior and opinion is one of the important parts in the field of sentiment analysis. It can help to establish strong association between user desire and viewpoint. It is a good way to explore fine or minute scope of improvement. Tweets can help in knowing the view of user against any incidence or events. Strong sentiment analysis can lead to improvement of current trend scenarios. For such reasons, sentiment analysis of user viewpoint becomes mandatory for all citizens.

Existing system provides good way to explore user sentiments but still suffers with certain limitations: It considers words as individual effort and tokenize sentence in form of words. They state, "Future enhancement to this work might be to use n-gram classification rather than limiting to unigram" to overcome this issue. Existing solution has used unigram Naïve classifier, which considers word probabilities for both training and testing purposes. Sentiment analysis of whole sentence can help to reach more close to user viewpoint. At the time of classification, they consider individual word rather than actual sentence. They also suggest to use on n-gram Naïve classifier rather unigram. Authors do not consider ambiguity as serious problem and only concentrate to improve performance of sentiment analysis. Existing solution may be compromised for taunting tweets or indirect comments, which comes with positive words but used in negative purpose.

By using different algorithms, different phrases can be identified with positive or negative annotation based on frequency of occurrence of phrase and through it; its weight can be calculated based on phrases. Tweets are the online data of social site, which contains variety of flaws with the probability to hinder sentiment analysis. Limitations arise here is quality of opinion because of the right to freely post, meaningless contents due to online spammers. Another one is truth; only opinions are taken as truth, which can be as positive, negative, or neutral. Sentimental polarity is classified as sentence level and tweet level, where sentence level defines sentiments through positive and negative terms. Data required for sentence level needs truth facts to convey sentence, and this is significant to arise as the truth.

3 Proposed Methodology

Proposed work includes some modules in which flow of the work is explained through proposed architecture. Steps involved in proposed methodology are as following:

Step 1: Data Collection and Preparation:
Data can be collected either directly from the user or from the existing system. In proposed work, we have used Airline Tweet Dataset from Kaggle data repository. A link to detail description is cited below; https://www.kaggle.com/crowdflower/twitter-airline-sentiment#Tweets. csv.

Step 2: Data Pre-processing:

- Data Cleaning: Data cleaning is done to remove redundant data, irrelevant tweets, image short path, and unwanted link.
- Lemmatization: Stanford Lemmatize Library is used. Lemmatization refers to vocabulary and morphological words. It removes infected word and concentrates on words involved in dictionary.
- Tokenization: Here sensitive data is replaced with non-sensitive equivalent data and is represented as tokens.

Step 3: Proposed Estimation:

- Ambiguity-based polarity estimation: To check polarity of ambiguity word. Whether the word is positive, negative, partial positive, partial negative, or neutral. This work will initially classify tweets in three categories which are positive, negative, and ambiguous. Example of ambiguous tweets is cited below;
- Ambiguous tweet: "This is good airline service but comes with high tariff. They do not provide good service. Best part of this airline is they are always available."
- Positive tweet: "This one is the best airline comes with awesome services and facility. 100% recommendation for new once."
- Negative tweet: "Worst services experience. Air hostess don't know about their responsibilities."

This work will investigate ambiguous tweets and observe sentiments of every word. After observing sentiment for every individual, it also checks sentiment of previous and next word. In case of ambiguous word, it fixes the word polarity based on occurrence of previous and next whether they are positive or negative.

The complete tweets will be forwarded to Naïve Bayes classifier.

Step 4: Classifier:
Naïve Bayes classifier: It is used for large volume of dataset. Naïve Bayes classifier supposes that features are independent with values of given class labels. It will apply unigram and n-gram technique and concludes results based on both techniques.

Step 5: Tweets Classification:

- Sentiment calculation: Polarity of sentiments is calculated.
- Tweets feature extraction: Relevant features are extracted (Fig. 1).

```
1. Consider Airline Dataset as Source of Input
2. Input: Set variables rtweet_id, airline_senti-
   ment airline_sentiment_confidence, negatverea-
   son,     negatvereason_confidence     airline,
   airline_sentiment_gold, name, negativereason_-
   gold,   retweet_count,   text,   tweet_coord,
   tweet_created, tweet_location, user_timezone
3. Define Array Stopwords[] = {is, am, are, an, this}
4. Remove Stop words
5. Perform Lemmatization
6. Perform Tokenization
7. Set variable pre_word, post_word, curr_word
8. Check if (polarity[curr_word] == ambiguous)
9. Recheck if (polarity[pre_word] == positive)
   {Set: curr_word = positive}
```

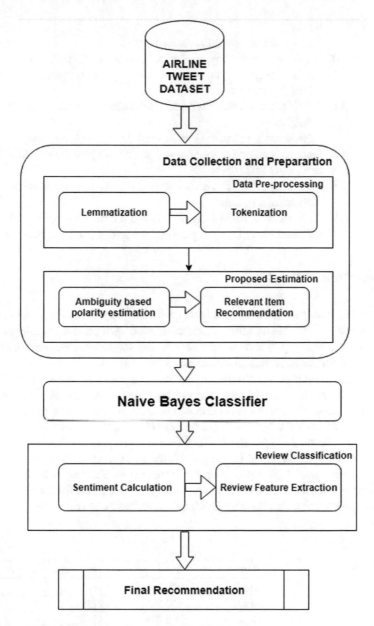

Fig. 1 Proposed architecture

```
      Else
      {Set: curr_word = negative}
  10. Replace word by polarity and estimate weight by
      individual    Tweets    for    ambiguous    Tweets
      classification
  11. If  (positive  Tweets  weight > negative  Tweets
      weight)
      {Tweets positive} else {Tweets negative}
  12. Set threshold value & Naive Bayes Classification
      and run unigram
  13. Set threshold value & Naive Bayes Classification
      and run n-gram
  14. Calculate average of 12 & 13 and classify tweets.
  15. Classification of positive & Negative Tweets
```

4 Experimental Analysis

The complete result examination has been performed for five different data sample. Proposed solution has been examined and compared with previous results to justify that proposed solution performs better than existing solutions. Here, size of different data samples has been shown in Table 1.

Headings. Headings should be capitalized (i.e., nouns, verbs, and all other words except articles, prepositions, and conjunctions should be set with an initial capital) and should, with the exception of the title, be aligned to the left. Words joined by a hyphen are subject to a special rule. If the first word can stand alone, the second word should be capitalized. The font sizes are given in Table 1.

Here are some examples of headings: "Criteria to Disprove Context-Freeness of Collage Languages," "On Correcting the Intrusion of Tracing Non-deterministic Programs by Software," "A User-Friendly and Extendable Data Distribution System," "Multi-flip Networks: Parallelizing GenSAT," and "Self-determinations of Man."

Table 1 Dataset

Title	Size	Total number of tweets
Input 1	100 KB	1325
Input 2	200 KB	2859
Input 3	300 KB	4256
Input 4	500 KB	7256
Input 5	1 MB	13,569

To observe the accuracy of proposed solution, precision and recall has been examined. Here, total tweets, recommended, and retrieved tweets have been calculated to estimate precision and recall parameters of proposed solution. Following formulas are used to estimate precision, recall, and f-score.

$$\text{Precision} = \frac{\text{Total Recommended Tweets}}{\text{Total Number of Tweets Retrieved}} \qquad (1)$$

$$\text{Recall/Accuracy} = \frac{\text{Total Recommended Tweets}}{\text{Total Number of Relevant Tweets Retrieved}} \qquad (2)$$

$$F\text{-Score} = 2 * ((\text{Precision} * \text{Recall})/(\text{Precision} + \text{Recall})) \qquad (3)$$

The complete performance observation is shown in Tables 2, 3, and 4 and with relevant graphs (Figs. 2, 3, 4, 5 and Table 5).

Table 2 Precision

Title	Precision
Input 1	0.74
Input 2	0.75
Input 3	0.77
Input 4	0.84
Input 5	0.87

Table 3 Recall/accuracy score

Title	Recall
Input 1	0.89
Input 2	0.91
Input 3	0.92
Input 4	0.93
Input 5	0.96

Table 4 F-score

Title	F-score
Input 1	0.79
Input 2	0.81
Input 3	0.82
Input 4	0.88
Input 5	0.90

Fig. 2 Precision comparison graph

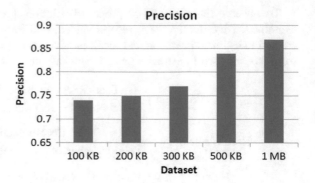

Fig. 3 Recall comparison graph

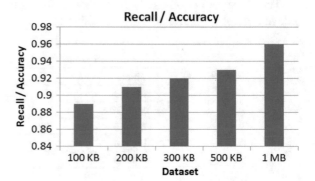

Fig. 4 *F*-score comparison graph

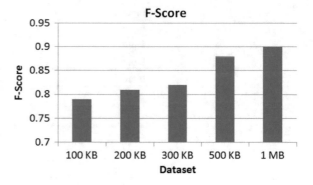

Fig. 5 Computation time

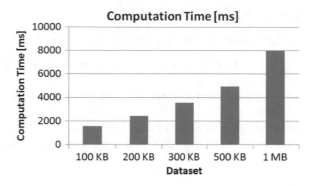

Table 5 Computation time

Title	Computation time (ms)
Input 1	1568
Input 2	2459
Input 3	3569
Input 4	4963
Input 5	8050

5 Conclusion

The complete work is based on Twitter Dataset for Airline Sentiments, where Airline Dataset is used as source input with analyzing and pre-processing the data to find sentiments of words. This work has proposed sentiment analysis model to observe positive and negative viewpoints of different users based on sentimental analysis approach. The complete work performs pre-processing and cleaning of data using Naïve Bayes classifier; the sentiments of airline tweets are estimated on this basis. By overcoming the unigram issue of existing work where single word tweet is taken, Naïve Bayes is used in proposed work and takes overall tweet to find number of words in tweet and their polarity.

Following conclusions have been drawn from complete experimental analysis.

1. Different size input gives different precision and recall but all can be scaled into single frame and shows growing nature.
2. Proposed algorithm performs better for high volume tweets in comparison with less numbers dataset.
3. Proposed solution is capable to classify ambiguous tweets and neutralize them according to their weight.
4. It implements unigram and *n*-gram technique to classify positive and negative tweets based on average value.
5. 0.79 as worst and 0.90 as best *f*-score have been observed which denote a great performance for ambiguous tweets.

Future Work

Following future work is predicted for proposed solution.

1. Proposed solution can be implemented using big data technology for large dataset.
2. Proposed algorithm can be used to classify different category tweets except airlines.
3. Proposed algorithm can be implemented with other classification techniques such as SVM and KNN to achieve better performance.

References

1. Dange, T., Bhalerao, P.: A novel approach for interpreting public sentiment variations on twitter: a tweets. Int. J. Sci. Res. (IJSR) 3(12) (2014). ISSN (Online): 2319-7064
2. Cambria, E., Hussain, A.: Sentic Computing: Techniques, Tools, and Applications. Springer, Dordrecht (2012)
3. Cambria, E., Olsher, D., Rajagopal, D.: SenticNet 3: a common and common sense knowledge base for cognition-driven sentiment analysis. In: AAAI. Quebec City, pp. 1515–1521 (2014)
4. Cambria, E., Schuller, B., Xia, Y., Havasi, C.: New avenues in opinion mining and sentiment analysis. IEEE Intell. Syst. 28(2), 15–21 (2013)
5. Turney, P.D.: Thumbs up or thumbs down?: Semantic orientation applied to unsupervised classification of Tweetss. In: Proceedings of the 40th Annual Meeting on Association for Computational Linguistics, pp. 417–424. Association for Computational Linguistics (2002)
6. Sehgal, D., Agarwal, A.K.: Sentiment analysis of big data applications using twitter data with the help of HADOOP framework. In: 5th International Conference on System Modeling & Advancement in Research Trends. IEEE (2016)
7. Mazhar Rathore, M., Paul, A., Ahmad, A.: Big data analytics of geosocial media for planning and real-time decisions. In: ICC SAC Symposium Big Data Networking Track. IEEE (2017)
8. Soni, J., Mathur, K.: Data mining and machine learning: design a generalized real time sentiment analysis system on tweeter data using natural language processing. Int. J. Eng. Adv. Technol. 8(6), 2139–2142 (2019)
9. Wawre, S.V., Deshmukh, S.N.: Sentiment classification using machine learning techniques. Int. J. Sci. Res. (IJSR) 5(4) (2016)
10. D'Andrea, A., Ferri, F., Grifoni, P., Guzzo, T.: Approaches, tools and applications for sentiment analysis implementation. Int. J. Comput. Appl. (0975–8887) 125(3) (2015)
11. Hailong, Z., Wenyan, G., Bo, J.: Machine learning and lexicon based methods for sentiment classification: a survey. IEEE (2014). https://doi.org/10.1109/wisa.2014.5. 978-1-4799-5727-9/14 $31.00
12. Turney, P.D., Littman, M.L.: Measuring praise and criticism: inference of semantic orientation from association. ACM Trans. Inf. Syst. 21(4), 315–346 (2003)
13. Ramakrishna Murthy, M., Murthy, J.V.R., Prasad Reddy, P.V.G.D.: Text document classification based on a least square vector support vector machines with singular value decomposition. Int. J. Comput. Appl. (IJCA) 27, 21–26 (2011)
14. Ramakrishna Murthy, M., Murthy, J.V.R., Prasad Reddy, P.V.G.D.: Statistical approach based keyword extraction aid dimensionality reduction. In: International Conference on Information Systems Design and Intelligent Application. Springer-AISC indexed by SCOPUS, ISI Proceeding DBLP etc., vol. 132, pp. 445–454 (2012). ISBN 978-3-642-27443-5

A Nonparametric Approach to the Prioritization of Customers' Online Service Adoption Dimensions in Indian Banks

Kishor Chandra Sahu and J. Kavita

Abstract Online banking service adoption depends on the Internet exposure and the customer's preparedness and individual's attitude. There may be numerous dimensions that can influence in the adoption of online banking services both Internet and mobile banking. However, it is very much essential to calculate the perception of the customers on what they are expecting from the service adoption. This paper proposes a methodology that prioritizes the parameters related to service adoption for the Indian banking industry grounded on a survey administered on the bank account holders. This paper examines the factors which influence the attitude of the customers in adopting the technologically driven banking modes for their day-to-day transactions. In this study, we have attempted to find the most important factors such as technology, perceived security, regulatory, customer orientation, attitude, perceived benefits and perceived service quality by following an exploratory factor analysis (EFA). The study tried to find the precedence list of factors by using nonparametric approaches such as ridit approach and Kruskal-Wallis method. The findings of the study will give the Indian banks in designing the attitude change strategies for wider adoption of their various online banking applications among their customer base.

Keywords Banking service adoption · Attitudinal factors · Exploratory factor analysis · Ridit analysis · Kruskal Wallis test

1 Introduction

In today's world, the financial institutions need to adopt the latest technologies and revisit the customer experience and at the same time need to drop their barriers to accept innovation and stay relevant in today's world which becoming more cus-

K. C. Sahu (✉) · J. Kavita
Department of Mathematics and Humanities, Mahatma Gandhi Institute of Technology, Hyderabad, Telangana 500030, India
e-mail: kishor.chandra.sahu@gmail.com

© Springer Nature Singapore Pte Ltd. 2020
K. S. Raju et al. (eds.), *Data Engineering and Communication Technology*, Advances in Intelligent Systems and Computing 1079, https://doi.org/10.1007/978-981-15-1097-7_18

tomer centric than ever before. With the ever-growing consumer expectations and adoption of the new technologies, the financial institutions are becoming more technology oriented. The innovation in the fintech industry has become a global phenomenon. Among the various areas, fintech adoption stands apart. India is one of the growing economies that have witnessed a massive growth in the fintech, driven by the clear regulatory reforms to guide the digital economy. The real impact of digital transformation comes from the adoption of open banking. The banks which were having restricted ecosystems are now adapting more flexible systems. As India leaps towards open digital economy, the banks are offering aadhar authentication, digital signature of the documents, unified payments interface and permission-based access fees.

In this age of radical advancement of technologies, the banking industry faces the challenge of their customers expecting a modernized banking experience. The banks are bringing new client offerings in a digital way as the digitalization and digital client interaction are becoming more influential. Changes in the consumer preference and technology drive the transition to a digital economy, and these factors are common to most of the banks. A new standard in consumer and digital experience has been created by the millennial. The future as predicted includes the role of the capability of the platforms and the application program interfaces. The providers of digital service will work as the change drivers and will create opportunities to serve their clients and segments better (KPMG report).

According to a global banking outlook report from EY [1], there is little doubt that technology can help banks tailor much better products and deliver improved customer experience. 85% of the banks consider digital transformation program implementation as a top business priority in the year 2018. After the global financial crisis, the global banking sector is considerably healthier now. The adoption of the finance technology providers for money transfer and payment services have raised to 50% in the year 2017 and understandably, more than 60% of the global banks expect to be a digital leader or maturing digitally in the year 2020. More than half of the banks in the world foresee their investment in technology increase by more than 10% in 2018.

Smartphone adoption in India getting boosted by the expansion of 4G/LTE networks and the Government of India initiatives such as Digital India, e-wallets, Make in India, smart cities. As per the last report from IAMAI, around 292 million people in India are using smartphone, and the smartphone owner base is expected to increase to 445 million by 2020 as the economy and the young population of India increases. This can be a growth enabler positively influencing the evolving knowledge economy.

Banking activities are digitized and automated as argued and the Internet banking is prevalent because of the benefits it provides such as convenience and comfort [2]. According to Moody [3], the banks are offering online banking in order to gain new retail customers.

As rightly put forward by Yousafzai and Yani-de-Soriano [4], the banking and finance world is not able to stop with the Internet revolution and banking institutions. If the customers fail to accept technology and use the capabilities there will not much return from the investments. They suggest the banks to make the customer feel secure by informing and make them understand how to protect their interests and the privacy.

Mobile banking offering has become a factor to differentiate while selecting a new bank. According to Juniper Research's "Mobile and online banking: developed and developing market strategies 2014–2019" report, it has been estimated that by 2019, the mobile banking users will exceed 1.75 billion [5].

At present, almost all kinds of bank transactions i.e. payments, transfer, enquiry etc. are using Internet. Adoption of Internet technology has changed the Indian banking industry. Indian banks have been offering mobile banking services over the last decade or so to offer seamless access to banking services. In India, mobile banking transaction has been on the rise in the last few years or so and as per KPGM, 2015; it is one of the highest rates of adoption in the world.

Online banking does not allow customers to speak to bank representatives in person. There have been numerous studies on the influential factors related to the use of online banking by the customers and its effect of privacy concerns on e-commerce variables.

2 Review of Literature

This section will discuss and argue the selection of the variables for empirical exploration. There are abundant existing literatures related to banking that evaluates the perceived service quality dimensions. In the area of services marketing, two important variables that impact the customer loyalty are the perceived service quality and customer satisfaction [6, 7].

Creation and retention of the satisfied customers are the yardsticks to measure the success of any service quality program [8]. A considerable amount of research has been published which shows that customer service quality dimensions and customer satisfaction are strongly related [9–11]. Major determinants of customer satisfaction are the quality dimensions, features, problems and recovery of service [12]. The role of perceived service quality in a high involvement industry like banking is immense. As the nature of customer relationship is changing and electronic banking becomes more prevalent than ever before, the service quality is measured in terms of technical support. Moreover, the customers evaluate banks based on their high-touch factors rather than high-tech factors. High-touch factors are becoming more relevant than the high-tech factors for the customers to evaluate the banks.

Jun and Cai [13] conducted a critical incident technique on the comments from the customers who use Internet banking to find out the significant attributes of quality related to the online banking products and services. Among the various dimensions identified, they suggested that the banks have to concentrate on the dimensions such as how to respond fast, the reliability, the access to the customers, and ease of use, transaction accuracy and product variety. Customer satisfaction is strongly dependent on the quality performance of these dimensions.

Majority of the previous research studies are focused on the Internet banking adoption from two aspects; a service provider and an employee perspective. The studies on the adoption of technologies by the user are important because the global banks are relying on the Internet technologies as the key achiever of customer satisfaction and efficiency for surviving in the competitive banking industry [14]. Service quality and service adoption are strongly related and that led to the conceptual foundation of the SERVQUAL model and the perceived quality model [15–17].

The technology adoption in service industries is becoming a trend very fast as the industry bodies are insisting the service providers to invest in technology so that they can secure their future in this electronic age. Numerous past studies have tried to explain the impact of technology on banking sector [18]. Much of the previous research has focused on measuring the customer service quality offered by the banking institutions [19, 20]. Hemmasi et al. [21] recommended that the banks need to concentrate in areas such as allowing the customers to verify accuracy of transactions, accessibility of outlets and a toll-free number, so that they can handle customer complaints, receive general feedback and improved security.

IT-based services and service quality dimensions are directly related whereas it is indirectly related to customer-perceived service quality and customer satisfaction. The customer preferences about traditional service, learnings during the use of IT services and the way they perceive the IT policies affect the way the customers assess the IT-based services. There is a large volume of published studies describing the role of IT in enhancing service quality, productivity and increase of revenue generation. All the service dimensions such as responsiveness, reliability and assurance are affected by the services which are IT based. These service dimensions in turn affect the customer satisfaction [22].

Whatever may be the technology, the customer service, technology usage easiness and reliability are the important service quality dimensions. These service dimensions impact positively on both customer satisfaction and customer loyalty [23].

Understanding the factors that influence the acceptance of new products will allow businesses to create a climate in which technological advances with real advantages can be embraced by a majority instead of just a few techno-savvy consumers. Most of the consumers will embrace the advantages of technology if they understand the factors influencing the new product acceptance. The probability of adoption increases and the risk with which e-banking is perceived becomes less if e-banking is easier, compatible and more observable [24].

Inclination to trust, guarantees and security of transactions and the content of word-of-mouth can predict the initial trust in the e-channel [25].

Trust and security are the two most important aspects associated with the rate with which the Internet banking is adopted. These insights were fetched from the theories like TAM, TRA and TPB. A study in Mauritius showed that perceived ease of use and perceived usefulness influence the Internet banking adoption. Behaviour in adopting new ITs and individual intentions is being explained effectively by TPB. Level of awareness of Internet banking, security and trust are also the factors that explain significantly the Internet banking adoption [26].

Similarly in India, when it comes to m-banking adoption intention of consumers, the factors impacting are consumer awareness, ease of use compatibility, privacy and security, and financial cost. Out of these, security, privacy and financial cost affect the adoption positively. As the customers seek suggestions from family members, social norms have also crucial factor that influences the mobile banking adoption [27].

The functional features of the mobile apps and enjoyment which the consumers get develop a positive attitude to adopt the mobile banking. They need to understand and evaluate the benefits they get and the risks in terms of security and privacy offered by a bank's technological infrastructure. It was found that an interesting feature called "fun" is important in influencing the adoption of m-banking. Fun is referred to the technology and consumers' innovativeness. However, security and privacy are the major reasons for consumers' not adopting mobile banking [28].

A study on the perceived risk and its dimensions in the Internet banking found that data security poses a serious challenge [29]. The dimensions identified by Marakarkandy et al. [30] are impacting the adoption of Internet banking. They are usage efficacy of the Internet banking, risk perceived, the trialability of the Internet banking, trust in the bank and self-efficacy. The study acknowledges the issue of the customer adoption of services. Another study on the Internet banking dimensions highlighted that financial risk, online banking awareness, time, performance risk, social risk, security and privacy affect the adoption [31].

Another study found integrity and credibility as trust dimensions and they positively influence perceived usefulness. Internet banking adoption and behavioural intention are affected as a result of the perceived usefulness [32].

Across the distribution channels which are technology based such as ATMs, Internet banking, phone banking trust is the most important factor for adoption [33]. A comprehensive attempt to understand the non-adoption of technological interfaces suggested that the aspects such as human interaction, risks and lack of trial are found to be inhibitors of adoption [34].

Although many factors in different scenarios can be listed that influence the adoption of digital banking service, this study will try to recognize and evaluate the relative significance of the various digital banking service adoption dimensions prepare the list of prioritized items so that the banks can draw some meaningful insights.

3 Research Methods

3.1 Research Design

Both exploratory and descriptive research methods have been used in this research study. Exploratory research has been employed to gain an understanding about the impact of relationship marketing on customer loyalty. The research study is descriptive as it includes the survey to collect the data of the survey. The variables/ dimensions of the online service adoption are dynamic and subjective in nature. The research process has been represented in a workflow diagram (Fig. 1).

The research study undertakes a combination of exploratory research and descriptive research. Exploratory research has been used in order to gain in-depth insights into the research area and comprehension of various issues related to relationship marketing and customer loyalty. The research study is descriptive as it includes the survey to collect the data of the survey. The variables/dimensions of the online service adoption are dynamic and subjective in nature. The research process has been represented in a workflow diagram (Fig. 1).

3.2 Sampling Design

The sample for the present study has been collected based on simple random sampling. Both online and offline users of various different private and public banks in two cities Hyderabad and Bhubaneswar (tier II cities) were surveyed. Eleven banks were considered for the survey. The respondents for the survey were bank account holders who were more than 18 years of age and the survey was conducted during August 2018 and November 2018.

The sample was selected randomly at the branches of banks, and the printed copy of the survey questionnaire was given with an orientation of the purpose of the

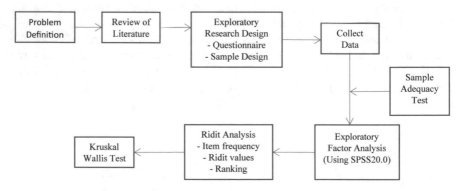

Fig. 1 Research process workflow

survey. The questionnaire was also translated in native languages Telugu and Odiya in addition to English for the convenience and easy understanding of the items in the questionnaire. Each of the respondents was given sufficient time to go through the questionnaire and fill the survey. Since this survey involves the human participants, informed consent form was filled and obtained from each of the participants before administering the questionnaire survey. The respondents were assured of the privacy of their information and were explained in detail the objectives of the study. They were also assured that the data gathered in this study were used only for this research and not for any commercial purpose, and the information collected were kept strictly confidential. Out of the 400 questionnaires distributed (including both offline and online professional survey form), 347 respondents' data were valid and complete. The response rate of 86.75% was significant to accept and proceed for the survey analysis.

The survey questionnaire encompassed a total of 29 items excluding socio-demographic and items on technology, perceived benefits, security and overall service quality. The questionnaire was translated from English to local languages for better understanding.

The items which are of Likert-type scale item contain five options which were assigned as follows: 1 being strongly disagree and 5 being strongly agree. The items of the questionnaire are listed in the Appendix section.

3.3 Data Analysis

3.3.1 Plan for Data Analysis

The study was divided into two phases. In the first phase, exploratory factor analysis was done on the variables identified for the purpose of this study. The second step involves ridit analysis where in the items are prioritized based on the ridit values computed.

3.3.2 Data Set

The data set contains few demographic details about the individuals and the details about digital banking service adoption factors such as technology, perceived security, regulatory, customer orientation, attitude, benefits and the service quality of the various banking services they are availing both online and offline. The variables other than the demographic variables were captured on the ordinal scale.

Data cleaning was done before the EFA was done to find the underlying structure. Data cleaning process included the variable identification and missing values treatment. The records with missing values were removed from the final data set for analysis.

3.3.3 Data Analysis

Exploratory factor analysis method was employed to identify the underlying relationships between the variables to be measured. The adequacy of the sample was checked using the Kaiser-Meyer-Olkin (KMO). The KMO test helps in identifying the level of variance in the variables which may be because of underlying factors. The Bartlett's test of sphericity was conducted to check whether the correlation matrix is an identity matrix or not. The result of the tests shows that the KMO and Bartlett's support the data set under study.

The KMO was found to be 0.812 which is a significant value. According to the Bartlett's test of sphericity, the p-value (significant) of $0.000 < 0.05$ indicates that the factor analysis is valid.

3.3.4 Varimax Rotation

To minimize the number of variables by distributing the loading across the variables varimax rotation method was used. Table 1 represents the factor scores of all the items identified for the study. Varimax rotation was used to do the factor analysis of the 29 items on the basis of principal component extraction. In comparison to the other orthogonal rotation methods, it was chosen as the preferred method of rotation. In varimax rotation, the results in all the coefficients are large or zero with a few intermittent values. Thus, the varimax rotation helps in associating each variable to a given factor when compared to other orthogonal rotation methods.

The rotated component matrix below shows the correlation of the variables or items with each of the extracted factors. Table 1 shows the highest loading of each variable in one factor. After selecting the highest values in each row, the 29 items were grouped into 7 components or factors.

The varimax rotation converged after 5 iterations and it resulted in 7 factors with eigenvalues of 7.894, 5.018, 4.134, 2.112, 1.823, 1.625 and 1.042. The total variance explained after the rotation was 72.992%. A variable, for which the communality is below 0.4, was considered to be struggled to load significantly on any factor. For the exploratory factor analysis, SPSS version 20.0 (IBM Corporation) was used.

The factors during the exploratory analysis fetched were labelled and their underlying construct considering the perceived digital services adoption items. The factors are labelled as customer oriented (Cronbach's $\alpha = 0.785$; five items), attitudinal (Cronbach's $\alpha = 0.815$; six items), perceived benefit (Cronbach's $\alpha = 0.802$; three items), technology factors (Cronbach's $\alpha = 0.853$; four items), perceived security (Cronbach's $\alpha = 0.842$; four items), regulatory factors (Cronbach's $\alpha = 0.807$; three items), perceived service quality (Cronbach's

Table 1 Rotated component matrix

Component							
	1	2	3	4	5	6	7
COF1	0.87						
COF2	0.75						
COF3	0.892						
COF4	0.888						
COF5	0.782						
ATT1		0.813					
ATT2		0.753					
ATT3		0.68					
ATT4		0.865					
ATT5		0.778					
ATT6		0.832					
PB1			0.779				
PB2			0.822				
PB3			0.709				
TF1				0.921			
TF2				0.934			
TF3				0.756			
TF4				0.806			
PS1					0.795		
PS2					0.797		
PS3					0.811		
PS4					0.833		
RF1						0.912	
RF2						0.907	
RF3						0.691	
PSQ1							0.679
PSQ2							0.872
PSQ3							0.888
PSQ4							0.903

Note Rotation converged in six iterations. The extraction method was principal component analysis and the rotation method was varimax with Kaiser normalization (Using SPSS 20.0)

$\alpha = 0.775$; four items). The overall Cronbach's alpha for the digital service adoption scale items was found to be 0.781 from we can infer that the questionnaire is capable of capturing the variables in a meaningful way. The factors of the online service adoptions thus found are shown in Fig. 2.

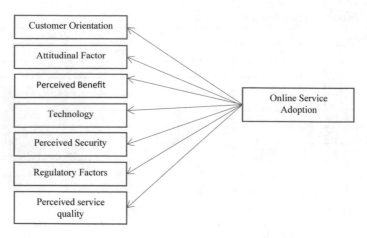

Fig. 2 Extracted factors

3.4 Ridit Analysis

Ridit is the acronym for "Relative to an identified distribution". In ridit analysis, probability transformation is done based on some empirical distributions which is taken as a reference class, it does not make any assumptions about the distribution of the population, and it is distribution free [35]. Ridit analysis does not make any assumption about the distribution of the population under study, hence distribution free. It has been applied to various behavioural and business management studies since it was first proposed by Bross [36].

In ridit analysis, the term ridit is the assigned weight to a response category which represents the probability of the category appearing in the reference distributions. Ridit analysis is used for ordered data but which cannot be put on interval scale. It is a statistical method to measure ordinal scale.

Ridit value is a number assigned to a particular response category of the variable that is equal to the proportion of individual in the reference class who have a lower value on that variable, plus one half of the proportion of individuals in category itself. A ridit value ranges from 0.00 to 1.00. Once the ridit values for each category of the independent variables have been computed, the individual scores transformed into the ridit value for the dependent variable. The average ridit value is calculated for a class rather than the proportion of the respondents giving each of the responses in the dependent variable.

The reference data set chosen was the digital service adoption at the banks in India. Table 2 shows the frequencies of the responses. The ridit values of the reference dataset for each item category are shown in the last row of Table 2. Table 3 shows the weights that are summed to derived ridit values and the corresponding ranking with those ridit scores.

Table 2 Item frequencies for reference data set

Variable	5	4	3	2	1	π
COF1	131	102	100	14	0	347
COF2	89	107	96	34	21	347
COF3	112	114	84	27	15	352
COF4	108	129	65	28	17	347
COF5	119	111	74	28	15	347
ATT1	100	111	92	29	15	347
ATT2	102	111	93	27	14	347
ATT3	107	112	87	27	14	347
ATT4	111	111	83	28	14	347
ATT5	115	119	77	20	16	347
ATT6	103	110	89	29	16	347
PB1	104	110	91	27	15	347
PB2	102	116	86	28	15	347
PB3	96	106	100	30	15	347
TF1	97	104	91	39	16	347
TF2	124	100	82	27	14	347
TF3	97	115	90	29	16	347
TF4	90	114	101	27	15	347
PS1	101	109	91	32	14	347
PS2	109	109	87	28	14	347
PS3	92	108	99	31	17	347
PS4	116	107	83	27	14	347
RF1	96	113	92	30	16	347
RF2	100	107	96	28	16	347
RF3	97	110	87	37	16	347
PSQ1	96	106	94	34	17	347
PSQ2	102	110	88	32	15	347
PSQ3	115	100	88	29	15	347
PSQ4	125	99	85	25	13	347
fj	3056	3180	2571	831	430	10068
1/2 fj	1528	1590	1285.5	415.5	215	
Fj	1528	4646	7521.5	9222.5	9853	
Rj	0.151768	0.461462	0.74707	0.916021	0.978645	

3.4.1 Computation of Ridit Value

In Table 3, the row for COF1, the value of 0.057 is derived from Table 2 by multiplying the frequency of 131 (Look for the row marked COF1 in Table 2) with the reference group ridit value of 0.15177 (look at the bottom row and 1st column of Table 2). Divide the result by the number of respondents i.e. 347 (last column of

Table 3 Ridit values computation and prioritization

Variables	5	4	3	2	1	RowSum	Rank	Upper bound	Lower bound
COF1	0.057	0.136	0.215	0.037	0.000	0.4452	1	0.451	0.439
COF2	0.039	0.142	0.207	0.090	0.059	0.5369	29	0.543	0.531
COF3	0.048	0.149	0.178	0.070	0.042	0.4880	8	0.494	0.482
COF4	0.047	0.172	0.140	0.074	0.048	0.4806	6	0.486	0.475
COF5	0.052	0.148	0.159	0.074	0.042	0.4752	5	0.481	0.469
ATT1	0.044	0.148	0.198	0.077	0.042	0.5083	18	0.514	0.503
ATT2	0.045	0.148	0.200	0.071	0.039	0.5032	15	0.509	0.497
ATT3	0.047	0.149	0.187	0.071	0.039	0.4938	12	0.500	0.488
ATT4	0.049	0.148	0.179	0.074	0.039	0.4883	9	0.494	0.483
ATT5	0.050	0.158	0.166	0.053	0.045	0.4722	3	0.478	0.466
ATT6	0.045	0.146	0.192	0.077	0.045	0.5046	16	0.510	0.499
PB1	0.045	0.146	0.196	0.071	0.042	0.5013	14	0.507	0.496
PB2	0.045	0.154	0.185	0.074	0.042	0.5002	13	0.506	0.494
PB3	0.042	0.141	0.215	0.079	0.042	0.5197	24	0.526	0.514
TF1	0.042	0.138	0.196	0.103	0.045	0.5247	27	0.530	0.519
TF2	0.054	0.133	0.177	0.071	0.039	0.4745	4	0.480	0.469
TF3	0.042	0.153	0.194	0.077	0.045	0.5108	20	0.517	0.505
TF4	0.039	0.152	0.217	0.071	0.042	0.5220	25	0.528	0.516
PS1	0.044	0.145	0.196	0.084	0.039	0.5090	19	0.515	0.503
PS2	0.048	0.145	0.187	0.074	0.039	0.4933	11	0.499	0.488
PS3	0.040	0.144	0.213	0.082	0.048	0.5268	28	0.533	0.521
PS4	0.051	0.142	0.179	0.071	0.039	0.4825	7	0.488	0.477
RF1	0.042	0.150	0.198	0.079	0.045	0.5147	22	0.520	0.509
RF2	0.044	0.142	0.207	0.074	0.045	0.5118	21	0.518	0.506
RF3	0.042	0.146	0.187	0.098	0.045	0.5188	23	0.525	0.513
PSQ1	0.042	0.141	0.202	0.090	0.048	0.5230	26	0.529	0.517
PSQ2	0.045	0.146	0.189	0.084	0.042	0.5071	17	0.513	0.501
PSQ3	0.050	0.133	0.189	0.077	0.042	0.4916	10	0.497	0.486
PSQ4	0.055	0.132	0.183	0.066	0.037	0.4720	2	0.478	0.466

Table 2). The weights from the five columns are then summed to get the ridit scores. Mathematically, the average ridit value will be 0.5. The items which have relatively more responses of 5 and 4 will tend to have a ridit value of less than 0.5 and the items which are having relatively more responses of 1 or 2 will have a ridit value greater than 0.5. Consequently, the lower is the ridit value; the higher is the priority of an item. So, we will assign the highest priority to the items with the lowest ridit value.

3.5 *Kruskal-Wallis Test*

The Kruskal-Wallis W was calculated to be 50.354. Kruskal-Wallis W adopts Chi square distribution with $(m - 1)$ degrees of freedom, wherein m represents the number of items. We found the W (**50.354**) to be significantly **greater** than the Chi-square χ^2 $(29 - 1) = $ **41.34**. From this we can imply that opinions about the items in the survey among the respondents are statistically not the same.

This is a nonparametric test and is used to determine significant differences between two or more groups of independent variables. It does not require the data to be normal, but instead the ranks of the data values are used for the analysis.

From Table 3 which represents the analysis of ridit ranking, of all the digital service adoption dimensions the customer orientation dimension was given more priority. Respondents have given the highest priority to items like the banks instructions about the usage of digital services followed by the overall quality of ban services provided to the customers and the online transfer facility provided by these banks.

The lowest priority ranking in the digital service adoption dimension was given to toll free lines for customer service, availability of various options in digital services, threat of virus attack.

This may be because of the inadequate awareness of the basic and core functional and risk and security factors involved in availing the digital banking services by the customers. It also indicates that the customers are not giving more priority to the provision of important augmented services. The banks' providing toll free lines, the security threats has also very little meaning for the customers. The other dimensions and their relative ranking are also found to be more or less symmetrical.

This means the groupings of the variables being done by factor analysis under each construct in a way justifies their rankings being done by ridit analysis. The overall ranking of the digital service adoption dimensions are shown in Table 3.

4 Conclusion and Managerial Implications

This research has been undertaken to understand the customer adoption of the digital banking solutions provided by the banks and the issues involved while adopting such services. Offering of the various digital services and capabilities should be based on the readiness of the customers' adoption and willingness to try those services. They are involved in identifying, evaluating and availing these various Internet and mobile banking services. Although service quality is an important issue, the banks in India are not doing much to know the preferred services for their account holders or customers.

In the Indian banking sector, there are several factors those drive the service adoption by the customers. Customer adoption of digital services is influenced by factors like customer orientation, attitude of the customers, perceived benefits, technology, perceived security, regulatory and perceived service quality. Banks would benefit from identifying the most important factors and prioritizing them in their promotion thereby speeding up the adoption of these digital services.

The study proposes an appropriate method to weigh and prioritize variables to manage better performance in the Indian banking setting. These constructs may help to focus the important areas to be improved in digital banking services. The ridit methodology was proposed for ranking the digital banking service adoption in the Indian banks with setting with seven dimensions for service adoption. Prioritization helps the banks to take better decision and customers in identifying the best service adoption characteristics that can in turn improve the providers' performance.

The current research suggests that the managers of the banks should identify those dimensions of the service adoption affecting customer satisfaction by developing a feedback system. The managers need to look into the top priority dimensions and ensure they match the customer's expectations. Future research can be replicated for different types of banks and customer segment wise in order to get more concrete insights. Also the relationship between the adoption priorities and the long-term customer values can be worth to explore.

The present study has some limitations. The ordinal response measures included in this research were the customers' perception based. So the data collected is not free of error due to predisposition or misinterpretation. Moreover, since the study has been carried out in two cities Hyderabad and Bhubaneswar, the perception based response of the respondents may vary from those of the other parts of India. A research study with better selection which can be a representative across India will give substantial insights.

Appendix

Online service adoption dimensions

Item	Variable
COF1	I need to be instructed on how to use online banking
COF2	Toll free lines are needed for customer service
COF3	I get the latest information about the new products and services offered by the bank through online banking applications
COF4	I believe that my bank educates its customers about the risk of any unauthorized transactions
COF5	I get adequate utility services (like tax, credit card bill payments, insurance premium) through the online banking applications
ATT1	My lifestyle influences me to use online banking
ATT2	I believe that I have control over the transactions conducted using online banking service
ATT3	Recommendations from family and friends motivate me to use online banking

(continued)

(continued)

Item	Variable
ATT4	I use online banking to manage my bank account
ATT5	I use online banking to transfer
ATT6	I believe online banking improves my payment and task efficiency
PB1	I find online banking easy to use
PB2	Online banking gives me the flexibility to conduct banking transactions 24×7
PB3	I find online banking to be more advantageous than traditional banking
TF1	I get adequate opportunities to try out different banking operations in online banking
TF2	I'm considering to subscribe to a mobile or internet banking service in the near future
TF3	I believe the data transfer speed over the network is an important factor for internet and mobile banking
TF4	I expect the easy navigation of online banking application
PS1	I trust online banking platform to perform transactions
PS2	While performing banking transactions, there is a risk of losing personal information
PS3	I feel online banking is prone to virus attack
PS4	Information related to the banking transactions will not be lost during an online transaction
RF1	My Bank's basic online banking system utilization fee is reasonable and economic
RF2	The system services, though include inter-bank and intra-bank account transfer fees, does not affect adoption intention
RF3	Inter-bank and intra-bank transaction fee is more advantageous than ATM and over-the-counter transactions
PSQ1	I believe that my bank processes my transactions accurately
PSQ2	I believe that my bank processes my transactions timely
PSQ3	I believe that online banking services provided by my bank are compatible for interbank transactions
PSQ4	The overall quality of online banking service offered by my bank is very important for all my banking needs

References

1. Pivoting towards an innovation led strategy: Global Banking Outlook 2018 [online]. Available at: http://www.ey.com/bankinginnovation (2018). Accessed 20 Nov 2018
2. Nor, K.M., Barbuta, N., Stroe, R.: A model for analysing the determinant factors of adoption e-banking services by Romanian customers. Available at: www.ecocyb.ase.ro/nr4%20eng/Radu%20Stroe.pdf (2010)
3. Moody, J.: Traditional banks gain edge with electronic banking. Available at: www.cendant.com/media/trends_information/trends_information.cgi/MarketingþServices/59 (2002)
4. Yousafzai, S., Yani-De-Soriano, M.: Understanding customer-specific factors underpinning internet banking adoption. Int. J. Bank Mark. **30**(1), 60–81 (2012)
5. Bhas, N.: Digital Banking—Mobile and Beyond, White Paper. Juniper Research, Basingstoke, Hampshire. Available at: www.juniperresearch.com/whitepaper/digital-banking-mobile-andbeyond (2014)

6. Oliver, R.L.: Satisfaction: A Behavioral Perspective on the Consumer. McGraw-Hill, New York, USA (1997)
7. Zeithaml, V.A., Bitner, J.M.: Services Marketing. McGraw-Hill, New York, USA (1996)
8. Yavas, U., Bilgin, Z., Shemwell, D.J.: Service quality in the banking sector in an emerging economy: a consumer survey. Int. J. Bank Mark. 15(6), 217–223 (1997). https://doi.org/10.1108/02652329710184442
9. Anderson, E.A., Sullivan, M.W.: The antecedents and consequences of customer satisfaction for firms. Mark. Sci. 12, 125–143 (1993)
10. Brown, T.J., Churchill Jr., G.A., Peter, J.P.: Improving the measurement of service quality. J. Retail. 69, 127–138 (1993)
11. Cronin, J.J., Taylor, S.A.: Measuring service quality: a re-examination and extension. J. Mark. 56(3), 55–68 (1992)
12. Levesque, T., McDougall, G.H.G.: Determinants of customer satisfaction in retail banking. Int. J. Bank Mark. 14(7), 12–20 (1996). https://doi.org/10.1108/02652329610151340
13. Jun, M., Cai, S.: The key determinants of Internet banking service quality: a content analysis. Int. J. Bank Mark. 19(7), 276–291 (2001). https://doi.org/10.1108/02652320110409825
14. Sharma, S.K., Govindaluri, S.M., Al Balushi, S.M.: Predicting determinants of internet banking adoption: a two-staged regression-neural network approach. Manag. Res. Rev. 38(7), 750–766 (2015). https://doi.org/10.1108/MRR-06-2014-0139
15. Parasuraman, A., Zeithaml, V.A., Berry, L.L.: A conceptual model of service quality and its implications for future research. J. Mark. 49(3), 41–50 (1985)
16. Shih, Y.Y., Fang, K.: Effects of network quality attributes on customer adoption intentions of internet banking. Total Qual. Manag. Bus. Excellence 17(1), 61–77 (2006)
17. Shostack, L.G.: Designing services that deliver. Harvard Bus. Rev. 62(1), 133–139 (1984)
18. Dabholkar, P.: Technology based service delivery. Adv. Serv. Mark. Manag. 3, 241–271 (1994)
19. Loudon, D., Della Bitta, A.: Consumer Behaviour: Concepts and Applications, 3rd edn. McGraw-Hill International, Singapore (1988)
20. Parasuraman, A., Zeithaml, V., Berry, L.: SERVQUAL: a multiple-item scale for measuring consumer perceptions of service quality. J. Retail. 64, 12–40 (1988)
21. Hemmasi, M., Strong, K., Taylor, S.: Measuring service quality for planning and analysis in service firms. J. Appl. Bus. Res. 10(4), 24–34 (1994)
22. Zhu, F.X., Wymer, W., Chen, I.: IT-based services and service quality in consumer banking. Int. J. Serv. Ind. Manag. 13(1), 69–90 (2002). https://doi.org/10.1108/09564230210421164
23. Ganguli, S., Roy, S.K.: Generic technology-based service quality dimensions in banking: impact on customer satisfaction and loyalty. Int. J. Bank Mark. 29(2), 168–189 (2011). https://doi.org/10.1108/02652321111107648
24. Kolodinsky, J.M., Hogarth, J.M., Hilgert, M.A.: The adoption of electronic banking technologies by US consumers. Int. J. Bank Mark. 22(4), 238–259 (2004). https://doi.org/10.1108/02652320410542536
25. Pamungkas, S., Kusuma, H.: Initial trust of customers and adoption of mobile banking: an empirical study from Indonesia. Ann. Univ. Petrosani Econ. 17(1), 223–234 (2017)
26. Juwaheer, T.D., Pudaruth, S., Ramdin, P.: Factors influencing the adoption of internet banking: a case study of commercial banks in Mauritius. World J. Sci. Technol. Sustain. Dev. 9(3), 204–234 (2012). https://doi.org/10.1108/20425941211250552
27. Shankar, A., Kumari, P.: Factors affecting mobile banking adoption behavior in India. J. Internet Bank. Commer. 21, 160 (2016)
28. Zhang, T., Lu, C., Kizildag, M.: Banking "on-the-go": examining consumers' adoption of mobile banking services. Int. J. Qual. Serv. Sci. 10(3), 279–295 (2018). https://doi.org/10.1108/IJQSS-07-2017-0067
29. Kesharwani, A., Tripathy, T.: Dimensionality of perceived risk and its impact on internet banking adoption: an empirical investigation. Serv. Mark. Q. 33(2), 177–193 (2012). https://doi.org/10.1080/15332969.2012.662461

30. Marakarkandy, B., Yajnik, N., Dasgupta, C.: Enabling internet banking adoption. J. Enterp. Inf. Manag. **30**(2), 263–294 (2017). https://doi.org/10.1108/JEIM-10-2015-0094
31. Hanafizadeh, P., Khedmatgozar, H.: The mediating role of the dimensions of the perceived risk in the effect of customers' awareness on the adoption of internet banking in Iran. Electron. Commer. Res. **12**(2), 151 (2012)
32. Ben Mansour, K.: An analysis of business' acceptance of internet banking: an integration of e-trust to the TAM. J. Bus. Ind. Mark. **31**(8), 982–994 (2016)
33. Dimitriadis, S., Kyrezis, N.: Does trust in the bank build trust in its technology-based channels? J. Financ. Serv. Mark. **13**(1), 28–38 (2008). https://doi.org/10.1057/fsm.2008.3
34. Patsiotis, A.G., Hughes, T., Webber, D.J.: An examination of consumers ' resistance to computer-based technologies. J. Serv. Mark. **27**(4), 294–311 (2013). https://doi.org/10.1108/08876041311330771
35. Fleiss, J.L., Chilton, N.W., Wallenstein, S.: Ridit analysis in dental clinical studies. J. Dent. Res. **58**(11), 2080–2084 (1979). https://doi.org/10.1177/00220345790580110701
36. Bross, I.: How to use ridit analysis. Biometrics **14**(1), 18–38 (1958)

Feasibility of Soft Real-Time Operations Over WLAN Infrastructure-Independent IoT Implementation by Enhancing Edge Computing

Sujanavan Tiruvayipati and Ramadevi Yellasiri

Abstract The subsequent generation of IoT devices must work on a multi-protocol architecture to facilitate M2M communication along with endpoint user interfacing to solve the network infrastructure dependencies accompanied by redundant data flow overhead. An ideological solution is proposed to facilitate a change while cutting down infrastructure cost and enhancing the current setups through proper implementation of edge computation. End devices cooperate with each other along with providing GUI and Internet to handsets; monitoring sensor information as well as issuing control signals.

Keywords IoT · Edge computing · Multi-protocol architecture · Infrastructure independent

1 Introduction

The existing IoT solutions need various investments which include procurement of sensors, actuators, edge computing devices, gateways, and network infrastructure setup. There could be some scenarios [1, 2] where the network infrastructure itself could accompany huge cost and maintenance. Such setups make sense when there is a high usability associated with network load, but there could be scenarios [3] where high usability is overseen by budget. Most of the soft real-time operations do not need high-performance network infrastructure, rather they need compact

S. Tiruvayipati (✉)
Department of Computer Science and Engineering, Maturi Venkata Subba Rao
Engineering College, Hyderabad, India
e-mail: sujanavan_cse@mvsrec.edu.in

R. Yellasiri
Department of Computer Science and Engineering, Chaitanya Bharathi
Institute of Technology, Hyderabad, India
e-mail: yramadevi_cse@cbit.ac.in

© Springer Nature Singapore Pte Ltd. 2020   223
K. S. Raju et al. (eds.), *Data Engineering and Communication Technology*,
Advances in Intelligent Systems and Computing 1079,
https://doi.org/10.1007/978-981-15-1097-7_19

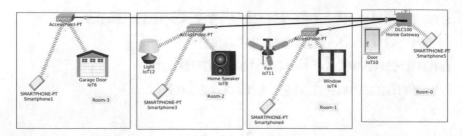

Fig. 1 Representation of the existing architecture of a simulated home IoT sample setup

solutions [4] using existing equipment. Home IoT setup is a forthcoming field of business where lower budget and maintenance are of a higher priority.

Most of the IoT end devices are becoming more pocket-friendly and easily programmable as the technology is advancing. Current IoT system structure (Fig. 1) is driven by only one among the major M2M protocols such as CoAP or MQTT [5] where the latter is more standardized and put to practice [6]. These M2M protocols require managing brokers as an intermediary computational device [7] along with a gateway. The gateway plays a vital role in expanding the structure by adding Internet and helps the user by providing a document interface that is accessible via a handset using any HTTP supported application, i.e., web browser.

The change proposed does have its own scope of operation and brings in major improvements by enhancing the functionality and features of edge computing. The ideology is to add broker features [8] to each edge device along with the ability of routing. The setup might not provide high performance, but could lead to one of a feasible solution [9–11] wherever required.

2 Proposed System Architecture

The architecture of the proposed solution does not have hardware changes, but some computational modules as a part of software need to be added as depicted in Figs. 2 and 3.

Existing systems at present contain sensors, actuators, and M2M broker–client. The proposed systems add on the following modules for enhancement.

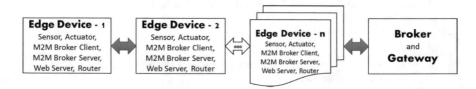

Fig. 2 Simplified architectural representation of the proposed system

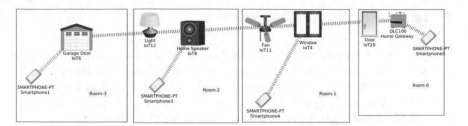

Fig. 3 Representation of the proposed architecture in a simulated home IoT sample setup

Edge M2M Broker–Server. It helps in the discovery of the neighboring devices instead of reaching the main broker.

Edge Web Server. Accessed through the built-in Wi-Fi module helps user interaction via a handheld device along with Internet connectivity fetched from the gateway.

Edge Router. This module makes communication more directional and effective.

The detailed architecture of the software part as seen in Fig. 4 displays parallel computation between local and global handles of the edge device. The local handle deals with maintaining neighbouring discovered nodes as well as processing requests over them. The global handle deals with handing over requests related to unknown edges to the next node as well as HTTP whose default handler is the gateway to provide Internet connectivity to the user's handheld device.

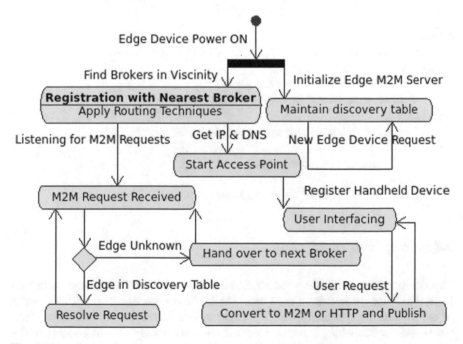

Fig. 4 State diagram for the overall flow of computational logic of the proposed solution

3 Implementation of Enhancement

The proposed ideological solution to be put to practice requires an ESP platform of the development boards in the production line 12F which work along with the Arduino IDE. The IDE provides several built-in libraries as well as support from GIT repositories. The major libraries to work with are HTTP web server and client [12], MQTT broker [13], painless mesh [14], and Wi-Fi repeater [15].

3.1 Pseudocode

To suit the needs of the computational architecture proposed as in Fig. 4, the following condensed pseudocode must be virtualized as per the requirements of the IDE and necessary conversions have to be made.

```
setup:                                 //Required Initializations
    start.WiFi_Station_Mode()
    send.broadcast(id,features)               //For Broker Discovery
    receive.address(ip,subnet,dns,gateway)
    start.WiFi_AP_Mode(ip,subnet,dns,gateway)
    start.WiFi_WebServer(80)



loop:
    listen.request(msg)                    // M2M or HTTP(UI)
    IF request is broadcast:
        receive.broadcast(id,features)
        broker.add_table_entry(id,features)
    ELSE
        broker.search_table_entries(id)
        IF id discovered:
            request.process(feature)
        ELSE
            send.request.next_hop(msg)//Next Broker
```

3.2 Setup and Integration

The devices must be arranged as seen in Fig. 3, where the four rooms numbered from 0 to 3 are isolated from each other which means the signals from the devices of the neighboring rooms can only be caught by one or two devices in another room successfully establishing a network backbone. User is only able to connect to the network backbone through any edge node in that particular room over a HTTP application in his or her handheld device. It is to be verified whether the edge M2M brokers do not miss any node undiscovered. Appropriate routing techniques employed should be able to provide Internet connectivity through the edge device by setting a path to the gateway.

4 Feasibility Study via Simulation

The test bed contained simulation of the projected method by developing appropriate scenarios in CISCO Packet Tracer version 7.1 and gathering the virtual serial outputs which print the various numbers of hops into a CSV file format for a number of continuous observations made over extensive intervals.

Scenario 1. Users issue control signal requests for devices in the same room.

Observation (Existing System). For any request issued from handheld device reaches the access point as a packet. Access point forwards the packet to the broker. Broker transfers the packet to the next access point (could be the same access point from where the packet arrived from) which is connected to the destined device. Access point forwards the packet to the device. A total of 4 hops are observed per request every time for any request.

Observation (Proposed System). Request issued from handheld device reaches the access point as a packet. Access point here is an edge computing device. If the request belongs to itself the action is taken, then and there else, it searches and transfers it to the destined device in the same room to take action. A minimum of 1 hop and a maximum of 2 hops are observed, averaging of 1.5 hops per request.

Scenario 2. Users issue control signal requests for devices in neighboring rooms.

Observation (Existing System). No change. Resultant as discussed in Scenario 1.

Observation (Proposed System). Request issued from handheld device reaches edge device which does not find the device in the room and forwards the request to the adjacent rooms where the request would be processed. A minimum of 2 hops and a maximum of 3 hops are observed, averaging of 2.5 hops per request.

Scenario 3. Users issue control signal requests for devices in faraway rooms. This scenario is also similar to users controlling the devices at home from another location (e.g., workplace) through the Internet.

Observation (Existing System). No change. Resultant as discussed in Scenario 1.

Observation (Proposed System). The resultant is based on the architectural design of the house. Relating how rooms are constructed to a tree structure solves the problem. A tree with depth one indicates all rooms can be reached by any edge device. A tree with depth two indicates two hops are required for a request to reach its destination room, and an extra hop might be required if the destination node is some other device in a room.

As the number of rooms increases and so does the number of users, the following formula helps to calculate the hops required.

$$H = nU \times nR \times (DT + 0.5)$$

where

H = Number of hops required by a request to reach its destination edge
nU = Number of users accessing the network
nR = Number of rooms equipped with the edge devices
DT = Depth of the architectural tree derived from the layout of rooms

Hence, for a soft real-time scenario, the total number of hops undergone by the network per second for all possible combinations of requests sent to various devices in the simulated home scenario can be calculated by the following formula.

$$H = \sum_{i=1}^{nU} \sum_{j=1}^{nR} i \times j \times (DT + 0.5)$$

Applying the above formula over the observations, an overall analysis can be made as represented as in Fig. 5. As the number of users and depth of the

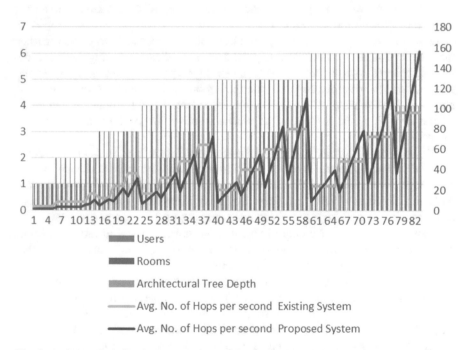

Fig. 5 A clustered combo chart prepared out of the observations. Primary y-axis maps variation in the number of users, rooms, and architectural tree depth; physical layout of rooms in a home. Secondary y-axis maps the average number of hops per second for the existing and proposed systems. The x-axis represents the average number of observations reflected per each instance of user request made over the network

architectural rooms' layout tree increases, the existing system takes lesser hops in comparison with the proposed system which requires hops.

5 Conclusion

The suggested solution outperforms the current implementations when there are a smaller amount of requests issued by the user along with architectural constraints over the physical layout of the rooms of a lower depth levels of the embraced tree structure. The existing architectures do have performance gain over the proposed enhancement, but when the cost in addition to maintenance is constraints, then the latter proves to be more feasible and effective. In order to facilitate transformation, the existing system needs to undergo computational enhancement which could be provided as system updates.

The feasibility resolution to be brought to soft real-time operations does require proper standardizations and would lead to the betterment of enriching M2M communication techniques as the years pass. These new features if introduced down the line will open up a new world of possibilities by further pushing the research to a greater extent.

References

1. Sen, S., Balasubramanian, A.: A highly resilient and scalable broker architecture for IoT applications. In: 2018 10th International Conference on Communication Systems & Networks (COMSNETS), Bengaluru, pp. 336–341 (2018)
2. Sun, X., Ansari, N.: Traffic load balancing among brokers at the IoT application layer. IEEE Trans. Netw. Serv. Manage. **15**(1), 489–502 (2018)
3. Guner, A., Kurtel, K., Celikkan, U.: A message broker based architecture for context aware IoT application development. In: 2017 International Conference on Computer Science and Engineering (UBMK), Antalya, pp. 233–238 (2017)
4. Pipatsakulroj, W., Visoottiviseth, V., Takano, R.: muMQ: a lightweight and scalable MQTT broker. In: 2017 IEEE International Symposium on Local and Metropolitan Area Networks (LANMAN), Osaka, pp. 1–6 (2017)
5. Veeramanikandan, M., Sankaranarayanan, S.: Publish/subscribe broker based architecture for fog computing. In: 2017 International Conference on Energy, Communication, Data Analytics and Soft Computing (ICECDS), Chennai, pp. 1024–1026 (2017)
6. Banno, R., Sun, J., Fujita, M., Takeuchi, S., Shudo, K.: Dissemination of edge-heavy data on heterogeneous MQTT brokers. In: 2017 IEEE 6th International Conference on Cloud Networking (CloudNet), Prague, pp. 1–7 (2017)
7. Tabatabai, S., Mohammed, I., Al-Fuqaha, A., Salahuddin, M.A.: Managing a cluster of IoT brokers in support of smart city applications. In: 2017 IEEE 28th Annual International Symposium on Personal, Indoor, and Mobile Radio Communications (PIMRC), Montreal, QC, pp. 1–6 (2017)
8. Kim, J., Hong, S., Yang, S., Kim, J.: Design of universal broker architecture for edge networking. In: 2017 International Conference on Networking, Architecture, and Storage (NAS), Shenzhen, pp. 1–2 (2017)

9. Jutadhamakorn, P., Pillavas, T., Visoottiviseth, V., Takano, R., Haga, J., Kobayashi, D.: A scalable and low-cost MQTT broker clustering system. In: 2017 2nd International Conference on Information Technology (INCIT), Nakhonpathom, pp. 1–5 (2017)
10. Espinosa-Aranda, J.L., Vallez, N., Sanchez-Bueno, C., Aguado-Araujo, D., Bueno, G., Deniz, O.: Pulga, a tiny open-source MQTT broker for flexible and secure IoT deployments. In: 2015 IEEE Conference on Communications and Network Security (CNS), Florence, pp. 690–694 (2015)
11. Li, W., Huang, H., Zhang, L.: A role-based distributed publish/subscribe system in IoT. In: 2015 4th International Conference on Computer Science and Network Technology (ICCSNT), Harbin, pp. 128–133 (2015)
12. ESP8266 Community Forum Projects and Libraries. https://github.com/esp8266
13. MQTT Broker library for ESP8266 Arduino. https://github.com/martin-ger/uMQTTBroker
14. Projects related to painlessMesh: a painless way to setup a mesh with ESP8266 and ESP32 devices. https://gitlab.com/painlessMesh (painlessMesh itself was based on easyMesh https://github.com/Coopdis/easyMesh)
15. ESP8266 WiFi repeater with ESP-12F (NodeMCU V3) and Arduino IDE. https://gitlab.com/forpdfsending/ESP8266-Wifi-Repeater

LANMAR Routing Protocol to Support Real-Time Communications in MANETs Using Soft Computing Technique

Anveshini Dumala and S. Pallam Setty

Abstract The collection of mobile wireless nodes that can self-organize in the absence of pre-installed infrastructure is claimed as mobile ad hoc network. Furthermore, nodes are open to enter or exit the network subsequently leading to dynamic topology. Nodes are equipped with limited energy resource. So carefulness in consuming the energy is required in MANET that affects the operational lifetime of the entire network. IETF draft of MANET suggests that the tuning of parameters with respect to dynamic network behaviour affects the overall protocol performance to a positive extent. This is called time series problem of LANMAR (LANdMARk) routing protocol and it is solved by dynamically calculating the configurable parameters (fisheye update interval) of the existing LANMAR using soft computing technique. However, few efforts are made to optimize the existing LANMAR routing protocol towards green-technology. In this paper, we created various network scenarios of LANMAR routing protocol using EXata simulator/emulator and MATLAB. Simulation results reveal that the proposed fuzzy-based LANMAR brought out better performance than the existing LANMAR.

Keywords Ad hoc network · LANMAR · Fisheye · FUI · Soft computing · Fuzzy logic · EXata

1 Introduction

Nodes in the MANET have the properties of moving, forming dynamic topology and are equipped with limited energy resource. The aim of this paper is to minimize the energy consumption. According to the IETF draft, the adaptation of protocol

A. Dumala (✉) · S. Pallam Setty
Department of Computer Science and Systems Engineering, AUCE(A), Andhra University, Visakhapatnam 530003, India
e-mail: danveshini@gmail.com

S. Pallam Setty
e-mail: drspsetty@gmail.com

© Springer Nature Singapore Pte Ltd. 2020
K. S. Raju et al. (eds.), *Data Engineering and Communication Technology*,
Advances in Intelligent Systems and Computing 1079,
https://doi.org/10.1007/978-981-15-1097-7_20

231

parameters according to the dynamic behaviour of the network is essential to attain the best performance of protocol. The notion of applying fuzzy logic is to obtain the optimal fisheye update interval (FUI) value in LANMAR. Based on the network structure, the routing protocols [1, 2] in MANETs are categorized into flat, hierarchical, and geographical routing protocols. In flat routing, all the nodes in the network work at the same level with same routing functionality. Flat routing is effective and simple for small networks but when the network size increases; it takes a pretty long time for routing data to reach nodes. This makes flat routing not suitable for scalable routing. Hierarchical routing protocols (e.g., LANMAR) are best suited for effective scalable routing in high mobile ad hoc networks.

In this paper, Sect. 2 presents literature regarding LANMAR routing protocol, Sect. 3 describes methodology used, and Sect. 4 demonstrates the simulation environment. Section 5 presents the experimental results on the application of a soft computing technique fuzzy logic, and Sect. 6 highlights the conclusions.

2 Literature Survey

LANMAR is one of the routing protocols that follow grouping of nodes method [3]. Here, in the below section, various literatures are presented how an energy-efficient LANMAR routing protocol can be implemented by using various ideas. Pei et al. [4, 5] presented a unique routing protocol called landmark ad hoc routing (LANMAR) that combines landmark routing and fisheye state routing (FSR) to reduce routing update overhead. Koushik [6] suggested a LANDMARK choosing process to save the energy with group mobility and concludes that the performance of MANET routing protocols is sensitive to scalability and mobility of network. Kumar [7] calculated the effect of group mobility model and random waypoint mobility model on LANMAR and OLSR. They carried out the simulation grounded analysis with constant traffic load for different network sizes. Venkatachalapathy and Ayyasamy [8] focused on QoS aware routing in MANETS using fuzzy. El-Hajj et al. [9] developed a fuzzy logic controller that takes residual energy, traffic, and mobility to elect cluster heads so that the network lifetime gets extended. Venkataramana and Setty [10] applied the fuzzy in ad hoc mobile network to find out the node traversal time with respect to the dynamic behaviour of the network.

3 Landmark Routing Protocol (LANMAR)

LANMAR adopts the idea of logical groups: moving as a crowd in a coordinated fashion. Every logical subnet of nodes (Fig. 1) has a leader (landmark) that gains the duty of dispatching the packets to a node in its subnet. Such LANDMARK header maintains subnet data [11, 12]. LANMAR utilizes FSR [13] as the local

Fig. 1 Logical subnet of nodes

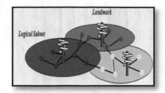

Fig. 2 Scope

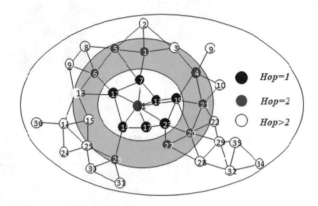

scope routing protocol where scope is measured in hop distance (Fig. 2). The landmarks locations are distributed by a distance-vector mechanism. Fisheye state routing protocol is used to forward the packets to the nodes within the range of a landmark by the landmark header. The distance-vector table maintains all the logical subnets addresses present in the network. The nodes which are far away from the scope of landmark are called as drifters. Always there is an exchange of the distance vectors of landmark nodes and the drifters by continuous periodical updates. LANMAR is a pro-active routing protocol which has the required data about nodes within range. Periodic monitoring of the network status helps in being updated and routing within scope. If a packet generator wants to direct a packet to the destination which is present in its scope then packet generator refers the local routing table and directs packet using FSR. A landmark is nominated dynamically. The fisheye (Fig. 3) captures the pixels with high detail which are nearer a focal point. The detail of the object shrinkages as the space between the object and focal point rises. Every node gradually reduces the updating speed as the destination moves away from the source. Therefore, node entries a scope is disseminated very often. Eventually, a large portion of entries in topology table are repressed, hence reducing network overhead. The timing parameters of LANMAR are presented in Table 1. LANMAR timing parameters values have worked well for high mobile large networks. The timing parameters should be administratively configurable for different network sizes and mobility speeds dynamically either experimentally determined values or dynamic adaptation.

Fig. 3 Fisheye

Table 1 LANMAR routing protocol timing parameters

Timing parameters	Default value
MINIMUM_MEMBER_THRESHOLD	8
APHA	1.3
LANDMARK_UPDATE_INTERVAL	4 s
NEIGHBOR_TIMEOUT_INTERVAL	6 s
MAXIMUM_LANDMARK_ENTRY_AGE	12 s
MAXIMUM_DRIFTER_ENTRY_AGE	12 s
FISHEYE_SCOPE (HOPS)	2
FISHEYE_UPDATE_INTERVAL	2 s
MAXIMUM_FISHEYE_ENTRY_AGE	6 s

4 Methodology

Simulation is used in our research. EXATA [14] is a virtual platform that permits to imitate communication networks. The parameters of simulation are listed below.

In this paper, the nodes are band together into four groups. Each group has equal no. of nodes as shown in Table 2. All the nodes are set to move in 'reference point group mobility' (RPGM) fashion. Figure 4 is a snapshot taken during simulation.

4.1 Computation of FB-FUI-LANMAR Routing Protocol

Fuzzy Logic: Multi-valued logic extension led to fuzzy logic. It follows approximate reasoning rather than exact solution. Instead of defining 0 or 1, the membership value in fuzzy logic describes the degree of belongingness to the fuzzy set.

Fuzzification is the conversion of crisp data into membership functions (MFs). MF is defined as a characteristic of fuzzy set which assigns every element to the membership value or the degree of membership, i.e. it maps each element in the

Table 2 Simulation parameters

Simulation parameters	Values
Number of nodes	20, 40, 60, 80, 100
Simulation area	1000×1000 m^2
Link	Wireless
Radio range	150 m
Item to send	512 bytes
MAC layer	IEEE 802.11
Antenna model	Omnidirectional
Data rate	2 mbps
Energy model	Generic
Pause time	0 s
Speed	10 m/s
Transport layer protocol	UDP
Routing protocol	LANMAR
Simulation time	900 s
Mobility	RPGM model

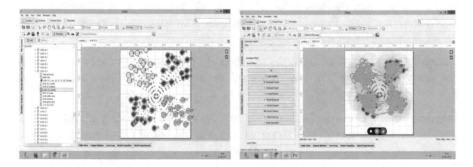

Fig. 4 Simulation environment of LANMAR protocol during simulation

fuzzy set to a membership value in [0, 1]. If X is a universe of discourse and $x \in X$ tn the fuzzy set A in X is defined as a set of ordered pairs, i.e. $A = \{(x, \mu_A(x))/ x \in X\}$ where $\mu_A(x)$ is membership function for fuzzy set A. We used Sugeno FIS that consists of two inputs: network size and mobility speed. The range of each parameter is as shown in Table 3. We consider two input parameters; one is network size and the mobility speed of a node. We considered the range of values of network size (I_1.) as [0, 20, 30, 40,…, 120] in number and mobility speed (I_2) as [0, 5, 10, …, 30] in metres per second. The triangular curve is a function of a vector, x, and depends on three scalar parameters a, b, and c, as given by

Table 3 Fuzzy inference system

Types of variables	FIS variables	Range					Types of membership functions
		Very low	Low	Medium	High	Very high	
Input 1	Network size (I_1)	0–40	20–60	40–80	60–100	80–120	TRIMF
Input 2	Mobility (I_2)	0–10	5–15	10–20	15–25	20–30	TRIMF
Output	Fisheye update interval (s)	2	3	3.5	4	5	Constant

$$f(x,a,b,c) = \begin{Bmatrix} 0, & x \leq a \\ \frac{x-a}{b-a}, & a \leq x \leq b \\ \frac{c-x}{c-b}, & b \leq x \leq c \\ 0, & c \leq x \end{Bmatrix}$$

The a and c parameters are at bottom and b parameter at the top of a triangle (Fig. 5). Where a and b and c represent the x_coordinate for capital triangle, x represents the real value (crisp value) from the private variable fuzzy universe of discourse. Function outputs ranging between 0 and 1 and represents the value of the degree of membership of x.

Fuzzy Inference System links the membership functions with the if-then rules to get fuzzy output (Fig. 6). The proposed system is decided by a maximum of 5 * 5 = 25 rules (Table 4).

De-fuzzification converts the fuzzy output into crisp output. Any ith rule, in this approach, is represented by 'If $(x_1$ is $A_1^i)$ and $(x_2$ is $A_2^i)$ and $(x_n$ is $A_n^i)$ then $y_i = a_0^i + a_1^i x_1 + a_2^i x_2 + \cdots + a_n^i x_n$'. Where $a_0, a_1, a_2, \ldots, a_n$ are the coefficients. The weights of the ith rule can be determined for a set of inputs $x_1, x_2, x_3, \ldots, x_n$ as

$$w^i = \mu_{A_1}^i(x_1) * \mu_{A_2}^i(x_2) * \cdots \mu_{A_n}^i(x_n) \tag{1}$$

Fig. 5 Triangular membership function

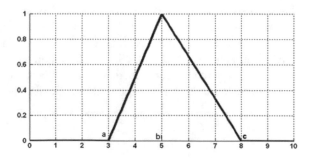

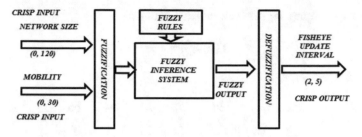

Fig. 6 Fuzzy inference system

Table 4 Fuzzy if-then rules

S. No.	Network size	Mobility	Fisheye update interval
1.	VL	VL	VH
2.	VL	L	VH
3.	VL	M	H
4.	VL	H	H
5.	VL	VH	M
6.	L	VL	VH
7.	L	L	H
8.	L	M	H
9.	L	H	M
10.	L	VH	M
11.	M	VL	M
12.	M	L	M
13.	M	M	M
14.	M	H	L
15.	M	VH	L
16.	H	VL	M
17.	H	L	L
IS.	H	M	L
19.	H	H	L
20.	H	VH	VL
21.	VH	VL	L
22.	VH	L	L
23.	VH	M	L
24.	VH	H	L
25.	VH	VH	VL

where $A_1, A_2, \ldots, A_n$ indicates membership function distributions of the linguistic hedges used to indicate the input variables. μ represents membership function value (Table 4).

The output of any ith rule can be expressed as:

$$y^i = f(I_1, I_2) = a_{ji}I_1 + b_{ki}I_2 \tag{2}$$

where j, k = 1, 2, 3, 4, 5

		b1	b2	b3	b4	b5
		VL	L	M	H	VH
a1	VL	y^1	y^2	y^3	y^4	y^5
a2	L	y^6	y^7	y^8	y^9	y^{10}
a3	M	y^{11}	y^{12}	y^{13}	y^{14}	y^{15}
a4	H	y^{16}	y^{17}	y^{18}	y^{19}	y^{20}
a5	VH	y^{21}	y^{22}	y^{23}	y^{24}	y^{25}

To calculate the output of FLC for I_1 = 70 and I_2 = 14 (Fig. 7).

(A) The input I_1 = 70 can be called either medium or high. Similarly, the input I_2 = 14 can be either low or medium.

(B) The triangular curve is a function of a vector 'x' that depends on three scalar parameters a, b, and c. Using the principle of similarity of a triangle, we have $\mu_M(I_1) = 0.5$, $\mu_H(I_1) = 0.5$, $\mu_L(I_2) = 0.2$, $\mu_M(I_2) = 0.8$. Then an aggregation of these fuzzy values into a single fuzzy output is given in a detailed rule set.

(C) For the input set, the following 4 rules can be fired out of 25 rules.

$$R_{12} : I_1 \text{ is M and } I_2 \text{ is L} \rightarrow y^{12}, \quad R_{13} : I_1 \text{ is M and } I_2 \text{ is M} \rightarrow y^{13}$$
$$R_{17} : I_1 \text{ is H and } I_2 \text{ is L} \rightarrow y^{17}, \quad R_{18} : I_1 \text{ is H and } I_2 \text{ is M} \rightarrow y^{18}$$

for each rule we get separate output y^i now we need to calculate the weight of each rule w^i.

(D) Now the weights for each of the above rules can be determined as follows

$$R_{12} : w^{12} = \mu_M * \mu_L = 0.5 * 0.2, \quad R_{13} : w^{13} = \mu_M * \mu_M = 0.5 * 0.8$$
$$R_{17} : w^{17} = \mu_H * \mu_L = 0.5 * 0.2, \quad R_{18} : w^{18} = \mu_H * \mu_M = 0.5 * 0.8$$

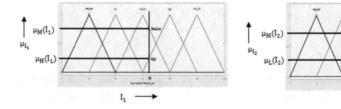

Fig. 7 Membership values of inputs

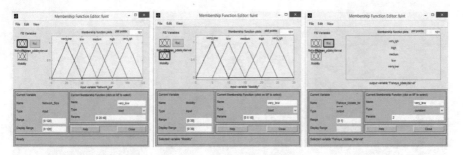

Fig. 8 Membership functions of the input and output parameters

(E) The fundamental sequential values for rule can be calculated as below

$$y^i = a_{ji}I_1 + b_{ki}I_2 \quad \text{where } j, k = 1, 2, 3, 4, 5$$

(F) Therefore, the output y of the controller can be determined as follows. The final output weighted average of all rules is calculated as

$$\frac{\sum_{i=1}^{N} W_i y_i}{\sum_{i=1}^{N} W_i} \quad \text{where } N = \text{No. of rules} \tag{3}$$

This FIS has 'network size' and 'mobility' as input parameters and 'fisheye update interval' as the output parameter (Fig. 8). The rule viewer gives picture of the designed FIS. Surface viewer is a 3D picture mapping inputs and output (Fig. 9).

5 Results and Analysis

Throughput (***bits/s***): number of bits sent in the network. Figure 10 shows 2271.93 bps for existing LANMAR (L) and 2493.99 bps for FB-LANMAR (FL).

Jitter (***sec***): It is the variation between maximum and minimum delay. Figure 11 presents 1.97425 ms for existing LANMAR (L) and 1.57525 ms for FB-LANMAR (FL).

End-to-End Delay (***sec***): The average time required for a data packet to reach the destination. Figure 12 presents 4.15133 ms for existing LANMAR (L) and 3.69061 ms for FB-LANMAR (FL). Thus, low delay in FB-LANMAR is witnessed.

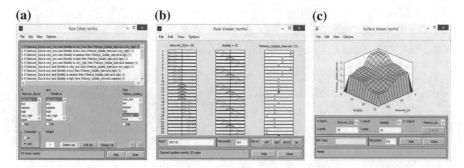

Fig. 9 **a** Fuzzy if-then rules, **b** rule viewer, **c** surface viewer

Fig. 10 Throughput

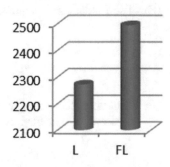

Fig. 11 Jitter

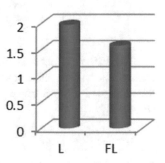

Control Overhead (***bytes***): Total number of bytes sent as control packets. Figure 13 shows 218764 bytes for existing LANMAR (L) and 2493.99 bytes for FB-LANMAR (FL). Control overhead is found to be less in FB-LANMAR.

Number of Control Packets (***CP***): Total number of control packets sent. Figure 14 presents 661.7 for existing LANMAR (L) and it is 475 for FB-LANMAR (FL). Thus, low number of control packets in FB-LANMAR is witnessed.

Energy Consumption (***EC***) ***in Transmit Mode*** (***TM***) (***MJ***): Energy consumed by a node to send data packet. Figure 15 shows 1.4205 MJ for existing LANMAR (L) and 0.651492 MJ for FB-LANMAR (FL). Energy consumed is less in FB-LANMAR.

Fig. 12 End-to-end delay

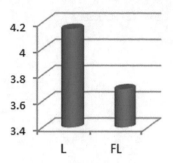

Fig. 13 Control overhead

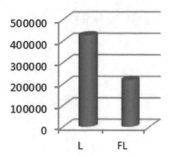

Fig. 14 No. of CP

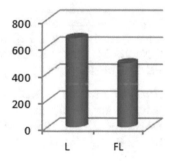

Fig. 15 EC in TM

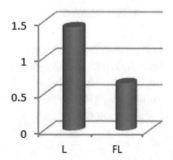

Fig. 16 EC in RM

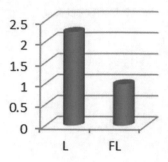

Fig. 17 EC in IM

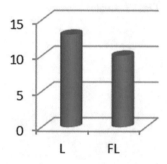

Energy Consumption (EC) in Receive Mode (RM) (MJ): Energy consumed by a node to receive packet. Fig. 16 shows 2.23182 MJ for existing LANMAR (L) and 0.986919 MJ for FB-LANMAR (FL). Energy consumed is minimized in FB-LANMAR.

Energy Consumption (EC) in Idle Mode (IM) (MJ): Node consumes power to listen to the traffic in the media. Figure 17 shows 12.9382 MJ for existing LANMAR (L) and 10.011 MJ for FB-LANMAR (FL). Energy consumed in IM is less in FB-LANMAR.

6   Conclusion

Dynamic values of fisheye update interval (FUI) are generated using soft computing technique taking network size and node mobility as inputs and FUI as output. Based on the proposed FB-LANMAR routing protocol, it is determined that the dynamic values of FUI showed substantial variations on the performance of LANMAR routing protocol.

References

1. Toh, C.K.: Ad Hoc Mobile Wireless Networks: Protocols and Systems. Prentice Hall Engle wood Cliff, NJ, pp. 297–333 (1987)
2. Ahmed, D.E.M., Khalifa, O.O.: A comprehensive classification of MANETs routing protocols. Int. J. Comput. Appl. Technol. Res. 6(3), 141–158 (2017)
3. Maan, F., Mazhar, N.: MANET routing protocols vs mobility models: a performance evaluation. In: Ubiquitous and Future Networks (ICUFN), pp. 179–184 (2011)
4. Pei, G., Gerla, M., Hong, X.: LANMAR: landmark routing for large scale wireless ad hoc networks with group mobility. In: Proceedings of IEEE/ACM MobiHOC 2000, Boston, MA (2000)
5. Pei, G., Gerla, M., Chen, T.-W.: Fisheye state routing: a routing scheme for ad hoc wireless networks. In: Proceedings of ICC 2000, New Orleans, LA (2000)
6. Koushik, C.P., Vetrivelan, P., Ratheesh, R.: Energy efficient landmark selection for group mobility model in MANET. Indian J. Sci. Technol. 8(26) (2015)
7. Kumar, S., Sharma, S.C., Suman, B.: Impact of mobility models with different scalability of networks on MANET routing protocols. Int. J. Sci. Eng. Res. 2(7) (2011)
8. Ayyasamy, A., Venkatachalapathy, K.: Context aware adaptive fuzzy based QoS routing scheme for streaming services over MANETs. Wireless Netw. 21, 421–430 (2014)
9. El-Hajj, W., Kountanis, D., Al-Fuqaha, A., Guizani, M.: A fuzzy-based hierarchical energy efficient routing protocol for large scale mobile ad hoc networks (FEER). In: IEEE ICC 2006 Proceedings, pp. 3585–3590 (2006)
10. Venkataramana, A., Setty, S.P.: Enhance the quality of service in mobile ad hoc networks by using fuzzy based NTT-DYMO. Wireless Pers. Commun. 95(3) (2017)
11. Gerla, M.: Landmark routing protocol (LANMAR) for large scale ad hoc networks. Internet Draft, draft-ietf-manet-lanmar-05.txt, work in progress (2003)
12. Tsuchiya, P.F.: The landmark hierarchy: a new hierarchy for routing in very large networks. Comput. Commun. Rev. 18, 35–42 (1988)
13. Gerla, M.: Fisheye state routing protocol (FSR) for ad hoc networks. Internet Draft, draft-ietf-manet-fsr-03.txt, work in progress (2002)
14. EXata Network Simulator/Emulator. http://www.scalable-networks.com

The SOP of the System with TAS Based Underlay CRN

Anshu Thakur and Ashok Kumar

Abstract The underlay cognitive radio is one of the authentic answers for spectrum utilization. The security of data in these networks is also correspondingly
important. In this paper, underlay cognitive radio network is considered with a
secondary network and eavesdroppers. The secondary users communicate with
secondary receiver in underlay cognitive radio network under the malicious attempt
of eavesdroppers. The multiple ratio combining (MRC) and secrecy combining
(SC) techniques are studied with transmit antenna selection (TAS) scheme to enrich
the secrecy performance of the system. The exact secrecy outage probability with
TAS is studied, and the results are validated by Monte Carlo simulation.

Keywords Secrecy outage probability · Cognitive radio network · Security ·
Spectrum scarcity

1 Introduction

With the advancement of wireless technology and increasing number of users, the
spectrum scarcity has become a bottleneck issue nowadays. To deal with this,
cognitive radio network (CRN) is one of the promising solutions [1]. CRN is an
intelligent network which programs itself dynamically and utilizes the spectrum
efficiently [2]. The sharing of spectrum is divided into underlay and overlay
approaches [3]. In the present work, we are considering the underlay spectrum
sharing; in this technique, the primary user (PU) and secondary user (SU) are
allowed to transmit over the same spectrum [4]. On the other hand, in overlay
approach, PU searches for the vacant spectrum for its data transmission, and
therefore, no interference is present to the primary users.

A. Thakur (✉) · A. Kumar
Department of Electronics and Communications Engineering,
NIT Hamirpur, Hamirpur, India
e-mail: anshuthakur@nith.ac.in

© Springer Nature Singapore Pte Ltd. 2020　　　　　　　　　　　　　　　　245
K. S. Raju et al. (eds.), *Data Engineering and Communication Technology*,
Advances in Intelligent Systems and Computing 1079,
https://doi.org/10.1007/978-981-15-1097-7_21

The spectrum sharing has leaded some serious security issues. Because of the security problem, a keen interest has been developed in providing data security without using the complex cryptographic techniques. The cryptographic techniques are complex and costly. Therefore, to deal with present security issue physical layer security is a promising solution. In this, the main channel is strengthened to provide the secrecy as compared to the legitimate receiver [5]. The security of the system with multiple antennas at transmitter, receiver and eavesdropper has been discussed in [6–9] papers. The multiple antennas have been considered in [6], and maximal ratio combining technique (MRC) was discussed to derive secrecy outage probability (SOP). The multiple eavesdroppers were considered in [7], and SOP was derived. To enhance the security of the system, transmit antenna selection (TAS) was discussed in [8] as well as [9]. The CRN with secure system is a new technology for the next-generation communication. The optimized secrecy with perfect and imperfect CSI and CRN was discussed in paper [10]. To maximize the secrecy rate, CRN with single eavesdropper is considered with relay selection technique in [11]. Physical layer security with underlay CRN with multiple antenna and passive eavesdroppers were considered in [12]. In passive eavesdropping, the secondary transmitter does not know anything about the channel of eavesdropper. This is a very complicated scenario, and in this, it is very difficult to achieve perfect secrecy. SOP is one the performance metrics to achieve the secrecy and is discussed in the paper.

In the present paper, the authors considered underlay CRN and necessary conditions are considered to secure the confidential information from eavesdroppers. The important parameter considered in this paper is TAS at secondary transmitter (BS). The TAS is an important aspect to reduce the complexity due to multiple antennas and improves the data rate. The strongest antenna is opted for data transmission at BS. This insures the better communication and secrecy between the secondary transmitter and receiver. In the present paper, multiple antennas at PU, eavesdropper and secondary receiver (IR) are considered. The maximal ratio combining (MRC) and selection combining (SC) techniques with TAS at BS are considered in our work.

2 System Model

In the present model, a wiretap cognitive channel is considered. The primary user (PU), base station (BS), information reciever(IR) and eaversdropper (eve) are having multiple antennas. The PU is having $1, \ldots, N_{pu}$ antenna, $1. \ldots, N_{bs}$ at BS, $1, \ldots, N_{eve}$ at eavesdropper and $1, \ldots, N_{IR}$ at IR. The system model is described in Fig. 1.

The primary as well as secondary channels are independent and identically distributed (iid) and having Rayleigh fading and $\{k_1\}_{i=1}^{n_b}$, $\{k_2\}_{j=1}^{n_e}$ and k_0 be the complex random variables having zero mean, and variance is Ω_1, Ω_2 and Ω_0. The average SNR of the channels is γ_0, γ_1 and γ_2 which be the average SNR for PU, IR and eavesdroppers, respectively. The perfect CSI is considered in the given paper,

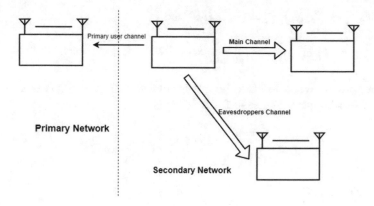

Fig. 1 System model

and strongest antenna is selected for transmission from BS to IR following TAS scheme. The selected antenna at BS maximizes the instantaneous SNR between BS and IR, and random SNR is opted out. The SNR value between the BS-IR and BS-eve is given by [12]:

$$\gamma_{\text{main}} = \max_{i=1,\ldots,n_{\text{IR}}} \left(\frac{P_{\text{bs}}}{N_0}\right) |k_1|^2 \quad \text{and} \quad \gamma_{\text{eve}} = \max_{i=1,\ldots,n_{\text{Ie}}} \left(\frac{P_{\text{bs}}}{N_0}\right) |k_2|^2 \tag{1}$$

where P_{bs} is the transmit power at BS, and N_0 is the variance of noise. For the reliable and secure communication, the transmitted power at BS is managed under an interference threshold power for reliable communication at PU. So the transmitted power at BS is given by:

$$P_{\text{bs}} = \min \left(\frac{I_{\text{P}}}{\max_{n=1,\ldots,n_{\text{p}}} \{k_1\}_{i=1}^{n_{\text{p}}}}, P_{\text{trans}}\right) \tag{2}$$

where P_{trans} represents the maximum transmitted power at BS. I_{p} is the interference power at PU, and k_1 is the coefficient of fading is the maximum SNR between the BS and eve. Parameters under consideration, the SNR can be further written as [13]:

$$\gamma_{\text{main}} = \min \left(\frac{\gamma_{\text{p}}}{x}, \gamma_0\right) y_{\text{main}} \tag{3}$$

$$\gamma_{\text{eve}} = \min \left(\frac{\gamma_{\text{p}}}{x}, \gamma_0\right) y_{\text{eve}} \tag{4}$$

In this $\gamma_p = (I_p/N_0)$, $\gamma_0 = (P_{trans}/N_0)$ and

$$X = |k_0|^2, y_{main} = \max_{i=1,\dots,n_{IR}} |k_1|^2, \quad \text{and} \quad y_{eve} = \max_{i=1,\dots,n_{eve}} |k_2|^2 \tag{5}$$

In our present work, MRC technique is studied at IR, and SC is used at PU as well as eavesdroppers. The PDF for SC [13] is:

$$f_x(x) = \sum_{l=0}^{n_p-1} \binom{n_p-1}{l} (-1)^l \frac{n_p}{\Omega_0} \exp^{-\frac{(l+1)}{\Omega_0}} \tag{6}$$

The CDF for SC is:

$$F_x(x) = \sum_{i=0}^{n_n} \binom{n_p}{i} (-1)^i \exp^{-\left(\frac{(n+1)x}{\Omega_0}\right)} \tag{7}$$

The PDF for the case of MRC is defined in [14]:

$$f_{\gamma_\varepsilon}(y) = \left(\frac{y^{n_{IR}} \exp^{-\left(\frac{y}{\gamma_1}\right)}}{(n_{IR}-1)!(\gamma_1)^{n_{IR}}} \right) \tag{8}$$

The CDF in the given case is as follows:

$$F_{\gamma_\varepsilon}(y) = 1 - \exp^{-\left(\frac{y}{\gamma_1}\right)} \sum_{l=0}^{n_{IR}} \frac{1}{l!} \left(\frac{y}{\gamma_1}\right)^l \tag{9}$$

3 Secrecy Outage Probability

The SOP for MRC combining scheme is derived. The passive eavesdropping is considered in present work. While considering passive eavesdropping, BS encodes the message at a constant rate R_s, without the CSI knowledge of Eve's channel. The secrecy capacity of the system is given in [14].

$$C_s = \begin{cases} C_{main} - C_{eve} = \log_2\left(\frac{1+\gamma_{main}}{1+\gamma_{eve}}\right) & \text{if} \quad \gamma_{main} > \gamma_{eve} \\ 0 & \text{if} \quad \gamma_{main} \leq \gamma_{eve} \end{cases} \tag{10}$$

The perfect secrecy is guaranteed when $R_s \leq C_s$. Contrarily, if $R_s > C_s$ the security of the system is compromised. The C_{main} is the channel capacity of main channel and C_{eve} is the channel capacity of eavesdropper's channel. $\varepsilon\gamma_\varepsilon = \gamma_{main} < 2^{R_s}(1 + \gamma_{eve}) - 1$ defines the capacity of main channel. The SOP is defined in [14]:

$$
\begin{aligned}
P_{out} &= P_r(C_s < R_s) \\
&= P_r(\gamma_{main} \leq \gamma_{eve}) + P_r(\gamma_{main} > \gamma_{eve}) \times P_r(C_s < R_s/\gamma_{main} > \gamma_{eve})
\end{aligned}
\tag{11}
$$

This equation can be further simplified as:

$$
P_{out} = \underbrace{\int_0^\psi \int_0^\infty F_{\gamma_{main}|X=x}(\varepsilon\gamma_\varepsilon) f_{\gamma_\varepsilon|X=x}(\gamma_\varepsilon) f_x(x) d\gamma_\varepsilon dx}_{P_{out_1}}
$$
$$
+ \underbrace{\int_\psi^\infty \int_0^\infty F_{\gamma_{main}|X=x}(\varepsilon\gamma_\varepsilon) f_{\gamma_\varepsilon|X=x}(\gamma_\varepsilon) f_x(x) d\gamma_\varepsilon dx}_{P_{out_2}}
\tag{12}
$$

Here, $f_{\gamma_\varepsilon|X=x}(\gamma_\varepsilon) f_x(\cdot)$ is PDF of γ_ε and conditioned on X. The CDF is $F_{\gamma_{main}|X=x}(\cdot)$, of γ_{main} which is conditioned on X. Here, $\gamma_1 = \Omega_1\gamma_0$ which represents maximum SNR between the BS and IR and $\gamma_2 = \Omega_2\gamma_0$ is the maximum SNR between the BS and eve.

After solving the equation, the P_{out_1} is

$$
P_{out_1} = \left(
\begin{aligned}
&1 - \sum_{l=0}^{n_{IR}} \sum_{r=0}^{l} \sum_{j=0}^{N_e} \sum_{i=0}^{N_{pu}} \binom{N_e - 1}{j} \binom{N_{PU}}{i} \binom{l}{r} \left(\frac{(-1)^{i+j}(2^{R_s})^r}{\gamma_2(\gamma_1)^l(l!)} \right) \exp\left(-\frac{(2^{R_s}-1)}{\gamma_1} \right) \\
&\times \left(\frac{(l-r)!(2^{R_s}-1)^{l-r}}{((J+1)/\gamma_2) + (2^{R_s}/\gamma_1))} \right) \times \left(\frac{\left(1 - \exp\left(\frac{-(i+1)}{\Omega_0} \right)\right)^\sigma}{(i+1)} \right)
\end{aligned}
\right)
\tag{13a}
$$

$$
P_{out_2} = \exp\left(\frac{-\sigma}{\Omega_0} \right) - \left(
\begin{aligned}
&\sum_{l=0}^{n_{IR}} \sum_{r=0}^{l} \sum_{j=0}^{N_e} \sum_{i=0}^{N_{pu}} \binom{N_e - 1}{j} \binom{N_{PU}}{i} \binom{l}{r} \frac{(-1)^{i+j}(2^{R_s})^r}{\gamma_2(\gamma_1)^l(l!)} \\
&\left(\frac{(l-k)!(2^{R_s}-)(\sigma)^{i+j-r}}{((J+1)/\gamma_2) + (2^{R_s}/\gamma_1))} \right) \left(\frac{1}{\frac{(2^{R_s}-1)}{\sigma\gamma_1} + \frac{1}{\Omega_0}} \right) \exp^{-\left(\frac{(2^{R_s}-1)}{\sigma\gamma_1} + \frac{1}{\Omega_0} \right)}
\end{aligned}
\right)
\tag{13b}
$$

3.1 Transmit Antenna Selection (TAS)

In case of TAS, only the best antenna is selected for data transmission instead of all the present antennas. The secrecy capacity in the present case is given as:

$$C_{\max}^{\text{TAS}} = \max_{\max \subset N_{\text{IR}}} (c_{\max}) \tag{14}$$

$$P_{\text{out}}^{\text{TAS}} = \Pr\left(\max_{\max \subset N_{\text{IR}}} (c_{\max}) < R_{\text{s}} \right) \tag{15}$$

4 Numerical Results

In the present work, the exact SOP is derived and the analytical results are verified with Monte Carlo simulation. The MRC scheme is adopted at IR, and SC is adopted at PU as well as eve. In the present section, the variance, i.e., $\Omega_0 = 1$ and the secrecy rate ($R_s = 1$ bit/s/Hz).

In Fig. 2, the plot is drawn between SOP and γ_1 (dB) for the various values of N_{IR} and σ. With the increasing number of antennas at IR, the secrecy diversity order also keeps on increasing which in return decreases the SOP. We also concluded that with the increase of σ, the SOP keeps on decreasing. This is due to relaxing the peak interference power.

In Fig. 3, the plot is drawn between SOP for TAS scheme and γ_1 (dB). The different numbers of antennas are selected at BS. From this, we have concluded that

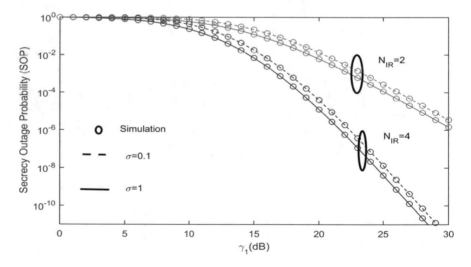

Fig. 2 Secrecy outage probability (SOP) versus γ_1 (dB) for $\gamma_2 = 10$ dB

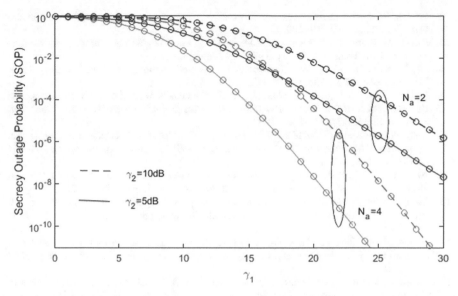

Fig. 3 Secrecy outage probability (SOP) versus $\gamma 1$ (dB) for $N_{\text{IR}} = N_{\text{PU}} = N_{\text{eve}} = 2$

as the number of antenna keeps on increasing, the SOP keeps on decreasing. The more the number of antenna at BS, the better is the secrecy of the system.

5 Conclusion

The underlay cognitive radio network with transmit antenna selection scheme is considered. The MRC and SC diversity combining schemes are also investigated with TAS scheme to increase the security of the system. The multiple antennas are considered at each block of the system. The more the number of antennas at secondary transmitter, the better is the secrecy of the system.

References

1. Goldsmith, A., Jafar, S., Maric, I., Srinivasa, S.: Breaking spectrum gridlock with cognitive radios: an information theoretic perspective. Proc. IEEE **97**(5), 894–914 (2009)
2. Shiang, H.P., Van der Schaar, M.: Distributed resource management in multihop cognitive radio networks for delay sensitive transmission. IEEE Trans. Veh. Technol. **58**(2), 941–953 (2009)
3. Pei, Y., Liang, Y.-C., Zhang, L., et al.: Secure communication over MISO cognitive radio channels. IEEE Trans. Wireless Commun. **9**(4), 1494–1502 (2010)

4. Haykin, S.: Cognitive radio: brain-empowered wireless communications. IEEE J. Sel. Areas Commun. **23**(2), 201–220 (2005)
5. Bloch, M., Barros, J., Rodrigues, M.R.D., McLaughlin, S.W.: Wireless information-theoretic security. IEEE Trans. Inf. Theory **54**(6), 2515–2534 (2008)
6. He, F., Man, H., Wang, W.: Maximal ratio diversity combining enhanced security. IEEE Commun. Lett. **15**(5), 509–511 (2011)
7. Prabhu, V.U., Rodrigues, M.R.D.: On wireless channels with m-antenna eavesdroppers: characterization of the outage probability and ε-outage secrecy capacity. IEEE Trans. Inf. Forensics Secur. **6**(3), 853–860 (2011)
8. Yang, N., Yeoh, P.L., Elkashlan, M., Schober, R., Collings, I.B.: Transmit antenna selection for security enhancement in MIMO wiretap channels. IEEE Trans. Commun. **61**(1), 144–154 (2013)
9. Alves, H., Souza, R.D., Debbah, M., Bennis, M.: Performance of transmit antenna selection physical layer security schemes. IEEE Signal Process. Lett. **19**(6), 372–375 (2012)
10. Zhang, J., Gursoy, M.C.: Secure relay beam forming over cognitive radio channels. In: 45th Annual Conference on Information Sciences and Systems, pp. 1–5. IEEE, Baltimore, MD, USA (2011)
11. Sakran, H., Shokair, M., Nasr, O., El-Rabaie, S., El-Azm, A.A.: Proposed relay selection scheme for physical layer security in cognitive radio networks. IET Commun. **6**(16), 2676–2687 (2012)
12. Elkashlan, M., Wang, L., Duong, T.Q., Karagiannidis, G.K., Nallanathan, A.: On the security of cognitive radio networks. IEEE Trans. Veh. Technol. **64**(8), 3790–3795 (2015)
13. Yang, N., Yeoh, P.L., Elkashlan, M., Schober, R., Collings, I.B.: Transmit antenna selection for security enhancement in MIMO wiretap channels. IEEE Trans. Commun. **61**(1) (2013)
14. Singh, A., Bhatnagar, M., Mallik, R.: Physical layer security of a multi-antenna based CR network with single and multiple primary users. IEEE Trans. Veh. Technol. **66**(12), 11011–11022 (2017)

Key Exchange and E-Mail Authentication Using Lagrange Interpolation

P. Lalitha Surya Kumari and G. Soma Sekhar

Abstract All corporate and government sectors like the aviation, banking, military, etc., need to determine genuineness of an authenticated user. This paper proposes an identity-based secure authentication system for e-mail security using Lagrange interpolation based on the concept of Zero-Knowledge Protocol. This protocol makes the user to prove to the server that he/she has password without sending any information, either encrypted password or clear text to the server and also without intervention of the third party (authentication server). In this protocol, one party allows the other party to confirm whether the statement is true without revealing any other information. Mutual identification of two users is done using this protocol, and it involves exchange of pseudorandom numbers to generate secret key or session key which in turn is used for encryption and decryption. Polynomial-based session key is generated using Lagrange interpolation by sender and receiver.

Keywords Authentication · Confidentiality · Key exchange · E-mail security · Zero-knowledge protocol · Session key · Lagrange interpolation

1 Introduction

Electronic mail (e-mail) is the commonly used system to transfer data between the users through the Internet. The basic e-mail components are mail server and clients. Mail servers are the applications that store, forward and deliver mail, whereas clients are the interfaces allow users to store, send, compose and read e-mail messages. Mail clients, mail servers and their supporting network infrastructure need security. Main security threats of e-mail security are denial-of-service (DoS) attacks and distribu-

P. Lalitha Surya Kumari (✉)
Koneru Lakshmaiah University, Hyderabad, Telangana, India
e-mail: vlalithanagesh@gmail.com

G. Soma Sekhar
Geethanjali College of Engineering and Technology, Hyderabad, Telangana, India
e-mail: somasekharonline@yahoo.co.in

© Springer Nature Singapore Pte Ltd. 2020
K. S. Raju et al. (eds.), *Data Engineering and Communication Technology*,
Advances in Intelligent Systems and Computing 1079,
https://doi.org/10.1007/978-981-15-1097-7_22

tion of secret information among unauthorized individuals. The organizations concerned in improving e-mail security can use this document. Mail transport standards ensure interoperability and reliability of all e-mail applications. The most widespread MTA transfer protocol is Simple Mail Transfer Protocol (SMTP) [1]. The most widely accepted protocol for e-mail delivery was Simple Mail Transfer Protocol. Main drawback of this protocol is lack of security features like authentication and privacy of sending party. SMTP servers were strategically and technologically changed to provide e-mail systems more secure which includes refusal of unauthorized servers, refusal of e-mail relaying, authentication of headers and protocols used e-mail security by limit the e-mail message size and filtering. Add-on security protocols are used to make email systems more secure. These protocols use validation or encryption standards or cryptographic techniques.

E-mail security provides consistency, confidentiality, authentication, non-repudiation and message integrity. The most common encryption techniques for secured e-mail include Pretty Good Privacy (PGP), Privacy-Enhanced Mail (PEM), Secure Multi-purpose Internet Mail Extensions (S/MIME) and GNU Privacy Guard (GPG). PEM requires trusting third party like Certificate Authority (CA), and hence, its acceptance is insignificant. GPG and PGP are using the concept of PKI system and limited to a small user society. The cryptographic services like message integrity, non-repudiation of sender, sender authentication, digital signatures and message security using encryption are provided by S/MIME. The main disadvantage of S/MIME is that at the time of usage of digital signature, both source and destination need to procure digital signature from certified CA. The strength of S/MIME is mainly depending on two things. They are public key infrastructure (PKI) and effectiveness of cryptographic algorithm. The other concerns related to encryption protocols like key distribution, key renewal and key management require suitable mail systems and skilled users. Zero-Knowledge Protocol can be used to overcome all the problems. Zero-Knowledge Protocol [2] provides services like confidentiality, message integrity, non-repudiation and authentication. The primary concern of any system is the authenticity of an individual. Hence, there is a need for the strong authentication between two parties. The following sessions discuss the proposed authentication protocol.

2 Related Work

Guo et al. [3] proposed an combined privacy-preserving identity verification scheme for complex Web services. It combines multiple interactions of component providers which are used for identity verification to form a single one concerning the user. They adopted zero-knowledge protocol so that it protects users from disclosure of confidential information. This approach can also decrease the computation time, without considering the number of component providers and identity attributes. Allam et al. [4] presented a Fiat–Shamir-like Zero-Knowledge identification scheme based on the elliptic curve discrete logarithm problem. This scheme is used with secret key exchange for subsequent conventional encryption and expanded to support mutual identification, for open network application. Mohamed et al. [5] analysed the various ways in which the phishing is achieved, the possible solutions, and the awareness along with some tips to be away from a victim of phishing attacks is discussed. Gokhale and Dasgupta [6] provide a new trust-based representation on secure groups and troups cryptographically. The verification of membership of the troup is done using the Zero-Knowledge Protocol and modular exponentiation. Each node of a group has its own identity inside a group, but they need not reveal their identity throughout verification. Hence, this particular trust model is a decentralized model and can be incrementally deployed in any network. In 2014, Yao and Zhao [7] developed deniable Internet key exchange (DIKE), a group of privacy-protecting authenticated DHKE protocols both in the identity-based system and also traditional PKI system. Asimi et al. [8] proposed a novel secured symmetric communication system build upon strong ZKP using session keys (SASK). In this, authentication is performed in two steps: the first is to recreate a virtual password and use this virtual password for assurance of confidentiality and the integrity of exchanged nonces through symmetric key encryption. The second is to generate a session key distributed between the Web server and the client to perform the symmetric encryption by using this session key. Kamal [9] proposed a key management scheme based on polynomial for both secure intra-group and inter-group communication. Both security and privacy are required for the exchange of key over the Internet. Suresh and Jagathy [10] in 2012, proposes a novel IBE-based secure e-mail system which utilize DNS as the infrastructure for exchange of key and a proxy service to carry out encryption/decryption on behalf of user and a secure key token or fingerprint authentication system for user authentication. Entity authentication and key distribution are central cryptographic problems in distributed computing. Ding [11] presented a broad overview of P2P computing topologies used, described architecture models and focused on content sharing networks and technologies.

Conventional Authentication Method of E-Mail using PGP

Figure 1 shows the following sequence of instructions to perform authentication:

1. The sender generates a message.
2. It uses SHA-1 to generate hash code of 160 bits length.

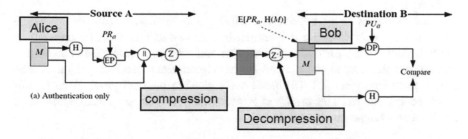

Fig. 1 E-mail authentication system using PGP

3. Sender uses private key along with RSA for the encryption of the hash code, and that result is added before the message.
4. The receiver decrypts the received message using public key of sender along with RSA algorithm and recover the hash code.
5. The receiver compares the decrypted hash code with newly generated hash code of the message. If both the two hash codes are matched, then the message is accepted as authentic.

3 Proposed Key Exchange Algorithm Using Lagrange Interpolation

Proposed protocol includes a method that performs the identification of two users among themselves. This authentication procedure involves the creation of a session key by exchange of pseudorandom numbers to verify the public keys over an open channel. This session key is used to encrypt or decrypt the message.

3.1 Generation of Session Key for Encryption or Decryption

Session key or private key is generated by exchanging data between sender and receiver. It is created using the concept of Lagrange interpolation. The polynomial generated is the session key to perform encryption and decryption for that session. It includes the following steps and showed in Fig. 2.

1. Generates a polynomial $P(x)$ by taking coordinates (X_i, Y_i) using Algorithm 1
2. The output of Algorithm 1 and pseudorandom number X send by the destination is used to generate another polynomial using Algorithm 2. This establishes connection between source and destination so that third party cannot intervene in between.
3. Destination generates polynomial using its own data and pseudorandom number X from source and compare both the polynomials using Algorithm 3.

   Procedure:
   Sender requests for a set of data to create a polynomial. Receiver responds with a set of data. Sender creates a polynomial by taking data and master secret key as input. Sender sends a pseudorandom number and asks receiver to create polynomial to compare the polynomial created by both sender and receiver are same. The Lagrange interpolation [12, 13]-based authentication protocol provides more efficient security service. This proposed protocol uses the concept of Zero-Knowledge Protocol and Lagrange interpolation.

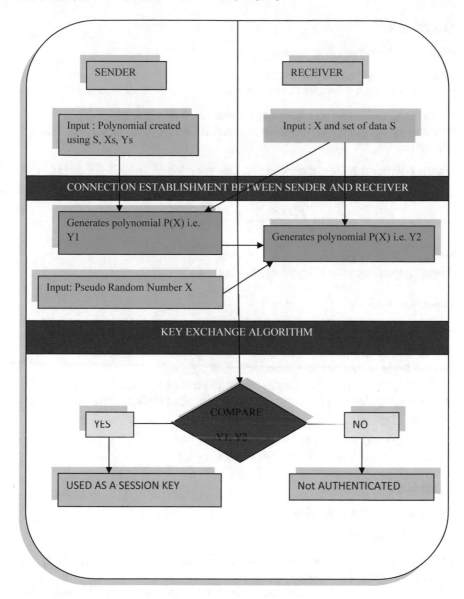

Fig. 2 Proposed key exchange algorithm

This novel mechanism includes the following algorithms:

1. Generation of polynomial by sender and receiver.
2. Establishing the connection between them.
3. Key exchange between sender and receiver.

Algorithm 1

Algorithm: Algorithm_Polynomial

This algorithm generates polynomial
Input: #(n) indicates the number of points, X

An Array $x_p[0, ..., n-1]$ represents x-coordinates of n points
An Array $y_p[0, ..., n-1]$ represents y-coordinates of n points X is a value used to calculate Y

Creates polynomial using the following Lagrange interpolation formula [14, 15]

$$p(x) = \sum_{i=1}^{m} y_i \prod_{j=1}^{m} \frac{(x - x_j)}{(x_i - x_j)} \tag{1}$$

We must consider pre-condition that $X_i \neq X_j$
Output: Algorithm returns Polynomial (PS(x)) #Y1

Algorithm 2

Algorithm: Connection_Establishment

This algorithm establishes connection between Source and Destination
Input: # (n) indicates the number of points, X_s, Y_s, X
An Array $x_p[0, ..., n-1]$ represents x-coordinates of a points
An Array $y_p[0,...,n-1]$ represents y-coordinates of a points X_s, Y_s represents secret key used by source

X is a value used to calculate Y
Output: Algorithm returns Polynomial (PD(x)) #Y2

Algorithm 3

Algorithm: Algorithm_KeyExchange

Compare Y1 and Y2

If both are same this procedure is repeated until the receiver (verifier) satisfies the sender is authorized (Prover)
After confirmation, Y1 is used for encryption and Y2 is used for decryption
Otherwise it is not authenticated.

4 Proposed Authentication Algorithm for E-Mail Security

Figure 3 shows the different steps followed in proposed authentication algorithm of E-Mail security. In Fig. 3, terms used are

M   Message
P(x)   Session Key
EP   Encryption using Polynomial using any symmetric key algorithm
DP   Decryption using Polynomial using any symmetric key algorithm

   Procedure:

1. This protocol creates polynomial through Lagrange interpolation based on Zero-Knowledge Protocol on both the sides, i.e. P(X) and Y1.
2. Message is encrypted using the shared secret or session key, i.e. Y1 using any symmetric key algorithm.
3. Original message is appended to the encrypted message (i.e. M + E(Y1,M)) and sent to destination.
4. Destination decrypts a message by using its own Polynomial Y2.
5. Compare the original message M sent by the source along with the message achieved through decryption by destination.
6. Both the messages will be same only when polynomials generated by both source and destination are same.
1. Perform encryption using any symmetric key algorithm on source-side
   EP represents encryption using polynomial

   P = EP (PS(x),M)// source encrypts
   append P(x) to P
   C = P‖P(x), i.e. EP(P(x),M) ‖P(x)
   Ciphertext C is sent to destination.

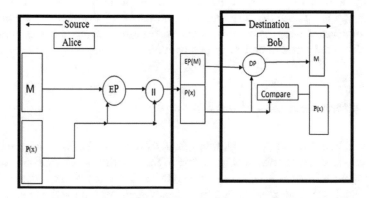

Fig. 3 Novel methodology used in e-mail security

2. Destination generates polynomial and decrypts to get original message
   Destination compares both the polynomials

   PS(x), PD(x) represents polynomials generated by source and destination
   respectively Compare (PS(x), PD(x))
   //DP represents Decryption using polynomial to get original message M
   M = DP (PD(x), C)
   Compares the Message M generated after decryption is compared with the
   original message. End of Algorithm.

5 Implemented Results of Key Exchange Algorithm Using NS2 in Peer-to-Peer Networks

The results are implemented using network simulator NS2 in peer-to-peer networks.
The procedure mentioned in Fig. 2 is applied using NS2 in peer-to-peer networks.
Implementation represents the following steps

1. Creation of topology of P2P network (peer to peer).
2. The terminal nodes of Peer A and Peer B exchange data in the form of packets to
   generate polynomials on both sides by following the steps of Fig. 2.
3. Performance of the P2P networks is defined by considering the packet delivery
   ratio, throughput and packet loss ratio.

5.1 Performance Based on Packet Delivery Ratio

From Fig. 4, it can be observed that as the time increases from 0.0 to 4.0 s significantly and the packet delivery ratio has also increased from 2 to 11 significantly
as time increases. X-axis and Y-axis indicate time in seconds and the number of
packets delivered through nodes, respectively.

5.2 Performance Based on Throughput

As the time increases significantly, throughput of proposed key exchange algorithm
also increases considerably. In Fig. 5, the highest performance ratio of proposed
key exchange scheme is 9. When the time interval is increased, more data will flow
and then leading to increased throughput, packet delivery ratio and decreases drops
in proposed authentication algorithm. X-axis and Y-axis indicate time in seconds
and the number of packets delivered through nodes, respectively.

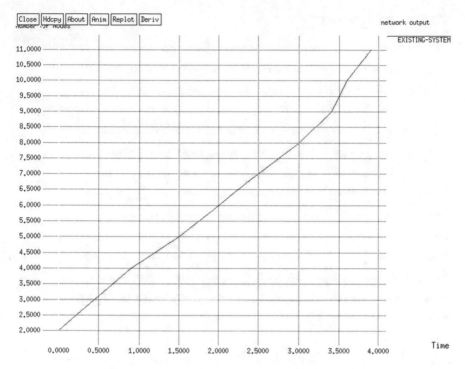

Fig. 4 Performance of packet delivery of peer-to-peer networks for key exchange

5.3 Performance Based on Packet Loss Ratio

From Fig. 6, it can be seen that the packet loss percentage decreases from 30 to 6 as the time varies from 0 to 160 s for proposed key exchange methodology. That means percentage of packet loss is decreasing as the time increases. That means performance of proposed key exchange algorithm is more efficient. X-axis indicates time in seconds, and Y-axis indicates the number of packets loss.

6 Security Analysis of Proposed Algorithm

6.1 Performance Analysis

It increases the performance due to less mathematical calculations, and these calculations require low storage capacity as it uses polynomials instead of modular arithmetic concept used in PKI algorithms. In addition to that, a third party cannot get any information and unable to impersonate client (source) to convince server (destination).

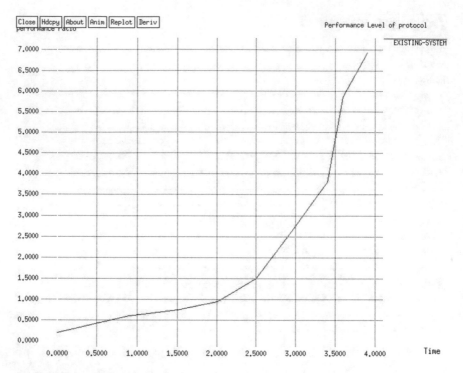

Fig. 5 Performance of throughput of peer-to-peer networks for key exchange

6.2 Cryptanalysis

1. Man-In-The-Middle Attack

   In proposed model, the sender never transmits secret key and hence the intruder does not get any chance to know any information about sender. Despite the usage of brute force method to find a secret key, the attacker could not break the security because public key (N) and random challenge question change for every communication.

2. Replay Attack

   Using replay attack previous message will be replayed by an attacker to prove himself as an authorized person to the destination (verifier). However, in this model, it will not prove the attacker as authenticate by replay attack as different challenge responses will be send by the verifier for every communication.

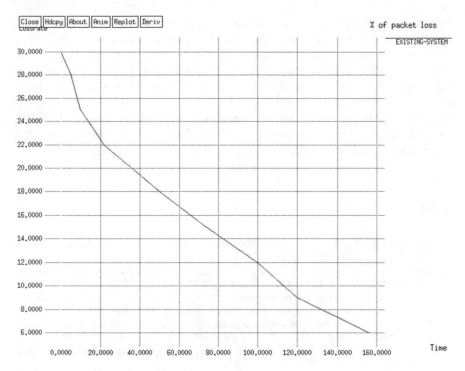

Fig. 6 Percentage of packet loss of peer-to-peer networks for key exchange

6.3 Cryptographic Strength

The cryptographic strength of ZKP is based on hard-to-solve problem. It is very difficult for a third party (intruder) to recognize the secret key because for each communication public key value changes. In addition to that, a random number will be generated the sender (prover) for every communication secret key. Change of public key by receiver (verifier) and generation of random number by sender (prover) make the attacker very difficult to breach the security.

6.4 Applications of Proposed Methodology

In areas like passports, PIN numbers, RFID tags, etc., where secret knowledge which is highly sensitive to disclose or needs to be verified, the proposed methodology can be used. This algorithm can also be used where exchange of keys is essential like in public key infrastructure (PKI), secure shell (SSH), secure socket layer (SSL)/Internet Protocol security (IPSec), transport layer security (TLS).

Some of the applications of ZKP [16] are verifiable outsourced computation, anonymous authentication, electronic voting and anonymous electronic ticketing for open transportation in which truthfulness needs to be forced without sacrificing privacy. Of late, there is a spurt in the amount of personal information that is being uploaded online pertaining to both corporate and government sectors. Hence, there is a need for efficient privacy-protecting systems that have become more and more important and a major focus of current research. This protocol can be a better alternative for a number of applications where there is no intervention of third party and also requires few mathematical calculations. Some of the real-world applications of the this algorithm would be Smart Cards, e-mail security, network authentications and key exchanges.

References

1. Tariq Banday, M.: Effectiveness and limitations of e-mail security protocols. Int. J. Distrib. Parallel Sys. (IJDPS). **2**(3) (2011)
2. Aronsson, H.A.: Zero knowledge protocols and small systems. In: Network Security Seminar (Tik-110.501). Helsinki University of Technology, Finland Fall (1995)
3. Guo, N., et al.: Aggregated privacy-preserving identity verification for composite web services. In: International Conference on Web Services, pp. 692–693 (2011)
4. Allam, A.M., et al.: Efficient zero-knowledge identification scheme with secret key exchange. In: 46th Midwest Symposium on Circuits and Systems, vol. 1, pp. 516–519 (2003)
5. Mohamed, G., et al.: E-mail phishing—an open threat to everyone. Int. J. Sci. Res. Publ. **4**(2), 1–4 (2014)
6. Gokhale, S., Dasgupta, P.: Distributed authentication for peer-to-peer networks. In: Proceedings of Symposium on Applications and the Internet Workshops, pp. 347–353 (2003)
7. Yao, A.C.C., Zhao, Y.: Privacy-preserving authenticated key-exchange over internet. Proc. IEEE Trans. Inf. Forensics Secur. **9**(1), 125–140 (2014)
8. Asimi, Y., et al.: Strong zero-knowledge authentication based on the session keys. Int. J. Netw. Secur. Appl. (IJNSA) **7**(1), 51–66 (2015)
9. Kamal, A.: Cryptanalysis of a polynomial based key management scheme for secure group communication. Int. J. Netw. Secur. **15**(1), 68–70 (2013)
10. Suresh, K.B., Jagathy, R.V.P.: A secure email system based on IBE, DNS and proxy service. J. Emer. Trends Comput. Inf. Sci. **3**(9), 1271–1276 (2012)
11. Ding, C.H., Nutanong, S., Buyya, R.: Peer-to-peer networks for content sharing. In: Grid Computing and Distributed Systems Laboratory, Department of Computer Science and Software Engineering, The University of Melbourne, Australia
12. Elia, M.: Polynomials and cryptography. Department of Electronics, Politecnico di Torino (2011)
13. Kovács, A., Kovács, L.I.: The Lagrange interpolation formula in determining the fluid's velocity potential through profile grids. University Politechnica of Timişoara, Romania, Research Institute for Symbolic Computation, Johannes Kepler University, Linz, Austria
14. Kovacevic, M.A.: A simple algorithm for the construction of Lagrange and Hermite interpolating polynomial in two variables, 2000 Mathematics Subject Classification. Primary 41A05; Secondary 65D05 (2007)
15. Shahsavaran, A., Shahsavaran, A.: Application of Lagrange interpolation for nonlinear integro differential equations. Appl. Math. Sci. **6**(18), 887–892 (2012)
16. Meiklejohn, S., Erway, C.C., Küpçü, A., Hinkle, T., Lysyanskaya, A.: ZKPDL: a language-based system for efficient zero-knowledge proofs and electronic cash

Author Profiles Prediction Using Syntactic and Content-Based Features

T. Raghunadha Reddy, M. Srilatha, M. Sreenivas and N. Rajasekhar

Abstract In digital forensics, the forensic analysts raised the major questions about the details of the author of a document like identity, demographic information of authors and the documents which were related these documents. To answer these questions, the researchers proposed a new research field of stylometry which uses the set of linguistic features and machine learning algorithms. Information extraction from the textual documents has become a popular research area in the last few years to know the details of the authors. In this context, author profiling is one research area concentrated by the several researchers to know the authors' demographic profiles like age, gender, and location by examining their style of writing. Several researchers proposed various types of stylistic features to analyze the style of the authors writing. In this paper, the experiment was performed with combination of syntactic features and content-based features. Various machine learning classifiers were used to evaluate the performance of the prediction of gender of reviews dataset. The proposed method achieved best accuracy for profiles prediction in author profiling.

Keywords Gender prediction · Author profiling · PDW model · Syntactic features · Content-based features

T. Raghunadha Reddy (✉)
Department of IT, Vardhaman College of Engineering, Hyderabad, India
e-mail: raghu.sas@gmail.com

M. Srilatha
Department of CSE, VR Siddhartha Engineering College, Vijayawada, India
e-mail: srilatha.manam@gmail.com

M. Sreenivas
Department of IT, Sreenidhi Institute of Science and Technology, Hyderabad, India
e-mail: mekala.sreenivas@gmail.com

N. Rajasekhar
Department of IT, Gokaraju Rangaraju Institute of Engineering and Technology, Hyderabad, India
e-mail: n.rajasekhar@griet.ac.in

© Springer Nature Singapore Pte Ltd. 2020   265
K. S. Raju et al. (eds.), *Data Engineering and Communication Technology*,
Advances in Intelligent Systems and Computing 1079,
https://doi.org/10.1007/978-981-15-1097-7_23

1 Introduction

The Internet is increasing rapidly with the huge amount of text day by day through reviews, blogs, documents, tweets, and other social media content. The researchers need the automated tools to process the information which is dynamic in nature. In this process, sometimes it is necessary to identify the owner who has created the text or document. Authorship analysis is one such area paying attention by the many researchers to find the author details of the text [1]. Authorship analysis is a process of reaching to a conclusion by understanding the characteristics of a piece of written document and thereby analyzing the author whose roots are coming from stylometric, which is a linguistic research field. The characteristics of the texts play an important role in the procedure of the authorship analysis. In general, the authorship analysis is categorized into three categories namely authorship attribution, authorship verification, and authorship profiling [2].

Every author has their own writing style but by looking at the writing style one can predict certain profiling characteristics of the text. The following are the profiling characteristics to analyze writing styles of the authors and their age, gender, location, personality traits, native language, occupation, educational background, etc. [3]. The writing style of a human being will never change throughout his lifetime. That is the reason the style of writing of author is same either he tweets, writes in a blog or writes in a document.

There is an important phenomenon observed while understanding the writing styles of male and female. It has been witnessed that the female writing style is different from male writing style, and female are more expressive and involve emotionally in their writings. The female writing expresses both positive and negative comments on an object or a person. Researchers have given another inference on male writing wherein, the male tends to narrate new stories and focus on what had happened but whereas the female expresses how they felt [1].

Koppel et al. assumed [4] that men used more determiners and quantifiers in their writings and women writings contain more pronouns by analyzing different types of corpuses. The female authors are interested in the topics like shopping, beauty, jewelry, and kitty party. In contrast to that male authors' interested in topics related to technology, sports, women, and politics. In the past, some of the researchers found that more number of prepositions [1, 4] was found in female author's writings than the male author's. In general, the writing style is defined as a set of grammar rules, words of choice, and clubbed with the selection of topics. In another observation [5], females use more adjectives and adverbs and talk on shopping and wedding styles and male author's writings consisting of technology and politics.

In this paper, we addressed predicting the profiling characteristic of gender of the authors from reviews dataset. The existing researchers extracted different set of features to predict the author profiles. In this work, two types of features were used with PDW (Profile specific Document Weighted) model for gender prediction of the authors using different classifiers.

This work is planned in six sections. The existing works in author profiling are analyzed in Sect. 2. The characteristics of dataset of the reviews were represented in Sect. 3. The PDW model was discussed in Sect. 4. Section 5 explains the experimental results PDW model with combination of content-based features and syntactic features. Section 6 concludes this work with future scope.

2 Related Works

To allocate predefined set of classes to text documents, classification process uses a set of features and variety of machine learning procedures. Since there is no direct mechanism to process the raw text, each document is specified in the form of vector. To identify the importance of each feature in a given document, traditional feature frequency measure is used. As numerous researchers addressed, frequency of a feature is not sufficient to discover the significance of different features. To overcome this difficulty, an information retrieval-based measure TF-IDF is used to find the feature weight based on its frequency and the number of text documents that contains these feature in the given corpus [6]. A variety of weight measures are proposed by many researchers to determine the weight of these features. The count of these features and its associated weight measures along with different machine learning algorithms involved in the prediction accuracies of the author profiles in author profiling.

In the experiments of Dang Duc [7], 298 features such as character-based and word-based features were extracted from approximately 3500 web pages of 70 Vietnamese blogs and concluded that word-based features are more intended to predict the gender than the features based on characters. Stylometric features are extracted from 1000 blog posts of 20 bloggers by many Greek researchers which also included most frequent n-grams such as word n-grams and character n-grams. They observed that character n-grams and word n-grams with more length are producing good accuracy in gender prediction when experimented with SVM.

The experiment of Soler and Wanner [8] on opinion blogs of New York Times corpus, features with different combinations were tried. These combinations include sentence-based, word-based, character-based, syntactic features, and dictionary-based features for prediction of gender, and the combination of all these features achieved good accuracy. A considerable amount of drop in the accuracy is noticed with the application of bag-of-words approach on 3000 words having most TF-IDF values.

Koppel et al. [9] experimented with British National Corpus (BNC), from which 566 text documents were collected, 1081 features were extracted and good accuracy is achieved when predicting the gender. Argamon et al. [10] composed a corpus from various blogs which includes different authors (approximately 19,320), by using both stylistic features and content-based features the accuracy is calculated. Argamon, S. et al. also concluded that stylistic features are contributing more in predicting and discriminating the gender.

Palomino-Garibay Alonso et al. stated that the character level n-grams with the value of n ranging from 2 to 6 were producing better accuracies over the languages like Spanish and Dutch with the combination of abovementioned features. In order to classify the styles of writing of the different authors, some researchers suggested different composition of features like readability, semantic, syntactic, structural, character-based, and lexical features [11]. Estival et al. [12] experimented with approximately 9800 emails and collected 689 features and different types of classification algorithms, and they observed that SMO algorithm produced good accuracy in predicting the gender.

3   Dataset Characteristics

In this work, the dataset contains 4000 English reviews of different hotels and it was gathered from www.TripAdvisor.com. The dataset for gender profile was balanced, i.e., both female and male profile groups include 2000 documents each.

The researchers used different types of performance measures such as recall, precision, accuracy, and F1-measure were used to determine the efficiency of their proposed system in the approaches of author profiling problem. In this paper, accuracy measure is used to compute the performance of the system. In this context, the accuracy is defined as the ratio among the number of test documents was predicted their author gender correctly and the total number of test documents considered for testing [13].

4   PDW Model

In our view of knowledge, few researchers concentrated on different representations of text in author profiling. At present, bag-of-words (BOW) model is used by most of the researchers to represent the document vectors in author profiling. In BOW model, the document vector values are computed by the frequency of content/ stylistic features. The existing document vector representation techniques face many problems. To overcome those difficulties and improving the prediction accuracy of author profiles, Raghunadha Reddy et al. [14] proposed a new approach named PDW which uses the relationship among the documents to author profile group and the terms to documents. In this work, we used the combination of content-based features such as most frequent terms and syntactic features like most frequent POS n-grams to represent the document.

Figure 1 shows the model of PDW approach. In this model, the training dataset consists of 4000 reviews of male and female, and the dataset is balanced that is each gender group contains 2000 documents. The dataset is used for extracting content-based and syntactic features. In content-based features extraction, two preprocessing methods like stop word removal and stemming were used to prepare

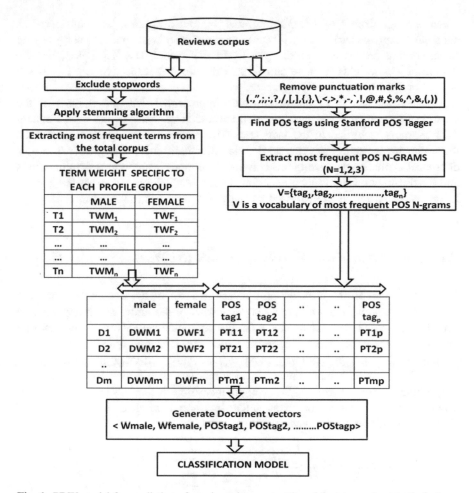

Fig. 1 PDW model for prediction of gender using content-based features and syntactic features

the data for efficient features extraction. Once the data is cleaned then extract the most frequent terms from the entire training dataset. After extraction of most frequent terms, each document is represented with these terms as document vector. In this model, term weight measure was used to compute the terms weight. Each term's importance is measured by the weight of the term against each gender group. These term weights of individual gender group were used to compute the weight of the documents by using document weight measure. The document weight is also computed against to each gender group.

In syntactic features extraction, first remove the punctuation marks to clean the text for efficient features extraction. We used Stanford POS tagger to identify the part of speech of words. The most frequent POS n-grams were extracted from the dataset as syntactic features. In this paper, the most frequent POS n-grams where n range from 1 to 3 (POS unigrams, bigrams, trigrams) was considered as syntactic features. {tag1, tag2, ..., tag$n$} is a set of POS tags, {male, female} is a set of sub-profile groups in a gender profile. In this model, DWM$_m$, DWF$_m$ are the document D_m weights in male and female documents, respectively. PT$_{mp}$ is the weight of POS tag 'p' in document 'm.'

Finally, the vectors of documents were represented with the weights of the document computed in the previous step and the frequencies of the most frequent POS n-grams identified in the later step. The machine learning algorithms used these document vectors to generate the classification model. The gender of a new document is predicted by using this classification model. The term weight measures and the document weight measures influence the prediction accuracy of the gender. In this paper, we used a supervised term weight measures proposed in [15].

4.1 Supervised Term Weight Measure (STWM)

The general idea is the term which appears in most of the documents has least power to distinguish from different classes to that of the terms which appears in a single document or few documents. Also the computed weights play a crucial role in assigning the class labels. The terms which are appearing only in single document or in few documents carry most representative information called unique terms. The terms with more concentrated inter-class distribution have strong distinguishing power. Therefore, this weight measure allocates weight to the terms according to their inner document, inter-class and intra-class distributions [15]. Equation (1) shows the STWM.

$$
\begin{aligned}
W(t_i, d_k \in \text{PG}_p) = \frac{\text{tf}(t_i, d_k)}{\text{DF}_k} &\times \frac{\sum_{x=1, d_x \in \text{PG}_p}^{m} \text{tf}(t_i, d_x)}{1 + \left(\sum_{y=1, d_y \notin \text{PG}_p}^{n} \text{tf}(t_i, d_y)\right)} \\
&\times \frac{\sum_{x=1, d_x \in \text{PG}_p}^{m} \text{DC}(t_i, d_x)}{1 + \left(\sum_{y=1, d_y \notin \text{PG}_p}^{n} \text{DC}(t_i, d_y)\right)}
\end{aligned}
\tag{1}
$$

In this measure, tf (t_i, d_k) represents the no. of times the term t_i appeared in document d_k, DF$_k$ represents total number of terms in a document d_k.

4.2 Document Weight Measure (DWM)

In this thesis, the document weight against to author groups is calculated by using a DWM proposed by Raghunadha Reddy et al. [16]. The DWM is represented in Eq. (2).

$$W(d_k, \mathrm{PG}_p) = \sum_{t_i \in d_k, d_k \in \mathrm{PG}_p} \mathrm{TFIDF}(t_i, d_k) * W(t_i, \mathrm{PG}_p) \qquad (2)$$

In this measure, $W(d_k, \mathrm{PG}_p)$ is the document weight of d_k in PG_p profile group. This DWM is a product of TF-IDF weight of a term and the weight of a term against to author group computed in previous step of the model.

5 Experimental Results

5.1 PDW Model with Content-Based Features and Syntactic Features

The major steps in PDW model are finding suitable features, identifying suitable sub-profiles, computing feature weights and document weights, and document representation. Table 1 presents the gender prediction accuracies of PDW model, when content-based and syntactic features were used to represent the vector of document and different types of machine learning algorithms such as Naïve Bayes Multinomial (NBM), Simple Logistic (SL), Logistic (LOG), IBK, Bagging (BAG), and Random Forest (RF) were used. In this work, the experiment performed with combination of 8000 most frequent terms and the number of POS n-grams varies with an interval of 500 from 500 to 3000 for computing the document weight.

The PDW model produced highest accuracy of 93.25% for prediction of gender when Random Forest classifier used 8000 most frequent terms and 3000 most frequent POS n-grams. The Random Forest classifier achieved best accuracy when compared with other classifiers. It was observed that the accuracies were increased when the number of features was increased.

Table 1 The accuracies of PDW model when content-based features and syntactic features were used as features

Classifier/number of features	NBM	SL	LOG	IBK	BAG	RF
8000 terms + 500 POS n-grams	84.51	78.63	80.54	70.65	64.56	87.65
8000 terms + 1000 POS n-grams	86.74	79.74	81.71	71.78	65.74	88.89
8000 terms + 1500 POS n-grams	87.20	81.32	83.32	73.94	67.91	90.12
8000 terms + 2000 POS n-grams	89.76	82.87	84.89	74.09	68.87	91.56
8000 terms + 2500 POS n-grams	90.94	84.21	86.54	75.74	69.41	92.71
8000 terms + 3000 POS n-grams	91.36	84.89	87.08	76.36	71.34	93.25

6 Conclusions

In this work, a PDW model was experimented with combination of syntactic and content-based features. The PDW model with the combination of both content-based features and syntactic features obtained 93.25% for gender prediction. It was observed that with the constant number of content-based features, the accuracy of gender prediction is increased when the number of POS n-grams was increased. It was also observed that the syntactic features influence is more when compared with content-based features to improve the accuracy of gender.

References

1. Koppel, M., Argamon, S., Shimoni, A.: Automatically categorizing written texts by author gender. Literary Linguist. Comput. 401–412 (2003)
2. Raghunadha Reddy, T., Vishnu Vardhan, B., Vijayapal Reddy, P.: A survey on author profiling techniques. Int. J. Appl. Eng. Res. **11**(5), 3092–3102 (2016)
3. Raghunadha Reddy, T., Vishnu Vardhan, B., Vijayapal Reddy, P.: N-gram approach for gender prediction. In: 7th IEEE International Advanced Computing Conference, Hyderabad, Telangana, pp. 860–865, 5–7 Jan 2017
4. Koppel, M., Schler, J., Bonchek-Dokow, E.: Measuring differentiability: unmasking pseudonymous authors. J. Mach. Learn. Res. **8**, 1261–1276 (2007)
5. Newman, M.L., Groom, C.J., Handelman, L.D., Pennebaker, J.W.: Gender differences in language use: an analysis of 14,000 text samples. Discourse Process. **45**(3), 211–236 (2008)
6. Pradeep Reddy, K., Raghunadha Reddy, T., Apparao Naidu, G., Vishnu Vardhan, B.: Term weight measures influence in information retrieval. Int. J. Eng. Technol. **7**(2), 832–836 (2018)
7. Dang Duc, P., Giang Binh, T., Son Bao, P.: Author profiling for vietnamese blogs. Asian Language Processing, 2009 (IALP '09), pp. 190–194 (2009)
8. Company, J.S., Wanner, L.: How to use less features and reach better performance in author gender identification. The 9th edition of the Language Resources and Evaluation Conference (LREC), 26–31 May 2007
9. Schler, J., Koppel, M., Argamon, S., Pennebaker, J.: Effects of age and gender on blogging. In: Proceedings of AAAI Spring Symposium on Computational Approaches for Analyzing Weblogs, vol. 6, pp. 199–205, Mar 2006
10. Argamon, S., Koppel, M., Pennebaker, J.W., Schler, J.: Automatically profiling the author of an anonymous text. Commun. ACM **52**(2), 119 (2009)
11. Palomino-Garibay, A., Camacho-Gonzalez, A.T., Fierro-Villaneda, R.A., Hernandez-Farias, I, Buscaldi, D., Meza-Ruiz, I.V.: A random forest approach for authorship profiling. In: Proceedings of CLEF 2015 Evaluation Labs (2015)
12. Estival, D., Gaustad, T., Pham, S.B., Radford, W., Hutchinson, B.: Author profiling for English emails. In: 10th Conference of the Pacific Association for Computational Linguistics (PACLING, 2007) (2007)
13. Raghunadha Reddy, T., Vishnu Vardhan, B., Vijayapal Reddy, P.: Author profile prediction using pivoted unique term normalization. Indian J. Sci. Technol. **9**(46) (2016)
14. Raghunadha Reddy, T., Vishnu Vardhan, B., Vijayapal Reddy, P.: Profile specific document weighted approach using a new term weighting measure for author profiling. Int. J. Intell. Eng. Syst. **9**(4), 136–146 (2016)

15. Sreenivas, M., Raghunadha Reddy, T., Vishnu Vardhan, B.: A novel document representation approach for authorship attribution. Int. J. Intell. Eng. Syst. **11**(3), 261–270 (2018)
16. Raghunadha Reddy, T., Vishnu Vardhan, B., Vijayapal Reddy, P.: A document weighted approach for gender and age prediction. Int. J. Eng. –Trans. B: Appl. **30**(5), 647–653 (2017)

An Investigation of the Effects of Missing Data Handling Using 'R'-Packages

Sobhan Sarkar, Anima Pramanik, Nikhil Khatedi and J. Maiti

Abstract The data in real-world scenario often consists of missing values which leads to difficulty in analysis. Though there has been an emergence of various algorithms handling the issue, it is, in fact, troublesome to implement from industry perspective. Based on this issue, the 'R'-software with various available packages for missing data handling can be a fruitful solution, which is hardly reported in any previous study. Hence, the availability of such packages demands analysis to compare their performances and check their suitability for a given dataset. A comparative study is performed using the 'R'-packages, namely missForest, Multivariate imputation by chained equations (MICE), and AMELIA-II. Two classifiers, support vector machine (SVM) and logistic regression (LR), are used for prediction. The packages are compared with regard to imputation time, effects on variance, and the efficiency. The experimental results reveal that the performances depend on the dataset size and the percentage of missing values in data.

Keywords Missing value · MICE · missForest · AMELIA II · Time for imputation · Accuracy of imputation · Variance

1 Introduction

Missing attribute values are often found in various datasets. They can occur in all types of studies including randomized controlled trials, cohort studies, casecontrol studies, and clinical registries due to a variety of reasons. The presence of missing values affects the data quality and henceforth, the final outcome [2, 15]. Handling missing data requires knowledge of the types of missing data mechanisms [5].

S. Sarkar (✉) · A. Pramanik · J. Maiti
Department of Industrial & Systems Engineering, IIT Kharagpur,
Kharagpur, India
e-mail: sobhan.sarkar@gmail.com

N. Khatedi
Department of Mechanical Engineering, IIT Kharagpur, Kharagpur, India

© Springer Nature Singapore Pte Ltd. 2020
K. S. Raju et al. (eds.), *Data Engineering and Communication Technology*,
Advances in Intelligent Systems and Computing 1079,
https://doi.org/10.1007/978-981-15-1097-7_24

Missing data is usually categorized into three types, namely missing completely at random (MCAR), missing at random (MAR), and missing not at random (MNAR). MCAR occurs when the missing value of an attribute is independent of the attribute value itself as well the values of other attributes, i.e., it does not depend on either missing or observed data. No systematic difference could be found between attributes with observed and unobserved values. MAR is only related to observed data, where the value is decided by the value of another attribute. Hence, correlations in the observed data could be used to recover the missing data. MNAR is dependent on the value of the concerned attribute itself, i.e., missing value depends on the value that would have been observed, but is currently missing. Hence, missing data cannot be recovered by utilizing the relationships in observed data since extra information is associated with it [1, 2, 10, 14]. Previous studies have indicated that inappropriate handling of these missing data types leads to estimation problems, including bias in parameters and standard error estimates, and reduction in classification power.

Therefore, the objective of the paper is to investigate the utility of the available three 'R'-packages, namely missForest [13], Multivariate imputation by chained equations (MICE), and AMELIA-II, on the two benchmark datasets, namely 'BNG heart statlog' and 'Poker hand'. The performances of the algorithms are checked by varying both data sizes (10,000 and 15,000 samples) and % of missing values (10, 20, and 30%). Two classifiers, logistic regression (LR) and support vector machine (SVM), are used for prediction task. The three performance metrics, namely imputation time, accuracy variance percentage (AVP), and variance decrease percentage (VDP), are used.

The rest of the paper is organized as follows: in Sect. 2, related works are presented. The proposed methodology is discussed in Sect. 3. In Sect. 4, the experimental results are discussed, and finally in Sect. 5, the conclusion with scopes for future works is presented.

2 Review of Literature

With the need to take care of missing data, multiple methods have been proposed ranging from deletion to imputation. Deletion can be classified into full deletion, deletion by list, or complete case analysis where all records containing one or more missing values are removed or specific deletion where records with a certain minimum number of missing values are deleted [6, 8, 14]. Imputation techniques aim on estimating the missing values on the basis of the observed features. The value which is imputed may be any of the measures of central tendency or an assumed value of the concerned attribute, or it could use a predicted value using some predictive models [9]. However, they have been criticized as they do not reflect uncertainty in the data. They might produce bias in the analysis. For numerical data, mean or median imputation is commonly used, while mode imputation is used with nominal data. Imputation can be classified as single imputation (SI) and multiple imputation (MI). Single imputation uses a single plausible value to substitute the missing items. Multiple imputation substitutes the missing values with plausible values multiple times, say m times, creating m imputed datasets. Each imputed dataset is analyzed individually and the results are then combined using Rubin's rule Rubin [11] multiple. While multiple imputation has a lot more advantages over single imputation, it also requires more resources [4]. Other methods to impute missing values include regression models, K-nearest neighbor (KNN), expectation maximization (EM) imputation [7], or as shown in Farhangfar [3] novel, etc. In these methods, existing information is used to develop a model which is used to predict values to impute in place of missing values.

Although there are several algorithms designed for missing value imputation, there usage remains limited due to the difficulty in coding the algorithms. Also, based on the aforementioned discussion, it is found that most of the studies have used different techniques consuming more resources and time. However, no study has been reported so far that addresses the exploration of the utilities of available 'R'-packages in taking care of missing data. Thus, the contribution of the study is to explore the utility of the available 'R'-packages to solve the missing value problems in data. This is investigated in two benchmark datasets varying their sizes and % missing values. In addition, this type of analysis is new as this investigation has been carried out for a case study of a steel plant.

3 The Proposed Methodology

The study investigated the effects of the size of the dataset and the missing value % in it on the accuracy of the imputation and the imputation time. Figure 1 shows the flowchart of the strategy of the methodology used.

To carry out the study, two large datasets, namely 'poker_hand'[1] and 'BNG heart statlog'[2] are considered. Each dataset has been divided into two random subsets of size 10,000 and 15,000 rows or samples. Additionally, a test set of size 25,000 rows has been created. Each subset has been introduced with 10, 20 and 30% missing values using prod NA() function of R. The missing values of each of these subsets have been imputed using the three R-packages, missForest, MICE, and AMELIA-II. Both of these datasets contain an attribute 'class.' The 'poker hand' dataset has 10 categories, among which records with classification 2–9 are very less in comparison to records with classification 0 and 1. Therefore, only the

---

[1]Source: https://archive.ics.uci.edu/ml/datasets/Poker+Hand.

[2]Source: https://archive.ics.uci.edu/ml/datasets/Statlog+%28Heart%29.

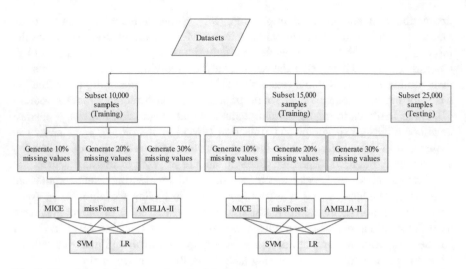

Fig. 1 Flowchart of the proposed methodology

records with classification 0 and 1 have been targeted. In case of 'BNG heart statlog' dataset, class attribute contains only two classifications, whether heart disease being 'present' or 'absent.' It has been converted into binary form to get 0 for 'absent' and 1 for 'present.' To check the accuracy of the imputation algorithms, two predictive models, SVM and LR, have been developed for each subset with the target attribute, 'class.' Each classifier has been trained on two different sets of training sets (10,000 and 15,000 samples) with varying % of missing values, i.e., 10, 20, and 30%. After training, the classifiers have been used for prediction on the 'class' of the test set.

4 Experimental Details

4.1 Experimental Setup and Datasets Used

The above study is carried out using R-Studio of Version 1.1.463 on a laptop with memory 8 GB, processor Intel(R) Core(TM) i7-4510U CPU @ 2.60 GHz, and 64-bit Windows 7 operating system. Two different datasets are used, namely 'BNG heart statlog' and 'poker hand.' 'BNG heart statlog' contains 14 attributes, namely a1, a2, a3, a4, a5, a6, a7, a8, a9, a10, a11, a12, a13, and class. It has 1,000,000 entries. On the other hand, 'poker hand' contains attributes, namely s1, c1, s2, c2, s3, c3, s4, c4, s5, c5, and class. This dataset has 800,000 entries. These two datasets are selected because they are freely available and large in size.

4.2 Performance Metrics

The performance metrics used in this study are: (a) time taken by the 'R'-package to carry out missing value imputation, (b) imputation accuracy based on the predictions made by models on test sets, and (c) variance of numerical attributes of the imputed data subsets as compared to the original data subsets. Two measures, namely accuracy variance percentage (AVP) and variance decrease percentage (VDP) are used for the evaluation of imputation performance, which are described below.

(a) *Accuracy variance percentage (AVP)*—It is defined as the ratio of the difference in the predicted accuracies of model using original data subset and imputed data subset to the predicted accuracy of model built using original data subset with no missing values. It is expressed as the following Eq. (1).

$$AVP = \frac{|(\text{Original\_accuracy} - \text{Imputed\_accuracy})|}{\text{Original\_accuracy}} \times 100\% \qquad (1)$$

where Original_accuracy is the accuracy of prediction when the model is built using the original training data subset with no missing values, and Imputed_accuracy implies the accuracy when the model is built using the same training data subset but after generating missing values and subsequent imputation.

(b) *Variance decrease percentage (VDP)*—It is based on the fact that the variance in any feature or variable of imputed data subset should not be too different than the variance of that feature or variable in the original data subset. It is expressed as the following Eq. (2).

$$VDP = \frac{|(\text{Variance\_original} - \text{Variance\_imputed})|}{\text{Variance\_original}} \times 100\% \qquad (2)$$

where Variance_original denotes the variance in the attribute of the original training data subset, and Variance_imputed implies the variance in the same attribute of the same training data subset but after generating missing values and subsequent imputation.

Table 1 Imputation time recorded of algorithms using different training subsets and % of missing values width = 0.48

Data size	Data set	Missing %	MICE (min)	missForest (min)	AMELIA-II (min)
10,000	BNG	10	7.202	12.749	6.885
10,000	BNG	20	8.404	9.708	7.552
10,000	BNG	30	9.144	7.430	7.995
15,000	BNG	10	15.627	29.428	12.856
15,000	BNG	20	19.662	21.533	15.669
15,000	BNG	30	22.101	17.562	16.445
10,000	Poker	10	5.443	12.798	4.885
10,000	Poker	20	5.607	9.859	5.005
10,000	Poker	30	6.152	7.994	5.996
15,000	Poker	10	10.314	30.935	9.554
15,000	Poker	20	12.088	22.454	11.447
15,000	Poker	30	12.713	16.989	12.336

4.3   Results and Discussion

In this section, the results of imputation time, predictive accuracy, and variance (AVP and VDP) are discussed.

4.3.1   Analysis of the Time Taken

For all the imputed data subsets of size 10,000 and 15,000, each subset with 10, 20, and 30% of missing values, the imputation time using MICE, missForest, and AMELIA-II was calculated. Table 1 provides with the imputation time for all the datasets with different % of missing values. The salient observations from this experiment are: (a) When MICE is used, the imputation time increases with an increment in the missing value percentage in a dataset. The time taken for imputation is increased as the size of dataset gets increased; (b) for missForest, the time of imputation reduces with an increase in the missing value percentage in a dataset. Here, the time for imputation is increased on increasing the size of dataset; (c) AMELIA-II consumes more time as % of missing values are increased and the size of each dataset is increased; and (d) considering the total time for all the imputations on all datasets, AMELIA-II is faster than other algorithms.

4.3.2   Predictive Accuracy Analysis

The predicted accuracy of models trained on imputed data subsets were compared with those trained on original data subsets using AVP. Table 2 shows the trends in

Table 2 AVP recorded of algorithms using different training subsets and % of missing values width = 0.41

Data size	Dataset	Missing value	Package	LR	SVM
10,000	BNG	10	MICE	0.082	0.032
10,000	BNG	20	MICE	0.059	0.059
10,000	BNG	30	MICE	0.191	0.123
10,000	BNG	10	missForest	0.063	0.049
10,000	BNG	20	missForest	0.041	0.056
10,000	BNG	30	missForest	0.091	0.074
10,000	BNG	10	AMELIA-II	0.077	0.028
10,000	BNG	20	AMELIA-II	0.067	0.049
10,000	BNG	30	AMELIA-II	0.120	0.114
15,000	BNG	10	MICE	0.055	0.059
15,000	BNG	20	MICE	0.014	0.059
15,000	BNG	30	MICE	0.064	0.127
15,000	BNG	10	missForest	0.014	0.040
15,000	BNG	20	missForest	0.064	0.048
15,000	BNG	30	missForest	0.036	0.060
15,000	BNG	10	AMELIA-II	0.043	0.035
15,000	BNG	20	AMELIA-II	0.024	0.019
15,000	BNG	30	AMELIA-II	0.054	0.036
10,000	Poker	10	MICE	0.045	0.020
10,000	Poker	20	MICE	0.082	0.020
10,000	Poker	30	MICE	0.030	0.031
10,000	Poker	10	missForest	0.035	0.031
10,000	Poker	20	missForest	0.062	0.047
10,000	Poker	30	missForest	0.070	0.054
10,000	Poker	10	AMELIA-II	0.028	0.029
10,000	Poker	20	AMELIA-II	0.059	0.044
10,000	Poker	30	AMELIA-II	0.061	0.053
15,000	Poker	10	MICE	0.052	0.035
15,000	Poker	20	MICE	0.062	0.045
15,000	Poker	30	MICE	0.070	0.051
15,000	Poker	10	missForest	0.025	0.021
15,000	Poker	20	missForest	0.030	0.042
15,000	Poker	30	missForest	0.056	0.061
15,000	Poker	10	AMELIA-II	0.021	0.020
15,000	Poker	20	AMELIA-II	0.027	0.039
15,000	Poker	30	AMELIA-II	0.047	0.061

Table 3 VDP recorded of algorithms using different training subsets and % of missing values

Size	Data	Algorithm	Missing value (%)	Field 1	Field 2	Field 3	Field 4	Field 5
10,000	BNG	MICE	10	0.456	3.564	0.050	0.134	0.811
10,000	BNG	MICE	20	1.335	6.703	0.450	0.219	1.738
10,000	BNG	MICE	30	0.097	9.431	1.581	1.518	2.028
10,000	BNG	missForest	10	8.381	7.741	9.074	8.694	8.182
10,000	BNG	missForest	20	16.332	16.149	18.159	18.819	15.918
10,000	BNG	missForest	30	23.046	19.794	28.631	28.355	20.256
10,000	BNG	AMELIA-II	10	8.214	7.014	9.045	7.995	8.044
10,000	BNG	AMELIA-II	20	15.225	16.557	17.556	17.859	12.119
10,000	BNG	AMELIA-II	30	22.891	18.446	12.155	15.220	15.778
15,000	BNG	MICE	10	0.472	3.529	0.451	0.402	0.198
15,000	BNG	MICE	20	0.411	6.159	0.461	0.144	0.686
15,000	BNG	MICE	30	0.376	10.527	1.407	0.498	0.845
15,000	BNG	missForest	10	7.350	8.106	9.756	9.126	7.965
15,000	BNG	missForest	20	15.375	14.872	18.262	18.602	14.314
15,000	BNG	missForest	30	23.589	21.103	26.839	26.813	22.429
15,000	BNG	AMELIA-II	10	6.889	7.556	8.889	8.364	6.908
15,000	BNG	AMELIA-II	20	14.556	13.885	17.556	17.448	13.604
15,000	BNG	AMELIA-II	30	22.669	21.445	26.448	25.948	21.406
10,000	Poker	MICE	10	0.948	0.852	1.159	0.306	0.943
10,000	Poker	MICE	20	1.939	1.155	1.467	0.177	0.143
10,000	Poker	MICE	30	2.661	1.711	0.072	2.274	1.518
10,000	Poker	missForest	10	9.594	8.742	9.181	9.324	9.589
10,000	Poker	missForest	20	18.704	18.241	18.500	18.134	18.219
10,000	Poker	missForest	30	26.121	25.012	26.440	24.312	24.923
10,000	Poker	AMELIA-II	10	8.506	6.362	9.112	8.325	5.556
10,000	Poker	AMELIA-II	20	17.448	17.529	17.225	17.055	17.889
10,000	Poker	AMELIA-II	30	23.998	21.057	21.556	24.368	22.550
15,000	Poker	MICE	10	1.095	0.405	0.168	1.577	1.381
15,000	Poker	MICE	20	1.238	1.283	0.270	1.283	1.912
15,000	Poker	MICE	30	1.126	0.274	1.728	3.280	1.864
15,000	Poker	missForest	10	9.328	9.427	9.742	9.466	9.141
15,000	Poker	missForest	20	18.847	18.052	17.901	18.231	17.845
15,000	Poker	missForest	30	26.273	25.257	24.753	24.934	24.652
15,000	Poker	AMELIA-II	10	5.226	7.889	8.445	8.996	8.779
15,000	Poker	AMELIA-II	20	12.556	12.556	12.559	15.226	15.448
15,000	Poker	AMELIA-II	30	15.445	19.998	20.504	23.557	23.556

AVP for both the algorithms used. The salient observations from this experiment are: (i) The predictive accuracy reduces as there are more and more missing values in the dataset for all cases with some aberrations, (ii) SVM performs better than logistic regression when the missing value % is small, and (iii) In general, AVP is increased on increasing % of missing values.

4.3.3 Variance Analysis

To check how the imputation affects the variance of the variable columns, variance analysis has been used. The variance of the attributes of imputed data subsets have been compared to the corresponding attribute of the original data subset for both the datasets using variance decrease percentage. Since the dataset named 'poker hand' contains only five numeric variables or features, only the initial five numeric features were considered from the 'BNG_heart_statlog' dataset so that comparison becomes easier. Field 1 is representative of c1 and a1 feature of 'poker hand' and 'BNG_heart_statlog' dataset, respectively. Field 2 represents c2 and a3 attribute. Field 3 represents c3 and a4. Field 4 represents c4 and a5. Field 5 represents c5 and a8. Table 3 provides the VDP for the attributes of the imputed datasets. From the analysis, some key insights are obtained. They are: (i) the variances of each attribute for all data subsets reduce with increase in the missing value % for both datasets and (ii) it shows an increase in VDP with the increase in % of missing values for all the algorithms, thus exhibiting a linear trend with some aberrations.

5   Conclusions

This study explored the effects of imputation of missing values using three R-packages, namely missForest, MICE, and AMELIA-II, on the performance of predictive models trained using datasets after the missing values were imputed. The reduction in variance for imputed datasets has also been investigated. For the time taken for imputations, AMELIA-II outperforms others in almost all the cases. The predictive accuracy is reduced as the percentage of missing values is increased in the datasets for both the models. SVM outperforms LR in cases with a lower percentage of missing values. Although this paper studied the performance of only three packages in R, there is a future scope for more extensive study on more imputation packages with optimization packages [12] available and correlating them with the models developed to get a better performance.

References

1. Allison, P.D.: Handling missing data by maximum likelihood. In: SAS Global Forum, vol. 2012. Statistical Horizons, Havenford, PA (2012)
2. Batista, G.E., Monard, M.C., et al.: A study of k-nearest neighbour as an imputation method. HIS **87**(251–260), 48 (2002)
3. Farhangfar, A., Kurgan, L.A., Pedrycz, W.: A novel framework for imputation of missing values in databases. IEEE Trans. Syst. Man Cybern.-Part A Syst. Hum. **37**(5), 692–709 (2007)
4. Fichman, M., Cummings, J.N.: Multiple imputation for missing data: making the most of what you know. Organ. Res. Methods **6**(3), 282–308 (2003)
5. Horton, N.J., Lipsitz, S.R.: Multiple imputation in practice: comparison of software packages for regression models with missing variables. Am. Stat. **55**(3), 244–254 (2001)
6. Jonsson, P., Wohlin, C.: An evaluation of k-nearest neighbour imputation using likert data. In: 10th International Symposium on Software Metrics, 2004. Proceedings, pp. 108–118. IEEE (2004)
7. Karmaker, A., Kwek, S.: Incorporating an em-approach for handling missing attribute-values in decision tree induction. In: Fifth International Conference on Hybrid Intelligent Systems, 2005. HIS'05, pp. 6–pp. IEEE (2005)
8. Kumutha, V., Palaniammal, S.: An enhanced approach on handling missing values using bagging k-nn imputation. In: International Conference on Computer Communication and Informatics (ICCCI), 2013, pp. 1–8. IEEE (2013)
9. Malarvizhi, M.R., Thanamani, A.S.: K-nearest neighbor in missing data imputation. Int. J. Eng. Res. Dev. **5**(1), 5–7 (2012)
10. Rubbin, D., Little, R.: Statistical analysis with missing data (1987)
11. Rubin, D.B.: Multiple imputation after 18+ years. J. Am. Stat. Assoc. **91**(434), 473–489 (1996)
12. Sarkar, S., Lohani, A., Maiti, J.: Genetic algorithm-based association rule mining approach towards rule generation of occupational accidents. In: International Conference on Computational Intelligence, Communications, and Business Analytics, pp. 517–530. Springer (2017)
13. Sarkar, S., Vinay, S., Raj, R., Maiti, J., Mitra, P.: Application of optimized machine learning techniques for prediction of occupational accidents. Comput. Oper. Res. **106**, 210–224 (2019)
14. Scheffer, J.: Dealing with missing data (2002)
15. Tsai, C.F., Chang, F.Y.: Combining instance selection for better missing value imputation. J. Syst. Softw. **122**, 63–71 (2016)

Performance Assessment of Multiple Machine Learning Classifiers for Detecting the Phishing URLs

Sheikh Shah Mohammad Motiur Rahman, Fatama Binta Rafiq,
Tapushe Rabaya Toma, Syeda Sumbul Hossain
and Khalid Been Badruzzaman Biplob

Abstract In the field of information security, phishing URLs detection and prevention has recently become egregious. For detecting, phishing attacks several anti-phishing systems have already been proposed by researchers. The performance of those systems can be affected due to the lack of proper selection of machine learning classifiers along with the types of feature sets. A details investigation on machine learning classifiers (KNN, DT, SVM, RF, ERT and GBT) along with three publicly available datasets with multidimensional feature sets have been presented on this paper. The performance of the classifiers has been evaluated by confusion matrix, precision, recall, F1-score, accuracy and misclassification rate. The best output obtained from Random Forest and Extremely Randomized Tree with dataset one and three (binary class feature set) of 97% and 98% accuracy accordingly. In multiclass feature set (dataset two), Gradient Boosting Tree provides highest performance with 92% accuracy.

Keywords Phishing · Malicious URLs · Anti-Phishing · Phishing detection

S. S. M. M. Rahman (✉) · F. B. Rafiq · T. R. Toma · S. S. Hossain · K. B. B. Biplob
Department of Software Engineering, Daffodil International University, Dhaka, Bangladesh
e-mail: motiur.swe@diu.edu.bd

F. B. Rafiq
e-mail: fatama.swe@diu.edu.bd

T. R. Toma
e-mail: toma.swe@diu.edu.bd

S. S. Hossain
e-mail: syeda.swe@diu.edu.bd

K. B. B. Biplob
e-mail: khalid@daffodilvarsity.edu.bd

© Springer Nature Singapore Pte Ltd. 2020
K. S. Raju et al. (eds.), *Data Engineering and Communication Technology*,
Advances in Intelligent Systems and Computing 1079,
https://doi.org/10.1007/978-981-15-1097-7_25

1 Introduction

Attackers are getting attracted by phishing which is a criminal activities-based scheme. Phishing is designed to convince users into exposing their sensitive information such as credit card information, pin numbers, login credentials and personal information. The collected information through phishing is used in gaining financial access and committing fraudulent activities [1, 2]. Phishing performed by sending malicious Uniform Resource Locator (URL) in e-mail or using other social communication medium. Typically, a malicious link or URL is being sent by cybercriminals to victims within a message in e-mails, private chat message, blogs, and forums as well as on banners which are representing legitimate sources. Phishing attacks exceed annually 3 billion dollar financial losses based according to the recent estimation [3]. There are three ways by which phishing can occur [4]. One, by making a duplicate web interface as like as other trusted Web sites so that users will submit the sensitive information which is defined as web-based phishing [5]. Two, by sending e-mails to victims which is defined as e-mail-based phishing in which web-based phishing can involve [5]. Another way to inject malicious code in a legitimate Web site and a malicious software will be installed on the user's system if users visit the site which is known as Malware-based phishing [5]. Web-based phishing is being considered in this research work to accurately identify phishing URLs by identifying the most relevant subset of features using multiple machine learning classifiers. The main contributions of this paper can be stated as follows:

(i) In this presented work, the performance of multiple machine learning classifiers as well as the ensemble methods has been investigated.

(ii) The machine learning classifiers K-Nearest Neighbors (KNN), Decision Tree (DT) and Support Vector Machine (SVM) as well as the ensemble classifiers Random Forest (RF), Extremely Randomized Tree (ERT) and Gradient Boosting Tree (GBT) are trained for classifying the phishing URL and evaluated.

(iii) A details assessment and evaluation of different classifiers have been presented in this paper perspective of multiple datasets which will help others to determine the classifiers during the development of anti-phishing intelligent tools.

(iv) F1-score, precision, recall, confusion matrix and accuracy are used to evaluate the effectiveness of classification algorithms.

(v) Misclassification rate has been also calculated.

The organization of the rest of the paper is constructed as: Related works have been presented those are related to our research works in Sect. 2. In Sect. 3, the methodology and the dataset information which is used to investigate have been described. Experiments, evaluation parameter and results have been described and discussed in Sect. 4. Finally, Sect. 5 concludes with a future steps.

2 Literature Review

A real-time, anti-phishing system has been proposed by Sahingoz, O. K., Buber, E., Demir, O., and Diri, B. which uses seven different classification algorithms along with different feature sets. They found that 97.98% accuracy rate for detection of phishing URLs by Random Forest algorithm with NLP-based features [6]. Based on Uniform Resource Locator (URL) features another anti-phishing system named PHISH-SAFE that has proposed with more than 90% accuracy in detecting phishing Web sites using SVM classifier [7]. An approach based on minimum time to detect phishing URL using classification technique is described by Gupta, S., and Singhal, A. They claimed that random tree is a good classification technique in comparison with others [8]. A different methodology has been proposed to detect phishing Web sites by Parekh et al. [9]. In their model, there started with parsing then classified with heuristic classification of data. Random Forest was chosen for classification and found that the accuracy level of Random Forest to be the highest around 95%. Another anti-phishing technique based on a weighted URL tokens system by extracting identity keywords from a query webpage has been presented in one study. A search engine was invoked to pinpoint the target domain name using the identity keywords as search terms and to determine the legitimacy of the query webpage. The model was proposed by Tan and Chiew [10] which was able to detect phishing webpages without using conventional language-dependent key-words extraction algorithms. Zouina and Outtaj [11] have presented a phishing detection approach which is lightweight claimed by them. The approach was completely based on the URL (Uniform Resource Locator) with only six URL features. 95.80% recognition rate was produced by their proposed system Ahmed and Abdullah [12] worked on a detection technique of phishing Web sites based on checking Uniform Resource Locators (URLs) of webpages. In their final result, it was shown that phishing webpages can be detected by PhishChecker with accuracy of 0.96, and the false-negative rates not more than 0.105.

3 Methodology

3.1 Dataset Used

For this assessment, three publicly available datasets have been used. One dataset (D1) has 11,055 different Web sites in which 4,898 are phishing and 6,157 are legitimate Web sites where 30 features are considered to classify [13]. Other dataset (D2) has 1,353 records of different Web sites in which 548 Web sites are legitimate, 702 are phishing, and 103 suspicious where only ten features are considered to identify [14]. For dataset one, 1 as legitimate and −1 as phishing are considered. 1 as legitimate, 0 as suspicious and −1 as phishing have been considered from dataset two. Another recently updated dataset (D3) has been used which consists of

10,000 different Web sites information in which 5,000 phishing and 5,000 legitimate webpages. Total 48 features were extracted from those webpages. This dataset was downloaded within (January–May), 2015 and (May–June), 2017. D1 and D2 considered only binary type features or attributes but D3 considered binary, discrete, categorical and continuous types of attributes. The features or attributes used in D1, D2 and D3 are described in [15, 16, 18], respectively.

3.2 Used Classifiers for Assessment

Multiple classifiers have been assessed and evaluated in this investigation. As labeled datasets are being used, supervised machine learning classifiers are considered as well as ensemble methods during the implementation. The classifiers used in this work are: K-Nearest Neighbors Classifier (KNN) [17], Decision Tree Classifier (DT) [18], Support Vector Machine (SVM) [19], Random Forest Classifier (Ensemble Method—RF), Extremely Randomized Tree (Ensemble Method—ERT) and Gradient Boosting Classifier (Ensemble Method—GBT) [20, 21].

3.3 The Architecture of Experimented Approach

This subsection will depict the overview of the architecture of implemented approach in Fig. 1. To evaluate the classifiers, a set of dataset being load from a CSV file and from that file the features have been loaded to split for train the classifiers. 65% data has been used to train the classifiers and then test on 35% data whether the classifier can identify or not.

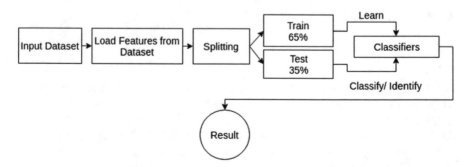

Fig. 1 The procedure of the experiment

4 Results Analysis and Discussion

4.1 Environment Setup

The experiment has been conducted on a machine which configuration is Intel(R) Core(TM) i5-6500 CPU @ 3.20 GHz processor, 64-bit PC along with 16 GB RAM. Ubuntu 18.04.1 LTS (Bionic Beaver) is the operating system. A well-established programming language named python has been used along with scikit-learn [17], SciPy, panda and numpy packages.

4.2 Evaluation Parameter

In this study, mostly binary decision has been focused which classifies the data as either phishing or legitimate. The decision of the classifiers can be expressed using a structure called confusion matrix. Confusion matrix consists of four attributes (Table 1).

Precision, recall, F1-score and accuracy have been used to evaluate the performance of classifiers. Some metrics are used to evaluate the assessment which is as outlined in Table 2.

4.3 Result Analysis and Discussion

This section has three parts: Case Study #1, Case Study #2 and Case Study #3. These studies based on three observations; the performance assessment for dataset one (D1) is the first observation, then for dataset two (D2) is another observation and final observation is the performance evaluation for dataset three (D3).

Table 1 Structure of confusion matrix

	Predicted label	
Actual label	True negatives (TN)	False positives (FP)
	Identified phishing as phishing correctly	Identified legitimate as phishing
	False negative (FN)	True positives (TP)
	Identified phishing as legitimate correctly	Identified the legitimate as legitimate correctly

Table 2 Evaluation parameters for assessment of the classifiers [19]

Evaluation parameter	Description	Formula
Accuracy	The ratio of correctly classified data to total data	Accuracy = (TP + TN)/ (TP + TN + FP + FN)
Precision	The measurement of the exact value of the result provided by the classifier	Positive precision = TP/ (TP + FP) Negative precision = TN/ (TN + FN)
Recall	The measurement of the thoroughness of the result provided by classifier	Positive recall = TP/ (TP + FN) Negative recall = TN/ (TN + FP)
F1-score	The harmonic mean of exactness and completeness	F1-score = (2 * Precision * Recall)/(Precision + Recall)
Misclassification Rate	Known as "Error Rate"	Misclassification Rate = (FP + FN)/ (TP + TN + FP + FN) or 1— Accuracy

4.3.1 Case Study #1

Figure 2 represents the confusion matrix generated from the classifiers from which the evaluation parameters accuracy, precision, recall, F1-score and error rate have been calculated. As the D1 has 11055 URLs and 65% means total around 7185 URLs used for training and rest 3870 URLs used to be tested where 1700 URLs are phishing and 2170 are legitimate. It has been found that for phishing identification total 1579, 1636, 1570, 1633, 1643 and 1592 number of URLs have been correctly detected and total 121, 64, 130, 67, 57 and 108 number of URLs are misclassified by KNN, DT, SVM, RF, ERT and GBT, respectively. From legitimate identification, KNN, DT, SVM, RF, ERT and GBT correctly identified total 2060, 2081, 2103, 2137, 2116 and 2090 number of URLs and the total number of misclassified URLs are 110, 89, 67, 33, 54 and 80, respectively. It has been obtained that to identify the phishing URLs ERT performs better and RF performs better to identify the legitimate URLs.

Figure 3 depicts the precision, recall and F1-score where also it can be figured out in transparent that in case of phishing (red color) ERT and for legitimate (green color) RF provides better performance. It has been recapitulated that RF performs better to identify the legitimate and ERT for phishing URLs.

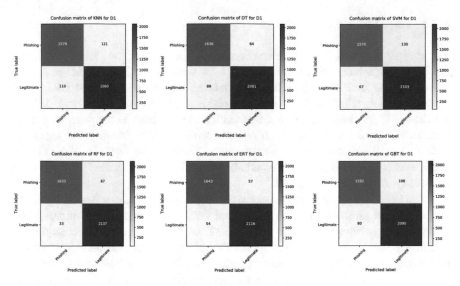

Fig. 2 The confusion matrix of the classifiers of D1

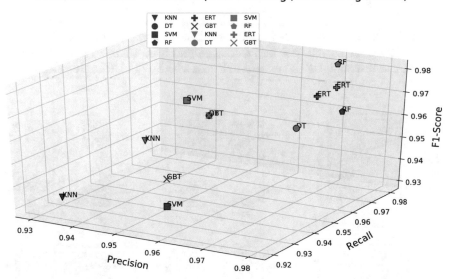

Fig. 3 Precision, recall and F1-score of the classifiers for D1

4.3.2 Case Study #2

The confusion matrix from the classifiers has depicted in Fig. 4. Since the D2 has phishing, suspicious and legitimate URLs as well as 1353 URLs in total, 65% means total around 879 URLs used for training and for test rest 474 URLs have used. In which where 254 URLs are phishing, 32 are suspicious and 188 are legitimate. It has been found that to identify the phishing along with suspicious URLs GBT performs better and SVM performs better to identify the legitimate URLs but SVM has poor performance rather than RF to classify the suspicious URLs. RF has almost equal ability to identify legitimate URLs and more to identify suspicious than SVM.

From Fig. 5 which depicts the precision, recall and F1-score, it can be visualized that in case of phishing and suspicious (red color) GBT and for legitimate (green color) SVM performs better. In sum, GBT in case of multiclass classification performs better to detect the phishing and suspicious URLs. On the other hand, in multiclass detection based on overall evaluation, RF provides better performance to classify legitimate URLs along with suspicious URLs. So, here RF wins the race with SVM.

4.3.3 Case Study #3

Figure 6 is the visualization of generated confusion matrix from the classifiers. The D3 has 10000 URLs and 65% means total around 6500 URLs have been used for training and rest 3500 URLs are used to detect where 1744 URLs are phishing and

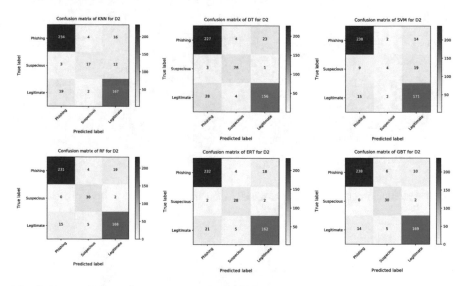

Fig. 4 The confusion matrix of the classifiers of D2

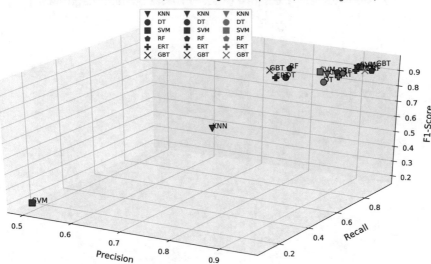

Fig. 5 Precision, recall and F1-score of the classifiers for D2

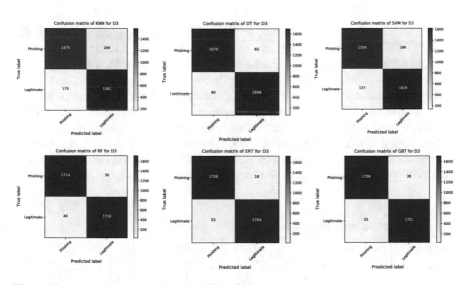

Fig. 6 The confusion matrix of the classifiers of D3

1756 are legitimate. From the confusion matrix, it has been stated that ERT identified maximum number of phishing URLs in summed 1726 and RF detected 1716 of legitimate URLs. The total number of misclassified URLs by ERT and RF are same in general.

Precision - Recall - F1-Score (Red: Phishing , Green: Legitimate)

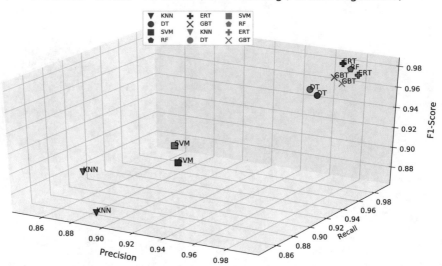

Fig. 7 Precision, recall and F1-score of the classifiers for D3

Figure 7 represents the precision, recall and F1-score where also it can be figured out that ERT for phishing (red color) and RF for legitimate (green color) provides better performance. It has been recapitulated that ERT and RF perform better and almost same to identify the legitimate and phishing URLs.

The performance comparison in case of accuracy and misclassification rate has been tabulated in Table 3. For binary class identification, it has been observed that Random Forest (RF) and Extremely Randomized Tree (ERT) both are fitted well with accuracy 97% and 98% with D1 and D3 accordingly along with 3% and 2% misclassification rate. On the other hand, for multiclass detection Gradient Boosting Tree (GBT) played well with 92% accuracy and 8% error rate.

Table 3 Evaluation parameters for assessment of the classifiers

Classifiers	Accuracy (D1)	Error rate (D1)	Accuracy (D2)	Error rate (D2)	Accuracy (D3)	Error rate (D3)
KNN	0.94	0.06	0.88	0.12	0.87	0.13
DT	0.96	0.04	0.87	0.13	0.96	0.04
SVM	0.95	0.05	0.87	0.13	0.91	0.09
RF	0.97	0.03	0.91	0.09	0.98	0.02
ERT	0.97	0.03	0.89	0.11	0.98	0.02
GBT	0.95	0.05	0.92	0.08	0.97	0.03

5 Conclusion

In order to develop anti-phishing tools or system, this investigation will lead the basement. It can be determined that Random Forest and Extremely Randomized Tree perform better for binary classification-based detection tools. They provide 97% and 98% accuracy with dataset one and two accordingly. Another finding is that latest dataset with more features in total 48 provides better learning base to the learning algorithms where number of URLs was less then dataset one. On the other side, Gradient Boosting Tree performed well in multiclass dataset with 92% accuracy where Random Forest also performs with nearest accuracy of 91%. In recapitulate, the performance of ensemble methods especially Random Forest algorithm provides a strong basement to anti-phishing system development. In the future, this study will continue to develop a real-time anti-phishing system.

References

1. Abutair, H.Y., Belghith, A.: Using case-based reasoning for phishing detection. Procedia Comput. Sci. **109**, 281–288 (2017)
2. Tan, C.L., Chiew, K.L.: Phishing website detection using URL-assisted brand name weighting system. In: 2014 International Symposium on Intelligent Signal Processing and Communication Systems (ISPACS), pp. 054–059. IEEE (2014)
3. Shirazi, H., Bezawada, B., Ray, I.: KnOw Thy Domaln Name: unbiased phishing detection using domain name based features. In: Proceedings of the 23rd ACM on Symposium on Access Control Models and Technologies, pp. 69–75. ACM (2018)
4. Hutchinson, S., Zhang, Z., Liu, Q.: Detecting phishing websites with random forest. In: International Conference on Machine Learning and Intelligent Communications, pp. 470–479. Springer, Cham (2018)
5. Dong, Z., Kapadia, A., Blythe, J., Camp, L.J.: Beyond the lock icon: real-time detection of phishing websites using public key certificates. In: 2015 APWG Symposium on Electronic Crime Research (eCrime), pp. 1–12. IEEE (2015)
6. Sahingoz, O.K., Buber, E., Demir, O., Diri, B.: Machine learning based phishing detection from URLs. Expert Syst. Appl. **117**, 345–357 (2019)
7. Jain, A.K., Gupta, B.B.: PHISH-SAFE: URL features-based phishing detection system using machine learning. In: Cyber Security: Proceedings of CSI 2015, pp. 467–474. Springer, Singapore (2018)
8. Gupta, S., Singhal, A.: Dynamic classification mining techniques for predicting phishing URL. In: Soft Computing: Theories and Applications, pp. 537–546. Springer, Singapore (2018)
9. Parekh, S., Parikh, D., Kotak, S., Sankhe, S.: A new method for detection of phishing websites: URL detection. In: 2018 Second International Conference on Inventive Communication and Computational Technologies (ICICCT), pp. 949–952. IEEE (2018)
10. Tan, C.L., Chiew, K.L.: Phishing webpage detection using weighted URL tokens for identity keywords retrieval. In: 9th International Conference on Robotic, Vision, Signal Processing and Power Applications, pp. 133–139. Springer, Singapore (2017)
11. Zouina, M., Outtaj, B.: A novel lightweight URL phishing detection system using SVM and similarity index. Human-Centric Comput. Inf. Sci. **7**(1), 17 (2017)

12. Ahmed, A.A., Abdullah, N.A.: Real time detection of phishing websites. In: 2016 IEEE 7th Annual Information Technology, Electronics and Mobile Communication Conference (IEMCON), pp. 1–6. IEEE (2016)
13. Mohammad, R.M., Thabtah, F., McCluskey, L.: Intelligent rule-based phishing websites classification. IET Inf. Secur. **8**(3), 153–160. IEEE (2014)
14. Abdelhamid, N., Ayesh, A., Thabtah, F.: Phishing detection based associative classification data mining. Expert Syst. Appl. **41**(13), 5948–5959 (2014)
15. Tan, C.L.: Phishing dataset for machine learning: feature evaluation. Mendeley Data, v1 (2018). http://dx.doi.org/10.17632/h3cgnj8hft.1
16. Thaker, M., Parikh, M., Shetty, P., Neogi, V., Jaswal, S.: Detecting phishing websites using data mining. In: 2018 Second International Conference on Electronics, Communication and Aerospace Technology (ICECA), pp. 1876–1879. IEEE (2018)
17. Scikit-learn: Machine Learning in Python. https://scikit-learn.org/stable/. Last accessed 11 Nov 2018
18. Chiew, K.L., Tan, C.L., Wong, K., Yong, K.S., Tiong, W.K.: A new hybrid ensemble feature selection framework for machine learning-based phishing detection system. Inf. Sci. (2019)
19. Rahman, S.S.M.M., Rahman, M.H., Sarker, K., Rahman, M.S., Ahsan, N., Sarker, M.M.: Supervised ensemble machine learning aided performance evaluation of sentiment classification. In: Journal of Physics: Conference Series, Vol. 1060(1), p. 012036. IOP Publishing (2018)
20. Rahman, S.S.M.M., Saha, S.K.: StackDroid: Evaluation of a multi-level approach for detecting the malware on android using stacked generalization. In International Conference on Recent Trends in Image Processing and Pattern Recognition, pp. 611–623. Springer, Singapore (2018)
21. Rana, M.S., Rahman, S.S.M.M., Sung, A.H.: Evaluation of tree based machine learning classifiers for android malware detection. In International Conference on Computational Collective Intelligence, pp. 377–385. Springer, Cham (2018)

Parallel Queuing Model in a Dynamic Cloud Environment-Study of Impact on QoS: An Analytical Approach

Shahbaz Afzal, G. Kavitha and Shabnum Gull

Abstract Stabilizing the Quality of Service (QoS) currently being viewed as the main challenge in cloud computing environment and can be achieved either by employing an efficient queuing model or robust scheduling and allocation policies with load balancing features or combination of both. The QoS attributes are analyzed in a parallel queuing model featuring a pure analytical approach and extend enhancement to our previous work. The aim of the paper is to study behavior of hybrid parallel queuing model under conditions of three time varying input parameters: arrival rate, service rate, and number of servers on the response variables: queue length, waiting time, response time, and server utilization factor. With the change in these input parameters, the impact on response variables is analyzed to arrive at a valid conclusion.

Keywords Quality of service · Cloud computing · Parallel queuing model · Input parameters · Response variables · Arrival rate · Service rate · CSP · CSC · SLA

1 Introduction

With the advances in Internet technologies, cloud computing has emerged as a potential Internet-based utility during current technological era having capability of providing services to a vast collection of scalable users. It has provision of providing both hardware and software utilities together for developmental platforms

S. Afzal (✉) · G. Kavitha
Department of Information Technology, B. S. Abdur Rahman Crescent Institute of Science and Technology, Chennai 600048, India
e-mail: shabazafzaldar@gmail.com

G. Kavitha
e-mail: gkavitha78@gmail.com

S. Gull
Freelance Researcher, Chennai, India
e-mail: gullshabnum786@gmail.com

© Springer Nature Singapore Pte Ltd. 2020   297
K. S. Raju et al. (eds.), *Data Engineering and Communication Technology*,
Advances in Intelligent Systems and Computing 1079,
https://doi.org/10.1007/978-981-15-1097-7_26

and testing tools with infrastructure as a service (IaaS), software as a service (SaaS) and platform as a service (PaaS), while Internet-as a service (iaaS) being a backbone. From technological perspective, there are three primary actors on the scene, the cloud service provider (CSP), the cloud service consumer (CSC), and the Internet while many actors behind the scene in backend [1–11]. Internet being the communication medium, CSP provides necessary services in the form of rented scalable virtual machines to the CSC at some nominal price on guaranteed levels of QoS as agreed between the two stakeholders in a proper documented service level agreement (SLA).

The techniques of virtualization make it possible to transform a single user physical machine into a multiple of shared multiuser virtual machines with variant of computing configurations in terms of operating system, memory, CPU power, number of cores, storage, SSD, operating frequency, network bandwidth, and much more, hence enabling better resource utilization with small wastages [12–14].

The present era of 4G and 5G communication technologies also played a major role in pushing the various profit and non-profit organizations and individual users into cloud making it the buzzword among the present business technologies. On the other hand with the growth of Internet of things (IoT), e-commerce, e-health, e-governance, e-education, and other Internet-enabled online services, big data, and data analytics, voluminous amount of data is being generated to be processed and stored making cloud computing the only versatile and cost-effective option of handling such huge dimensions of data. The phenomena are further made complex by Moore's law-based increase in the number of smart devices, laptops, tablets, and other data processing devices. The cloud users and data-generating devices are increasing at an exponential rate while the computing resources at cloud providers side increase in a linear way leaving in between unparallel diverging gaps which cannot be bridged.

Technically, when a huge number of user tasks requesting service are arriving at provider, there is a direct consequence on the deliverable Quality of Service in terms of performance, deadline, cost, efficiency, profit, etc., for both provider and consumer. To preserve QoS according to SLA keeping in mind provider's turnover and other provider associated metrics, it is essential to go for an efficient system model. This can be achieved by either developing an efficient queuing model or an efficient scheduling and allocation algorithm (with or without load balancing features) or combination of both. While queuing models deal with enhancement of queuing metrics before actual execution of user tasks [15–19], scheduling and allocation algorithms deals with augmentation of scheduling metrics once the user tasks are sent for execution on virtual machines. The scheduling and allocation algorithms fall under the categories of task scheduling, task allocation, resource scheduling, resource allocation, VM scheduling, and VM allocation approaches [20–26]. The present paper focuses on the queuing model approach to improving QoS in cloud computing. This paper does not deal with scheduling and allocation part of reinforcing the QoS metrics in cloud computing. From queuing model perspective, the generic cloud computing architecture is based on the queuing theory model.

The objective of current work is to analyze effect on QoS in a hybrid parallel queuing model implemented through the joint use of M/M/1: ∞/FCFS and M/M/s: N/FCFS as proposed by Shahbaz and Kavitha [27]. In a hybrid parallel queuing model, an infinite global queue is enforced at the datacenter level followed by finite local parallel queues at each of the virtual machines.

The rest of paper is structured as follows. Section 2 discusses literature survey, and Sect. 3 presents brief overview about the queuing theory. The hybrid parallel queuing model-based cloud computing is reviewed in Sect. 4. Section 5 provides data collection. The computational results are provided in Sect. 6. Section 7 discusses the research findings while Sect. 8 concludes our research work and points out future work.

2 Literature Survey

Improving QoS parameters in cloud computing has been a marathon among the researchers since its birth to ensure performance, efficiency, fault tolerance, scalability, dynamism, utilization, robustness, and economy. Despite a lot of milestones been reached, no one has broken a leg and still the race is on. However, on the one hand giant leaps have been made to improve these parameters through scheduling strategies and allocation policies; on the other hand, little attention was paid toward the queuing side to improve these parameters. The concern of the paper is to have birds view on the queuing side of the cloud computing hopping the scheduling part which is not discussed in this paper.

Shahbaz and Kavitha analyzed the behavior of parallel queuing model with a strictly increasing arrival rate as a special case to observe the impact on QoS parameters [27]. It is evident when arrival rate is below the service rate (keeping other input parameters constant), and there is no deviation in waiting time and response time which is limited by parallel queues; however, as arrival rate overshoots service rate, waiting time and response time show a tremendous increase. In contrast with the existing queuing models, the analytical observations proved it much more efficient in terms of achieving better QoS. The paper does not consider impact on QoS under dynamic input conditions which is more realistic. Jordi Vilaplana et al. studied impact on waiting time and response time of tasks in queue by varying the mean arrival rate 'λ' with time keeping the mean service rate 'μ' constant. Queuing theory and open Jackson's networks were selected as the benchmark to guarantee a desired level of performance. The results show that as the arrival rate increases the response time also increases proportionally. However, the response time decreases as the number of servers increase, and the decrease in value of response time is achieved at lower values of servers after which it stabilizes quickly. The authors also identified the limitation of their model which is 'servers as the bottlenecks.' The limitation is eliminated in our work by employing parallel queues [15]. Kirsal et al. proposed an analytical modeling and performability analysis for cloud computing using queuing system. The QoS measurements are determined using the Markov reward model (MRM). The model consists of 'C' number of servers which allow 'n' requests at a time 't.' The buffer capacity of server is assumed to be infinite. In the proposed model, the failure and repair of the datacenter are considered. The paper suffers from a weakness of contemplating single scenario and single output parameter of cloud computing-based queuing model in which effect on mean queue length is studied while increasing the arrival rate of user tasks keeping the service rate ($\mu = 500$ requests/sec) constant along with the number of servers which is fixed at 10 [28]. Eisa M. et al. proposed the enhancing cloud computing scheduling-based queuing model to improve the QoS by employing single queue multiple server model to reduce the queue length, waiting time and to increase the server utilization rate [29]. Sowjanya T. S. et al. proposed a queuing theory approach to reduce waiting time in cloud computing. The model used to find the waiting time with variation in arrival and service rates, and the experiments were conducted only on one and two servers, respectively, using M/M/1 and M/M/2 queuing models [30]. Suneeta Mohanty et al. applied queuing approach to reduce waiting time in cloud computing by employing M/M/c model and Erlang's model. Results were conducted on different values of arrival rate (λ), service rate (μ), and number of servers (s) [31]. Suresh Varma et al. calculated the effect on mean number of packets, queue length, throughput, and mean delay assuming constant arrival and service rates at different time instances in one experiment while changing only the arrival rate in second experiment keeping other two parameters constant [32].

3 Queuing Theory Fundamentals

A queuing framework consists of collection of customers to be served on service facilities or vice versa and the same is applicable to cloud computing where a multitude of user tasks arrive at the virtualized computing servers for their culmination. The queuing models are portrayed in terms of arrival process, service mechanism, and queue discipline where arrival process governs the incoming pattern of cloud tasks with time, service mechanism deals with way of servicing, and exit from the service system. The queue discipline stipulates the arrangement of tasks from queue for execution. In most of the real-time situations, arrival process usually follows the exponential distribution sometimes Poisson or Erlang distributions [28, 33–35]. The queuing model for cloud computing depicted in this work assumes that the arrival process follows a Poisson distribution under the assumption that net inflow of task arrivals in a cloud computing system at any given point of time is independent of the number of arrivals that have already occurred prior to the beginning of the time interval [35].

4 Methodology of Hybrid Parallel Queuing Model-Based Cloud Computing

The cloud computing model with respect to the queuing model is represented by a hybrid parallel queuing model consisting of one global queue and multiple parallel local queues as depicted in Fig. 1 [27, 36]. Ideally, the global queue is considered to have infinite queuing capacity and corresponds to the datacenter controller in the cloud model. The global queue is implemented using the M/M/1: ∞/FCFS, symbols having the usual meanings as in Kendall notations. The characteristics of the queuing model deployed at datacenter controller are governed by equations of M/M/1: ∞/FCFS [35].

The user tasks from datacenter controller are sent to global scheduler for scheduling in virtual machines. Before actual scheduling, global scheduler dispatches tasks to local queues maintained by each of the virtual machines. From local queues, the tasks are scheduled to actual computing resources for execution by the local schedulers. Each of local queues is implemented by adopting the M/M/s: N model, where N is the maximum finite capacity of each local queue in terms of cloud tasks.

After implementing analytical approach, following queuing characteristics are obtained in the hybrid parallel model.

The average queue length of each of the local queue is given by

$$L_q = \frac{(s\rho)^s \rho}{(s! s^n)(1-\rho)^2} P_o \left[1 - \rho^{N-s+1} - (1-\rho)(N-s+1)\rho^{N-s}\right]. \tag{1}$$

The expected number of tasks in the system is given

$$L_s = L_q + \left(\frac{\lambda}{s\mu}\right)(1 - P_N) \tag{2}$$

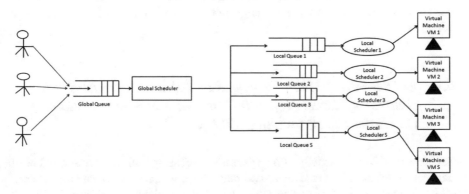

Fig. 1 Hybrid parallel queuing model-based cloud computing

The expected waiting time in the system is calculated as

$$W_s = \frac{L_s}{\lambda/s(1 - P_N)} \qquad (3)$$

The expected waiting time in the queue is calculated as

$$W_q = W_s - \frac{1}{s\mu} \qquad (4)$$

The server utilization rate of the system is calculated as

$$\rho = \frac{\lambda}{s\mu} \qquad (5)$$

5 Data Collection

The data collected in this paper has not been gathered from any real-time data source or from any cloud provider because of the diverse objectives. The data collected is assumed data meant for experimental observation only to study the behavior of hybrid parallel queuing model under varying conditions.

6 Computational Results

The results have been generated and tested for 22 different possible scenarios in cloud computing environment for a given number of tasks (n) and maximum number of tasks allowed in the system (N) where the values of n and N are incremented at each observation level. Further each condition is tested for two possibilities when mean service rate is greater or equal to mean arrival rate and vice versa.

6.1 At Increasing Mean Service Rate (μ) and Increasing Number of Servers (s) with Constant Mean Arrival Rate (λ) Under the Assumption that μ ≥ λ

The task completion capacity of the system is increasing with respect to number of parallel virtual machines at static arrival rate inferring mean service rate is greater or equal to mean arrival rate. The effect on output variables is plotted in Fig. 2.

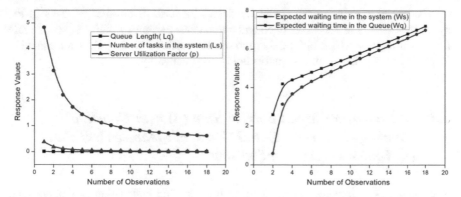

Fig. 2 Response values at increasing μ and s with constant λ at $\mu \geq \lambda$

The queue length is maintained at zero level, number of tasks in system decreases monotonously, and the server utilization factor also decreases as portrayed in Fig. 2. The expected waiting time in the system and the expected waiting time in the queue which according to Fig. 2 should decrease behave conversely and increases. This is because at each observation point, n and N are elevated in the model.

6.2 At Increasing Mean Service Rate (μ) and Increasing Number of Servers (s) with Constant Mean Arrival Rate (λ) Under the Assumption that μ ≤ λ

With these conditions when mean service rate is less or equal to arrival rate, server utilization factor drops rapidly, the queue length attain steady state at zero level, the number of tasks in the system goes beyond zero and achieves negative values while expected waiting time in the system and expected waiting time in queue increase drastically as presented in Fig. 3, respectively.

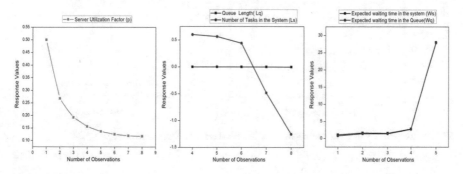

Fig. 3 Response values at increasing μ and s with constant λ at $\mu \leq \lambda$

The negative values of L_s radically depict increase in number of tasks in the system which can never be serviced and approaches to minus infinity as service rate falls continuously which ultimately give rise to increased waiting time approaching to plus infinity. Hence, this scenario holds no practicality.

6.3 At Increasing Mean Service Rate (μ) with Constant Number of Servers (s) and Constant Mean Arrival Rate (λ) Under the Assumption that μ ≤ λ

The mean service rate keeps increasing with other two inputs being kept at constant state assuming mean service rate is less or equal to mean arrival rate. Figure 4 interprets almost zero server utilization factor and zero queue length with the number of tasks in the system increases negative exponentially. Figure 4 further illustrates that waiting times go toward positive exponential. Under this circumstance, the queuing model is also of no real-time use.

6.4 At Increasing Mean Service Rate (μ) with Constant Number of Servers (s) and Constant Mean Arrival Rate (λ) Under the Assumption that μ ≥ λ

Figure 5 shows response values of output parameters under the assumption that mean service rate is greater or equal to mean arrival rate. It is evident from graphs that queue length is zero, and server utilization factor is close to zero while the number of tasks in the system shows the increasing trend. The system and queue

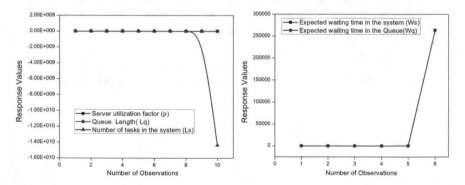

Fig. 4 Response values at increasing μ with constant s and constant λ at $\mu \leq \lambda$

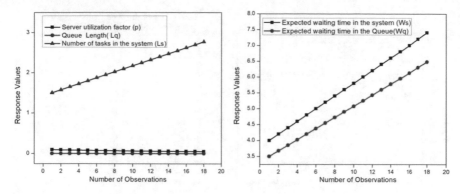

Fig. 5 Response values at increasing μ with constant s and constant λ at $\mu \geq \lambda$

waiting times also increase linearly. The graph further holds the evidence that here waiting time is directly proportional to number of tasks in the system.

6.5 At Increasing Mean Service Rate (μ) and Increasing Mean Arrival Rate (λ) with Constant Number of Servers (s) Under the Assumption that $\mu \geq \lambda$

The situation discusses the impact on QoS when mean service rate and mean arrival rate are both increasing while the number of servers is being fixed with service rate greater or equal to arrival rate. Figure 6 shows that system behaves exactly with one discussed in Sect. 6.4.

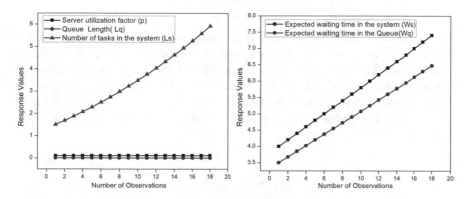

Fig. 6 Response values at increasing μ and λ with constant s at $\mu \geq \lambda$

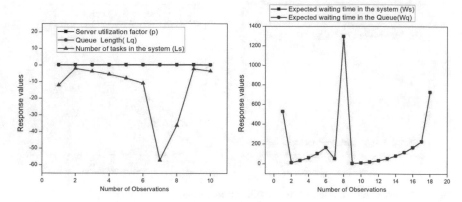

Fig. 7 Response values at increasing μ and λ with constant s at $\mu \leq \lambda$

6.6 At Increasing Mean Service Rate (μ) and Increasing Mean Arrival Rate (λ) with Constant Number of Servers (s) Under the Assumption that μ ≤ λ

It is evident from Fig. 7 that the number of tasks in the system oscillates between negative values likewise the waiting time also shows unusual behavior corresponding to the input variables. The queuing model under these conditions does not guarantee any practicality.

6.7 At Increasing Mean Arrival Rate (λ) and Increasing Number of Servers (s) with Constant Mean Service Rate (μ) Under the Assumption that λ ≥ μ

The section describes the phenomena at increasing mean arrival rate and number of servers while mean service rate being kept static such that mean arrival rate is greater or equal to mean service rate. The number of tasks in the system decreases following a decaying exponential curve with no queue formed and server utilization factor tending to zero. At the same time, system waiting time achieves steady state while queue waiting time attains logarithmic growth, respectively, as characterized in Fig. 8.

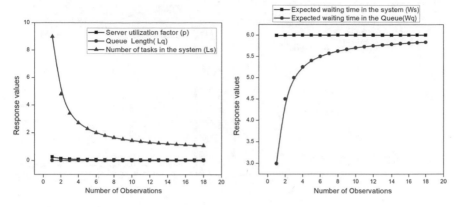

Fig. 8 Response values at increasing λ and s with constant μ at $\lambda \geq \mu$

6.8 At Increasing Mean Arrival Rate (λ) and Increasing Number of Servers (s) with Constant Mean Service Rate (μ) Under the Assumption that $\lambda \leq \mu$

As illustrated in Fig. 9, the queuing model under these factors perform exactly the dupe of Sect. 6.7 as long as mean arrival rate is less than or equal to mean service rate.

6.9 At Increasing Mean Arrival Rate (λ) with Constant Number of Servers (s) and Constant Mean Service Rate (μ) Under the Assumption that $\lambda \leq \mu$

The queuing model is viewed at increasing 'λ' at constant 's' and constant 'μ' assuming arrival rate is less or equal to service rate. The experimental observations

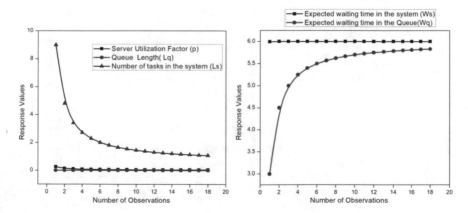

Fig. 9 Response values at increasing λ and s with constant μ at $\lambda \leq \mu$

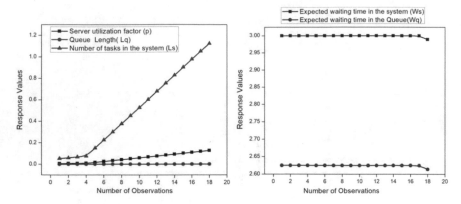

Fig. 10 Response values at increasing λ with constant s and μ at $\lambda \leq \mu$

show a linear increase in number of tasks in system, slight increase in server utilization factor with queue length being maintained at zero level, respectively. On the other hand, the waiting times show a steady state which later falls down as shown in Fig. 10.

6.10 At Increasing Mean Arrival Rate (λ) with Constant Number of Servers (s) and Constant Mean Service Rate (μ) Under the Assumption that $\lambda \geq \mu$

The server utilization factor and queue length are stabilized at zero level while queue length approaches zero making its way to fourth quadrant. Meanwhile waiting times acquire exponential values tending to infinity corresponding to the number of tasks in the system as illustrated in Fig. 11. The inferences drawn in favor of this action of queue model under these conditions suggests it to be impractical.

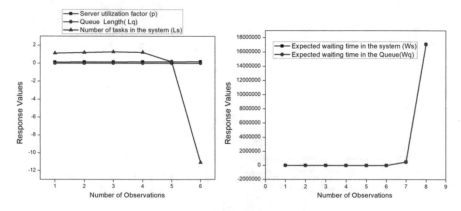

Fig. 11 Response values at increasing λ with constant s and μ at $\lambda \geq \mu$

6.11 At Increasing Mean Arrival Rate (λ) with Constant Number of Servers (s) and Decreasing Mean Service Rate (μ) Under the Assumption that $\lambda \geq \mu$

Under these aspects, the number of tasks achieves higher negative values approaching to negative infinity after some point with no queue formation and no server utilization factor. Concurrently, the waiting times achieve higher positive values approaching to positive infinity as plotted in Fig. 12 again rendering the model unfit for practical application.

6.12 At Increasing Mean Arrival Rate (λ) with Constant Number of Servers (s) and Decreasing Mean Service Rate (μ) Under the Assumption that $\lambda \leq \mu$

Figure 13 interprets number of tasks increase gradually until it gains a uniform slope with queue length being stabilized at zero and server utilization slightly grows above the zero level. Meanwhile the waiting times follows a decreasing linear slope with converging nature between the two.

6.13 At Decreasing Mean Arrival Rate (λ) and Increasing Number of Servers (s) with Constant Mean Service Rate (μ) Under the Assumption that $\lambda \leq \mu$

The queue length is maintained at zero level with the number of tasks decrease monotonically while the server utilization factor follows negative exponential curve

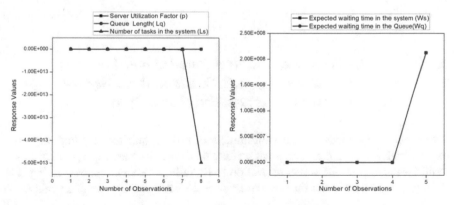

Fig. 12 Response values at increasing λ with constant s and decreasing μ at $\lambda \geq \mu$

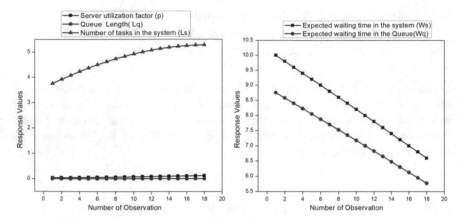

Fig. 13 Response values at increasing λ with constant s and decreasing μ at $\lambda \leq \mu$

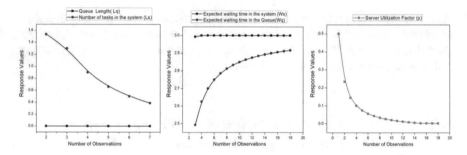

Fig. 14 Response values at decreasing λ and increasing s with constant μ at $\lambda \leq \mu$

with the value tending to zero. At the same time, queue waiting time achieves a logarithmic curve while the system waiting time achieves a constant line as depicted in Fig. 14.

6.14 At Decreasing Mean Arrival Rate (λ) and Increasing Number of Servers (s) with Constant Mean Service Rate (μ) Under the Assumption that $\lambda \geq \mu$

The queue is not formed at all with number of tasks acquiring very high-negative exponential values at the same point waiting times acquire very high-positive exponential values while the server utilization factor falls gradually as shown in Fig. 15. The queuing model under these circumstances does not possess any practicality.

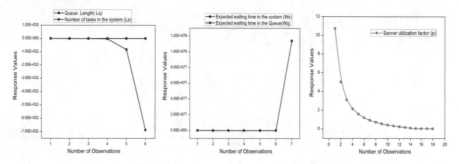

Fig. 15 Response values at decreasing λ and increasing s with constant μ at $\lambda \geq \mu$

6.15 At Decreasing Mean Arrival Rate (λ) with Constant Number of Servers (s) and Constant Mean Service Rate (μ) Under the Assumption that λ ≤ μ

The system under conditions of decreasing λ at constant s and constant μ with arrival rate less than or equal to service rate shows no formation of queue length, linear decrease in number of tasks in the system and server utilization factor which later on goes uniform. On the other hand, the waiting times attain an instant uniform straight line parallel to horizontal axis with a handsome gradient between the two as interpreted in Fig. 16.

6.16 At Decreasing Mean Arrival Rate (λ) with Constant Number of Servers (s) and Constant Mean Service Rate (μ) Under the Assumption that λ ≥ μ

The parallel queuing model under these factors attains infinity values for number of tasks in the system and waiting times with no formation of queue as in Fig. 17.

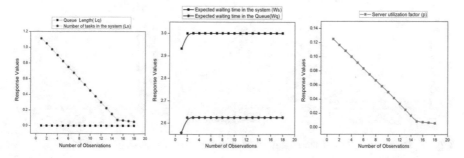

Fig. 16 Response values at decreasing λ with constant s and constant μ at $\lambda \leq \mu$

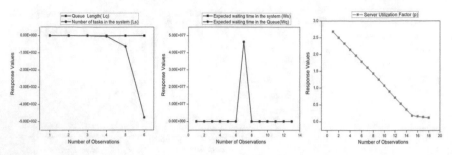

Fig. 17 Response values at decreasing λ with constant s and constant μ at $\lambda \geq \mu$

6.17   At Decreasing Mean Arrival Rate (λ) with Constant Number of Servers (s) and Decreasing Mean Service Rate (μ) Under the Assumption that $\lambda \leq \mu$

With this scenario, queue is not formed and the number of tasks in the system decreases progressively toward zero level while the waiting times follow the decreasing linear plot with the two trying to coincide with each other. The server utilization factor acquires a parabolic path as shown in Fig. 18.

6.18   At Decreasing Mean Arrival Rate (λ) with Constant Number of Servers (s) and Decreasing Mean Service Rate (μ) Under the Assumption that $\lambda \geq \mu$

The queue is not formed with number of tasks falling moderately toward zero level trying to cross horizontal axis while the waiting times show a sharp decrease in the values. The server utilization factor initially increases continuously, then suddenly falls and again increases as shown in Fig. 19.

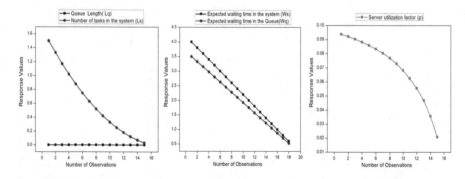

Fig. 18 Response values at decreasing λ with constant s and decreasing μ at $\lambda \leq \mu$

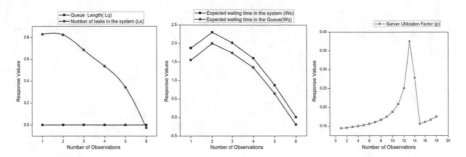

Fig. 19 Response values at decreasing λ with constant s and decreasing μ at $\lambda \geq \mu$

6.19 At Decreasing Mean Service Rate (μ) and Increasing Number of Servers (s) with Constant Mean Arrival Rate (λ) Under the Assumption that μ ≥ λ

The queue length is maintained at zero level with the number of tasks following inverse parabolic path while the queue waiting time follows parabolic curve, and the system waiting time follows a decreasing linear plot. The server utilization factor also follows an inverse parabolic curve as shown in Fig. 20.

6.20 At Decreasing Mean Service Rate (μ) and Increasing Number of Servers (s) with Constant Mean Arrival Rate (λ) Under the Assumption that μ ≤ λ

When the parallel queuing system bears above pattern of parameters, queue is not formed and the number of tasks in the system attains values in fourth quadrant of the coordinate system with waiting times oscillate between high and low values while the server utilization factor embraces an inverse bell curve as shown in Fig. 21.

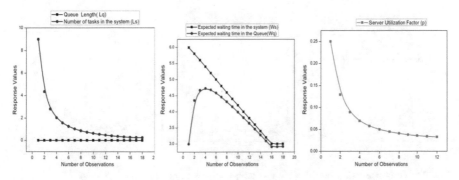

Fig. 20 Response values at decreasing μ and increasing s with constant λ at $\mu \geq \lambda$

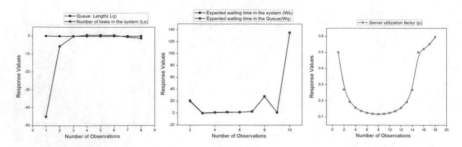

Fig. 21 Response values at decreasing μ and increasing s with constant λ at $\mu \leq \lambda$

6.21 At Decreasing Mean Service Rate (μ) with Constant Number of Servers (s) and Constant Mean Arrival Rate (λ) Under the Assumption that μ ≤ λ

The queue is not formed and the number of tasks fall down to linear negative values with waiting times show a constant trend first and then achieve sharp linear rise while server utilization factor shows parabolic growth as illustrated in Fig. 22.

6.22 At Decreasing Mean Service Rate (μ) with Constant Number of Servers (s) and Constant Mean Arrival Rate (λ) Under the Assumption that μ ≥ λ

The parallel queuing system is observed at decreasing μ, constant s, and constant λ with the mean service rate greater or equal to mean arrival rate where it is found that no queue of tasks is formed and the number of tasks in the system decreases linearly until it achieves uniform path. Meanwhile waiting times also decrease linearly which later on take a constant value and the server utilization factor increases progressively reaches a peak value and maintains that peak as shown in Fig. 23.

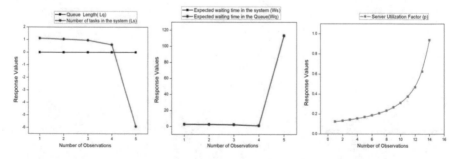

Fig. 22 Response values at decreasing μ with constant s and constant λ at $\mu \leq \lambda$

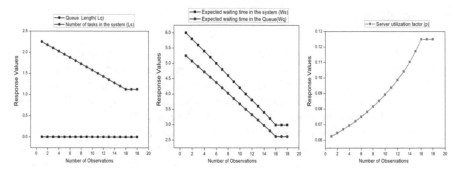

Fig. 23 Response values at decreasing μ with constant s and constant λ at $\mu \geq \lambda$

7 Research Findings

In this research work, twenty-two different possible cases have been studied and analyzed in a parallel queuing model proposed in our previous work. The following observations have been outlined in this paper:

I. The numerical value of the server utilization factor lies in the range of 0–1; however, at times under abnormal conditions when the number of tasks in the system goes in negative quadrants, the value of 's' may go out of its range.

II. One of the significant observations established from experimental evidences validates that any queuing model in general and parallel queuing model in particular holds practicality if and only if the mean service rate is greater or equal to the mean arrival rate.

III. From the twenty-two cases examined in this work, it is evident that under eight special cases discussed in Sects. 6.2, 6.3, 6.6, 6.10, 6.11, 6.14, 6.16, and 6.21 the parallel queuing model does not hold any practicality.

IV. Five cases have been found where there is a direct relationship between the number of tasks in the system and waiting times as reviewed in Sects. 6.4, 6.5, 6.17, 6.18, and 6.22. Further out of these, four hold linear relationship with the two mentioned parameters and they are cases discussed in Sects. 6.4, 6.5, 6.17, and 6.22.

V. Also, five cases have been identified which follow an inverse relationship between the number of tasks in the system and waiting times as figured in 6.7, 6.8, 6.10, 6.11, and 6.14.

8 Conclusion

From experimental results and research findings, it is concluded that in a parallel queuing model the queue is not formed at all so the user tasks do not have to wait for longer times, thus enabling higher response rates improving performance of the cloud system. The peak of waiting times does not go very high usually shows a decrease in the trend; however, it may also achieve the steady state at lower values considered for best cases only in the paper. Further, it is observed that parallel queuing model holds best when service rate is greater or equal to arrival rate and mentions the best possible cases where higher QoS is achieved. The experimental observations lead us that parallel queuing model is best to achieve QoS in terms of waiting time, queue length, server utilization, and response time.

References

1. Mishra, S.K., Sahoo, B., Parida, P.P.: Load balancing in cloud computing: a big picture. J. King Saud Univ.—Comput. Inf. Sci. (2018)
2. Pradhan, P., Behera, P.K., Ray, B.N.B.: Modified round robin algorithm for resource allocation in cloud computing. Procedia Comput. Sci. **85**, 878–890 (2016)
3. Reddy, V.K., Rao, B.T., Reddy, L.S.S.: Research issues in cloud computing. Global J. Comput. Sci. Technol. (2011)
4. Buyya, R., Broberg, J., Goscinski, A.M. (Eds.) Cloud Computing: Principles and Paradigms, vol. 87. Wiley (2010)
5. Liu, F., Tong, J., Mao, J., Bohn, R., Messina, J., Badger, L., Leaf, D.: NIST Cloud Computing Reference Architecture 500-292 2011. NIST Special Publication 35 (2011)
6. Abrishami, S., Naghibzadeh, M.: Deadline-constrained workflow scheduling in software as a service cloud. Sci. Iran. **19**(3), 680–689 (2012)
7. Celesti, A., et al.: Virtual machine provisioning through satellite communications in federated cloud environments. Future Gener. Comput. Syst. **28**(1), 85–93 (2012)
8. Jose Moura, D.H.: Review and analysis of networking challenges in cloud computing. J. Netw. Comput. Appl. (2015)
9. Marston, S., et al.: Cloud computing—the business perspective. Decis. Support Syst. **51**(1), 176–189 (2011)
10. Voorsluys, W., etal.: Introduction to cloud computing. Cloud Computing, pp. 1–41. Wiley (2011)
11. Chang, V.: An overview, examples and impacts offered by emerging services and analytics in cloud computing. Int. J. Inf. Manag. (2015)
12. Jain, N., Choudhary, S.: Overview of virtualization in cloud computing. In: Symposium on Colossal Data Analysis and Networking (CDAN), pp. 1–4, IEEE, Mar 2016
13. Alouane, M., El Bakkali, H.: Virtualization in cloud computing: no hype vs hyperwall new approach. In: 2016 International Conference on Electrical and Information Technologies (ICEIT), pp. 49–54, IEEE, May 2016
14. Rimal, B.P., Choi, E., Lumb, I.: A taxonomy and survey of cloud computing systems. In: Fifth International Joint Conference on INC, IMS and IDC, 2009, NCM'09, pp. 44–51, IEEE, Aug 2009
15. Vilaplana, J., Solsona, F., Teixidó, I., Mateo, J., Abella, F., Rius, J.: A queuing theory model for cloud computing. J. Supercomput. **69**(1), 492–507 (2014)

16. Khazaei, H., Mišić, J., Mišić, V.B.: Performance analysis of cloud computing centers using m/g/m/m+r queuing systems. IEEE Trans. Parallel Distrib. Syst. **5**, 936–943 (2011)
17. Takagi, H.: Queueing Analysis, Vol. 1: Vacation and Priority Systems. North Holland, Amsterdam (1991)
18. Subramani, V., Kettimuthu, R., Srinivasan, S., Sadayappan, S.: Distributed job scheduling on computational grids using multiple simultaneous requests. In: Proceedings of the 11th IEEE International Symposium on High Performance Distributed Computing HPDC-11 2002, pp. 359–366. IEEE (2002)
19. Gong, L., Sun, X.H., Watson, E.F.: Performance modeling and prediction of nondedicated network computing. IEEE Trans. Comput. **9**, 1041–1055 (2002)
20. Cho, K.M., Tsai, P.W., Tsai, C.W., Yang, C.S.: A hybrid meta-heuristic algorithm for VM scheduling with load balancing in cloud computing. Neural Comput. Appl. **26**(6), 1297–1309 (2015)
21. Li, K., Xu, G., Zhao, G., Dong, Y., Wang, D.: Cloud task scheduling based on load balancing ant colony optimization. In: 2011 Sixth Annual ChinaGrid Conference, pp. 3–9. IEEE
22. Kapur, R.: A workload balanced approach for resource scheduling in cloud computing. In: 2015 Eighth International Conference on Contemporary Computing (IC3), pp. 36–41, IEEE, Aug 2015
23. Ajit, M., Vidya, G.: VM level load balancing in cloud environment. In: 2013 Fourth International Conference on Computing, Communications and Networking Technologies (ICCCNT), pp. 1–5, IEEE, July 2013
24. Ibrahim, A.H., Faheem, H.E.D.M., Mahdy, Y.B., Hedar, A.R.: Resource allocation algorithm for GPUs in a private cloud. Int. J. Cloud Comput. **5**(1–2), 45–56 (2016)
25. Milani, A.S., Navimipour, N.J.: Load balancing mechanisms and techniques in the cloud environments: systematic literature review and future trends. J. Netw. Comput. Appl. **71**, 86–98 (2016)
26. Kalra, M., Singh, S.: A review of metaheuristic scheduling techniques in cloud computing. Egypt. Inf. J. **16**(3), 275–295 (2015)
27. Shahbaz, A., Kavitha, G.: Impact on QoS in cloud computing employing a parallel queuing model with strictly increasing arrival rate-an analytical approach. Int. J. Innovation Eng. Res. Manag. **5**(2) (2018). ISSN 2348-4918, ISO 2000-9001 certified, E
28. Kirsal, Y., Ever, Y.K., Mostarda, L., Gemikonakli, O.: Analytical modelling and performability analysis for cloud computing using queuing system. In: 2015 IEEE/ACM 8th International Conference on Utility and Cloud Computing (UCC), pp. 643–647, IEEE, Dec 2015
29. Eisa, M., Esedimy, E.I., Rashad, M.Z.: Enhancing cloud computing scheduling based on queuing models. Int. J. Comput. Appl. **85**(2) (2014)
30. Sowjanya, T.S., Praveen, D., Satish, K., Rahiman, A.: The queueing theory in cloud computing to reduce the waiting time. Int. J. Comput. Sci. Eng. Technol. **1**(3)
31. Mohanty, S., Pattnaik, P.K., Mund, G.B.: A comparative approach to reduce the waiting time using queuing theory in cloud computing environment. Int. J. Inf. Comput. Technol. (2014). ISSN 0974-2239
32. Varma, P.S., Satyanarayana, A., Sundari, M.R.: Performance analysis of cloud computing using queuing models. In: 2012 International Conference on Cloud Computing Technologies, Applications and Management (ICCCTAM), pp. 12–15, IEEE, Dec 2012
33. Sharma, J.K.: Operations Research: Theory and Applications. Macmillan (1997)
34. Saaty, T.L.: Elements of queueing theory: with applications, p. 423. McGraw-Hill, New York (1961)
35. Taha, H.A. Operations Research: An Introduction. Macmillan (1992)
36. Afzal S, Kavitha G.: A hybrid multiple parallel queuing model to enhance QoS in cloud computing. Int. J. Grid High Perform. Comput. **12**(1), 18–34

Analysis of Efficient Classification Algorithms in Web Mining

K. Prem Chander, S. S. V. N. Sharma, S. Nagaprasad, M. Anjaneyulu
and V. Ajantha Devi

Abstract Education is a tremendous and critical concern, in current years the amount of data stored in academic database is developing rapidly. The protected database consists of hidden data about the development of the scholar's performance and behavior. The ability to assess the faculties overall performance annually could be very critical in academic environments. Web content mining is the method of extracting useful data from the contents of web pages. Content records are the collection of data that a web page is designed to hold. Content data consist of text, photograph, audio, video, hyperlink or dependent data which includes lists or tables. From this method, the scholars' behavior is analyzed by using the Internet log files, which holds the data such as staffs searched web pages and their history was extracted for analysis. The analysis of staffs' behavior in the learning and teaching environment is based on log files created on the server during the course of interaction between learners and the electronic syllabus. This study emphasizes on concept of various classification algorithms, for processing the data available in the web. There are numerous algorithms for classification, among those three major algorithms such as Naïve Bayes algorithm, Support Vector Machine algorithm

K. Prem Chander · M. Anjaneyulu
Department of CS, Dravidian University, Kuppam, Andhra Pradesh, India
e-mail: kpc.1279@gmail.com

M. Anjaneyulu
e-mail: anjan.lingam1@gmail.com

S. S. V. N. Sharma
Department of CSE, Vaagdevi College of Engineering, Warangal, India
e-mail: ssvn.sarma@gmail.com

S. Nagaprasad (✉)
Department of CS, Tara Government Degree College, Sangareddy, Telangana, India
e-mail: nagkanna80@gmail.com

V. Ajantha Devi
Department of Computer Application, Guru Nanak College, Chennai, Tamil Nadu, India
e-mail: ajay.press@gmail.com

© Springer Nature Singapore Pte Ltd. 2020   319
K. S. Raju et al. (eds.), *Data Engineering and Communication Technology*,
Advances in Intelligent Systems and Computing 1079,
https://doi.org/10.1007/978-981-15-1097-7_27

(SVM) and finally Artificial Neural Network algorithms (ANN) were considered for analysis. These three algorithms were used for analyzing the web log files, where the data to be handles resides. The results of all the three algorithms have been clearly defined, in order to choose the efficient classification algorithm.

1 Introduction

Internet Services are available in many of the educational institutions for their working staff members and students which is one of the major research areas which provide a new medium of communication [1]. In this research, the researcher strives to get better insight into how Internet utilization behaviors of academicians in colleges contribute in promoting students' skills, as staffs play a key role in sharing the knowledge among the students which in turn improves their academic skills as well as extracurricular activities.

2 Web Server Log

One or more log files automatically created and maintained when user navigates the pages of the Web site to access the information [2]. This file is called an Internet server log or a web log. The log files keep a lot of information about each user's access to the web server. Each line or entry in the web log represents a request for a resource. The different web server supports different log format [3]. The exact content of an entry varies from log format to log format. Nearly all the server log file contains information such as Visitor's IP address, access date and time, URL of page accessed, the status code returned by the server, bytes transferred to the user, etc.

3 Methodology Implementation Process

During the implementation process, collected raw web log files by the web mining techniques are preprocessed by normalization method to extract useful data from the raw web log files where the unwanted and duplicate raw files are removed [4]. To discover the pattern of these data, the researcher applies a predictive model which is nothing data mining technique. This predictive model is taken in which three algorithms are implemented [5] (Fig. 1).

- Naïve Bayes
- Support Vector Machine
- Artificial Neural Network.

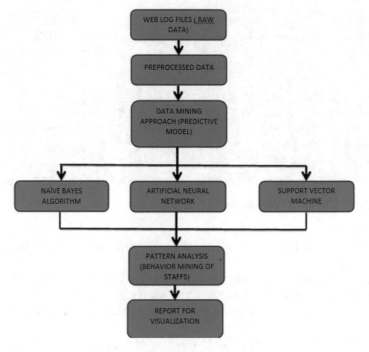

Fig. 1 Flowchart for the proposed system workflow

From these three algorithms, the discovered patterns are analyzed and filter out the useless rules and thereby extract the interesting rules and patterns. Finally, a text report is generated from that the researcher can analyze the behavior of staffs [6].

3.1 Screenshot for Web Log File

Figure 2 is a part of web log file of the first 11 staffs where SRNR is nothing but the serial number of the staffs. Here, the serial number is given according to the Internet Protocol (IP) address.

The age of the staff is obtained from the college database which is available in that particular college [7]. Likewise, age, family income, region, location and engineering stream are taken from the college database which is available in that particular college. And also three recently accessed Internet date of that particular semester for every staff is taken and its corresponding day was also obtained. What Web site the particular staff is surfing is collected and what kind of Web site is also collected. Time session is also collected [8].

Figure 3 is a part of web log file where the day of that top used Internet date. What Web site the particular is surfing is collected and what kind of Web site is also

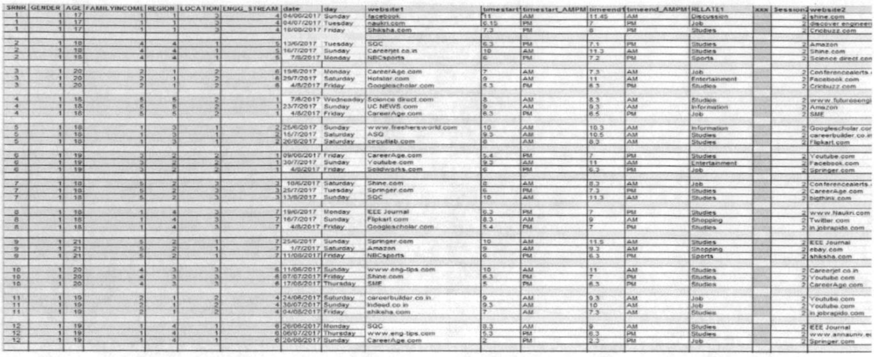

SRNR	GENDER	AGE	FAMILYINCOME	REGION	LOCATION	ENGG_STREAM	date	day	website1	timestart1	timestart_AMPM	timeend1	timeend_AMPM	RELATE1	xxx	Session2	website2
1	1	17	1	1	3	4	04/06/2017	Sunday	facebook	11	AM	11.45	AM	Discussion		2	shine.com
1	1	17	1	1	3	4	04/07/2017	Tuesday	naukri.com	6.15	PM	7	PM	Job		2	discover engineeri
1	1	17	1	1	3	4	18/08/2017	Friday	Shiksha.com	7.3	PM	8	PM	Studies		2	Cricbuzz.com
2	1	18	4	4	1	5	13/6/2017	Tuesday	SQC	6.3	PM	7.1	PM	Studies		2	Amazon
2	1	18	4	4	1	5	16/7/2017	Sunday	Careerjet.co.in	10	AM	11.3	AM	Studies		2	Shine.com
2	1	18	4	4	1	5	7/8/2017	Monday	NBCsports	6	PM	7.2	PM	Sports		2	Science direct.con
3	1	20	2	1	2	6	19/6/2017	Monday	CareerAge.com	7	AM	7.3	AM	Job		2	Conferencealerts.
3	1	20	2	1	2	6	29/7/2017	Saturday	Hotstar.com	9	AM	11	AM	Entertainment		2	Facebook.com
3	1	20	2	1	2	6	4/8/2017	Friday	Googlescholar.com	5.3	PM	6.3	PM	Studies		2	Cricbuzz.com
4	1	18	5	5	2	1	7/6/2017	Wednesday	Science direct.com	8	AM	8.3	AM	Studies		2	www.futuresengi
4	1	18	5	5	2	1	23/7/2017	Sunday	UC NEWS.com	9	AM	9.3	AM	Information		2	Amazon
4	1	18	5	5	2	1	4/8/2017	Friday	CareerAge.com	6.3	PM	6.5	PM	Job		2	SME
5	1	18	1	3	1	2	25/6/2017	Sunday	www.freshersworld.com	10	AM	10.3	AM	Information		2	Googlescholar.cor
5	1	18	1	3	1	2	15/7/2017	Saturday	ASQ	9.3	AM	10.5	AM	Studies		2	careerbuilder.co.i
5	1	18	1	3	1	2	20/8/2017	Saturday	circuitlab.com	8	AM	8.3	AM	Studies		2	Flipkart.com
6	1	19	3	2	2	1	09/06/2017	Friday	CareerAge.com	5.4	PM	7	PM	Studies		2	Youtube.com
6	1	19	3	2	2	1	30/7/2017	Sunday	Youtube.com	9.3	AM	11	AM	Entertainment		2	Facebook.com
6	1	19	3	2	2	1	4/8/2017	Friday	Solidworks.com	6	PM	6.3	PM	Job		2	Springer.com
7	1	18	5	2	3	3	10/6/2017	Saturday	Shine.com	8	AM	8.3	AM	Job		2	Conferencealerts.
7	1	18	5	2	3	3	25/7/2017	Tuesday	Springer.com	6	PM	7.3	PM	Studies		2	CareerAge.com
7	1	18	5	2	3	3	13/8/2017	Sunday	SQC	10	AM	11.3	AM	Studies		2	bigthink.com
8	1	18	1	4	3	7	19/6/2017	Monday	IEEE Journal	6.3	PM	7	PM	Studies		2	www.Naukri.com
8	1	18	1	4	3	7	16/7/2017	Sunday	Flipkart.com	8.3	AM	9	AM	Shopping		2	Twitter.com
8	1	18	1	4	3	7	4/8/2017	Friday	Googlescholar.com	5.4	PM	7	PM	Studies		2	in.jobrapido.com
9	1	21	5	2	1	7	25/6/2017	Sunday	Springer.com	10	AM	11.5	AM	Studies		2	IEEE Journal
9	1	21	5	2	1	7	1/7/2017	Saturday	Amazon	9	AM	9.3	AM	Shopping		2	ebay.com
9	1	21	5	2	1	7	11/08/2017	Friday	NBCsports	6	PM	6.3	PM	Sports		2	shiksha.com
10	1	20	4	3	3	6	11/06/2017	Sunday	www.eng-tips.com	10	AM	11	AM	Studies		2	Careerjet.co.in
10	1	20	4	3	3	6	07/07/2017	Friday	Shine.com	6.3	PM	7	PM	Studies		2	Youtube.com
10	1	20	4	3	3	6	17/08/2017	Thursday	SME	5	PM	6.3	PM	Studies		2	CareerAge.com
11	1	19	2	1	2	4	24/06/2017	Saturday	careerbuilder.co.in	9	AM	9.3	AM	Job		2	Youtube.com
11	1	19	2	1	2	4	30/07/2017	Sunday	Indeed.co.in	9.3	AM	10	AM	Job		2	Youtube.com
11	1	19	2	1	2	4	04/08/2017	Friday	shiksha.com	7	AM	7.3	AM	Studies		2	in.jobrapido.com
12	1	19	1	4	1	6	26/06/2017	Monday	SQC	8.3	AM	9	AM	Studies		2	IEEE Journal
12	1	19	1	4	1	6	06/07/2017	Thursday	www.eng-tips.com	5.3	PM	6.3	PM	Studies		2	www.annauniv.e
12	1	19	1	4	1	6	20/08/2017	Sunday	CareerAge.com	2	PM	2.3	PM	Job		2	Springer.com

Fig. 2 Staffs' web log file of Web site 1

collected. Time session is also collected. Likewise, every session is collected in this web log [9].

Figure 4 is a part of web log file where the day of that top used Internet date. What Web site the particular is surfing is collected and what kind of Web site is also collected. Time session is also collected. Likewise, every session is collected in this web log.

Figures 2, 3 and 4 describe the screenshots of web log files of ten staffs of a college which comprise staffs age, engineering stream, number of times the Internet being used by the staffs, what type of site is being used by the staffs, at what time

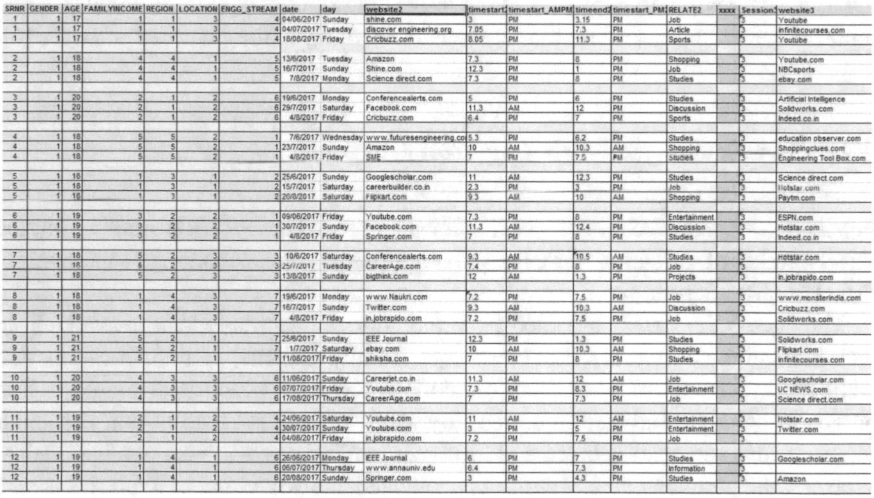

SRNR	GENDER	AGE	FAMILYINCOME	REGION	LOCATION	ENGG_STREAM	date	day	website2	timestart2	timestart_AMPM	timeend2	timestart_PM2	RELATE2	xxxx	Session3	website3
1	1	17	1	1	3	4	04/06/2017	Sunday	shine.com	3	PM	3.15	PM	Job		3	Youtube
1	1	17	1	1	3	4	04/07/2017	Tuesday	discover engineering.org	7.05	PM	7.3	PM	Article		3	infinitecourses.com
1	1	17	1	1	3	4	18/08/2017	Friday	Cricbuzz.com	8.05	PM	11.3	PM	Sports		3	Youtube
2	1	18	4	4	1	5	13/6/2017	Tuesday	Amazon	7.3	PM	8	PM	Shopping		3	Youtube.com
2	1	18	4	4	1	5	16/7/2017	Sunday	Shine.com	12.3	PM	1	PM	Job		3	NBCsports
2	1	18	4	4	1	5	7/8/2017	Monday	Science direct.com	7.3	PM	8	PM	Studies		3	ebay.com
3	1	20	2	1	2	6	19/6/2017	Monday	Conferencealerts.com	5	PM	6	PM	Studies		3	Artificial Intelligence
3	1	20	2	1	2	6	29/7/2017	Saturday	Facebook.com	11.3	AM	12	PM	Discussion		3	Solidworks.com
3	1	20	2	1	2	6	4/8/2017	Friday	Cricbuzz.com	6.4	PM	7	PM	Sports		3	Indeed.co.in
4	1	18	5	5	2	1	7/6/2017	Wednesday	www.futuresengineering.co	5.3	PM	6.2	PM	Studies		3	education observer.com
4	1	18	5	5	2	1	23/7/2017	Sunday	Amazon	10	AM	10.3	AM	Shopping		3	Shoppingclues.com
4	1	18	5	5	2	1	4/8/2017	Friday	SME	7	PM	7.5	PM	Studies		3	Engineering Tool Box.com
5	1	18	1	3	1	2	25/6/2017	Sunday	Googlescholar.com	11	AM	12.3	PM	Studies		3	Science direct.com
5	1	18	1	3	1	2	15/7/2017	Saturday	careerbuilder.co.in	2.3	PM	3	PM	Job		3	Hotstar.com
5	1	18	1	3	1	2	20/8/2017	Saturday	Flipkart.com	9.3	AM	10	AM	Shopping		3	Paytm.com
6	1	19	3	2	2	1	09/06/2017	Friday	Youtube.com	7.3	PM	8	PM	Entertainment		3	ESPN.com
6	1	19	3	2	2	1	30/7/2017	Sunday	Facebook.com	11.3	AM	12.4	PM	Discussion		3	Hotstar.com
6	1	19	3	2	2	1	4/8/2017	Friday	Springer.com	7	PM	8	PM	Studies		3	Indeed.co.in
7	1	18	5	2	3	3	10/6/2017	Saturday	Conferencealerts.com	9.3	AM	10.5	AM	Studies		3	Hotstar.com
7	1	18	5	2	3	3	25/7/2017	Tuesday	CareerAge.com	7.4	PM	8	PM	Job		3	
7	1	18	5	2	3	3	13/8/2017	Sunday	bigthink.com	12	AM	1.3	PM	Projects		3	in.jobrapido.com
8	1	18	1	4	3	7	19/6/2017	Monday	www.Naukri.com	7.2	PM	7.5	PM	Job		3	www.monsterindia.com
8	1	18	1	4	3	7	16/7/2017	Sunday	Twitter.com	9.3	AM	10.3	AM	Discussion		3	Cricbuzz.com
8	1	18	1	4	3	7	4/8/2017	Friday	in.jobrapido.com	7.2	PM	7.5	PM	Job		3	Solidworks.com
9	1	21	5	2	1	7	25/6/2017	Sunday	IEEE Journal	12.3	PM	1.3	PM	Studies		3	Solidworks.com
9	1	21	5	2	1	7	1/7/2017	Saturday	ebay.com	10	AM	10.3	AM	Shopping		3	Flipkart.com
9	1	21	5	2	1	7	11/08/2017	Friday	shiksha.com	7	PM	8	PM	Studies		3	infinitecourses.com
10	1	20	4	3	3	6	11/06/2017	Sunday	Careerjet.co.in	11.3	AM	12	AM	Job		3	Googlescholar.com
10	1	20	4	3	3	6	07/07/2017	Friday	Youtube.com	7.3	PM	8.3	PM	Entertainment		3	UC NEWS.com
10	1	20	4	3	3	6	17/08/2017	Thursday	CareerAge.com	7	PM	7.3	PM	Job		3	Science direct.com
11	1	19	2	1	2	4	24/06/2017	Saturday	Youtube.com	11	AM	12	AM	Entertainment		3	Hotstar.com
11	1	19	2	1	2	4	30/07/2017	Sunday	Youtube.com	3	PM	5	PM	Entertainment		3	Twitter.com
11	1	19	2	1	2	4	04/08/2017	Friday	in.jobrapido.com	7.2	PM	7.5	PM	Job		3	
12	1	19	1	4	1	6	26/06/2017	Monday	IEEE Journal	6	PM	7	PM	Studies		3	Googlescholar.com
12	1	19	1	4	1	6	06/07/2017	Thursday	www.annauniv.edu	6.4	PM	7.3	PM	Information		3	
12	1	19	1	4	1	6	20/08/2017	Sunday	Springer.com	3	PM	4.3	PM	Studies		3	Amazon

Fig. 3 Staffs' web log file of Web site 2 (Continuation)

SRNR	GENDER	AGE	FAMILYINCOME	REGION	LOCATION	ENGG_STREAM	date	day
1	1	17	1	1	3	4	04/06/2017	Sunday
1	1	17	1	1	3	4	04/07/2017	Tuesday
1	1	17	1	1	3	4	18/08/2017	Friday
2	1	18	4	4	1	5	13/6/2017	Tuesday
2	1	18	4	4	1	5	16/7/2017	Sunday
2	1	18	4	4	1	5	7/8/2017	Monday
3	1	20	2	1	2	6	19/6/2017	Monday
3	1	20	2	1	2	6	29/7/2017	Saturday
3	1	20	2	1	2	6	4/8/2017	Friday
4	1	18	5	5	2	1	7/6/2017	Wednesday
4	1	18	5	5	2	1	23/7/2017	Sunday
4	1	18	5	5	2	1	4/8/2017	Friday
5	1	18	1	3	1	2	25/6/2017	Sunday
5	1	18	1	3	1	2	15/7/2017	Saturday
5	1	18	1	3	1	2	26/8/2017	Saturday
6	1	19	3	2	2	1	09/06/2017	Friday
6	1	19	3	2	2	1	30/7/2017	Sunday
6	1	19	3	2	2	1	4/8/2017	Friday
7	1	18	5	2	3	3	10/6/2017	Saturday
7	1	18	5	2	3	3	25/7/2017	Tuesday
7	1	18	5	2	3	3	13/8/2017	Sunday
8	1	18	1	4	3	7	19/6/2017	Monday
8	1	18	1	4	3	7	16/7/2017	Sunday
8	1	18	1	4	3	7	4/8/2017	Friday
9	1	21	5	2	1	7	25/6/2017	Sunday
9	1	21	5	2	1	7	1/7/2017	Saturday
9	1	21	5	2	1	7	11/08/2017	Friday
10	1	20	4	3	3	6	11/06/2017	Sunday
10	1	20	4	3	3	6	07/07/2017	Friday
10	1	20	4	3	3	6	17/08/2017	Thursday
11	1	19	2	1	2	4	24/06/2017	Saturday
11	1	19	2	1	2	4	30/07/2017	Sunday
11	1	19	2	1	2	4	04/08/2017	Friday
12	1	19	1	4	1	6	26/06/2017	Monday
12	1	19	1	4	1	6	06/07/2017	Thursday
12	1	19	1	4	1	6	20/08/2017	Sunday

SRNR	timestart	timestart_AMPM	timeend3	timesend_AMPM	RELATE3	xxxx	Session4	website4	timestart	timestart_A
1	5	PM	5.3	PM	Program		4	IEEE Journals	6.3	PM
1	8.05	PM	8.3	PM	Studies		4			
1	11.52	PM	11.55	PM	Entertainment		4			
2	8.3	PM	8.3	PM	Studies		4	CareerAge.com	9.4	PM
2	2.2	PM	3.3	PM	Sports		4	Science direct.com	4	PM
2	8.2	PM	8.5	PM	Shopping		4	NBCsports	9	PM
3	6.3	PM	7	PM	Studies		4	Engineers Edge.com	8	PM
3	3	PM	3.3	PM	Job		4	www.Naukri.com	5	PM
3	7.3	PM	8.3	PM	Job		4	Youtube.com	8.4	PM
4	8.3	PM	7	PM	Studies		4			
4	10.4	AM	11	AM	Shopping		4	Twitter.com	2	PM
4	8	PM	8.4	PM	Studies		4			
5	1	PM	1.3	PM	Studies		4	Twitter.com	3	PM
5	3.3	PM	4	PM	Entertainment		4	Cricbuzz.com	5.3	PM
5	11	AM	11.3	AM	Shopping		4	ESPN.com	2.3	PM
6	8.3	PM	9	PM	Studies		4	Shoppingclues.com	9.3	PM
6	3.3	PM	4.3	PM	Entertainment		4	TimesJob.com	5.3	PM
6	8.3	PM	8.6	PM	Job		4			
7	11	AM	1	PM	Entertainment		4			
7							4			
7	3.3	PM	4.3	PM	Job		4	ebay.com	5	PM
8	8	PM	8.3	PM	job		4			
8	1	PM	2	PM	Sports		4	Googlescholar.com	7	PM
8	6	PM	6.3	PM	Studies		4	Paytm.com	8.4	PM
9	3	PM	3.3	PM	Studies		4	SOC	5.3	PM
9	1	PM	1.3	PM	Shopping		4	Youtube.com	3.3	PM
9	8.3	PM	9.3	PM	Studies		4	education observer.com	10	PM
10	2	PM	3	PM	Studies		4			
10	8.4	PM	9	PM	Information		4	NBCsports	8.2	PM
10	6	PM	9	PM	Studies		4			
11	2.3	PM	3.3	PM	Entertainment		4	ebay.com	5	PM
11	5.3	PM	6	PM	Discussion		4			
11							4			
12	7.3	PM	8	PM	Studies		4			
12							4			
12	5	PM	5.3	PM	Shopping		4	in.jobrapido.com	6	PM

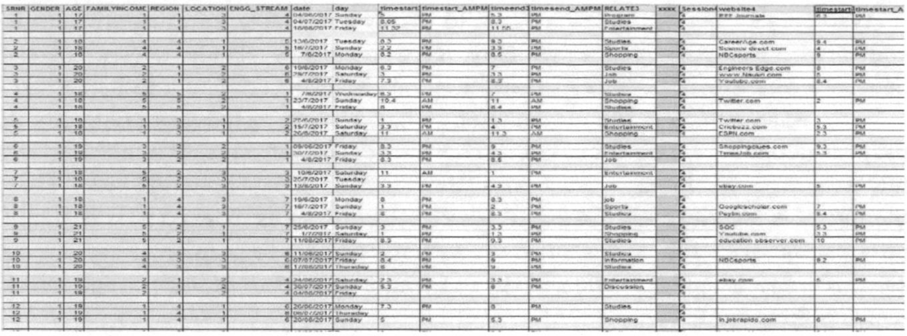

Fig. 4 Staffs' web log file of Website 3 (Continuation)

the staffs started to use the Internet access, at what time and till end of using of Internet access [10]. These web log files are processed one because they are mined based on the criteria what the researcher need for the research.

4 Web Usage Mining

Web mining is the usage of data mining methods to automatically determine and extract data from web files and services.

There are three common classes of data that can be found out by web mining:

- Web interest, from server logs and web browser pastime tracking.
- Web graph, from hyperlinks among pages, human beings and other data.
- Web content material, for the data found on Web pages and inside of files (Fig. 5).

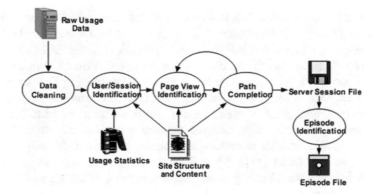

Fig. 5 Process of web mining

Web content mining is the method to discover valuable information from the content of a web page [11]. Essentially, the web content includes several kinds of data such as textual, video, metadata image, audio and in addition to hyperlinks.

Web usage mining is the presentation of data mining techniques to identify the usage patterns from web data [12]. Data were collected from user's interface with the web, e.g., web/proxy server logs, user queries and registration data. Usage mining tools identify and calculate the user behavior, so as to assist the designer to progress the Web site, to fascinate the visitors, or to give consistent users a personalized and adaptive service [13]. The whole procedure of web usage mining includes three steps:

- Data collection and data preprocessing,
- Knowledge discovery or (pattern mining),
- Knowledge application.

5 Experimental Analysis—Web Mining

Web mining is the use of data mining techniques to automatically discover and extract information from web documents and services [14].

There are three general classes of information that can be discovered by web mining:

- Web activity—activity tracking between server logs and web browser.
- Web graph—links between pages, people and other data.
- Web content—the data found on web pages and documents.

5.1 Data Preprocessing

In data mining and statistical data analysis, before models can be built or algorithms can be used data needs to be prepared [15]. In this context, preparing the data means transforming them prior to the analysis so as to minimize the algorithm's job. Often, the motive will be to modify the data so that the hypotheses, on which the algorithms are based, are verified, while at the same time preserving their information content intact. One of the most basic transformations is normalization [16].

The word normalization is used in many situations, with different, but associated, meanings. Fundamentally, normalizing means transforming so as to render normal. When data are seen as vectors, normalizing means transforming the vector so that ith as unit norm [17]. When data are thought of as random variables, normalizing means transforming to normal distribution. When data are hypothesized to be normal, normalizing means transforming to unit variance.

5.2 Implementation of Naïve Bayes Algorithm

Naïve Bayes is a basic method for building classifiers: models that allocate class labels to hassle instances, represented as vectors of characteristic values, in which the class labels are derived from a few finite sets [18]. It is not a solitary algorithm for education, such classifiers; however, an own family of algorithms grounded totally on a general principle: All Naïve Bayes classifiers anticipate that the cost of a selected feature is independent of the price of another function, given the class variable [19] (Fig. 6).

In this coding part, how the web log files are imported, how the Internet is used by the staffs, how the Web sites are classified into 16 categories and how the Naïve Bayes algorithm works according to the dataset are explained [20].

From the above result, it is shown that, from the category of 16 Web sites, a study-based Web site largely used by the staffs. X-axis—top 2 most searched Web site category for first 32 staffs and Y-axis—histogram of usage.

Figure 7 explains the Behavior Mining of staffs by using Naïve Bayes algorithm [21]. Here, the Web sites used by the staffs are classified into 16 categories. From this algorithm, staffs spent their time, largely on an article and career drives are identified. Figure 8 divides the 16 categories into three groups such as co-curriculum, extracurriculum and others. In Naïve Bayes, the output shows that the staffs spent most of the time on co-curriculum.

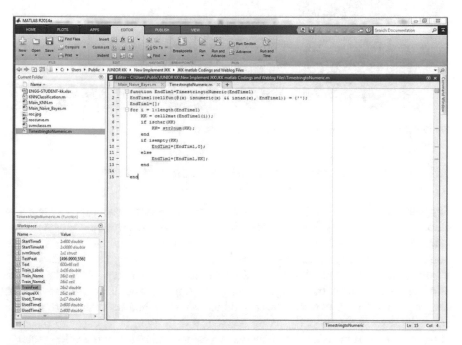

Fig. 6 Naïve Bayes algorithm coding

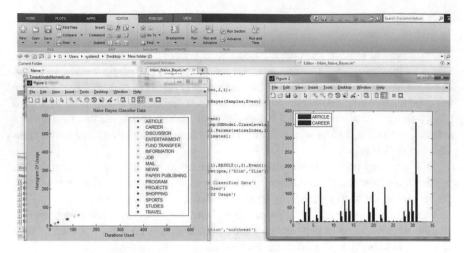

Fig. 7 Staffs' Behavior Mining output by using Naïve Bayes

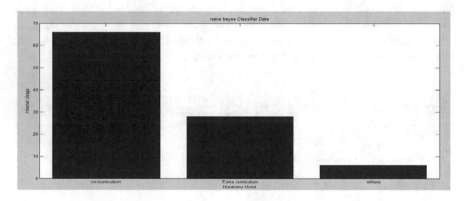

Fig. 8 Staffs' Behavior Mining output by using Naïve Bayes

5.3 Implementation of the Support Vector Machine

Support Vector Machine (SVM) is a powerful supervised classifier and correct learning approach [22]. From the statistical idea, it was derived and advanced by using Vapnik in 1982. It yields a success classification result in various utility domains, as an example, medical diagnosis, text categorization, face popularity and bioinformatics. The kernel controls the empirical risk and type capacity with a purpose to maximize the margin among the training and decrease the proper charges [23]. The remarkable benefit of the SVM technique over conventional strategies, except the properly set up theoretical definition, is its capacity of operating with high dimensional characteristic vectors without dropping the generalization performance [23].

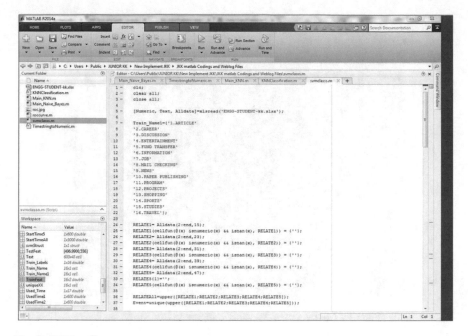

Fig. 9 SVM coding

Figure 9 shows the engineering staffs' data are related to each other in terms of Internet usage and academic performance and are explained in the form of coding. The above coding explains show the staffs' data are represented and are done in the form of coding [24].

Figure 10 shows the output of Support Vector Machine. Here, the 16 Web site category is grouped into three categories, i.e., co-curricular, extracurricular and others. From this figure, it is clear that the staffs spend most of the time on other factors.

5.4 Implementations of ANN Algorithm

ANNs are nothing but a data processing model which works similar to the brain's neural structure. ANNs are considered as nonlinear statistical modeling tools; here, the complex bonds between input and output are modeled clearly and also that the information passes through the networks will affect the arrangement of the ANNs, because it changes the neural structure depends on that input and output [24].

Figure 11 shows the engineering staffs' data are related to each other in terms of Internet usage and academic performance and are explained in the form of coding. The above coding explains show the staffs' data are represented and are done in the form of coding [25]. Here, the Web sites which are used by the staffs are classified

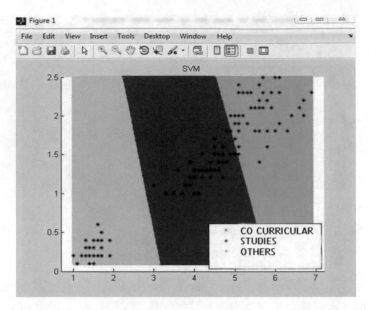

Fig. 10 Staffs' Behavior Mining output by using Support Vector Machine

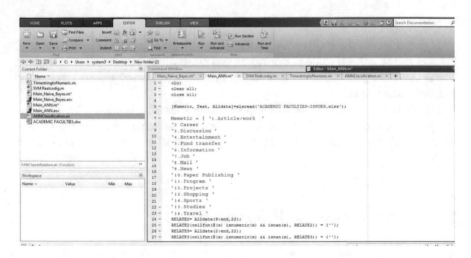

Fig. 11 ANN algorithm coding

into 16 categories. From these 16 categories, the majority of the staffs frequently used Internet access for the study purposes.

In Fig. 12, the 16 categories are divided into three categories such as co-curriculum, extracurriculum and others. In K-nn, the staffs spend most of the time on extracurricular activities.

Fig. 12 Staffs' Behavior Mining output by using ANN algorithm

5.5 Result Finding

Our research is beneficial for the researchers and also for the educational institutions in order to find out the knowledge level of academicians, who plays a very important role in enhancing the student's skills, which in turn improves the status of specific institutions or colleges in society [25]. It supports to analyze Internet usage for educational purposes that overcome the major limitations of both manual qualitative analysis and large-scale computational survey method. Our research can inform educational administrators, practitioners and other relevant decision-makers gain further understanding of faculties or staff's efficiency toward academy. With respect to the traditional classroom environment, faculty and instructional designers have successfully utilized Internet for various types of academic activities that include collaborative learning, inquiry-based learning and reflective learning [26]. Depends on the responses, it is apparent that by means of social media in distance learning environments allowed for increased collaboration, communication and interaction.

The social networking sites are a highly principal medium of communiqué and entertainment. Nowadays all age groups from kids to old age were more attracted toward use of social networking Web sites. This may be as a result of the massive benefits that these Web sites furnish including higher entry to persons around the world, immediate messaging, video calling, entry to quite a lot of products and services of many firms and manufacturers and rather more. With this technology, the progress of technological expertise and its accessibility has enabled rapid expansion and reputation of social networking Web sites [27]. With these improved technology in the world, it becomes the quite necessary to analyze the behavior in terms of social network usage. This includes the present study to focus mainly upon the institution scholars in the Indian context. University scholars are the mass users of these Web sites and for that reason it turns into important to analyze them in relation to their social networking site usage.

Hence, to have a clear view, it is highly important to have the abilities of present reviews and articles with regard to social networking Web sites. As a result, this aids in higher method of the reward study and a proper working out of the various opinions existential. The entire studies stated within the overview give a plethora of views about social networking Web sites and effective role of the present study. It is noteworthy that no longer many researchers have been undertaken related to interpersonal relationships with university scholars and social networking sites in the Indian context. Therefore, it is some distance more principle to research it in the Indian context.

6 Conclusion

In this research, the researcher used Naïve Bayes classifiers, Support Vector Machine and ANN algorithms to categorize the web log which reflects the behavior of academicians in colleges. In the Naïve Bayes classifier, a comparative technique is utilized to predict the probability of various classes considering different characteristics. This calculation is generally utilized as a part of content characterization and with the Web site having different categories. This result is analyzed by using ROC curve.

In the Support Vector Machine, 16 Web site categories are grouped into three groups, namely co-curricular, extracurricular and another group. In SVM, 60 staffs' data are converted into 100%. 43% of staffs are using the Internet for co-curricular usage, 37% of staffs are using the Internet for extracurricular other than the academic programs. 20% of staffs are using the Internet for supplementary purposes. From this analysis, it is clear that the staffs spend most of the time on additional factors, other than co-curricular and extracurricular.

In Artificial Neural Network Algorithm, frequent hit Web sites have been identified. The study of analysis shows that staffs were engaged in using the Internet for paper publishing, articles, conference, etc. in order to assist the student's knowledge, therefore most of the time they spend for the purpose of extracurricular activities. However, the prediction of best classifier is still a major task for this research. So, with the help of Receiver Operating Characteristic (ROC) Curve, the prediction of best classifier has been done.

References

1. Tanasa, D. et al.: Advanced Data Preprocessing for Inter Sites Web Usage Mining. IEEE Computer Society (2012)
2. Cooley, R., Srivastava, J., Mobasher, B.: Web mining: information and pattern discovery on the world wide web. In: Proceedings of the Ninth IEEE International Conference on Tools with Artificial Intelligence (ICTAI '97) (2014)
3. Cooley, R., Srivastava, J., Mobasher, B.: Data preparation for mining: discovery and applications of usage pattern from web data. SIGKDD Explor. 1(2), 12–23 (2016)
4. Zhou, B., Hui, S.C., Fong, A.C.M.: Discovering and visualizing temporal-based web access behavior. In: Proceedings of the 2015 IEEE/WIC/ACM International Conference on Web Intelligence (WI '05) (2015)

5. Zhou, B., Hui, S.C., Fong, A.C.M.: An effective approach for periodic web personalization. In: Proceedings of the 2016 IEEE/WIC/ACM International Conference on Web Intelligence (WI '06)
6. Suneetha, K.R., Krishnamoorthy, K.R.: Identifying user behavior by analyzing web server access log file. IJCSNS Int. J. Comput. Sci. Netw. Secur. 9(4) (2017)
7. Peng, Y.Q., Xiaq, G.X., Lin, T.: Predicting of user's behavior based on matrix clustering. In: Proceedings of the Fifth IEEE Conference on Machine Learning and Cybernetics, Dalian, 13–16 Aug 2016
8. Jalali, M., Mustapha, N., Mamat, A., Sulaiman, N.B.: A new classification model for online predicting users "future movements" 2015. IEEE, 978-1-4244, 6/08
9. EL-Halees, A.: Mining students data to analyze learning behavior: a case study. Int. J. Comput. Appl. (IJCA J.) 22, Article 4
10. Merceron, A., Yacef, K.: Educational data mining: a case study. In: Proceeding of the 2015 Conference on Artificial Intelligence in Education: Supporting Learning through Intelligent and Socially Informed Technology, pp. 467–474
11. Song, A.-B., Liang, Z.-P., Zhao, M.-X., Dong, Y.-S.: Mining web log data based on key path. In: Proceedings of the First International Conference on Machine Learning and Cybernetics, Beijing, 4–5 Nov 2012
12. Vaarandi, R.: A data clustering algorithm for mining patterns from event logs. In: Proceedings of the 3rd IEEE Workshop on IP Operations and Management, pp. 119–126, IEEE Press, Kansas City, MO, USA, Oct 2003
13. Chen, X., Vorvoreanu, M., Madhavan, K.: Mining social media data for understanding students' learning experiences. IEEE Trans. Learn. Technol. 7(3) (2014)
14. Olivera, G., Zita, B., Sasa, B.: Students behavior on social media sites a data mining approach. In: 2013 IEEE 11th International Symposium on Intelligent Systems and Informatics (SISY), Subotica, Serbia, 26–28 Sept 2013
15. Patel, P., Mistry, K.: Classification of student's e-learning experiences' in social media via text mining. IOSR J. Comput. Eng. (IOSR-JCE) 17(3), 81–89 (2015). Department of CSE, PIET, Vadodara, India. e-ISSN: 2278-0661,p-ISSN: 2278-8727, Ver. VI
16. Pearson, E.: All the world wide web's a stage: the performance of identity in online social networks. First Monday 14(3), 1–7 (2016)
17. Fu, Y.J.: Data mining: tasks, techniques and applications. IEEE Potentials 16(4), 18–20 (2014)
18. Bennett, D., Demiritz, A.: Semi-supervised support vector machines. Adv. Neural Inf. Process Syst. 11, 368–374 (2016)
19. Zhang, Y., Xiao, B.F.L.: Web page classification based on a least square support vector machine with latent semantic analysis. In: Proceedings of the 5th International Conference on Fuzzy Systems and Knowledge Discovery, vol. 2, pp. 528–532 (2014)
20. Dehghan, S., Rahmani, A.M.: A classifier-CMAC neural network model for web mining. In: Proceedings of the IEEE/WIC/ACM International Conference on Web Intelligence and Intelligent Agent Technology, vol. 1, pp. 427–431 (2015)
21. Feng, Z., Huyou, C.: Research and development in web usage mining system-key issues and proposed solutions: a survey. Mach. Learn. Cybern. 2, 986–990 (2002)
22. Dou, S., Jian-Tao, S., Qiang, Y., Zheng, C.: A comparison of implicit and explicit links for web page classification. In: Proceedings of the 15th International Conference on World Wide Web, pp. 643–650. Edinburgh, Scotland (2016)
23. Oskouei, R.J., Chaudhary, B.D.: Internet usage pattern by female students: a case study. In: ITNG, Seventh International Conference on Information Technology, pp. 1247–1250 (2010)
24. Srivastava, J., Cooley, R., Deshpande, M., Tan, P.-N.: Web usage mining: discovery and applications of usage patterns from web data. SIGKDD Explor. 1(2), 1–12 (2000)
25. Shahabi, C., Kashani, F.B.: A framework for efficient and anonymous web usage mining based on client-side tracking. In: Proceedings WEBKDD 2001: Mining Web Log Data across All Customer Touch Points, LNCS, vol. 2356, pp. 113–144. Springer-Verlag (2002)

26. Berendt, B., Mobasher, B., Nakagawa, M., Spiliopoulou, M.: The impact of site structure and user environment on session reconstruction in web usage analysis. In: Proceedings WEBKDD 2002: Mining Web Data for Discovery Usage Patterns and Profiles, LNCS, vol. 2703, pp. 159–179. Springer-Verlag (2002)
27. Menaka, P., Prathimadevi, A.: A Survey on Web Mining and Its Techniques (2015)

RETRACTED CHAPTER: Students' Performance Prediction Using Machine Learning Approach

Srinivasu Badugu and Bhavani Rachakatla

Abstract This report aims to cut back the manual procedures concerned within the performance analysis and analysis of scholars, by automating the method right from retrieval of results to pre-processing, segregating, and storing them into information. We additionally expect to perform examination on immense measures of information viably and encourage simple recovery of different sorts of data identified with understudies' execution. We give a degree to build up to information stockroom wherein, we can apply information mining methods to perform different sorts of examinations, making a learning base and use it further, for forecast purposes.

Keywords Data analytics · Performance analysis · Decision tree · Machine learning methods · Regression and correlations

1 Introduction

Any instructive establishment endeavors to give quality training to its understudies along these lines creating alumni of extraordinary quality who exceed expectations in scholastics, viable learning, and so forth. In this procedure, they take different activities to cultivate value education and survey themselves in different ways. Also, understudies' scholarly execution stands an essential measurement in their appraisal.

The original version of this chapter was retracted. The retraction note to this chapter is available at https://doi.org/10.1007/978-981-15-1097-7_82

S. Badugu (✉)
Stanley College of Engineering and Technology for Women, Hyderabad, India
e-mail: srinivasucse@gmail.com

B. Rachakatla
Vistex Asia Pacific Pvt Ltd, Hyderabad, India
e-mail: bhavanirachakatla96@gmail.com

The data identified with understudies' execution is generally put away in spreadsheets, records, and so forth and is investigated to extricate valuable data. However, the companies have spent their time developing themselves, settling for certain options in order to achieve better results. They base their choices on the investigation done utilizing the customary strategies.

Information mining in advanced education is an ongoing rising examination field, and this territory of research is picking up notoriety on account of its possibilities to instructive organizations and has altered the customary basic leadership. Kekane et al. and Venkatesan and Selvaragini [1, 2] it tends to be made to figure the correct principles to infer insightful choices.

This conventional methodology toward understudies' execution examination more often than not does not include any mining procedures, however, is an aftereffect of the different totals done on the understudies information put away in spreadsheets, records, and so on. An after investigation report may include less outcomes like the no. of passed understudies, no. of disappointments, the most astounding level of imprints, and so on. Be that as it may, these conventional strategies and other database devices are not all that powerful in removing different sorts of required data. The issue emerges when we need to dissect the understudies execution in progressively conceivable ways and furthermore stretch out the examination to discover affiliations, connection between the different elements influencing an understudies act. Additionally, such a sort of an examination may include a great deal of difficult work as well. Thus, there is a requirement for a powerful instrument which conquers these constraints and furthermore helps to anticipate understudies' execution.

We propose a framework which applies information mining procedures on the understudies result database gotten by cleaning, coordinating, and changing the understudy result information and help us set up a learning base. Further analysis of the information obtained can be used to predict the future execution of the understudies. This early forecast encourages us distinguish understudies who are probably going to imitate comparative conduct in the coming long periods of study, which may assist the understudy with knowing the zones to be concentrated on, the regions of progress before close by. Likewise we intend to stretch out the framework to discover affiliations and relationships between different variables influencing an understudies' act. A better estimate of understudies' execution in instructive establishments is one way to deal with achieve higher quality in training framework wherein the instructive foundations base their choices on substantial raw numbers in the instruction part in settling on better and canny choices. We should execute the above framework for our school (Osmania University).

Our arrangements of targets recognized in building up the above framework stay as pursues:

- Build up a framework that computerizes the procedure directly from extraction of results to putting away the outcome information into the database along these lines decreasing the manual work included.

- Build up a framework that performs different examinations and conglomerations utilizing data mining strategies on immense outcome informational indexes and gives us the ideal outcome.
- Additionally make a knowledge base from the outcome database and concentrate the information through data mining methods, giving us a degree to foresee their future execution.
- To enable the instructive establishments to settle on better choices for the advancement of understudy dependent on substantial statistical data points.

2 Literature Survey

From the sooner studies on analyzing students' performance, Kekane et al. [1] have tried to produce the areas of improvement and generate a score card for the scholar. Venkatesan and Selvaragini [2] have all over that among C4.5 classification rule performed well whereas analysing students' performance. Sai Baba et al. [3] have used classification techniques for analysis and even have enclosed alternative attributes like students' tenth, 12th and B. Tech. first year results for a lot of accuracies.

Chew Li Sa et al. [4] projected a system named student performance analysis system (SPAS) to remain track of students' finish within the college of technology and knowledge Technology (FCSIT). Their projected system offered a clairvoyant system that is able to predict the students' performance in course "TMC1013 System Analysis and Design" that in turn assists the lecturers from system department to identify students that area unit foretold to possess unhealthy performance in that course. Chew Li Sa et al. [4] reserves everywhere that their project concentrates on the event of a system for student performance analysis. Associate info mining technique, classification rule is applied throughout this project to verify the prediction of the scholar performance in course "TMC1013 System Analysis and Design" is possible. The foremost contribution of the SPAS is that it assists the lecturers in conducting student performance analysis. The system helps lecturers classify the students who are expected to fail in the "TMC1013 Program Analysis and Development" course.

Singh et al. [5] proposed to use Data Mining techniques to extract and evaluate faculty results. The principle behind exploitation processing is to cluster college performances on varied criteria subject to the sure constraints and to boot extracting the dependencies among the parameters which might facilitate finding purposeful associations between them. These associations in turn facilitate to identify new patterns for deciding.

Umamaheswari and Niraimathi [6] projected to categorize the scholars into grade order all told their education studies exploring the varied socio-demographic variables (age, gender, name, socio-economic class grade, higher category grade, degree proficiency and further data or talent, etc.) and examine to what extent these factors helps to categorize students in ordering to rearrange for the accomplishment method. Umamaheswari and Niraimathi [6] used bunch, association rules, classification, and outlier detection to judge the scholars' performance.

Venkatesan and Selvaragini [7] presupposed to traverse the data mining techniques that area unit used for the advance of students' performance and in addition establish the foremost effective suited structure of curriculum for this atmosphere. The survey analyses regarding the use of classification algorithms ID3 and C4.5 for the scholar performance analysis system and have analyzed data set containing data regarding students, like gender, marks scored at intervals the board examinations of classes X and XII, marks and rank in entrance examinations and results in history of the previous batch of students; Applied ID3 (Iterative Dichotomiser 3) and C4.5 classification algorithms on this data and expected the and individual performance of freshly admitted students in future examinations. Venkatesan and Selvaragini [7] all over that among the choice of classification algorithms, the performance of C4.5 performs well in method the student's data.

3 Proposed System

Overcoming the constraints of the sooner methodologies, we tend to propose associate in nursing finish to finish system which will majorly facilitate USA create intelligent selections by analyzing large sets of information with the assistance of information mining techniques for our school, i.e., Osmania University at the side of fulfilling alternative objectives mentioned within the introduction.

In Osmania University, there are two examinations (internal and external) command for a combined total of a hundred marks (twenty-five marks for internal examination and seventy-five marks for external examination) for a theory subject and a combined total of seventy-five marks (twenty-five marks for internal examination and fifty marks for external examination) for a practical/lab. Our proposed system has two major tasks involved in it.

  i. Classification
  ii. Prediction.

In achieving our end-to-end system, we start the data extraction and processing task which involves challenges as in for our university the results are displayed as a web page [8]. We can facilitate to download the results from online portals using a web crawler. Then apply pre-processing result and extract information and combine them into one structured form. We can group them further as semester wise results and proceed further with data storage creating a result database [8]. Automated processes are once the process of extracting the results from web portals, deleting noise, grouping and storing them into a database [1].

In Classification, we propose to build a classification model making use of suitable classification technique to categorize the student's result data into categorical class labels like pass as *P* and fail as *F* in an examination. One instance is dividing whole class of 60 students (say of our college) result into two classes "All clear students (pass)" and "students who have backlog (fail)." Likewise we can also classify them into subject-wise pass or fail or categorize each students' result as pass or fail. Here, we ought to use rule-based classification technique.

In prediction, we propose to build a prediction model making use of suitable data mining technique to predict students' future performance in examination. We take each students performance in internal and external examinations and train the prediction model to predict students' external marks based on his/her performance in the internal examination. Also we try to predict if a student might pass or fail in the external examination based on his/her performance in internal examination.

4 Results and Discussion

Task 2—Prediction

In this task, we ought to predict the students' future performances in examination using data mining techniques. Firstly, we try to find the statistical relationship between students' performance in internal examinations and external examinations (EM) and a relationship between students' internal performance and final result (pass/fail) in external examinations. Our aim is to evaluate a linear relationship between internal marks and external marks, internal marks and total marks (Fig. 1).

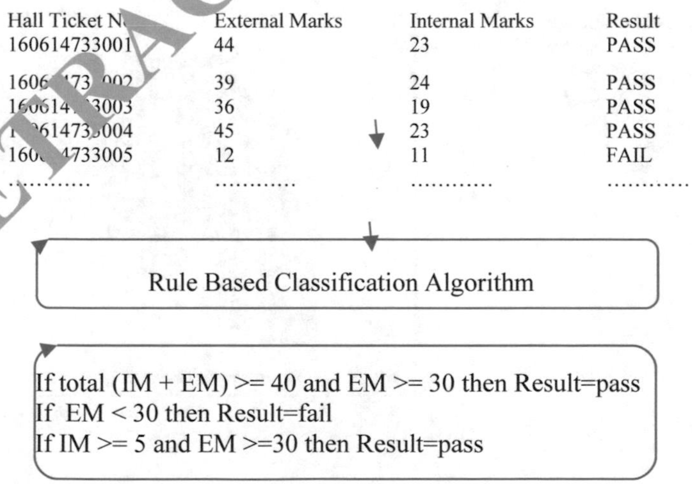

Hall Ticket No.	External Marks	Internal Marks	Result
160614733001	44	23	PASS
1606 73 002	39	24	PASS
160614 3003	36	19	PASS
61473 004	45	23	PASS
160 4733005	12	11	FAIL
…………	…………	…………	…………

Rule Based Classification Algorithm

If total (IM + EM) >= 40 and EM >= 30 then Result=pass
If EM < 30 then Result=fail
If IM >= 5 and EM >=30 then Result=pass

Fig. 1 Rule-based classification system

The goal is to search for an operation that will pull a linear relationship between a group of Predictor options and the dependent variable value.

Linear regression could be a applied math approach for modeling relationship between a variable and a given set of freelance variables. Multivariate analysis could be a variety of prognostic modeling technique that investigates the link between a dependent (target) and experimental variable (s) (predictor). It indicates the strength of impact of multiple independent variables on a variable [9]. It calculates the best-fit line for the determined information by minimizing the total of the squares of the vertical deviations from every information to the road. As a result of the deviations area unit first square, when added, there is no canceling out between positive and negative values [10, 11].

$$Y = bx + a \tag{1}$$

We now use the above formula to calculate a and b as follows

$a = (n\Sigma xy - \Sigma x \Sigma y)/(n\Sigma x^2 - (\Sigma x))$ and $b = (1/n)(\sum y - a \sum x)$

Estimated coefficients for the student's internal marks and external marks (Fig. 2):

$b = -0.000124964745481293$
$a = 1.7425213483574113.$

Estimated coefficients for student's internal marks and total marks (Fig. 3):

$b = -0.00012496474546708214$
$a = 2.74252134835741$ (Table 1).

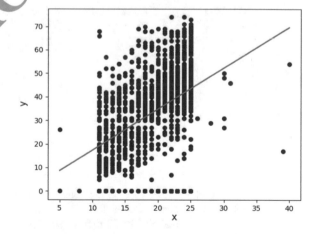

Fig. 2 Regression line for the relation between student's internal marks and external marks

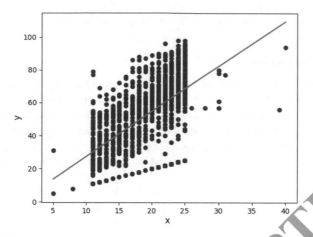

Fig. 3 Regression line for the relation between student's internal mark and total marks

Table 1 Prediction external and total marks using internal marks

Internal marks	Predicted external marks	Internal marks	Predicted external marks
5	8	7	19
10	17	10	27
13	22	3	32
15	26	13	41
18	31	17	46
20	24	20	54
23	40	23	63
25	43	25	68

5   Conclusion and Future Work

Conclusion

We primarily aim to simplify the process right from collecting of results from their respective home gateways till depositing them into a consequence data base. And conjointly to modify the then advanced operations like finding the mixture of a student, the general pass and fail share in an exceedingly subject, etc., and analyze the scholar performance in numerous potential ways in which. Conjointly derive out info sort of a students' stronger and weaker areas, establish the set of scholars United Nations agency would like special attention, etc., with more information aggregations, establishing a knowledge warehouse. Such a sort of machine-driven system can ease the work of the tutorial establishments and facilitate them have a much better situation of the factors touching a students' performance. And conjointly helps the establishments create selections supported valid facts and figures to attain higher results.

Recommendations for Future Work

We could any extend our plan to develop a prediction system creating use of machine learning algorithms and techniques which will predict a students' performance supported his/her previous log, find associations and correlations between varied factors poignant students' performance, etc., includes parameters apart from marks like group action, responsiveness of a student category |in school| at school and alternative class space parameters and facilitate establishments build intelligent selections.

References

1. Kekane, S., Khairnar, D., Patil, R., Vispute, S.R., Gawande, N.: Automatic student performance analysis and monitoring. Int. J. Innovative Res. Comput. Commun. Eng. (2016)
2. Venkatesan, E., Selvaragini, S.: A study on the result based analysis of student performance using data mining techniques. Int. J. Pure Appl. Math. (2017)
3. Sai Baba, Ch. M.H., Govindu, A., Raavi, M.K.S., Som. etty, V.P.: Student performance analysis using classification techniques. Int. J. Pure Appl. Math. (2016)
4. Chew Li Sa, bt. Abang Ibrahim, D.H., Hossain, E.D., bin Hossin, M.: Student performance analysis system (SPAS). In: 2014 the 5th International Conference on Information and Communication Technology for the Muslim World, ICT4M 2014 (2015). https://doi.org/10.1109/ict4m.2014.7020662
5. Singh, C., Gopal, A., Mishra, S.: Extraction and Analysis of Faculty Performance of Management Discipline from Student Feedback Using Clustering and Association Rule Mining Techniques. IEEE (2011)
6. Umamaheswari, K., Niraimathi, S.: A study on student data analysis using data mining techniques. Int. J. Adv. Res. Comput. Sci. Softw. Eng. (2013)
7. Venkatesan, E., Selvaragini, S.: A study on the result based analysis of diabetes data set using mining techniques. Int. J. Pure Appl. Math. **116**, 319–325 (2017)
8. Rachakatla, B., Srinivasu, B., Laxmi, C.P., Thasleem, S.: Students' performance evaluation and analysis manager's. J. Softw. Eng. (JSE) **13**(2), 29–36 (2018)
9. Chatterjee, S., Hadi, A.S.: Regression Analysis by Example. Wiley (2015)
10. Kaytez, F., Cengiz Taplamacioglu, M., Cam, E., Hardalac, F.: Forecasting electricity consumption: a comparison of regression analysis, neural networks and least squares support vector machines. Int. J. Electr. Power Energy Syst. **67**, 431–438 (2015)8
11. Veronese, N., Cereda, E., Stubbs, B., Solmi, M., Luchini, C., Manzato, E., Sergi, G., et al.: Risk of cardiovascular disease morbidity and mortality in frail and pre-frail older adults: results from a meta-analysis and exploratory meta-regression analysis. Ageing Res. Rev. **35**, 63–73 (2017)

Bio-Inspired Scheme of Killer Whale Hunting-Based Behaviour for Enhancing Performance of Wireless Sensor Network

C. Parvathi and Suresha Talanki

Abstract Optimization is one of the best alternative solutions for addressing majority of the computational challenges associated with the wireless sensor network (WSN). At present, the successful implications of bio-inspired approaches are significantly proven to solve diversified complex problems in wireless network. However, after reviewing the existing literatures, it was found that yet there is bigger scope to enhance the capability of bio-inspired approach in the area of WSN. Therefore, this paper introduces a novel bio-inspired algorithm called as killer whale hunting (KWH) targeted for optimizing the data aggregation process with highly improved network sustenance capability. The formulation of the algorithm is designed on the basis of social and cognitive behaviour of killer whale which has a distinct style of hunting its prey. Using analytical modelling, the proposed approach was found to offer better energy efficiency and sufficient data forwarding capability in contrast to existing energy-efficient bio-inspired techniques.

Keywords Energy · Bio-inspired algorithm · Optimization · Data aggregation · Wireless sensor network

1 Introduction

Wireless sensor network (WSN) has established its true identity in wireless networking technologies at present times owing to its cost-effective communication performance [1]. There are various literatures that report of increased usage of WSN in different sets of commercial applications; however, there are a lot of differences

C. Parvathi (✉)
Department of Computer Science & Engineering, Sri Venkateshwara College of Engineering, Bengaluru, India
e-mail: pvc3119@gmail.com

S. Talanki
Sri Venkateshwara College of Engineering, Bengaluru, India
e-mail: suresha_rec@rediffmail.com

© Springer Nature Singapore Pte Ltd. 2020
K. S. Raju et al. (eds.), *Data Engineering and Communication Technology*,
Advances in Intelligent Systems and Computing 1079,
https://doi.org/10.1007/978-981-15-1097-7_29

between information obtained from research-based literatures and actual commercial products and services of WSN in reality. Irrespective of such increasing discussion of adoption of WSN in Internet of things (IoT) in present time, it should be known that even a 10% of such products cannot be seen in real-time application in much abundance. There are various technical reasons behind this contradiction, viz. (i) a sensor node does not have high computational capability that restricts the node to carry out increased number of processes, (ii) the battery life of the sensor is highly important factor which is practically challenging to save for longer network operation, and (iii) existing routing techniques are meant to perform discrete set of operation and hence fail to encapsulate maximum networking issues/demands. Various literatures, e.g. [2–4], will offer evidence that there has been enough research work being carried out towards energy-related problems. However, none of such approaches are proven to offer full-proof solution towards energy problems in different scenarios of WSN. There are various optimization-based studies carried out to improve the performance of the WSN, e.g. neural network-based [5], fuzzy logic-based [6], genetic algorithm-based [7], game theory-based [8]. All these optimization approaches are capable of identifying the problems to higher accuracy, and hence they can perform good optimization over a defined set of problems. However, such approaches are highly unsuitable for any sensory application that demands, e.g., fault-tolerant capture of event, (ii) faster relaying of information, (iii) highly reliable data with lesser data correlation and (iv) increased scalability. Therefore, in search of a better form of optimization, it was found that bio-inspired algorithm has a significant contribution towards optimization [9]. There are significant advantages of using bio-inspired approaches, e.g. evolutionary technique, swarm-based technique and ecology-based technique. There are various reasons to believe that bio-inspired algorithms are highly suitable for solving network-related parameters, viz. (i) a network has a definitive set of behaviour to be characterized that can be modelled using various living organisms very easily, (ii) any principle of uncertainty can be mapped to a problem space of the bio-inspired algorithms using non-complex steps of optimization, and (iii) solutions towards network-based problems could be solved by discrete social and cognitive behaviour of bio-inspired algorithms. The best part of the usage is that bio-inspired algorithms are highly flexible in the form of its implementation by higher degree of fine-tuning capabilities that are quite complex in any other form of optimization-based solution in existing times. Therefore, the potential of bio-inspired algorithm is discussed in proposed system where the prime target is to ensure that network lifetime is significantly retained to highest extent with assurance of optimal data delivery performance in WSN. The proposed paper discusses two algorithms where the first algorithm is responsible for showcasing the novel bio-inspired algorithm design on the basis of hunting behaviour of killer whale while the second algorithm is responsible for further optimizing the performance suitable for both static and mobile nodes in WSN. The idea of the proposed manuscript is mainly to discuss a cost-effective and novel bio-inspired algorithm to offer better network lifetime and node performance in WSN. Section 1.1 discusses the existing literatures where different techniques are discussed for energy efficiency and bio-inspired algorithms followed by discussion of research problems in Sect. 1.2 and proposed solution in Sect. 1.3. Section 2 discusses algorithm building strategy followed by discussion of algorithm implementation in Sect. 3. Result analysis is discussed in Sect. 4. Finally, the conclusive remarks are provided in Sect. 5.

1.1 Background

Our prior study has surveyed existing approaches about energy efficiencies in WSN with more updates on this paper [10]. It has been observed that there are various bio-inspired-based approaches in order to perform optimization. The most recent study carried out by Yang and Yoo [11] has hybridized the ant colony optimization as well as genetic algorithm for route configuration. Alamouri et al. [12] have implemented swarm-based optimization based on cognitive behaviour of wolf and whale for solving the localization problems in WSN. The literatures have also witnessed implementation of bio-inspired technique on advanced applications of WSN as seen in study of Hamrioui and Lorenz [13]. Pheromone-based optimization approach was presented by Verma [14] in order to incorporate reliability factor associated with the routes. Work of Atakan and Akan [15] have presented discussion of bio-inspired algorithm for improving communication over cross-layers in vehicular network integrated with sensors. The study also claims to offer controlled delay and enhanced lifetime. Usage of epidemic theory for propagation of data is discussed in the study of Byun and So [16]. Similar authors [17] have used the bio-inspired approach for ensuring better system stability to claim significant saving of energy as well as controlled delay. Study of conventional ant colony optimization was carried out by Wang et al. [18] considering the presence of mobile base stations for improving the communication of smaller network with energy efficiency. Study emphasizing the potential of using bio-inspired protocol in networking was carried out by Charalambous and Cul [19], where a different network behaviour has been investigated. Improvement of communication in WSN has been carried out by Saleem et al. [20] emphasizing the security aspect too. The study claims of lowered value of delay and energy consumption. Gao et al. [21] have used a mathematical model called as Physarum Solver to address the shortest route issues in WSN. Bitam et al. [22] have studied security aspect with respect to its implications using bio-inspired algorithm. Senel et al. [23] have addressed the localization problems for rectifying the defective node in WSN using spider web concept of bio-inspired algorithm. Song et al. [24] have addressed the problems of coverage Physarum optimization. Connectivity between bio-inspired algorithms with social-driven networking has been presented by Duan et al. [25]. Security problems associated with identification of attacker were also found to be solved by bio-inspired algorithms as seen in the work of Fu et al. [26]. Hybridizing of bacterial foraging and particle swarm optimization was reported to solve localization problem in WSN as claimed in the study of Kulkarni and Venayagamoorthy [27]. Concept of aquatic swarm was reported to assist in location-based services in WSN as discussed by Viera et al. [28]. Similar problems have been also addressed by Reid et al. [29]. Usage of ant colony optimization was reported in the study of Ribeiro and Castro [30] for addressing energy problems along with coverage problems in WSN. Therefore, there are various bio-inspired schemes reported to be benefited from WSN in different perspectives. The next section outlines the research issues associated with it.

1.2   Research Problem

The significant research problems are as follows:

- The novel evolutions of cognitive behaviour of living organisms explored in the area of WSN with respect to improvement in energy related problems are very less in number.
- Existing approaches are highly iterative for which energy benefits are obtained but at the cost of computational complexity in WSN.
- Bio-inspired algorithm towards ensuring faster data aggregation process with higher data quality is yet to be explored.
- A balanced bio-inspired approach to be maintained between energy consumption and other data forwarding performance is yet to be seen with benchmarking.

Therefore, the problem statement of the proposed study can be stated as 'constructing a novel and innovative cognitive behaviour of bio-inspired algorithm to leverage the overall performance of sensor to improve network lifetime in highly cost-effective manner'. The next section discusses solution.

1.3   Proposed Solution

The proposed work is a continuation of our prior study towards the goal of achieving energy efficiency in WSN [31]. This part of the study is meant for accomplishing dual objectives, i.e. (i) to extend the efficiency of framework by incorporating autonomous decision-making routing protocols for better formulation of routes using a novel bio-inspired approach and (ii) to consider both static and mobile nodes for assessing routing performance in order to optimize further lifetime of the network. The schematic diagram of the proposed system is shown in Fig. 1.

In order to implement proposed KWH, the considered bio-inspired parameters are as follows, viz. (i) click interval: corresponding to biological click of killer whale to locate prey, it is considered as a control message to trace the node with high data content that is advertised by every node in this proposed system, (ii) sensing range of node, (iii) fraction of transmitted click to total clicks required to trace the prey, (iv) active communication in the form of two control messages,

Fig. 1 Implementation process for proposed bio-inspired optimization

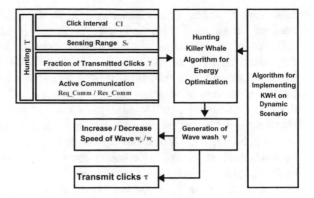

i.e. request message and response message, and (v) hunting time, i.e. total time interval required to successfully hunt the prey (or trace the node with higher degree of information). It should be noted that a target node never gets to read the content of data but it can only check the size of data borne by its neighbouring node. It was done for security reasons. The proposed algorithm uses probability theory to map the bio-inspired parameters with the technical specifications of a sensor node. The KWH algorithm processes to generate wave wash as an outcome, which is a phenomenon that all the active cluster heads must perform communication with the base station at same time (which is similar to the scenario of generating attack on the prey by different killer whales located at different distances from the prey). The communication by all the active cluster heads to base station is carried out by controlling the rate of data transfer; i.e. cluster head located at remote position is allocated faster rate, while cluster head that is at the vicinity of base station is allocated with the lower rate in order to ensure that all active cluster heads perform a highly synchronized data aggregation with base station. The novelty of the proposed system is that it not only positively affects energy conservation but also reduces the computational time required in process of data aggregation, i.e. lesser delay. The next section elaborates about the algorithm design and implementation followed by result analysis.

2 Designing Killer Whale Hunting (KWH) Algorithm

The concept of KWH is descriptively elaborated in this phase; however, a gist of the mapping concept can be as follows: the concept of the term 'killer whale' is mapped with that of sensor nodes in WSN, and the term 'prey' is assumed as data packet. As the study intends to capture as much as data possible by the aggregator node and follow the same path or route for the purpose of forwarding the unique data to base station, the study also deploys the swarm theory of clicks and phenomenon used by killer whale to find out its prey. The study considers clicks equivalent to route request beacon while as the acknowledged route when the beacons are assumed to hit the target and retraced it path back to killer whale. The prime cognitive behaviour of killer whale considered in the current work is that of the establishment of the route using forwarding condition. The work also considers the establishment of communication links from one to another aggregator node with the higher number of neighbour nodes in possession. It also plays the prime role in the work, where the possible aggregator node communicates using multiple hops about the uniqueness or redundancies in their aggregated data before even forwarding the aggregated data directly to the base station in general. Deploying the swarm behaviour proposed, the sensor can eventually substantiate the redundancies in their data and thereby minimize the possibilities of retransmission of aggregated data directed to the base station. Therefore, proposed killer whale swarm optimization mitigates redundancies and retransmission and thereby contributes to energy conservation in wireless sensor network.

In order to understand the mapping principles of the considered work and issues associated with it, consider Fig. 2 with events of data aggregation in WSN. We consider three different swarms of killer whale in Fig. 1, where one of the members inside the pod is assumed to have highest level of physical capability (or energy). Figure 2 also shows availability of two preys. As the work considers the swarm element in aquatic environment, hence both the killer whales and its prey are in mobile condition with respect to time. It can be also seen that there is availability of two preys at two different distances from three swarms of killer whale. Using the social behaviour of killer whale, it can be said that killer whales perform click phenomenon to send the signals for checking the distance of the prey from them. Hence, it can be assumed that all the three different swarms of killer whales are attempting for capturing their prey from three different directions. The killer whale always moves in pods, and they are always connected to each other when they attempt for hunting its prey. This instance will mean that by performing technique, the killer whale has the capability to distinguish, which one is prey or their member of other pods or some non-living objects in the aquatic environment. Moreover, due to constant communication with other pods, killer whales know the exact location of their prey with less exercising of procedure. This cognitive characteristic of killer whale equally assists the killer whales to consume less energy that may be required to perform or swim towards the prey in long routes. Hence, the cognitive behaviour of killer whale assures two prospects, e.g. (1) killer whales do not need to perform much extra effort for hunting, and (2) killer whales are in communication with other pods. This behaviour is considered for the proposed work in wireless sensor network with an aim of achieving effective data aggregation. As discussed in the prior chapters, that proposed work does not use the concept of mobility of any sensor nodes or even base station, so hypothetical concept of killer whales is acquired keeping in mind for the adopted optimization technique for better data aggregation in WSN. It is already known that communication protocols in WSN employ retransmission schemes as the wireless transmission channels are lossy. Therefore, retransmission schemes are applied to boost the network performances. The next section discusses the algorithm implementation.

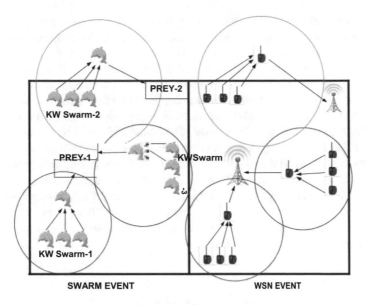

Fig. 2 Mapping of killer whale swarm events in WSN

3 Algorithm Implementation

The first stage of algorithm implementation is related to setting up a network suitable for KWH behavioural study: the proposed technique assumes a larger-scale WSN system where all the sensor nodes are widely distributed in random fashion. Mobility of the nodes is not considered as in previous implementation of technique. Consecutively, every sensor node being distributed in the simulation area is assumed to have uniform communication range and initial energy (in Joules) of all the nodes is same. However, rate of energy drainage by every node differs after the simulation round is initiated. Once the node depletes its energy, it eventually dies and there is no possibility of that node to be acting as aggregator node in future rounds, quite logically. It is also assumed that every sensor node has their position information as well as information related to their residual energy. The network is also assumed to be free from any interference or fading issues. Finally, the base station is also considered as statically located and has no limitation of power supply. The second stage of the algorithm construction is more about designing nodes. The assumptions done on the design of the nodes are as follows: (i) each killer whale represents a sensor node, and coordinate of the killer whales is state position of the sensor node. However, interesting question here lies is how coordinates of killer whales can be mapped by the term state in WSN. Some of the literature study tells that the usage of the term 'state' has been frequently used in the mapping principles of swarm intelligence with WSN. For example, Haberman and Sheppard [32] discussed that conventional routing protocols for wireless networks typically need the state messages to be broadcasted using flooding techniques throughout the network (global information) for each node to make informed and correct path selection that are optimal. According to the author, the term 'state' will mean solution, strategies or attributes of a function to be optimized in SI. The author has used the concept of 'state' in their work that represents the current power level of each neighbour's battery and the hop count to the destination through each neighbour as a single objective measure. However, the usage of the term 'state' differs from the objectives of the study. The study performed by Dorigo et al. [33] highlights states as sending and receiving of messages by swarm elements in WSN. The author has also analysed around 1010 states for designing robot swarm intelligence. In the proposed study, the term state will represent two properties of node communication, data receiving state ((E_{RX})-amount of energy required to perform receiving of data in process) and transmitting state ((E_{TX})-amount of energy required to perform sending the beacons in click generation). The brief discussion of the proposed algorithm is as follows.

3.1 Killer Whale Hunting Algorithm for Energy Optimization

The algorithm takes the input of all sensors (Line-1) but selects only the sensors which has more proportion of data within itself using distance between them (Line-2 and Line-3). A tree is constructed on the basis of wave matrix in order to initiate generation of wave wash process by all the selected sensors considering input arguments of position of sensor node (n_x, $_y$) and sensor node with more data, i.e. (p_x, $_y$) (Line-4). The next step of proposed KWH is about performing clustering operation by a simple processing of generating clustering identity, i.e. cluster head node arbitrarily (Line-8). After the cluster is formulated, the sensors start transmitting clicks τ with respect to their hunting time (T) (Line-10). A unique message is acquired among all the communicating nodes on the basis of click interval (CI), sensing range (S_r), fraction of transmitted clicks (γ) and duration of click exchange (θ) (Line-11). Two objective functions $f_1(x)$ and $f_2(x)$ are constructed that is meant for maximizing transmission of such unique message (Line-12) and joint maximization of speed control of wave (i.e. w^+ and w^- to represent increase and decrease speed of wave, respectively). The wave speed is basically used to ensure that all the sensor nodes generate the wave from different positions at same time such that it could hit the prey at same time by increasing or decreasing its velocity. The steps of the algorithm are as follows:

Killer Whale Hunting Algorithm for Energy Optimization

Input: n (number of sensors)
Output: f_1/f_2(Multi-objective Optimization)
Start
1. **For** i =1: n
2.   **If** $d < S_r$
3.   filter n
4. $\psi \xrightarrow{} \text{graph}(W_{mat}[n_{x,y}, p_{x,y}])$
5.   **End**
6.   **For** j=1: n_c
7.   **If** $k < n_c$
8. $C_{id} \xrightarrow{} \text{arb}[(k\text{-}1).n / n_c]$
9.   **End**
10. $\tau \xrightarrow{} C_{id}(T)$
11. $\text{msg} \xrightarrow{} \text{unique}(CI, S_r, \gamma, \theta)$
12. $f_1(x) \xrightarrow{} \text{argmax}(\text{msg})$
14. $f_2(x) \xrightarrow{} (\text{argmax}(w^+) \,\&\&\, \text{argmax}(w^-))$
15.   **End**
16. **End**
End

3.2 Algorithm for Implementing KWH on Dynamic Scenario

This algorithm is responsible for making the KWH algorithm compatible to be implemented for both conditions of static and mobile nodes. It is imperative that mobility induces more energy consumption and hence it is restrained using proposed KWH algorithm by slightly fine-tuning it. For this reason, the algorithm constructs a protocol that is meant for convergence point of local optima (L_{prot}) and global optima (G_{prot}) (Line-2). An explicit energy modelling function $f_3(E)$ was carried out using second-order radio energy model in order to compute energy dissipation (Line-3). For all the static nodes, the algorithm primarily checks for all the nodes that has energy dissipation rate higher than normal states (Line-5) followed by implying KWH algorithm. For all the mobile nodes, the algorithm computes all the draining nodes whose energy consumption is more than certain cut-off value (Line-9) followed by implying KWH algorithm (Line-10). The final step of the algorithm is to perform optimization of network performance by ensuring maximum level of energy conservation both in static state and in mobile state (Line-11). The algorithm emphasizes more energy optimization on mobility state as it drains more energy than static states. The steps of the algorithm are as follows:

Algorithm for Implementing KWH on Dynamic Scenario

Input: n (sensors), Th (Threshold)
Output: f_3 (optimize node performance)
Start
1. **For** $i = 1: n$
2.   Prot $\rightarrow$ (L_{prot}, G_{prot})
3.   Apply $f_3(E)$
4.   **If** $n=n_{stat}$
5.   identify d_{node}
6.   Apply algorithm-1
8.   **Else**
9.   Select n_{mob} $\rightarrow$ d_{node}>Th
10.   Apply algorithm-1
11. $f_3(x)$ $\rightarrow$ $\arg_{max}(n_{stat}, n_{mob})$
12. **End**
End

4   Result Analysis

The proposed system emphasizes introducing a novel bio-inspired algorithm with major target to achieve energy efficiency. The study outcome is analysed over standard benchmarked configuration of MEMSIC nodes along with the following numerical setting of γ (fraction of transmitted clicks) = 0.15, θ (duration of click exchange) = 0.03, Req_comm/Res_comm (active communication for request and response) = 0.12, w^+ (increased speed of wave) = 0.17, w^- (decreased speed of wave) = 0.68, length of frame = 1 s, size of data = 20 bytes. The study outcome is assessed with respect to multiple performance parameters, e.g. energy, delay, throughput, and is also compared with work carried out by Celestino et al. [34] and Zhou et al. [35]. For identification, we use BEEC to represent work of Celestino et al. [34] and IPSO as work of Zhou et al. [35].

From the above numerical outcomes, it can be seen that proposed system offers significant benefits with respect to residual energy (Fig. 3) and end-to-end delay (Fig. 4). The above outcomes are a direct representation of proposed bio-inspired algorithm implemented, i.e. KWH only. A closer look into the outcome shows that

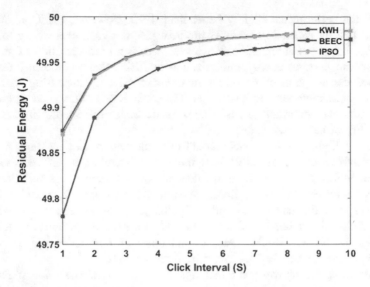

Fig. 3 Comparative analysis of residual energy

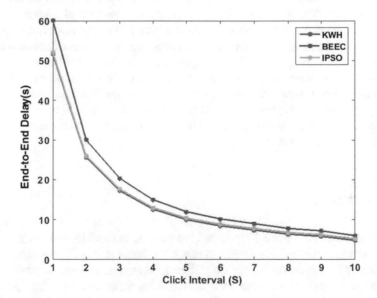

Fig. 4 Comparative analysis of delay

there is no significant difference between the outcome of proposed system and IPSO; it is because the optimization carried out by implementing improved particle swarm optimization by Zhou et al. [35] offers similar pattern of convergence in order to explore energy-efficient clusters in WSN along with exploring shortest route. Hence, both the algorithms behave nearly in same manner, but yet KWH exceeds the performance slightly better than IPSO. However, BEEC algorithm is implemented by enhancing LEACH protocol [36] designed exclusively to handle multihop- based communication. However, there is no incorporation of any function that can perform energy minimization in BEEC that significantly results in increased consumption of resources in order to cater up increasing traffic with increase of click intervals of the sensors. Therefore, KWH offers good performance optimization where further optimization is carried out by considering different states of mobility of the sensor node.

Figure 5 highlights the benchmarked outcome where it offers distinct performance analysis of energy and throughput for a discrete state of node mobility. A closer look into Fig. 5a, b shows that proposed system offers significant performance improvement by retaining maximum alive nodes in mobility state as compared to static state of the nodes. Similarly, it can be seen that IPSO offers better lifetime in mobility state as compared to static state whereas performance degrades for BEEC. Similar trend of outcomes of residual energy as well as throughput can be seen in Figs. 5c, d, e, f, respectively.

Apart from the above-mentioned outcomes, it was also seen that proposed system offers 87% of faster response time in performing data aggregation as compared to existing system of BEEC and IPSO. The prime reason behind this is that proposed KWH algorithm offers extremely less number of iterative steps in order to compute the best local and global solution that not only result in faster processing time but directly contribute towards achieving lesser dependencies on different resources for performing data aggregation in WSN. Another reason is that proposed system offers multi-objective optimization designed on ensuring satisfaction of different objective functions in order to ensure that more energy saving could be rendered during mobility condition of node. Hence, the implementation of proposed KWH offers more practicality in operating over uncertain environment as compared to existing system.

5 Conclusion

The evolution of bio-inspired algorithm has proven its capability to solve complex mathematical problems. However, existing bio-inspired algorithms have certain characteristics, e.g. dependencies on defining convergence point in advance, and elite outcomes only appear after load of iterations, dependencies of computational resources, etc. They could be very efficient for other cases of network but definitely not for WSN which has highly resource-constrained sensor nodes. Therefore, proposed KWH offers competitive edge over other bio-inspired techniques owing to its following contribution towards WSN as follows: (i) it is highly reduced iterative operation and does not have any form of memory dependencies, (ii) it is more practical in operation as it considers temporal factor of every instance during data

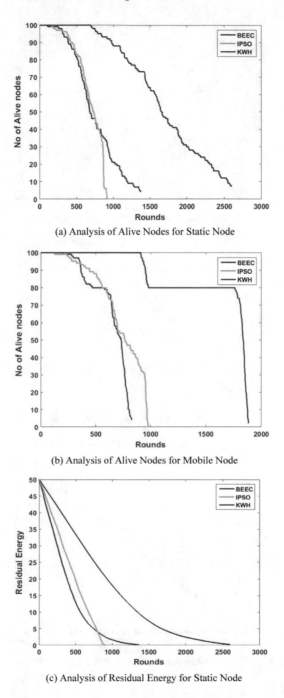

(a) Analysis of Alive Nodes for Static Node

(b) Analysis of Alive Nodes for Mobile Node

(c) Analysis of Residual Energy for Static Node

Fig. 5 Benchmarked outcomes on multiple scenarios

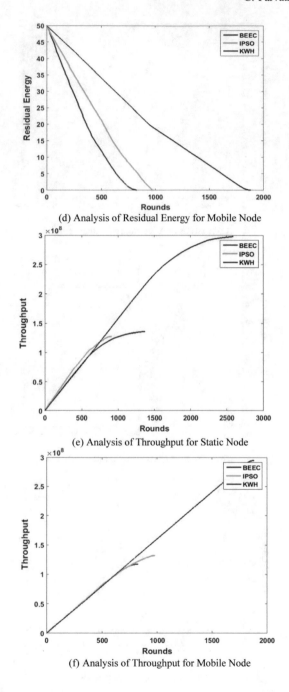

(d) Analysis of Residual Energy for Mobile Node

(e) Analysis of Throughput for Static Node

(f) Analysis of Throughput for Mobile Node

Fig. 5 (continued)

aggregation, (iii) it also offers a complete scheduling process of increasing/
decreasing allocation of transmittance energy in order to ensure higher volume of
unique data is forwarded to base station.

References

1. Sghaier, N., Mellouk, A.: Energy Efficient Design of Wireless Sensor Networks. Elsevier-ISTE Press Limited (2018)
2. Manikannu, J., Nagarajan, V.: A survey of energy efficient routing and optimization techniques in wireless sensor networks. In: 2017 International Conference on Communication and Signal Processing (ICCSP), Chennai, pp. 2075–2080 (2017)
3. Bhushan, B., Sahoo, G.: A comprehensive survey of secure and energy efficient routing protocols and data collection approaches in wireless sensor networks. In: 2017 International Conference on Signal Processing and Communication (ICSPC), Coimbatore, pp. 294–299 (2017)
4. Ali, N.F., Said, A.M., Nisar, K., Aziz, I.A.: A survey on software defined network approaches for achieving energy efficiency in wireless sensor network. In: 2017 IEEE Conference on Wireless Sensors (ICWiSe), Miri, pp. 1–6 (2017)
5. Sahagun, M.A.M., Dela Cruz, J.C., Garcia, R.G.: Wireless sensor nodes for flood forecasting using artificial neural network. In: 2017 IEEE 9th International Conference on Humanoid, Nanotechnology, Information Technology, Communication and Control, Environment and Management (HNICEM), Manila, pp. 1–6 (2017). https://doi.org/10.1109/hnicem.2017.8269462
6. The, P.T., Manh, V.N., Hung, T.C., Dien Tam, L.: Improving network lifetime in wireless sensor network using fuzzy logic based clustering combined with mobile sink. In: 2018 20th International Conference on Advanced Communication Technology (ICACT), Chuncheon-si Gangwon-do, Korea (South), pp. 1–1 (2018)
7. Shehab, A., Elhoseny, M., Sahlol, A.T., Aziz, M.A.E.: Self-organizing single-hop wireless sensor network using a genetic algorithm: longer lifetimes and maximal throughputs. In: 2017 IEEE International Conference on Intelligent Techniques in Control, Optimization and Signal Processing (INCOS), Srivilliputhur, pp. 1–6 (2017)
8. Wu, X., Tang, Y.Y., Fang, B., Zeng, X.: An efficient distributed clustering protocol based on game-theory for wireless sensor networks. In: 2016 7th International Conference on Cloud Computing and Big Data (CCBD), Macau, pp. 289–294 (2016)
9. Barocio, E., Regalado, J., Cuevas, E., Uribe, F., Zúñiga, P., Torres, P.J.R.: Modified bio-inspired optimisation algorithm with a centroid decision making approach for solving a multi-objective optimal power flow problem. IET Gener. Transm. Distrib. 11(4), 1012–1022 (2017)
10. Parvathy, C., Suresha: Existing routing protocols for wireless sensor network-a study. Int. J. Comput. Eng. Res. 4(7) (2014)
11. Yang, Q., Yoo, S.J.: Optimal UAV path planning: sensing data acquisition over IoT sensor networks using multi-objective bio-inspired algorithms. IEEE Access 6, 13671–13684 (2018)
12. Alomari, A., Phillips, W., Aslam, N., Comeau, F.: Swarm intelligence optimization techniques for obstacle-avoidance mobility-assisted localization in wireless sensor networks. IEEE Access. https://doi.org/10.1109/access.2017.2787140
13. Hamrioui, S., Lorenz, P.: Bio inspired routing algorithm and efficient communications within IoT. IEEE Netw 31(5), 74–79 (2017)
14. Verma, V.K.: Pheromone and path length factor-based trustworthiness estimations in heterogeneous wireless sensor networks. IEEE Sens. J. 17(1), 215–220 (2017)

15. Atakan, B., Akan, O.B.: Bio-inspired cross-layer communication and coordination in sensor and vehicular actor networks. IEEE Trans. Veh. Technol. **61**(5), 2185–2193 (2012)
16. Byun, H., So, J.: Node scheduling control inspired by epidemic theory for data dissemination in wireless sensor-actuator networks with delay constraints. IEEE Trans. Wireless Commun. **15**(3), 1794–1807 (2016)
17. Byun, H., Son, S., Yang, S.: Biologically inspired node scheduling control for wireless sensor networks. J. Commun. Netw. **17**(5), 506–516 (2015)
18. Wang, J., Cao, J., Li, B., Lee, S., Sherratt, R.S.: Bio-inspired ant colony optimization based clustering algorithm with mobile sinks for applications in consumer home automation networks. IEEE Trans. Consum. Electron. **61**(4), 438–444 (2015)
19. Charalambous, C., Cui, S.: A biologically inspired networking model for wireless sensor networks. IEEE Netw. **24**(3), 6–13 (2010)
20. Saleem, K., Fisal, N., Al-Muhtadi, J.: Empirical studies of bio-inspired self-organized secure autonomous routing protocol. IEEE Sens. J. **14**(7), 2232–2239 (2014)
21. Gao, C., Yan, C., Adamatzky, A., Deng, Y.: A bio-inspired algorithm for route selection in wireless sensor networks. IEEE Commun. Lett. **18**(11), 2019–2022 (2014)
22. Bitam, S., Zeadally, S., Mellouk, A.: Bio-inspired cybersecurity for wireless sensor networks. IEEE Commun. Mag. **54**(6), 68–74 (2016)
23. Senel, F., Younis, M.F., Akkaya, K.: Bio-inspired relay node placement heuristics for repairing damaged wireless sensor networks. IEEE Trans. Veh. Technol. **60**(4), 1835–1848 (2011)
24. Song, Y., Liu, L., Ma, H., Vasilakos, A.V.: A biology-based algorithm to minimal exposure problem of wireless sensor networks. IEEE Trans. Netw. Serv. Manage. **11**(3), 417–430 (2014)
25. Duan, D., Yang, L., Cao, Y., Wei, J., Cheng, X.: Self-organizing networks: from bio-inspired to social-driven. IEEE Intell. Syst. **29**(2), 86–90 (2014)
26. Fu, B., Xiao, Y., Liang, X., Philip Chen, C.L.: Bio-inspired group modeling and analysis for intruder detection in mobile sensor/robotic networks. IEEE Trans. Cybern. **45**(1), 103–115 (2015)
27. Kulkarni, R.V., Venayagamoorthy, G.K.: Bio-inspired algorithms for autonomous deployment and localization of sensor nodes. IEEE Trans. Syst. Man Cybern. Part C (Appl. Rev.) **40**(6), 663–675 (2010)
28. Vieira, L.F.M., Lee, U., Gerla, M.: Phero-trail: a bio-inspired location service for mobile underwater sensor networks. IEEE J. Sel. Areas Commun. **28**(4), 553–563 (2010)
29. Reid, A., Uttamchandani, D., Windmill, J.F.C.: Optimization of a bio-inspired sound localization sensor for high directional sensitivity. In: 2015 IEEE SENSORS, Busan, pp. 1–4 (2015)
30. Ribeiro, L.B., de Castro, M.F.: BiO4SeL: a bio-inspired routing algorithm for sensor network lifetime optimization. In: 2010 17th International Conference on Telecommunications, Doha, pp. 728–734 (2010)
31. Parvathi, C., Suresha: AEOC: a novel algorithm for energy optimization clustering in wireless sensor network. In: Springer-Computer Science On-line Conference, pp. 216–224 (2017)
32. Haberman, B.K., Sheppard, J.W.: Overlapping particle swarms for energy-efficient routing in sensor networks. ACM-Digital Library. J. Wirel. Netw. **18**(4), 351–363 (2012)
33. Dorigo, M., Birattari, M.: Swarm Intelligence: 7th International Conference. Springer Science & Business Media (2010)
34. da Silva Rego, A., Celestino, J., dos Santos, A., Cerqueira, E.C., Patel, A., Taghavi, M.: BEE-C: a bio-inspired energy efficient cluster-based algorithm for data continuous dissemination in wireless sensor networks. In: 2012 18th IEEE International Conference on Networks (ICON), Singapore, pp. 405–410 (2012)

35. Zhou, Y., Wang, N., Xiang, W.: Clustering hierarchy protocol in wireless sensor networks using an improved PSO algorithm. IEEE Access **5**, 2241–2253 (2017)
36. Heinzelman, W., Chandrakasan, A., Balakrishnan, H.: Energy-efficient communication protocols for wireless microsensor networks. In: Proceedings of the 33rd Hawaaian International Conference on Systems Science (HICSS) (2000)

Fingerprint Cryptosystem Using Variable Selection of Minutiae Points

Mulagala Sandhya, Mulagala Dileep, Akurathi Narayana Murthy and Md. Misbahuddin

Abstract In recent years, many template protection methods for fingerprints have been developed. We present a method for constructing a bio-cryptosystem for fingerprints by selecting neighbors of a minutia point variable. To select a variable number of neighbors for each minutia point, we construct a Delaunay triangulation for the minutiae points in the fingerprint image. We derive a bit string from the neighbors in the Delaunay triangulation. This bit string is protected using fuzzy commitment scheme. Here, we use convolution coding to encrypt the bit string. The experimental tests conducted on FVC 2002 databases, yield satisfactory results, thereby proving the viability of the proposed method.

Keywords Template protection · Fuzzy commitment · Cancelable biometrics

1 Introduction

Biometrics is recognizing a person using her physiological/behavioral characteristics/traits. It is used in many applications including authentication, security, access control and surveillance. Traditional authentication, e.g. passwords or tokens, subjected to theft or loss made security systems search for alternatives. Biometrics provided a solution to this issue because they are permanent and are

M. Sandhya (✉)
Department of CSE, National Institute of Technology, Warangal, India
e-mail: msandhya@nitw.ac.in

M. Dileep
Department of ECE, Vishnu Institute of Technology, Bhimavaram, India

A. N. Murthy · Md. Misbahuddin
Department of ECE, Osmania University, Hyderabad, India

© Springer Nature Singapore Pte Ltd. 2020
K. S. Raju et al. (eds.), *Data Engineering and Communication Technology*,
Advances in Intelligent Systems and Computing 1079,
https://doi.org/10.1007/978-981-15-1097-7_30

difficult to counterfeit [1]. Fingerprint is mostly used and well-researched biometric trait. If an attacker tries to copy a biometric system, she has to get the biometric sample to make a trial as a spoof. In the security applications, a biometric template is transformed to a new one, called as cancelable biometrics. A one-way transformation is performed on the biometric data during the enrollment phase. During authentication, the transformation used during enrollment is applied to the query's biometric. The matching module does its function in the transformed domain [2].

The paper is arranged as follows: The related work is given in Sect. 2. Section 3 provides a detailed explanation of proposed work, algorithm and flowchart. Section 4 discusses the experiments conducted and their analysis. The conclusion is given in Sect. 5.

2 Related Work

Wang et al. [3] use partial Hadamard transform to protect binary representation of fingerprint templates. It preserves the stochastic distance between binary vectors after the transformation. Gomezbarrer et al. [4] use Bloom filter-based protection in weighted feature level fusion for face, iris, fingerprint, and finger vein. Sandhya and Prasad [5] developed a score level fusion method for generating cancelable templates of fingerprints. Fierrez et al. [6] provided a review on applying multiple classifier systems to biometrics.

Wu et al. [7] given FVHS key generation algorithm from finger vein biometrics. This paper gave new avenue by the combination of machine learning, cryptography and biometric technologies, a feature vector can be mined by using the biometrics space which can be stabilized into a fixed number sequence in a higher-dimensional space. Dhall et al. [8] demonstrate cryptanalysis of an image encryption scheme based on the 1D chaotic system. Anees et al. [9] generate a secure cryptographic key from facial features by the equalized local binary pattern. Sandhya and Prasad [10] developed a cancelable cryptosystem for fingerprints using convolution coding for encoding bit string generated from fingerprint features.

3 Proposed Method

The following steps are used in variable selection of minutiae points:

1. Construction of Delaunay triangulation from the minutiae points
2. For each minutia point, find neighbors in the Delaunay triangulation
3. Map neighbors to a 3D array and derive bit string

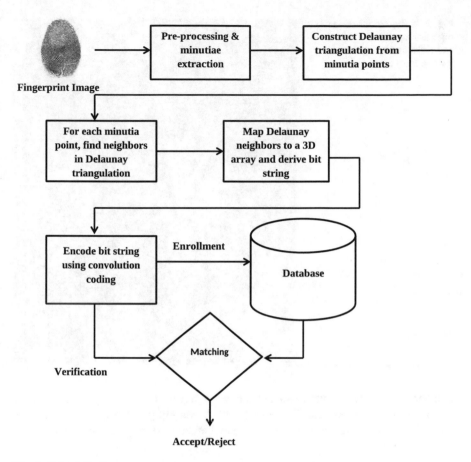

Fig. 1 Schematic diagram

4. Use convolution coding to encode the bit string
5. Matching.

The flow chart is given in Fig. 1. The detailed Algorithm 1 gives the step-by-step procedure of the proposed method.

3.1 Construction of Delaunay Triangulation

The fingerprint image is preprocessed and minutiae points are extracted. Let, $M_i = (x_i, y_i, \theta_i)_1^k$ represents the minutiae points extracted from a fingerprint image, k is the

Fig. 2 A fingerprint image
representing minutia points
and Delaunay triangulation
formed from them

number of minutiae points extracted, (x_i, y_i) represent the x- and y-coordinates of
minutia point, θ_i represents the orientation of minutia point. Figure 2 refers to a
fingerprint image, its minutiae points and Delaunay triangulation formed from
minutia points.

3.2   Finding Neighbors in Delaunay Triangulation

From Fig. 2, the neighbors in of each minutia point are found from the Delaunay
triangulation. This is as shown in Fig. 3. Now, the neighbors are mapped to 3D
array as shown in Fig. 4.

Algorithm 1 CRYPT $(x_i, y_i, \theta_i, A_1, c_x, c_y, c_z, n, k, \lambda)$: To compute Delaunay
neighbor structures and generate secure template

Input: Minutiae locations (x_i, y_i)
Orientation of minutiae points θ_i
Predefined 3D Array A_1 divided into cells of size c_x, c_y, c_z

Fig. 3 Neighbors of minutia *r* in a Delauany triangulation

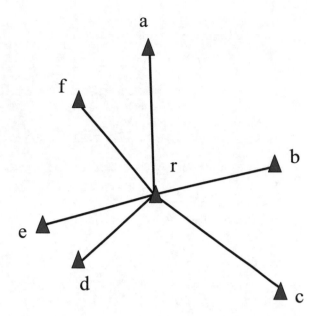

Fig. 4 3D array to which neighbors are mapped

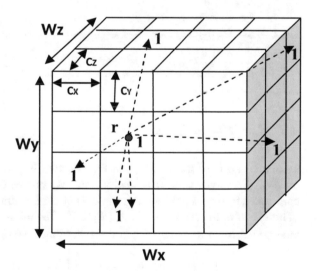

Output: Normalized match score

1: **begin**
2: $t \leftarrow$ number of minutiae points in the fingerprint image;
3: Construct Delaunay triangulation net from the fingerprint minutia;
4: Initialize minutia counter $i \leftarrow 0$;
5: **while** $i \leq t$ **do**
   //*Find neighbor minutiae in Delaunay triangulation net for each minutia point*
6: $DNS(i) \leftarrow$ Neighboring minutiae of i in Delaunay triangulation net;
7: Quantize $DNS(i)$ by cell sizes c_x, c_y, c_z, map to 3D array A_1;
8: Generate bit string B_s;
9: Encrypt B_s wit convolution coding to et Codeword c
10: $e \leftarrow b \oplus c$;
   //*Hash value generation*
11: $h \leftarrow \text{SHA} - 1(c)$;
   //*Store encrypted template e*
12: $Encrypt\_temp(i) \leftarrow e$;
13: $i \leftarrow i + 1$;
14: **end while**
15: Match score generation;
16: **return** match score;
17: **end**

3.3 Deriving a Bit String

Take a 3D array of size W_x, W_y and W_z shown in Fig. 4, where W_x ranges from 0 to max(x), W_y ranges from 0 to max(y), and W_z ranges from 0 to 2π. This array is partitioned into cells of size c_x, c_y and c_z [11]. The central minutia is placed at center of first layer of 3D array, i.e., $(W_x/2, W_y/2, 0)$. The minutiae around it are rotated and then transformed to $(x_i^1, y_i^1, \theta_i^1)$ in the following manner:

$$\begin{bmatrix} x_i^1 \\ y_i^1 \end{bmatrix} = \begin{bmatrix} \cos\theta_r & -\sin\theta_r \\ \sin\theta_r & \cos\theta_r \end{bmatrix} \begin{bmatrix} (x_i - x_r) \\ -(y_i - y_r) \end{bmatrix} + \begin{bmatrix} W_x/2 \\ W_y/2 \end{bmatrix} \tag{1}$$

$$\theta_i^1 = \begin{cases} \theta_i - \theta_r & \text{if } \theta_i \geq \theta_r \\ 2\pi + \theta_i - \theta_r & \text{if } \theta_i < \theta_r \end{cases} \tag{2}$$

Quantization by the cell sizes c_x, c_y, c_z is as follows:

$$\begin{bmatrix} x_i^{11} \\ y_i^{11} \\ \theta_i^{11} \end{bmatrix} = \begin{bmatrix} x_i^1/c_x \\ y_i^1/c_y \\ \theta_i^1/c_z \end{bmatrix} \qquad (3)$$

$(x_i^{11}, y_i^{11}, \theta_i^{11})$ is mapped to a 3D array. If one or more $(x_i^{11}, y_i^{11}, \theta_i^{11})$ occupy a cell, its value is treated as 1 otherwise 0 as in Fig. 4. Visiting the cells in a sequential manner produces a bit string.

3.4 Encoding and Decoding

A codeword c is selected from random error-correcting codes. A key is generated randomly and encoded by convolution code, say c. An XOR operation is applied between b and c, represented as an encrypted template $e = b \oplus c$. A hash value is generated from c is computed using Secure Hash Algorithm (SHA-1), say $h(c)$. This $h(c)$ is stored with e, termed as helper data. This process is called encoding phase.

A query image is presented during decoding phase, from which the binary feature vector b^1 is obtained. b^1 is XOR'd with 'e,' $e^1 = b^1 \oplus e$. The hash value of c^1 is computed. The two hashes are compared to make an accept/reject decision and to decrypt e.

4 Results and Discussion

We use FVC 2002 databases [12] for testing the proposed method. The minutiae points are extracted from fingerprint images using Neurotechnology VeriFinger SDK [13]. FVC 2002 database contains 800 fingerprint images collected from 100 users. Each finger of a user has eight samples. The first sample of each user is used for training, the second sample is used for testing our method. The EER value is used for performance evaluation.

Table 1 EER and d^l values

Cell sizes of 3D array			FVC 2002					
			DB1		DB2		DB3	
c_x	c_y	c_z	EER	d^l	EER	d^l	EER	d^l
10	10	10	3.56	2.98	3.56	2.67	11.51	1.35
10	**10**	**15**	**2.96**	**3.13**	**1.63**	**3.52**	**6.78**	**2.35**
20	20	15	4.43	3.81	5.34	4.06	8.55	3.71
20	20	20	3.16	4.46	2.98	4.41	8.38	3.45

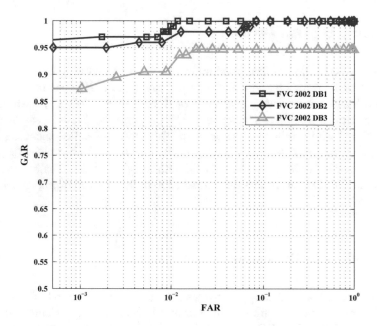

Fig. 5 ROC curves

4.1 Accuracy

We take different cell sizes, c_x, c_y and c_z, and test the proposed method. The EER obtained by varying cell sizes is represented in Table 1. The cells with size $c_x = 10$, $c_y = 10$ and $c_z = 15$, shows optimal EER value (low).

The ROC curves obtained are shown in Fig. 5. From the figure, it is shown that the GAR is better for FVC 2002 DB1 than that of DB2 nd DB3. The EER obtained for the three databases is given in Fig. 6. Score distributions are plotted in Fig. 7

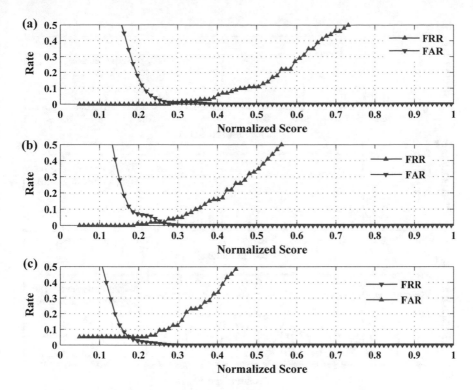

Fig. 6 EER obtained for FVC 2002 **a** DB1 **b** DB2 **c** DB3

5   Performance Comparison

Table 2 depicts the comparison of EER value over FVC 2002 databases. Our method shows low EER compared to [10, 14, 15] for FVC 2002 databases. Our method has shown low EER value for FVC2002 DB1 compared to [16].

6   Conclusion

We presented bio-cryptosystem for securing fingerprint templates. A bio-cryptosystem for fingerprints is developed by variable selection of minutiae points. The use of convolution code reduces EER. The method proposed in this paper assures the requirements of template protection schemes. The experimental evaluation conducted proves that the proposed method achieves better accuracy as well as security.

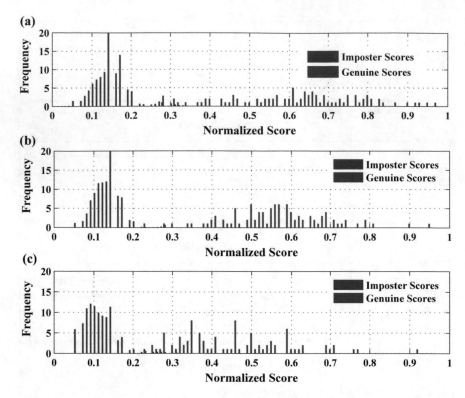

Fig. 7 Genuine and imposter distribution for FVC 2002 **a** DB1 **b** DB2 **c** DB3

Table 2 EER comparison

Method	FVC 2002		
	DB1	DB2	DB3
Yang et al. [15]	5.93	4	–
Jin et al. [16]	4.36	1.77	–
Sandhya et al. [10]	4.38	3.14	6.98
Proposed method	2.96	1.63	6.78

References

1. Ratha, N., Connell, J., Bolle, R.: An analysis of minutiae matching strength. In: Audio- and Video-Based Biometric Person Authentication, Lecture Notes in Computer Science, vol. 2091, pp. 223–228. Springer, Berlin, Heidelberg (2001)
2. Ratha, N., Chikkerur, S., Connell, J., Bolle, R.: Generating cancelable fingerprint templates. IEEE Trans. Pattern Anal. Mach. Intell. **29**(4), 561–572 (2007)

3. Wang, S., Deng, G., Hu, J.: A partial Hadamard transform approach to the design of cancelable fingerprint templates containing binary biometric representations. Pattern Recogn. **61**, 447–458 (2017). https://doi.org/10.1016/j.patcog.2016.08.017; http://www.sciencedirect.com/science/article/pii/S0031320316302291
4. Gomez-Barrero, M., Rathgeb, C., Li, G., Ramachandra, R., Galbally, J., Busch, C.: Multi-biometric template protection based on bloom filters. Inf. Fusion **42**, 37–50 (2018)
5. Sandhya, M., Prasad, M.V.: Multi-algorithmic cancelable fingerprint template generation based on weighted sum rule and t-operators. Pattern Anal. Appl. 1–16 (2016)
6. Fierrez, J., Morales, A., Vera-Rodriguez, R., Camacho, D.: Multiple classifiers in biometrics, Part 1: Fundamentals and review. Inf. Fusion **44**, 57–64 (2018). https://doi.org/10.1016/j.inffus.2017.12.003; http://www.sciencedirect.com/science/article/pii/S1566253517307285
7. Wu, Z., Tian, L., Li, P., Wu, T., Jiang, M., Wu, C.: Generating stable biometric keys for flexible cloud computing authentication using finger vein. Inf. Sci. 433–434, 431–447 (2018). https://doi.org/10.1016/j.ins.2016.12.048; http://www.sciencedirect.com/science/article/pii/S0020025516322769
8. Dhall, S., Pal, S.K., Sharma, K.: Cryptanalysis of image encryption scheme based on a new 1d chaotic system. Sig. Process. **146**, 22–32 (2018). https://doi.org/10.1016/j.sigpro.2017.12.021; http://www.sciencedirect.com/science/article/pii/S0165168417304346
9. Anees, A., Chen, Y.P.P.: Discriminative binary feature learning and quantization in bio-metric key generation. Pattern Recogn. **77**, 289–305 (2018). https://doi.org/10.1016/j.patcog.2017.11.018; http://www.sciencedirect.com/science/article/pii/S0031320317304739
10. Sandhya, M., Prasad, M.V.: Cancelable fingerprint cryptosystem based on convolution coding. In: Advances in Signal Processing and Intelligent Recognition Systems, pp. 145–157. Springer (2016)
11. Lee, C., Kim, J.: Cancelable fingerprint templates using minutiae-based bit-strings. J. Netw. Comput. Appl. **33**(3), 236–246 (2010)
12. http://bias.csr.unibo.it/fvc2002: Fingerprint Verification Competition. Accessed 30 Jan 2014
13. http://www.neurotechnology.com: Neurotechnology VeriFinger SDK. Accessed 11 Dec 2013
14. Yang, W., Hu, J., Wang, S., Stojmenovic, M.: An alignment-free fingerprint bio-cryptosystem based on modified voronoi neighbor structures. Pattern Recogn. **47**(3), 1309–1320 (2014)
15. Yang, W., Hu, J., Wang, S., Yang, J.: Cancelable fingerprint templates with Delaunay triangle-based local structures. In: Cyberspace Safety and Security, Lecture Notes in Computer Science, vol. 8300, pp. 81–91. Springer International Publishing (2013)
16. Jin, Z., Lim, M.H., Teoh, A.B.J., Goi, B.M.: A non-invertible randomized graph-based ham-ming embedding for generating cancelable fingerprint template. Pattern Recogn. Lett. **42**, 137–147 (2014)

Multi-secret Sharing Scheme Using Modular Inverse for Compartmented Access Structure

Abdul Basit, V. Ch. Venkaiah and Salman Abdul Moiz

Abstract Secret sharing scheme is a cryptographic primitive or a method for increasing the security of crucial information or data. It is used to share a secret among a set of participants, such that specific sets of participants can uniquely reconstruct the secret by pooling their shares. In this paper, we have proposed a new multi-secret sharing scheme for compartment access structure. In this access structure, the set of participants is partitioned into different compartments. The secret can be obtained only if the threshold number of participants from each of the compartments reconstruct their compartment secret, and participate in recovering the actual secret. The proposed scheme uses Shamir's scheme first to retrieve partial secrets and combines them to form the requested secret. The scheme can also verify whether the retrieved secret is valid or not. Security analysis of the scheme is carried out and showed that the scheme resists both the insider as well as outsider attacks. Our proposed scheme is simple and easy to understand as we have used only the modular inverse concept.

Keywords Secret sharing · Multi-secret · Compartmented access structure · Modular inverse

1 Introduction

Information security has grown much more since electronic communication is used in our daily life. The cryptographic secret key, which is used for securing the information, is shared among a set of players by a dealer in the distribution process. The sharing is done in such a way that by pooling specific sets of shares, the secret can be reconstructed. Initially, secret sharing schemes are proposed to solve

A. Basit (✉) · V. Ch. Venkaiah · S. A. Moiz
School of Computer & Information Sciences, University of Hyderabad,
500046 Hyderabad, India
e-mail: abdulmcajh@gmail.com

© Springer Nature Singapore Pte Ltd. 2020
K. S. Raju et al. (eds.), *Data Engineering and Communication Technology*,
Advances in Intelligent Systems and Computing 1079,
https://doi.org/10.1007/978-981-15-1097-7_31

problems that arise in application such as safeguarding cryptographic keys. Later, several useful applications have been identified in various cryptographic protocols. Nowadays, it is widely used in many interesting applications including multiparty computations, threshold cryptography, generalized oblivious transfer protocol, missile launching, and biometric authentication system.

Secret sharing schemes that share multiple secrets are known as multi-secret sharing schemes. In a multi-secret sharing scheme, participants keep only a single share that corresponds to many secrets shared using the scheme. The dealer uses a public information board to publish the necessary public values for the reconstruction of the secrets. The participants use pseudo shares for the reconstruction of multiple secrets. These pseudo shares are calculated using the original share and public information. Reconstruction of a secret does not reveal any information about the remaining secrets that have not been reconstructed. There is a multistage secret sharing scheme, in which multiple secrets are revealed stage by stage with each secret revealed in one stage. In a single-stage secret sharing scheme, all the secrets are revealed in the single stage of the protocol.

In literature, several access structures have been introduced. Some of them are the generalized access structure, the threshold (t, n) access structure, level-ordered access structure, and the multipartite access structure. The multipartite access structure is further divided into compartmented and hierarchical access structure.

Definition 1 *Generalized access structure*
Generalized access structure applies to cases wherein the group of permissible subsets of U may be any group of elements of 2^U having the monotonicity property; U is the set of participants, and 2^U is the power set U.

If an access structure is monotone, then whenever a set is qualified its superset is also qualified. Similarly, if a set is unauthorized, then its subset is also unauthorized. A set that can reconstruct the secret is called a qualified set.

Definition 2 *(Threshold (t, n) access structure)*
The (t, n) threshold access structure is an essential class of access structure consisting of n participants where an authorized set consists of any t or more than t participants. An authorized set can recover the secret. An unauthorized set belongs to less than t participants. An unauthorized cannot recover the secret. A formal definition of the (t, n) threshold access structure is as follows:

$$\Gamma = \{A \in 2^u : |A| \geq t\} \text{ and } \bar{\Gamma} = \{A \in 2^u : |A| < t\},$$

where 2^u is the power set of the set of participants.

Definition 3 *(Hierarchical access structure)*
All n participants in a hierarchical access structure are divided into k disjoint levels $L_1, L_2, \ldots, L_k$. These levels form a hierarchical structure. The level i consists of n_i participants and any t_i or more participants at ith level can reconstruct the secret. Whenever the number of participants at ith level is less than the t_i say r_i, then the

remaining $t_i - r_i$ participants can get from higher levels. A formal definition of the hierarchical access structure is as follows:

$$\Gamma = \left\{ V \subseteq U : \left| V \cap \left(\cup_{j=1}^{i} L_j \right) \right| \geq t_i, i \in \{1, 2, \ldots, k\} \right\}$$

where $U = \{P_1, \ldots, P_n\}$ denote the set of participants.

Definition 4 (*Compartmented access structure*)
In a compartmented access structure, all the n participants are divided into k disjoint compartments $C_1, C_2, \ldots, C_k$. The access structure consists of subsets of participants containing at least t_i from C_i for $i = 1, 2, \ldots, k$ and a total of at least t_0 participants. Formal definition of compartmented access structure is defined as follows:

$$\Gamma = \{ V \subseteq U : |V \cap C_i| \geq t_i \text{ for all } i \in 1, \ldots, k \text{ and } |V| \geq t_0 \}$$

where $U = \cup_{i=1}^{k} C_i, C_i \cap C_j = \emptyset$ for all $1 \leq i < j \leq k$ and $t_0 \geq \sum_{i=1}^{k} t_i$

1.1 Motivation and Contribution

Security is a major challenge in the digital storage and transmission of data. Secret sharing protocols provide solutions to several security problems including secure key management, distributed access control, secure distributed storage and transmission, and secure multiparty computation. This paper mainly contributes to the development of a secret sharing protocol and also explores its uses in typical areas of application. In this paper, we proposed a multi-secret sharing scheme for the compartmented access structure by using modular inverse. The reconstructed secret can be verified whether it is correct or not, and no participant in the compartment knows other compartment secrets. The proposed secret sharing scheme for compartmented access structure applies Shamir scheme to recover compartment secret and combines them into the requested secret. The scheme is ideal as well as perfect. Our proposed scheme can be applied in the following situation: Suppose an organization has a confidential project. To make work easier and for on-time delivery, the management divides the project into different modules. Each module is given to different team, where each team consists of a set of employees. In order to achieve the project's confidentiality, they can use any encryption scheme, but applying an encryption scheme and having the key for each employee is a time-consuming and costly process with the risk of key losses. Therefore, distribute the key to the employee of each team by using the proposed scheme instead of any encryption scheme.

2   Related Work

Secret sharing systems have been extensively studied in the literature over the last few decades. Shamir and Blakley proposed a secret sharing scheme for the first time. The Blakley scheme is based on the geometric idea that uses the concept of intersecting hyperplanes, while Shamir's scheme is based on polynomial interpolation of Lagrange. Later on, many researchers contributed to the area of secret sharing by proposing several secret sharing schemes [1–4].

These schemes are intended to share only one secret. In order to share multiple secrets concept of multi-secret sharing scheme and the associated terminology comes into the secret sharing. A multi-secret sharing scheme (MSSS) is an extension of the secret sharing scheme of Shamir [1]. There are several multi-secret sharing schemes in the literature. Dawson et al. [5] was the first to propose a multi-secret sharing scheme, in which many secrets can be shared to each participant at once using a single share. Karnin et al. [6] introduced a multiple secret sharing scheme where many secrets can be reconstructed simultaneously by a threshold number of participants. Franklin et al. [7] used a method in which a single secret sharing scheme based on a polynomial has been modified to a scheme in which many secrets are hidden in a single polynomial. In [8], Yang et al. proposed two different schemes based on Shamir's secret sharing scheme. Chien et al. [9], introduced a realistic multi-secret sharing scheme based on systematic block codes. Endurthi et al. [10], have developed a new secret sharing scheme called CRT-based multistage multi-secret sharing scheme. Afterward, many new multi-secret sharing systems are subsequently developed using various mathematical tools and techniques [8, 11–14].

The line of work was started by Simmon [15], who proposed two classes of access structures, the compartmented access structure and the multilevel access structure for ideal secret sharing schemes. An access structure is the family of authorized subsets of participants who can recover the secret. Many access structures were proposed to cater to various application scenarios. Research community contributed to the secret sharing area by designing many secret sharing schemes that realize different access structures including multilevel (hierarchical) access structure, level ordered access structure and compartmental access structure. Simmons [15] proposed the concept of compartmented secret sharing using geometric techniques. Brickell [2] proposed a more general family called a lower bounded compartmented access structure. Tassa and Dyn [16] introduced a new access structure known as compartmented access structures with lower limit. Tassa and Dyn also devised a secret sharing scheme with bivariate interpolation polynomial for these access structures. Farras et al. [4] characterized some new ideal compartmented access structures, such as the upper and lower bounded compartmented access structure. Brickell and Ghodosi have suggested an ideal scheme [17] for a compartmented access structure by using Shamir secret sharing scheme twice, first to obtain partial secrets and secondly to consolidate them into the required secret. Iftene [3] suggested a compartmented scheme primarily based on the

Chinese Remainder Theorem. Nojoumian et al. [18] suggested a new hierarchical sequential secret sharing scheme, which differs from the existing hierarchical secret sharing schemes. Basit et al. [12] introduced a multi-secret sharing scheme based on polynomials for hierarchical access structure. In 2016 Singh et al. [13] proposed a sequential secret sharing scheme realizing hierarchical access structure for the case of multiple secrets. In 2018 Tentu et al. [19] introduced new multi-secret sharing scheme for level ordered access structure. Many multi-secret sharing schemes exist in the literature [20–24] realizing various access structures.

Since 1979, literature has seen various secret sharing schemes. Shamir and Blakley proposed their secret sharing scheme independently. Secret sharing schemes can be classified based on the mathematical tools and techniques they use, as presented in Table 1. Another categorization of the secret sharing schemes can be in accordance with the number of secrets they shared as presented in Table 2. Secret sharing schemes can have many properties. In this paper, we introduced a new multi-secret sharing scheme that uses the modular inverse. To date, there is no multi-secret scheme for compartment access structures using modular inverse.

Outline of the Paper, The rest of the paper is organized as follows: Sect. 3 presents some of the preliminaries used to build the proposed scheme. We presented our proposed multi-secret sharing scheme for compartmented access structure in Sect. 4. Detailed example of the proposed scheme is given Sect. 5. Security analysis of the scheme is shown in Sect. 6 and in Sect. 7, conclusions and future work are addressed.

Table 1 Various mathematical techniques in secret sharing scheme

Vector-space-based	Blakley
Polynomial-based	Shamir; Ghodosi; Basit; Yang; Franklin; He; ours
CRT-based	Mignotte; Asmuth-Bloom; Endurthi; Iftene; Hsu; Singh; Harn
Matrix-projection-based	Bai; Wang
Systematic block codes based	Chien
Super increasing sequence	Harsha; Basit
Finite geometry	Duari

Table 2 Single and multi-secret sharing schemes

Single secret sharing scheme	Multi-secret sharing scheme
Shamir; Brickell; Ghodosi	Chien; Shao
Asmuth-Bloom; Blakley; Mignotte	Singh; Franklin
Iftene; Farras; Ghodosi	Tentu; Yang; He
Jackson; Harsha	Zhang; Bai; Ours

3 Background and Preliminaries

In this section, we present briefly basic concepts on modular inverse and Shamir secret sharing scheme, which are necessary to understand the rest of our work.

3.1 Modular Inverse

Consider a tuple $c = (c_1, c_2, c_3, \ldots, c_m)$ such that the elements in the tuple are pairwise co-prime. Then there exists an unique tuple $L = (l_1, l_2, l_3, \ldots, l_m)$ such that $\frac{l_1}{c_1} + \frac{l_2^1}{c_2} + \frac{l_3}{c_3} + \cdots + \frac{l_m}{c_m} + \frac{1}{c}$ is always an integer.

A proof of this statement is as follows:

Let $c_1, c_2, c_3, \ldots, c_m$ be a set of positive pairwise co-primes and $c = c_1 * c_2 * c_3 * \cdots * c_m$. That is,

$$\Rightarrow \gcd(c_i, c_j) = 1 \quad \forall i \neq j$$

$$\Rightarrow \gcd\left(\frac{c}{c_i}, c_i\right) = 1 \quad \forall i$$

$$\Rightarrow \gcd\left(\frac{-c}{c_i}, c_i\right) = 1 \quad \forall i$$

$$\Rightarrow \exists l_i | l_i \equiv \left(\frac{-c}{c_i}\right)^{-1} \bmod c_i \quad \forall i$$

$$\Rightarrow l_i\left(\frac{-c}{c_i}\right) \equiv 1 \bmod c_i \quad \forall i$$

$$\Rightarrow l_i\left(\frac{c}{c_i}\right) \equiv -1 \bmod c_i \quad \forall i$$

$$\Rightarrow l_i\left(\frac{c}{c_i}\right) + 1 \equiv 0 \bmod c_i \quad \forall i$$

$$s = l_1\left(\frac{c}{c_1}\right) + l_2\left(\frac{c}{c_2}\right) + l_3\left(\frac{c}{c_3}\right) + \cdots + l_n\left(\frac{c}{c_m}\right) + 1$$

$$\Rightarrow s \equiv l_i\left(\frac{c}{c_i}\right) + 1 \equiv 0 \bmod c_i, \forall i$$

$$\Rightarrow s \equiv 0 \bmod c. \text{ That is, } l_1\left(\frac{c}{c_1}\right) + l_2\left(\frac{c}{c_2}\right) + l_3\left(\frac{c}{c_3}\right) + \cdots + l_n\left(\frac{c}{c_m}\right) + 1$$

$$\equiv 0 \bmod c$$

$$\Rightarrow \exists k | l_1\left(\frac{c}{c_1}\right) + l_2\left(\frac{c}{c_2}\right) + l_3\left(\frac{c}{c_3}\right) + \cdots + l_n\left(\frac{c}{c_m}\right) + 1 = kc$$

$$\Rightarrow \exists k | \left(\frac{l_1}{c_1}\right) + \left(\frac{l_2}{c_2}\right) + \left(\frac{l_3}{c_3}\right) + \cdots + \left(\frac{l_n}{c_m}\right) + \frac{1}{c} = k$$

3.2 Shamir's Secret Sharing Scheme

The Shamir scheme consists of two phases. It is based on the Lagrange polynomial interpolation. Let the number of participants be n, the threshold be t, the secret be S and let $F_P = GF(P)$ be a Galois field with P elements.

I. Share Generation Phase:

Dealer chooses a polynomial $F(x)$ randomly of degree $(t-1)$ as $F(x) = a_0 + a_1 \times x + a_2 \times x^2 + \cdots + a_{t-1} \times x^{t-1}$ mod P, where $S = a_0$ is the secret. All the coefficients $a_i, i = 1, \ldots, t-1$ are from Galois finite field $F_P = GF(P)$, where P is prime and $P > S$. Dealer calculates n shares, $y_i = F(x_i)$ mod P; where $i = 1, 2, \ldots, n$ and x_i is the public identity for the participant i. The dealer then securely distributes (x_i, y_i) as the ith participant share, for $i = 1, 2, \ldots, n$

II. Share Reconstruction Phase:

Any t or more than t participants can reconstruct the secret S by using the Lagrange polynomial interpolating formula as follows:

$$f(x) = \sum_{i=1}^{t} y_i \times l_i(x)$$

where

$$l_i(x) = \prod_{k=1, k \neq i}^{t} \frac{x - x_k}{x_i - x_k} \quad \text{mod } P.$$

4 Proposed Scheme

4.1 Overview of the Proposed Scheme

In this paper, we introduced a new multi-secret sharing scheme for the compartment access structure, In the beginning, the dealer has n number of participants $P_1, P_2, \ldots, P_n$, which is partitioned into k different compartments $C_1, C_2, \ldots, C_k$ and the dealer chooses a set of k co-prime integer $c_1, c_2, \ldots, c_k$ as the compartment secrets for the k compartments. The dealer calculates partial secrets s_i using compartment secrets $c_i, 1 \leq i \leq k$ and add all the partial secrets to make pseudo secret S. Then dealer computes shift values Z_i by adding pseudo secret S into actual secrets S_i, i.e., $Z_i = S + S_i$ and make it public. The dealer now applies the Shamir's secret sharing scheme to distribute shares of the compartment secrets c_i to the participants of corresponding compartments C_i. While reconstructing the secret, the participants of the corresponding compartment first apply Lagrange interpolation and recover the corresponding compartment secrets c_i. They compute partial secrets s_i at corresponding compartments and add all partial secrets s_i to get pseudo secret S. Then to recover actual secrets S_i, Subtract pseudo secrets from the corresponding shift values Z_i, which are public values. That is, $S_i = Z_i - S$.

4.2 Algorithm

Algorithm 1 Initialization:

1: Let $\mathcal{P} = \{P_1, P_2, P_3, \ldots, P_n\}$ be the set of n participants partitioned into k disjoint compartment.
2: Let $\{C_1, C_2, \ldots, C_k\}$ be the k compartments.
3: Every compartment C_i is associated with (t_i, n_i) access structure, where n_i is the total number of participants and t_i is the threshold value for compartment C_i, $i \in \{1, 2, \ldots, k\}$.
4: Choose l secrets i.e. choose $S_1, S_2, \ldots, S_l$.
5: Choose k pairwise co-prime integers i.e. choose $c_1, c_2, \ldots, c_k$ s.t. $\mathrm{GCD}(c_i, c_j) = 1, \forall i \neq j$ as the compartmented secret.
6: Select large prime p.
7: id_j^i denoted as the identifier for the participant j at the compartment i, where $1 \leq j \leq n_i$ and $1 \leq i \leq k$.

Algorithm 2 Share Distribution

1: Compute $c = c_1 \times c_2 \times \cdots c_k$.
2: Compute the pseudo secret $S = \sum_{i=1}^{k} s_i, s_i = \left(\frac{-c}{c_i}\right)^{-1} \bmod c_i$
3: Compute shift value $Z_i = S + S_i$, where $1 \leq i \leq l$.
4: for $i = 1 \rightarrow k$ do

- Choose $t_i - 1$ positive integers randomly from $F_p, a_1^i, a_2^i, \ldots, a_{t_i-1}^i$, and let $a_0^i = c_i$
- Construct the Polynomial of degree $t_i - 1$,

$$F(x) = a_0^i + a_1^i * x + \cdots + a_{t_i-1}^i * x^{(t_i-1)}$$

- Compute $\mathrm{Shr}_j^i = F\left(id_j^i\right) \bmod p$ where, $1 \leq j \leq n_i$

5: Distribute shares Shr_j^i to the respective participants via secure channel and make c and Z_i values public.

Algorithm 3 Share Reconstruction

1: Authorized subset of any t_i or more than t_i participants using their ID's id_j^i and shares Shr_j^i from compartment C_i reconstruct compartmented secret $c_i, 1 \leq i \leq k$, by computing Lagrange interpolation formula as follows:

2: for $i = 1 \rightarrow k$ do

$$- \quad c_i = \sum_{j=1}^{t} \text{shr}_j^i \prod_{k=1, k \neq j}^{t} \frac{\text{id}_k^i}{\text{id}_k^i - \text{id}_j^i} \bmod p$$

$$- \quad \text{Compute partial secret } s_i = \left(\frac{-c}{c_i}\right)^{-1} \bmod c_i$$

5: Calculate pseudo secret $S = \sum_{i=1}^{k} s_i$

6: Compute Secret $S_i = Z_i - S, 1 \leq i \leq l$.

Algorithm 4 Secret Verification

1: $k = \sum_{i=1}^{l} \left\{ s_i \times \left(\frac{c}{c_i}\right) \right\} + \frac{1}{c}$

2: if k is any integer.

3: secret is correct.

4: else secret is wrong.

5 Example

The proposed scheme with small artificial parameters is described here.

1. *Initialization*:

   - Suppose total number of participants $n = 15$. These participants are partitioned into three disjoint compartments C_1, C_2, C_3 such that threshold values and number of participants of the corresponding compartment as (t_1, n_1) is (2, 4), (t_2, n_2) is (3, 6) and (t_3, n_3) is (3, 5).
   - Choose 4 secrets i.e. $S_1 = 11, S_2 = 13, S_3 = 15, S_4 = 17$.
   - Let 3 pairwise co-prime integers be $c_1 = 143, c_2 = 35, c_3 = 6$. These are the compartmented secret.
   - Let prime $p = 59$.
   - ID's of the 1st compartment participants be

$id_1^1 = 1, id_2^1 = 2, id_3^1 = 3, id_4^1 = 4$,

2nd compartment ID's be

$id_1^2 = 5, id_2^2 = 6, id_3^2 = 7, id_4^2 = 8, id_5^2 = 9, id_6^2 = 10$ and

3rd compartment ID's be

$id_1^3 = 11, id_2^3 = 12, id_3^3 = 13, id_4^3 = 14, id_5^3 = 15$,

2. *Distribution*:

- Compute $c = c_1 \times c_2 \times c_3 = 143 \times 35 \times 6 = 30,030$
- Calculate pseudo secret

$$S = \sum_{i=1}^{k} \left(\frac{-c}{c_i} \right)^{-1} \mod c_i = 70$$

- Compute shift values $Z_i = S_i + S$ for $1 \le i \le l$;
- $Z_1 = 11 + 70 = 81, Z_2 = 13 + 70 = 83$,
  $Z_3 = 15 + 70 = 85, Z_4 = 17 + 70 = 87$.
- At compartment C_1, $t_1 = 2$, $n_1 = 4$, $c_1 = 143$

  • $F(x) = c_1 + 3 * x = 143 + 3 * x$
  • Compute shares $Shr_1^1 = 146, Shr_2^1 = 149, Shr_3^1 = 1, Shr_4^1 = 4$

- At compartment C_2, $t_2 = 3$, $n_2 = 6$, $c_2 = 35$

  • $F(x) = c_2 + 3 * x + 4 * x^2 = 35 + 3 * x + 4 * x^2$
  • Compute shares $Shr_1^2 = 150, Shr_2^2 = 46, \ Shr_3^2 = 101, Shr_4^2 = 13, Shr_5^2 = 84, Shr_6^2 = 12$

- At compartment C_3, $t_3 = 3$, $n_3 = 5$, $c_3 = 6$

  • $F(x) = c_3 + 2 * x + 3 * x^2 = 6 + 2 * x + 3 * x^2$
  • Compute shares $Shr_1^3 = 89, Shr_2^3 = 9, \ Shr_3^3 = 86, Shr_4^3 = 18, Shr_5^3 = 107$.

3. *Reconstruction*:

- Any t_i or more than t_i participants from an authorized set. Using their ID's and shares Shr_j^i each compartment reconstructs compartmented secret c_i by computing Lagrange interpolation formula.
- **Compartment secret reconstruction at compartment C_1**

  • Suppose participant 2nd and 3rd want to reconstruct the secret by sharing their ID's and shares $Shr_2^1 = 149, Shr_3^1 = 1$ respectively.
  • Apply Lagrange polynomial interpolation on (2, 149), (3, 1) and get

$$c_1 = \sum_{j=1}^{2} \mathrm{Shr}_j^1 \prod_{k=1, k\neq j}^{2} \frac{\mathrm{id}_k^1}{\mathrm{id}_k^i - \mathrm{id}_j^1} \bmod p$$

$$c_1 = 149 \times \left(\frac{3}{3-2}\right) + 1 \times \left(\frac{2}{2-3}\right) \bmod 151$$

$$= 445 \quad \bmod 151 = 143$$

- $\therefore c_1 = 143$
- Compute partial secret $s_1 = \left(\frac{-c}{c_1}\right)^{-1} \bmod c_1$
- Compute partial secret

$$s_1 = \left(\frac{-30,030}{143}\right)^{-1} \bmod 143 = 32$$

– **Compartment secret reconstruction at compartment C_2**

- Suppose participant 1st, 2nd, and 3rd want to reconstruct the secret by sharing their ID's and shares $\mathrm{Shr}_1^2 = 150, \mathrm{Shr}_2^2 = 46, \mathrm{Shr}_3^2 = 101$ respectively.
- Apply Lagrange polynomial interpolation on (5, 150), (6, 46), (7, 101) and get

$$c_2 = \sum_{j=1}^{3} \mathrm{shr}_j^2 \prod_{k=1, k\neq j}^{3} \frac{\mathrm{id}_k^2}{\mathrm{id}_k^2 - \mathrm{id}_j^2} \quad \bmod p$$

$$c_2 = 150 \times \left(\frac{6}{6-5}\right) \times \left(\frac{7}{7-5}\right) + 46 \times \left(\frac{5}{5-6}\right)$$

$$\times \left(\frac{7}{7-6}\right) + 101 \times \left(\frac{5}{5-7}\right) \times \left(\frac{6}{6-7}\right) \bmod 151$$

$$= 3055 \bmod 151 = 35$$

- $\therefore c_2 = 35$
- Compute partial secret $s_2 = \left(\frac{-c}{c_2}\right)^{-1} \bmod c_2$
- Compute partial secret

$$s_2 = \left(\frac{-30,030}{35}\right)^{-1} \bmod 35 = 33$$

– **Compartment secret reconstruction at compartment C_3**

- Suppose participant 2nd, 3rd, and 4th want to reconstruct the secret by sharing their ID's and shares $\mathrm{Shr}_2^3 = 9, \mathrm{Shr}_3^3 = 86, \mathrm{Shr}_4^3 = 18$ respectively
- Apply Lagrange polynomial interpolation on (12, 9), (13, 86), (14, 18) and get

$$
c_3 = \sum_{j=1}^{3} \mathrm{shr}_j^3 \prod_{k=1, k \neq j}^{3} \frac{id_k^3}{id_k^3 - id_j^3} \mod p
$$

$$
c_3 = 9 \times \left(\frac{13}{13 - 12}\right) \times \left(\frac{14}{14 - 12}\right) + 86 \times \left(\frac{12}{12 - 13}\right)
$$

$$
\times \left(\frac{14}{14 - 13}\right) + 18 \times \left(\frac{12}{12 - 14}\right) \times \left(\frac{13}{13 - 14}\right) \mod 151
$$

$$
= -12,225 \mod 151 = 6
$$

- $\therefore c_3 = 6$
- Compute partial secret $s_3 = \left(\frac{-c}{c_3}\right)^{-1} \mod c_3$
- Compute partial secret
  $s_3 = \left(\frac{-30,030}{6}\right)^{-1} \mod 6 = 5$

– Calculate pseudo secret $S = \sum_{i=1}^{k} s_i = 32 + 33 + 5 = 70$
– Compute secret $S_1 = Z_1 - S = 81 - 70 = 11, S_2 = Z_2 - S = 83 - 70 = 13, S_3 = Z_3 - S = 85 - 70 = 15, S_4 = Z_4 - S = 87 - 70 = 17$

6 Security Analysis

In this section, we discuss security analysis of the proposed multi-secret sharing scheme for compartmented access structure assuming that the dealer is honest and the communication channels between two connecting nodes are secure, so that information cannot leak to the non-authenticating node. Consequently, we have discussed the security analysis for the outside adversaries and inside adversaries.

Following are the possible attacks:

– Unauthorized set of participants tries to recover the secret s_i at compartment i.
– Less than k compartments participants tries to recover the secret S_i.
– Outside adversary tries to recover the secret S_i using public values Z_i.

Lemma 1 *Unauthorized set of participants tries to recover the secret s_i at compartment i.*

Proof Partial secret s_i can be got only when c_i is known and c_i is distributed using Shamir's (t_i, n_i) threshold scheme. Less than t_i participants can not obtain c_i, which is essential to get s_i. Therefore, unauthorized set of participants can not get s_i.

Lemma 2 *Less than k compartments participant tries to recover the secret S_i.*

Proof The secret S_i can be recovered by subtracting pseudo secret S from available public values Z_i. However, pseudo secret S is sum all the k compartmented secrets. Therefore, less than k compartment participants can not recover the secret S_i.

Lemma 3 *Outside adversary tries to recover the secret S_i using public values Z_i.*

Proof Recover of the secret S_i requires public value Z_i and pseudo secret S. S can be computed only by authorized participants of all the compartments. Hence, outside adversary cannot obtain the secret S_i using public values Z_i only.

6.1 Privacy

The secrets can be recovered by an authorized group of participants; while an unauthorized group of participants cannot recover the secrets. In the proposed scheme no one knows others compartment secret. As every compartment uses Shamir secret sharing to distribute the shares to its participants, at least t or more than t participants must cooperate in order to get the compartment secret. Therefore no outside adversary can cheat to get the compartment secret.

7 Conclusion and Future Work

In this paper, we presented a new multi-secret sharing scheme that realizes the compartmented access structure by using the concept of modular inverse. It can check whether the reconstructed secret is correct or not, and no compartment participant knows another compartment secret. The proposed secret sharing scheme for the compartmented access structure uses the Shamir scheme in order to get the partial secrets and combines them into the actual secret. The secret can only be obtained if the threshold number of participants in each compartment reconstructs their compartment secret, and participates in the recovery of the secret. Security analysis is carried out for both inside and outside adversaries on possible attacks. The proposed scheme can be improved by using for the many secrets only one polynomial in each compartment.

References

1. Shamir, A.: How to share a secret. Commun. ACM **22**(11), 612–613 (1979)
2. Blakley, G.R.: Safeguarding cryptographic keys. AFIPS, vol. 48, pp. 313–317 (1979)
3. Brickell, E.F.: Some ideal secret sharing schemes. In: Workshop on the Theory and Application of Cryptographic Techniques, pp. 468–475, Springer, Berlin, Heidelberg, 10 Apr 1989
4. Iftene, S.: General secret sharing based on the chinese remainder theorem with applications" in e-voting. Electron. Notes Theor. Comput. Sci. **186**, 67–84 (2007)
5. Farras, O., Padro, C., Xing, C., Yang, A.: Natural generalizations of threshold secret sharing. IEEE Trans. Inf. Theory. **60**(3):1652–64 (2004)
6. Dawson, E., Donovan, D.: The breadth of Shamir's secret-sharing scheme. Comput. Secur. **13**(1), 69–78 (1994)
7. Karnin, E., Greene, J., Hellman, M.: On secret sharing systems. IEEE Trans. Inf. Theory **29**(1), 35–41 (1983)
8. Franklin, M., Yung, M.: Communication complexity of secure computation. In: Proceedings of the Twenty-Fourth Annual ACM Symposium on Theory of Computing, pp. 699–710, ACM, 1 July 1992
9. Yang, C.-C., Chang, T.-Y., Hwang, M.-S.: A (t, n) multi-secret sharing scheme. Appl. Math. Comput. **151**(2), 483–490 (2004)
10. Chien, H.-Y., Jan, J.-K., Tseng, Y.-M.: A practical (t, n) multi-secret sharing scheme. IEICE Trans. Fundam. Electron. Commun. Comput. Sci. **83**(12), 2762–2765 (2000)
11. Endurthi, A., Chanu, O.B., Tentu, A.N., Ch Venkaiah, V.: Reusable multi-stage multi-secret sharing schemes based on CRT. 15–24 (2015)
12. Wang, K., Zou, X., Sui, Y.: A multiple secret sharing scheme based on matrix projection. In: 2009 33rd Annual IEEE International Computer Software and Applications Conference, COMPSAC'09, vol. 1, pp. 400–405, IEEE, 20 July 2009
13. Basit, A., Chaitanya Kumar, N., Ch Venkaiah, V., Moiz, S.A., Tentu, A.N., Naik, W.: Multi-stage multi-secret sharing scheme for hierarchical access structure. In: 2017 International Conference on Computing, communication and automation (IC-CCA), pp. 557–563. IEEE (2017)
14. Singh, N., Tentu, A.N., Basit, A., Ch Venkaiah, V.: Sequential secret sharing scheme based on Chinese remainder theorem. In: 2016 IEEE International Conference on Computational Intelligence and Computing Research (ICCIC), pp. 1–6. IEEE (2016)
15. Zhang, T., Ke, X., Liu, Y.: (t, n) multi-secret sharing scheme ex-tended from Harn-Hsu's scheme. EURASIP J. Wirel. Commun. Netw. **2018**(1), 71 (2018)
16. Simmons, G.J.: How to (really) share a secret. In: Conference on the Theory and Application of Cryptography, pp. 390–448. Springer, New York, NY (1988)
17. Tassa, T., Dyn, N.: Multipartite secret sharing by bivariate interpolation. J. Cryptol. **22**(2), 227–58 (2009)
18. Ghodosi, H., Pieprzyk, J., Safavi-Naini, R.: Secret sharing in multilevel and compartmented groups. Information Security and Privacy, pp. 367–378. Springer (1998)
19. Nojoumian, M., Stinson, D.R.: Sequential secret sharing as a new hierarchical access structure. J. Internet Serv. Inf. Secur. **5**(2), 24–32 (2015)
20. Tentu, A.N., Basit, A., Bhavani, K., Ch Venkaiah, V.: Multi-secret sharing scheme for level-ordered access structures. In: International Conference on Number-Theoretic Methods in Cryptology, pp. 267–278. Springer, Cham (2017)
21. Wang, X., Xiang, C., Fu, F.W.: Secret sharing schemes for compartmented access structures. Crypt. Commun. **9**(5), 625–635 (2017)
22. Selcuk, A.A., Yilmaz, R.: Joint compartmented threshold access structures. IACR Cryptol ePrint Arch. **2013**, 54 (2013)

23. Duari, B., Giri, D.: An ideal and perfect (t, n) Multi-secret sharing scheme based on finite geometry. In: Information Technology and Applied Mathematics, pp. 85–94. Springer, Singapore (2019)
24. Fathimal, P.M, Arockia Jansi Rani, P.: Threshold secret sharing scheme for compartmented access structures. Int. J. Inf. Secur. Privacy (IJISP) **10**(3), 1–9 (2016)
25. Ito, M., Saito, A., Nishizeki, T.: Secret sharing scheme realizing general access structure. Electron. Commun. Jpn. (Part III: Fundam. Electron. Sci.) **72**(9), 56–64 (1989)

FM Broadcast Audio Signal Analysis Using Time–Frequency Distribution Function

Kartik Patel and Paawan Sharma

Abstract For any product, the end customer is at highest priority with regard to available user options. Presently, commercial radio broadcast contains more number of advertisements instead of content. The paper presents a solution to detect advertisement in between duration for which songs being played. With the recent advancements in the area of audio processing and analysis, the analysis of songs and advertisements is made by studying real-time data collected from various radio broadcasts. After examining each particular characteristics of decomposed signal in time, frequency, time–frequency, and statistical domains for different audio clippings, it is observed that there is a pattern between the songs and advertisements. This data can be used for the classification of songs and advertisements. After correct classification, a software-based routine can be made to automatically switch to another station in case of advertisement being played.

Keywords Audio signal processing · FFT · Wavelet analysis · Commercial audio · Time–frequency distribution

1 Introduction

Audio signal processing is an emerging field as soon there will be a need for content discovery and indexing applications ability which will be required by the system to automatically analyze and classify the signal [1]. Analysis of signals is the basic fundamental step for classifications, as classification between song and advertisement can help in solving problems faced by the people such as skimming of repeated advertisements on radio broadcast. The proposed work deals with

K. Patel (✉) · P. Sharma
Department of CE and ICT, School of Technology, Pandit Deendayal Petroleum University, Gandhinagar, India
e-mail: kartikpatel8088@gmail.com

P. Sharma
e-mail: Paawan.sharma@sot.pdpu.ac.in

© Springer Nature Singapore Pte Ltd. 2020
K. S. Raju et al. (eds.), *Data Engineering and Communication Technology*,
Advances in Intelligent Systems and Computing 1079,
https://doi.org/10.1007/978-981-15-1097-7_32

identification of song or advertisement, i.e., it classifies an audio either as song or advertisement. In this paper, analysis is performed on a large number of real-time short audio frame samples. However, the audio signals contain a monotonous and a similar pattern at different frequencies and amplitudes. In fact, it is necessary to better characterize and bifurcate each component of audio signals through its statistical data for the analysis. Thus, research has been carried out with three key steps simultaneously by analyzing each short audio frame in: (a) frequency domain (b) time domain (c) time–frequency domain, and (d) statistical (AI) domain [2].

Each of the domain plots of song/advertisement is carefully examined, and various calculations have been carried out. First, frequency-domain plot is constructed by setting specific and appropriate algorithm, size, function, and axis. Similarly, time-domain and time–frequency-domain graphs are plotted for all the sample short audio frames. Then according to the category, the graphs are differentiated and on keen observation several variations at different amplitudes and different times are noted which suggested a similarity among particular category signals.

However, the varying statistical sets of the graphs had a far more pronounced effect on the classification accuracy. But it is not easy to extract the similarities of the signal. This approach of classification can reduce the computational complexity which can further help in the research field for a particular signal [3].

It has been observed that the classification of songs/advertisements can also be achieved through keywords as there are distinct and unique words which are repeatedly or are only used in advertisements. The paper discusses each step in detail and proposed analysis and conclusion.

2   Research Methodology

Beginning with the review of the existing research papers based on audio signal processing. The literature review of many articles and papers on different types of audio classifications using several methods such as short-time Fourier transform (STFT), neural network, and support vector machine (SVM) are extensively used in this research field to solve related issues and real-time problems [4]. From review of literature, it is observed that these extensively used methods increase computational complexity and the advent of ubiquitous advertisement especially in radio broadcast should be particularly taken into consideration [5]. In order to solve the task in real time, the need for classification of advertisement/songs for radio using audio signal processing is required, thereby formulating a problem [6]. After that, both categories were individually investigated and found out some similarities which could be used for the classification (Fig. 1).

The idea of short-time Fourier transform [7] is also used for the study of sinusoidal frequency and to obtain local properties of each audio samples. Then, the STFT of a function f with respect to g is defined as follows:

Fig. 1 Analysis flow

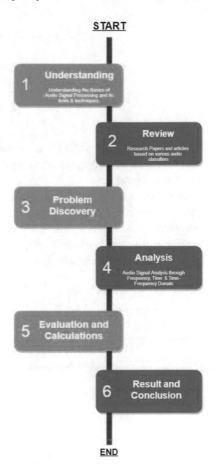

$$V_g f(x, w) = \int \mathbb{R}^d f(t)\, g(t - x) e^{-2\pi i t \cdot \omega} \mathrm{d}t, \quad \text{for } x, \omega \in \mathbb{R}^d$$

$$S_\omega(t, \tau) = s(\tau)\omega(\tau - t),$$
$$F_s^\omega(t, f) = F_{\tau \to f}\{s(\tau)\omega(\tau - t)\}$$

$F_s^\omega(t, f)$ is called short-time Fourier transform (STFT).

Further, calculating spectrogram using STFT, i.e., computing the squared magnitude of the short-time Fourier transform of signal for a window width ω,

$$\text{spectrogram}(t, \omega) = |\text{STFT}(t, \omega)|^2$$

$$S_s^\omega(t,f) = \left| F_s^\omega(t,f) \right|^2$$

$$= \left| F_{\tau \to f} \{ s(\tau)\omega(\tau - t) \} \right|^2$$

$$= \left| \int_{-\infty}^{\infty} s(\tau)\omega(\tau - t)e^{-j2\pi f\tau} d\tau \right|^2$$

A block diagram of STFT is shown below, where nfft is the length of the DFT, *nonoverlap* is the number of samples the two frames overlap, while window is weighting vector.

After evaluation in each domain, several attempts were made to formulate a common pattern. Comparisons were made between all the steps, i.e., time domain, frequency domain, and time–frequency domain. At last, specific characteristics of signals in frequency-domain plot are utilized for the operation and performance which is represented in result and conclusion.

Block diagram STFT

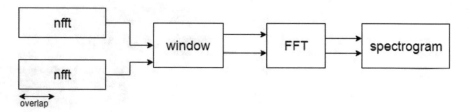

$$\text{Spectrogram}(t, \omega) = |\text{STFT}(t, \omega)|^2$$

3   Time Domain

Time-domain signal plots depict variation of amplitude with time; oscilloscope is the most commonly used to visualize signals in time domain. Python Library such as EEMD (ensemble empirical mode decomposition) which works on the basic principle of EMD (empirical mode decomposition) is a method of breaking down of signals without leaving time domain. These components form a complete and nearly orthogonal basis for the original signal, i.e., intrinsic mode functions (IMF) which can also be used to optimize the analysis of signals.

First, data collected is converted into raw data which optimizes analysis. Time-domain plots of sample raw data of real-time short, i.e., 25–30 s, audio are shown below which are formulated by using Scilab 6.0.1 open-source software; below signal plots are built by assigning measurement scale as dB and rate in Hz for each of the following.

Code used to plot:

```
wavread("SCI/modules/sound/filename.wav","size")
[y,Fs,bits]=wavread("SCI/modules/sound/filename.wav");Fs,bits
subplot(2,1,1)
plot2d(y(1,:))//first channel
subplot(2,1,2)
plot2d(y(2,:))//second channel
```

Observing its extracted features and characteristics individually using Scilab and EEMD, no resemblance or likeness was found. The amplitude of nonstationary signals gives no remark of how the frequency content of a signal changes with time. Hence, time-domain signal processing cannot be used to classify (Figs. 2 and 3).

4 Frequency Domain

Frequency domain represents variation of signals over a range of frequencies; theoretically, signals are composed of many sinusoidal signals with different frequencies. Thereby, analysis using frequency domain is much simple as we can figure out the key points than examining each variation which occurs in time-domain analysis. Spectrum analyzer is a tool commonly used to visualize signals in frequency domain.

Raw data in suitable form is used to plot frequency domain plots of sample real-time short, i.e., 25–30 s, audio as formulated below using audacity open-source software, by assigning the following specific features:

- Algorithm: Spectrum
- Size: 1024
- Function: Hamming window
- Axis: Linear frequency (Figs. 4 and 5).

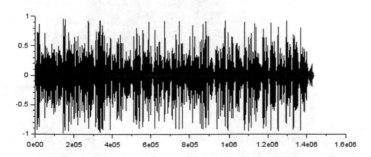

Fig. 2 Time-domain plot of sample advertisements

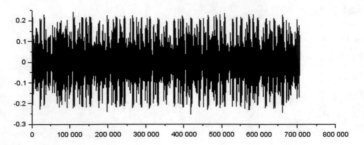

Fig. 3 Time-domain plot of sample song

Fig. 4 Frequency-domain
plot of sample advertisement

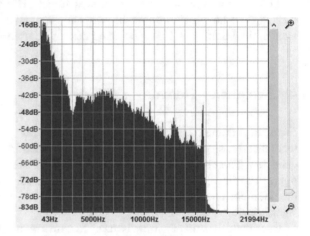

Fig. 5 Frequency-domain
plot of sample song

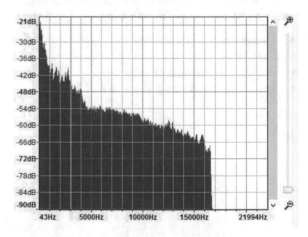

The frequency variation usually contains crucial information about the signal and its study of applications, but the magnitude spectrum cannot. Hence, frequency-domain signal processing cannot be used.

5 Time–Frequency Domain

Time–frequency domain is often complex-valued field over time and frequency, where the modulus of the field represents either amplitude or energy density, and the argument of field represents phase.

One of the basic forms of time–frequency analysis is the short-time Fourier transform (STFT), in which it provides some temporal information and some spectral information. These days, analysis is extended from basic to some other approaches such as Gabor transform (GT) and Wigner distribution function (WDF) [8].

Similarly, Python Audio Analysis-based library is pytftb, which includes signal generation files and processing files including the time–frequency distributions and other related processing functions. Converted raw data is used to plot time–frequency-domain plots of sample real-time short, i.e., 25–30 s, audio as shown below which are formulated by using audacity open-source software; these signal plots are known as spectrogram, built by assigning the following characteristics:

Scale

- *Scale: Linear*
- *Minimum frequency* (Hz): 0
- *Maximum frequency* (Hz): 8000.

Colors

- *Gain* (dB): 20
- *Range* (dB): 80
- *Frequency gain* (dB/dec): 0.

Algorithm

- *Algorithm: frequencies*
- *Window size*: 1024
- *Window type: Hanning*
- *Zero padding factor:* 1 (Figs. 6 and 7).

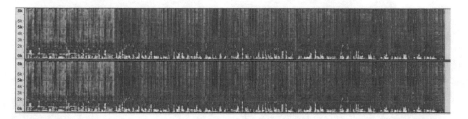

Fig. 6 Time–frequency plot of sample advertisement

Fig. 7 Time–frequency plot of sample song

These time–frequency representation and analysis of signal contains multiple time-varying frequencies. After careful examination, time–frequency domain by adjusting various algorithms, window types, scales, and their different combinations, still no clues or resemblance was found. Hence, time–frequency-domain plots are not viable for classifying the audio.

6 Conclusion

This study of audio signal processing carried out in time domain, frequency domain, and time–frequency domain for the possible classification of audio samples is conducted. A number of different tools and techniques in each domain have been reviewed for the analysis of various sample short audios. With regard to the study, the results of different analysis and examination of various related article data [9, 10] confirmed that varying audio signals cannot be classified only based on their time-domain, frequency-domain, or time–frequency-domain characteristics. Thereby, it suggests the use of MFCC or LPCC for classification of audio as song and advertisement along with these techniques [11]. Therefore, further research might be of interest on this subject.

Acknowledgements This research was supported by School of Technology, Pandit Deendayal Petroleum University, for allowing me to boost my knowledge. I sincerely thank my entire Computer Science and Engineering Department for significantly contributing behind my knowledge in the audio signal processing domain. I would also like to thank our Head of Department, Prof. T. P. Singh, for his encouragement for me to publish a research paper.

References

1. Lachambre, H., Ricaud, B., Stempfel, G., Torrésani, B., Wiesmeyr, C., Onchis-Moaca, D.: Optimal window and lattice in Gabor transform. Application to audio analysis. In: 2015 17th International Symposium on Symbolic and Numeric Algorithms for Scientific Computing (SYNASC), pp. 109–112, IEEE, Sept 2015

2. Saggese, A., Strisciuglio, N., Vento, M., Petkov, N.: Time-frequency analysis for audio event detection in real scenarios. In: 2016 13th IEEE International Conference on Advanced Video and Signal Based Surveillance (AVSS), pp. 438–443, IEEE, Aug 2016
3. Lim, T.Y., Yeh, R.A., Xu, Y., Do, M.N., Hasegawa-Johnson, M.: Time-frequency networks for audio super-resolution. In 2018 IEEE International Conference on Acoustics, Speech and Signal Processing (ICASSP), pp. 646–650, IEEE, Apr 2018
4. Tzanetakis, G., Cook, P.: Musical genre classification of audio signals. IEEE Trans. Speech Audio Process. **10**(5), 293–302 (2002)
5. Quatieri, T.F., McAulay, R.J.: Audio signal processing based on sinusoidal analysis/synthesis. In: Applications of Digital Signal Processing to Audio and Acoustics, pp. 343–416. Springer, Boston, MA (2002)
6. Lu, L., Zhang, H.J., Jiang, H.: Content analysis for audio classification and segmentation. IEEE Trans. Speech Audio Process. **10**(7), 504–516 (2002)
7. Zanardelli, W.G., Strangas, E.G., Aviyente, S.: Identification of intermittent electrical and mechanical faults in permanent-magnet AC drives based on time–frequency analysis. IEEE Trans. Ind. Appl. **43**(4), 971–980 (2007)
8. Boashash, B. Time-Frequency Signal Analysis and Processing: A Comprehensive Reference. Academic Press (2015)
9. Radhakrishnan, R., Divakaran, A., Smaragdis, A.: Audio analysis for surveillance applications. In: IEEE Workshop on Applications of Signal Processing to Audio and Acoustics 2005, pp. 158–161, IEEE, Oct 2005
10. Liu, Z., Wang, Y., Chen, T.: Audio feature extraction and analysis for scene segmentation and classification. J. VLSI Sig. Process. Syst. Sig. Image Video Technol. **20**(1–2), 61–79 (1998)
11. Li, D., Sethi, I.K., Dimitrova, N., McGee, T.: Classification of general audio data for content-based retrieval. Pattern Recogn. Lett. **22**(5), 533–544 (2001)

A Sentiment Analysis Based Approach for Understanding the User Satisfaction on Android Application

Md. Mahfuzur Rahman, Sheikh Shah Mohammad Motiur Rahman, Shaikh Muhammad Allayear, Md. Fazlul Karim Patwary and Md. Tahsir Ahmed Munna

Abstract The consistency of user satisfaction on mobile application has been more competitive because of the rapid growth of multi-featured applications. The analysis of user reviews or opinions can play a major role to understand the user's emotions or demands. Several approaches in different areas of sentiment analysis have been proposed recently. The main objective of this work is to assist the developers in identifying the user's opinion on their apps whether positive or negative. A sentiment analysis based approach has been proposed in this paper. NLP-based techniques Bags-of-Words, N-Gram, and TF-IDF along with Machine Learning Classifiers, namely, KNN, Random Forest (RF), SVM, Decision Tree, Naive Byes have been used to determine and generate a well-fitted model. It's been found that RF provides 87.1% accuracy, 91.4% precision, 81.8% recall, 86.3% F1-Score. 88.9% of accuracy, 90.8% of precision, 86.4% of recall, and 88.5% of F1-Score are obtained from SVM.

Keywords NLP · TF-IDF · Sentiment analysis · Machine learning · Mobile apps review

Md. M. Rahman (✉)
Division of Research, Daffodil International University, Dhaka, Bangladesh
e-mail: mahfuzur.web@daffodilvarsity.edu.bd

S. S. M. M. Rahman
Department of Software Engineering, Daffodil International University, Dhaka, Bangladesh
e-mail: motiur.swe@diu.edu.bd

S. M. Allayear
Department of Multimedia and Creative Technology, Daffodil International University,
Dhaka, Bangladesh
e-mail: drallayear.swe@diu.edu.bd

Md. F. K. Patwary
Institute of Information Technology, Jahangirnagar University, Savar, Bangladesh
e-mail: patwary@juniv.edu

Md. T. A. Munna
National Research University Higher School of Economics, Moscow, Russia
e-mail: tahsir.se@gmail.com

K. S. Raju et al. (eds.), *Data Engineering and Communication Technology*,
Advances in Intelligent Systems and Computing 1079,
https://doi.org/10.1007/978-981-15-1097-7_33

1   Introduction

Sentiment Analysis (SA) broadly involves data mining processes and techniques to extract insights, however, it is famous for Opinion Mining (OM) or emotion AI. Though it has different definition and uses but mostly it refers to the use of Natural Language Processing (NLP), text analysis, computational linguistics, and biometrics. It systematically identifies, extracts, quantifies, and studies affective states and subjective information. People's sentiment and emotion takes place towards certain entities. There are various categories that people use their sentiment for various purposes. Sophisticated categorization of a huge number of recent articles is provided by one survey. The related field to SA includes emotion detection, building resources, and transfer learning [1].

Machine learning approach and lexicon-based approach to evolve the problem of sentiment classification can be categorized. Machine learning and lexical need to enhance sentiment classification performance [2].

Application programming interfaces (APIs) are provided by many social media sites to prompt data collection and analysis by researchers and developers. For example, Twitter has three different versions of APIs available such as the REST API, the Streaming API, and the Search API. For collecting the status data and user information developers should use the REST API; with the Streaming API developers able to gather Twitter content in real-time, whereas the Search API allows developers to query specific Twitter content. Moreover, developers can mix those APIs to create their own applications. Hence, sentiment analysis seems to have a strong fundament with the support of massive online data [3].

Subjective detection, sentiment prediction, aspect-based sentiment summarization, text summarization for opinions, etc. are main field of sentiment research. Negative or positive opinions are predicting the polarity of text. The field of Natural Language Processing (NLP) has discovered the concept of sentiment analysis [4].

Classifying an opinionated document express as a positive or negative opinion and classifying a sentence or a clause of the sentence as subjective or objective has been studied in one paper [5]. The reviewers determine the product by positive or negative review. Positive or negative opinion is called sentence-level sentiment classification.

Due to the rapid growth of online shop, most of the people are getting attracted to buy products which are needed through online. On that case, customer reviews on the products getting importance for online merchants. So that, they can measure the satisfaction rate of their customers and can make decision [6]. The main contributions of this paper are listed as follows:

1. A model has been proposed to make a decision and be financially profitable for android apps developers using sentence-level sentiment analysis.
2. More focus on android apps reviews which will help developers to know about the real scenario of the developed application in market.

3. Multiple Machine learning classifiers has been evaluated to find out the classifier which can be fitted well in the proposed model.
4. The result of the evaluation provides a strong basis for building effective tools for Sentiment Analysis on the reviews from mobile application users.
5. The effectiveness of classification algorithm is evaluated in terms of accuracy score, precision, recall and F1-score.

To analyze overall sentiment of United States people, to determine the 2016 United States presidential election, there were many tweets collected from United States people. These tweets are collected from not only United States people but also other foreign countries [7].

The organization of this paper is structured as follows: Sect. 2 describes the related works done in this relevant field. The proposed model has been constructed and described in Sect. 3. The experiments, result analysis, and performance evaluation have been briefly discussed, in Sect. 4, which explains result and discussion. Section 5 concludes the paper with a possible future step.

2 Literature Review

Recently, sentiment analysis is one of the big areas for researchers because of increasing social media networks, blogger sites, different types of forum, feedback of products. This section provides an overview of previous studies regarding sentiment analysis.

Kanakaraj [8] proposed an idea to increase the accuracy of classification by semantics and Word Sense Disambiguation (WSD) in NLP (Natural Language Processing Techniques). Ensemble classification is analyzed to mined text information in sentiment analysis. Ensemble classifiers outperformed traditional machine learning classifiers by 3–5%.

Purchasing online products are preferred by maximum customers. Kamalpurkar [9] proposed a technique, classifying negative or positive reviews using machine learning based approach based on feature. Based on user's feedback, users choose to purchase products.

Using two product's datasets, namely, Nokia Lumia 1020 and Apple iPhone 4s, Venkata Rajeev [10] determines analysis combination on four parameters: star rating average, polarity rating, reviews per month, and helpful score. Apple iPhone 4s is better than Nokia Lumia 1020 to choosing the product. Star rating average 3.98, polarity rating 0.51, reviews per month 39, helpful score 78%, and grand score 8.03.

Mining products based on feature by customer's negative or positive opinion. Hu [6] proposed a SentiWordNet-based algorithm and divides the opinion analysis tasks into three steps: identifying opinion sentences and their polarity, mining the features that are customers' opinion, and removing the incorrect features.

Hu [6] gathered the customers' comments on electronic products from Amazon.com and C|net.com. They have considered the reviews of five products during their experiment such as Digital Cameras, DVD player, mp3 player, and cellular phone in quantity 2,1,1,1, and 1, respectively. During their work, they focused on only the features of the product mentioned by the reviewer. After that, they applied different techniques to mine the features. As a result, the obtained maximum 80% of recall and 72% of precision on average.

Individual performance is improved by the different types of sentiment approaches and consists of three combinations: ranking algorithms for scoring sentiment features as bi-grams and skip-grams extracted from annotated corpora, a polarity classifier based on a deep learning algorithm and a semi-supervised system founded on the combination of sentiment resources. TASS competition is evaluated based on general corpus average [11].

With statistical significance test, five text categorization methods are reported by a controlled study. Five methods are used, namely, Support Vector Machine (SVM), K-Nearest Neighbor(KNN), Neural Network (NNet), LLSF (Linear Least Squares Fit), and Naive Byes (NB). Dealing with a skewed category and performance as function of the training-set category frequency are focused on robustness of these methods. Yang [12] claimed that when the number of positive training instances per category are small, SVM, KNN, and LLSF significantly outperform.

Nicholas [13] investigated cross-domain data to detect polarity of sentiment analysis which data collected from YouTube. It represents bag-of-words feature.

Most significant challenges in sentiment evaluation are focused. It explains techniques and access of sentiment analysis objection. It represents challenges of sentiment [14].

Murty [15] explained cross-domain text classification techniques impotences and strengths. Basic knowledge right from the beginning of text classification methods are provided by cross-domain text classification algorithms.

Murty [16] proposed a new algorithm named LS-SVM which is efficient scheme for documents classification. LS-SVM is enhancing accuracy and retrenchment dimensionality of large text data with Singular Value Decomposition.

3 Proposed Model

In this section, the proposed model has been presented and described in details.

3.1 Data Collection

Reviews are collected from google play store using Web-based Crawler. Then collected positive reviews (PR) are labeled as 1 and for negative reviews (NR) are labeled as 0.

3.2   Pre-processing

For balancing the dataset, in this step, the total number of PR and total number of NR have been checked. After that, a balanced dataset has been generated from equal number of PR and NR. The following tasks have been also done as a part of pre-processing of the data:

- Steaming: Steaming is the process of reducing a word to its word stem that affixes to suffixes and prefixes or to the roots of words known as a lemma. Stemming is important in natural language understanding and NLP.
- Remove stopwords: The process has been carried out by removing frequently used stopwords (preposition, irrelevant words, special character, etc).

3.3   Bag-of-Words (BoW) Model Certain

The bag-of-words model is widely used in feature extraction procedures of Natural Language Processing (NLP) for a sentence or a document. Bag-Of-Words Model has been generated which includes following steps.

- Words Vocabulary: If there is a word that occurred 20 times in the corpus, it will be counted only once in the word's vocabulary. During defining the words vocabulary, every word will ensure the uniqueness of that presence.
- Sentence Vector and Manage Vocabulary: Each word from vocabulary will be scored in binary which is known as word scoring in NLP. This process has been conducted by applying n-gram to minimize the large size of vocabulary. The output of this process is (for one sentence).
- Feature Extraction: TF-IDF has been used to extract the features from the vectorized vocabulary. TF-IDF stands for Terms Frequency Inverse and Document Frequency which can define the importance of words corpus datasets.

3.4   Extracted Feature Vector

As an output of BoW, a vector containing the extracted and minimized features has been found in this stage.

3.5 Splitting

In this stage, the dataset has been split in ratio of 67% for train the model and the rest 33% for test.

3.6 Machine Learning Algorithm

In this stage, machine learning algorithms can be applied for training and test. In this research work, K-Nearest Neighbors (KNN), Random Forest (RF), Support Vector Machine (SVM), Decision Tree (DT), and Naive Byes (NB) classification techniques have been applied to classify the sentiment and evaluate the performance of the classifiers. As a result, a decision can be made whether which algorithm performs better with proposed model (Fig. 1).

4 Environment Setup and Experiment

The experiments and evaluation of the classifiers in proposed approach has been discussed in this section.

4.1 Environmental Setup

A desktop computer in configuration with Intel Core i5 Processor, 8 GB DDR3 RAM has been used during the experiment. Windows 10 64 bit was the installed Operating System. The implementation of the model in coding has been done using python 3.5 programming language along with the packages of scikit-learn, pandas, scipy, numpy, etc.

4.2 Dataset Used

"Mobile App Review" dataset collected by a web Crawler has been used [17]. The dataset contains 20 k reviews where 10 k data are positive and 10 k data are negative. Every app has about 40 reviews. The reviews are classified into positive and negative reviews and ignored neutral reviews. For positive review and negative review, node value are 1 and 0, respectively.

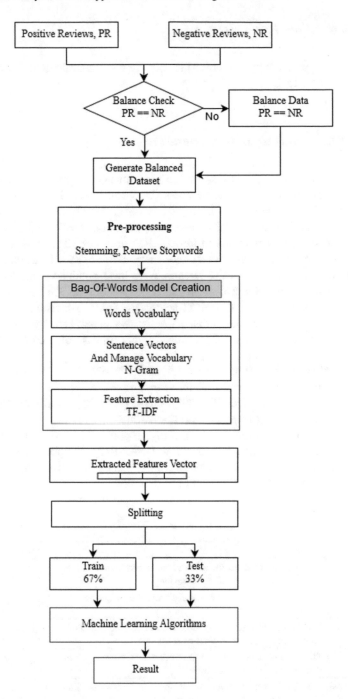

Fig. 1 Architecture of proposed model

4.3 Evaluation Parameter

There are some parameters or metrics [18–20] are used to evaluate the performance of classifiers (Table 1).

4.4 Result Analysis and Evaluation

Table 2 represents the comparison of the performance of proposed model along with KNN, Random Forest, SVM, Decision Tree, and Naive Byes classifiers. KNN provides 62.4% of accuracy, 70.2% of precision, 42.9% of recall and 53.3% f1-score. 87.1, 91.4, 81.8, 86.3% of accuracy, precision, recall, and f1-score have been found respectively for random forest. SVM provides 88.9, 90.8, 86.4, 88.5% of accuracy, precision, recall and f1-score respectfully. 85.4% accuracy, 87.3% precision, 82.2% recall and 84.7% f1-score have been obtained from decision tree classifiers. The results obtained from Naive Bayes are 60.3% accuracy, 78.4% precision, 28.4% recall and 41.7% f1-score. Maximum precision score obtains 91.4% by Random Forest classifiers. SVM is also given better performance for Recall and F-score 86.4% and 88.5%, respectively. In case of accuracy, SVM has been performed better with the proposed.

Figure 2 depicts the comparison chart of the performance of precision from proposed model obtained from KNN, Random Forest (RF), Support Vector Machine (SVM), Decision Tree (DT) and Naive Byes (NB) classifiers. The highest Precision score 91.4% by Random Forest classifier.

Figure 3 depicts the comparison chart of the performance of Recall from proposed model obtained from KNN, Random Forest (RF), Support Vector Machine (SVM), Decision Tree (DT) and Naive Byes (NB) classifiers. The highest Recall is given by Support Vector Machine and score 86.4%.

Table 1 Evaluation parameters

Parameters	Definition	Formula
Precision	The measurement exact value of the result provided by the classifier is defined as precision	$\frac{TP}{TP+FP}$
Recall	The thoroughness of the result provided by classifier measured by recall	$\frac{TP}{TP+FN}$
F1-score	The harmonic mean of exactness and completeness is known as F1-Score as well as F-Score	$\frac{2*Precision*Recall}{Precision+Recall}$
Accuracy	The ratio of accurately classified data to total number of data is defined as accuracy	$\frac{TP+TN}{TP+TN+FP+FN}$

TP (True Positive): The reviews which are positive and also classified as positive; FP (False Positive): False Positive represent the reviews are positive but classifier classified as negative; TN (True Negative): True Negative represents the reviews are negative and also classifier classified as negative; FN (False Negative): False Negative represent the reviews are negative but classifier classified as negative

Table 2 Comparison of experimented classifiers

Classifiers	Accuracy (%)	Precision (%)	Recall (%)	F-1 score (%)
KNN	62.4	70.2	42.9	53.3
RF	87.1	**91.4**	81.8	86.3
SVM	**88.9**	90.8	**86.4**	**88.5**
DT	85.4	87.3	82.2	84.7
NB	60.3	78.4	28.4	41.7

Fig. 2 Comparison of classifier performance for Precision

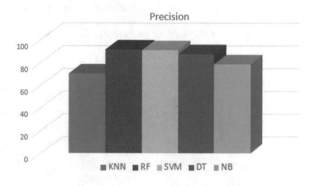

Fig. 3 Comparison of classifier performance for recall

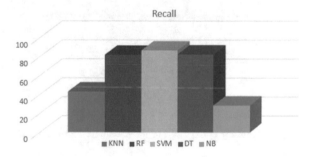

Figure 4 depicts the comparison chart of the performance of F-1 Score from proposed model obtained from KNN, Random Forest (RF), Support Vector Machine (SVM), Decision Tree (DT) and Naive Byes (NB) classifiers. Support Vector Machine is the highest F-1 score.

The final accuracy obtained from different classifiers have been depicted in Fig. 5. It's been found that Support Vector Machine (SVM) provide better performance in terms of accuracy and proved as a strong candidate to be used in the proposed model. SVM provides 88.90% of accuracy.

To recapitulate, it's been claimed that the proposed approach can be a strong basis of sentiment analysis with SVM classifier.

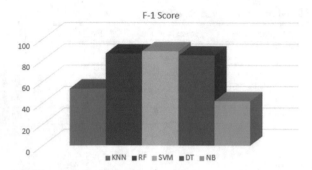

Fig. 4 Comparison of classifier performance for F-1 Score

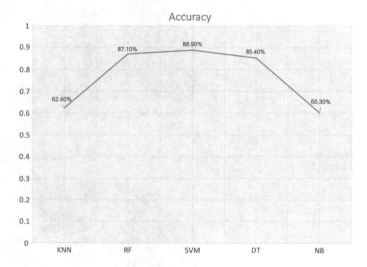

Fig. 5 Accuracy comparison of the classifiers on the proposed model

5 Conclusion

Sentiment analysis of mobile application users has been explained as well as been proposed a model. The proposed model has been broadly discussed and evaluated. Sentence-level sentiment classification has been focused on this research work. Multiple machine learning classifiers are also being assessed on the model. From Table 2 and Fig. 2, it can be concluded that the model can be a strong basement for building effective tools for Sentiment Analysis. Maximum accuracy has been found from SVM which is 88.9%. Even in case of recall and f1-score, it's been found that SVM performs better with the proposed model. But in case of precision, Random forest classifiers provide maximum precision which is 91.4%.

As a future step, the proposed model will be enhanced based on "Ensemble Feature Selection Scheme" and will be investigated.

References

1. Medhat, W., Hassan, A., Korashy, H.: Sentiment analysis algorithms and applications: A survey. Ain Shams Eng. J. **5**(4), 1093–1113 (2014)
2. Ghosh, M., Sanyal, G.: Performance assessment of multiple classifiers based on ensemble feature selection scheme for sentiment analysis. Appl. Comput. Intell. Soft Comput. (2018)
3. Twitter: Twitter apis. https://dev.twitter.com/start. Accessed 10 Jan 2019 (2014)
4. Saifee, V., Jay, T.: Applications and challenges for sentiment analysis: a survey. Int. J. Eng. Res. Technol. (IJERT) **2**
5. Liu, B.: Sentiment analysis and subjectivity. Handb. Nat. Lang. Process. **2**, 627–666 (2010)
6. Hu, M., Liu, B.: Mining opinion features in customer reviews. In AAAI, vol. 4, issue No. 4, pp. 755–760 (2004, July)
7. Agrawal, A., Hamling, T.: Sentiment analysis of tweets to gain insights into the 2016 US election. Columbia Undergraduate Sci. J. **11** (2017)
8. Kanakaraj, M., Guddeti, R.M.R.: Performance analysis of Ensemble methods on Twitter sentiment analysis using NLP techniques. In: 2015 IEEE International Conference on Semantic Computing (ICSC), pp. 169–170. IEEE (2015, February)
9. Kamalapurkar, D., Bagwe, N., Harikrishnan, R., Shahane, S., Gahirwal, M.: Sentiment analysis of product reviews. Int. J. Eng. Sci. Res. Technol. **6**(1), 456–460 (2017)
10. Rajeev, P.V., & Rekha, V.S.: Recommending products to customers using opinion mining of online product reviews and features. In: 2015 International Conference on Circuits, Power and Computing Technologies [ICCPCT-2015], pp. 1–5. IEEE (2015)
11. Martinez-Cámara, E., Gutiérrez-Vázquez, Y., Fernández, J., Montejo-Ráez, A., Munoz-Guillena, R.: Ensemble classifier for Twitter Sentiment Analysis (2015). Available at: http://wordpress.let.vupr.nl/nlpapplications/files/2015/06/WNACP-2015_submission_6.pdf
12. Yang, Y., Liu, X.: A re-examination of text categorization methods. In: Proceedings of the 22nd Annual International ACM SIGIR Conference on Research and Development in Information Retrieval, pp. 42–49. ACM (1999, August)
13. Cummins, N., Amiriparian, S., Ottl, S., Gerczuk, M., Schmitt, M., Schuller, B.: Multimodal bag-of-words for cross domains sentiment analysis. In: 2018 IEEE International Conference on Acoustics, Speech and Signal Processing (ICASSP), pp. 4954–4958. IEEE (2018, April)
14. Hussein, D.M.E.D.M.: A survey on sentiment analysis challenges. J. King Saud Univ.-Eng. Sci. **30**(4), 330–338 (2018)
15. Murty, M.R., Murthy, J.V.R., Reddy, P.P., Satapathy, S.C.: A survey of cross-domain text categorization techniques. In: 2012 1st International Conference on Recent Advances in Information Technology (RAIT), pp. 499–504. IEEE (2012, March)
16. Murty, M.R., Murthy, J.V.R., PVGD, P.R.: Text document classification based-on least square support vector machines with singular value decomposition. Int. J. Comput. Appl. **27** (7):21–26 (2011)
17. Android App Review Dataset, https://github.com/amitt001/Android-App-Reviews-Dataset. Accessed 10 Jan 2019
18. Rahman, S.S.M.M., Rahman, M.H., Sarker, K., Rahman, M.S., Ahsan, N., Sarker, M.M.: Supervised ensemble machine learning aided performance evaluation of sentiment classification. J. Phys. Conference Ser. **1060**(1), 012036 (2018). (IOP)
19. Rahman, S.S.M.M., & Saha, S.K.: StackDroid: Evaluation of a multi-level approach for detecting the malware on android using stacked generalization. In: International Conference on Recent Trends in Image Processing and Pattern Recognition, pp. 611–623. Springer, Singapore. (2018)
20. Rana, M.S., Rahman, S.S.M.M., & Sung, A.H: Evaluation of tree based machine learning classifiers for android malware detection. In: International Conference on Computational Collective Intelligence, pp. 377–385. Springer, Cham. (2018)

ECG Arrhythmia Detection
with Machine Learning Algorithms

**Saroj Kumar Pandey, Vineetha Reddy Sodum, Rekh Ram Janghel
and Anamika Raj**

Abstract Arrhythmia is one of the major causes of deaths across the globe. Almost
17.9 million deaths are caused due to cardiovascular diseases. This study has been
conducted to classify heartbeats into two classes, the one with regular heartbeat and
the other having irregular heartbeat. The dataset that is used here has been collected
from California University at Irvine Machine Learning Data Repository. First of all,
the dataset is pre-processed in which the data normalization is performed and the
missing values are removed. Following the previous step, feature selection is performed by Principal Component Analysis (PCA). Then, 8 classifiers are applied on
the various data splits. Finally, Accuracy, Sensitivity, and Specificity are calculated.
The maximum accuracy of **89.74%** is obtained using SVM and Naïve Bayes after
applying feature selection method on 90-10 data split.

Keywords Arrhythmia · PCA · SVM · Naïve bayes

1 Introduction

The major cause of deaths in the world according to the report of WHO are
cardiovascular diseases (CVDs): In 2016, an approx of 18 million people died
because of cardiac-related diseases, which reflect 31% of all deaths globally. 85%
of the deaths among these 31% are due to heart attack and stroke. The leading cause
of CVDs is long-term effect of cardiac arrhythmias. Arrhythmias occur when the

S. K. Pandey (✉) · V. R. Sodum · R. R. Janghel · A. Raj
National Institute of Technology, Raipur, Chhattisgarh, India
e-mail: sarojpandey23@gmail.com

V. R. Sodum
e-mail: vineetharedy007@gmail.com

R. R. Janghel
e-mail: rrjanghel.it@nitrr.ac.in

A. Raj
e-mail: nmkraj50@gmail.com

© Springer Nature Singapore Pte Ltd. 2020
K. S. Raju et al. (eds.), *Data Engineering and Communication Technology*,
Advances in Intelligent Systems and Computing 1079,
https://doi.org/10.1007/978-981-15-1097-7_34

Fig. 1 Normal ECG signal

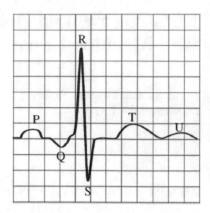

electrical signals to the heart that coordinate heartbeats don't work properly. Arrhythmia can be referred to a kind of disorder in which heartbeat is irregular, too fast or too slow [1].

The electrocardiogram is a significant diagnostic tool that is used to assess and monitor the electrical activities and muscular functions of the human heart. These heart activities result in creating some waves called P-QRS-T waves [2]. Some arrhythmias do exist with no sort of symptoms. For those arrhythmias that exhibit symptoms, some of the following may be symptoms: dizziness, breathlessness and noticeably rapid, strong, or irregular heartbeat due to agitation. There are several causes for arrhythmia, which may include diabetes, mental stress, smoking and many others [3] (Fig. 1).

Arrhythmias can be broken down into following types: Slow heartbeat, Fast heartbeat and Irregular heartbeat, early heartbeat which are scientifically termed as Bradycardia, Tachycardia, Flutter or fibrillation, premature contraction respectively. Not all arrhythmias are serious, but there are chances that it can lead the individual to cardiac arrest or stroke [4]. In general, for any person having a healthy heart, the heart rate of 60–100 beats per min should exist while relaxing [5].

Organization of the rest of the paper is as follows: Sect. 2 presents the related work. Section 3 presents the proposed methodology. The experimental analysis can be viewed in the Sect. 4. Conclusion is given in the Sect. 5.

2 Related Work

Since many years, several different classification models for arrhythmia detection are being proposed by Researchers. Correlation-based feature selection (CFS) with linear forward selection search was applied along with incremental back propagation neural network for better results in the classification process [6].

Another research is carried out using Multilayer perceptron (MLP) network model with static backpropagation algorithm to classify arrhythmia cases into

Fig. 2 Block diagram of the proposed methodology

normal and abnormal classes [7]. Another work used support vector machines (SVM) and K-nearest neighbour (KNN) as classification algorithms and sequential forward search selection as feature selection method. This setup used KNN and SVM alternatively in the wrapper part to obtain the best feature subset. Further, 20 fold cross-validation was performed to improve the further results [8]. A new approach was introduced to examine and calculate the performance of arrhythmia classification model using a generalized feedforward network (GFNN). The proposed classifier is trained using static backpropagation algorithm to classify arrhythmia patient cases into normal and abnormal classes [9].

Random forests have been made use for classification of the cardiac arrhythmia for the purpose of designing a computer-assisted diagnosis (CAD) system. It is based on the resampling strategy which was put forward for improving the classification systems that have been specially designed for detecting arrhythmia [10]. In [11], an intelligent system for classification was introduced in which multiple C5.0 decision tree models along with boosting process have been used to improve the accuracy of the model. In another work, a novel technique for automated analysis of electrocardiogram signal records, based on constructing a forest of binary decision trees was introduced. Each individual binary decision tree has been trained to identify a unique disease using medically important feature pertaining to that disease [12]. In [13], neural network is used to predict cardiac arrhythmias (Fig. 2).

3 Proposed Methodology

3.1 Removal of Missing Values and Data Normalization

There are a total of 279 features, among those 206 are linear and the remaining are nominal. Among all the subjects, there are 32 with missing information, So discarding all those, we have a net of only 392 instances, This is because we delete rows with all 0's and also columns with all 0's. There are 182 columns in total, out of which 173 are numerical ones and remaining is categorical. Among all these attributes, only 162 are used for training and testing. For the purpose of differentiating healthy and unhealthy Cardiac, classify it in one of the 2 groups. Class 1 implies Normal ECG data; Class 0 implies Arrhythmia. In this study, the dataset's feature values have a large numeric range for distinct attributes. In such scenarios, there may exist a greater impact of the features with larger numeric range on the other features. Data normalization will be done in order to overcome this impact [14]. Enhancing or improving the performance is the major motive of data normalization. For original dataset values, centering and scaling techniques have been adopted for data normalization, with the help of which the numerical stability of the proposed system can be improved.

3.2 *Principal Component Analysis*

It is a statistical method used to reduce the number of variables in a dataset. It does so by lumping highly correlated variables together. The dimensionality of dataset can be minimized using Principal Component Analysis. The various steps involved in PCA are as follows [15]:

(i) Standardize: standardize the scale of the data.
(ii) Calculate covariance: Calculate the covariance matrix for the data.

$$\mathrm{Cor}(X, Y) = \frac{\mathrm{Cov}(X, Y)}{\sigma_X, \sigma_Y} \tag{1}$$

(iii) Deduce Eigens: In the process of converting the data into new axes, the original data is multiplied by Eigenvectors, which indicates the direction of the new axes.

As per the definition of Eigenvalue and Eigenvector:

$$[\mathrm{Covariance\,matrix}] \cdot [\mathrm{Eigenvector}] = [\mathrm{Eigenvalue}] \cdot [\mathrm{Eigenvector}] \tag{2}$$

(iv) Re-orient data: Since the direction of the new axes is indicated by the Eigenvectors, the original data is multiplied by the Eigenvectors for re-orienting our data onto the new axes.

$$\mathrm{sc} = [\mathrm{Orig.data}] \cdot [\mathrm{eigenvector}] \tag{3}$$

(v) Plotting the re-oriented data: The rotated data plotted.
(vi) Bi-Plotting: PCA is incomplete without a bi-plot. We standardize the axes on the same scale, and add the arrows to reflect the original variables.

3.3 Proposed Classification Model

The pre-processed data will be passed onto the next phase. Here, the normalized data is classified to identify whether there is any disease. Partition of the filtered and normalized dataset into 2 distinct groups i.e., training and testing datasets will be done prior to classification process [16, 17]. A short description of these classifiers is given in the following subsections.

Multilayer Perceptron: Multilayer perceptron classifier is an artificial neural network, of feedforward type [18]. This classifier comprises a minimum of 3 layers of nodes. They are Input layer, Output layer, and Hidden layer. This Multilayer perceptron classifier makes use of a supervised learning technique called back-propagation for training purpose. [7].

Naïve Bayes: The basis of the Naïve Bayes (NB) Classifier is the "Bayesian theorem" and it is appropriately suited during high dimensionality of the inputs. This algorithm constructed the model that is a set of probabilities. Each member of this set corresponds to the probability that a specific feature appear in the instances of class c. Given a new instance, basing on the multiplication result of the conditional probabilities of the individuals for the feature values in the instance, the probability is estimated by the classifier, that the instance belongs to a particular class [19].

Decision Tree: A Decision tree is a flowchart like tree structure, where each and every individual internal node indicates a test on an attribute, each branch reflects an outcome of the test, and each terminal node holds a class label [20].

K-Nearest Neighbours: This classification technique is supervised one with desired computational speed and is based on mathematics, simple theory. KNN gives the sense that the nearest neighbours in any plane are expected that, they are of same kind i.e. they come under same class. In KNN, the class assignment is done by labelling the test sample with the majority class among the K neighbours. K is always ≥ 1 [8].

Random Forest: Random forests, which are also known as random decision forests, are learning methods for many tasks like regression, classification, etc. [15]. It gives performance better than a single individual tree, since it's an ensemble classifier that contains many decision trees [21]. Regression and classification both can be performed using Random forest classifier.

AdaBoost: AdaBoost is the short term for "Adaptive Boosting". Freund and Schapire proposed this first practical boosting algorithm in the year 1996. It mainly targets on the problems of classification and then sets its goal to convert a set of weak classifiers into a strong one [22].

Long Short-Term Memory: This is a method similar to recurrent neural networks. It is like the basic RNN which has recurrent unit to be used across time where the context value is obtained from result of the hidden node [23].

Support vector machine: The goal of the support vector machine technique is to discover a hyperplane in an N-dimensional space, that which distinctly classifies the data points. For the separation of any two classes of data points, we can choose many possible hyperplanes. Discovering a plane with maximum margin is our aim. Maximizing the margin distance provides some reinforcement so that future data points can be classified with more confidence [8].

4   Results

UCI Machine learning repository's arrhythmia dataset, which is composed of 452 samples have been made use for experimenting purpose in this research. All the samples were classified into a total of 16 distinct classes, where the first class refers to normal scenarios and the rest refer to various kinds of arrhythmias. For each and every individual sample, there are a total of 279 attributes. The first four attributes comprise of 'general information' like weight, age, etc., while the rest of the attributes are extracted from electrocardiogram (ECG) signals that have been recorded by a standard twelve lead recorder including P-Q-R-S-T waves' information.

Especially in the scenario of detecting the disease, the cost of wrongly classifying vary in severity, whereas accuracy assigns equal costs for all errors including false negative and false positive errors [12]. The proposed methodology's performance is evaluated for each category of ECG signal. It is found by calculating false positive, true positive, true negative and false negative. Sensitivity, Specificity, and Accuracy can be calculated as

$$S_E = \frac{T_P * 100}{T_P + F_N} \tag{4}$$

$$S_P = \frac{T_N * 100}{T_N + F_P} \tag{5}$$

$$\text{Accuracy} = \frac{(T_N + T_P) * 100}{T_N + F_P + T_P + F_N} \tag{6}$$

The results of different classifiers are shown in the tables given below.

Table 1 shows that performances of SVM and Naïve Bayes with and without applying PCA on 90-10 data split with respect to accuracy are best i.e. 89.74% with PCA. We can observe that sensitivity of random forest has reached to a maximum of 100%. The maximum performance is shown by Naïve Bayes with an accuracy of 89.74% without PCA. Its performance is not affected by applying PCA.

Table 1 Classifiers' performance for 90-10 data split with and without PCA

Classifiers	With PCA			Without PCA		
	Accuracy	Sensitivity	Specificity	Accuracy	Sensitivity	Specificity
MLP	82.05	80.00	75.00	84.61	70.00	78.78
RF	87.17	**100.00**	70.58	87.17	100.00	70.58
NB	**89.74**	70.00	80.00	**89.74**	70.00	80.00
SVM	**89.74**	70.00	80.00	74.35	50.00	82.75
KNN	84.61	40.00	87.87	76.92	30.00	90.00
DT	79.48	80.00	74.19	82.05	80.00	75.00
AdaBoost	87.17	80.00	76.47	87.17	80.00	76.47
LSTM	79.48	100.00	33.33	74.35	70.00	78.78

4.1 Comparison with State of the Art

For the evaluation of the proposed method, its result is compared with the state-of-the-art methods for the UCI arrhythmia dataset. The performance of the proposed method in comparison to the recent methods is shown in Table 2. It is clearly deduced that the proposed method shows much better performance in comparison to state-of-the-art methods. Malay, Mitra and Samantha's [6] work on 68-32 data split with incremental back propagation neural network (IBPLN) and Levenburg—Marquardt (LM) achieved an accuracy of 87.71%.

Jadhav et al. [10] worked on 90-10 data split achieving the accuracy of 78.89% by applying modular neural network. Batra, Anish [19] applied SVM on 75-25 data split to achieve an accuracy of 84%.

Table 2 Comparing the proposed classifiers with other methods

Method	Train-test split	No. of features selected	Accuracy (%)
Proposed	**90-10**	**102**	**89.74**
IBPLN +LM [13]	68-32	18	87.71
Modular NN [14, 15]	90-10	198	78.89
SVM [8]	75-25	60	84
Naïve Bayes [18]	70-30	205	70.50
KNN [5]	20 fold CV	148	73.80
MLPNN [6]	3 fold CV	–	88.24
NN [7]	90-10	79	76.67

5 Conclusion and Future Scope

This research paper presents a binary class classification of arrhythmic beats where the beats are either regular or irregular. This research uses 8 classifiers which are SVM, MLP, Naive Bayes, Random forest, Decision tree, KNN, LSTM, and AdaBoost. First of the pre-processing of data is performed in which missing values are removed from the dataset and further the data is normalized to bring uniformity. Later feature selection is applied on the dataset using Principal Component Analysis (PCA). Further the 8 classifiers are applied where SVM and Naive Bayes show the highest accuracy of 89.74% in the 90-10 data split after applying PCA followed by sensitivity of 70% and 80% respectively and specificity of 80 and 76.47% respectively whereas Random forest and LSTM displays the maximum sensitivity up to 100%. The potential of SVM, AdaBoost, Naïve Bayes, and Random forest can be used for classifying arrhythmia with much better results in the future and help the medical supervisors assist in their medical practice which will ultimately help patients to get specific cure for the heartbeat irregularities.

References

1. Mondejar-Guerra, V., et al.: Heartbeat classification fusing temporal and morphological information of ECGs via ensemble of classifiers. Biomed. Signal Process. Control **47**, 41–48 (2019)
2. https://medlineplus.gov/arrhythmia.html
3. https://en.wikipedia.org/wiki/Heart_arrhythmia
4. http://www.heart.org/en/health-topics/arrhythmia/about-arrhythmia
5. https://www.biotronik.com/en-us/patients/health-conditions/cardiac-arrhythmia
6. Mitra, M., Samanta, R.K.: Cardiac arrhythmia classification using neural networks with selected features. Procedia Technol. **10**, 76–84 (2013)
7. Jadhav, S.M., Nalbalwar, S.L., Ghatol, A.: Artificial neural network based cardiac arrhythmia classification using ECG signal data. In: 2010 International Conference on Electronics and Information Engineering (ICEIE), vol. 1. IEEE (2010)
8. Niazi, K., Khan, A., et al.: Identifying best feature subset for cardiac arrhythmia classification. Science and Information Conference (SAI), 2015. IEEE (2015)
9. Jadhav, S., et al.: Performance evaluation of generalized feedforward neural network based ECG arrhythmia classifier. IJCSI Int. J. Comput. Sci. Issues **9**, 379 (2012)
10. Namsrai, E., et al.: A feature selection-based ensemble method for arrhythmia classification. J. Inf. Process. Syst. **9.1**, 31–40 (2013)
11. Elsayyad, A., Nassef, A.M., Baareh, A.K.: Cardiac arrhythmia classification using boosted decision trees. Int. Rev. Comput. Softw. **10**, 280–289 (2015)
12. Bin, G., et al.: Detection of atrial fibrillation using decision tree ensemble. In 2017 Computing in Cardiology (CinC). IEEE (2017)
13. Umale, V., et al.: Prediction and classification of cardiac arrhythmia using ELM. Int. Res. J. Eng. and Technol. (IRJET) **3** (2016)
14. Jadhav, S.M., Nalbalwar, S.L., Ghatol, A.A.: ECG arrhythmia classification using modular neural network model. In 2010 IEEE EMBS Conference on Biomedical Engineering and Sciences (IECBES). IEEE (2010)

15. Abdi, H., Williams, L.J.: Principal component analysis. Wiley Interdisc. Rev.: Comput. Stat. **2.4**, 433–459 (2010)
16. Gupta, V., Srinivasan, S., Kudli, S.S.: Prediction and classification of cardiac arrhythmia (2014)
17. Fazel, A., Algharbi, F., Haider, B.: Classification of cardiac arrhythmias patients. In: CS229 Final Project Report (2014)
18. Raut, R.D., Dudul, S.V.: Arrhythmias classification with MLP neural network and statistical analysis. In: ICETET'08. First International Conference on Emerging Trends in Engineering and Technology, 2008. IEEE (2008)
19. Samad, S., et al.: Classification of arrhythmia. Int. J Electr. Energy **2.1**, 57–61 (2014)
20. Soman, T., Bobbie, P.O.: Classification of arrhythmia using machine learning techniques. WSEAS Trans. Comput. **4.6** 548–552 (2005)
21. Batra, A., Jawa, V.: Classification of arrhythmia using conjunction of machine learning algorithms and ECG diagnostic criteria. Train. J. (1975)
22. https://towardsdatascience.com/boosting-algorithm-adaboost-b6737a9ee60c
23. Assodiky, H., Syarif, I., Badriyah, T.: Arrhythmia classification using long short-term memory with adaptive learning rate. EMITTER Int. J. Eng. Technol. **6.1**, 75–91 (2018)
24. Kohli, N., Verma, N.K.: Arrhythmia classification using SVM with selected features. Int. J. Eng. Sci. Technol. **3**(8), 122–131 (2011)
25. Pandey, S.K., Janghel, R.R.: ECG arrhythmia classification using artificial neural networks. In Proceedings of 2nd International Conference on Communication, Computing and Networking. Springer, Singapore, 2019

Innovative Sensing and Communication Model to Enhance Disaster Management in Traffic

K. S. Sandeep Sagar and G. Narendra Kumar

Abstract Traffic management is the major problem and challenge these days especially in metropolitan cities. The congestion of vehicles at the traffic signal junction can lead to delay in movement of emergency vehicles, viz., ambulance, etc., a person struggling for life, not planned properly, congestion lead to loss of life. The project proposes a solution wherein the drones at each signal junction monitor the emergency vehicles movement and clear the path such that there are no other vehicles in the way during emergency vehicle movement. Emergency vehicle selects the shortest path to destination, sends signals to all the drones in the designated path using Global System for Mobile Communication (GSM). The drones at signal junction takes over control from central traffic control system and gives back to the control system once the emergency vehicle attains seamless signal. The next drone at next signal junction is ready for the operation and continues till the emergency vehicle reaches destination.

Keywords GSM · WSN · Drone · NS-2

1 Introduction

In developing countries, traffic congestion is the problem and a challenge. Traffic congestion is the root cause for various problems like accidents, traffic jam, and traffic rule violation. This has adverse effect on human lives too. Due to congestion, an emergency vehicle like ambulance may not reach hospital on time. The problem of traffic congestion can be solved using drones at each signal and they continuously monitor the movement of vehicles. Before starting off the ambulance driver

K. S. Sandeep Sagar (✉) · G. Narendra Kumar
Department of Electronics and Communication Engineering, U V C E,
Bangalore University, Bangalore, Karnataka, India
e-mail: sandeepsagarks@gmail.com

G. Narendra Kumar
e-mail: gnarenk@yahoo.com

© Springer Nature Singapore Pte Ltd. 2020   419
K. S. Raju et al. (eds.), *Data Engineering and Communication Technology*,
Advances in Intelligent Systems and Computing 1079,
https://doi.org/10.1007/978-981-15-1097-7_35

finds out the shortest path to hospital using Dikshtra's algorithm and sends signals to all the drones in that route using GSM.

Whenever there is an ambulance arriving a signal drone immediately takes control from the Traffic Control System and turns signal into blue color indicating that there is an ambulance arriving and all other roads are blocked [1]. The vehicles in the road where ambulance is moving automatically clear the lane for the smooth movement of ambulance. Once the ambulance crosses the signal, the drone gives back control to the Traffic Control System and the next drone will be ready for the operation. Due to congestion of traffic in peak time is horrible. To control this traffic wireless sensor network (WSN) is used for controlling congestion.

2 Related Work

The problem traffic management for emergency vehicles has been an issue since long time and is of great concern. Some already proposed systems in the area are as explained, Said Kafumbe proposed Improved Traffic Clearance for Emergency Vehicles. Here, every individual vehicle is provided with Radio Frequency Identification (RFID) tag. Once the machine is incoming at the junction, the RFID reader reads the RFID and communicates to the traffic controller [2] at the junction to show ON the inexperienced lightweight till the machine passes and altogether alternative ways red lightweight is turned ON.

Yong-Kul Ki proposed Traffic Clearance for Ambulance Services [3]. Here, an android application is developed which will help the ambulance to reach the hospital and help the traffic police clear the traffic. The proposed system makes use of GPS system to track the location of the ambulance which will give clear picture of the traffic hub/policeman for clearing the traffic.

B. Janani Saradha et al. proposed Advanced Traffic Clearance System for machine Clearance victimization RF-434 Module during this system the machine is fitted with a RF transmitter and controlled by the microcontroller [4]. The road intersections the traffic signal stand, the RF receiver in the stand receives the RF signal and automatically switches the traffic signal to green thus making the ambulance pass through the road intersections.

Sahil Mirchandani, et al., proposed Intelligent Traffic Control System for Emergency Vehicle. Here, ZigBee transmitter module is placed in each emergency vehicle. ZigBee receiver module is placed at traffic junction. The switch will be turned ON when the vehicle is used for emergency purpose. This will send signal through the ZigBee transmitter to the ZigBee receiver [5]. As the signal is sent to the traffic post, the red light will turn into green light. After crossing the traffic signal ZigBee module automatically turns OFF and green light will turn into red light.

3 Performance Evaluation of Protocols

3.1 Ad Hoc On-demand Distance-Vector

Ad Hoc On-Demand Distance-Vector (AODV) is a routing protocol designed for wireless and mobile unexpected networks. This protocol establishes routes to destinations on-demand and supports each unicast and multicast routing.

The AODV protocol creates paths between nodes as long as they are on demanded by the supply nodes [6]. Consequently, AODV is associated with the degree on-demand formula and will not yields to any added traffic for communication. AODV uses sequence of numbers to create freshness in the route and they are self-starting and loop-free in addition with scaling to various mobile nodes.

3.2 Destination-Sequenced Distance-Vector Routing

Destination-Sequenced Distance-Vector Routing (DSDV) is a table-driven routing subject for inadvertent mobile networks, reinforced the Bellman–Ford algorithmic rule. The impact of the algorithmic rule was to unravel the drawback of loop routing. For every entry within the routing table comprises a variety of sequences, the area of the sequence numbers unit is usually not enduring the link present. The amount is produced by the destination, and as well as the electrode must channelize successive update with this diversity [6]. Routing data is spread among nodes by producing full junk yard occasionally and smaller additional advanced updates regularly.

3.3 Intelligent Transportation System

An Intelligent Transportation System (ITS) is a complex application that, while not embodying intelligence in and of itself. It proposes to supply advanced services about entirely different modes of transport and traffic management [6].

3.4 Unmanned Aerial Vehicle

An Unmanned Aerial Vehicle (UAV), is also called as Drone, is associated with the Nursing craft while not an entity's pilot aboard [7]. In the nursing pilotless craft system, the UAVs square measure a part of Associate that represent a UAV, a ground-based controller, and communication system. The flight of UAVs could operate with wide-ranging degrees of self-sufficiency, either inferior to device by an entity's operator or separately by on-board.

3.5   NS-2

Network Simulator-2 (NS-2) is a well-known network simulation tool for its architecture which is appropriate for extensions interfacing with new simulation modules. It is implemented using IEEE 802.11 protocol [8] and has been extended to comprise other parameters to simulate the IEEE 802.11p. NS-2 also structures as a simple model for the illustration of reflections, refraction and shadowing effects affected by buildings.

4   Proposed Performance of Vehicles

The proposed work aims at clearing traffic at the signal junctions. Initially, the emergency vehicle finds shortest path using Dijkstra's algorithm. Using GSM the emergency vehicle sends information to the corresponding drones in the selected path. As the emergency vehicle is approaching a signal the drone at the particular signal take over the control of the traffic control system and monitors accordingly as shown in Fig. 1. Once the emergency vehicle crosses the particular signal the control taken by the drone is given back to the traffic control system and the signals are set normally. The next the drone is ready for the operation. This continues till the emergency vehicle reaches the destination.

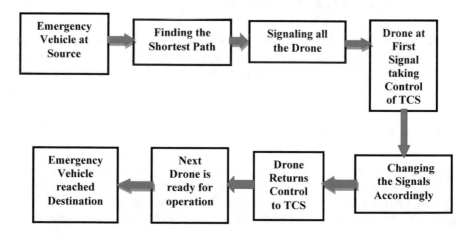

Fig. 1 Overview of proposed work

4.1 Signaling All the Drones

A Global System for Mobile communications (GPS) is installed in the emergency vehicle and drones at every signal. After finding the shortest path, the emergency vehicle sends a message to all drones in that path using GSM. The message is received by the drones using GSM.

The communication between two GSM modules can be achieved in 3 ways:

Using SMS messages, using a Data Call, using GPRS.

4.2 Drone Taking Control of Traffic Control System

GPS is installed in the drone which will continuously track the location of the emergency vehicle. When the emergency vehicle is approaching the signal junction, the drone takes control over the traffic control system (TCS). This can be done by communication between drone and TCS using GSM.

4.3 Changing the Signals Accordingly

Once the drone takes control over the Traffic Control System (TCS), the drone changes the signal lights accordingly. This is achieved by using Intelligent Transport System (ITS). When the emergency vehicle arrives at the traffic signal junction, all other signal lights will turn into red except the one with emergency vehicle. The signal light of the path in which emergency vehicle is present will turn into blue indicating the arrival of emergency vehicle. When the signal light turns blue, all the vehicles in that path move aside allowing the emergency vehicle move smoothly.

4.4 Drone Gives Back Control to TCS

Once the emergency vehicle crosses the junction, the drone gives back the control to the Traffic Control System (TCS) and the signals operate normally. This is done by communicating between GSM's in then drone and Traffic Control System (TCS).

4.5 Next Drone is Ready for the Operation

The drone at the next signal will be continuously monitoring the movement of emergency vehicle using GPS and will be ready for operation. This procedure continues until the emergency vehicle reaches the destination.

5 Development of Green Corridor

The Development of Green Corridor for Emergency Vehicles requires various hardware components:

- Drones at every traffic signal junction.
- Global System for Mobile communication (GSM) module in drone, emergency vehicle and Traffic Control System (TCS).
- Global Positioning System (GPS) in the drone and emergency vehicle.
- Intelligent Transport System (ITS) in the drone.

Initially, the emergency vehicle is at the source and to reach the destination it has to cross N number of signal junctions. At the signal junction, the drone takes control from the Traffic Control System (TCS) and operates the signal lights as shown in Fig. 2. The emergency vehicle crosses the junction smoothly and the drone gives back the control to TCS. This is repeated for N times. The emergency vehicle at the source selects the shortest route to the destination and sends message to all the drones in the selected path. The messages are sent to the drones using GSM and these messages are received by the GSMs in the drones. The GPS in the emergency vehicle continuously updates its location to the GPS in the drone.

As the emergency vehicle approaches the signal junction, the drone at the junction communicates with the TCS using GSM to GSM communication as shown in Fig. 3. The TCS hands over the control to drone. The drone sets all the traffic signal lights into red color except for the emergency vehicle's path. The traffic

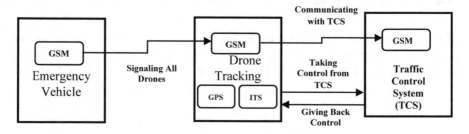

Fig. 2 Schematic representation

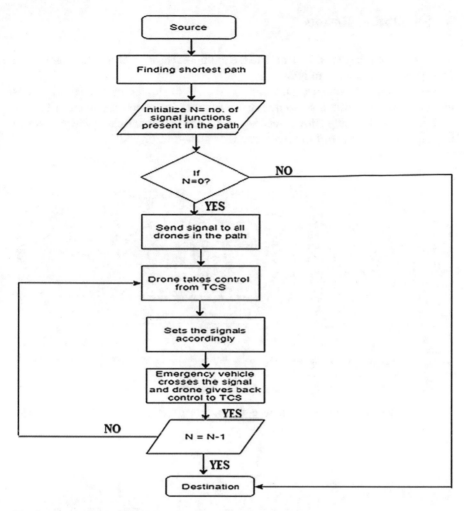

Fig. 3 Development of green corridor for emergency vehicles

signal light of the path in which emergency vehicle is moving is set to blue color, indicating the arrival of emergency vehicle. This process is done by Intelligent Transport System (ITS). The vehicles in the path of emergency vehicle move aside making green corridor for smooth movement of ambulance.

After the emergency vehicle crosses the traffic signal junction, using GSM to GSM communication the drone gives back control to TCS and the traffic signal lights operate normally. In case, multiple emergency vehicles arrive at the junction, all the traffic signal lights of the paths in which emergency vehicles are arriving will be set to blue and the same process repeats.

6 Simulation Results

In Experimental Analysis: To simulate the multiple scenarios of wireless network, a TCL program is written in NS2.

The emergency vehicles transforming signals to drones before vehicle starts moving towards destination, so that drone acquires indication is shown in Fig. 4.

Traffic signal turning Blue when the emergency vehicle reaching towards signal for creating to give path to move forward as shown in Fig. 5.

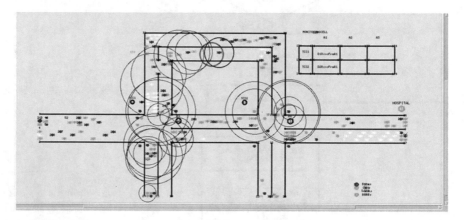

Fig. 4 Snapshot of emergency vehicle sending signals to all drones in the path before leaving

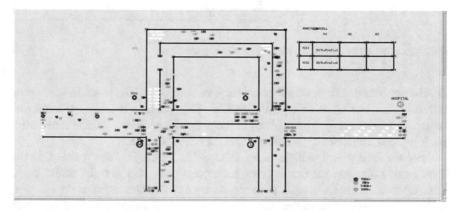

Fig. 5 Snapshot of signal light turning blue on arrival of emergency vehicle

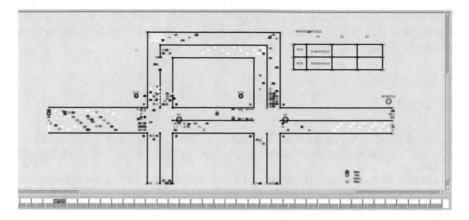

Fig. 6 Snapshot of vehicles making green corridor as emergency vehicle arrives

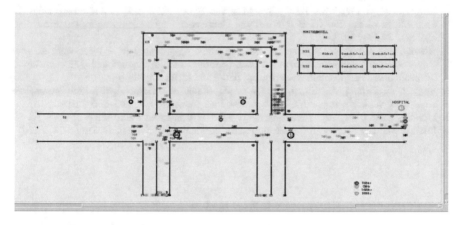

Fig. 7 Snapshot of two emergency vehicles moving freely

Traffic signal turning Green when the emergency vehicle reaching towards signal, hence the vehicle moves towards destination easier as shown in Fig. 6.

The emergency vehicles moving freely towards destination, after vehicle transient the signal drove takings control of TCS is shown in Fig. 7.

7 Conclusion

The Development of Green Corridor for Emergency Vehicles is simulated using Network Simulator 2. The duration for reaching the ambulance to nearest hospital by short duration. The emergency vehicles can reach destination in shortest time.

No disturbances in the route for emergency vehicles. The system lifetime is high compared to IoT. Initial cost of implementation is high. The emergency vehicles will be reaching without any delay and in perfect time.

Reference

1. Narendra Kumar, G., Smitha Shekar, B., Palani, P., Soumya, M. Murali, G.: VANET based congestion control using Wireless Sensor Networks (WSN). IJCSI Int. J. Comput. Sci. **10**(3) (2013)
2. Kafumbe, S.: iRovers: real-time unmanned four-wheel iot vehicles for air and noise pollution monitoring. IEEE ASET, Abu Dhabi, UAE (2018)
3. Ki, Y.-K.: Accident detection system using image processing and Mdr. IJCSNS **7**(3), 35–39 (2007)
4. Janani Saradha, B., Vijayshri, G., Subha, T.: Intelligent traffic signal control system for ambulance using rfid and cloud. In: IEEE Computing and Communications Technologies (ICCCT), Chennai (2017, February)
5. Mirchandani, S., Wadhwa, P.: Optimized rescue system for accidents and emergencies. In: Proceedings of the 2nd IEEE International Conference on Communication and Electronics Systems, 22 Mar 2017.
6. Abdoos, M., Mozayani, N., Bazzan, A.L.C.: Traffic light control in non-stationary environments based on multi agent Qlearning. In: Proceedings of 14th International IEEE Conference on Intelligent Transportation Systems, pp. 580–1585 (2011, October)
7. Sharma, S., Pithora, A., Gupta, G., Goel, M., Sinha, M.: Traffic light priority control for emergency vehicle using RFID. Int. J. Innov. Eng. Technol. **2**(2), 363–366 (2013)
8. L'hadi, I.: Rifai, M., Alj, Y.S.: An energy-efficient wsn-based traffic safety system. In: 5th International Conference on Information and Communication System, Jordan (2014, April)

Prediction of Student's Educational Performance Using Machine Learning Techniques

B. Mallikarjun Rao and B. V. Ramana Murthy

Abstract Educational data mining indicates an area of research in which the data mining, machine learning, and statistics are applied to predict information from academic environment. Educating is an act of imparting or acquiring knowledge to/ from a person formally engaged in learning and developing their innate quality. Over the years, the data mining techniques are being applied to academics to find out the hidden knowledge from educational datasets and other external factors. Previous research has been done to identify the elements that change the performance, and these elements can be termed as emotional and external factors. One's performance can be affected by factors such as not attending classes, diversion, remembrance, physical or mental exhaustion due to exertion, sentiments, surroundings, pecuniary, and pressure from family members. This research effort is on external factors and organizational elements. For teachers to foretell the future of a student is very useful and it identifies a student with his performance. In this research paper, external factors are studied and investigated and implemented using XGBoost classifier for predicting the student's performance.

Keywords Educational Data Mining (EDM) · Classification · XGBoost · Boosting · Stratified K-fold · Prediction

1 Introduction

As the data increases, managing the data is always a concern. Since the educational data pertaining to students, course contents, timetables, classrooms, and staff increases enormously, to manage this data also becomes a concern to the authorities.

B. Mallikarjun Rao (✉)
Research Scholar, Rayalaseema University, Kurnool, India
e-mail: bmrao2002@gmail.com

B. V. Ramana Murthy
Stanley College of Engineering, Abids, Hyderabad, India
e-mail: drbvrm@gmail.com; drbvrm@stanley.edu.in

© Springer Nature Singapore Pte Ltd. 2020   429
K. S. Raju et al. (eds.), *Data Engineering and Communication Technology*,
Advances in Intelligent Systems and Computing 1079,
https://doi.org/10.1007/978-981-15-1097-7_36

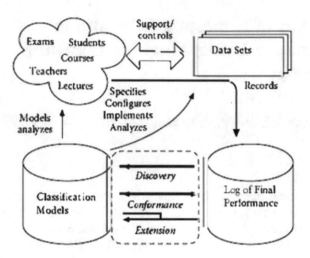

Fig. 1 EDM model workflow

The exponential increase of data in education has given a potential to a new field of data mining, known as educational data mining [1] (EDM).

Data mining [2] in academic field is an integrative study and known as educational data mining (EDM). EDM generally explicates data generated by any type of system which supports instructional methods of knowledge acquired or in institutions who provides a design for teaching. The EDM process workflow is given in Fig. 1.

The fundamental purpose of educational data mining is to declare student's performance and assess a learner. The performance and assessing a learner issues are well extensively connected with academic environment.

EDM has come forth to view as an autonomous investigative area in recent years for data mining researchers over the terrene from various and associated field of education which attempts to give knowledge and skills based on interaction, and also investigate cognitively how humans learn. The statistical and mathematical techniques have been applied to data like performance, responses to internal and external stimuli, modules, and mental traits, and such data was taken from third-party tools.

This research paper is organized into six sections. Section 1 gives introduction about EDM process. Section 2 is about the literature review. Section 3 examines the data mining process. In Sect. 4, the research work is examined. Section 5 discusses the results, and Sect. 6 presents the conclusion and the future work.

2 Literature Review

A lot of studies on EDM have been done and are proven to be beneficial. The study by Han and Kamber [2] shows how data mining power to produce results and performance. A case study builds a model based on data using the decision tree algorithm to show how it affects the low academic performance. Many prototypes based on ID3 decision tree algorithm were generated. Their work gave the model [3] based on the data as one of the accurate models. The learner's or student's performance in the course of study is indicator of his achievement.

The second predicting performance in higher courses of university, humanities department [4] developed a model to predict the performance in their course of learning based on the previous year marks. Their work reduces the dropout rates by predicting student's performance before registration of the course. Two experiments were supervised using the rule-generation algorithm; The first experiment used grades in English courses and mathematics courses, which gave rules with accuracy of 62.75%. The second experiment used student's grades only in two English courses, which gave four rules with accuracy of 67.33%. These results convey that the student's work [5] in English courses will have the effect on their performance in the programming course.

In another application of EDM [6] in predicting behavioral patterns [7] which gave a system of classification for the students and to predict the uncontrolled behavior with the help of different parameters other than academic details.

An evaluation of student's performance using data mining techniques [8] provides an overview of the data mining techniques that have been used to predict student's performance.

Dr. Tajuniza [9] gave analysis of the elements affecting learning that aids the prediction of learning performance. They used map-reduce concepts for predicting the student performance.

Baker [1] analyzed the trends in the area of EDM and focused on the prediction and the scope of work using the existing models.

Galathiya [10] examined the classification techniques with decision tree algorithms using feature selection, and a new concept of feature engineering is evolved.

Bharadwaj and Pal [11] investigated features like students present in the class, slip test, seminars, and internal marks, to predict the performance at the end of the semester.

Ho and Tin Kam [12] investigated techniques like random subspace which is referred to as attribute or feature bagging technique, and it reduces the correlation between different elements by training them on random feature samples. Ho analyzed how bagging and random feature projection aid in giving accurate gain under various conditions.

Methodology

The dataset has been generated programmatically and used as a stand-in for test datasets of production or relating to data, to validate mathematical models and to train machine learning models. Moreover, the data can be used to identify benchmarks which can be later used to conduct quantitative performance testing and systematic comparisons of approaches which also allows for determining statistical significance of the findings.

KDD is a process in which we used to extract useful and meaningful information by doing various actions on the dataset. Data mining [13] and KDD are often used interchangeably to serve the purpose of mining the dataset. The data mining is considered as a step in the knowledge discovery, and the KDD is the overall process of mining. Knowledge discovery in databases involves:

(1) Selection of data.
(2) Preprocessing of data.
(3) Transformation of data.
(4) Data mining.
(5) Interpretation and evaluation of data.

3   Data Mining Process

In modern educational system, the performance of student is influenced by emotional and external factors and they have to be investigated and analyzed. An instance of motivation leads to enthusiasm and then to success. Right evaluation of talented student can help the teacher to assess the student to perform better. All the learners require proper ambience at learning place and also at home. The factors like economic conditions and background of parents also produce an effect on the achievement of the student as they are helpless to get appropriate study and learning. In this research work, we consider external and academic factors, which help the instructor to recognize the features or elements that are associated with the good, average, and bad learners and take proper action to improve their performance.

A. *Data Preparation*

The resource set used in this research paper was programmatically created rather by the real-world events. Initially, the dataset size is very small and we can go for large datasets after applying different models. In this step, data is prepared for any errors and missing values and the resultant can be stored as a complete dataset.

B. *Data Selection and Analysis*

In data selection [4], the required factors were selected. A few derived variables were selected, and other variables were derived from the dataset. The forecasting and response factors were extracted from the dataset of Fig. 2. In general, all institutions conduct several exams and test the theoretical and practical knowledge of the students in that subject. This assessment helps the learners to get the better understanding of that particular course. The subject marks are given by the institution and must be scored by each and every student in order to continue the program. It is compulsory for the student to attain that score based on which he/she can be considered to be passed. This information gives student's performance during the entire semester and before the final exams. The aim of this study is to forecast the accomplishment or prediction of the performance of students in final exams based on their previous performance in that semester. This study will also provide the information about how much a student is learning from these exams and how well one has performed. Each semester consists of five core subjects. Two midsems and two assignment tests are being conducted for each subject during the entire semester. A sample of original dataset is given in Fig. 2; Figure 3 is the preprocessed dataset; Table 1 gives the attributes of dataset; Table 2 gives statistical values of dataset; and Table 3 gives the predicted final values.

The variables taken from the dataset are of categorical and numeric, and final examination performance is the response variable. Initially, the attribute Id is shown in the original dataset and it is a unique identifier and it need not be used in the training of the model.

Description of each attribute is shown below:

(1) Gender—Student's gender (nominal: M → Male, F → Female)
(2) Location-Student's location(nominal: U—Urban, R—Rural)
(3) Student ID: Identification of student (numeric: 0–10,000)
(4) Stage—Student level and belongs (nominal: 'lowerlevel', 'MiddleSchool', 'HighSchool')
(5) Previous year mark—Marks secured by the student last year (numeric: 0–100)
(6) Section—Classroom section (nominal: 'A', 'B', 'C')
(7) Stream—Course topic (nominal: 'CSE', 'IT', 'META', 'CIVIL', 'MECH', 'ECE', 'CHEM', 'BIO TECH', 'EIE', 'OTHERS')
(8) Semester—College year semester (nominal: 'First', 'Second')
(9) Relation: Relation with student and responsible for student (nominal: 'Mother', 'Father', 'Guardian')
(10) Raised hand—Student raising his/her hand in classroom (numeric: 0–100)
(11) Visited resources—Student visits a course content (numeric: 0–100)
(12) Viewing announcements—Student checking new announcements (numeric: 0–100)
(13) Discussion groups—Number of times student participated in group discussion (numeric: 0–100)
(14) Parent answering survey—Survey answered by parent (nominal: 'Yes', 'No')
(15) Parent college satisfaction—Parent satisfied with college or not (nominal: 'Yes', 'No')
(16) Student absence days—Number of days student is absent (nominal: above-7, under-7)
(17) Parent income—Income of parent of each student (nominal: High, Middle, Low)

ID	Location	Gender	Section	Stream	Semester	Relation	Raisedhands	VisitedResources	AnnouncementsView	Discussion	ParentAnsweringSurvey	Income_Level	ParentscollegeSatisfaction	StudentAbsenceDays	Parent_Education
0	7922	Urban	M	A	IT	F	Father	15	16	2	20	Yes	Low	Good	Under-7	Y
1	7809	Urban	M	A	IT	F	Father	20	20	3	25	Yes	Low	Good	Under-7	Y
2	2322	Urban	M	A	IT	F	Father	10	7	0	30	No	Low	Bad	Under-7	N
3	7386	Urban	M	A	IT	F	Father	30	25	5	35	No	Low	Bad	Under-7	N
4	6324	Urban	M	A	IT	F	Father	40	50	12	50	No	Low	Bad	Under-7	Y
5	3718	Urban	F	A	IT	F	Father	42	30	13	70	Yes	Low	Bad	Under-7	Y
6	8032	Urban	NaN	A	ECE	F	Father	35	12	0	17	No	Middle	Bad	Under-7	N
7	6411	Urban	M	A	ECE	F	Father	50	10	15	22	Yes	Middle	Good	Under-7	Y
8	4849	Urban	F	A	ECE	F	Father	12	21	16	50	Yes	Middle	Good	Under-7	Y
9	5478	Urban	F	B	IT	F	Father	70	80	25	70	Yes	Middle	Good	Under-7	Y

Fig. 2 Sample of original dataset

Fig. 3 Preprocessed dataset

Data Cleaning Steps:

Checking if any data is missing

ID	0
Location	3
Gender	3
Section	0
Stream	0
Semester	0
Relation	0
Raisedhands	0
VisitedResources	0
AnnouncementsView	0
Discussion	0
ParentAnsweringSurvey	0
IncomeLevel	0
ParentscollegeSatisfaction	0
StudentAbsenceDays	0
ParentEducation	0
AssignmentSubmitted	0
PreviousYearMark	0
Result	0

Table 1 Attributes of dataset

Feature	Description
ID	Student Id
Location	Location of student
Gender	Male or female
Section	Classroom student belongs to A, B, C
Stream	Course topics like CSE, ECE, etc.
Semester	First semester or second semester
Relation	Relation with student like father or mother or guardian
Raised hands	Number of times the student raises his/her hand
Visited resources	How many times the student visits a course content
Announcement's view	Number of times the student checks the new announcements
Discussion	How many times the student participates in group discussion
Parent answering survey	Parent participation of surveys
Income level	Income of parent
Parents college satisfaction	Parent satisfied with college or not
Student absent days	Number of days student is absent
Parent education	Parent education
Assignment submitted	Assignment submitted or not
Previous year mark	Last year's mark
Result	Final result

Table 2 Statistical values of dataset

	ID	Raisedhands	VisitedResources	AnnouncementsView	Discussion	PreviousYear_Mark
count	480.000000	480.000000	480.000000	480.000000	480.000000	480.000000
mean	5767.816667	46.775000	54.797917	37.918750	43.283333	26.452083
std	2631.313209	30.779223	33.080007	26.611244	27.637735	9.571498
min	1008.000000	0.000000	0.000000	0.000000	1.000000	10.000000
25%	3640.750000	15.750000	20.000000	14.000000	20.000000	19.000000
50%	5837.500000	50.000000	65.000000	33.000000	39.000000	26.000000
75%	8100.500000	75.000000	84.000000	58.000000	70.000000	33.000000
max	9946.000000	100.000000	99.000000	98.000000	99.000000	50.000000

Table 3 Predicted final values as result

Assignment_Submitted	PreviousYear_Mark	Result
Y	34	M
N	29	M
N	18	L
N	23	L
N	24	M
N	24	M
N	29	L
N	26	M
N	23	M
N	23	M

(18) Is Parent Educated?—Parent's education (nominal: Y → Yes, N → No)

(19) Assignment submitted (nominal: Y → Yes, N → No)

(20) Previous years mark—Last year marks (numeric: 0–100)

(21) Result— Final result (nominal: H → High/Excellent, M → Above Average, L → Below Average).

Dataset as a structure is divided into finite elements to prepare for data analysis, which is a form of data transformation, and also the process of minimizing the values for a given continuous attribute. It is done by reducing the attribute value into intervals.

Each attribute for the research paper is as follows and is shown in Table 1:

Data mining is the most important step which extracts interesting patterns existing in the transformed data. Considering the goal of the study and the knowledge that should be extracted, proper mining tasks are chosen. The classification [14, 15] aims to predict certain results based on the given set of inputs.

The values of count, mean, min, max, and standard deviation with 25%, 50%, and 75% split levels are given below.

The different classification algorithms are used in machine learning process, and for the prediction [15] of learner's performance, we have used XGBoost classifier. The data cleaning is done on raw data for any missing values and repeated values.

The data is replaced for missing values (categorical variables) with their mode. We can also choose to remove the rows where some values are null. But that would lead to information loss.

Feature 'Location' has '2' unique categories.
Feature 'Gender' has '2' unique categories.
Feature 'Section' has '3' unique categories.
Feature 'Stream' has '10' unique categories.
Feature 'Semester' has '2' unique categories.
Feature 'Relation' has '2' unique categories.
Feature 'ParentAnsweringSurvey' has '2' unique categories
Feature 'Income_Level' has '3' unique categories.
Feature 'ParentscollegeSatisfaction' has '2' unique categories.
Feature 'StudentAbsenceDays' has '2' unique categories.
Feature 'Parent_Education' has '2' unique categories.
Feature 'Assignment Submitted' has '2' unique categories.
Feature 'Result' has '3' unique categories.

XGBoost Classifier

This model works on the small-to-medium structured form of data (pattern) and uses a sequential technique which works on the principle of ensemble, which combines a set of weak learning model features, in which trees are grown one after another, and reduces the calculations if any and are made in subsequent iterations. The cross-validation is done by stratified sampling method, which refers to a method of dividing the dataset into groups, and a sample is drawn from each group.

Stratified K-fold validation is used to validate the model by providing the train and test split at the user level for the data and the features where the folds are given as percentage of samples of each class. After this it gives the best features for the model and their by the validation of the selected model.

4   Research Work

Different solutions have been given for educational data mining, but classification algorithms have been acknowledged as important in the area of prediction of [16] the student performance. It focuses on the relevant attributes through feature selection methods. In Sect. 4, an outline has been given for the feature selection and dataset is taken to be 80 for training dataset as X_Train and Y_Train and 20% for the testing dataset as X_Test and y_test. The y_test is the final predicted value.

A.   **Feature Selection**   [17]

In the dataset, there may be many related and unrelated attributes present in the data to be mined [3]. The unrelated, missing, and different data formats are to be removed or filled with appropriate values, and irrelevant attributes are to be removed. Different mining algorithms will not perform well with large datasets, and therefore, feature selection techniques need to be applied before any kind of mining algorithm is applied. The main objective of feature selection is to avoid overfitting and improve the performance of the model.

Every machine learning model depends on the data that is used to train it, and the data used for training has to be clean and free of unnecessary data points. Hence, we need to ensure there is less noise in data. Moreover, there is no hard and fast rule to include all that features in our model. There are many statistical tests based on filter methods and wrapper methods to undertake the process of feature selection. This process can also help in reducing the training evaluation time and also improve accuracy of the model. It also has an added advantage in interpretation of our model.

B.   **Bootstrap Aggregation**

Bootstrap aggregating is also called as boosting or bagging is an ensemble method designed to improve the accuracy of algorithms used in classification models and reduce the variance and help to avoid overfitting.

In this research paper, feature selection is made with bootstrap aggregation or bagging, which works on the appropriate features through boosting method.

5   Results and Discussion

In this paper, factors with values more than 0.20 were considered and feature variables for prediction with high values are shown.

From the table, parent education, visited resources, and assignment submitted have given high priority over other features. After the above features, income has taken priority, and then, working of parents, medium of instruction, and style of learning have given next priority.

The values and the features like education of mother and stream, which makes the students to get low grades in their examinations, but if we consider impact from source, institution helps students to achieve good results.

Finally, this research helps stakeholders and teachers, who give high preference to after-school hours of study and house income, which improve their performance or the student's performance.

The parent has to give more choice to features like location and after-school hours study. If they do not give high priority to emotional features like working parents and household income, then the students and their parents are suggested to give full choice to features relating to institute. But if they give priority to the above features, most of the learners will get good marks in their studies. From the above discussion, it is clear that the research study will help the teachers to improve the performance of the students.

6 Conclusion and Future Work

The performance forecasting or prediction will help teachers and parents to concentrate on improving student's performance, and this helps research study to improve the current model or to select good models among the classifier algorithms for predicting the student's performance.

The results in Table 3 show the results. This work will also help to identify the students who perform low and why they need special attention. For future work, we would like to refine our work by taking more number of example datasets or features and come up with more accuracy and with other techniques to help students in their educational careers.

As a conclusion, we had met our objective of predicting the performance of students by the XGBoost classifier and it was implemented in Python 3.7.1. This study gave an accuracy of 100% with a time of 1.2 s, and the time varies for large datasets. XGBoost classifier is evaluated as the best algorithm for predicting student's performance.

References

1. Baker, R.S.J.D., Yacef, K.: The state of educational data mining in 2009: a review and future visions. J. Edu. Data Min. **1**(1) (2009, Fall)
2. Han, J., Kamber, M., Pei, J.: Data Mining: Concepts and Techniques. Morgan Kaufmann, 3rd edn. (2012). ISBN 978-0-12- 381479-1
3. Yadav, S.K.: Data mining: a prediction for performance improvement of engineering students using classification. WCSIT **2** (2012)
4. Badr, G., Algobail, A., Almutairi, H., Almutery, M.: Predicting students performance in university courses: a case study and tool in KSU maths department. Procedia Comput. Sci. **82**, 80 (2016)

5. Suresh, B.K.: Mining educational data to analyze student's performance. Int. J. Adv. Comput. Sci. Appl. **2**(N0), 6 (2011)
6. Romero, C., Ventura, S.: Data mining in education. WIREs Data Min. Knowl. Disc. **3**(1), 12–27 (2013)
7. Elakia, G., Aarthi, N.J.: Application of data mining in educational database for predicting behavioural patterns of the students. IJCSIT **5**(3), 4649–4652 (2014)
8. Altujjar, Y., Altamimi, W., Al-TuraikiAl-Razgan, M.: Predictingcriticalcourses affecting students performance: a case study. Procedia Comput. Sci. **82**, 65–71 (2016)
9. Tajunisha, N., Anjali, M.: Predicting student performance using MapReduce. IJECS **4**(1) (2015)
10. Galathiya, A.S., Ganatra, A.P.: Classification with an improved decision tree algorithm. IJECS **46** (2012)
11. Bharadwaj, B.K., Pal, S.: Data mining: a prediction for performance improvement using classification. IJCSIS **9** (2011)
12. Ho, T.K.: Random decision forests. In: Proceedings of the 3rd International Conference on Document Analysis and Recognition, Montreal, Canada, pp. 278–282, 14 Aug 1995
13. Pujari, A.K.: Data mining techniques. ISBN 8173713804
14. Shahiri,A., M., Husain, W., Rashid, N.A.: A review on predicting students performance using data mining techniques. Procedia Comput. Sci. **72**, 414–422 (2015)
15. Pandey, U.K., Pal, S.: Data mining: a prediction of performer or underperformer using classification. Int. J. Comput. Sci. Inf. Technol. (IJCSIT) **2**(2), 686–690 (2011). ISSN:0975-9646
16. Galathiya, A., Ganatra, A.P.: Classification with an improved decision tree algorithm. IJECS **46** (2012)
17. Beniwal, S., Arora, J.: Classification and feature selection techniques in data mining, IJERT **1** (6) (2012)
18. Ho, T.K.: The random subspace method for constructing decision forests. IEEE Transactions on Pattern Analysis and Machine Intelligence (1998)

A Novel Approach for Authorship Verification

P. Buddha Reddy, T. Murali Mohan, P. Vamsi Krishna Raja
and T. Raghunadha Reddy

Abstract The Internet is increasing with huge amount of textual data. The crimes also increased in the Internet along with textual data. The authorship analysis is one important area attracted by the several researchers to reduce the problems raised through the text in the Internet. Authorship verification is one type of authorship analysis which is used to verify an author by checking whether the textual document is written by the disputed author or not. The accuracy of authorship verification majorly depends on the features that are used for distinguishing the style of writing followed in the documents. In the previous works of authorship verification, the researchers proposed various types of stylistic features to distinguish the authors writing style. The researchers analyzed that the performance of authorship verification was poor when the stylistic features were used alone in the experiment. In this work, a new approach is proposed for authorship verification where the content-based features were used in the experiment. The term importance is computed by using term weight measure, and these term weights were used to calculate the document weight. The document weights of training document and document weights of test documents were compared to verify the test document. The proposed approach accuracy is good when compared with state-of-the-art existing approaches for authorship verification.

P. Buddha Reddy (✉)
Department of CSE, Vardhaman College of Engineering,
Shamshabad, Telangana, India
e-mail: buddhareddy.polepelli@gmail.com

T. Murali Mohan
Department of CSE, Swarnandhra Institute of Engineering and Technology,
Narsapuram, West Godavari, Andhra Pradesh, India
e-mail: drtmm512@gmail.com

P. Vamsi Krishna Raja
Department of CSE, Swarnandhra College of Engineering and Technology,
Narsapuram, West Godavari, Andhra Pradesh, India
e-mail: drpvkraja@gmail.com

T. Raghunadha Reddy
Department of IT, Vardhaman College of Engineering, Shamshabad, Telangana, India
e-mail: raghu.sas@gmail.com

© Springer Nature Singapore Pte Ltd. 2020   441
K. S. Raju et al. (eds.), *Data Engineering and Communication Technology*,
Advances in Intelligent Systems and Computing 1079,
https://doi.org/10.1007/978-981-15-1097-7_37

Keywords Authorship verification · Term weight measures · Document weight measure · Mean · Accuracy · Authorship analysis

1 Introduction

The exponential growth of data spreading across the Web may shape in different forms, and it may result in fraudulent, manipulated, morphed, unidentified, and stolen data. A wrong message can get viral and reach to the people very fast. The prime concern of the researchers, forensic department, news authorities, and government is to find the fraudulent data. In this process, in pursuit of truth, finding the author of the text is highly required. Authorship analysis is one such area paying attention by the researchers to find the information of the authors of the text. The text characteristics play vital role in the procedure of finding the authors in authorship analysis [1].

Authorship analysis is a technique of reaching a conclusion by understanding the characteristics of a piece of written document and thereby analyzing the author whose roots are coming from stylomeric, which is a linguistic research field [1]. In the present research field, the authorship analysis utilizes the knowledge of the text statistics and the researchers are using the machine learning techniques in the process of identifying the author. According to Zheng et al. [2], there are three subfields that researchers focus on the field of authorship analysis, i.e., authorship identification (authorship attribution), authorship profiling, and similarity detection. Similarly, Koppel et al. summarized [2] the problems that authorship analysis aims to solve into three scenarios—authorship identification, authorship verification problem, and author profiling problem. In general, the author ship analysis is made as three categories namely authorship attribution, author profiling, and authorship verification [3]. In this work, a new approach is proposed for authorship verification.

Authorship verification approaches were important in various applications [4]. In cyberspace application, digital documents were used as a proof to prove whether the suspect is a criminal or not by analyzing their documents. If the suspect authors are unknown, i.e., there is no suspect, this is generally recognized as an authorship identification problem. However, there are also some situations when the detection of the author is not necessary, i.e., it is enough just to know if the document in dispute was written by the suspected author or not. This is a problem faced by many forensic linguistic experts which is called as authorship verification problem [5].

The procedure of authorship verification is displayed in Fig. 1. As explained in Fig. 1, the unknown text 'd' is to be assigned to an author by looking at various scripts of the author. That means in the figure $d_1, d_2, \ldots d_n$ are the known text of an author, and we are verifying whether the unknown text 'd' belongs to the author or not.

This paper is organized in five sections. The previous works in authorship verification are discussed in Sect. 2. The corpus characteristics were presented in Sect. 3. The proposed approach for authorship verification is described in Sect. 4. Section 5 concludes this work with proper future directions.

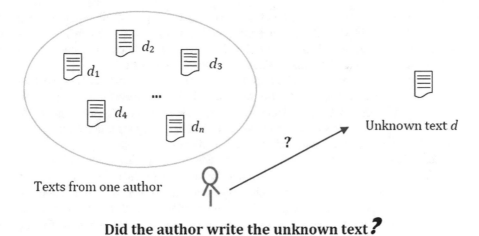

Texts from one author

Unknown text d

?

Did the author write the unknown text ?

Fig. 1 Procedure of authorship verification

2 Literature Survey

Authorship analysis is a technique of analyzing the anonymous chunks of text document and predicting the author based on the stylometry which is a linguistic research area that involves different statistical approaches and machine learning techniques [6]. Authorship verification is a process of verifying the several written text documents and finding out whether these textual documents were written by particular author or not, but it is not involving the identification of the author [5].

Vanessa et al. proposed [7] a new technique called unmasking approach. This method is helpful in enhancing the feature quality when building the classifiers. In this work, they experimented with 399 Spanish language features, 538 English language features, and 568 Greek language features. These features comprised of both coherent and stylometry features. The experimental results have shown that higher accuracy is achieved with English and Spanish documents and less accuracy was achieved for Greek documents.

Bobicev proposed [8] a system, which depends on statistical n-gram model called prediction by partial matching (PPM) for automatic detection of author when a text document is given with corpus which contains known authors and a small training set. The training corpus contains 30 authors; for each author, 100 posts are extracted and corpus is normalized to 150–200 words to maintain the uniformity. The experimental results are exhibiting that when the length of the document is increased, the accuracy measure (F-score) is not increased.

In the experiments of Vilarino et al. [9], the documents were represented with a vector graph-based features and lexical and syntactic features. Graph-based features were extracted by using data mining tool called subdue. Classification model is built by using SVM. This experiment is carried out by using lexical–syntactic features which include phrase-level features, such as punctuation marks, stop words, word suffixes, and POS trigrams, and character-based features such as combination vowels and their permutation. It showed that the run time complexity is more when compared to other submissions for a competition.

A profile-based approach called common n-gram (CNG) method was used by van Dam [10] which implements the normalized distance measure among imbalance text and short text. Every document is represented with character n-grams to implement CNG method. For Spanish and English language documents, this method showed a good accuracy but failed for Greek language.

Compression-based dissimilarity measure was used by Veenman et al. [11] to calculate the compression distance among the text documents. Researchers used three approaches for author verification task such as two-class classifications in compression prototype, nearest neighbor with compression distances, space and bootstrapped document samples. This experiment got the best accuracy for the authorship verification task over all the submissions in the competition of PAN 2013.

A new machine learning technique was proposed by Bartoli et al. [12] which uses the combination of linguistic features. Many features were extracted like n-grams of character, word, POS, sentence lengths, word lengths, POS tag n-grams, word richness, text shape n-grams, sentence lengths n-grams, and punctuation n-grams. In PAN 2015 conference, this technique ranked first in authorship verification.

General impostor's technique was implemented by Seidman [13]. In this technique, the similarity among the known document and number of external documents is calculated. This approach stood first in competition. Frequent significant features were extracted by Petmanson [14] such as nouns, verbs, punctuations, and beginning words of lines or sentences, and they calculated the Matthews correlation coefficient (MCC) for each pair of extracted features by using principle component analysis (PCA).

3 Corpus Characteristics

Since 2009, several authorship analysis tasks have been organized including author attribution, author profiling, author clustering, and author obfuscation in PAN competition. In three consecutive competitions of PAN (2013, 2014, and 2015), a shared task in authorship verification was conducted and attracted the participation of multiple research teams. PAN organizers built new benchmark corpora covering several languages (English, Dutch, Spanish, and Greek), genres (essays, novels, reviews, newspaper articles) and provided an online experimentation framework for software submissions and evaluation [15].

Each PAN corpus comprises a set of authorship verification problems, and within each problem, a set of labeled documents and exactly one unlabeled (or unknown) document are given. PAN participants should provide a binary answer

Table 1 Corpus characteristics of PAN 2015 competition authorship verification

Features	Training Data	Testing Data
Number of problems	100	500
Number of documents	200	1000
Average number of known documents	1.0	1.0
Average words per document	366	536

and a score in [0, 1] indicating the probability of a positive answer (0 means it is certain that the unknown and known textual documents are not written by the particular author and 1 means the opposite). In this work, the experimentation was performed on PAN 2015 competition authorship verification dataset. Table 1 shows the characteristics of the PAN 2015 competition corpus for authorship verification.

4 Proposed Approach

In this work, a new approach is proposed to verify the author of a new document in authorship verification. The procedure of new approach is displayed in Fig. 2. In this approach, first, the documents of a suspected author are analyzed. Two pre-processing approaches like stop word removal and stemming were identified to prepare the data further analysis [16]. The stop word removal technique removes the words which are not having distinguished power in the classification. Stemming converts the words into root forms to reduce the number of distinct words. After performing preprocessing approaches, the content-based features (most frequent terms) were extracted from the corpus.

The documents were represented with these terms in the form document vectors. In vector space model, every document vector contains the numerical weights of features or terms extracted from the dataset. Text classification mainly depends on the term weight in the document, but the procedure of calculating the term weight certainly affects the accuracy of classification [17]. Conventional term weight measures binary, term frequency (TF), and TF-IDF were proposed in various domains such as information retrieval and text categorization. The binary measure gives values 0 or 1, which denotes the absence or presence of a term with in a document, respectively. The term frequency (TF) assigns weights to the terms. Depending upon the frequency of a term in a given document, the term frequency (TF) measure assigns the weights to all the terms. Higher weight values were assigned to terms those occur frequently in the document when compared with the terms which are occurred rarely in the document. In general, the rare terms are having strong text discriminating power compared to frequent terms [18].

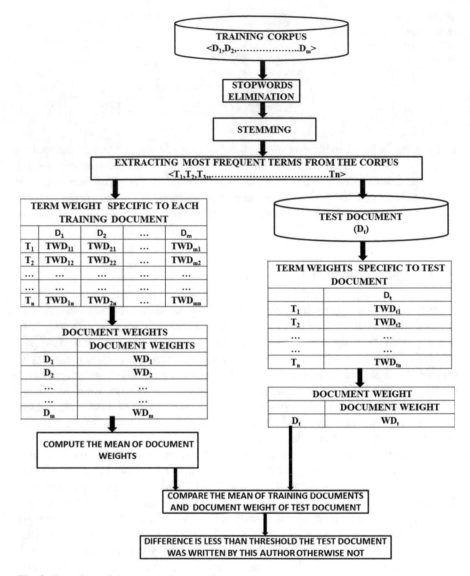

Fig. 2 Procedure of the proposed approach

In this work, a non-uniform distributed term weight (NDTW) measure is used [19] to calculate the term weight that was extracted in the previous step. The Eq. (1) represents the NDTW measure.

$$w(t_i, D_k) = \text{Log}(\text{TOTF}_{ti}) - \sum_{k=1}^{m} \left(\frac{tf(t_i, D_k)}{\text{TOTF}_{ti}} \text{Log} \left[\frac{1 + tf(t_i, D_k)}{1 + TOTF_{ti}} \right] \right) \tag{1}$$

where $tf(ti, D_k)$ is the term t_i frequency in d_k document. TOTF_{ti} represents the total occurrence of the term t_i in all the documents of the entire corpus. In the proposed approach, TWD_{mn} is the term T_n weight in D_m document.

The next step in the proposed approach calculates the weight of the documents. The document weight measure [20] is represented in Eq. (2).

$$w(D_k) = \sum_{i=1}^{n} \text{TF}(T_i, D_k) * \text{TWD}_{ik} \tag{2}$$

Calculate the mean value of the weights of the documents using the Eq. (3).

$$\text{Mean}(A_1) = \frac{\sum_{i=1}^{m} \text{WD}_i}{m} \tag{3}$$

where Mean (A_1) is the mean value of suspected author A_1, WD_i is the weight of document D_i, and m is the number of documents in the corpus of A_1.

For a test document, compute the document weight by performing all the steps that were performed on the training dataset. Finally, the test document weight and mean value of an author are compared to verify whether suspected author has written the test document or not. The difference between mean value and test document weight is less than a threshold value; then, the document is written by that suspected author; otherwise, the document was not written by the suspected author.

In this work, the experimentation was carried out with the threshold value as 5 and it is observed that our approach obtained 89.35% accuracy for authorship verification. Here, accuracy is the ratio between the number of test documents verified correctly to the total number of test documents used for experimentation.

5 Conclusions and Future Scope

In this work, a new approach was proposed for authorship verification. In this approach, we used a term weight measure to assign more suitable weights to the terms and these weights of the terms were used to calculate the document weight. The document weights were used to verify whether the suspected author has written the test document or not. We obtained 89.35% accuracy for author verification.

In future, we are planning to propose a new weight measures for computing weights of term and document to improve the accuracy of authorship verification. It was also planning to apply our approach to multiple languages corpuses.

References

1. Raghunadha Reddy, T., Vishnu Vardhan, B., Vijayapal Reddy, P., A survey on author profiling techniques. Int. J. Appl. Eng. Res. **11**(5), 3092–3102 (2016)
2. Zheng, R., Li, J., Chen, H., Huang, Z.: A framework for authorship identification of online messages: writing style features and classification techniques. J. Am. Soc. Inf. Sci. Technol. **57**(3), 378–393 (2006)
3. Argamon, S., Koppel, M., Pennebaker, J.W., Schler, J.: Automatically profiling the author of an anonymous text. Commun. ACM **52**(2), 119–123 (2009)
4. Bhanu Prasad, A., Rajeswari, S., Venkanna Babu, A., Raghunadha Reddy, T.: Author verification using rich set of linguistic features. Proc. Adv. Intell. Syst. Comput. **701**, 197–203 (2018)
5. Koppel, M., Schler, J., Argamon, S.: Computational methods in authorship attribution. J. Am. Soc. Inf. Sci. Technol. **60**(1), 9–26 (2009)
6. Raghunadha Reddy, T., Vishnu Vardhan, B., Vijayapal Reddy, P.: Author profile prediction using pivoted unique term normalization. Indian J. Sci. Technol. **9**(46) (2016)
7. Feng, V.W., Hirst, G.: Authorship verification with entity coherence and other rich linguistic features. In: Proceedings of CLEF 2013 Evaluation Labs (2013)
8. Bobicev, V.B.: Authorship detection with PPM. In: Proceedings of CLEF 2013 Evaluation Labs (2013)
9. Vilariño, D., Pinto, D., Gómez, H., León, S., Castillo, E.: Lexical-syntactic and graph-based features for authorship verification. In: Proceedings of CLEF 2013 Evaluation Labs (2013)
10. van Dam, M.: A basic character N-gram approach to authorship verification. In: Proceedings of CLEF 2013 Evaluation Labs (2013)
11. Veenman, C.J., Li, Z.: Authorship verification with compression features. In: Proceedings of CLEF 2013 Evaluation Labs (2013)
12. Bartoli, A., Dagri, A., De Lorenzo, A., Medvet, E., Tarlao, F.: An author verification approach based on differential features. In: Proceedings of CLEF 2013 Evaluation Labs (2015)
13. Seidman, S.: Authorship verification using the impostors method. In: Proceedings of CLEF 2013 Evaluation Labs (2013)
14. Petmanson, T.: Authorship identification using correlations of frequent features. In: Proceedings of CLEF 2013 Evaluation Labs (2013)
15. Gollub, T., Potthast, M., Beyer, A., Busse, M., Pardo, F.M.R., Rosso, P., Stamatatos, E., Stein, B.: Recent trends in digital text forensics and its evaluation—plagiarism detection, author identification, and author profiling. In: Proceedings of the 4th International Conference of the CLEF Initiative, pp. 282–302 (2013)
16. Raghunadha Reddy, T., Vishnu Vardhan, B., Vijayapal Reddy, P.: Profile specific document weighted approach using a new term weighting measure for author profiling. Int. J. Intell. Eng. Syst. **9**(4), 136–146 (2016)
17. Sreenivas, M., Raghunadha Reddy, T., Vishnu Vardhan, B.: A novel document representation approach for authorship attribution. Int. J. Intell. Eng. Syst. **11**(3):261–270 (2018)
18. Swathi, C., Karunakar, K., Archana, G., Raghunadha Reddy, T.: A new term weight measure for gender prediction in author profiling. In: Proceedings in Advances in Intelligent Systems and Computing, vol. 695, pp. 11–18 (2018)
19. Dennis, S.F.: The design and testing of a fully automated indexing-searching system for documents consisting of expository text. In: Schecter, G. (ed.) Informational Retrieval: A Critical Review, pp. 67–94. Thompson Book Company, Washington D.C. (1967)
20. Raghunadha Reddy, T., Vishnu Vardhan, B., Vijayapal Reddy, P.: A document weighted approach for gender and age prediction. Int. J. Eng. Trans. B: Appl. **30**(5), 647–653 (2017)

A System for Efficient Examination Seat Allocation

Charitha Tuniki, Vanitha Kunta and M. Trupthi

Abstract Examination seating arrangement is a major issue faced by an institution. A university consists of a large number of students and classrooms, which makes it difficult for the university authorities to design seating arrangement manually. The proposed solution provides an efficient set of algorithms for examination seating allocation problems. It also gives the best combination of rooms to be utilized for the examination and dynamically organizes seating based on the orientation of the room and students. The described solution encompasses methods to address some common issues like eliminating students who are ineligible to write one or more examinations and adding the students who are retaking the examinations. The presented system is made examiner-friendly such that the user can swiftly get a perfect seating arrangement based upon the above cases without manually excluding the ineligible students and rearranging the system. Excel sheets of the classroom view are generated automatically along with the room number and capacity of the classroom. These ready-to-go sheets can be printed and used by the examiners.

Keywords Examination seat allocation · Room optimization · Dynamic · Algorithms · Ready-to-go sheets

C. Tuniki (✉) · V. Kunta · M. Trupthi
Chaitanya Bharathi Institute of Technology, Hyderabad, India
e-mail: charithathuniki@gmail.com

V. Kunta
e-mail: vanithakunta2406@gmail.com

M. Trupthi
e-mail: mtrupthi_it@cbit.ac.in

© Springer Nature Singapore Pte Ltd. 2020
K. S. Raju et al. (eds.), *Data Engineering and Communication Technology*,
Advances in Intelligent Systems and Computing 1079,
https://doi.org/10.1007/978-981-15-1097-7_38

1 Introduction

Examination hall seating arrangement is an important strategy for avoiding malpractice and conducting a fair examination. University authorities find it difficult to allocate students into examination halls due to high number of courses and departments. In the worst case, a student may even sit adjacent to other students belonging to same class even if there is place to arrange them away from each other. These kinds of seating arrangements may cause trouble to the invigilators and also the university management.

The main criteria for examination seating arrangement are to properly arrange the students in an efficient and cost-friendly way with minimum utilization of resources. For this, the collection of appropriate data such as total number of courses, students, availability of rooms, and room orientation is needed. The aim of the proposed system is to utilize all the available rooms optimally and allocate the students into rooms in a structured and systematized manner. The room optimization algorithm described later provides a method to utilize the available rooms to the maximum by eliminating the extra rooms. Further, taking the utilized rooms into consideration, the students are seated according to their IDs such that students belonging to the same course do not sit in the each other's vicinity.

The excel sheets representing the examination hall generated for each room provide a clear view of the room number, capacity of the room, and allotment of each student represented by his ID number. The examination authorities can easily know the capacity of each room and thus make use of it to assign invigilators and know the number of examination scripts to be issued per classroom.

2 Related Work

Examination hall seating arrangement has widely been researched for a number of years. The methodology proposed by Alam [1] considered each room to be of equal seating capacity, and Chaki and Anirban [2] proposed a system where students belonging to the same course were allowed to sit in the nearest row, but not in the nearest column. Gokila and Dass [3] put forward a GUI implementation which showcases the range of student IDs allotted to each class and generates a report to the admin. Room penalty cost model is formulated by Ayob and Malik [4] to minimize the students shifting between the rooms allotted to them when there are continuous examinations in a single day.

PHP- and Java-based application is created by Sowmiya et al. [5] where the students are arranged in seats behind each other, considering the last digits of their roll numbers. PriyaDharshini et al. [6] suggested a GUI where the students can register and locate their classroom for examination. This research manually updates the examination hall information prior releasing it to the students. The research done by Chandewar et al. [7] arranges students behind each other using a dynamic array. A software is created by Prabnarong and Vasupongayya [8] which allows the examination invigilator to choose their scheduled time, and based on the room type and capacity, each subject is assigned to each available room. Constraints regarding the distance between rooms and dividing the courses across several rooms are addressed by Kahar and Kendall [9]. Graphs and networks are used by de Werra [10] to provide arranging of courses and examinations in a university.

3 Existing System

The existing manual system is slow and takes a lot of human effort. The rate of errors made by a human is higher than that made by a computer. The criteria where the work is split among many people to decrease workload creates a havoc.

The traditional way of examination hall seating arrangement involves collection of huge data about the students and rooms and processing it manually to arrange students in the classroom.

The limitations of existing system are as follows:

- It may lead to missing of few students, allotting a student multiple times, etc.
- Students ineligible for examination are removed, and seating rearrangement is done manually.
- Less efficiency.
- Hectic work and time taking.

4 Proposed Method

Our research aims to aid the management of the universities to solve the problem of examination seating plan. It is sensitive and efficient enough to work even with small number of courses and less rooms. The students' and classrooms' data is collected and is operated on with the algorithms to create an efficient seating arrangement. This system is built to be effective and scalable. The reference model that is used in the proposed method is shown in Fig. 1.

5 Methodology

(A) *System Description*

In this research, the algorithms are designed for three phases. They are as follows:

Phase I: The first phase deals with collection of input data such as total number of rooms available, the room orientation, and total number of students attending the

5001	8007
7005	6003
5002	8008
7006	6004
5003	8009
7007	6005

5004	8010
7008	6006
5005	9001
7009	6007
5006	9002
7010	6008

5007	9003
8001	6009
5008	9004
8002	6010
5009	9005
8003	7001

5010	9006
8004	7002
6001	9007
8005	7003
6002	9008
8006	7004

Fig. 1 Reference model

examination. In this phase, we mark the rooms that are to be used for examination, hence eliminating the extra rooms.

Phase II: The second phase consists of two algorithms. In first algorithm, a set of students of each course are allocated to a particular room. That is, each room is filled with a batch of students who belong to different courses. In the second algorithm of this phase, we allocate the ID numbers of the students to each room, with reference to the first algorithm in this phase.

Phase III: In the third phase, we place the students according to their ID numbers in the room. This arrangement is done by the algorithm in an efficient manner such that no two students writing examination of same course are placed in the proximity of each other.

In Fig. 2, the phases of the discussed methodology are represented using a flowchart.

(B) *Algorithms*

Phase I: Algorithm 1.1

In this algorithm, we mark rooms for utilization using their capacities and the number of students attending the examination. Extra seats are calculated using the total capacity of rooms and total students. These extra seats are further used in the algorithm to choose the optimal rooms for seating arrangement.

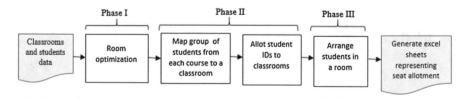

Fig. 2 Flowchart depicting the phases of proposed methodology

Parameters used in algorithm:

roomCount= total number of rooms

roomCapacity[roomCount]= list of capacities of each room in sorted order

totalCapacity= $\sum_{i=0}^{roomCount}$ roomCapacity[i]

totalStudents= total number of students attending the examination

utilizedRooms[]= list containing the indices of rooms that should be used for examination

ignoredRooms[]= list containing the indices of rooms that should be ignored for examination

The pseudocode for Algorithm 1.1 is given in Fig. 3.

Inputs: roomCapacity[], roomCount, totalCapacity, totalStudents
Output: utilizedRooms[], ignoredRooms[]

1. Initialize extraSeats to the difference of totalCapacity and totalStudents
2. Declare utilizedRooms, ignoredRooms are empty list
3. Initialize roomIndex to one less than roomCount
4. Function RoomOptimization(extraSeats, roomIndex, ignoredRooms)
   a. While extraSeats is less than roomCapacity[roomIndex] and roomIndex is greater than or equal to zero
      i. Decrement roomIndex by one
      ii. If roomIndex is less than zero
           Return
         End If
      End While
   b. Add roomIndex in ignoredRooms
   c. Deduct roomCapacity[roomIndex] from extraSeats
   d. If extraSeats is greater than or equal to roomCapacity[0]
      i. Decrement roomIndex by one
      ii. If roomIndex is greater than or equal to zero
           value = Room Optimization a(extraSeats, roomIndex, ignoredRooms)
         End If
      End If
   e. Set outValue to ignoredRooms
   f. Return outValue
5. If ignoredRooms is not empty
   a. For every room
      i. If a room is not in ignoredRooms
           Add that room in utilizedRooms
         End If
      End For
   End If
6. Else
           Add all rooms in utilizedRooms

Fig. 3 Pseudocode of Algorithm 1.1

Phase II: Algorithm 2.1

In this algorithm, we calculate the number of students from each course to be accommodated in a particular room. This is calculated by the formula 1 given below:

$$\text{Students from a course in a room} = \frac{\text{Total capacity of students from the course}}{\text{Total number of utilized rooms}} \quad (1)$$

The pseudocode for Algorithm 2.1 is given in Fig. 4.

List of parameters for both algorithm 2.1 and calculation:

coursesCapacity[]= capacity of students in each course.
utilizedRoomsCapacity[]= capacity of utilized rooms in sorted order.
roomCoursesCapacity[[]]= number of students from each course in a particular room.
courseCapacityNotFilled[]= number of students of a course who are not allocated.
roomEmptySeats[]= number of empty seats in each room.

Description:

Step 1:

If the capacity of a particular room is sufficient enough to accommodate a group of students belonging to a course (calculated using formula 1), then those students are mapped to that classroom. Else, if the capacity is not sufficient enough, then the appropriate number of students are accommodated and rest are directed to the next room.

Step 2:

We calculate the number of empty seats in each room to further reference it in Step 4.

Step 3:

We calculate the number of students who are not accommodated into any of the rooms. It will be further referenced in Step 4.

Step 4:

The students who are not yet accommodated (calculated in Step 3) are assigned to the classrooms having empty seats (calculated in Step 2). The algorithm terminates if all the students from each course are assigned to classrooms.

Algorithm 2.2

This algorithm is designed to allot student IDs of each course to a room.

List of parameters for both algorithm and calculation:

Inputs: utilizedRooms[], coursesCapacity[]
Output: roomCoursesCapacity[[]], roomCourseIDs[[]]
1. For each room in utilizedRooms
    a. For each course in the list of Courses
        i. Set j to coursesCapacity[course] // total utilizedRooms
        ii. If utilizedRoomsCapacity[room] is greater than or equal to j
                Add j to roomCoursesCapacity[room]
                Subtract j from utilizedRoomsCapacity[room]
                        End If
      iii. Else
                If utilizedRoomsCapacity[room] is greater than or equal to 0
                    Add utilizedRoomsCapacity[room] to
                    roomCoursesCapacity[room]
                    Set utilizedRoomsCapacity[room] to 0
                        End If
            Else
                Add 0 to roomCoursesCapacity[room]
2. For each room in utilizedRooms
    a. If sum of roomCoursesCapacity in a room is greater than or equal to
        utilizedRoomsCapacity of that room
        i. Add 0 to roomEmptySeats of that room
    b. Else
        i.   Find difference between (utilizedRoomsCapacity of a room and sum
        of roomCoursesCapacity in a room)
        ii. Add calculated difference to roomEmptySeats of that room
3. For each course in the list of Courses
    a. If coursesCapacity of a course is equal to the sum of that course
    capacity in all the rooms assigned to that course
        i. Add 0 to courseCapacityNotFilled of that room
    b. Else
        i.   Find difference between (coursesCapacity of a course and sum of that
        course capacity in all the rooms assigned to that   course)
        ii. Add calculated difference to courseCapacityNotFilled of that room
4. For each course in the list of Courses
    a. If courseCapacityNotFilled of course is not equal to 0
        i. For each room in utilizedRooms
            i.a.   If roomEmptySeats of room is not equal to 0
                If roomEmptySeats of room is greater than or equal
                to courseCapacityNotFilled of course
                    Subtract courseCapacityNotFilled of course
                    from roomEmptySeats of room
                    Add courseCapacityNotFilled of course to
                    roomCoursesCapacity of room
                    Set courseCapacityNotFilled of course to 0
                Else
                    Subtract roomEmptySeats of room from
                    courseCapacityNotFilled of course
                    Add roomEmptySeats of room to
                    roomCoursesCapacity of room
                    Set roomEmptySeats of room to 0
            i.b.   If courseCapacityNotFilled of course is equal to 0
                break

Fig. 4 Pseudo code of Algorithm 2.1

```
Inputs: roomCoursesCapacity[[]], studentIDs[[]]
Output: roomCourseIDs[[]]
1.           For each room in utilizedRooms
     a.     For each course in the list(Courses)
            i.        Set k to roomCoursesCapacity[room][course]
            ii.       For n =1 to k
                      Add studentIDs[course][n] to roomCourseIDs[room]
                           End For
            iii.      While k is greater than 0
                      remove studentIDs[course][0] from studentIDs[course]
                      Decrement k by 1
                           End While
                      End For
            End For
```

Fig. 5 Pseudocode of Algorithm 2.2

studentIDs[]= student IDs of each course. It is computed with reference to starting and ending IDs including the IDs to be added and excluding the IDs of ineligible students.

roomCourseIDs[[]]= student IDs of each course who are accommodated in each room.

Description:

The capacity of students from a course in each room is considered, and the IDs of the students are assigned to each classroom.

The pseudocode for Algorithm 2.2 is given in Fig. 5.

Phase III:

This phase uses the XlsxWriter available in Python. XlsxWriter helps to visually represent the student's IDs in an excel sheet and creates an easy view of seat allocation to the invigilator.

Description:

The students are arranged as shown in the process diagram Fig. 6. We assume that each bench can accommodate two students.

Step 1: Each student is allocated in an alternative manner along the left side of the bench, starting from first row in the first column, as represented by (1) in Fig. 6.
Step 2: At the last column, after the filling of last possible row (6) on left side of bench, the next students are placed starting from right side of the bench in the second row of first column (7).

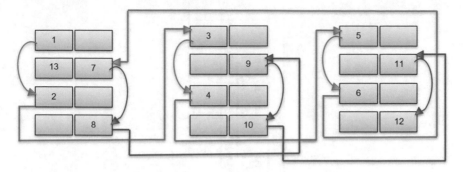

Fig. 6 Process diagram of Phase III

Step 3: In the last column, after the filling of last possible row (12) on the right side of bench, the next students are placed starting from the left side of bench in the second row of first column (13) until the last possible row is filled.

Step 4: After the completion of Step 3, the students are filled starting from the right side of the bench in the first row of second column.

6   Results

Table 1 depicts the input data about the classrooms and the students' courses.

Three rooms with IDs 301, 302, and 303 are considered. Each room has 5 rows and 5 columns. The capacity of each room is 50. Four different courses are considered with each course having a student capacity of 30.

(i)   Initially, the extra seats are calculated, which is difference between the total capacity available and total number of students.

(ii)   Algorithm 1.1 is run to optimize room's utilization. The extra rooms are further not taken into consideration.

(iii)   The number of students from a course to be allotted in a room is calculated using the formula (1). Thus, 10 members from each course are allocated to every room.

(iv)   We recheck if all students are allocated in the rooms using Algorithm 2.1.

(v)   The students' IDs of each course are allocated to rooms using Algorithm 2.2.

(vi)   The allocation of seats takes place according to process diagram of Phase III described in Fig. 6.

(vii)   The student IDs are color coded, with each color representing a course. In the displayed result, we represent the four courses using four different colors.

Table 1 Input data

S. No.	Rooms input			Students input					Any ID to be deleted? (Y/N)	Any ID to be added?(Y/N)
	ROWS	Columns	Room ID	Course name	Course capacity	Starting Student's ID	Ending Student's ID			
1	5	5	301	C	30	160116737001	160116737030		Y,160116737025	Y,160116737032
2	5	5	302	C++	30	160115737001	160115737030		N	N
3	5	5	303	Python	30	160117737001	160117737030		N	N
4	–	–	–	Java	30	160118737001	160118737030		N	N

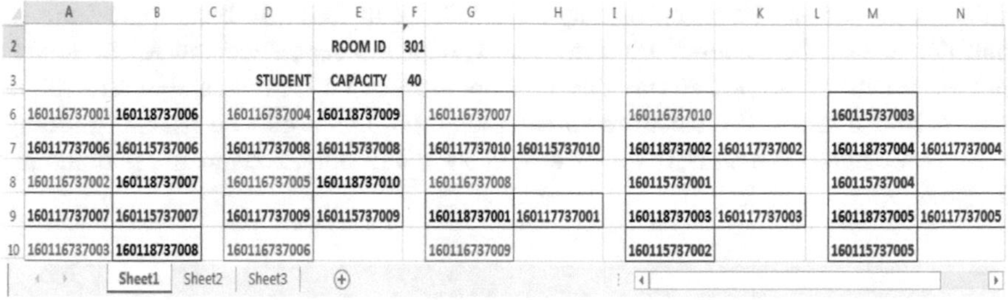

	A	B	C	D	E	F	G	H	I	J	K	L	M	N
2					ROOM ID	301								
3				STUDENT	CAPACITY	40								
6	160116737001	160118737006		160116737004	160118737009		160116737007			160116737010			160115737003	
7	160117737006	160115737006		160117737008	160118737008		160117737010	160115737010		160118737002	160117737002		160118737004	160117737004
8	160116737002	160118737007		160116737005	160118737010		160116737008			160115737001			160115737004	
9	160117737007	160115737007		160117737009	160115737009		160118737001	160117737001		160118737003	160117737003		160118737005	160117737005
10	160116737003	160118737008		160116737006			160116737009			160115737002			160115737005	

Sheet1 Sheet2 Sheet3 (+)

Fig. 7 Sheet 1 representing classroom 301

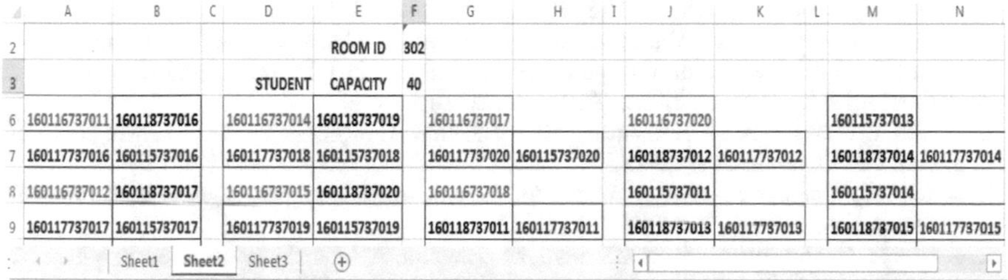

	A	B	C	D	E	F	G	H	I	J	K	L	M	N
2					ROOM ID	302								
3				STUDENT	CAPACITY	40								
6	160116737011	160118737016		160116737014	160118737019		160116737017			160116737020			160115737013	
7	160117737016	160115737016		160117737018	160115737018		160117737020	160115737020		160118737012	160117737012		160118737014	160117737014
8	160116737012	160118737017		160116737015	160118737020		160116737018			160115737011			160115737014	
9	160117737017	160115737017		160117737019	160115737019		160118737011	160117737011		160118737013	160117737013		160118737015	160117737015

Sheet1 **Sheet2** Sheet3 (+)

Fig. 8 Sheet 2 representing classroom 302

(viii) Excel sheets are generated separately for each classroom in a single excel workbook. Each sheet consists of the classroom ID and the total students' capacity in the room. Figures 7 and 8 show the results obtained in the excel sheets.

(ix) The third room-303 would be arranged similar to the rooms 301 and 302 as shown in Figs. 7 and 8.

7 Conclusion

This system has overcome the disadvantages of the existing system. It has the advantages of using less human effort, time, reduced rate of errors, and user-friendly. The system has been tested on various constraints in which students are removed or extra students are added. Each room's capacity and total students' capacity are taken into consideration, and extra rooms are eliminated efficiently. The results displayed in excel sheets can be easily interpreted by anyone.

The proposed system can be further enhanced by adding a few more constraints like seat arrangement for session-based examinations. This project can also be implemented as a mobile application which would be student-friendly so that students can know which room they are allocated in through this application. The hall tickets can be generated automatically with the seat allocated to the student printed on it. These algorithms can be improved to be more scalable and robust. Apart from limiting the software to examinations, this could be useful to arrange members attending a conference or a workshop and allot beds to the patients in a hospital.

References

1. Alam, A.F.: Automatic seating arrangement tool for examinations in universities/colleges. Int. J. Eng. Appl. Sci. Technol. **1**(4). ISSN No. 2455-2143 (2016)
2. Chaki, P.K., Anirban S.: Algorithm for efficient seating plan for centralized exam system. International Conference on Computational Techniques in Information and Communication Technologies (ICCTICT), IEEE: 07514601 (2016)
3. Gokila, R., Dass, R.A.: Examination hall and seating arrangement application using PHP. Int. J. Eng. Sci. Comput. **8**(2) (2018)
4. Ayob, M., Malik, A.: A new model for an examination-room assignment problem. Int. J. Comput. Sci. Netw. Secur. **11**(10) (2011)
5. Sowmiya, S., Sivakumar, V., Kalaimathi, M., Kavitha, S.V.: Automation of exam hall seating arrangement. Int. J. Adv. Res. Ideas Innov. Technol. **4**(2):1048–1052 (2018)
6. PriyaDharshini, S., SelvaSudha, M., Anithalakshmi, V.: Exam cell automation system. Int. J. Eng. Sci. Comput. **7**(3) (2017)
7. Chandewar, D., Saha, M., Deshkar, P., Wankhede, P., Hajare, S.: Automatic seating arrangement of university exam. Int. J. Sci. Technol. Eng. **3**(09). ISSN: 2349-784X (2017)
8. Prabnarong, T., Vasupongayya, S.: Examination management system: room assignment and seating layout. World Academy of Science, Engineering and Technology. Int. J. Comput. Inf. Eng. **7**(7) (2013)
9. Kahar, M.N.M., Kendall, G.: The examination timetabling problem at Universiti Malaysia Pahang. MIT Press and McGraw-Hill
10. de Werra, D.: An introduction to timetabling. Eur. J. Oper. Res. **19**(2):151–162 (1985)

Clustering-Based Blockchain Technique for Securing Wireless Sensor Networks

T. C. Swetha Priya and A. Kanaka Durga

Abstract Security is one of the major issues in WSNs. Due to the emergence of the WSNs, new requirements such as accessibility to services, data consistency, and network lucidity have come into picture. Although many protocols have been developed, they satisfy only a few requirements in managing security. This paper proposes a blockchain technology-based energy-efficient clustering approach for WSNs to provide security to the transaction records by including cryptographic techniques. It uses multiple collaborative base stations to reduce the network malfunctioning due to the compromised base stations. The proposed method achieves accessibility to services, data consistency, and network lucidity.

Keywords Wireless sensor network · Blockchain · Base station · Sensor node · Cluster head · Policy · Transaction

1 Introduction

1.1 Wireless Sensor Network

A WSN consists of sensor nodes and base station. A sensor node captures the information about the conditions such as the temperature, traffic, and moisture. The sensors are implanted in an area that has to be monitored. Sensor nodes send the collected information to the central controller known as base station (BS).

T. C. Swetha Priya (✉) · A. Kanaka Durga
IT Department, Stanley College of Engineering and Technology for Women,
Hyderabad, Telangana, India
e-mail: tcswethapriya@stanley.edu.in

A. Kanaka Durga
e-mail: drakanakadurga@stanley.edu.in

© Springer Nature Singapore Pte Ltd. 2020
K. S. Raju et al. (eds.), *Data Engineering and Communication Technology*,
Advances in Intelligent Systems and Computing 1079,
https://doi.org/10.1007/978-981-15-1097-7_39

1.2   Clustering

Clustering [1] is the partitioning of the network into several subnetworks each consisting of associated nodes and a CH for each subnetwork that was created. Each node in the cluster gathers data and transmits that data to its CH which sends the data to the BS [2]. Such type of clustering can extend the lifetime of network.

1.3   Blockchain

A blockchain is a collection of blocks. Each block is associated with an ID of its own and the ID of previous block to establish a link between the corresponding blocks and the transactions. The links continue until it reaches the first block. The hash value of the transactions in the respective block is added to the following blocks in the chain. The changes made to data in a block affect the following blocks in the chain [3, 4]. This method guarantees consistency.

1.4   Key Establishment

The protocol creates a group key between the sensor nodes. It includes distribution of keys, key agreement, key updation, and distribution of group keys [5].

2   Literature Review

In this section, a brief analysis of the existing systems is discussed.

2.1   Energy-Balancing Multiple-Sink Optimal Deployment in Multi-hop WSNs

In a multi-hop WSN, most of the energy is spent in transmission of data. This method uses an optimal multi-sink algorithm in which sinks are placed in an optimal manner by forming several disjoint clusters. This method provides accessibility to data and services but does not provide consistency of data. It does not include any method to identify the compromised nodes [6, 2].

2.2 Hexagonal Cell-Based Data Dissemination

This scheme uses a hexagonal grid-type structure with multiple base stations [7]. It enforces mobility. If there is similar data from multiple paths, then the routing paths do not know whether the data from different base stations is trustworthy or not. There is no aggregation of similar data which leads to inconsistency. If a base station is compromised, the associated data and accessibility to services are also lost.

2.3 Weight-Based Energy-Balancing Unequal Clustering

It is based on an unequal formation of clusters [8, 1]. The CH is randomly chosen based on probabilistic model. Here, after cluster formation, a weight is assigned to a node based on degree of a node. This scheme provides stability in the consumption of energy by distributing the load [9]. It does not provide security to the data from the compromised nodes.

2.4 Fuzzy Logic-Based Cluster Head Selection

In this method, fuzzy logic [10] is used to select the CH and determine the size of cluster and data processing. This method has less communication overhead. It causes degradation in the lifetime of a network with the computation logic and lack in security.

3 Proposed System

The proposed system presents an efficient method to provide security to WSN and addresses the accessibility of services, consistency, and network lucidity issues. In the proposed system, there is no interruption in transmission even if some of the base stations get compromised by adopting the cooperative base station's scheme. Each base station maintains a blockchain (BC) to provide consistency of the stored data and network lucidity. It allows user access to the status of permitted nodes. A user can also verify the reliability of the data when multiple base stations in the network share the same data using the blockchain technology [4, 11].

3.1 Components of the Proposed System

3.1.1 Sensor Node

The sensor nodes are dispersed over an area of monitoring. They provide their IDs to join the network. These IDs are used to generate pairwise keys and group keys used for communication between two clusters. If any data is requested by a base station, immediately the sensor nodes will gather the data and send to CH which will pass it to BS.

3.1.2 Base Station (BS)

It provides IDs to nodes, selects the CH, distributes group keys to the clusters, and gathers information from sensors. It also controls the access to the data from hacker nodes. The proposed method uses multiple collaborative BSs [2, 7] that supervise all the data using blockchain technology.

3.1.3 Cluster Head (CH)

From the sensor nodes present in the network, the BS will choose the CH based on the available energy of a node, the nearest BS, etc.

3.1.4 Users (U)

The BS provides data access to the users. The BS manages a record of all the nodes that are accessible. The proposed method uses two protocols: (i) access and (ii) monitor. Access protocol refers to the retrieval of records of data from the sensor nodes. Monitor protocol retrieves the status of the neighboring sensor nodes.

3.1.5 Transactions (T)

In the proposed method, the blockchain method is used. It considers the message that goes to or comes from BS as a transaction. A transaction is represented as shown in Table 1.

All the messages that are sent between the nodes, users, and the BSs are considered as transactions. The first field in the transaction indicates the previous transaction number, and it is used to link the corresponding transactions as a linked list for easy retrieval of data. The second field represents the number of a transaction. The number of transaction is increased by 1 for each new transaction. The third field indicates the ID of nodes. The fourth field specifies the type of the

Table 1 Representation of a transaction

Previous transaction	Transaction number	Node ID (s)	Transaction type	Sig. Req	Data
			0 = Genesis		
			1 = Store		
			2 = Access		
			3 = Monitor		
			4 = Update		

transaction. There are 5 types of transactions: Genesis transaction is the first transaction that initializes to be committed as entity. Store transaction is used when the data is sent to the BS. Access transaction is used for effective retrieval of data. Monitor transaction is used to acquire the status of the node. Update transaction is used to update the status of the withdrawn nodes and to clear the credentials assigned to the left node. The fifth field indicates the user's signature for unique identification. The sixth field consists of the data present in the transaction.

3.1.6 Blockchain

A block is a set of transactions. Each transaction is filled in the empty space available in the blockchain, and once the block is filled, the transactions have to be entered into the new block. Each block consists of a block header which is the hash value of the previous blocks. So such type of creation of chains for the blocks ensures the consistency of the data. The representation of blockchain [4] is shown in Fig. 1.

3.1.7 Policies

The permissions given to the nodes and users are policies. The policies are saved in the block structure. Any modifications in the network structure leads to the creation of new policies. The representation of a policy is shown in Table 2.

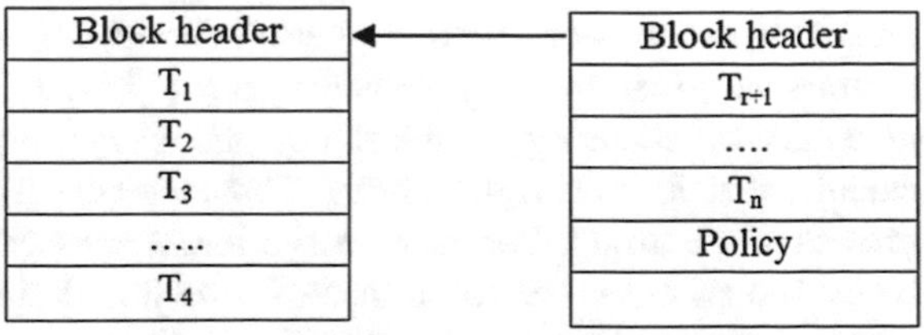

Fig. 1 Blockchain structure

Block header
T_1
T_2
T_3
…..
T_4

Block header
T_{r+1}
….
T_n
Policy

Table 2 Representation of policy

Requester	Request for	Device ID(s)	Action
	...	...	
U_h	Access	⟨List of node IDs⟩	Allow
...	...	...	
$N_1...N_h$	Update	⟨List of BS IDs⟩	Allow

3.2 Phases of the Proposed System

The proposed system consists of seven phases: initializing, creation of group keys, revocation of nodes, updation of keys, storing, accessing of data, and monitoring process.

3.2.1 Initializing and Creation of Group Keys

The BS creates a matrix and loads the ith column of matrix to node N_i. Then, the transaction is saved into the block by the BS. This transaction is exchanged to the other BSs. Initially, as there are no transactions, the type of this transaction is genesis transaction, so the structure of the transaction is filled as $T = \langle 0, I, N_i, 0, 0, G_i||A_i \rangle$. The BS concatenates G_i and A_i. For providing security, signature field is added to T and encryption is done using the public key K+. Once the other BSs receive the message, the decryption process is done using the private key and then with the BS's public key to extract T. After each node's initializing process, a pointer is created to the list of policies. This list includes the user list and the action associated with each rules. The initializing and creation of group keys process is shown in Fig. 2.

The details provided by the BS during the initializing process will be used in creation of pairwise keys [5]. But these keys have to be created for each node present in a cluster. CH will compute the key and broadcast a message. Once it reaches the cluster member, they decrypt the message to get the key. But this process will consume more energy and time. So to avoid this, in the proposed system, a group key is established for each cluster instead of generating keys for every cluster member. The group key provides security to the nodes in communication of messages with other nodes. After this, the BS is responsible for the formation of clusters and transmission of group keys to all the nodes in the clusters present in the network with the permission of CH. After this, for monitoring the permissions given for each node, the policies have to be updated. This group key is generated by choosing a random key and encrypting the key with the K which is calculated as $K = G_i. A_i$ for node i. Then, the keys formed are sent to all the nodes present in the group. For each encryption, a transaction has to be entered into the block and an entry has to be made for each node in the cluster. After successful entry of each transaction into the block, the transactions have to be committed. For each new transaction, the transaction number has to be incremented by 1.

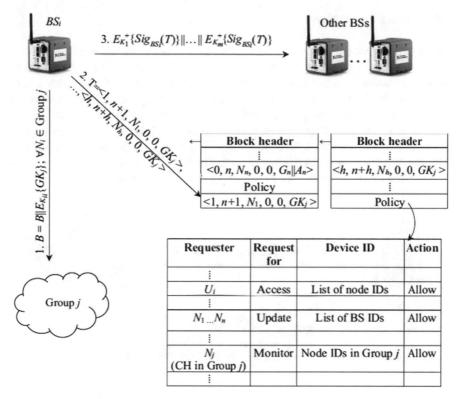

Fig. 2 Initializing and creation of group keys

The changes made each time to the network, i.e., addition or removal of nodes, has to be updated to provide security.

3.2.2 Revocation of Nodes

This phase helps in the detection of malicious nodes. A malicious node always tries to enter into the network by revealing the important information about the nodes or sometimes reveals the secret group keys or tries to introduce corrupted data into the block Chain. So, it is necessary for a BS to identify and separate them from the network so as to avoid malicious activities. So, the revocation of nodes has to be done at that point of time. For revocation of a node, the BS records the IDs of all revoked nodes and forms a list of revoked nodes, namely R. Then, the policies in the block structure has to be modified and these revoked nodes are not allowed any access to the network [3, 11]. After modification, the updation of keys phase has to be initiated to all the nodes except R (list of revoked nodes). In this manner, the malicious nodes are separated from the network.

3.2.3 Updation of Keys

This phase is initiated when the BS identifies the malicious activities in the network or when a malicious node is identified or in the formation of new clusters. For providing newness in the keys that are used, this phase has to be repeated periodically to generate updated group keys. The process of updating the keys is shown in Fig. 3. Each update that is made to the key has to be entered into the block in the form of a transaction. This type of transaction is called update, so the transaction type has to be written as 4. But the updated key has to be given a different name than the one that is used in the last update. Then, the updated keys are exchanged between the clusters and the nodes present in the clusters except the nodes that are

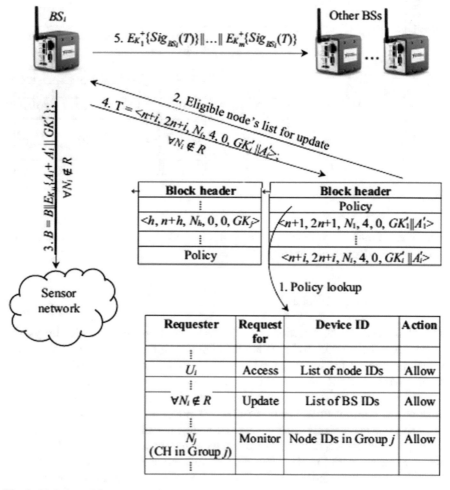

Fig. 3 Updation of keys

revoked. For the updation of keys, there is no need of computing the signature values for the transactions.

3.2.4 Storing and Accessing of Data

As the nodes gather information, the data has to be sent to the CH first and it forwards the data it received to the BS. The BS checks policy table to know whether it has access to view the data or not. If the BS has the access to monitor, then it successfully generates a transaction in the block using the transaction type as 3 (for monitor) and commits the current transaction and adds it into the present block in the block chain and this transaction has to be shared to all the reachable BSs. Then, the data sent by nodes has to be encrypted using the key K. Then, this encrypted information has to be forwarded to the BS. The BS before decrypting verifies the policies available in the block, and on successful verification of policy, it can decrypt the information and save this transaction to the current block. Then, the BS sends the updated copy of the transaction in the block to other BSs which in turn will maintain a copy of this block in their blockchain. The process of storing and accessing of the data is illustrated in Fig. 4.

Accessing the data means getting the data from the blockchain. For accessing the data [2, 12], a two-step process is followed. First, the user U will send a query about the data that he wants to the BS to which he belongs to. This request is sent along with the user signature for verification of user identity. BS checks the ID presented by the user, and if he is an authenticated user, the BS will retrieve the data, put its signature, and transfer the data to the user.

3.2.5 Monitoring Process

The last phase in the proposed system helps in viewing the requested user's status in the network. Here, the request can come to a BS about knowing the node that has access to a particular user. If the user wants to know the present state of a node, verification is done on the list of accepted node IDs in the network. Then, on successful verification, a check is made on whether they have monitor access in the block or not. If it has permission, then the BS requests the node about the current position. Once the node receives the request, it performs sensing and sends its status in an encrypted form to ensure security to the BS. Then, the BS adds its own signature and sends the status to the requested user. Then, immediately, an entry is made about the transaction in the block in a secret manner [11].

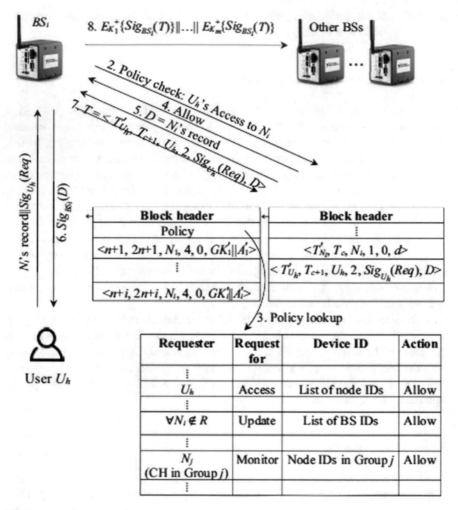

Fig. 4 Storing and accessing of data

4   Conclusion

The proposed system provides security to the WSN using blockchain technology. It can also resolve some issues like accessibility to services, consistency of data, and network lucidity. This scheme uses multiple collaborative BSs to continue the normal operation without interruption even if some of the BSs are compromised due to some malicious activity. It provides security and consistency of data using blockchain-based cryptographic techniques for the data records. It also improves the energy efficiency by using clustering approach. Because of the use of the blockchain technique, extra resources are also not needed in the proposed approach.

5 Future Work

As the nodes start sending more data, the transactions will increase and there will be a growth in the size of blockchain. So, in future, the proposed research work can be extended to include a more cost-effective memory management method that uses a cloud which can be connected to the BSs.

References

1. Hari, U., Ramachandran, B., Johnson, C.: An unequally clustered multihop routing protocol for wireless sensor networks. In: 2013 International Conference on Advances in Computing, Communications and Informatics (ICACCI), pp. 1007–1011, Aug 2013
2. Deng, R., He, S., Chen, J.: An online algorithm for data collection by multiple sinks in wireless sensor networks. IEEE Trans. Control Netw. Syst. **5**(1), 93–104 (2018)
3. Zyskind, G., Nathan, O., Pentland, A.S.: Decentralizing privacy: using blockchain to protect personal data. In: Proceedings—2015 IEEE Security and Privacy Workshops, SPW 2015, pp. 180–184 (2015)
4. Yuan, Y., Wang, F.: Blockchain: the state of the art and future trends. Acta Automatica Sinica **42**(4), 481–494 (2016)
5. Rahman, M., Sampalli, S.: An efficient pair wise and group key management protocol for wireless sensor network. Wirel. Pers. Commun. **84**(3), 2035–2053 (2015)
6. Xia, H., Zhang, R., Yu, J., Pan, Z.: Energy Efficient routing algorithm based on unequal clustering and connected graph in wireless sensor networks. Int. J. Wirel. Inf. Netw. **23**(2), 141–150 (2016)
7. Reddy, N.G., Chitare, N., Sampalli, S.: Deployment of multiple base-stations in clustering protocols of wireless sensor networks. In: 2013 International Conference on Advances in Computing, Communications and Informatics (ICACCI), pp. 1003–1006, Aug 2013
8. Gong, B., Li, L., Wang, S., Zhou, X.: Multi hop routing protocol with unequal clustering for wireless sensor networks. In: International Colloquium on Computing, Communication, Control, and Management (ISECS), vol. 2, pp. 552–556. IEEE (2008)
9. Zaman, N., Jung, L.T., Yasin, M.M.: Enhancing energy efficiency of wireless sensor network through the design of energy efficient routing protocol. J Sens. **2016**, 16 (2016). Article ID 9278701
10. Gajjar, S., Talati, A., Sarkar, M., Dasgupta, K.: FUCP: fuzzy based unequal clustering protocol for wireless sensor networks. In: 2015 39th National Systems Conference (NSC), pp. 1–6, Dec 2015
11. Qiao, R., Dong, S., Wei, Q., Wang, Q.: Blockchain based secure storage scheme of dynamic data. Comput. Sci. **45**, 57–62 (2018)
12. Tang, B., Wang, J., Geng, X., Zheng, Y., Kim, J.-U.: A novel data retrieving mechanism in wireless sensor networks with path-limited mobile sink. Int. J. Grid Distrib. Comput. **5**, 133–140 (2012)

IoT-Based Agriculture Monitoring System

Kantamneni Raviteja and M. Supriya

Abstract Internet of things (IoT) plays a crucial role in smart agriculture. Smart farming is an emerging concept, as IoT sensors are capable of providing information about their agriculture fields. Our focus is to provide farmers with an IoT-based Web application for monitoring the agriculture fields and its conditions. With the arrival of open supply NodeMCU boards beside low-cost wet sensors, it is viable to make devices that monitor the temperature/humidity sensor and soil wet content associated, consequently irrigating the fields or the landscape as and when required. The projected system makes use of microcontroller NodeMCU platform and IoT that alert farmers to remotely monitor the standing of sprinklers placed on the agricultural farms by knowing the sensing element values, thereby making the farmers' work a lot of easier when considering different farm-related activities.

Keywords IoT · NodeMCU · Temperature/humidity sensor · Soil moisture sensor · DC motor · Android

1 Introduction

Digital Republic of India is a campaign passed by the Government of India to form this country a digitally authorized country. The aim of launching this campaign is to supply Indian voter's electronic government services by reducing the work. Information and communications technology (ICT) is deeply tangled in digital banking, insurance, e-commerce, diversion, e-health, e-education, knowledge process outsourcing (KPOs), information technology-enabled services (ITES), fashionable producing, transportation, agriculture, and lots of such sectors these days.

K. Raviteja (✉) · M. Supriya
Department of Computer Science and Engineering, Amrita School of Engineering,
Amrita Vishwa Vidyapeetham, Bengaluru, India
e-mail: ravikantamneni94@gmail.com

M. Supriya
e-mail: m_supriya@blr.amrita.edu

Our focus is on agriculture sector, as it plays a very vital role in the growth of Indian economy. The growth of agricultural sector can be visually seen from Fig. 1. With rising population, there is a desire for inflated agricultural production. So as to support larger production in farms, the need of the amount of water utilized in irrigation conjointly rises. Currently, agriculture accounts 83% of the whole water consumption in India. Unplanned use of water unwittingly ends up in wastage of water. This means that there is a high priority to develop systems that stop water wastage while not imposing pressure on farmers.

The major factors affecting agriculture are:

(1) **Irrigation**: Correct irrigation ensures correct growth of crops and therefore the modification within the water level will even ruin the complete cultivation [1].
(2) **Soil conditions**: The kind of soil plays a very important role in agriculture. The soil has got to be chosen in line with the crop kind. The aspects of soil resembling its pH scale level, moisture, temperature, etc., all play an enormous role in irrigation.
(3) **Climatic condition**: The climate ought to be appropriate for the crop to induce high yield.

A system monitor could be a hardware or computer code element used to monitor resources and performance in an information technology. Monitoring systems are responsible for controlling the technology used by a business (hardware, networks and communications, operating systems, or applications, among others) in order to analyze their operation and performance, and to detect and alert about possible errors [2].

The Internet of things (IoT) is the system of physical tendency, transport, residence, and different gadgets enclosed with physics, computer program, detector, rotary actuator, and property which allows this stuff to attach, accumulate, and interchange facts as shown in Fig. 2. Internet of things might be a spread of high technology that ports data from whole totally assorted detectors and makes one thing to hitch the online to interchange data [3].

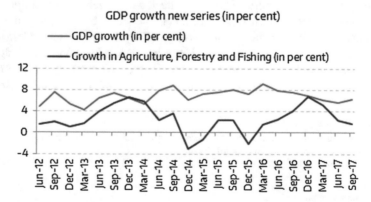

Fig. 1 Growth of agriculture. *Source* Centre for Monitoring Indian Economy (www.equity master.com)

Fig. 2 Internet of things

Our focus is to provide farmers with an IoT-based Web application for monitoring the agriculture fields and its conditions. With the arrival of open supply NodeMCU boards beside low-cost wet sensors, it is viable to make devices that monitors the soil wet content associated, consequently irrigating the fields or the landscape as and when required. The projected system makes use of microcontroller NodeMCU platform and IoT that alert farmers to remotely monitor the standing of sprinklers placed on the farm by knowing the sensing element values, thereby creating the farmers' work a lot of easier as they will consider different farm activities. The technology can be a simple card involuntary usage, with four selections. This includes electric motor standing, wetness, melting point, and status values. The electric motor standing predicts the current standing of the field.

2 Literature Survey

Agriculture Land Suitability Evaluator (ALSE) has been proposed to test a selection of styles of land to seek out the top-notch land for some of varieties of flowers with the aid of the usage of inspecting surrounding parameters [4]. ALSE approves the standardization of the framework for characterizing geo-environmental conditions like temperature, clay, spoiling, wave, topographical, etc., that are relevant for manufacturing of main plants (e.g., mangifera, buffon, watermelon, orange, and sodium). A cellular computing-based framework for agriculturists proposed in [5] developed an application called AgroMobile which can be used by the farmers for cultivation, promotion, and assessment of yield. Here, the cellular is used to analyze the sickness in crops by using methods of image processing. Jong et al. [6] proposed cloud-based completely wellness assertion and farm animal observance system (DFLMS) throughout where the detector networks have been used to collect records and manages simply about it. Farm Management Information System (FMIS)-based complete methodology has been proposed which comprehends the accuracy of agriculture necessities for data structures through Web-based approach. Nickola et al. [7] modified the FMIS to analyze high power needs of agriculturist to boost name techniques and their related practicality. Moreover, they record that identity of technique used for preliminary analysis of patron goals is obligatory for correct mode of FMIS. Hu et al. [8] proposed WASS (Internet-based agricultural help machine) and viewed the functionalities (information, cooperative work, and pick-out aid) and capabilities of WASS.

The literature survey focuses on the various QoS parameters and addresses few of them to achieve an efficient agriculture monitoring system [9, 10]. However, none of them presents a mechanism for self-management of resources which may lead to customer dissatisfaction. The parameter-wise summary of the survey has been presented in Table 1 for easiness. The proposed system is a single QoS-aware cloud-based completely automatic information device which allocates and also manages the agriculture-based resources effectively.

3 System Architecture

Figure 3 shows the overall diagram of IoT primarily based agriculture monitoring system which holds three sensors that are hooked up to the controller, and sensed data from these detectors are sent to the android application [11, 12].

Table 1 Summary of literature survey

Agriculture system	Mechanism	QoS parameters	Domains	Resource management	Big data
ALSE [4]	Non-autonomic	Yes (suitability)	Soil	No	No
AgroMobile [5]	Voluntary	Favorable (quality)	Yield	Unfavorable	No
DFLMS [6]	Voluntary	Unfavorable	Yield	Favorable	Unfavorable
FMIS [7]	Willing	Unfavorable	Pest and crop	Unfavorable	No
WASS [8]	Voluntary	Unfavorable	Productivity	No	No
Present system (Aaas)	Involuntary	Favorable (price, instance, development, latency, throughput)	Yield, atmosphere, clay, pest, plant food and supply	Favorable	Favorable

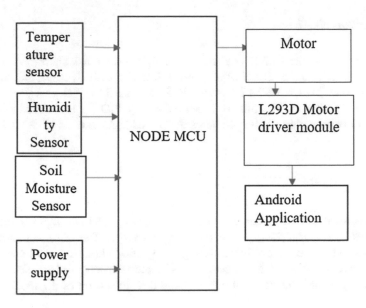

Fig. 3 Block diagram of IoT-based agriculture monitoring system

3.1 DHT11 Sensor

DHT11 is a humidity and temperature sensor, which generates calibrated digital output. DHT11 can be interfaced with any microcontrollers like Arduino, Raspberry Pi, etc. It provides high reliability, and sensor consists of three predominant components, namely a resistive kind humidity sensor, an NTC (poor temperature coefficient) thermistor (to degree the temperature), and an 8-bit microcontroller, which converts the analog alerts from each sensor into digital signal. The range of voltage for VCC pin is 3.5–5.5 V. A 5 V supply might be used for high-quality output. The information from the statistics out pin is a serial virtual statistic.

3.2 Soil Moisture

The soil moisture is used to measure the water content in the soil. Once the soil has water scarcity, the module output is at high level; else, the output is at low level. This tool reminds the user to water their plant life and in addition shows the moist content material of soil on the display screen. It has been loosely applied in agriculture, especially in land irrigation. Specifications of this sensor include working voltage: 5 V, working current: <20 mA, interface type: analog, and working temperature: 10–30 °C.

3.3 NodeMCU

Node Microcontroller Unit (NodeMCU) is an open source IoT platform. It includes firmware which runs on the ESP8266 Wi-fi SoC Espressif Systems, and hardware which is based on the ESP-12 module. ESP8266 has 17 GPIO pins (0–16); however, one can only use 11 of them, because 6 pins (GPIO 6–11) are used to connect the flash memory chip. Figure 4 shows the pin configuration of NodeMCU.

3.4 DC Motor

A DC motor is any of a category of rotary electrical machines that convert direct present-day electrical energy into mechanical energy. The most common types remember on the forces produced through magnetic fields. Nearly all sorts of DC motors have some internal mechanism, both electromechanical or electronic, to periodically change the route of cutting-edge drift in section of the motor.

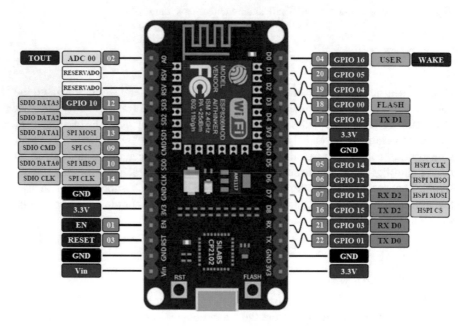

Fig. 4 NodeMCU

3.5 L293D Motor Driver Module

The L293D device is quadruple excessive cutting-edge 1/2-H motive force. The L293D is designed to provide bidirectional force up to 600 mA with a voltage from 5 to 36 V. L293D adapter board can be used as twin DC motor motive force or bipolar stepper motor motive force. This device is beneficial in robotics application, bidirectional DC motor controller, and stepper motor motive force. L293D includes the output clamping diodes for protection.

Figure 5 represents the flowchart of the entire system.

Farmers start to utilize a variety of monitoring and managed system in order to increase the yield. With the help of automation, agricultural parameters like temperature, humidity, and soil moisture are monitored and managed in the device and in turn can assist the farmers to improve the yield.

This device affords wearable and required level of water for the farmhouse and is kept way from water wastage. Once the moisture stage in the soil reaches below threshold value, then gadget mechanically switches the motor ON. When the water degree reaches normal degree, the motor robotically switches OFF. The sensed parameters and current status of the motor will be displayed on mobile device.

This agricultural mechanism is meant to find most useful answer to the water crisis. It has a sizeable influence on guaranteeing the atmosphere friendly use of freshwater assets as properly as guaranteeing the effectuality and firmness of the agribusiness. With further development within the discipline of IoT anticipated within the returning years, these structures are often larger economical, faster, and fewer dearer.

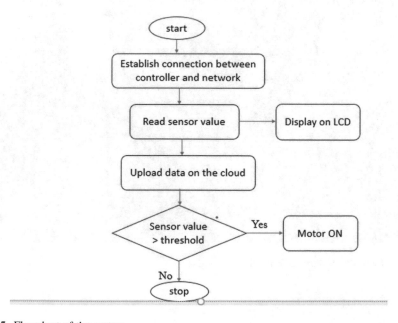

Fig. 5 Flowchart of the system

4   Results and Discussions

It has an influence on the atmosphere friendly use of freshwater assets properly as well guaranteeing the effectuality and firmness of the agribusiness. Further, considering the growth and use of IoT in the coming years, these structures are often larger, economical, and faster.

DHT 11 sensor is connected to Pin A0 of NodeMCU. Soil moisture pin is connected to pin D0 of NodeMCU. Pin A5 is connected to output of L293D module. Pin A9 is connected to power supply of L293D module. Power supply pin is connected to Pin D9 of NodeMCU. The hardware setup of the above architecture named agricultural monitoring system is shown in Fig. 6.

This graph in Fig. 7 shows temperature values displayed on user android application by using ThingSpeak app by using the cloud server. X-axis shows date-wise details, and y-axis shows the temperature monitored using the sensor at the agricultural site. By switching on the hardware system, the data on the LCD screen of the hardware prototype get transmitted to cloud via Wi-fi and then get retransmitted to the android application on the mobile.

The graph in Fig. 8 shows humidity values displayed on user android application by using ThingSpeak app. Here, x-axis shows date, and y-axis shows the humidity measure. Similarly, Figs. 9 and 10 show the motor status and the soil moisture displayed on the app.

Fig. 6 Prototype of agriculture monitoring system

Fig. 7 Temperature values displayed on user android application

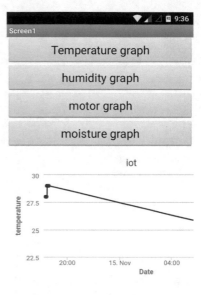

Fig. 8 Humidity values displayed on user android application

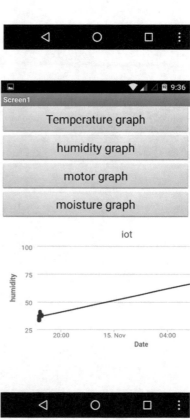

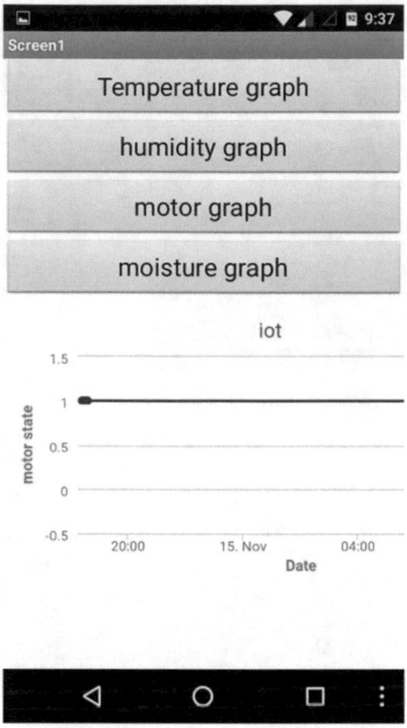

Fig. 9 Motor status displayed on user android application

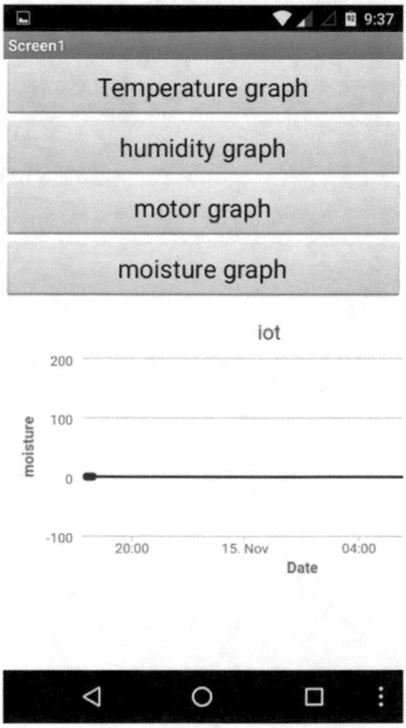

Fig. 10 Soil moisture values displayed on user android application

5   Conclusion and Future Scope

With additional development of IoT expected within the forth coming years, these systems may be additionally economical, plenty quicker, and fewer costlier. With the high technology of provision, the humanity is also reduced. In the future, this system can be created as associate smart system, wherever within the device predicts human actions, precipitation pattern, time to reap, animal intruder in the area and communicating the records. Through advanced science, IoT can be applied so that agricultural device can be made independent of human operation and in flip first rate and big extent yield can be obtained.

References

1. Prabha, R., Sinitambirivoutin, E., Passelaigue, F., Ramesh, M.V.: Design and development of an IoT based smart irrigation and fertilization system for chilly farming. Paper presented at the 2018 International Conference on Wireless Communications, Signal Processing and Networking, Wisp NET 2018 (2018). https://doi.org/10.1109/WiSPNET.2018.8538568
2. Akhil, R., Gokul, M.S., Menon, S., Nair, L.S.: Automated soil nutrient monitoring for improved agriculture. Paper presented at the Proceedings of the 2018 IEEE International Conference on Communication and Signal Processing, ICCSP 2018, pp. 688–692 (2018). https://doi.org/10.1109/ICCSP.2018.8524512
3. Aruul Mozhi Varman, S., Baskaran, A.R., Aravindh, S., Prabhu, E.: Deep learning and IoT for smart agriculture using WSN. Paper presented at the 2017 IEEE International Conference on Computational Intelligence and Computing Research, ICCIC 2017 (2018). https://doi.org/10.1109/ICCIC.2017.8524140
4. Sheikh, R., Mohamed Sharif, A.R.B., Amira, F., Ahmad, N.B., Lasiandra, S.K., Sodom, M.A.M.: "Agriculture Land Suitability Evaluator." (ALSE): A decision and planning support tool for tropical and subtropical crops. Comput. Electron. Agric. **93**, 98–110 (2013). https://doi.org/10.1016/j.compag.2013.02.003
5. Prasad, S., Peddoju, S.K., Ghosh, D.: AgroMobile: a cloud-based framework for agriculturists on mobile platform. Int. J. Adv. Sci. Technol. **59**, 41–52 (2013)
6. Jong, S., Jong, H., Kim, H., Yuen, H.: Cloud computing based livestock monitoring and disease forecasting system. Int. J. Smart Home **7**(6), 313–320 (2013). https://doi.org/10.14257/ijsh.2013.7.6.30
7. Nickola, R., Silone, I., Oscine, K.: Software architecture for farm management information systems in precision agriculture. Comput. Electron. Agric. **70**(2), 328–336 (2010). https://doi.org/10.1016/j.compag.2009.08.013
8. Hu, Y., Quant, Z., Yao, Y.: Web-based agricultural support systems. In: Proceeding of the Workshop on Web-based Support Systems, pp. 75–80 (2004)
9. Gill, S.S., Bunya, R.K.: IoT based agriculture as a cloud and big data service. J. Organ. End User Comput. **29**(4) (2017)
10. Khanna, A., Kaur, S.: Evolution of internet of things (IoT) and its significant impact in the field of precision agriculture. Comput. Electron. Agric. **157**, 218–231 (2019)
11. Pillai, P., Supriya, M.: Real time CO_2 monitoring and alert system based on wireless sensor networks. In: Intelligent Systems Technologies and Applications. Advances in Intelligent Systems and Computing, Kochi, India, 10–11 Aug, pp. 91–103 (2015)
12. Joshi, V.B., Goudar, R.H.: IoT-based automated solution to irrigation: An approach to control electric motors through android phones (2019). https://doi.org/10.1007/978-981-10-8639-7_33

A Review on Automatic Glaucoma Detection in Retinal Fundus Images

Shahistha, K. Vaidehi and J. Srilatha

Abstract Glaucoma is a very dangerous disease which is increasing alarmingly year by year. Often, people do not realize the severity or who is affected and also do not get to know the signs. The ganglia cells present in the retina of eye are affected due to the glaucoma, and it will lead to loss of eyesight. The main reason behind this disease is the increase of the intra-ocular pressure inside the eye which also damages the optic nerve. There are many types of glaucoma out of which two of them draw the attention which are namely open-angle glaucoma and angle-closure glaucoma. Initially, glaucoma shows no clear symptoms. Later on, the disease progresses that the eye vision becomes obscure. Thereby, glaucoma should be diagnosed early so as to prevent the loss of eye vision. If the manual examination of the eye images is done, it will be time consuming as well as correctness will depend on the proficiency of the experts. A vibrant tool nowadays is automatic detection of glaucoma which helps in preventing and detecting the most dangerous eye disease, glaucoma. For clear examination, the fundus images are collected from the fundus camera. The methodologies which are reviewed in this paper have certain benefits and flaws. With the help of this study, it will be easy to determine the best methodology for detecting glaucoma automatically.

Keywords Glaucoma · Intra-ocular pressure · Open-angle glaucoma · Angle-closure glaucoma

Shahistha (✉) · K. Vaidehi · J. Srilatha
Department of Computer Science and Engineering, Stanley College of Engineering and Technology, Abids, Hyderabad 500001, India
e-mail: shahistha02@gmail.com

K. Vaidehi
e-mail: kvaidehi@stanley.edu.in

J. Srilatha
e-mail: jsrilatha@stanley.edu.in

© Springer Nature Singapore Pte Ltd. 2020
K. S. Raju et al. (eds.), *Data Engineering and Communication Technology*,
Advances in Intelligent Systems and Computing 1079,
https://doi.org/10.1007/978-981-15-1097-7_41

1 Introduction

The glaucoma word is originated from the Greek ancient word which means 'clouded or blue-green hue' apparently to describe anyone who has an enlarged cornea or a person who is developing cataract inside the eye, the two of which due to the enduring intra-ocular pressure inside the eye. It is known as 'silent thief of sight' as the blindness occurs progressively for a long stretch of time and the indications are unseen in earlier stages. Figure 1 shows the healthy eye and glaucoma-affected eye images. This disease is the highest cause of blindness in the whole world. It is the major cause behind loss of sight among working population. Our India alone hosts the 20% of all the glaucoma cases among global ratio. Approximately, 79 million people are likely to be affected in the year 2020 which is estimated by WHO.

Computational techniques have great impact on the field of medicine. These techniques help the medical practitioners to diagnose any abnormality in advance and provide fruitful treatment. Retinal image analysis has been an ongoing area of research. Automated retinal image analysis aids the ophthalmologists in detecting abnormalities in the retinal structures which includes the optic disc and blood vessels, thus diagnosing sight-threatening retinal diseases such as glaucoma and retinopathy. It is a disease that infects the optic nerve as well as optic nerve cells that causes blindness. It is caused due to an increase in inter-ocular pressure. An eye liquid known as 'aqueous' unceasingly flows in the eye. The aqueous liquid generates heaviness in the sense pressure on the core shallow surface in the eye.

In any human eye, the intra-ocular pressure is between 14- and 20-mm Hg. If the pressure is between the 20- and 24-mm Hg, then it is suspected for glaucoma image, whereas if it exceeds the 24-mm Hg then it is a mandatory glaucoma image. In the good and fit eyes, all the liquids flow in balancing way. Internally, one end of the eye is for producing liquid and the other end is for draining. This balancing fluid flow maintains the inter-ocular pressure (IOP), constant in the eye. But in glaucoma,

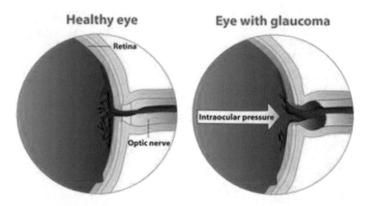

Fig. 1 Healthy eye and glaucoma-affected eye

this fluid balance is not upheld properly. As a outcome, it causes an increase in IOP, resulting in the damage of optic nerve. Due to increase in IOP, the cup dimensions begin to increase which consequently increases the optic cup-to-optic disc ratio.

Glaucoma is characterized by increased IOP inside the eye leading to variations in the optic disc and optic nerve. It does not reveal its symptoms until advanced stage. Hence, regular screening of the patients is required to identify this disease, thus demanding huge labour, time and proficiency. Thus, computational methods are sought for analysis.

Fundus is a Latin term, automatically referring to a portion of organ opposite from its opening. Fundus image of an eye is the snap of the interior surface of the eye which contains retinal blood vessels, macula, fovea, optic disc and optic cup. Since glaucoma affects optic disc and optic cup by altering cup-to-disc ratio and rim-to-disc ratio, proper segmentation of both the features is important for glaucoma detection.

As the changes in the profile of optic disc act as a biomarker for the onset of the disease, optic disc is segmented through image processing techniques. Optic disc is the brightest part portrayed as oval structure in the retinal fundus image. It encompasses optic cup, which is the brightest central part, optic rim, the surrounding pale part and the blood vessels. Glaucoma is a group of diseases that result in damage of eye's optic nerve which cause blindness and vision loss. This is a kind of disease which could be hereditary. Enormous pressure of the eye's internal surface distresses the brain.

Figure 2 shows the optic nerve image. The optic nerve carries information from eye to brain. It consists of 1 million nerve fibres that connect the retina of the eye to the brain tissues. The retina is a light and delicate tissue inside the eye. Nerve fibre gets damaged, creates the blind spots and also causes blindness. Maintaining healthy optic nerve is essential for proper vision. The pressure arises inside the eye because the aqueous liquid which normally flows in and outside of the eye cannot drain properly. This pressure is causing the optic nerve damage which is resulting in

Fig. 2 Optic nerve image

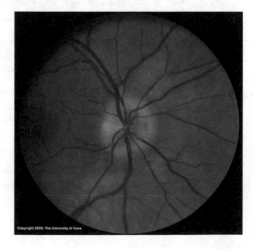

glaucoma. The threatening thing about this disease is it shows no signs like pain. If it is left without diagnosing, then the person starts losing their vision and feels like vision through a tunnel.

2   Types of Glaucoma

Primary open-angle glaucoma: The aqueous fluid which circulates in the anterior part of the eye is jammed from moving out of the eye's drainage system. This type of utmost glaucoma is also known as wide-angle glaucoma. It happens due to partial blockage of drainage channel in which pressure increases slowly because the fluid cannot drain properly. It does not get noticed until central vision is affected.

- Progression: Rate of progress of the disease is very slow, so the patient only gets to know when there is complete vision loss.
- Causes: As no symptoms can be seen till the vision is lost, it is termed as sneak thief of sight.
- Symptoms: It will continue for so many years and progress sometimes without the visible loss of vision.

Angle-closure glaucoma: The drainage canals are blocked by the outer edge of the iris blocking the flow of aqueous fluid. Due to the sudden and full blockage of aqueous, it is also called acute angle glaucoma. Loss of vision occurs quickly due to the rising pressure. It is because of the slender drainage angle and thin and droopy iris.

- Progression: This is aching and sudden that is the sole cause there is a loss of vision at a significant rate.
- Causes: This is a much more incomparable form of glaucoma that requires prompt medical treatment as it grows very swiftly.
- Symptoms: Symptoms of this are vomiting, extreme eye and head annoyance, concealed vision.

Normal-tension glaucoma: This type of glaucoma also called normal-tension or low-tension glaucoma occurs to those who have an eye pressure of 21–22-mm Hg, vision loss occurs, and optic nerve is damaged.

Ocular hypertension: It is a type of glaucoma where only the pressure is high rather than normal eye pressure and does not possess any other signs of glaucoma.

Secondary glaucoma: It is the type of glaucoma that results from alternative eye illness. For example, if anyone is having eye injury or someone is on steroid therapy, then they can suffer from this type of glaucoma.

Congenital glaucoma: It is an infrequent type of eye disease that progresses in newborns and teenagers and can also be hereditary.

Childhood glaucoma: It is the condition where disease starts in childhood.

3 Literature Review

The existing approaches towards glaucoma diagnosis are concisely presented here. Generally, the process of glaucoma detection involves the abstraction of optic disc and optic cup followed by elicitation of its properties such as CD ratio and ISNT ratio to distinguish normal images from glaucoma-affected images.

In 2014, Alexandra Guerre and Paul Miller worked on automatic glaucoma detection. The data set was captured from the stereoscopic camera which contained 29 images out of which 14 are normal eye images whereas 15 are glaucomatous eyes. Initially, pre-processing is done through the green channel extraction and then a low-pass filter is applied to decrease the fine grain noise. Then, histogram equalization is done to confirm consistency among the images. Then, Otsu thresholding method is applied to identify several different image regions. Feature extraction is done through the CDR which can be obtained by localizing the highest pixel and lowest pixel values in vertical axis for rim and cup. Classification is done through the linear kernel of support vector machine.

In 2014, Rajendra Acharya and Prabhakar Nayak worked on glaucoma detection with 510 database images from Kasturba Medical College, Manipal, all images of size 2588 * 1958 which are resized to 240 * 180 pixels. Gabor transformation is done to extract texture features. Various feature extractions are namely mean, variance, skewness, kurtosis, Shannon entropy and Kapoor entropy. Principal component analysis is done to rescue the data dimensionality. High rank features are selected and then applied support vector machine classifier achieved 93.10%.

In 2015, Ayush Agarwal and Malay Kishore Dutta worked on automatic detection of glaucoma through the database collected from Venu Hospital, New Delhi, and achieved an accuracy of 90%. In their work, input is RGB images and their ROI was optic disc instead of processing the whole region. ROI is calculated by passing image through window of $r * r$ dimension, and coordinates of optic disc are obtained. Then, optic disc is segmented through the Otsu thresholding technique using only the R channel. Then, optic cup is segmented through extracting mean and standard deviation. Then, a histogram is drawn by taking the tonal variation on x-axis and pixel values on y-axis. Addition of mean and standard deviation gives the intensity level in the grey region that points to highest number of pixels. Neuroretinal rim is calculated by the thickness of rim. Neural rim-to-disc ratio is calculated by ratio of rim area in inferior, and temporal regions-to-the total disc area ratio is taken as rim-to-disc ratio. RBF kernel is used for classification of the images; if the CDR exceeds 0.3, then it is said to be a glaucomatous eye. Then, RDR is calculated; if it is less than 0.4, then it is said to be glaucomatous eye.

In 2015, Abhishek Dey et al. detected glaucoma in retinal fundus images from the databases, Susrut Eye Foundation & Research Centre, Kolkata; first, the images are pre-processed by converting to greyscale, resizing, noise removal and improval of image contrast, and principal component analysis is used for extracting features, by subtracting the mean, then calculating the covariance matrix, then calculating the eigenvectors and eigenvalues, choosing the particular components and forming a feature vector. They are fed into support vector machine for classifier which helps in distinguishing normal images and glaucoma-affected images by achieving an accuracy of 96%.

In 2016, Shishir Maheshwari and Ram Bas Pilari worked on automatic detection of glaucoma using wavelet transform because empirical wavelet transform is signal-dependent decomposition technique. In this work, two databases are used. Publicly available database which has 250 normal images and 250 glaucoma images is used. Another private database from Kasturba Medical College is used. Two-dimensional empirical wavelet transform is signal decomposition. Correntropy is nonlinear kernel-based method used for feature extraction. Feature selection is done to using Students t-test algorithm which is highly discriminative. Classification is done using least square support vector machine which achieved an accuracy of 98.33% using RBF kernel.

In 2016, Sughitharani and Geetha Ramani worked on glaucoma detection. At first, the green channel is extracted and then the blood vessel is segmented by median filtering, image compliment, discrete wavelet transform using symmlet filter. Then, blood vessel is extracted. The optic disc is segmented through the three algorithms, namely K-means, Haar wavelet and histogram analysis. Then, the optic disc is segmented as the one through maximum voting. Certain features are selected, then classification is done through the various classifiers which includes random tree, and reduced error pruning tree performance of the classifier is evaluated through the tenfold cross-validation where ninefold training data and onefold as testing data-rich achieved an accuracy of 96.4%.

In 2016, Anushika Singh and Malay, Kishore Dutta worked on automatic glaucoma detection; initially, the fundus images are taken from the database Venu Eye Centre, New Delhi, in which the images are of JPEG format of image size 2540 * 1696. Blood vessel in-painting is done, optic disc is segmented, and wavelet features are extracted without manual intervention. Optic disc is segmented through the bit-plane analysis and calculated using double windowing method. Features are selected to improve the accuracy prominent feature selection using attribute selection, and various classifiers like random forest, KNN, Naïve Bayes, ANN and SVM are used. In SVM, kernel function which has the best accuracy is selected for classification.

In 2018, Sumayya Pathan and Preetham Kumar worked on computer-based automated glaucoma detection method using RIM-ONE version 3 which consists of fundus image database of 124 retinal images. Initially, blood in-painting is done by median filtering or Laplacian filtering. Feature extraction includes colour features, and textual features are extracted. Colour features include mean, standard deviation, variance, skewness, entropy. Textual features include GLCM representation square matrix. Each element in GLCM gives the number of occurrences of pixel with its neighbour. Support vector machine is used for classification which achieved accuracy of 92%.

M. Madhusudhan et al. used the public database MESSIDOR which addresses the various image processing techniques to diagnose the glaucoma based on the

CDR evaluation of pre-processed fundus images. Pre-processing is done by vessel removal by in-painting the images. Illumination correction is done by morphological operation and normalization. Features are extracted using the Snakes: active contour method. Then, segmentation is done by the region growing algorithm. These algorithms are applied on fundus images, and then results are compared. The actual value of the cup-to-disc ratio lies between 0.3 and 0.5. The sensitivity and specificity for these algorithms are found to be very favourable and achieved an accuracy of 92%.

Gayathri Devi et al. detected glaucoma from retinal images (http://www. manipal.edu) from the database. The textual characteristics obtained from retinal images are used for this classification. The system is using discrete wavelet transform to extract dissimilar features which are obtained from the three filters symmlets Daubechies and biorthogonal wavelet filters. For classification, classifiers like least square support vector machine, random forest, dual sequential minimal optimization, Naive Bayes and artificial neural networks are used. The energy signatures found from 2-D discrete wavelet transform are used for detecting glaucoma. SVM achieved an accuracy of 92%, for random forest 91%, for Naïve Bayes 92% and for ANN 98%.

Maila Claro and Leonardo Santon worked on the detection of glaucoma with the three databases, namely DRISHTI, DRIONS-DB and RIM-ONE. Initially, the red channel is extracted and region of interest is derived by identifying more pixel intense value and localization based on threshold. Feature extraction is done by GLCM features which include homogeneity, correlation, energy, contrast. Classification is done by the random forest which achieved an accuracy of 85%.

Fouzia Khan and Shoaib A. Khan collected the database from the Hamilton eye institute which contained 50 images, another dataset which has 19 images from the fried rich alexander university, another database from MESSIDOR which has 18 images are used. Pre-processing is done by resizing the image. Feature extraction is done by converting to greyscale, and then calculate the mean and finally set a threshold value. Classification is done by cup-to-disc ratio features; if the disc has CDR greater than 0.5, it violates ISNT rule. Normal disc obeys ISNT rule which means CDR value less is than 0.5; if tie occurs, then disc is considered to be suspected for glaucoma. This paper achieved an accuracy of 94%.

Syed Sibte Raza Abidi et al. used the public database which contained 125 images. Pre-processing is done by the moment methods. Then, feature extraction is performed by the subset selection strategy that combines filter and wrapper models. For Classification, 75% data is used for training and 25% is used for testing in support vector machine which achieved an accuracy of 79%.

Suraya Mohammed et al. used the database collected from the Manchester Royal Eye Hospital. In this paper unlike other papers, no pre-processing is done. Feature extraction is done through Binary Robust Independent Elementary Features (BRIEF). Classification is done through the support vector machine using linear radial basis kernel achieving an accuracy of 84%.

Fig. 3 Block diagram for
automatic glaucoma detection

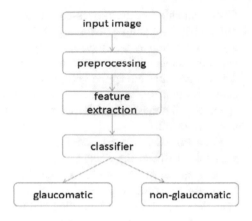

4 Methodology

The general block diagram for the automatic glaucoma detection is shown Fig. 3. The following are the different steps involved.

5 Comparison of Related Works

See Tables 1 and 2.

6 Conclusion

In this review paper, different techniques are studied which are tangled in detecting glaucoma disease automatically. Mainstream of blindness worldwide is due to the sight-threatening disease. Glaucoma with the methodologies presented in the paper, it is necessary to develop cost-effective automated technique, so that glaucoma disease is detected precisely. These procedures will be supportive for underdeveloped nations. In the near future, precise glaucoma detection with the less cost may help in benefitting every class of people. Once this disease is properly diagnosed, then we may avoid total blindness. In view of the fact of the visual system disparity and maintaining posture in any individual, people suffering from glaucoma should contemplate themselves at huge risk of falls, and are recommended to take the essential precautions to help prevent from glaucoma.

Table 1 Comparison between related works for glaucoma detection using fundus images

S. No	Authors	Title and database	Pre-processing	Feature extraction	Classification	Accuracy
1.	[1]	Automatic analysis of digital retinal images for glaucoma detection DB: http://www.optic-disc.org/	Low-pass wavelet filter Histogram Equalization	Otsu adaptive thresholding and cup-to-disc ratio Neuroretinal rim width variation	Support vector machine	90%
2.	[2]	Decision support system for the glaucoma using gabor transformation DB: Kasturba Medical College	Filtering	t-test space algorithm	Support vector machine	93%
3.	[3]	A novel approach to detect glaucoma in retinal fundus images using cup-disk and rim-disk ratio DB: Venu Eye Centre	Windowing method	Otsu thresholding technique for optic disc segmentation Histogram equalization Cup-to-disc ratio Rim–to-disc ratio	Support vector machine RBF kernel	90%
4.	[4]	Automated glaucoma detection using support vector machine classification method DB: public database	Normalization Colour conversion	Principal component analysis	Support vector machine	96%
5.	[5]	Automated diagnosis of glaucoma using empirical wavelet transform and correntropy features extracted from fundus images Public DB: http://medimrg.webs. ull.es/Private DB: Kasturba Medical College	RGB extraction	Students t-test algorithm Two-dimensional empirical wavelet transform	Least square support vector machine	98%

(continued)

Table 1 (continued)

S. No	Authors	Title and database	Pre-processing	Feature extraction	Classification	Accuracy
6.	[6]	Automatic detection of glaucoma in retinal fundus images through image processing and data mining techniques DB: publicly available database	Median filtering and image compliment	Border filtering, K-means, histogram equalization, Haar wavelet. Border suppression	Support vector machine	96%
7.	[7]	Image processing-based automatic diagnosis of glaucoma using wavelet features of segmented optic disc from fundus image DB: Venu Eye Centre, New Delhi	Blood vessel in-painting	Optic disc is segmented using bit-plane analysis and calculated using double windowing method Evolutionary attribute selection	Artificial neural network and support vector machine	89% and 94%
8.	[8]	Image processing techniques for glaucoma detection DB: MESSIDOR	Vessel removal by in-painting	Snakes active contour method	Multi-thresholding technique Region growing algorithm	92%
9.	[9]	The role of color and texture features in glaucoma detection DB: RIM-ONE version 3	Blood vessel in-painting by median filtering and Laplacian filtering	Colour features and texture features GLCM square matrix	Support vector machine	92%
10.	[10]	Glaucoma detection from retinal images DB: http://www.manipal.edu	Smoothening using filters like median filter	Wavelet transform using symmlet filters, Daubechies, biorthogonal	Support vector machine and artificial neural network	92% and 98%
11.	[11]	Image processing techniques for glaucoma detection DB: MESSIDOR	Resized and converted to greyscale	Edge detection using canny filter	Cup-to-disc ratio and rim-to-disc ratio are calculated. Globally accepted value is 0.3	94%

(continued)

Table 1 (continued)

S. No	Authors	Title and database	Pre-processing	Feature extraction	Classification	Accuracy
12.	[12]	Automatic glaucoma detection based on optic disc segmentation and texture feature extraction DB: DRISHTI, DRIONS-DB, RIM-ONE	Red channel extraction and region of interest demarcation	GLCM	Random forest	85%
13.	[13]	Automated optic nerve analysis for diagnostic support in glaucoma DB: publicly available database	Moment methods	Feature subset selection strategy	Support vector machine	79%
14.	[14]	Characterizing glaucoma using texture DB: Manchester Royal Eye Hospital	Not done	Binary Robust Independent Elementary Features (BRIEF)	Nonlinear radial basis kernel	84%
15.	[17]	Joint OD and OC segmentation on deep network DB: ORIGA database	Multiscale input layer	U-shaped convolutional network	Side output layer is used as classifier	
16.	[18]	Automated glaucoma diagnosis using deep and transfer learning	Images are cropped	VGG19 and Inception Resnet V2	Tuned classifier	92%
17.	[19]	Detection of glaucoma eye disease DB: DRIONS and HRF	Green channel extraction	CNN	Soft max linear classifier	99%
18.	[20]	Automated glaucoma detection using clustering and Ellipse fitting	Green channel extraction	K-means clustering	CDR value 0.3, then glaucomatous image	61%

Table 2 Comparison between merits and demerits of the related works

S. No	Author	Title	Merits	Demerits
1.	[1]	Automatic analysis of digital retinal images for glaucoma detection	Higher accuracy in the presence of various stages of glaucoma	Does not work accurately on the largest training data samples
2.	[2]	Decision support system for the glaucoma using gabor transformation	Online and offline techniques are performed	Tenfold has lower variance
3.	[3]	A novel approach to detect glaucoma in retinal fundus images using cup-disk and rim-disk ratio	Technique makes the method vigorous and operative	Only few parameters are used
4.	[4]	Automated glaucoma detection using support vector machine classification method	Instead of ROI, entire images are processed. All of the observations are used for both training and validation	Only RBF kernel of SVM is used
5.	[5]	Automated diagnosis of glaucoma using empirical wavelet transform and correntropy features extracted from fundus images	Preserves both statistical information and temporal information	Not tested for huge database. Only few images are taken
6.	[6]	Automatic detection of glaucoma in retinal fundus images through image processing and data mining techniques	Blood vessels are also extracted, and optic cup is also extracted through three various algorithms	Manual intervention is required to determine the optic disc
7.	[7]	Image processing-based automatic diagnosis of glaucoma using wavelet features of segmented optic disc from fundus image	No manual intervention is required It is done automatically	First-order discrimination is used because higher-order discrimination costs more
8.	[8]	Image processing techniques for glaucoma detection	Cup-to-disc ratio and rim-to-disc ratio If that exceeds 0.3 then suspected	Snakes is not efficient with this technique
9.	[9]	The role of color and texture features in glaucoma detection	Helps in mass screening	Avoids segmentation of anatomical structures of the retina

(continued)

Table 2 (continued)

S. No	Author	Title	Merits	Demerits
10.	[10]	Glaucoma detection from retinal images	Works very efficiently with the data set	Multi-resolution is not possible
11.	[11]	Detection of glaucoma using retinal fundus images	Computational cost is less, i.e. 1.42 s	Blood vessel is not segmented
12.	[12]	Automatic glaucoma detection based on optic disc segmentation and texture feature extraction	Reduced processing time	Could have achieved more accuracy by using a different classifier
13.	[13]	Automated optic nerve analysis for diagnostic support in glaucoma	Reduced large input space by moment methods	Less accuracy is achieved
14.	[14]	Characterising glaucoma Using texture	Easy to implement	No pre-processing is done
15.	[17]	Joint OD and OC segmentation on deep network DB: ORIGA database	Jointly separated optic disc and optic cup	Accuracy level is not good
16.	[18]	Automated glaucoma diagnosis using deep and transfer learning	Achieved higher accuracy	Did not overcome overfitting problem
17.	[19]	Detection of glaucoma eye disease DB: DRIONS and HRF	Attained huge accuracy level of 99%	ROI is extracted manually
18.	[20]	Automated glaucoma detection using clustering and Ellipse fitting	Average time for classification is 984.37 ms	Very less accuracy

References

1. Del Rincón, A.G.J.M., Miller, P., Azuara Blanco, A.: Automatic analysis of digital retinal images for glaucoma detection
2. Acharya, U., Rajendra, et al.: Decision support system for the glaucoma using Gabor transformation. Biomed. Signal Process. Control **15** (2015)
3. Agarwal, A., et al.: A novel approach to detect glaucoma in retinal fundus images using cup-disk and rim-disk ratio. Bioinspired Intelligence (IWOBI), 2015 4th International Work Conference on. IEEE, 2015
4. Dey, A., Bandyopadhyay. S.K.: Automated glaucoma detection using support vector machine classification method. Br. J. Med. Med. Res. **11**(12) (2016)
5. Maheshwari, S., Pachori, R.B., Rajendra Acharya, U.: Automated diagnosis of glaucoma using empirical wavelet transform and correntropy features extracted from fundus images. IEEE J. Biomed. Health Inf.

6. Ramani, R.G., Sugirtharani, S., Lakshmi. B.: Automatic detection of glaucoma in retinal fundus images through image processing and data mining techniques. Int. J. Comput. Appl. **166**

7. Singh, A., et al.: Image processing based automatic diagnosis of glaucoma using wavelet features of egmented optic disc from fundus image. Comput. Methods Progr. Biomed. **124** (2016)

8. Madhusudhan, M., et al.: Image processing techniques for glaucoma detection. In: *International Conference on Advances in Computing and Communications.* Springer, Berlin, Heidelberg (2011)

9. Pathan, S., Kumar, P., Radhika M.P.: The role of color and texture features in glaucoma detection. In: *2018 International Conference on Advances in Computing, Communications and Informatics (ICACCI).* IEEE, 2018

10. Gayathri Devi, T.M., Sudha, S., Suraj, P.: Glaucoma detection from retinal images. In: *2015 2nd International Conference on Electronics and Communication Systems (ICECS).* IEEE, 2015

11. Khan, F., et al.: Detection of glaucoma using retinal fundus images. In: *Biomedical Engineering International Conference (BMEiCON), 2013 6th.* IEEE, 2013

12. Claro, M., et al.: Automatic glaucoma detection based on optic disc segmentation and texture feature extraction. CLEI Electron. J. **19**(2), 5–5 (2016)

13. Yu, J., et al.: Automated optic nerve analysis for diagnostic support in glaucoma. In: Proceedings 18th IEEE Symposium on Computer-Based Medical Systems, 2005. IEEE, 2005

14. Mohammad, S., Morris, D.T.: Texture analysis for glaucoma classification. In: 2015 International Conference on BioSignal Analysis Processing and Systems (ICBAPS). IEEE, 2015

15. DATABASE :http://www.manipal.edu

16. DATABASE: http://medimrg.webs.ull.es/

17. Fu, H., et al.: Joint optic disc and cup segmentation based on multi-label deep network and polar transformation. arXiv preprint arXiv:1801.00926 (2018)

18. Norouzifard, M., et al.: Automated glaucoma diagnosis using deep and transfer learning: proposal of a system for clinical testing

19. Abbas, Q.: Glaucoma-deep: detection of glaucoma eye disease on retinal fundus images using deep learning. Int. J. Adv. Comput. Sci. Appl. **8**(6), 41–45 (2017)

20. Sarkar, A., Sarkar, S.K., Nag, A.: Automated glaucoma detection from fundus images using clustering and Ellipse-fitting

IoT-Based Monitoring System for Safe Driving

Bulusu Sowjanya and C. R. Kavitha

Abstract Several studies are made on both physiological and psychological states of the driver. With the increase in the technology in today's world lead to the development of new devices. Driving is a complex activity that requires multi-level skills. Most of our driving skills will be improved by experience. Many people lost and even losing their lives because of this distracted, drunken, and rash driving due to lack of proper system. The main purpose of the IoT project is to design a system which will detect the drunken and drowsiness of the driver and provide safety by controlling the speed of the vehicle. GSM technology is used to alert the owner in case of drunken driving. Smart controlling of the headlight brightness is also involved in the project.

Keywords GSM · Smart controlling · Internet of things (IoT)

1   Introduction

Automobiles are one of the most convenient means of transport and with the increase in science and technology, the new advancements were added to these automobiles which not only developed the technical aspects of the vehicle but also concentrated on the safety features, and luxury. The use of these vehicles for transportation is rapidly increased from 193 million in 1970 to 800 million in the recent years [1]. Though these automobiles eased the living of the people, they have created many serious issues such as traffic, accidents, and pollution. The accidents are mainly due to inattention of the driver, over speed, alcohol consumption, misperception, error in decision making, drowsiness, and health issues like heart attack [2, 3].

B. Sowjanya (✉) · C. R. Kavitha
Department of Computer Science and Engineering, Amrita School of Engineering,
Amrita Vishwa Vidyapeetham, Bengaluru, India
e-mail: sowjibulusu@gmail.com

C. R. Kavitha
e-mail: cr_kavitha@blr.amrita.edu

© Springer Nature Singapore Pte Ltd. 2020   499
K. S. Raju et al. (eds.), *Data Engineering and Communication Technology*,
Advances in Intelligent Systems and Computing 1079,
https://doi.org/10.1007/978-981-15-1097-7_42

In today's world, there happens to occur lot of road accidents due to drunken and rash driving. Drunken driving is already a serious problem, which is likely to be added as one of the most noteworthy problems in the near prospect. Checking if the driver took alcohol or not is quite important in order to provide proper work and road safety [4]. The American Automobile Association Foundation for Traffic Safety (AAA FTS) has classified the various stages of inattention of the driver while driving which are: (1) attentive; (2) distracted; (3) looked but did not see; (4) sleepy; and (5) unknown. Driving is a complex activity that requires multi-level skills. Most of our driving skill will be improved by experience. Many people lost and even losing their lives because of this distracted, drunken, and rash driving due to lack of proper system [5].

The National Highway Traffic Safety Administration (NHTSA) conducted a survey over 12 months with 241 drivers. As per that report around 25% of the total road accidents are mainly due to inattention of the driver. From the 78% of the total accidents that came up 65% can be avoided if the driver had properly paid his attention towards the road [6]. From the official reports of Beijing, the traffic accidents are mainly due to drivers, vehicle, road, and weather and out of which accidents caused due to driver is the major factor, as this is the reason for 95% of the road accidents [7, 8]. Hence, it is quite very important to detect the abnormal driving by considering various factors and improve the traffic safety.

In general, we can divide the factors related to behaviour of driving into two groups. Firstly, variables determining physiological behaviour of the driver which may include heart rate, age, gender, and different environmental factors like traffic and weather conditions in that area. Secondly, different vehicle-related factors such as speed, throttle plate position, along with driver-related information such as intake of alcohol, drowsiness, and recklessness [7, 9].

India is having world's second-largest population with 1.3 billion. In average, around 20 million people every year are injured by the traffic accidents. It is a known fact that most of the people use private transport instead of public transport (mainly in India) [6]. Out of 48%, only 18% people use public transport and the remaining 30% use their own transport that is private vehicles. According to a survey conducted in India by Transport Research Wing, the percentage occurrence of accidents has been increased by 2.5% in one year. As per the report, every day around 1374 deaths occur due to road accidents. Among those 1374 deaths, 54.1% are in between the age of 15–34 years [10]. 52 billion US dollars are lost by India every year due to these road accidents. If proper care is taken regarding drunken drive, high beam of upfront vehicles, seat belt, drowsiness of the driver, then 65% of the total road accidents can be reduced in India.

The driver's behaviour as mentioned above (such as alcohol consumption, drowsiness, and recklessness) is the major cause of these fatal accidents on road which not only effect the co-drivers but also effect pedestrians sometimes and may even cause severe loss of both public and private properties along with loss of valuable lives. When the driver takes alcohol, sudden accelerating and decelerating will be done with a delayed response which determines a poor control of speed of the vehicle. The second factor drowsiness is mainly caused due to sleepiness of the driver but without alcohol consumption at all. Even the driver with drowsiness will not be able to control the vehicle properly like the drunken driver as his case also involves sudden rise and fall of the speed of the vehicle. The third case where the driver is reckless but is awake may result in sudden increase in the speed of the vehicle crossing speed limits and this may be because driver might be under some emotional stress [11].

The main aim of this project is to detect whether driver has consumed alcohol or not by monitoring different parameters like heartbeat, tilt angle of the head, amount of oxygen content in blood and accordingly, avoid false data and lock the speed of the vehicle if suspected by any reason that driver has consumed alcohol but not stop the ignition. The above step is provided so that once confirmed that the driver has drunk, the vehicle can be stopped safely without any accidents.

This paper is organized as follows. In Sect. 2, related works till now are discussed under the heading 'Literature Survey'. In Sect. 3, architecture and design of the system are described under the heading 'System Design'. In Sect. 4, implementation of the system is explained. Section 5 deals with experimental results and Sect. 6 gives conclusion and future enhancements.

2 Literature Review

This section provides an overview of the related works done in this field.

In paper [4], a process is designed for both alcohol detection and seat belt control system. This paper mainly focuses on providing protection to the driver by checking if the driver had seat belt or not along with monitoring and checking if the driver had consumed alcohol. In this project, the author mainly used Arduino micro controller and different sensors like alcohol sensor, IR sensor along with GSM, GPS systems.

In paper [7], the authors conducted a detailed study on the abnormal driving detection based on normalized driving behaviour analysis was done. AbnormIndex was proposed which is the abnormality index used to detect and evaluate the abnormal driving behaviour typically recklessness, fatigue, drunk, and lack of attention due to usage of phone. These were all simulated and then applied to AbnormIndex. System also helps in the development of an intelligent driving system in future as the AbnormIndex uses the onboard ordinary data of all the cars and will also be able to distinguish between the normal and abnormal driving under different abnormal behaviours. The drawback with this system is weather, steering, etc., are not taken into consideration in the driving model.

In paper [2], different sensors such as alcohol, passive infrared, MQ7 for alcohol detection and also monitors the pulse rate. Apart from these, the other method involved in this project is driver vigilance detection. An SMS alert will be sent to the registered users with GSM modem. Theft detection along with security is present in the hybrid model.

In paper [11], main focus was on drunk driving and drowsiness detection so that the accidents rate is reduced, they used an alcohol sensor, a Web camera for detection of drowsiness by monitoring the eyes, yawn, and face of the driver. Eye capturing is done to detect drowsiness and mouth capturing is done to detect yawn. If driver is found to be drowsy then alarm will be on and seat will vibrate alerting the driver.

In paper [3], Pughazendi N., Sathishkumar R., Balaji S., Sathyavenkateshwaren S., Subash Chander S. and Surendar V., mainly concentrated on detection and monitoring of alcohol and heat rate of the driver by providing smart technology using Internet of things. Monitoring on the heartbeat rate of the driver is done continuously with the respective sensor apart from the above-mentioned, this project also helps the driver to check if the desired destination can be reached with the available fuel in the vehicle. Results will be provided based on the destination set by the driver in the Google Maps and will even check the nearby fuel stations if the available fuel in the vehicle is not sufficient to reach the destination. With the help of ZigBee, whenever the driver is about to reach the traffic signal junction, the controller will alert if there is any change in the traffic signal.

In paper [9], Anuva Chowdhury, Rajan Shankaran, Manolya Kavakli, and Md. Mokammel Haque, mainly discussed about the measures available (as per the physiological signals) for the detection of drowsiness. With the help of this analysis, detailed information about alcohol detection through different physiological parameters that show drowsiness of the driver can be detected along with the respective measures advantages and disadvantages.

In paper [1], Youjun Choi, Sang Ik Han, Seung-Hyun Kong, and Hyunwoo Ko, mentioned about DSM system with and without vision sensors, steering angle sensors, and physiological sensors which are developed by major manufacturers of automobiles like Ford, Toyota, BMW, etc., to reduce number of fatal traffic accidents and few technologies are even got released which are in use and such are named as smart vehicles. The overall design of DSM systems based on different physiological sensors which are installed in the smart vehicles by various manufacturers of automobiles nowadays are introduced in this article.

This paper introduces a system which not only provides a different methodology for detection of alcohol without causing false alarms but also provides speed control of the vehicle by locking the maximum speed the vehicle has access to go. It will also alert the driver and surrounding vehicles to keep a safe distance through display and buzzer. This system includes different hardware components as alcohol sensor, MEMS sensor, heartbeat sensor, IR sensor, GSM, Buzzer, and LDR along with Arduino NANO. The alert message is sent with the help of GSM. Arduino C is the software used in the project. Proteus software is also used for simulation analysis.

Fig. 1 System block diagram

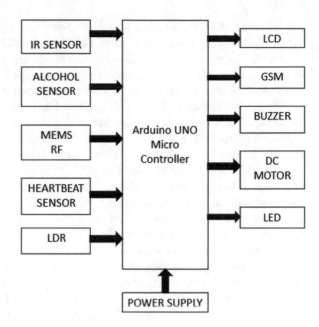

3 Proposed System

Figure 1 shows the block diagram of the proposed system of alcohol detection by different parameters and speed control.

The proposed system mainly helps in the detection of the alcohol by considering different parameters which are measured with the help of different sensors like alcohol, MEMS, heartbeat, and the counter is given for each sensor. The counter value is incremented only when the sensor detects the drunken and drowsy scenarios of the driver. For each sensor detection, the counter value is incremented, and the speed of the vehicle is locked accordingly and with the help of GSM technology, the necessary information is sent to the subscribed users and also surrounding vehicles are alerted with buzzer. The additional feature present in this is the headlight brightness, which is controlled depending upon the intensity of the light in the surrounding of the vehicle with the help of LDR sensor.

4 Implementation

4.1 GSM-Based Alerting

Initially, when the supply is given to the system, all other sensors will be activated but they remain in sleep state. The IR sensor is installed in the car in such a way that when a person enters the driver seat, the sensor reads. Once the IR sensor sends the output of its obstacle detection to micro controller, the controller sends the activation signal to the remaining sensors like alcohol, MEMS, and heartbeat sensors. This methodology is mainly used to save the power and also the life of the sensor so that the sensors will not be undergoing wear and tear because of their continuous readings. The alert message will be sent to the owner that the vehicle is started.

The alcohol sensor will detect if the driver has consumed alcohol or not through his breath. Once the alcohol sensor detects the alert, then a display will be given to the surrounding vehicles to warn them that the driver is drunk so that they maintain the safe distance to avoid any accidents. Buzzer is triggered alerting the driver. The speed of the vehicle will be locked to 70% of the available speed automatically, so that even if the accelerator is driven by the driver the speed will not exceed the limit. An SMS alert will be sent to the owner of the vehicle and all other registered users along with ambulance and cops so that necessary action can take place. After detection of the readings from the alcohol sensor, the counter which is initially set to zero will be incremented to one.

Drowsiness detection can also be done along with alcohol detection in this project. If MEMS sensor detects any head tilting angle of the driver then the counter value which is previously increased to one because of the alcohol sensor will now be incremented by one again which makes the counter value now two.

As soon as the counter value becomes two, the speed of the vehicle now be locked to 30% of the total speed and will not allow the driver to increase the speed of the vehicle beyond 30%. A display alert from the back window of the vehicle will be given to the surrounding vehicles that the driver is drunk and drowsy, speed of the vehicle will be reduced automatically. This alert display will help the surrounding vehicles to keep their safe distance from the current vehicle with drunken driver so that accidents are prevented. A SMS alert will be sent to the owner along with cops and other subscribed users so that needful action can be taken. To alert the driver immediately, so that the control of the vehicle is not lost, buzzer will be triggered. Here, one of the notable point is that, the MEMS sensor must be worn on the head of the driver to measure the head tilt angle. For the feasibility purpose, the transmission of the data through sensor to controller is made wireless with the help of radio frequency (RF) transmitter and receiver. Where the sensor will be having the RF transmitter and controller will be connected to RF receiver. The receiver will be receiving the data only under the particular frequency at which the transmitter will be sending and the readings are successful from large distance also and transmitter and receiver need not face each other for this to happen.

The drivers heart rate is measured with the help of the heart beat sensor. Some threshold limits are set in the controller so that if the sensor reads the heart rate of the person beyond the set limits, then the counter value will be incremented automatically from two which was set because of MEMS, the counter value now will be incremented to three. As soon as this happens, an SMS alert will be sent to the owner and all other registered users along with cops. The speed of the vehicle will be now dropped to zero slowly alerting the driver again with buzzer and the surrounding vehicles. The vehicles surrounding will receive the information regarding the state of the driver and speed of the vehicle, as the speed is dropping the remaining vehicles must keep up the safe distance to avoid accidents.

4.2 Headlight Intensity Controlling

All the above features in the project mainly represent automation for preventing accidents by sudden stopping of the vehicle on the road and also help in preventing false alarms. In addition to the above-automated features, the project is also having smartness incorporated in it. The brightness of the headlights of the car can be controlled with the intensity of the light in the surrounding environment of the vehicle. This mainly has three modes, namely day mode, night mode, and less intensity mode.

Whenever the vehicle is in the bright day light, there is no need of the headlights of the car to be in on-state. So, during the day mode, the headlights will be in off-state. In other words, zero brightness. When the vehicle is moving in during night, the surrounding intensity of the light around the vehicle will be very less. As a result, the headlights will be in on-mode, with full brightness. Now, when the vehicle is moving at night, the drivers eyes will be blurred for some nick of time because of the brightness of the headlights from the vehicle coming in opposite direction. The driver will not be able to see the road properly during this period and which may sometimes lead to a very drastic accident causing lot of loss for both life and property. Under such condition, the LDR sensor which will be placed in front of the car comes to use. Depending upon the light intensity of the opposite vehicle headlight, the brightness of the headlights is reduced (but will not be off completely) so that both the vehicle drivers eyes are not blurred. This makes the road visible to them all the time as the headlights brightness will not be turned off completely but will be reduced to 30% till the opposite vehicle passes and later will again increase the brightness to 100%. This leads to safe and secure driving.

The hardware module of the project along with the MEMS sensor with its transmitter is as shown in Figs. 2 and 3, respectively.

5 Experimental Results

When the supply is given to the kit, the LCD will be displaying a welcome message as shown below in Figs. 4 and 5.

Initially IR sensor does not read any input but whenever a person appears in front of IR sensor, it reads input from the heat radiated from the human body. Once this sensor reads input, then the activation of the remaining sensors will take place and an SMS alert will be sent to owner regarding the same.

Fig. 2 Hardware module

Fig. 3 Hardware module of MEMS transmitter

Fig. 4 LCD display to the project

Fig. 5 LCD display with welcome message

When the alcohol sensor detects the readings from the driver breath, an alert signal will be given. Buzzer will be activated along with message will be sent to the subscribed users. The count in the program will be incremented by one and the speed of the vehicle will be dropped to 70% of the total speed. The output of the sensor through LCD display is as shown in Fig. 6.

After the increment of the counter to one, then the counter will be incremented again if the reading or the alert is detected from a different sensor but not from the alcohol sensor again.

If the MEMS sensor detects the drowsiness next, then similar to above detection, LCD display alert will be given with increment in the counter value to two. The speed of the vehicle will be dropped to 30% and SMS alert will be sent to the subscribed users. The output of the MEMS sensor when the tilt is detected is shown through LCD display as per Fig. 7.

After this for the detection of the heart rate, the device will ask to put your finger into the place and will calculate the heart rate. If the heart rate of the person is either above or below the prescribed value, as per the code an alert message will be displayed, and the buzzer will be triggered, and SMS will be sent to the subscribed users. Figure 8 shows the different steps it follows while calculating the heart rate of the person along with output displayed in LCD.

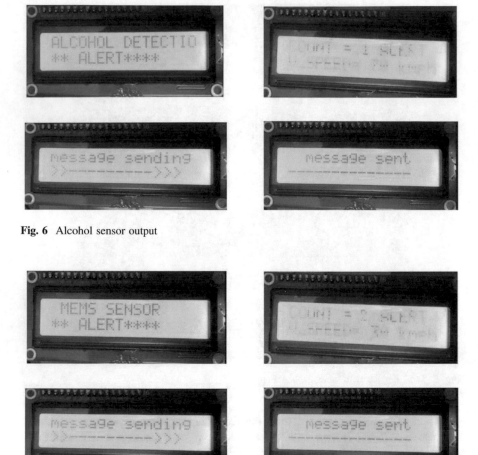

Fig. 6 Alcohol sensor output

Fig. 7 MEMS sensor output

Now, when the intensity of the light around the LDR sensor is too high similar to day mode then the headlights will be off-mode which will be displayed in the LCD sensor as follows in Fig. 9.

When the intensity of the surrounding light around the sensor starts decreasing then the headlights will be turned on and the brightness of the headlights will be a little less as the surroundings are not complete dark. Figure 10 shows the LCD display which will be showing the current brightness of the headlights.

Fig. 8 Heartbeat sensor output

Fig. 9 Day mode display

Fig. 10 Headlights with 30% brightness

Fig. 11 Head lights with 100% brightness

When the surroundings are completely dark then the brightness of the headlights will be 100% automatically without any human intervention and the same will be displayed in the LCD as shown below in Fig. 11.

SMS received by the owner will be as in Fig. 12.

After successful compilation and execution of the code, the same is dumped into the virtual controller in Proteus software after building the circuit in the software and execution of the Proteus circuit is done where the possible outputs are displayed in virtual terminal as in Figs. 13 and 14.

6 Conclusion and Future Enhancement

Different sensors are integrated, and a single module is developed which provides the smart way of avoiding accidents under drunken and drowsiness condition of the driver. Several measures are taken to avoid the false alarms of alcohol detection which mostly occurs due to the dependency of alcohol detection on a single sensor. Because of this continuous and often use of the single sensor happens which results in wear and tear of the sensor and its life diminishes and will be giving inaccurate.

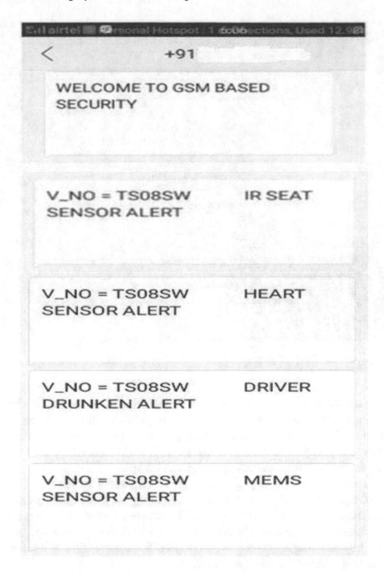

Fig. 12 SMS alerts sent to the subscribed user

In some cases, sensor will also be triggered because of some external interventions. All these problems are resolved in this project by the usage of different sensors like alcohol sensor, heartbeat sensor, and MEMS sensor.

This project also involves the process of saving power with the help of IR sensor. Only when the IR sensor detects the presence of person in the driver seat, it will trigger remaining sensors to active mode and they start working only then. With the detection of the drunken or drowsiness scenarios by each sensor, the speed of the vehicle will be locked so that the driver cannot go any speed further beyond

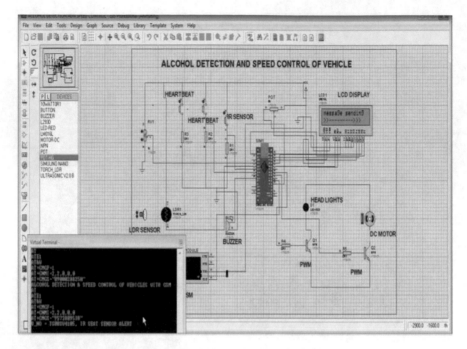

Fig. 13 Proteus virtual terminal output 1

the value set. Here, initially the speed will be locked to 70% and then to 30% with the increment of the counter value from 0 to 2, respectively. When counter is 3, then the vehicle will be stopped and throughout the process LCD and Buzzer is used to alert the surrounding vehicles and driver as well. GSM technology is also used to alert the subscribed users regarding driver condition.

In addition to the above process, headlight brightness of the vehicle is also controlled based on the light intensity of the opposite vehicles headlights so that the vision of the driver is not blurred so that accidents are avoided. The project is easily installed into any vehicle and is cost effective. Flexibility, portability, and scalability are key advantages present in this project. It has long life and the results are very accurate with low power consumption.

Enhancements can be done to this project by involving few extra features for security to prevent theft by using camera, finger print with various algorithms. The GSM technology used can be replaced by WI-FI technology which improves broadcast of the information to larger crowd with less expense. Ultrasonic sensors can be installed on four sides of the vehicles which will alert regarding the distance contraction between the vehicles and will try to avoid accidents.

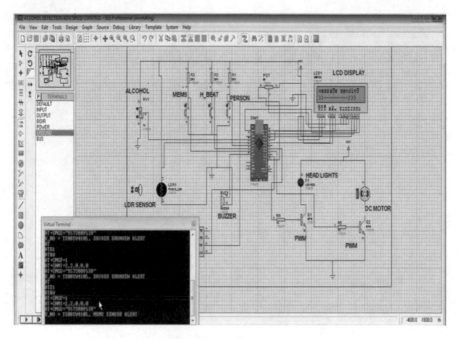

Fig. 14 Proteus virtual terminal output 2

References

1. Choi, Y., Han, S.I., Kong, S.-H., Ko, H.: Driver status monitoring systems for smart vehicles using physiological sensors—a safety enhancement system from automobile manufacturers. IEEE Mag. Sig. Process. Smart Veh. Technol. (2016)
2. Dhivya, M., Kathiravan, S.: Hybrid driver safety, vigilance and security system for vehicle. In: IEEE Sponsored 2nd International Conference on Innovations in Information Embedded and Communication Systems ICIIECS'15 (2015)
3. Pughazendi, N., Sathishkumar, R., Balaji, S., Sathyavenkateshwaren, S., Subash Chander, S., Surendar, V.: Heart attack and alcohol detection sensor monitoring in smart transportation system using internet of things. In: IEEE International Conference on Energy, Communication, Data Analytics and Soft Computing (ICECDS-2017) (2017)
4. Malathi, M., Sujitha, R., Revathi, M.R.: Alcohol detection and seat belt control system using Arduino. In: IEEE International Conference on Innovations in Information Embedded and Communication Systems (ICIIECS) (2017)
5. Parakkal, P.G., Sajith Variyar, V.V.: GPS based navigation system for autonomous car. In: IEEE International Conference on Advances in Computing, Communications and Informatics (2017)
6. Kodire, V., Bhaskaran, S., Vishwas, H.N.: GPS and ZigBee based traffic signal preemption. In: IEEE International Conference on Inventive Computational Technologies (2016)
7. Hu, J., Xu, L., He, X., Meng, W.: Abnormal driving detection based on normalized driving behaviour. IEEE Trans. Veh. Technol. **66**(8) (2017)
8. Vishal, D., Afaque, H.S., Bhardwaj, H., Ramesh, T.K.: IoT-driven road safety system. In: International Conference on Electrical, Electronics, Communication, Computer and Optimization Techniques (2017)

9. Chowdhury, A., Shankaran, R., Kavakli, M., Haque, M.M.: Sensor applications and physiological features in drivers' drowsiness detection: a review. IEEE Sens. J. **18**(8) (2018)
10. Sandeep, K., Kumar, P.R., Ranjith, S.: Novel drunken driving detection and prevention models using Internet of Things. In: International Conference on Recent Trends in Electrical, Electronics and Computing Technologies (2017)
11. Charniya, N.N., Nair, V.R.: Drunk driving and drowsiness detection. In: 2017 IEEE International Conference on Intelligent Computing and Control (I2C2) (2017)

Toward Secure Quantum Key Distribution Protocol for Super Dense Coding Attack: A Hybrid Approach

R. Lalu Naik, Seelam Sai Satyanarayana Reddy and M. Gopi Chand

Abstract The essential objectives of security, for example, verification, privacy, respectability and non-denial in correspondence systems are able to accomplish by safe input circulation. Quantum instruments are exceptionally safe methods for conveying mystery key because they are genuinely safe. QKDP is able to adequately counteract different assaults in Quantum Channel (QC) at the same time Traditional Cryptography (TC) is effective into confirmation plus check of mystery key. Through consolidating together QC plus TC safety of correspondences more than systems be able to exist utilized. Inside this manuscript, we suggest another toward secure quantum key appropriation convention for too thick coding assault. Since our plan be recognized because HQKDP.

Keywords Quantum cryptography · Active attacks · Passive attacks · Super dense coding attack

1 Introduction

Quantum Cryptography is to be verified utilizing security systems, for example, verification, classification, uprightness and non-renouncement. The current arrangements are of two sorts in particular customary or TC plus QC. Once more, the TCs are isolated keen on symmetric plus awry model. The safety of open key crypto-systems relies upon the computational multifaceted nature. Present be rejection assurance to the safety of it can't be broken down. Generally, novel option is QKC which depends on QC with the purpose of creating genuinely safe methods for key circulation instruments. In this manner, QKDP is able to forestall different assaults in QC as TC is effective inside confirmation plus check of mystery key.

R. Lalu Naik (✉) · M. Gopi Chand
Vardhaman College of Engineering, Information Technology, Shamshabad, India
e-mail: rlalunaik519@gmail.com

S. S. S. Reddy
Vardhaman College of Engineering, Computer Science and Engineering, Shamshabad, India

© Springer Nature Singapore Pte Ltd. 2020
K. S. Raju et al. (eds.), *Data Engineering and Communication Technology*,
Advances in Intelligent Systems and Computing 1079,
https://doi.org/10.1007/978-981-15-1097-7_43

In TC, instance stamp [1–3] plus confront reaction systems [3–6] is utilized. These instruments are particularly utilized in three gathering input dispersion conventions as investigated inside [1, 4, 5]. Established cryptography experiences the accompanying issues which are portrayed at this time. Least of two correspondence surroundings are essential at what time they employ confront reaction systems. Presumption of timepiece harmonization is necessary at what time they employ timestamp intended for input conveyance which is not appropriate within the sight of conceivable assaults and eccentric deferrals.

All QKDPs referenced above depend on either security confirmation or hypothetical plan or physical usage. Be that as it may, they are not customized for HQKDP. In this paper, we proposed a half and half convention named HQKDP for SDC which utilizes both established and quantum cryptography to guarantee secure key dissemination. Our commitments in this paper are the following: We planned a quantum key distribution convention named hybrid quantum key distribution so as to incorporate the customary plus QC.

2 Preliminaries

2.1 CC (Classical Cryptography)

Established cryptography has been approximate for a long time. Essentially, it is of two sorts to be specific symmetric cryptography and hilter kilter cryptography. The present is input distribution issue in private key at the same time as the quality of open key relies upon its computational multifaceted nature. The three party input conveyance conventions that appeared give increasingly secure correspondence over systems. The three party input appropriation conventions use instance stamp and confront reaction systems so as to forestall replay assaults. The disadvantage of established cryptography incorporates that at what time they employ confront reaction components they utilize additional correspondence round. At what time they employ time-stamps, they welcome the issue of clock harmonization because it anything but down to earth arrangement in reality. Moreover, the conventional cryptographic techniques cannot recognize detached assaults like listening in. To beat these disadvantages, quantum cryptography appeared.

2.2 QC (Quantum Cryptography)

QC depends on quantum mechanics which is unequivocally safe. QC dispenses with assaults, for example, replay and listening in. By utilizing this, it is conceivable to lessen no.of surrounding of correspondence at what time contrasted and established so as to utilize confront reaction instruments. Present be a lot of obtainable QKDPs [1–5] that utilize quantum material science so as to disperse open dialogs plus sitting enters inside provably safe design. Additional data on top of the down to earth use of QC be able to exist establish inside area.

2.3 Incentive to Mix Move Toward

This manuscript is gone for investigating unequivocally secure correspondences in excessively thick coding. As of late Hwang, Lee and Li [1, 2, 6] abused both cryptographic ideal models, for example, established plus QC intended for provably safe interchanges to avert play again, man-in-the-center plus latent assaults.

3 Proposed HQKDP for Super Dense Coding

This is a straightforward so far supervised request of essential quantum mechanics [4]. Super Dense Coding (SDC) includes two parties traditionally identified as Alice and Bob; they will likely convey some traditional data from Alice to Bob. Expect Alice is in responsibility for established bits of data which she needs to send Bob, yet is just permitted to send a solitary qubit to Bob. SDC lets us know expect Alice and Bob at first divide a couple of qubits in the ensnared state.

$$|\psi\rangle = (|00\rangle + |11\rangle)/\sqrt{2} \tag{1}$$

Alice is at first in responsibility for first qubit, while Bob has responsibility for second qubit. By sending the single qubit in her proprietorship to Bob, incidentally, Alice can convey two bits of traditional data to Bob. Here is the technique she employs.

- It she needs to drive the small piece threads '00' to move up and down after that she does not anything at all to her qubit $\sigma_0 = \begin{bmatrix} 1 & 0 \\ 0 & 1 \end{bmatrix}$.

- If she needs to drive the small piece thread '00' to move up and down after that she not anything at all to her qubit $\sigma_3 = \begin{bmatrix} 1 & 0 \\ 0 & -1 \end{bmatrix}$.

- If she needs to send '10', then she applies the quantum not gate, $\sigma_1 = \begin{bmatrix} 0 & 1 \\ 1 & 0 \end{bmatrix}$ to her qubit.

- If she needs to send '11', then she applies the $_i\sigma_2 = \begin{bmatrix} 0 & -i \\ i & 0 \end{bmatrix}$ gate to her qubit.

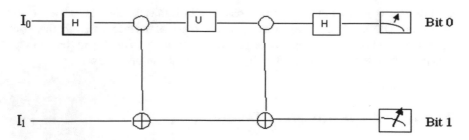

Fig. 1 Quantum circuit implements super dense coding

The four quantum entryways that are utilized here are pauli lattices [4, 5] σ_0, σ_1, $i\sigma_2$ and σ_3 mixes of them are reasonable as the Hadamard (H) task that Alice introduction on her half of the EPR matches or Bell states, in Fig. 1: quantum circuit executing very thick coding. The utilization of four distinct changes results in the four conditions of Bell [or] EPR states as in the accompanying conditions:

$$00 : |\psi\rangle \to (|00\rangle + |11\rangle)/\sqrt{2} \tag{2}$$

$$10 : |\psi\rangle \to (|00\rangle - |11\rangle)/\sqrt{2} \tag{3}$$

$$01 : |\psi\rangle \to (|01\rangle + |10\rangle)/\sqrt{2} \tag{4}$$

$$11 : |\psi\rangle \to (|01\rangle - |10\rangle)/\sqrt{2} \tag{5}$$

In Fig. 1, which has a Hadamard door pursued by a controlled NOT (CNOT) and changes the four computational premise states. As an open precedent, the Hadamard door takes the info and afterward the CNOT gives the yield state. First, the Hadamard change puts the top qubit in a wonderful location; this at that point goes about as a compose contribution to the CNOT, and the objective gets incorrect far up just when the sort out is 1.

Input $|00\rangle$ to $(|0\rangle + |1\rangle)|0\rangle/\sqrt{2} \to$ CNOT gives the yield state $(|00\rangle + |11\rangle)/\sqrt{2}$.
Information $|01\rangle$ to $(|0\rangle + |1\rangle)|1\rangle/\sqrt{2} \to$ CNOT gives the yield state $(|01\rangle + |10\rangle)/\sqrt{2}$.
Information $|10\rangle$ to $(|0\rangle - |1\rangle)|0\rangle/\sqrt{2} \to$ CNOT gives the yield state $(|00\rangle - |11\rangle)/\sqrt{2}$.
Information $|11\rangle$ to $(|0\rangle - |1\rangle)|1\rangle/\sqrt{2} \to$ CNOT gives the yield state $(|01\rangle - |10\rangle)/\sqrt{2}$.

Our proposed new convention of the HQKDP for SDC utilizes the maximally trapped Bell states (or) EPR states to encode the message bits unitary administrator (U) utilizing SDC that was referenced above and after that transmit them on quantum channels to the opposite side with less likelihood of the spying and more effectiveness by sending two established bits utilizing one (qubit) quantum bit.

3.1 Basic Design of HQKDP

$$\sigma_0 = \begin{bmatrix} 1 & 0 \\ 0 & 1 \end{bmatrix}, \sigma_1 = \begin{bmatrix} 0 & 1 \\ 1 & 0 \end{bmatrix}, i\sigma_2 = \begin{bmatrix} 0 & -i \\ i & 0 \end{bmatrix}, \sigma_3 = \begin{bmatrix} 1 & 0 \\ 0 & -1 \end{bmatrix} \quad (6)$$

Case-1, 2, 3, and 4 state changes after Alice's encoding on the fake qubits. The first column is the original states, second column is bit values, third column is combined operator (U), fourth column is Alice's encoding $|\Psi^0\rangle$ and last column is final output states. The first column and last column are Eve's entangled pairs. The second column is bit values of R_1 and B_1, respectively. The third column is the combined operation for Alice's encoding (Fig. 2; Table 1).

- In the event that the I-th bit estimations of 00 individually, as appeared in the second column on the off chance that 1, Alice's joined activity will be $(\sigma_0)|\varphi+\rangle$, and afterward the state will be changed into $|\varphi+\rangle$ after the encoding.

$$(\sigma_0 \otimes \sigma_0)|\Phi^+\rangle = \begin{bmatrix} 1 & 0 & 0 & 0 \\ 0 & 1 & 0 & 0 \\ 0 & 0 & 1 & 0 \\ 0 & 0 & 0 & 1 \end{bmatrix} * \sqrt{2} \begin{bmatrix} 1 \\ 0 \\ 0 \\ 1 \end{bmatrix} = \sqrt{2} \begin{bmatrix} 1 \\ 0 \\ 0 \\ 1 \end{bmatrix} = |\Phi^+\rangle \quad (7)$$

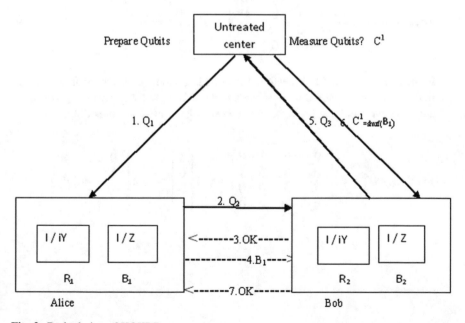

Fig. 2 Basic design of HQKDP

Table 1 Carrier bell state $|\Phi^+\rangle$

Original state	Bit values	Unitary operator (U)	Alice encoding ($|\Psi^0\rangle$)	Final output
Φ^+	00	σ_0	$(I\sigma_0 \otimes \sigma_0)\,|\Phi^+\rangle$	$|\Phi^+\rangle$
Φ^+	01	σ_1	$(\sigma_1 \otimes \sigma_0)\,|\Phi^+\rangle$	$|\Psi^+\rangle$
Φ^+	10	σ_3	$(\sigma_3 \otimes \sigma_0)\,|\Phi^+\rangle$	$|\Phi^-\rangle$
Φ^+	11	$i\sigma_2$	$(i\sigma_2 \otimes \sigma_0)\,|\Phi^+\rangle$	$|\Psi^-\rangle$

- In the event that the I-th bit estimations of 01 separately, as appeared in the third line on the off chance that 1, Alice's consolidated activity will be $(\sigma_1\sigma_0)|\varphi+\rangle$, and afterward the state will be changed into $|\psi+\rangle$ after the encoding.

$$(\sigma_1 \otimes \sigma_0)|\Phi^+\rangle = \begin{bmatrix} 0 & 0 & 1 & 0 \\ 0 & 0 & 0 & 1 \\ 1 & 0 & 0 & 0 \\ 0 & 1 & 0 & 0 \end{bmatrix} * \sqrt[1]{2} \begin{bmatrix} 1 \\ 0 \\ 0 \\ 1 \end{bmatrix} = \sqrt[1]{2} \begin{bmatrix} 0 \\ 1 \\ 1 \\ 0 \end{bmatrix} = |\Phi^+\rangle \qquad (8)$$

- In the event that the I-th bit estimations of 10 separately, as appeared in the third line on the off chance that 1, Alice's consolidated activity will be $(\sigma_1\sigma_0)|\varphi+\rangle$, and afterward the state will be changed into $|\psi+\rangle$ after the encoding.

$$(\sigma_3 \otimes \sigma_0)|\Phi^+\rangle = \begin{bmatrix} 1 & 0 & 0 & 0 \\ 0 & 1 & 0 & 0 \\ 0 & 0 & -1 & 0 \\ 0 & 0 & 0 & -1 \end{bmatrix} * \sqrt[1]{2} \begin{bmatrix} 1 \\ 0 \\ 0 \\ 1 \end{bmatrix} = \sqrt[1]{2} \begin{bmatrix} 1 \\ 0 \\ 0 \\ -1 \end{bmatrix} = |\Phi^-\rangle \qquad (9)$$

- In the event that the I-th bit estimations of 11 separately, as appeared in the third line on the off chance that 1, Alice's consolidated activity will bc $(\sigma_1\sigma_0)|\varphi+\rangle$, and afterward the state will be changed into $|\psi+\rangle$ after the encoding.

$$(i\sigma_2 \otimes \sigma_0)|\Phi^+\rangle = \begin{bmatrix} 0 & 0 & 1 & 0 \\ 0 & 0 & 0 & 1 \\ -1 & 0 & 0 & 0 \\ 0 & -1 & 0 & 0 \end{bmatrix} * \sqrt[1]{2} \begin{bmatrix} 1 \\ 0 \\ 0 \\ 1 \end{bmatrix} = \sqrt[1]{2} \begin{bmatrix} 0 \\ 1 \\ -1 \\ 0 \end{bmatrix} = |\Psi^-\rangle \qquad (10)$$

Table 2 Carrier bell state $|\Phi^-\rangle$

Original state	Bit values	Unitary operator (U)	Alice encoding ($|\Psi^0\rangle$)	Final output
$|\Phi^-\rangle$	00	σ_3	$(\sigma_3 \otimes \sigma_0)\,|\Phi^-\rangle$	$|\Phi^+\rangle$
$|\Phi^-\rangle$	01	$\sigma_1\sigma_3$	$(\sigma_1\sigma_3 \otimes \sigma_0)\,|\Phi^-\rangle$	$|\Psi^+\rangle$
$|\Phi^-\rangle$	10	σ_0	$(\sigma_0 \otimes \sigma_0)\,|\Phi^-\rangle$	$|\Phi^-\rangle$
$|\Phi^-\rangle$	11	$_i\sigma_2\sigma_3$	$(_i\sigma_2\sigma_3 \otimes \sigma_0)\,|\Phi^-\rangle$	$|\Psi^-\rangle$

- In the event that the I-th bit estimations of 00 separately, as appeared in the third line on the off chance that 1, Alice's consolidated activity will be $(\sigma_1\sigma_0)|\varphi+\rangle$, and afterward the state will be changed into $|\psi+\rangle$ after the encoding (Table 2).

$$(\sigma_3 \otimes \sigma_0)|\Phi^-\rangle = \begin{bmatrix} 1 & 0 & 0 & 0 \\ 0 & 1 & 0 & 0 \\ 0 & 0 & -1 & 0 \\ 0 & 0 & 0 & -1 \end{bmatrix} * \sqrt{2}\begin{bmatrix} 1 \\ 0 \\ 0 \\ -1 \end{bmatrix} = \sqrt{2}\begin{bmatrix} 1 \\ 0 \\ 0 \\ 1 \end{bmatrix} = |\Phi^+\rangle \quad (11)$$

- In the event that the I-th bit estimations of 01 separately, as appeared in the third line on the off chance that 1, Alice's consolidated activity will be $(\sigma_1\sigma_0)|\varphi+\rangle$, and afterward the state will be changed into $|\psi+\rangle$ after the encoding.

$$(\sigma_1\sigma_3\sigma_0) \otimes |\Phi^{-2}\rangle = \begin{bmatrix} 0 & 0 & -1 & 0 \\ 0 & 0 & 0 & -1 \\ 1 & 0 & 0 & 0 \\ 0 & 1 & 0 & 0 \end{bmatrix} * \sqrt{2}\begin{bmatrix} 1 \\ 0 \\ 0 \\ -1 \end{bmatrix} = \sqrt{2}\begin{bmatrix} 0 \\ 1 \\ 1 \\ 0 \end{bmatrix} = |\Psi^+\rangle \quad (12)$$

- In the event that the I-th bit estimations of 10 separately, as appeared in the third line on the off chance that 1, Alice's consolidated activity will be $(\sigma_1\sigma_0)|\varphi+\rangle$, and afterward the state will be changed into $|\psi+\rangle$ after the encoding.

$$(\sigma_0 \otimes \sigma_0)|\Phi^-\rangle = \begin{bmatrix} 1 & 0 & 0 & 0 \\ 0 & 1 & 0 & 0 \\ 0 & 0 & 1 & 0 \\ 0 & 0 & 0 & 1 \end{bmatrix} * \sqrt{2}\begin{bmatrix} 1 \\ 0 \\ 0 \\ -1 \end{bmatrix} = \sqrt{2}\begin{bmatrix} 1 \\ 0 \\ 0 \\ -1 \end{bmatrix} = |\Phi^-\rangle \quad (13)$$

- In the event that the I-th bit estimations of 11 separately, as appeared in the third line on the off chance that 1, Alice's consolidated activity will be $(\sigma_1\sigma_0)|\varphi+\rangle$, and afterward the state will be changed into $|\psi+\rangle$ after the encoding.

$$(_i\sigma_2\sigma_3\sigma_0) \otimes |\Phi^-\rangle = \begin{bmatrix} 0 & 0 & -1 & 0 \\ 0 & 0 & 0 & -1 \\ -1 & 0 & 0 & 0 \\ 0 & -1 & 0 & 0 \end{bmatrix} * \sqrt[1]{2} \begin{bmatrix} 1 \\ 0 \\ 0 \\ -1 \end{bmatrix} = \sqrt{2} \begin{bmatrix} 0 \\ 1 \\ -1 \\ 0 \end{bmatrix} = |\Psi^-\rangle$$

(14)

- In the event that the I-th bit estimations of 00 separately, as appeared in the third line on the off chance that 1, Alice's consolidated activity will be $(\sigma_1\sigma_0)|\varphi+\rangle$, and afterward the state will be changed into $|\psi+\rangle$ after the encoding (Table 3).

$$(\sigma_1 \otimes \sigma_0)|\Psi^+\rangle = \begin{bmatrix} 0 & 0 & 1 & 0 \\ 0 & 0 & 0 & 1 \\ 1 & 0 & 0 & 0 \\ 0 & 1 & 0 & 0 \end{bmatrix} * \sqrt[1]{2} \begin{bmatrix} 0 \\ 1 \\ 1 \\ 0 \end{bmatrix} = \sqrt[1]{2} \begin{bmatrix} 1 \\ 0 \\ 0 \\ 1 \end{bmatrix} = |\Phi^+\rangle \qquad (15)$$

- In the event that the I-th bit estimations of 01 separately, as appeared in the third line on the off chance that 1, Alice's consolidated activity will be $(\sigma_1\sigma_0)|\varphi+\rangle$, and afterward the state will be changed into $|\psi+\rangle$ after the encoding.

$$(\sigma_0 \otimes \sigma_0)|\Psi^+\rangle = \begin{bmatrix} 1 & 0 & 0 & 0 \\ 0 & 1 & 0 & 0 \\ 0 & 0 & 1 & 0 \\ 0 & 0 & 0 & 1 \end{bmatrix} * \sqrt[1]{2} \begin{bmatrix} 0 \\ 1 \\ 1 \\ 0 \end{bmatrix} = \sqrt[1]{2} \begin{bmatrix} 0 \\ 1 \\ 1 \\ 0 \end{bmatrix} = |\Psi^+\rangle \qquad (16)$$

Table 3 Carrier bell state $|\Psi^+\rangle$

Original state	Bit values	Unitary operator (U)	Alice encoding (Ψ^0) B	Final output
$|\Psi^+\rangle$	00	σ_1	$(\sigma_1\sigma_0)\,|\Psi^+\rangle$	$|\Phi^+\rangle$
$|\Psi^+\rangle$	01	σ_0	$(\sigma_0\sigma_0)\,|\Psi^+\rangle$	$|\Psi^+\rangle$
$|\Psi^+\rangle$	10	$_i\sigma_2$	$(_i\sigma_2\sigma_0)\,|\Psi^+\rangle$	$|\Phi^-\rangle$
$|\Psi^+\rangle$	11	σ_3	$(\sigma_3\sigma_0)\,|\Psi^+\rangle$	$|\Psi^-\rangle$

- In the event that the I-th bit estimations of 10 separately, as appeared in the third line on the off chance that 1, Alice's consolidated activity will be $(\sigma_1\sigma_0)|\varphi+\rangle$, and afterward the state will be changed into $|\psi+\rangle$ after the encoding.

$$(_i\sigma_2 \otimes \sigma_0)|\Psi^+\rangle = \begin{bmatrix} 0 & 0 & 1 & 0 \\ 0 & 0 & 0 & 1 \\ -1 & 0 & 0 & 0 \\ 0 & -1 & 0 & 0 \end{bmatrix} * \sqrt{2}\begin{bmatrix} 0 \\ 1 \\ 1 \\ 0 \end{bmatrix} = \sqrt{2}\begin{bmatrix} 1 \\ 0 \\ 0 \\ -1 \end{bmatrix} = |\Phi^-\rangle \quad (17)$$

- In the event that the I-th bit estimations of 11 separately, as appeared in the third line on the off chance that 1, Alice's consolidated activity will be $(\sigma_1\sigma_0)|\varphi+\rangle$, and afterward the state will be changed into $|\psi+\rangle$ after the encoding.

$$(\sigma_3 \otimes \sigma_0)|\Psi^+\rangle = \begin{bmatrix} 1 & 0 & 0 & 0 \\ 0 & 1 & 0 & 0 \\ 0 & 0 & -1 & 0 \\ 0 & 0 & 0 & -1 \end{bmatrix} * \sqrt{2}\begin{bmatrix} 0 \\ 1 \\ 1 \\ 0 \end{bmatrix} = \sqrt{2}\begin{bmatrix} 0 \\ 1 \\ -1 \\ 0 \end{bmatrix} = |\Psi^-\rangle \quad (18)$$

- In the event that the I-th bit estimations of 00 separately, as appeared in the third line on the off chance that 1, Alice's consolidated activity will be $(\sigma_1\sigma_0)|\varphi+\rangle$, and afterward the state will be changed into $|\psi+\rangle$ after the encoding (Table 4).

$$(\sigma_1\sigma_3\sigma_0)|\Psi^-\rangle = \begin{bmatrix} 0 & 0 & -1 & 0 \\ 0 & 0 & 0 & -1 \\ 1 & 0 & 0 & 0 \\ 0 & 1 & 0 & 0 \end{bmatrix} * \sqrt{2}\begin{bmatrix} 0 \\ 1 \\ -1 \\ 0 \end{bmatrix} = \sqrt{2}\begin{bmatrix} 1 \\ 0 \\ 0 \\ 1 \end{bmatrix} = |\Phi^+\rangle \quad (19)$$

- In the event that the I-th bit estimations of 01 separately, as appeared in the third line on the off chance that 1, Alice's consolidated activity will be $(\sigma_1\sigma_0)|\varphi+\rangle$, and afterward the state will be changed into $|\psi+\rangle$ after the encoding.

Table 4 Carrier bell state $|\Psi^-\rangle$

Original state	Bit values	Unitary operator (U)	Alice encoding (Ψ^0) B	Final output
$|\Psi^-\rangle$	00	$\sigma_1\sigma_3$	$(\sigma_1\sigma_3 \otimes \sigma_0)\,|\Psi^-\rangle$	$|\Phi^+\rangle$
$|\Psi^-\rangle$	01	σ_3	$(\sigma_3\sigma_0) \otimes |\Psi^-\rangle$	$|\Psi^+\rangle$
$|\Psi^-\rangle$	10	$_i\sigma_2\sigma_3$	$(_i\sigma_2\sigma_3\sigma_0) \otimes |\Psi^-\rangle$	$|\Phi^-\rangle$
$|\Psi^-\rangle$	11	σ_0	$(\sigma_0\sigma_0) \otimes |\Psi^-\rangle$	$|\Psi^-\rangle$

$$(\sigma_3\sigma_0) \otimes |\Psi^-\rangle = \begin{bmatrix} 1 & 0 & 0 & 0 \\ 0 & 1 & 0 & 0 \\ 0 & 0 & -1 & 0 \\ 0 & 0 & 0 & -1 \end{bmatrix} * \sqrt{2} \begin{bmatrix} 0 \\ 1 \\ -1 \\ 0 \end{bmatrix} = \sqrt{2} \begin{bmatrix} 0 \\ 1 \\ 1 \\ 0 \end{bmatrix} = |\Psi^+\rangle \quad (20)$$

- In the event that the I-th bit estimations of 10 separately, as appeared in the third line on the off chance that 1, Alice's consolidated activity will be $(\sigma_1\sigma_0)|\varphi+\rangle$, and afterward the state will be changed into $|\psi+\rangle$ after the encoding.

$$(_i\sigma_2\sigma_3\sigma_0) \otimes |\Psi^-\rangle = \begin{bmatrix} 0 & 0 & -1 & 0 \\ 0 & 0 & 0 & -1 \\ -1 & 0 & 0 & 0 \\ 0 & -1 & 0 & 1 \end{bmatrix} * \sqrt{2} \begin{bmatrix} 0 \\ 1 \\ -1 \\ 0 \end{bmatrix} = \sqrt{2} \begin{bmatrix} 1 \\ 0 \\ 0 \\ -1 \end{bmatrix} = |\Phi^-\rangle$$

$$(21)$$

- In the event that the I-th bit estimations of 11 separately, as appeared in the third line on the off chance that 1, Alice's consolidated activity will be $(\sigma_1\sigma_0)|\varphi+\rangle$, and afterward the state will be changed into $|\psi+\rangle$ after the encoding.

$$(\sigma_0 \otimes \sigma_0)|\Psi^-\rangle = \begin{bmatrix} 1 & 0 & 0 & 0 \\ 0 & 1 & 0 & 0 \\ 0 & 0 & 1 & 0 \\ 0 & 0 & 0 & 1 \end{bmatrix} * \sqrt{2} \begin{bmatrix} 0 \\ 1 \\ -1 \\ 0 \end{bmatrix} = \sqrt{2} \begin{bmatrix} 0 \\ 1 \\ -1 \\ 0 \end{bmatrix} = |\Phi^{-1}\rangle \quad (22)$$

4   Analysis

A Quantum framework whose states $|\psi\rangle$ is known intently is said to be in an unadulterated state. For this situation, the thickness administrator is essentially $p = |\psi\rangle\langle\psi|$. Something else, p is in a blended state; it is said to be a blend of the diverse unadulterated states in the troupe for p. In the activities, you will be approached to exhibit a straightforward paradigm for deciding if a state is unadulterated or blended: an unadulterated state fulfills tr($p2$) = 1, while a blended state fulfills tr($p2$) = 1. In the investigation of the above convention, it is funda-mental that the quantum channel utilized from Alice to bounce. This keeps an eve from getting to the quantum divert in one area and utilizing a similar strategy that ought to be utilized by sway to get unique message.

All the chime states utilized are unadulterated maximally snared states since in the event that we think about one of them; $|\psi\rangle = |00\rangle + |11\rangle/\sqrt{2}$, at that point its thickness lattice $p = |00\rangle\langle00| + |11\rangle\langle00| + |00\rangle\langle11| + |11\rangle\langle11|/2 = 1/2$. Since tr $(p2) = 1$, at that point this is an unadulterated state. The halfway follow over the first qubit is $p1 = 1/2$ since $\text{tr}((p1)2) = 1/2$ which is under 1, at that point the first qubit is in a blended state. On the off chance that the eve approaches quantum channel and makes an estimation, she gets $|0\rangle$ or $|1\rangle$ with likelihood 1/2 for each case.

5 Conclusions

In this paper, we examined the security components utilizing unadulterated maximally entrapped ringer states, for example, TC plus QC. Our investigation uncovered that QC can endure different sorts of assaults, for example, play again uninvolved assaults. We consolidated together the cryptographic techniques so as to use the upsides of together. We connected these instruments alongside a very thick coding so as to guarantee genuinely verify interchanges in excess of the system. The future plan is recognized because Hybrid QKDP. As we utilize 3 gatherings QKDP replica so as to decreases correspondence rounds to 3 other than diminishing the correspondence price and recollection utilization. Our expository investigation uncovered so as to the future methodology be able to give genuinely safe interchanges inside overly thick coding assault.

References

1. Li, G.: Efficient network authentication protocols: lower bounds and optimal implementations. Distrib. Comput. **9**(3), 131–145 (1995)
2. Wen, H.A., Lee, T.F., Hwang, T.: A provably secure three-party password-based authenticated key exchange protocol using Weil pairing. IEE Proc. Comm. **152**(2), 138–143 (2005)
3. Bellare, M., Rogaway, P.: Provably secure session key distribution: the three party case. In: Proceedings 27th ACM Symposium Theory of Computing, pp. 57–66 (1995)
4. Bennett, C.H., Brassard, G.: Quantum cryptography: public key distribution and coin tossing. In: Proceedings IEEE International Conference Computers, Systems, and Signal Processing, pp. 175–179 (1984)
5. Bennett, C.H.: Quantum cryptography using any two nonorthogonal states. Phys. Rev. Lett. **68**, 3121 (1992)
6. Hwang, W.Y., Koh, I.G., Han, Y.D.: Quantum cryptography without public announcement of bases. Phys. Lett. A **244**, 489–494 (1998)

RETRACTED CHAPTER: Data Transmission Based on Selection of Cluster Head Using M-RED Technique

Arjumand Sayeed, T. Nagalaxmi, P. Chandrasekhar and Satya Prasad Lanka

Abstract Wireless Sensor Networks are considered as basic technique for the social event of data; this will be executed in all organizations, for instance, helpful, opposition, auto-points of view, etc. Remote Sensor Networks generally include group head related to each and every other hub. The assurance of cluster head rule issue, in which the vitality essential, is careful in light of its propensity of social occasion the data's from neighbouring hubs. The group head needs the careful vitality so it can pass on the whole system information transmission to sink. For handling these issues, a novel calculation is proposed as Dynamic Energy Efficient Mid-point-Based Distance Aware figuring (DEE-M-DA). This is a vitality capable group decision part in the remote sensor arrangement. Basic fundamental is cluster head selection; this depends upon the standard of Midpoint-Residual Energy and Distance (M-RED) structure. K-medoid calculation is utilized to discover the mid-point of the hubs between the sink. This calculation is utilized to compel the division and send the information quickly. Smart information transmission has done among CH and sink. The proposed convention has been recreated utilizing NS-2 test structure and separated other existing conventions.

The original version of this chapter was retracted. The retraction note to this chapter is available at https://doi.org/10.1007/978-981-15-1097-7_82

A. Sayeed
Department of ECE, SCETW, Hyderabad, India
e-mail: arjumandsayeed137@gmail.com

T. Nagalaxmi (✉)
SCETW, Osmania University, Hyderabad, India
e-mail: tnagalaxmi@stanley.edu.in

P. Chandrasekhar
Osmania University, Hyderabad, India
e-mail: sekharpaidimarry@gmail.com

S. P. Lanka
SCETW, Hyderabad, India
e-mail: satyaprasadl@yahoo.com

Keywords Wireless sensor network · DEE-M-DA · M-RED · CH · Residual energy and distance · Clustering · K-medoids

1 Introduction

Wireless Sensor Networks (WSNs) can be defined as a self-configured and infrastructure-less wireless networks to monitor physical or environmental conditions, such as temperature, sound, vibration, pressure, motion or pollutants and to cooperatively pass their data through the network to a main location or sink where the data can be observed and analysed. WSN is a wireless network that consists of base stations and numbers of nodes (wireless sensors). These networks are used to monitor physical or environmental conditions like sound, pressure, temperature and co-operatively pass data through the network to a main location.

Nodes have been intended for coordinated correspondence among the diverse applications. Nodes have been structured depending on the applications, for example, the protection, therapeutic and shopper. Critical attributes of wireless sensor systems, as being self-ruling and involving little or small-scale devices, are accomplished to the detriment of strict accessible vitality-related impediments. To accomplish (to bring about result by effort; to bring to its goal or conclusion; to perform) such mindfulness, proper estimation of systems are required, empowering dependable and precise power utilization estimations of basic functionalities [1]. In sensor systems, nodes are versatile contingent upon their application necessities. Managing versatility can represent some considerable difficulties in protocol structure, especially, at the link layer [2].

Streamlining of vitality utilization has been a functioning exploration field for the most recent decades, and different methodologies have been investigated. Actually, a very much structured vitality utilization shows the establishment for creating and assessing a power board plot in system of vitality obliged gadgets, for example, WSN developed an optimal unified power control arrangements for vitality collecting WMSN furnished with photo-voltaic cells [3]. In numerous utilizations of WSN, the sink that is mounted on a versatile robot consistently moves over the observed territory to gather information from all sensor nodes [4]. Sensor network organizes the high thickness of nodes conveyance that will result in transmission crash and vitality dispersal of repetitive information. To determine the above issues, a vitality productive rest booking network with comparability measures to plan the network sensors into the dynamic or rest mode to decreases vitality utilization effectively [5]. The expansion of system lifespan turned into a fundamental matter in sensor networks [6].

Clustering instrument is one of the frameworks to save imperativeness in the WSNs. The data which is gathered from the cluster hub is passed to the sink for the further taking care of [7]. The choice criteria of the target work depend on the left over vitality, intra-cluster separate, node degree and head tally of the likely cluster heads [8]. Vitality collecting sources are different and include sun-powered, wind and vibratory sources [3]. Dynamic clustering calculation successfully wipes out contention among sensors and renders increasingly exact evaluations of target areas because of better quality information gathered and less collision incurred [8].

In the previous method, DEEDA algorithm, the essential standard is to pick the cluster head and energy consumption is high a direct result of more separation between Cluster head and sink. Here, we address the problem of DEEDA algorithm, by proposing the concept of a Mid-point-Residual Energy and distance from cluster head to sink (base station), named as M-RED. Here, mid-point calculation depends on K-medoids method. It is utilized to discover mid point of network nodes between the sink. The choice of the cluster head is done, and energy utilization is differentiated, and different calculations like LEACH, MLEACH and DEEDA algorithms send the data rapidly. Fast data transmission has done between CH and SINK. The proposed M-RED TECHNIQUE PROTOCOL is implemented using NS-2 test system and compared with other existing protocols.

The remaining of the paper composed as pursues. Sect. 2 gives a back-ground on LEACH technology, Sect. 3 explains the existing DEEDA algorithm; Sect. 4 explains the DEE-M-DA algorithm and Proposed Approach based on M-RED Technique; Sect. 5 gives NS-2 simulation results and parameters comparison results; and at last at Sect. 6, conclusions are drawn.

2 Literature Survey

A couple of different levelled steering conventions have been proposed for remote sensor arrangements over the most recent couple of years. A large number of them presented in multi-hop inter-cluster communication dealing with increment in the network lifetime, such as LEACH protocol [9]. Ring routing protocol, and DEEDA algorithms, etc. All these algorithms are used as energy-efficient algorithms. LEACH is the primary protocol of various levelled steerings which proposed information combination; it is of achievement centrality in clustering routing protocol. Various levelled protocols are utilized to decrease energy utilization by conglomerating information and to diminish the transmissions to base sink. LEACH protocol is a TDMA-based MAC design. The principle used is to expand network lifespan by bringing down levels of energy. LEACH protocol contains two rounds; each round will consist of cluster setup stage and steady phase. LEACH will not give any abstraction about the quantity of heads within the cluster in the network system. By using this process, clusters are divided randomly, due to this energy consumption is increased, and network life will be decreased. A multi-hop chart-based technique for a energy proficient steering (MH-GEER) convention intends to convey energy utilization between clustering at a fair rate and along these lines expand network life expectancies. MULTI-HOP GRAPH based approach for energy efficient routing protocol using k-meansalgorithm. MH-GEER deals with node clustering and inter-cluster multi-hop routing selection. The clustering stage is based upon the concentrated development of clusters, and the dispersed determination of cluster goes to that of low-energy versatile clustering chain of command (LEACH) [9]. (MSIEEP) to reduce the energy openings. To draw out WSN lifetime, the current clustering plans are equipped towards homogeneous WSN [10]. To increment the system lifetime, another protocol has been come into existence that is a Ring Routing protocol. This protocol is suitable for time sensitive applications. This structure is planned effectively; it will be available and effectively reconfigurable. The structure prerequisite of our protocol is to moderate the normal hotspot issue seen in hierarchical routing route, and it will limit the information revealing postponement by considering the different portability parameters of the mobile sink. One of the disadvantages of this strategy is more information traffic is concentrating towards the sink [11].

DEEDA algorithm will outflank as energy utilization instrument for the choice of cluster head. The cluster head assurance relies upon the RED (Residual Energy and Distance) measures. DEEDA algorithms chips away at two diverse phases, i.e. The Phase I manages picking head of the cluster by utilizing the respective rules that depends on energy and distance. For the picking of choice heads, rank framework is executed. Stage II manages the synchronization between the head of the cluster with alternate nodes dependent on the distance [7].

3　Dynamic Energy Efficient Mid-Point Based Distance Aware (DEE-M-DA) Algorithm

In previous method (DEEDA), the determination of cluster head depends on two phases and the cluster head assurance relies upon the RED (Residual Energy and Distance) measures. DEEDA algorithms done at two diverse phases, i.e. The Phase I manages picking head of the cluster by utilizing the respective rules that depends on energy and distance. For the picking of choice heads, rank framework is executed. Stage II manages the synchronization between the head of the cluster with alternate nodes dependent on the distance [7] and energy utilization is high because of more distance between cluster head and sink. Here, we address the problem of DEEDA algorithm, by proposing the concept of a Mid-point-Residual Energy and distance from cluster head to sink (base station), named as M-RED. Here, mid-point calculation depends on K-medoids method. It is utilized to discover the mid-point of the nodes between the sink using Dynamic Energy Efficient Mid-point based Distance Aware (DEE-M-DA) for the energy adequate cluster choice components in the Sensor Networks. During clustering process, the nodes will compare their information with each other and designate the cluster heads. Once the cluster head is selected, then cluster head node intimates its selection to all neighbouring nodes, and remaining nodes join as member. The difference between clustering algorithms is distance and energy. The K-medoids calculates distance between the sensor nodes and sink node and chooses cluster heads.

Cluster Head Selection

Though there are various routing protocols based on previous analysis, LEACH is the guide for sensor network protocols. Hierarchical routing aims at making sure less energy is used by sensor nodes by using less number of multi-hops in a cluster and to gather data and fuse it in a way to degrade the count of transmitted messages moving to sink as the formation of cluster is dependent on energy saved by sensors and its adjacency to cluster head. Such routing is basically given for wireless networks and is in preference to scalability and effective communication which is used in WSN. Here, nodes with higher energy are used for processing and transferring information while nodes with low energy are used for sensing the occurrence of target. The system performance as scalability, lifespan and energy usage is given by clusters and the tasks performed by them. Transmission of data regularly will use more energy hence the task of CH is given to other sensor nodes at regular intervals as to have uniform distribution of energy. LEACH performs in two different stages: steady phase where data is sent to BS and time taken to transmit is greater than setup phase so as to minimize overhead.

Nodes residual energy level calculation taken based on

$$E(\text{resd}) = \sum_{i=0}^{n} E(\text{init}) - E(\text{cons}) \tag{1}$$

where $E(\text{resd})$ = residual energy of nodes; $E(\text{init})$ = initial energy provided to each node; $E(\text{cons})$ is consumed energy of nodes.

The distance can be calculated using Euclidean distance formula $\sqrt{(x_i - x_j)^2 + (y_j - y_i)^2}$.

Post calculations of M-RED technique based on Eqs. 1 and 2, the nodes elect the CHs among them which spread information to non-CHs for formation of cluster. Non-CH chooses a CH that can come to expend the less energy for communication. Cluster head choice is explained in the Algorithm 1. This algorithm is utilized to minimize the distance and send the data rapidly. Fast data transmission can be done between CH and SINK. The proposed protocol has been mimicked utilizing NS-2 simulator.

Algorithm 1: Cluster Head Selection-CHS Based on M-RED

Step1: choose first medoids

1.1 Use Euclidean distance as a divergence measure computes distance at each pair as follows:

$$\left\{ d_{ij} = \sqrt{\sum_{a=1}^{p} (X_{ia} - X_{ja})^2}\, i = 1, \ldots, n;\ j = 1, \ldots, n \right\} \quad (1)$$

X_{ia} ———— $\rightarrow$ first point (x-topology) representation
X_{ja} ———— $\rightarrow$ second point (y-topology) represenation
a– $\rightarrow$ initial value
p— > ending value
d_{ij}- $\rightarrow$ euclidean distance

1.2 Compute P_{ij} for making an initial guess at the centers of the clusters.

$$\left\{ P_{ij} = \frac{d_{ij}}{\sum_{l=1}^{n} d_{il}}\, i = 1, \ldots, n;\ i = 1, \ldots, n \right\} \quad (2)$$

P_{ij} – $\rightarrow$ medoid point

1.3 At each object, calculate $\sum_{l=1} d_{ij}(j = 1, \ldots, n)$ and arrange them in ascending order. Select objects as initial group medoids which are having the minimum value.

1.4 Set to the nearest medoid for each object.

1.5 Determine the summation of distance from all objects to their medoids i.e., the current optimal value.

Step2: Find out new medoids

The current medoid minimizes the total distance to other objects in its cluster is replaced by the object in each cluster.

2.1 To the nearest new medoid, each object is assigned.

2.2 Decide the total of separation from all articles to their new medoids which is known as the new ideal esteem. On the off chance that the ideal esteem is equivalent to the past one, at that point stop the calculation else, return to the Step 2.

4   Implementation Process

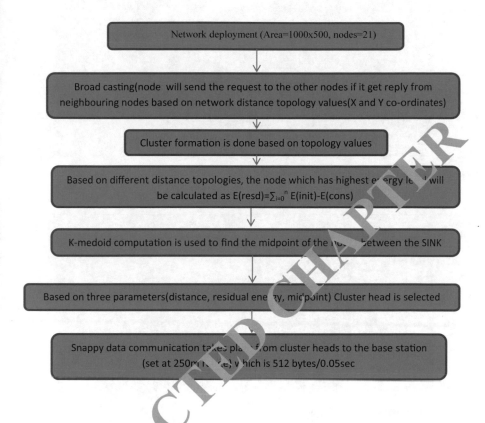

5   Experimental Results

Here, the numbers of sensor nodes are considered as 21 which are arbitrarily distributed over a 1000×500 m^2 field. In NS-2 simulator, we accept no such gap exists in the detecting field, and fixed sensors are the same in their abilities. We accept that the sink is located in the mid-point of network.

Figures 1 and 2 shows all nodes placed in network deployment. All nodes are displayed based on topology values, and all properties of Network Analysis Module must be mentioned. Twenty one nodes have taken for the formation of the WSN, considering 21 sensor nodes which are arbitrarily distributed over a 1000×500 m^2 field, in which one of the nodes is taken as base station (sink). Initially, each node will have 100% energy. The broadcasting occurs throughout the network. Here, broadcasting occurs for communication purpose. All nodes should be involved in this process. All the nodes will send route request if it gets the route reply then the

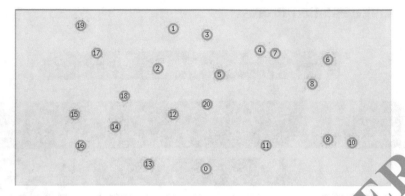

Fig. 1 Network deployment

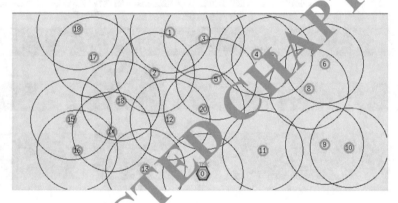

Fig. 2 Broadcasting process in network

data transmission will be started from node to node, and one of the nodes will be taken as base station (SINK).

Figures 3 and 4 shows node-to-node distance will be calculated and based on this cluster formation will be started which is based on the topology values. All cluster members participate in routing process. In this, CBR indicates traffic protocol and supports for data communication.

By using three parameters, i.e. energy, distance and mid-point, the node which has high energy, least distance and least mid-point when compared with all other nodes in a cluster, will be selected as cluster head as shown in Fig. 5. In Fig. 6, data communication starts from cluster head to sink node. By using k-medoid algorithm (see Algorithm 1), the data will be transmitted rapidly from cluster head to sink.

Nearest cluster head sends the information to base station and also send the data to base station. Snappy data transmission takes place between the cluster heads and base station (set at 250 m range), which is 512 bytes/0.05 s as shown in Figs. 7 and 8.

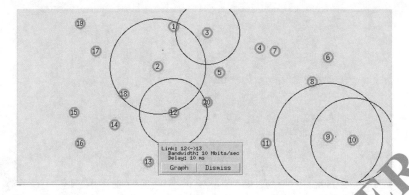

Fig. 3 Cluster formation starts

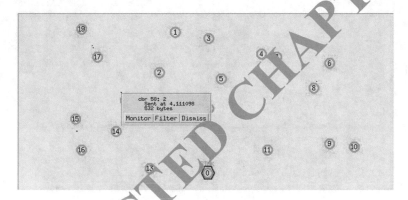

Fig. 4 Data transmission between cluster member and head

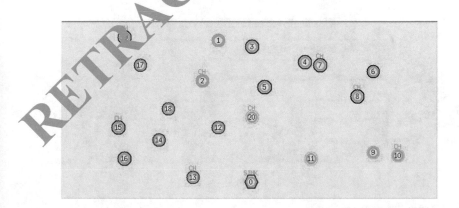

Fig. 5 All cluster heads in network

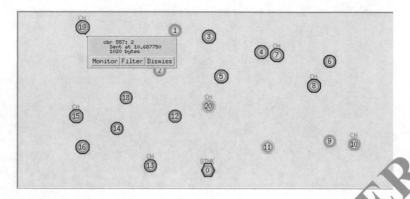

Fig. 6 Data communication starts from cluster head

Fig. 7 Sink receives the data from hop nodes

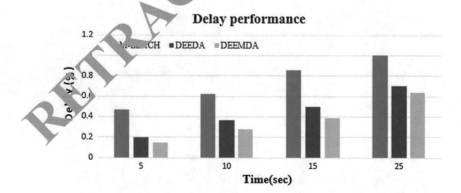

Fig. 8 Delay performance

Experimental results and appropriate analysis have been done on different parameters like delay,energy consumption, routing overhead and throughtput. Delay represents end2end delay, based on simulation time versus delay. The performance of DEE-M-DA mechanism reduces delay time between communication nodes compared with DEEDA and M-LEACH algorithm.

Analysis on energy consumption based on $E(\text{resd}) = \sum_{i=0}^{n} E(\text{init}) - E(\text{cons})$ Eq. (1) was been carried out and shown in Fig. 9 between simulation time versus energy. The performance of DEE-M-DA mechanism reduces the energy consumption compared with DEEDA algorithm and M-LEACH method.

Experimental results for routing overhead are shown in Fig. 10 between simulation time versus routing overhead. The performance of DEE-M-DA mechanism improves overhead compared with DEEDA algorithm and M-LEACH method.

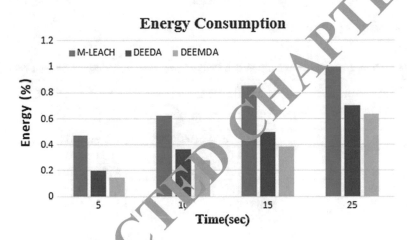

Fig. 9 Energy consumption

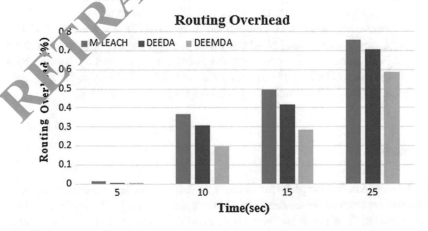

Fig. 10 Routing overhead

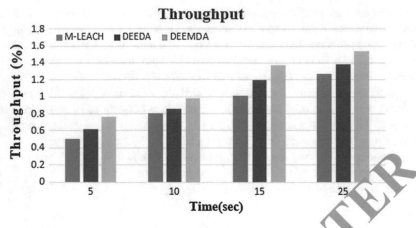

Fig. 11 Network performance

Table 1 Comparison between DEEDA and DEE-M-DA parameters gathered by sensor nodes

Parameter	DEEDA values	DEE-M-DA values
Transmission rate	–	512 bytes/0.05 s
Radio range	–	250 m
Packet size	512 and 1024 bytes	512 and 1024 bytes
Delay performance	At 25 s 0.7%	At 25 s 0.62%
Energy consumption	At 25 s 0.75%	At 25 s 0.61%
Routing overhead	At 25 s 0.7%	At 25 s 0.59%
Throughput	At 25 s 1.4%	At 25 s 1.58%

Analysis is carried out to find the throughput as shown in Fig. 11, simulation time versus throughput. The performance of DEE-M-DA algorithm improves the throughput compare to DEEDA algorithm and M-LEACH method.

Table 1 includes parameters comparison between DEEDA and DEE-M-DA gathered by sensor nodes using NS-2 test framework. So, as to improve scheduling for cluster head selection, we accept that the information gathered by sensor nodes is the referral tolerant information. Here, we used 512 and 1024 bytes for each data transmission process. The simulation of our network process is 25 s.

6   Conclusion

DEE-M-DA algorithm outperforms in terms of data transmission from cluster head to base station from other existing algorithms named as DEEDA and M-LEACH. The proposed method is used to transmit accurate data between cluster head to sink. Here, mid-point calculation is helpful to optimaze position of the sink and cluster heads. This algorithm is used to increment the network lifespan. The output performance improved, and fast data transmission has been achieved between cluster head and the base station.

References

1. Antonopoulos, C., Prayati, A., Stoyanova, T., Koulamas, C., Papadopoulos, G.: Experimental evaluation of a WSN platform power consumption. In: IPDPS 2009—Proceedings of the 2009 IEEE International Parallel and Distributed Processing Symposium, 2009
2. Dong, Q., Dargie, W.: A survey on mobility and mobility-aware MAC protocols in WSN. IEEE Commun. Surv. Tutorials **15**(1), 88–100 (2013)
3. Koulali, M.A., Kobbane, A., El Koutbi, M., Tembine, H., Ben-Othman, J.: Dynamic power control for energy harvesting wireless multimedia sensor networks. Eurasip J. Wirel. Commun. Netw. **2012**, 1–8 (2012)
4. Tashtarian, F., Yaghmaee Moghaddam, M.H., Sohraby, K., Effati, S.: On maximizing the lifetime of wireless sensor networks in event-driven applications with mobile sinks. IEEE Trans. Veh. Technol. **64**(7), 3177–3189 (2014)
5. Wan, R., Xiong, N., Loc, N.T.: An energy-efficient sleep scheduling mechanism with similarity measure for wireless sensor networks. Human-centric Comput. Inf. Sci. **8**(1) (2018)
6. Singh, B., Lobiyal, D.K.: A novel energy-aware cluster head selection based on particle swarm optimization for wireless sensor networks. Human-centric Comput. Inf. Sci. **2**(1), 1–18 (2012)
7. Kumar, M.V.M., Chaparala, A.: Dynamic energy efficient distance aware protocol for the cluster head selection in the wireless sensor networks. In: RTEICT 2017—2nd IEEE International Conference on Recent Trends in Electronics, Information & Communication Technology Proceedings, vol. 2018, Jan, pp. 147–150 (2017)
8. Chen, W.-P., Hou, J., Sha, L.: Dynamic clustering for acoustic target tracking in wireless sensor networks. In: ICNP 2003
9. Rhim, H., Tamine, K., Abassi, R., Sauveron, D., Guemara, S.: A multi-hop graph-based approach for an energy-efficient routing protocol in wireless sensor networks. Human-centric Comput. Inf. Sci. **8**(1) (2018)
10. Javaid, N., et al.: An energy-efficient distributed clustering algorithm for heterogeneous WSNs. Eurasip J. Wirel. Commun. Netw. **1**, 2015 (2015)
11. Tunca, C., Isik, S., Donmez, M.Y., Ersoy, C.: Ring routing: an energy-efficient routing protocol for wireless sensor networks with a mobile sink. IEEE Trans. Mob. Comput. **14**(9), 1947–1960 (2014)

Machine Translation Evaluation: Manual Versus Automatic—A Comparative Study

Kaushal Kumar Maurya, Renjith P. Ravindran, Ch Ram Anirudh
and Kavi Narayana Murthy

Abstract The quality of machine translation (MT) is best judged by humans well versed in both source and target languages. However, automatic techniques are often used as these are much faster, cheaper and language independent. The goal of this paper is to check for correlation between manual and automatic evaluation, specifically in the context of Indian languages. To the extent automatic evaluation methods correlate with the manual evaluations, we can get the best of both worlds. In this paper, we perform a comparative study of automatic evaluation metrics—BLEU, NIST, METEOR, TER and WER, against the manual evaluation metric (*adequacy*), for English-Hindi translation. We also attempt to estimate the manual evaluation score of a given MT output from its automatic evaluation score. The data for the study was sourced from the Workshop on Statistical Machine Translation WMT14.

Keywords Machine translation (MT) · MT evaluation · Manual metrics · Automatic metrics

1 Introduction

Machine translation (MT) deals with the conversion of natural language texts from one language to another using computers. Developing techniques to adequately judge the quality of machine translation has been a major concern in the MT

K. K. Maurya (✉) · R. P. Ravindran · C. R. Anirudh · K. N. Murthy
School of Computer and Information Sciences, University of Hyderabad, Hyderabad, India
e-mail: kaushalmaurya94@gmail.com

R. P. Ravindran
e-mail: rpr@uohyd.ac.in

C. R. Anirudh
e-mail: ramanirudh28@gmail.com

K. N. Murthy
e-mail: knmuh@yahoo.com

© Springer Nature Singapore Pte Ltd. 2020
K. S. Raju et al. (eds.), *Data Engineering and Communication Technology*,
Advances in Intelligent Systems and Computing 1079,
https://doi.org/10.1007/978-981-15-1097-7_45

research community. The main concern is whether the meaning of the source sentence is properly preserved in the target sentence. Therefore, the quality of machine translation output is best judged by a human, well versed in both source and target languages. However, like any other cognitive task, manual evaluation of MT is tedious, time-consuming, expensive and can be inconsistent. The general tendency, therefore, is to look for automatic techniques for evaluation. However, automatic evaluation of translation is hard, as computers are incapable of judging the meaning directly. Automatic evaluation metrics try to indirectly capture the meaning by comparing MT output with professional translations of the source sentence, called *reference translations*. Automatic techniques mainly rely on string comparisons [5], and it is well known that they fail to objectively judge the meaning conveyed in all cases. However, automatic techniques, being fast and consistent, can be used to track the progress in MT system development on a fixed data set. Automatic metrics, in spite of their limitations, are also used today by the MT community to judge and compare the performance of MT systems. Correlation studies on automatic and manual metrics show that automatic metrics can be useful in practice [5], although it is hard to adequately judge the quality of translations.

Adequacy and *fluency* are two of the widely used manual evaluation metrics today. *Adequacy* measures how much of the information in the source sentence is preserved in the translation [28]. *Fluency* measures how good the translation is with respect to the target language quality in terms of intelligibility and flow [28]. One or more human annotators score the output produced by an MT system using these metrics. On the other hand, automatic metrics like BLEU [23] and NIST [10] score the MT output based on lexical similarity with reference translations. Lexical similarity is computed through n-gram statistics and therefore is sensitive to word ordering and synonymy. Other automatic metrics like TER [25] and METEOR [1] try to address these issues to a certain degree.

It would be great if we can exploit the advantages of automatic evaluation and at the same time get a better feel for the actual quality of translations. The goal of this paper is to check for correlation between manual and automatic evaluation, specifically in the context of Indian languages. To the extent automatic evaluation methods correlate with the manual evaluations, we can get the best of both worlds. In this paper, we perform a comparative study of automatic evaluation metrics— BLEU, NIST, METEOR, TER and WER [26], against the manual evaluation metric (*adequacy*), in the context of English-Hindi translation. We also attempt to estimate the manual evaluation score of a given MT output from its automatic evaluation score. The data for the study was sourced from the Workshop on Statistical Machine Translation WMT14 [3].

2 Literature Survey

Intelligibility and *fidelity* were the first two manual evaluation metrics used by Automatic Language Processing Advisory Committee (ALPAC) [6]. In early 1990s, Advanced Research Projects Agency (ARPA) proposed three manual evaluation metrics, viz. *adequacy, fluency* and *comprehension* [28] in different MT evaluation campaigns. Few more extended criteria such as *suitability, interoperability, reliability, usability, efficiency, maintainability* and *portability* were discussed by King et al. [18]. A task-oriented metric was developed by White et al. [29] which can be used to judge whether an MT system is suitable for a given task such as publishing, gisting or extraction. Farrús et al. [12] and Costa-jussà [8] proposed an objective method for manual evaluation which takes various linguistic aspects like orthography, morphology, syntax and semantics into account. They provide guidelines to classify the MT output errors. In recent evaluation campaigns of WMT [4], *segment ranking* is used where judges rank the sentences from different systems according to their quality. This gives a relative scoring between the MT systems, and to choose the best MT system. This cannot be used to judge the quality of MT output as such. One of the problems with manual evaluation is that different human evaluators may disagree on the scores for the same MT output. The results may be subjective and irreproducible. Also, human evaluators are expensive in terms of money and time.

The core idea behind automatic evaluation is: *"the closer the machine generated translation is to a professional human translation, the better it is"* [23]. Automatic evaluation techniques use lexical similarity measures like the edit distance and overlap in n-gram sequences for measuring the closeness between machine output and a reference translation. Translation Error Rate (TER) and Word Error Rate (WER) [26] are edit-distance-based metrics, and BLEU, NIST, METEOR, etc., are n-gram sequence-based metrics. Edit distance is concerned with the calculation of minimum number of edit operations required to transform a given translation into a reference translation. WER finds the proportion of *insertions, substitutions* and *deletions* in the output with respect to reference translation. Hence, the higher the WER, the lower is the performance of MT system. It counts reordering between words as *deletions* and *insertions*, increasing the WER score. TER overcomes this problem, by also considering *shift* in addition to the above three operations.

BLEU is one of the most widely used automatic metrics today. The BLEU score is calculated by taking the product of the geometric mean of modified n-gram precision scores with brevity penalty. Brevity penalty penalizes the score if the output sentence is shorter than the reference sentence. The drawback of BLEU is that it gives equal weightage to all words and it fails if exact n-grams are not present in the reference translations [5].

NIST is a modification of BLEU. It uses arithmetic mean of n-gram matching precision scores instead of geometric mean. This is to avoid the nullification of the score if one or more n-grams do not match with the reference. The precision scores are weighted by information weights that heavily weigh infrequently occurring n-grams. Further, this approach modifies the brevity penalty so that small variations in sentence length do not affect the score much.

METEOR [1] is based on the matching of unigrams, matching of morphological variants based on their stems and matching synonyms. METEOR requires linguistic resources such as morphological analyzers for stemming and WordNet [21] for matching synonyms. For Indian languages, Gupta et al. [17] proposed an automatic evaluation metric *METEOR-Hindi* for Hindi as target language which is a modified version of original METEOR (uses Hindi-specific language resources).

Giménez et al. [16] proposed a metric that involves various linguistic features from shallow syntactic similarity, dependency parsing, shallow semantic similarity and semantic roles. They later extended it to include discourse representation structures also [15].

Gautam et al. [14] proposed another automatic metric called "LAYERED," based on three layers of NLP: lexical, syntactic and semantic. In lexical layer, BLEU is considered as the baseline metric. Syntactic layer focuses on reordering of sentences. Semantic layer uses features from a dependency parse of the sentence.

More recently proposed, COBALTF [13] uses target *language models* (LM)-based features to measure fluency of candidate translations. Features have been classified as adequacy-based and fluency-based features. Adequacy features are based on counts of words and n-grams aligned in the target and reference translations. Fluency-based features rely on the LM probability of candidate translation and reference translation, linguistic features like POS information, percentage of content/function words, etc.

3 Setup of the Experiments

3.1 Source of Data

All data required for our experiments were taken from the English-Hindi translation task of the Workshop on Statistical Machine Translation, 2014 edition (WMT14) [3]. In this translation task, participants were required to translate a shared test set. WMT14 hosted ten translation tasks—between English and each of Czech, French, German, Hindi and Russian in both directions. The more recent editions of WMT have not included the English-Hindi translation task; we could therefore source data only from WMT14. The test set for the English-Hindi translation task in WMT14 had 2507 English sentences (source) along with corresponding translations in Hindi (reference). As manual evaluation is expensive, we have restricted our studies to a subset of this whole data. We have made a random selection of 450 source-reference pairs from the 2507 source-reference pairs available. For each of the 450 source-reference pairs, we selected corresponding system-outputs from three MT systems, out of a total of 12 systems that competed in the English-Hindi translation task in WMT14. Thus, we have a total of $450 \times 3 = 1350$ triples where each triple is <source, reference, system-output>. The three MT systems we have chosen are: online-B, IIT-BOMBAY(IIT-B) [11] and MANAWI-RMOOVE [27] which were ranked in English-Hindi task in WMT14 as 1st, 5th, and 9th, respectively. Our choice of systems was to ensure a representative range of MT output quality. The IIT-B system is a factored SMT system. MANAWI-RMOOVE system is an improvement over the MOSES toolkit [20], and it uses Multi-Word expression and Named-Entity recognition. Online-B system is an online machine translation service that was anonymized by WMT14, for which translations were collected by the WMT organizing committee.

3.2 Choice of Metrics

Manual evaluation metrics are supposed to capture two different aspects of translation quality: *Adequacy* and *Fluency*. There can be fluent translations that may not be adequate and there may be adequate translations that are not fluent. Preservation of meaning being the most essential requirement in translation, adequacy is more important. We use adequacy alone for all our manual evaluations. Each sentence is assigned a score ranging from 1 to 5 based on the criteria mentioned in Table 1 [19].

We use the following metrics for automatic evaluation: BLEU [23], NIST [10], METEOR [1], TER [25] and WER [26]. These were chosen as their open implementations are available on the Internet and also they do not require elaborate linguistic resources. METEOR-Hindi is tailor-made for Hindi as a target language, but an open implementation is not available and we were not able to source it from the authors. METEOR allows inclusion of linguistic resources as modules, but only a few of those are openly available for Hindi. Therefore, in our experiments, we have used only the unigram module and the synonymy module through the Hindi-WordNet [22].

Automatic evaluation scores for all the five metrics mentioned above were computed on the entire test corpus (1350 sentences). Both manual and automatic evaluations are performed at segment level. A segment is a unit of translation which is usually one or a few sentences [10].

Segment-level scores are computed using the script `mtevalv13a.pl`[1] which is a part of Moses tool kit and was used in WMT [4]. We used the tool `meteor-1.5`[2] [9] for computing METEOR scores. TER and WER scores were obtained using the tool `tercom.7.25`[3] which implements the original idea of TER proposed by Snover et al. [25]. TER and WER are error metrics with values

---

[1]https://github.com/moses-smt/mosesdecoder/blob/master/scripts/generic/mteval-v13a.pl.
[2]http://www.cs.cmu.edu/ ~ alavie/METEOR/.
[3]http://www.cs.umd.edu/ ~ snover/tercom/.

Table 1 Manual evaluation: adequacy

Scores	Adequacy
5	All meaning is preserved
4	Most meaning is preserved
3	Much meaning is preserved
2	Little meaning is preserved
1	None of the meaning is preserved

ranging from 0 to 100. Higher scores indicate worse quality. We subtract the TER and WER scores from 100 for making them consistent with other metrics.

3.3 Manual Evaluation Setup

Manual evaluation was done by nine bilingual annotators. None of our annotators are professional translators, and they are graduate-level students with Hindi as their first language and English their second language during their studies. The evaluation experiment is set up as follows: (1) Each annotator will annotate 300 system-outputs in two rounds with 150 sentences in each. (2) Each will get equal proportions from all three MT systems. (3) No two annotators will get same system-outputs from the same MT systems. (4) Every system-output will be annotated by exactly two annotators (for getting inter-annotator agreement).

Before the actual evaluations were conducted, a pilot run with two annotators was carried out to get a fair understanding of common mistakes and difficulties the annotators would face during evaluation. The findings of the pilot run are as follows: (1) Annotators were sometimes inconsistent in the way they judge the sentence—sometimes they look at the source sentence first, sometimes they look at the reference translations, sometimes they directly judge the output. (2) The annotators were using the English word to fill the meaning gap in case of untranslated words. (3) A few annotators did not take any breaks during the evaluation process. This can cause fatigue.

Before the actual evaluation, we discussed the above points with all nine annotators and gave the following instructions that would help them do a fair and consistent evaluation. (1) Read the source sentence and the reference translation before scoring system-output. (2) Untranslated words should simply be treated as untranslated words. (3) At least one break was made mandatory to avoid fatigue and boredom during the evaluation. For statistics regarding the manual evaluation refer to Table 2.

Table 2 Statistics of manual evaluation

Min./Max./Avg. Time taken per annotator	50/250/82 min
Min./Max./Avg. Time per sentence (overall)	0.33/1.66/0.54 min

Table 3 Interpretation of k-values for inter-annotator agreement

Kappa	Agreement
<0	Less than chance agreement
0.01–0.20	Slight agreement
0.21–0.40	Fair agreement
0.41–0.60	Moderate agreement
0.61–0.80	Substantial agreement
0.81–0.99	Almost perfect agreement
1	Perfect agreement

Table 4 K-values obtained for measuring inter-annotator agreement

MT system	#Sentences	k-values
Online-B	450	0.2366
IIT-Bombay	450	0.2327
MANAWI-RMOOVE	450	0.2821
All systems	1350	0.2884

Inter-Annotator Agreement: Inter-annotator agreement scores are a measure of reliability of manual evaluation. We measure the pairwise inter-annotator agreement for each system as well as for the whole data using kappa coefficient (k) [7].

$$k = \frac{P(A) - P(E)}{1 - P(E)} \tag{1}$$

where $P(A)$ is the proportion of times the annotators agree and $P(E)$ is proportion of times they would agree by chance. The range of k value lies between 0 and 1 where 1 indicates perfect agreement and 0 no agreement. Interpretation of the kappa coefficient (k) is shown in Table 3. Results (k-values) are shown in Table 4.

From Table 4, it can be concluded that there is a fair inter-annotator agreement in all cases. We take these manual evaluations as reliable.

As mentioned earlier, each segment is annotated by two annotators. Average of adequacy scores from both annotators is considered as final manual evaluation score [19].

3.4 Correlation: Manual Versus Automatic

To find the segment-level correlation between automatic and manual evaluation, we use the Pearson's rho (ρ) correlation coefficient and Kendall's tau (τ) rank correlation coefficient.

Pearson's Correlation Coefficient: Pearson's correlation coefficient [24] ρ is given by:

Table 5 Interpretation of
Pearson's ρ correlation
coefficient

Correlation	Negative	Positive
Small	−0.29 to −0.10	0.10–0.29
Medium	−0.49 to −0.30	0.30–0.49
Large	−1.00 to −0.50	0.50–1.00

$$\rho = \frac{\sum_{i=1}^{n} (H_i - \overline{H})(M_i - \overline{M})}{\sqrt{\sum_{i=1}^{n} (H_i - \overline{H})^2}\sqrt{\sum_{i=1}^{n} (M_i - \overline{M})^2}} \tag{2}$$

where H_i is the manual evaluation score of segment i. M_i is the automatic evaluation score of segment i. $\overline{H}$ and $\overline{M}$ are the average of manual and automatic scores, respectively. Range of ρ lies between −1 and +1 where $\rho = 1$ counts as perfect correlation, $\rho = 0$ is total independence between two evaluation scores, and $\rho = -1$ indicates very strong negative correlation. Interpretation of ρ value is given in Table 5.

Kendall's τ Rank Correlation Coefficient: The advantage of Kendall's τ rank correlation coefficient over Pearson's ρ is that it does not assume a normal distribution of data. But it needs the data to be ranked. We order the sentences based on their adequacy scores and use them to calculate Kendall's tau. We used a variant of Kendall's tau [2] called Kendall's tau b:

$$\tau_b = \frac{n_c - n_d}{\sqrt{(n_0 - n_1)(n_0 - n_2)}} \tag{3}$$

where $n_0 = n(n-1)/2$, $n =$ number of segments, $n_1 = \sum_i t_i(t_i - 1)/2$, $n_2 = \sum_j u_j(u_j - 1)/2$, $n_c =$ number of concordant pairs, $n_d =$ number of discordant pairs, $t_i =$ number of tied values in the ith group of ties for the first quantity and $t_j =$ number of tied values in the jth group of ties for the second quantity.

For a given set of manual score and automatic score pairs, $(x_1, y_1), (x_2, y_2), \ldots, (x_n, y_n)$, any pair of scores, (x_i, y_i) and (x_j, y_j) such that $i \neq j$, are said to be concordant if $x_i > x_j$ and $y_i > y_j$; or if both $x_i < x_j$ and $y_i < y_j$. They are said to be discordant if $x_i < x_j$ and $y_i > y_j$; or if $x_i > x_j$ and $y_i < y_j$. The pair is a tie, if $x_i = x_j$ and $y_i = y_j$.

4   Results and Analysis

4.1   Correlation: Manual Versus Automatic

Manual versus automatic evaluation scores for the five automatic metrics are given as separate scatter plots in Fig. 1. Each plot contains 1350 data points corresponding to 1350 system-outputs. The x-axis in each plot gives the automatic metric

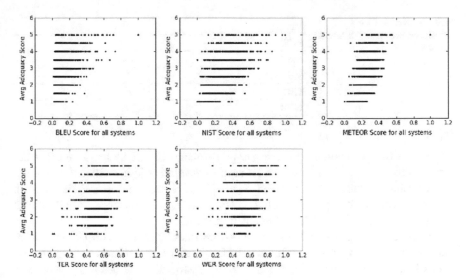

Fig. 1 Automatic metric score versus average adequacy score

score for a given MT system-output, and the y-axis gives the corresponding adequacy score. The plot gives a fair idea about the correlation between automatic and manual evaluation scores. It is evident that the correlation is weak as there is a substantial spread in the data points. The spread is minimum for METEOR; therefore, we can infer that METEOR scores agree the most with human judgments. However, to quantify the correlation we test the same using Pearson's and Kendall's correlation coefficients.

Tables 6 and 7 show segment-level Pearson's correlation coefficients and Kendall's tau correlation coefficients, respectively. Both the correlation coefficients suggest that METEOR correlates the best with manual evaluation scores.

4.2 Distribution: Manual Versus Automatic

One of the goals of this study is to see if manual evaluation score for a given MT output can be reliably estimated given its automatic evaluation score. Automatic

Table 6 Pearson's ρ correlation coefficient for different metrics

Metrics	ρ-value
BLEU	0.401
NIST	0.481
METEOR	0.513
TER	0.384
WER	0.345

Table 7 Kendall's τ correlation scores for different metrics

Metrics	τ-value
BLEU	0.287
NIST	0.336
METEOR	0.361
TER	0.269
WER	0.219

metrics are used to compare the relative performance of MT systems. But it is not clear whether automatic metrics can be used to absolutely judge translation quality. The correlation study above showed that METEOR has the best correlation with human judgments at segment level. Therefore, we now consider METEOR scores to estimate segment-level manual evaluation scores.

First, we observe how adequacy scores are distributed in the METEOR score range of 0.0–1.0. For this, we bin the METEOR scores into 10 bins each having an interval of 0.1. For each bin, we depict the distribution of adequacy scores as histograms in Fig. 2. The histogram shows how many segments scored adequacy scores of 1–5 in each interval of METEOR scores. Note that the range of y-axis for the bins is not normalized and is therefore different for each bin.

A majority of segment scores fall in the METEOR score interval of 0.2–0.4 (C, D in Fig. 2). The distribution of adequacy scores is very regular in the METEOR interval 0.1–0.5 (B–E in Fig. 2), with the central trend of adequacy score shifting positively with the METEOR scores. Bins in the interval 0.6–0.9 (G in Fig. 2) are empty as no segments received a METEOR score in this interval. Also, the first bin and the last bin have very little data points to make any valid conclusions.

This distribution study shows that there is some statistical evidence for the correlation between automatic scores and adequacy scores in the range 0.1–0.6 of METEOR scores. In order to get a rough estimate of manual scores for a given range of METEOR scores, we calculate 95% confidence interval [4] for the mean of adequacy scores in each bin. A 95% confidence interval for a given bin does not indicate that there is a 95% probability that the mean for a given bin lies within the interval. Instead, it tells us that the calculated interval will include the true mean for a given bin with a probability of 95%. Table 8 lists the most probable (95% confidence interval) range of adequacy scores for a given interval of METEOR score.

---

[4]We used the tconfint_mean function available in the statsmodels package in Python.

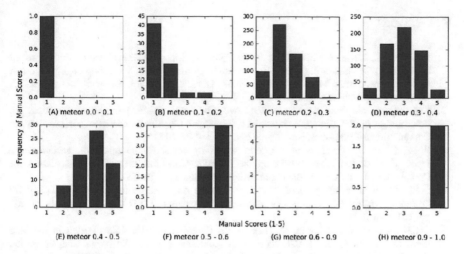

Fig. 2 Distribution of manual scores for each interval of METEOR scores

METEOR scores	Manual scores
0.0–0.1	NA
0.1–0.2	1.48–1.88
0.2–0.3	2.52–2.66
0.3–0.4	3.11–3.26
0.4–0.5	3.73–4.12
0.5–0.6	4.56–5.0
0.6–0.9	NA
0.9–1.0	5.0–5.0

Table 8 95% Confidence interval for the mean of manual scores for each bin of METEOR scores

5 Conclusions and Future Work

In this paper, we have presented an empirical study to compare automatic machine translation evaluation metrics with the manual evaluation metric *Adequacy*. We see that automatic metrics have a weak correlation with *adequacy*. However, among the various metrics considered, METEOR correlates best with *adequacy*. We see that for METEOR scores in the 0.1–0.6 interval, we can get a somewhat better idea of the adequacy scores. Thus, the quality of MT can be estimated from METEOR scores in certain situations. Our data did not contain METEOR scores in the interval 0.6–1.0, although we used the MT system that performed best in the WMT14 [3]. A more thorough study including other carefully selected language pairs and MT paradigms including rule-based and neural MT systems will be useful.

References

1. Banerjee, S., Lavie, A.: METEOR: an automatic metric for MT evaluation with improved correlation with human judgments. In: Proceedings of the ACL Workshop on Intrinsic and Extrinsic Evaluation Measures for Machine Translation and/or Summarization, vol. 29. University of Michigan, Ann Arbor, pp. 65–72 (2005)
2. Bojar, O. et al.: Findings of the 2013 workshop on statistical machine translation. In: Proceedings of the Eighth Workshop on Statistical Machine Translation. Association for Computational Linguistics, Sofia, Bulgaria, pp. 1–44 (2013)
3. Bojar, O. et al.: Findings of the 2014 workshop on statistical machine translation. In: Proceedings of the Ninth Workshop on Statistical Machine Translation. Association for Computational Linguistics. Baltimore, Maryland, USA, pp. 12–58 (2014)
4. Bojar, O. et al. Findings of the 2016 conference on machine translation (WMT16). In: Proceedings of the First Conference on Machine Translation (WMT), vol. 2. Berlin, Germany, pp. 131–198 (2016)
5. Callison-Burch, C., Osborne, M., Koehn, P.: Re-evaluation the role of Bleu in machine translation research. In: EACL-2006: 11th Conference of the European Chapter of the Association for Computational Linguistics. Trento, Italy, pp. 249–256 (2006)
6. Carroll, J.B.: An experiment in evaluating the quality of translations. Mech. Transl. Comput. Linguist. 9(3-4), 55–66 (1966)
7. Cohen, J.: A coefficient of agreement for nominal scales. Educ. Psychol. Measur. 20(1), 37–46 (1960)
8. Costa-jussà, M.R., Farrús, M.: Towards human linguistic machine translation evaluation. Digit. Scholarsh. Humanit. 30(2), 157–166 (2015)
9. Denkowski, M., Lavie, A.: Meteor universal: language specific translation evaluation for any target language. In: Proceedings of the Ninth Workshop on Statistical Machine Translation. Baltimore, Maryland, USA, pp. 376–380 (2014)
10. Doddington, G.: Automatic evaluation of machine translation quality using n-gram co-occurrence statistics. In: Proceedings of the Second International Conference on Human Language Technology Research. Morgan Kaufmann Publishers Inc. San Diego, California, pp. 138–145 (2002)
11. Dungarwal, P. et al.: The IIT Bombay Hindi-English translation system at WMT 2014. In: Proceedings of the Ninth Workshop on Statistical Machine Translation. Baltimore, Maryland, USA, pp. 90–96 (2014)
12. Farrús, M., Costa-jussà, M.R., Popović Morse, M.: Study and correlation analysis of linguistic, perceptual, and automatic machine translation evaluations. J. Assoc. Inf. Sci. Technol. 63(1), 174–184 (2012)
13. Fomicheva, M. et al.: CobaltF: a fluent metric for MT evaluation. In: Proceedings of the First Conference on Machine Translation: Volume 2, Shared Task Papers, vol. 2, pp. 483–490 (2016)
14. Gautam, S., Bhattacharyya, P.: LAYERED: metric for machine translation evaluation. In: Proceedings of the Ninth Workshop on Statistical Machine Translation. Baltimore, Maryland, USA, pp. 387–393 (2014)
15. Giménez, J., Màrquez, L.: A smorgasbord of features for automatic MT evaluation. In: Proceedings of the Third Workshop on Statistical Machine Translation. StatMT '08. Association for Computational Linguistics, Columbus, Ohio, pp. 195–198 (2008). ISBN: 978-1-932432-09-1
16. Giménez, J., Màrquez, L.: Linguistic features for automatic evaluation of heterogenous MT systems. In: Proceedings of the Second Workshop on Statistical Machine Translation. StatMT '07. Association for Computational Linguistics, Prague, Czech Republic, pp. 256–264 (2007)
17. Gupta, A., Venkatapathy, S., Sangal, R.: Meteor-Hindi: automatic MT evaluation metric for hindi as a target language. In: Proceedings of ICON-2010: 8th International Conference on Natural Language Processing. Macmillan Publishers, Kharagpur, India (2010)

18. King, M., Popescu-Belis, A., Hovy, E.: FEMTI: creating and using a framework for MT evaluation. In: Proceedings of MT Summit IX. New Orleans, USA, pp. 224–231 (2003)
19. Koehn, P.: Statistical Machine Translation. Cambridge University Press (2009). Chap. 8
20. Koehn, P. et al.: Moses: open source toolkit for statistical machine translation. In: Proceedings of the 45th Annual Meeting of the ACL on Interactive Poster and Demonstration Sessions. Association for Computational Linguistics. Prague, Czech Republic, pp. 177–180 (2007)
21. Miller, G.A.: WordNet: a lexical database for English. Commun. ACM **38**(11), 39–41 (1995)
22. Narayan, D. et al.: An experience in building the indo WordNet-a WordNet for Hindi. In: First International Conference on Global WordNet. Mysore, India (2002)
23. Papineni, K. et al.: BLEU: a method for automatic evaluation of machine translation. In: Proceedings of the 40th Annual Meeting on Association for Computational Linguistics. Philadelphia, pp. 311–318 (2002)
24. Pearson, K.: On the criterion that a given system of deviations from the probable in the case of a correlated system of variables is such that it can be reasonably supposed to have arisen from random sampling. London, Edinburgh, Dublin Philos. Mag. J. Sci. **50**(302), 157–175 (1900)
25. Snover, M. et al. A study of translation edit rate with targeted human annotation. In: Proceedings of the 7th Conference of the Association for Machine Translation in the Americas, "Visions for the Future of Machine Translation". Cambridge, Massachusetts, USA, pp. 223–231 (2006)
26. Su, K.-Y., Wu, M.-W., Chang, J.-S.: A new quantitative quality measure for machine translation systems. In: Proceedings of the 14th Conference on Computational linguistics, vol. 2. Association for Computational Linguistics, Nantes, pp. 433–439 (1992)
27. Tan, L., Pal, S.: Manawi: using multi-word expressions and named entities to improve machine translation. In: Proceedings of the Ninth Workshop on Statistical Machine Translation. Baltimore, Maryland, USA, pp. 201–206 (2014)
28. White, J., O'Connell, T., O'Mara, F.: The ARPA MT evaluation methodologies: evolution, lessons, and future approaches. In: Technology Partnerships for Crossing the Language Barrier: Proceedings of the First Conference of the Association for Machine Translation in the Americas. Columbia, Maryland, USA, pp. 193–205 (1994)
29. White, J.S., Taylor, K.B.: A task-oriented evaluation metric for machine translation. In: Proceedings of Language Resources and Evaluation Conference, LREC-98, vol. 1. Granada, Spain, pp. 21–27 (1998)

A Structural Topic Modeling-Based Machine Learning Approach for Pattern Extraction from Accident Data

Sobhan Sarkar, Suvo Gaine, Aditya Deshmukh, Nikhil Khatedi
and J. Maiti

Abstract In occupational accident analysis, the application of machine learning is still unexplored. In addition, the presence of unstructured text data makes the analysis really difficult. Therefore, the aim of the paper is to utilize the information within accident texts using structural topic model (STM) and predict loss time injury (LTI) and non-LTI. Random forest (RF) has been used in this study for prediction as well as rule extraction. The performance of RF has also been compared with that of support vector machine (SVM), and k-nearest neighbor (KNN). The experimental results reveal that RF outperforms other classifiers in terms of accuracy. Moreover, a set of interpretable nine rules are extracted exploring the causes of the occurrence of both LTI and non-LTI.

Keywords LTI/non-LTI prediction · STM · SVM · KNN · RF · Rule extraction

S. Sarkar (✉) · S. Gaine · J. Maiti
Department of Industrial & Systems Engineering, IIT Kharagpur, Kharagpur, India
e-mail: sobhan.sarkar@gmail.com

S. Gaine
e-mail: suvogaine786@gmail.com

J. Maiti
e-mail: jhareswar.maiti@gmail.com

A. Deshmukh
Department of Metallurgical & Materials Engineering, IIT Kharagpur, Kharagpur, India
e-mail: adityadeshmukh281997@gmail.com

N. Khatedi
Department of Mechanical Engineering, IIT Kharagpur, Kharagpur, India
e-mail: nkhatedi@gmail.com

© Springer Nature Singapore Pte Ltd. 2020   555
K. S. Raju et al. (eds.), *Data Engineering and Communication Technology*,
Advances in Intelligent Systems and Computing 1079,
https://doi.org/10.1007/978-981-15-1097-7_46

1 Introduction

In any organization, occupational injuries and accidents are severe problems. As per the International Labor Organization (ILO), almost 2.3 million labors are suffering every year and it leads to death because of diseases and accidents, the count of the fatal accidents reaches approximately 3.6 lakh every year. Any injury incident occurs because of the existence of a series of factors. If the factors are known, then the prediction of accidents can be done. In addition, contribution by several causal factors in accidents can be quantified using the predictive models. For analyzing occupational accidents, the predictive models may be either machine learning (ML) or statistical learning based on system safety [4, 17]. ML outperforms the classical statistical learning in estimating future incidents, which is used in many applications-based domains like engineering, health industry, etc. Usefulness of ML techniques according to explanatory capacity and predictive power has been shown in studies done in this domain. Various ML techniques like Bayesian network [8], k-nearest neighbor (KNN), support vector machine (SVM) [15], neural networks [16], decision tree (DT) [11], random forest (RF), etc., have been utilized with [10] or without parameter tuning using proactive or reactive data [12, 13, 16, 18] but RF gives the best result in terms of performance among them. Some important issues in the occupational analysis have been outlined after reviewing previous studies, which are as follows: (i) Previous studies focus more either on categorical data or on continuous data, or both categorical and continuous or using only textual data; (ii) none of the studies has shown the rule extraction from accident data; (iii) though the analysis of accident using machine learning approach has been done in different domains, in steel industry domain no studies are reported. Taking into account the issues in accident literature, our research endeavors to contribute as follows: (i) Handling both categorical and unstructured text data for accident analysis; (ii) use of several ML classifiers (SVM, RF, and KNN) for predicting LTI and non-LTI type incident categories; (iii) analysis of the important predictor variables for the occurrence of events and rule extraction from RF for analysis of accidents; and (iv) Prepared model is endorsed by a case study of a steel plant's safety data.

2 Methodology

In this section, our proposed methodology has been described briefly and in Fig. 1, the total proposed methodological flowchart is given. Three phases used in this study include: (i) *Data preprocessing phase*: Three important tasks, i.e., feature inclusion, missing value imputation, and feature selection are done. Our primary dataset has 2885 event records with 17 attributes; among them, six attributes have very low% of missing values. Missing values are imputed using random forest. Two new attributes are developed from present textual information using topic modeling. Finally, feature importance is calculated and important features are selected for model building using Boruta algorithm. The dataset generated using all above actions has 2884 event records and 15 attributes without having any value missing. (ii) *Model fitting and prediction phase*: Three important ML classifier (SVM, RF, and KNN) techniques have been applied and have been validated by 10-fold cross-validation. Then, the best model based on efficiency is taken and the prediction has been done using that model. (iii) *Rule extraction phase*: crucial rules are extracted from the best classifier (i.e., RF). Important techniques in these phases are described in below subsections.

2.1 Data Preprocessing

In this study, structural topic model (STM) as a topic modeling tool is used for converting unstructured texts into structured topic form. STM is a generative model which begins with document, topic, and word distributions. In STM, a topic is a combination of words and each word has a certain probability of belonging to a particular topic and a document is defined as a combination of topics which means each document can be composed of many topics. Roberts et al. [7] show that STM executes far well than other traditional topic modeling techniques like LDA [8, 16], LSI, etc. For a detailed description, readers are referred to [7]. Except this, other preprocessing, such as missing data handling and inconsistency removal, are also performed [14].

2.2 Predictive Algorithms and Rule Extraction

In this study, three popular algorithms, namely SVM, KNN, and RF, have been used. SVM has the ability to work as universal comparable functions like multi-variate for getting expected accuracy. Primarily, it has been made to do regression works, but nowadays it is used in classification problems owing to its good results. For further study, readers may refer to [5]. KNN is one of the earliest and easiest ML algorithms for classification problem. However, it sometimes gives nice results and in areas, where it is properly linked with domain knowledge, it has extremely advanced its performance [1]. RF is very useful for feature selection by evaluating the significance of any attribute. Interdependencies between features are not considered in RF. RF is acknowledged as the best model as it gives great accuracy in prediction. A detailed description of this algorithm can be found in [2].

In this study, the performance metrics, namely accuracy, precision, recall, F1-score, and geometric mean (G-mean), are used. Proper definitions of these metrics can be found in [16]. In this step, the rules are extracted using random forest algorithm for predicting 'LTI' or 'Non-LTI.' Using the 'inTrees' framework proposed by Deng et al. [3], the rule extraction is done. Once the rules are extracted, rule efficiency is calculated and the rules based on the maximum efficiency are selected.

3   A Case Study

The dataset with accident records is collected from a steel plant in India. Our dataset contains a total of 2885 records of incidents with 28 attributes. After preprocessing, there remains a set of 12 attributes. In the set of attributes, 'Month,' 'Division,' 'Incident category,' 'Primary cause of accident,' 'Brief description of incident,' 'Event leading to incidents,' 'Working condition,' 'Observation types,' 'Employee types,' 'Serious process incident scores,' 'Equipment damage score,' and 'Incident types' are used. Among them, 'Incident category' with two categories, 'LTI' and 'non-LTI,' is used as response attribute and rest are used predictor attributes.

4   Results and Discussion

4.1   Attribute Generation

In our data, both categorical and text data type attributes are present there. In our study, we have generated features for text data.

In our study, STM-based topic modeling has been used to categorize two text attributes namely 'Event Leading to Incident' and 'Brief Description of Incident' into ten and twelve topics, respectively, according to semantic coherence of the topics. From the analysis, two categorical attributes namely 'event' and 'description' are generated, respectively. Tables 1 and 2 shows top five words in each topic for 'description' and 'event' attribute, respectively. From top five words, we can roughly understand the core issues of each topic. Figures 2 and 3 show topic

Table 1 Top five words in each topic for 'description' attribute

Topic	Brief description of incident	Top five words in each topic
Topic 1	Desc 1	Hand, finger, right, left, middle
Topic 2	Desc 2	Face, drill, roof, coal, bolt
Topic 3	Desc 3	Duty, shift, road, bike, fall
Topic 4	Desc 4	Gate, yard, driver, dumper, load
Topic 5	Desc 5	Door, open, head, box, cable
Topic 6	Desc 6	Pipe, worker, fall, hit, level
Topic 7	Desc 7	Crane, job, plate, person, ground
Topic 8	Desc 8	Valve, eye, hammer, line, hot

(continued)

Table 1 (continued)

Topic	Brief description of incident	Top five words in each topic
Topic 9	Desc 9	Road, side, bike, injury, near
Topic 10	Desc 10	Right, leg, left, slip, injury
Topic 11	Desc 11	Cut, wire, hit, coil, hand
Topic 12	Desc 12	First, aid, plant, immediate, hospital

Table 2 Top five words in each topic for 'event' attribute

Topic	Event leading to incident	Top five words in each topic
Topic 1	Event 1	First, wire, aid, remove, coil
Topic 2	Event 2	Time, plate, wear, weld, cover
Topic 3	Event 3	Fall, cycle, open, piece, face
Topic 4	Event 4	Drive, person, careless, unsafe, proper
Topic 5	Event 5	Work, gas, condition, clean, belt
Topic 6	Event 6	Hand, got, cut, finger, injury
Topic 7	Event 7	Road, sudden, bike, fall, side
Topic 8	Event 8	Slip, pipe, bolt, roof, hold
Topic 9	Event 9	Machine, hammer, rod, surface, operation
Topic 10	Event 10	Hit, material, handle, floor, move

frequencies for 'description' and 'event' attribute, respectively. Topics with high frequency imply the chance of occurrence of those events is high.

4.2 Predictive Analysis and Rule Generation

In our study, we have used RF, SVM, and KNN algorithms for prediction of incident category. Results of our analysis reveal that the RF algorithm gives the best result in term of accuracy. Apart from accuracy, other performance measures are also computed and reported in Table 3. From these results, it can be seen that RF is the best classifier. There were 4478 rules extracted (length ≤ 6) from the first 100 trees in RF. After rule generation, only nine rules have been finally selected with a minimum length of two (refer to Table 4). Rule 1 can be interpreted as if it is the

Table 3 Model comparison based on accuracy, precision, recall, F1-score, and G-mean

Model	Accuracy (%)	Precision (%)	Recall (%)	F1 score (%)	G-mean (%)
RF	91.90	95.77	94.97	95.36	95.37
SVM	89.08	93.68	93.09	93.38	93.38
KNN	85.65	96.72	87.21	91.72	91.84

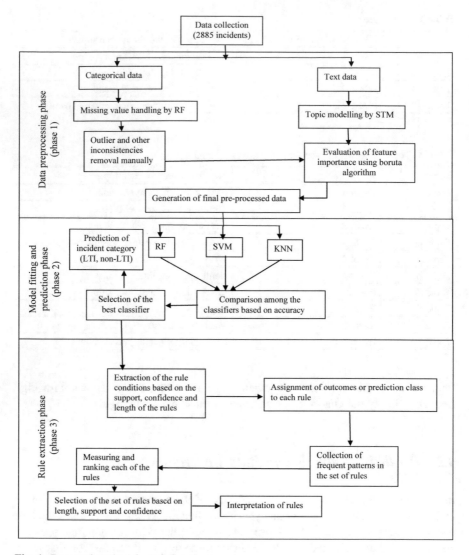

Fig. 1 Proposed methodological flowchart

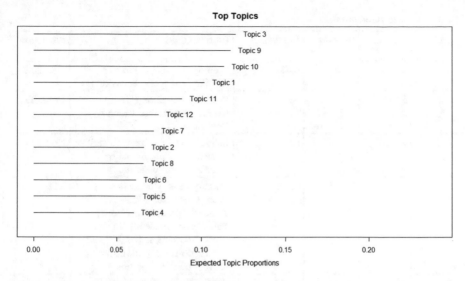

Fig. 2 Topic frequencies for 'description' attribute

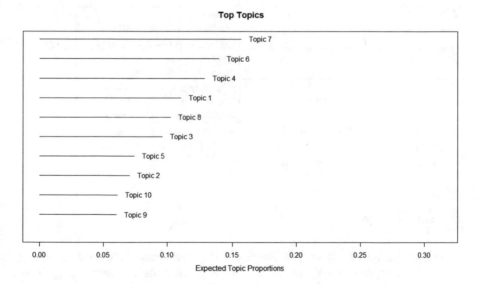

Fig. 3 Topic frequencies for 'event' attribute

contractor and incident type is behavioral then it is more likely to happen non-LTI type accident. Similarly, rule 6 can be interpreted as if the condition is unsafe and description of the event is either desc 4 or desc 7, then it is more likely to happen LTI. Similarly, all the other rules can also be interpreted.

Table 4 Rule summarization

Rule no.	Length	Support	Confidence	Rule condition	Prediction
1	2	0.007	0.837	Employee type ('contractor') and incident type ('behavior')	Non-LTI
2	2	0.005	0.893	Working condition ('group working') and observation type ('unsafe act')	Non-LTI
3	2	0.004	0.9	Incident type ('behavior') and observation ('unsafe act by other', 'unsafe condition')	Non-LTI
4	2	0.004	0.778	Primary cause ('crane dashing', 'energy isolation', 'medical ailment', 'process incidents', 'rail', 'road incident', 'run over', 'working at height') and observation type ('unsafe act by other', 'unsafe condition')	LTI
5	2	0.003	0.875	Employee type ('contractor') and events ('event 1', 'event 10')	Non-LTI
6	2	0.003	0.772	Observation type ('unsafe condition') and description ('desc 4', 'desc 7')	LTI
7	2	0.003	0.654	Employee type ('contractor') and incident type ('process')	LTI
8	2	0.003	0.598	Observation type ('unsafe act', 'unsafe condition') and employee type ('employee')	LTI
9	2	0.002	0.692	Working condition ('group working') and employee type ('contractor')	Non-LTI

5 Conclusions

In our study, prediction models based on ML have been developed to predict incident categories in the organization. Three efficient classifiers, namely RF, SVM, and KNN, have been used in this study. The findings of analysis revealed that RF algorithm performs best among others in term of accuracy. Some main interpretations from our study are: if observation type is either 'unsafe act' or 'unsafe condition,' then there are more chances that injury type will be 'LTI' and if injury potential score is high, then it has more chance that injury type will 'non-LTI.' Like several other studies, even this study has its limitations. In the data preprocessing phase, a lot of work customized for the given dataset needs to be performed to clean the dataset and make it ready for efficient analysis. The dataset used by us accounts very less number of accident records. It is advisable to use a larger dataset for the analysis to obtain a much more accurate model. Further, development of an autonomous decision support system [9] or vision-based safety system [6] can be done which may predict the incident categories and also be able to assist in making smart decisions according to the rules extracted from models. Other heuristic methods such as simulated annealing, Tabu search, and genetic algorithms can also be used to experience the better accuracy of classifiers and find sets of rules.

References

1. Belongie, S., Malik, J., Puzicha, J.: Shape matching and object recognition using shape contexts. Technical Reports, California Univ San Diego La Jolla Dept of Computer Science and Engineering (2002)
2. Breiman, L.: Random forests. Mach. Learn. **45**(1), 5–32 (2001)
3. Deng, H., Runger, G., Tuv, E., Bannister, W.: Cbc: an associative classifier with a small number of rules. Decis. Support Syst. **59**, 163–170 (2014)
4. Gautam, S., Maiti, J., Syamsundar, A., Sarkar, S.: Segmented point process models for work system safety analysis. Saf. Sci. **95**, 15–27 (2017)
5. Kecman, V.: Support vector machines—an introduction. In: Support Vector Machines: Theory and Applications, pp. 1–47. Springer (2005)
6. Pramanik, A., Sarkar, S., Maiti, J.: Oil spill detection using image processing technique: An occupational safety perspective of a steel plant. In: Emerging Technologies in Data Mining and Information Security, pp. 247–257. Springer (2019)
7. Roberts, M.E., Stewart, B.M., Airoldi, E.M.: A model of text for experimentation in the social sciences. J. Am. Stat. Assoc. **111**(515), 988–1003 (2016)
8. Sarkar, S., Baidya, S., Maiti, J.: Application of rough set theory in accident analysis at work: a case study. In: 2017 Third International Conference on Research in Computational Intelligence and Communication Networks (ICRCICN), pp. 245–250. IEEE (2017)
9. Sarkar, S., Chain, M., Nayak, S., Maiti, J.: Decision support system for prediction of occupational accident: a case study from a steel plant. In: Emerging Technologies in Data Mining and Information Security, pp. 787–796. Springer (2019)
10. Sarkar, S., Lohani, A., Maiti, J.: Genetic algorithm-based association rule mining approach towards rule generation of occupational accidents. In: International Conference on Computational Intelligence, Communications, and Business Analytics, pp. 517–530. Springer (2017)
11. Sarkar, S., Patel, A., Madaan, S., Maiti, J.: Prediction of occupational accidents using decision tree approach. In: India Conference (INDICON), 2016 IEEE Annual, pp. 1–6. IEEE (2016)
12. Sarkar, S., Pateshwari, V., Maiti, J.: Predictive model for incident occurrences in steel plant in india. In: 2017 8th International Conference on Computing, Communication and Networking Technologies (ICCCNT), pp. 1–5. IEEE (2017)
13. Sarkar, S., Verma, A., Maiti, J.: Prediction of occupational incidents using proactive and reactive data: a data mining approach. In: Industrial Safety Management, pp. 65–79. Springer (2018)
14. Sarkar, S., Vinay, S., Maiti, J.: Text mining based safety risk assessment and prediction of occupational accidents in a steel plant. In: 2016 International Conference on Computational Techniques in Information and Communication Technologies (ICCTICT), pp. 439–444. IEEE (2016)

15. Sarkar, S., Vinay, S., Pateshwari, V., Maiti, J.: Study of optimized SVM for incident prediction of a steel plant in india. In: India Conference (INDICON), 2016 IEEE Annual, pp. 1–6. IEEE (2016)
16. Sarkar, S., Vinay, S., Raj, R., Maiti, J., Mitra, P.: Application of optimized machine learning techniques for prediction of occupational accidents. Comput. Oper. Res. (2018)
17. Singh, K., Raj, N., Sahu, S., Behera, R., Sarkar, S., Maiti, J.: Modelling safety of gantry crane operations using petri nets. Int. J. Injury Control Saf. Promot. 24(1), 32–43 (2017)
18. Verma, A., Chatterjee, S., Sarkar, S., Maiti, J.: Data-driven mapping between proactive and reactive measures of occupational safety performance. In: Industrial Safety Management, pp. 53–63. Springer (2018)

Emotion Detection Framework for Twitter Data Using Supervised Classifiers

Matla Suhasini and Badugu Srinivasu

Abstract "The task of emotion detection usually involves the analysis of text. Humans show universal consistency in identifying emotions however shows an excellent deal of variation between individuals in their abilities." We have detected the emotion for Twitter messages as they provide rich ensemble of human emotions. We have used machine learning algorithms namely Naive Bayes (NB) and k-nearest neighbor algorithm (KNN) to detect the emotion of Twitter message and then classify the Twitter messages into four emotional categories. We also made a comparative study of two supervised machine learning algorithms; the eager learning classifier (NB) performed well when compared with lazy learning classifier (KNN).

Keywords Tweets · Emotion detection · Natural language processing · Machine learning

1 Introduction

The rise of social media has attracted significant interest in sentiment analysis such as emotion detection and opinion mining. Unlike more formal methods of communication, microblog posts (hereafter, "tweets") frequently reflect the author's opinions and emotional states [1]. A large number of Web sites provide a long list of topics varying from politics to entertainment [2]. It has been noted that tweets commonly tend to fall in one among two totally different categories: users that microblog concerning themselves and people that use microblog primarily to share

M. Suhasini (✉) · B. Srinivasu
SCETW, Hyderabad, India
e-mail: msuhasini26@gmail.com

B. Srinivasu
e-mail: srinivasucse@gmail.com

© Springer Nature Singapore Pte Ltd. 2020 565
K. S. Raju et al. (eds.), *Data Engineering and Communication Technology*,
Advances in Intelligent Systems and Computing 1079,
https://doi.org/10.1007/978-981-15-1097-7_47

information [3, 4]. Twitter can be considered as a large repository that includes a rich ensemble of emotions, sentiments and moods [5].

A number of analyses are targeted on specific events [3] like the study targeted on response to the death of celebrity or about a political election. Using publically available on-line information to perform emotion analyses considerably reduces the efforts and time required to administer large-scale public surveys and questionnaires [3, 6, 7, 8]. Researchers also noticed that tweets usually convey pertinent information about the user's emotional states [9]. Reason behind choosing the Twitter is it covers a very broad range of topics which leads to a large range of potential word interactions [10]. Twitter so has become a robust platform with several kinds of information from worldwide breaking news to buy merchandise reception [3, 11]. Emotion analysis can be applied to all kinds of text, but certain domains like psychology and modes of communication tend to have more overt expressions of emotions than others [12] like Richard, Dan Roth, Virginia, Michal, Alena [13–17] worked on fairy tales.

Section 2 provides related work with emotion detection and the use of Twitter as a sentiment corpus with different learning algorithms and how NLP techniques helped in text categorization. Section 3 about the system. Section 4 analyzes the results. Section 5 includes conclusion and the future work.

2   Literature Survey

Authors Suhasini and Srinivasu proposed a method called two-step approach to detect and classify the Twitter messages. The two-step approach was based on two approaches, one was rule-based approach (RBA) and the other was machine learning approach. They have compared the accuracy of both the approaches, stated that machine learning approach performed well when compared with rule-based approach as they have fed less error data to learning model [3]. Authors have described an easy way on how to create labeled data using knowledge-based approach, which in turn can be used for classification. Knowledge-based approach detects the emotion of the tweet and classifies the Twitter message under acceptable emotional class [18].

Apoorv Agarwal et al. led work on sentiment analysis on Twitter data. Their data was a random sample of streaming tweets. They have used manually annotated Twitter data; manual annotation is a tedious task when we want to go for large number of tweets. Sometimes, this may restrict in choosing the larger datasets because annotation takes much time. Unlike their approach, our approach is not a manual annotation; we have labeled the tweets with the help of knowledge-based approach. Their approach is restricted to the sentiment level, but we have classified the tweets into finer-grained level of emotion [19].

A. Montejo-R´aez et al. proposed a new approach to resolve the evaluation of posts. This polarity detection drawback was resolved by combining SentiWordNet scores over the word—Internet graph. Few issues remained open in their work like handling negated phrases; stopwords were ignored. We have addressed the negated phrases in our work by considering all the negation carrying words in the seed, as negated phrases which have the capability of turning the tweet into opposite side. Their work was limited to polarity finding of a tweet, whereas our work is not restricted to polarity but to find the emotion of tweet [20]. Akshi Kumar and Teeja Mary Sebastian proposed and investigated a paradigm to mine the sentiment [21]. Purver investigated a different approach via distant by using standard markers of emotional content among the texts themselves as a surrogate for specific labels [22]. Tanaka et al. used Japanese-style emoticons as classification labels [23]. The study led by Maryam Hasan et al. [2] to model emotional states utilized the well-established Circumplex model. Using hash-tags as labels, their methodology trained supervised classifiers to detect multiple classes of emotion on potentially huge data sets with no manual effort by using the weka tool [2]. Naaman created three independent annotators, which manually coded a sample of public tweets and found nine representative categories [24]. Swati Agarwal and Ashish Sureka's worked on how to tackle the problems faced by Intelligence Bureau and Security Informatic agents [25].

Jasy Liew Suet Yan classified the emotions into 28 finer-grained emotional categories. The class distribution became more unbalanced with finer-grained emotion classes. The most frequent category was happiness (13%) while the least frequent category was jealously (0.09%) [5, 26, 27]. Pearl and Steyvers focused on eight emotional categories [28]. Aamera Z. H. Khan, Mohammad Atique and V. M. Thakare proposed an augmented lexicon-based method specific to the Twitter data, which was applied to perform sentiment analysis [29, 30]. Some researchers applied a lexical approach to identify emotions in text. For example, Strapparava and Choudhury created an outsized lexicon [9, 31].

3 System

3.1 Our Twitter Corpus

The dataset is obtained from sentiment140 [32]. Though the dataset has different attributes like tweet, label, etc., we are taking only text attribute, i.e., tweet. In the dataset, their purpose was to spot the sentiment; sentiment says whether or not the tweet is positive, however, the proposed system is not restricted to a pair of classes of emotion. Instead, it identifies the finer level of emotion [3, 18]. In our work, we have classified the tweets into four different categories, namely C1 (Happy-Active), C2 (Happy -Inactive), C3 (Unhappy-Active) and C4 (Unhappy -Inactive).

3.2 Pre-Processing

Language utilized in Twitter has some distinctive characteristics, which have been removed because they do not provide relevant information for the detection process [20]. These specific characteristics are:

 (i) **Retweeting**: Rewriting the tweet which is interesting. Old tweet and the new tweet are separated [20].
 (ii) **Mentions**: Mentions are generally called as usernames. When a user wants to refer or to comment or to reply to another user, he or she introduces the person with Mention. A Mention can be recognized easily with the help of @ symbol.
   Input Example: Tweet1: @canny I AM Sooooo much Excited really Excited really wowwww!!
   Output Example: Tweet1: I AM Sooooo much Excited really Excited really wowwww!!

In the above example @canny and @silentdew are removed. To remove usernames we have used regular expression (RE). The word starts with @ symbol will be removed.

 (iii) **Links**: Twitter messages include Web links. In our approach, we are not analyzing the urls since they do not provide any useful information, so we have eliminated them from all tweets using RE.
   Input Example: Tweet1: i really love you http://tinyurl.com/mf88dz
   Output Example: Tweet1: i really love you

In the above example, the url http://tinyurl.com/mf88dz has been removed. To remove url, we have created a pattern "http://" so wherever the pattern matches in the tweet, that part will be removed.

 (iv) **Removal of new lines**: Some Twitter messages may extend to two or three different lines, so there is a need to remove new line symbols.
   Input Example: Tweet1: I AM Sooooo much Excited really Excited really wowwww!!
   Output Example: Tweet1: I AM Sooooo much Excited really Excited really wowwww!!

In the above example new line character (\n) is removed.

 (v) **Numeric**: Some tweets may contain numeric data [0–9]; we are not bothered about integer type text, so integers are removed using regular expression since integers do not play any role in detection process.
   Input Example: Tweet 1: 5 hot days in a row i am not happy very sad
   Output Example: Tweet 1: hot days in a row i am not happy very sad

In the above example, numeric 5 has been removed.

 (vi) **Special characters**: Tweets may contain special characters such as (*,/,., > ,! etc....) can be stripped of using strip() function.

Input Example: Tweet1: I AM Sooooo much Excited really
Excited really wowwww!!
Output Example: Tweet1: I AM Sooooo much Excited really
Excited really wowwww

In the above example symbol ! has been removed.

(vii) **Lower case**: The tweets are converted to lower case.
Input Example: Tweet1: I AM Sooooo much Excited really
Excited really wowwww!!
Output Example: Tweet1: i am sooooo much excited really
excited really wowwww

(viii) **Standardization:** The text should be standardized, social media contains
shortcuts, and the necessary text should be standardized to full form. For
this, we have maintained a predefined list.
Input Example: Tweet1: i am soooo much excited really
excited really wowwww
Output Example:Tweet1: i am so much excited really
excited really wow

In the above example, in tweet1, wowwww and soooo are converted to wow and
so.

3.3 Labeled Data Creation

We have created the labeled data using rule-based approach in our previous work
by taking Russell's Circumplex model as reference [18]. Russell's Circumplex says
that emotions are circulated into two-dimensional area, containing pleasure (valence) and activation (arousal) dimensions [2, 33, 34] (Fig. 1).

3.3.1 Negated Words

Negated words are the negation carrying words which can show the other side of
the tweet.

Example *Tweet*: today i am not happy

The above tweet is actually a sad emotion carrying tweet, but if we ignore "not"
before happy, then the tweet may be classified as happy emotion carrying class
which is a misclassification. Misclassification rate increases if we ignore the
negated words, so to reduce error rate, special care has to be taken for negation
carrying words. For this purpose, we have maintained a separate list for negated
words (Table 1).

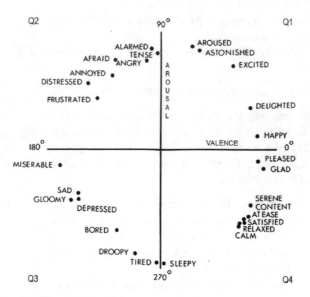

Fig. 1 J. A. Russell's Circumplex model

Table 1 Sample negated words

Do	Have	Could	Was	Does	Would	Should	Cannot
not	not	not	not	not	not	not	

3.4 Methodology

As we are working on text classification problem, we would like to explore machine learning algorithms. Supervised machine learning algorithms namely Naive Bayes and k-nearest neighbor are used for the classification process.

3.4.1 Experiment *1*

In this, we discuss Naive Bayes algorithm. Naive Bayes is based on Bayes theorem. In the testing phase, a new tweet (never seen by the model) is passed to the model. With the previously provided knowledge, the Bayes algorithm is able to classify the tweets. In training phase, the machine is categorizing the tweet into one among the category provided [3]. It picks up the hypothesis with maximum probability.

$$P\left(\frac{\text{label}}{\text{features}}\right) = P(\text{label}) * \frac{P\left(\frac{\text{features}}{\text{label}}\right)}{P(\text{features})} \tag{1}$$

where
P(label) = prior probability of a label, P(features/label) = prior probability that a given feature set is being classified as a label, P(features) = prior probability that a given feature set is occurred, P(label/features) = posterior probability.

3.4.2 Experiment 2

In this, we discuss k-nearest neighbor algorithm. The model is trained with 80% of the tweets, and remaining 20% is used for testing. Here, we have considered the similarity distance between test data point and train data point. So, every test data point is compared with all training data points for the similarity measure. We have used cosine similarity to measure the distance. After measuring, the distance scores are assigned to the tweets. Scores are sorted in descending order to know the maximum similar distance. We have considered the label of the tweet which is having highest score, and the same label is assigned to an unseen tweet.

$$\cos \theta = \frac{\vec{a} \cdot \vec{b}}{\|\vec{a}\| \cdot \|\vec{b}\|} \tag{2}$$

4 Experimental Results

We have used one supervised eager learning classifier (Naive Bayes) and one supervised lazy learning classifier (K-Nearest Neighbor).

4.1 Experiment 1

Experiment 1 provides the test results of supervised eager learning classifier(Naive Bayes).For training, we have used 80% of the tweets, and for testing, we have used 20% of the tweets.

Table 2 Confusion matrix, precision, recall and F-score of Naïve Bayes

$N = 1000$	C1	C2	C3	C4	Precision	Recall	F-measure
C1	201	34	14	13	0.73	0.76	0.74
C2	40	154	18	17	0.7	0.67	0.68
C3	18	19	221	35	0.79	0.75	0.76
C4	15	12	25	164	0.71	0.75	0.72

Fig. 2 *F*-measure of Naïve
Bayes

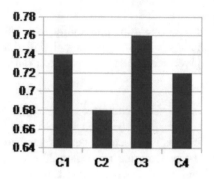

Observation 1: Observed the following F-scores for Naïve Bayes (Table 2; Fig. 2).

4.2 Experiment 2

Experiment 2 provides the test results of supervised lazy learning classifier (K-Nearest Neighbor). We have used 5000 examples out of which 80% are used for training, and 20% are used for testing. We have used the cosine similarity measure to measure the distance. The distances are sorted in descending order. Our k values are 1, 2, 3, 5 and 10.

We are considering the first appeared label as the final label for an unseen tweet in cases $k = 5, 10$.

Observations

Observation 1: Observed the following values when $k = 1$ (Table 3).
Observation 2: Observed the following values when $k = 2$ (Table 4).
Observation 3: Observed the following values when $k = 3$ (Table 5).
Observation 4: Observed the following values when $k = 5$. We are ignoring the second label if it has equal priority as first label in the case $k = 5$ (Table 6).
Observation 5: Observed the following values when k = 10. We are ignoring second and third labels if they have equal priority as first label in the case $k = 10$ (Tables 7, 8; Fig. 3).

Table 3 Confusion matrix, precision, recall and *F*-score of *K*-NN when $k = 1$

$K = 1$							
$N = 1000$	C1	C2	C3	C4	Precision	Recall	F-measure
C1	112	78	38	22	0.5	0.44	0.46
C2	53	102	41	27	0.45	0.45	0.45
C3	24	29	147	98	0.48	0.49	0.48
C4	32	16	75	106	0.41	0.46	0.43

Table 4 Confusion matrix, Precision, Recall and F-score of K-NN when $k = 2$

$K = 2$							
$N = 1000$	C1	C2	C3	C4	Precision	Recall	F-measure
C1	118	76	37	21	0.54	0.47	0.51
C2	49	107	39	27	0.48	0.48	0.48
C3	21	24	156	97	0.56	0.52	0.53
C4	28	14	74	112	0.43	0.48	0.45

Table 5 Confusion matrix, precision, recall and F-score of K-NN when $k = 3$

$K = 3$							
$N = 1000$	C1	C2	C3	C4	Precision	Recall	F-measure
C1	139	71	25	17	0.62	0.55	0.58
C2	48	114	37	23	0.54	0.51	0.52
C3	11	14	181	91	0.59	0.6	0.59
C4	23	9	62	135	0.5	0.58	0.53

Table 6 Confusion matrix, precision, recall and F-score of K-NN when $k = 5$

$K = 5$							
$N = 1000$	C1	C2	C3	C4	Precision	Recall	F-measure
C1	117	76	37	22	0.55	0.46	0.5
C2	49	107	39	26	0.48	0.48	0.48
C3	18	24	158	98	0.5	0.53	0.51
C4	27	15	77	110	0.42	0.48	0.44

Table 7 Confusion matrix, precision, recall and F-score of K-NN when $k = 10$

$K = 10$							
$N = 1000$	C1	C2	C3	C4	Precision	Recall	F-measure
C1	110	80	37	23	0.49	0.44	0.46
C2	54	101	41	27	0.44	0.45	0.44
C3	25	28	146	99	0.48	0.48	0.48
C4	33	17	75	104	0.41	0.45	0.42

Table 8 Different K values with corresponding F-score of KNN

K	F-score
1	0.45
2	0.49
3	0.55
5	0.48
10	0.45

Fig. 3 *F*-score for different *K* values of KNN

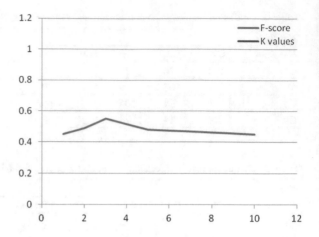

In the above graph, *X*-axis represents the values of *k* (where *k* = 1, 2, 3, 5, 10), and *Y*-axis represents the *F*-score for corresponding *k* values. From the graph, we could say that the optimum value for *k* is 3.

4.3 Comparison of Two Learning Models

We have compared both the algorithms and found the results as in the following Table 9.

The below figure represents the performance of Naïve Bayes and *k*-nearest neighbor algorithms (Fig. 4).

Table 9 Comparison of learning models

Model	Accuracy (%)
Naïve Bayes	72.60
KNN	55.50

Fig. 4 Performance of learning models

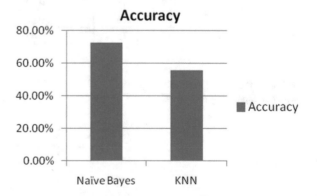

In the above graph, X-axis represents the two learning models (NB and KNN), and Y-axis represents the accuracy of the models in terms of percentage.

5 Conclusion and Future Work

We have compared both the results and found that Naive Bayes outperformed well when compared with K-NN. This has been a major topic of study in psychology. In the future, we would prefer to take several other models and multi-model forms of data.

References

1. Roberts, K., Roach, M.A., Johnson, J., Guthrie, J., Harabagiu, S.M.: EmpaTweet: annotating and detecting emotions on Twitter. In: Proceedings of the Eighth International Conference on Language Resources and Evaluation (LREC), ROBERTS12.201.L12-1059 (2012)
2. Hasan, M., Rundensteiner, E., Agu, E.: Emotex: detecting emotions in Twitter messages. Ase BigData/SocialCom/CyberSecurity Conference, 27–31 May 2014
3. Suhasini, M., Badugu, S.: Two step approach for emotion detection on Twitter data. Int. J. Comput. Appl. (0975 – 8887) **179**(53) (2018)
4. Bollen, J., Mao, H., Pepe, A.: Modeling public mood and emotion: Twitter sentiment and socio-economic phenomena. In: International AAAI Conference on Weblogs and Social Media (ICWSM'11) (2011)
5. Thelwall, M., Buckley, K., Paltoglou, G.: Sentiment in Twitter events. J. Am. Soc. Tavel Model. Simul. Design. (2007)
6. Diener, E., Seligman, M.E.P.: Beyond money: toward an economy of well-being. In: Psychological Science in the Public Interest. American Psychological Society (2004)
7. Diener, E.: Assessing Well-Being: The Collected Works of Ed Diener, vol. 3. Springer (2009)
8. Diener, E., Lucas, R.E., Oishi, S.: Subjective Wellbeing: The Science of Happiness and Life Satisfaction
9. De Choudhury, M., Counts,S., Gamon, M.: Not all moods are created equal! Exploring human emotional states in social media. In: Sixth International AAAI Conference on Weblogs and Social Media (ICWSM'12) (2012)
10. Lund, K., Burgess, C.: Producing high-dimensional semantic spaces from lexical co-occurrence. Behav. Res. Methods Instrum. Comput. **28**(2), 203–208 (1996)
11. Go A., Bhayani, R., & Huang L., Twitter Sentiment Classification Using Distant Supervision. Retrieved (2014)
12. Mohammad, S.M.: #Emotional tweets. In: First Joint Conference on Lexical and Computational Semantics, pp. 246–255, Canada (2012)
13. Alm, C.O., Sproat, R.: Emotional sequencing and development in fairy tales, pp. 668–674. Springer (2005)
14. Alm, C.O., Roth, D., Sproat, R.: Emotions from text: machine learning for textbased emotion prediction. In: Proceedings of the Joint Conference on HLT–EMNLP, Vancouver, Canada (2005)
15. Francisco, V., Gervás, P.: Automated mark up of affective information in english texts. In: Text, Speech and Dialogue, vol. 4188, pp. 375–382. Lecture Notes in Computer Science. Springer, Berlin, Heidelberg (2006)

16. Genereux, M., Evans, R.P.: Distinguishing affective states in weblogs. In: AAAI-2006 Spring Symposium on Computational Approaches to AnalysingWeblogs, pp. 27–29, Stanford, California (2006)
17. Neviarouskaya, A., Prendinger, H., Ishizuka, M.: Compositionality principle in recognition of fine-grained emotions from text. In: Proceedings of the Proceedings of the Third International Conference on Weblogs and Social Media (ICWSM-09), pp. 278–281, San Jose, California (2009)
18. Badugu, S., Suhasini, M.: Emotion detection on Twitter data using knowledge base approach. Int. J. Comput. Appl. (0975 – 8887) **162**(10) (2017)
19. Agarwal, A., Xie, B., Vovsha, I., Rambow, O., Passonnea, R.: Sentiment analysis of Twitter data. In: Proceedings of the Workshop on Language in Social Media (LSM), pp. 30–38, Portland, Oregon (2011)
20. Montejo-Ráez, A., Martínez-Cámara, E., Martín-Valdivia, M. T., Urena-Lopez, L. A.: Random Walk Weighting over SentiWordNet for Sentiment Polarity Detection on Twitter. In: Proceedings of the 3rd Workshop on Computational Approaches to Subjectivity and Sentiment Analysis, pp. 3–10, Republic of Korea (2012)
21. Kumar, A., Sebastian, T.M.: Sentiment analysis on Twitter. Int. J. Comput. Sci. **9**(4). ISSN:1694-0814 (2012). (Online)
22. Purver, M., Battersby, S.: Experimenting with distant supervision for emotion classification. In: Proceedings of the 13th Conference of the European Chapter of the Association for Computational Linguistics, pp. 482–491, Avignon, France (2012)
23. Tanaka, Y., Takamura, H., Okumura, M.: Extraction and classification of facemarks with kernel methods. In: Proceedings of IUI (2005)
24. Naaman, M., Boase, J., Lai, C.-H.: Is it really about me? Message content in social awareness streams. In ACM Conference on Computer Supported Cooperative Work (2010)
25. Agarwal, S., Sureka, A.: Using KNN and SVM based one-class classifier for detecting online radicalization on Twitters, ICDCIT, LNCS 8956, pp. 431–442, Switzerland (2015). https://link.springer.com/chapter/10.1007/978-3-319-14977-6_47
26. Liew, J.S.Y., Turtle, H.R.: Exploring fine-grained emotion detection in Tweets. In: Proceedings of NAACL-HLT, pp. 73–80, Association for Computational Linguistics, California (2016)
27. De Choudhury, M., Gamon, M., Counts, S., Horvitz, E.: Predicting depression via social media. In: International AAAI Conference on Weblogs and Social Media (ICWSM'13). The AAAI Press (2013)
28. Pearl, L., Steyvers, M.: Identifying emotions, intentions, and attitudes in text using a game with a purpose. In: Proceedings of the NAACLHLT Workshop on Computational Approaches to Analysis and Generation of Emotion in Text, Los Angeles, California (2010)
29. Golder, S., Loke, Y.K., Bland, M.: Meta-analyses of adverse effects data derived from randomized controlled trials as compared to observational studies: methodological overview. PLoS Med **8**(5), e1001026 (2011). https://doi.org/10.1371/journal.pmed.1001026
30. Khan, A.Z.H., Atique, M., Thakare, V. M.: Combining lexicon-based and learning-based methods for Twitter sentiment analysis. Int. J. Electron. CSCSE. ISSN:2277-9477 (2015)
31. Strapparava, C., Mihalcea, R.: Learning to identify emotions in text. In: Proceedings of the ACM Symposium on Applied computing. ACM, pp. 1556–1560 (2008)
32. http://sentiwordnet.isti.cnr.it
33. Russell, J.A.: A circumplex model of affect. J. Pers. Soc. Psychol. **39**, 1161–1178 (1980)
34. Olson, D.H., Sprenkle, D.H., Russell, C.S.: Circumplex model of marital and family system: I. Cohesion and adaptability dimensions, family types and clinical applications, vol. 18, Issue No. 1, pp. 3–28. Wiley Online Library (1979)

Multiple Object Detection Mechanism Using YOLO

G. A. Vishnu Lohit and Nalini Sampath

Abstract Object detection is a computer technology which relates with image processing and computer technology. There are many cases where in a given situation there is a need for faster object detection. For example, consider a traffic scenario or a case of natural disaster. In such areas, the detection of humans or specified objects becomes difficult. In such cases, there is a need for the better and quicker detection mechanism. Thus, in this paper, we proposed the YOLO detection technique. This technique uses the concept of deep neural networks and helps in the faster recognition of humans and objects.

Keywords IDLE · YOLO · IoU · UAV · NMS

1 Introduction

Humans will look at a picture once and instantly know what objects are present in the image, in what position they are, and how the interaction of objects has been done. The human visual system is speed in detecting the objects. Along with the speed if we also add conscious, we are able to perform many complex tasks. Similarly, to detect objects or humans in disasters, we require speed and accuracy. So, we select such algorithms for object detection that would allow robots and drones to detect the objects and enable the relative devices to perform its action and convey real-time scene information to human users. In the present scenario, the current detection systems repurpose the classifiers to perform detection. For object detection, a classifier is considered for that object, and it is evaluated across various locations and scale of the test image. For example, the systems like deformable part

G. A. Vishnu Lohit (✉) · N. Sampath
Department of Computer Science and Engineering, Amrita School of Engineering, Amrita Vishwa Vidyapeetham, Bengaluru, India
e-mail: gangaraju.vishnu@gmail.com

N. Sampath
e-mail: s_nalini@blr.amrita.edu

© Springer Nature Singapore Pte Ltd. 2020
K. S. Raju et al. (eds.), *Data Engineering and Communication Technology*,
Advances in Intelligent Systems and Computing 1079,
https://doi.org/10.1007/978-981-15-1097-7_48

model use the sliding window method where the classifier is applied over the full image with equal spacing of the locations. One of the approaches is the usage of R-CNN which uses a regional proposal. Initially, the given input image is resized, then it is passed through the convolutional network, and finally non-maximum suppression is performed. But, there are certain problems using this method. One such problem is that R-CNN takes a large time for the network training. The time taken for each test image is around 47 s. Hence, it cannot be implemented for real-time applications like disaster management and in traffic scenarios. The algorithm used in this method is the selective search algorithm which is a fixed algorithm, since there is no learning that leads to the generation of bad region proposals.

There are various methods which generate the bounding boxes in an image, and later a classifier is applied to the bounding box. After the process of classification, bounding boxes are refined and duplication is eliminated. Since the pipelines are complex, they are slow and difficult to optimize because separate training is required for each individual component. The object detection is then reframed as a problem of single regression starting from image pixels till the class probabilities. With the help of our proposed concept, we scan the picture completely to detect different objects present in the picture. With the help of YOLO, the complete pictures are trained and optimize the detection performance. The advantage of using this method is that the speed of the base network is around 45fps. This helps in streaming a video in less than 25 ms latency. The precision of YOLO is double than that of the mean average precision of other real-time systems. In this project, a webcam is used for the demo.

2 Related Work

Recent works have been conducted on object detection. A recent work was conducted on intelligent rescuer robot for detecting victims accurately in natural disasters [1]. This project describes a robot rescue team which uses its designed systems and helps in detection of victims when natural disasters occur. A 360° camera with virtual reality technology is used. The controlling of robot is done with the help of software by creating a package, and a joystick is used. The sensor data is displayed in another package.ag

Another method on the object detection was [2] worked on real-time, cloud-based object detection for unmanned aerial vehicles. The priority of this project is to detect an object in real time with the help of aerial vehicles at smaller computational time. The computation is an off-board computation while low-level object detection is performed by using CNN and R-CNN approach. Object detection is performed with the help of Faster R-CNN. R-CNN is an approach for object detection. With the combination of both the classifier and fast proposal mechanism, it is performed. The results are computed as follows. The experiments were conducted with three sets of data which help to show that the proposed approach performs successfully in a realistic but controlled environment. With the help of first set of data, the experiment is conducted and the accuracy of the deep network-based object detectors is tested. The first set of data is the images captured by the aerial vehicle. The second set of experiments, another performance parameter which is speed of the cloud-based object detection approach is evaluated. In the final set, the verification process of drone searching for a target object in an indoor environment is conducted, and it is verified. The verification is done as a simple simulation process in the field of hazard management in the form of a search and rescue or surveillance application. An analysis [3] on disaster management in India has been performed, and analysis is done based on five principles.

A project on Disaster Monitoring using unmanned aerial vehicles and deep learning [4] was done. The major objective of this project focused on the state of the artwork related to the use of deep learning techniques for disaster monitoring and identification. Moreover, the potential of this technique in identifying disasters automatically, with high accuracy, by means of a relatively simple deep learning model is demonstrated. As a deep learning model, the VGG engineering, pre-prepared with the ImageNet2 picture dataset is utilized. VGG comprises one of the numerous fruitful structures, which scientists may use to begin assembling their models as opposed to the beginning without any preparation. The images of natural disasters are taken from the Internet. The images from the Internet are considered in such a way that only images taken from the drone are used. Based on this dataset, the training of deep learning architecture is performed in order to classify automatically aerial photos according to the image group considered. The combination 80-20 (i.e., for splitting training and testing data) showed the best results in the experiment conducted. The time taken for the processing of training set was 20 min while the time taken for the processing of test set was around 5 min.

Hence, in most detection mechanisms, either CNN or virtual reality technology is used. The major drawback of these technologies is that the computation time is high and requires high processing GPU. YOLO also requires high processing GPU but the time of computation is very less and gives us much faster results. There are various other methods such as the security system using UAV to detect humans while harassing women [5], and indentifying the accident vehicles in the traffic scenario. The other works also include the usage of virtual 3D technology [1, 6] and also the general case studies on disaster management [7]. Furthermore, the usage of YOLO the traffic scenarios is studied [8, 9]. The co-author helped in giving a detailed analysis of different detection techniques using deep learning techniques in order to select the best and fastest detection techniques.

3 Implementation

YOLO stands for You Only Look Once. It is used for object detection. The object detection is done with the help of features that are learned with the help of a deep convolutional neural network. With the help of a convolution layer, the prediction is done and the convolutions are 1×1. YOLO is a fully convolutional network, and it has various skip connections and upsampling layers. Pooling is not used in YOLO, and the convolutional layer of stride 2 is performed. This helps in the downsampling of the feature maps. This is done in order to avoid the loss of low-level features.

YOLO does not vary with the size of the input image since it is a fully convolutional network. However, in practice, we use constant input size due to various problems which occur when we implement the algorithm. One of the problems among all the problems is that if the processing of images is done in batches all the images should have a constant height and width. This is required for concatenation of multiple images into a large batch. For the interpretation of the output, the features must be obtained by the convolutional layer. The resultant of the above step is then given as an input to the classifier for performing the detection.

Generally, convolutions of 1×1 are used by YOLO, and size of both the prediction map and feature map should be same. In YOLO v3 which is used in our project, the interpretation of the prediction map is done in such a way that the fixed number of bounding boxes is predicted by each cell. In neural networks, the term to describe the unit is called as a cell. In terms of the technical words, it is called as neuron.

If we consider in terms of depth, then the entries of a feature map can be calculated with the help of the formula given by ($B \times (5 + C)$). The number of bounding boxes per each cell is represented by B. $5 + C$ stands for the attributes for each of the bounding box. This helps in identifying the coordinates of the center and its dimensions. It also helps in the identification of confidences per bounding box and in general predicts three bounding boxes for every cell.

In a feature map, every cell of a map is expected to predict an object with the help of the bounding boxes if the object's center falls in the receptive field of the same cell. This can be done with the help of training YOLO in such a way that a single bounding box is responsible for object detection. In order to achieve it first, we must identify the cells to which this bounding box belongs to. Hence, the given image is divided into a grid having dimensions which is same as that of the feature map.

The example shown below consists of an image given as input. The size of the given image is 416×416, and stride of the network is 32. As mentioned, the feature map will have the dimensions of 13×13. The input image is then divided into 13×13 cells (Fig. 1).

Then, the cell which contains the center of the truth box of the given object is selected as the required one for the object prediction.

As shown in the figure, the red cell is the 7th cell in the 7th row on the grid. Then, we assign that cell on the feature map which is responsible for the detection of the dog in the image.

If we apply the log space transforms to the predefined bounding boxes, then we can obtain the dimensions of the bounding box, i.e., its width and the height. These predefined bounding boxes can also be called as anchor boxes.

Fig. 1 Prediction feature
map of an image

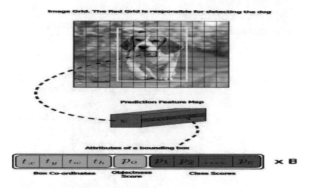

In the current version of YOLO which we are using in our project, there are three anchors. Hence, we can predict three bounding boxes for each cell. Later, the above-mentioned transformation method is applied to the anchors in order to find the prediction. The flowchart of the proposed work is shown (Fig. 2).

Intersection Over Union

IoU is called as intersection over union. Whenever a bounding box intersects with the anchor box, then IoU computes the size of intersection to the overall size of the union. For better analysis of object detection, IoU should be greater than or equal to 0.5. If the value of IoU falls below 0.5, then those boxes are not considered for the computation of object detection.

Making Predictions

The formula shown below gives the description of how the output network is transformed for obtaining the predictions of the bounding boxes.

$$b_x = \sigma(t_x) + c_x$$
$$b_y = \sigma(t_y) + c_y$$
$$b_w = p_w e^{t_w}$$
$$b_h = p_h e^{t_h}$$

Here, b_x and by represent the coordinates of the center of the grid cell x, y. The remaining coordinate b_w represents the width and b_h represents the height of bounding box, respectively. The coordinates t_x, t_y, t_w, and t_h shown in the above formulae are the output of the network. The coordinates c_x and c_y are the coordinates of the grid. The dimensions of the anchors for the box are represented by p_w and p_h, respectively. With all these values, we determine the values of the bounding box.

Fig. 2 Flowchart of the
proposed work

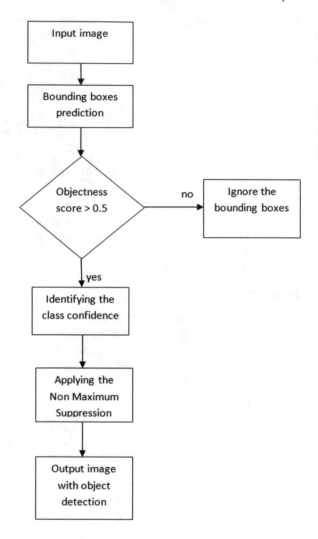

Center Coordinates

By using a sigmoid function, we can predict the center coordinates. By the help of this function, the output will lie between 0 and 1. The absolute coordinates of the center of the bounding box cannot be predicted.

For a brief explanation, let us consider image 1. Let the prediction values for obtaining the center be (0.4, 0.7) and the top left coordinates be (6, 6). Then, the coordinates of the center are (6.4, 6.7).

Let the predicted values be greater than one. Let us assume it as (1.3, 0.6). Then, the resultant center coordinates are (7.3, 6.6). From the observation, the center is the 8th cell of the 7th row which is the right side of the red cell. This violates the proposed theory because the assumption was that the red box is responsible for the object prediction. Hence, the center should lie in the red cell only.

With the help of the sigmoid function, we can avoid the problem by keeping the output values in the range of 0 to 1.

Objectness Score

The probability of an object that lies inside the bounding box is called as objectness score. This value should be equal to one or nearer to one for the red grid and the neighboring grids. If the grids are at the corner, then the objectness score should be close to 0. In order to interpret the value as probability, the sigmoid function is applied.

Class Confidences

If the detected object in an image belongs to a particular class, it can be shown in terms of probability value. These probability values are called as class confidences. Before v3, YOLO used to Softmax the class scores.

However, in v3, the usage of the sigmoid function took place instead of Softmax. This is because of that the assumption in Softmaxing is that the class scores are independent to each other. In other words, we can say that if the class of an object is found then that object does not belong to the other classes. This condition satisfies for the usage of COCO database. However, this condition does not hold true when we define classes like persons and ladies. Hence, sigmoid is used in determining the confidence values.

Output Processing

Consider an input. Let the size of the input be 416×416. The number of the bounding boxes is calculated by using the formula mentioned above, and we get 10,647 bounding boxes. Consider image 1 in this case, the input image consists of only a single object which is a dog. In order to reduce the detections from 10,647 bounding boxes to 1 bounding box, we apply the concept of thresholding. Thresholding is done with the help of using Object Confidence. In the first step, the boxes are filtered based on their objectness score calculated. The boxes which are not useful for further, i.e., having scores below a given threshold value are ignored. In general, we keep the threshold value to 0.5.

Non-maximum Suppression

After the removal of bounding boxes, there can be other boxes, which are duplicated, present in the feature map. In order to remove those duplicate boxes, we apply the concept of NMS. The main intention of NMS is to rectify the problem of many detections of the same picture.

4   Results

The size of the input image is considered as 416 × 416, the threshold value for confidence is considered as 0.5, and the threshold value for non-maximum suppression is considered as 0.4. The dataset used in the proposed work is the COCO dataset. It is large-scale object detection and captioning dataset. It consists of 80 classes with 80,000 training images and 40,000 of validation images.

The results are captured in two phases. In the first phase, the captured image is given as the input. The outputs are shown below.

Figure 3 is a captured image which is given as an input, and the objects detected in the image are a car and a motorcycle. These are already predefined classes in the COCO dataset. Even if the motorcycle is crashed, it can still be detected. This helps majorly in the application of disaster management.

Consider Fig. 4. This image was also captured and given as an input. With the help of YOLO, we can also identify the persons and other object. Here, a person is defined as a class. Hence, we can identify multiple persons in a single image.

Fig. 3 Object detection of a car and a motor cycle

Fig. 4 Object detection of persons and their bicycles

Fig. 5 Human detection by its body part

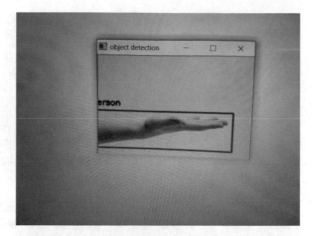

Fig. 6 Human detection with the help of streaming

In Fig. 5 a single human body part is enough for the identification of the person. It does not give the gender discrimination but we can identify it as a human body. With the help of this example, we can show that with the help of this technique we can detect the humans in natural disasters. As described above, the images 1, 2, 3 were taken from the Internet source.

In the second phase with the help of the camera, we can stream the data live and detect the objects and humans. In order to do so, we used the webcam of the laptop for live streaming and detect the objects. The inference time measured is also displayed. The dataset used in object detection while streaming is the same COCO dataset.

Figure 6 represents the screenshot of detecting a person with the help of live streaming using webcam as mentioned above. In Fig. 6, a person is sat in front of the webcam, and the streaming is done. With the help of YOLO, we can identify it as a person with the inference time as 447.07 ms.

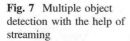

Fig. 7 Multiple object detection with the help of streaming

Figure 7 is another example for object detection with the help of streaming the camera. Different objects such as chair, mouse, and laptop are detected. The inference time is also displayed which is 542.35 ms. The images 4 and 5 are the screenshots of the live video streaming. This detection technique is performed with the help of a NVIDIA 1050 graphic processor unit.

5 Conclusion and Future Scope

Object detection with the help of YOLO software is implemented, and the results are shown. With the help of object detection using YOLO, we can obtain the required result with a decent accuracy and less time. As shown in Figs. 1 and 3, we can identify humans and objects even when they are damaged or injured. This technique is also helpful even with the live streaming where many objects are detected in a single frame with small inference time.

For future work, we try to connect a camera to a quadcopter and try to stream the data so that for longer distances which are out of visible range we try to detect the objects and humans. We will also connect a navigation device to the drone and also try to identify the location of the humans.

References

1. Pouransari, A., Pouransari, H., Inallou, M.M.: Intelligent rescuer robot for detecting victims accurately in natural disasters. In: 2nd International Conference on Knowledge-Based Engineering and Innovation (KBEI) (2015)
2. Lee, J., Wang, J., Crandall, D., Šabanović, S., Fox, G.: Real-time, cloud-based object detection for unmanned aerial vehicles. In: First IEEE International Conference on Robotic Computing (IRC) (2017)

3. Mohanan, C., Menon, V.: Disaster management in India—an analysis using COBIT 5 principles. In: 6th Annual IEEE Global Humanitarian Technology Conference, GHTC (2016)
4. Kamilaris1, A., Prenafeta-Boldú, F.X.: Disaster monitoring using unmanned aerial vehicles and deep learning (2017)
5. Mohan, R., Raj, C.V. Aswathi, P., Bhavani, R.R.: UAV based security system for prevention of harassment against woman. In: International Conference on Intelligent Computing, Instrumentation and Control Technologies (ICICICT) (2017)
6. Lee, S., Har, D., Kum, D.: Drone-assisted disaster management: finding victims via infrared camera and lidar sensor fusion. In: 3rd Asia-Pacific World Congress on Computer Science and Engineering (APWC on CSE) (2016)
7. Youngjib, H., Han, K.K., Lin, J.J., Golparvar-Fard, M.: Visual monitoring of civil infrastructure systems via camera-equipped Unmanned Aerial Vehicles (UAVs): a review of related works (2016)
8. Tao, J., Wang, H., Zhang, X., Li, X., Yang, H.: An object detection system based on YOLO in traffic. In: 6th International Conference on Computer Science and Network Technology (ICCSNT) (2016)
9. Prasath, N.K., Sampath, N., Vasudevan, S.K.: Dynamic reliable algorithm for IoT based systems. In: 2017 International Conference on Signal Processing and Communication (ICSPC) (2017)

Prediction of Remaining Useful Life of an End Mill Using ANSYS

Venkateswara Rao Mudunuru and Saisumasri Komarraju

Abstract Remaining useful life (RUL) is the amount of time remaining for a mechanical component to perform its functional capabilities before failure. There are several prognostics prediction methodologies and techniques available to determine the RUL of mechanical components. In this paper, we are applying a combination of analytical and model-based approaches to generate a tool life prediction equation for a two-flute end mill. We performed static and dynamic analysis on a two-flute end mill in ANSYS Workbench with and without crack. With the tool life results that are obtained from fatigue loading for an end mill with crack, we generated design of experiments by various techniques like Central Composite Design, Optimal Space Filling Design, Box–Behnken Design, Sparse Grid Initialization, and Latin Hypercube Design. We modeled tool life equation for the experimental data using regression techniques.

Keywords Remaining useful life · Design of experiments · Two-flute end mill · Prediction · ANSYS · Regression · Prediction

1 Introduction

Milling is defined as a machining operation in which work is fed past a rotating tool with multiple cutting edges. In traditional milling operations, axis of the cutting tool rotation is perpendicular to the feed. Milling creates planar surfaces and based on the cutter path or shape, other geometries can be obtained. The cutting tool is sometimes referred as cutter. End milling is one of the milling method, where the diameter of the cutter is smaller than the width of the workpiece to be machined. In the mechanical workshops, the experts have commonly utilized cutters from various providers and used their experience and knowledge for opting cutting conditions and the ideal opportunity for changing the cutting tool. Regardless of the consid-

V. R. Mudunuru (✉) · S. Komarraju
University of South Florida, Tampa, FL, USA
e-mail: vmudunur@mail.usf.edu

© Springer Nature Singapore Pte Ltd. 2020
K. S. Raju et al. (eds.), *Data Engineering and Communication Technology*,
Advances in Intelligent Systems and Computing 1079,
https://doi.org/10.1007/978-981-15-1097-7_49

erable number of guidelines and encounters accessible, it is hard to make a dependable forecast of the cutting tool life; this was clearly insufficient for utilizing the devices effectively and with ideal effectiveness. Cutting tool providers promote their products as exceedingly proficient machining tools. The success rate may vary in practice [1]. Even though the experts use their knowledge and experience before changing the tool after using it for a particular time, it is very important to evaluate the RUL of the tool while being used, since it has impacts on the maintenance of tool's performance. RUL estimation is vital in the administration of tool reuse and reuse that affects energy utilization. There are several prognostics prediction methodologies and techniques available to determine the RUL of mechanical components. These methodologies include model-based methodology, knowledge-based, analytical-based, hybrid- and statistics-based. Analytical-based RUL estimation methodology represents the physical failure technique of the tool. Tool fracture modes such as crack propagated by fatigue, flank wear, and corrosion of surfaces can be studied and modeled using the methodology [2]. As the name suggests, statistical-driven approaches estimate the RUL by application of statistical techniques like fitting model [3]. Data collected for these methods is called as direct control monitoring data [4]. These methodologies can also be considered as indirect methods as these are based on the optimizing the machining parameters [5].

2 Literature Review

Meidert et al. [6] developed a software to calculate damage parameters and predict lifetime from results of nonlinear FEM die stress analysis. Liang et al. [7] applied finite element method to analyze precision forming process of the speed-reducer shaft of auto starter. This research acts as a guiding role on the simulation of the cold forging and the design of the die set. Horita et al. [8] performed fatigue analysis on forging die by conducting repeated forging. Further, they proposed a new simulation method for estimating lifetimes of the forging die. Gohil [9] analyzed the closed die forging process of AISI 1016 by simulation. Authors studied the variation in forging by introducing variation in stress, strain, force, temperature, etc. Desai et al. [10] investigated failure analysis of HSS grade steel used as a punch tool during cold forging process.

3 Methodology

3.1 Static and Dynamic Analysis in Workbench

The basic assumptions made for the static component analysis are that the symmetrical machining process is in consistent state, the cutting tool is an inflexible body, chip is generated in a continuous manner, and the material of the workpiece is elastic–viscoplastic. A static analysis ascertains the impacts of consistent loading conditions on a component, while disregarding latency and damping impacts, for example, those brought about by time-changing loading [11]. In this analysis, a rigidity matrix is determined for every component as per the given determinations. A rigidity matrix for the entire system is developed by summing up rigidity matrix of each element of the component. The result of this analysis discovers the areas of maximum stress and displacements on the component. These areas render to decide the most important zones to be utilized in the dynamic analysis. The initial step of dynamic analysis is the modal analysis. Since the behavior of tool should be linear for this method, any non-linearity, for example, versatility and contact components are disregarded regardless of whether they are characterized. The assumptions made for this analysis are that the model has stable hardness and unbending body motions, Damping is substantial just for the damped mode extraction strategy and is overlooked for the others, forces are not time fluctuating. Removals, weights, and temperatures are not considered [12]. The point of harmonic analysis of a cutting tool is to apply forces on the structure from specific locations at specific intervals and to decide the response of these areas to the forces. As the aftereffect of this analysis is that, the cutting edge is displaced more in the y-axis. Even then, we consider displacements in all directions for optimal results. At the point, when the driving frequency of the resultant force concurs with one of the common frequencies of the cutting apparatus structure, the device breaks. The assumptions made for this analysis are that all forces must be sinusoidal time fluctuating with same frequency. Even for this analysis, linear behavior is accepted. Non-transient impacts are not determined [13].

3.2 Design of Experiments

Design of experiments (DOE) or experimental structure is the plan of any data gathering practices where variety is available, regardless of whether under the full control of the experimenter or not. A system for structuring tests was proposed by Ronald A. Fisher. DOE is a grouping of tests in which changes are made to the parameters of a framework and the consequences for reaction factors are estimated [14]. DOE is applicable to physical procedures and to simulation models. Anyway, in measurements these terms are typically utilized for controlled tests. DOE is a powerful technique for augmenting the measure of data collected from an examination while limiting the measure of information to be gathered. Factorial trial plans examine the impacts of a wide range of components by differing them at the same time as opposed to changing just a single factor at any given moment. Factorial plans enable estimation of the affectability to each factor and furthermore combined impact of at least two components [15]. Figure 1 gives the picture of workflow of this research work.

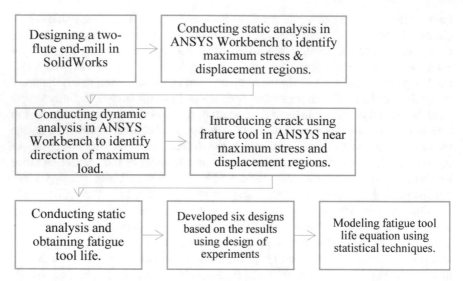

Fig. 1 Workflow of the simulation process

3.3 Framework Adopted for RUL Modeling

This paper is divided into four steps. In the first step, we designed two-flute end mill in SolidWorks. All the analysis part is done using ANSYS Workbench. In our next step, we conducted static and dynamic analysis (Modal and Harmonic), of a two-flute end mill without introducing the crack into the design. We performed static analysis to understand the behavior of end mill under static loading and to identify the maximum stress and displacement regions in the design. In this analysis, along with areas of maximum stress and displacements, we also studied the end mill life under fatigue loading conditions. To study the dynamic behavior of the cutting tool, we performed modal and harmonic analysis. Modal analysis helped us to find out natural frequencies of the end mill along with its mode shapes. Based on the results obtained from modal analysis, we performed harmonic analysis to identify the direction of maximum displacement because of varied loads at intervals. We repeated step-2 after introducing a crack in the design of the end mill. In the third step, we developed design of experiments for life of end mill with crack. DOE is developed by following ANSYS Design Exploration module documented technique [16]. We have developed DOE using various techniques including Central Composite Design (CCD), Optimal Space filling Design (OSD), Box–Behnken Design (BBD), Sparse Grid initialization Design (SGD), and Latin Hypercube Design (LHD). Results helped us to assess the relationship of each input parameter with the output parameter through correlation scatter plots, response surfaces, sensitivity charts. In the final step, we modeled tool life equation by applying both first-order and second-order regression techniques. In some cases, the input factors exhibit some curvature of the cutting parameters. Accordingly, it is helpful to consider second-order polynomials to demonstrate this investigation. This helps us to study the second-order effect of each contributing factor to the output parameter.

4 Computational Conditions for RUL Modeling

4.1 Design Parameters

Factors affecting tool life are geometry of the cutting tool, machining parameters, cutting fluid, and workpiece being machined. Theoretically, main factors that contribute to tool life are cutting speed, feed rate, depth of cut, and rake angle. The conventional tool life equation framed by Taylor suggests that the cutting tool life depends on the velocity of the cutting tool along with cutting tool material constants. In our paper, we used, High-speed steel (HSS) two-flute end mill. The end mill has an overall diameter of 0.75″ with length of cut of 2″, overall length of 2.357″, and minimum mill diameter of 0.1875″, maximum mill diameter of 0.375″, and a helix angle of 15°. Our experimental analysis for a two-flute end mill made up of HSS milling AISI-H13 tool steel. Based on the cutting tool and workpiece material combination, the operational machining parameters are as follows: Cutting speeds vary between 250 and 800 SFM, cutting feed ranges from 0.002 to 0.006 FPT, and a depth of cut of 0.06–0.125″.

4.2 Crack Dimensions

We initiated a semi-elliptical crack with major radius of 0.005″ and minor radius of 0.002″ with a largest counter radius of 0.02″. The static analysis parameters are as follows: spindle speed—4800 RPM, feed rate—38.4 IPM, and depth of cut—0.07″.

5 Results and Discussion

5.1 Results of Static and Dynamic Analysis

After performing static analysis of end mill without fracture, maximum principal stress values and strain values are recorded and are given in Table 1. We initiated crack in the area closer to the maximum principal stress region to analyze the life. A figure of maximum stress regions is given in Fig. 2 (ANSYS image).

Table 1 Results of static and dynamic analysis

Analysis type	Equivalent stress (psi)	Principal stress (psi)		Principal strain (psi)	
		Max.	Min.	Max.	Min.
Static	4.4625	5.7595	1.8831	$1.5742 \times e^{-7}$	$-1.0651 \times e^{-9}$
Dynamic	16,283	8765.9	53.773	0.00039	$-5.5491 \times e^{-8}$

Fig. 2 ANSYS—static structural image with maximum principal stress

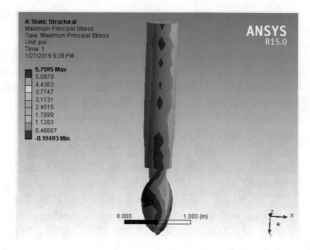

6 Design of Experiments Results

Table 2 has the results obtained after performing design of experiments using the parameters applied for static analysis under fatigue loading.

6.1 Estimation of the Model Parameters

The regression parameters of the developed models are estimated by the method of least squares estimates. The three parameters of the first-order predictive model are

Table 2 Results of ANSYS design of experiments (DOE)

DOE	Number of design points	Mean RUL	|d|
BBD	13	16.16	3.41
CCD	15	15.94	3.29
LHD	15	15.72	3.17
OSD	15	15.72	3.17
SGD	7	15.69	3.14

d = variation in Stress and Strain life of the tool

Table 3 Parameter estimates and measures of performance of models

DOE	Coefficients of parameters			F	R^2 (%)	Adj R^2 (%)	RMSE
	b_1	b_2	b_3				
BBD	0.001	-0.056^*	-0.092	17.75	85.5	80.7	1.33
CCD	0.001	0.048^*	-0.088	24.81	87.1	83.6	1.02
LHD	0.001	-5.519^*	-0.103	135.78	97.4	96.7	0.37
OSD	0.001	2.771^*	-0.089	45.539	92.5	90.5	0.63
SGD	0.001	0.007^*	-0.102	9.19	90.2	80.4	0.97

estimated for all the designs considered in the study. The estimates of β coefficients (b_1, b_2, b_3) of all the models along with the model validation statistics including F-statistic, R^2, adjusted R^2 (Adj. R^2), and root mean square error (RSME) are presented in Table 3. The coefficient of depth of cut, b_2, for all the models was insignificant with high p-values and hence eliminated from the model. The R^2 and adjusted R^2 values of Latin Hypercube Design model are high when compared with the other models and the RMSE is very low. The residual plot has a close to linear trend and the attributable variable cutting speed and feed rate also fall in a close to straight line as shown in Fig. 3. With these qualities, we conclude that Latin Hypercube Design has a better performance and the details of the cutting conditions and results are in Table 4. Using these simulation results of Latin Hypercube DOE using ANSYS, the following prediction model for remaining useful life, RUL, (in cycles) is developed using SPSS and SAS Enterprise Miner.

$$\ln (\text{RUL}) = 16.63 + 0.001(\text{cutting speed}) - 0.103(\text{feed rate}) \qquad (1)$$

Applying antilog to Eq. (1) will provide the predictive tool life (in cycles) as a function of cutting speed and feed rate. We further evaluated the linear regression models using ANOVA to check the adequacy. ANOVA result for LHD is given in Table 5. The F-ratio obtained is significant for all the models and all the models are adequate with 95% confidence interval.

7 Conclusion

This goal of this research work is to take a diversion from traditional methodology of estimating remaining useful life in a mechanical workshop environment. We designed and developed a two-flute end mill made of HSS using SolidWorks to evaluate its remaining useful life. Later we imported the design to ANSYS and simulated the results to identify the RUL of the tool by imposing the required lab conditions. We employed various design of experiment techniques available in ANSYS and recorded the results for each design. These simulated results are considered to develop multiple linear regression models to predict the remaining

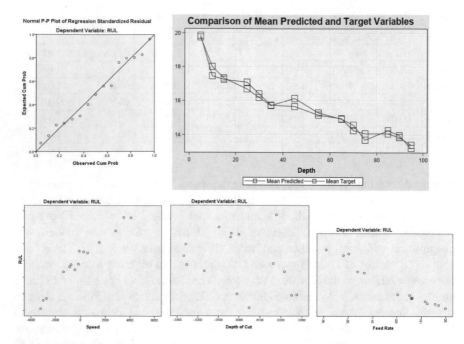

Fig. 3 (Top left) residual plot of RUL; (top right) comparison of mean predicted and target variables; (below) partial regression plots of speed; depth of cut and feed Rate parameters compared with RUL

Table 4 Cutting conditions and results of latin hypercube design

Trail	Spindle speed (RPM)	Depth of cut (IN)	Feed rate (IN/MIN)	RUL
1	7200	0.1185	38.4	18.00324
2	9866.667	0.1055	78.93333	15.28798
3	11,466.67	0.070833	48.92444	19.73685
4	9333.333	0.062167	79.64444	14.91183
5	8266.667	0.088167	83.76889	13.96595
6	7733.333	0.0795	53.61778	16.17962
7	12,000	0.114167	76.8	16.69918
8	6133.333	0.0665	71.96444	13.21696
9	6666.667	0.096833	39.11111	17.31163
10	10,400	0.075167	94.29333	14.56695
11	5600	0.122833	59.73333	13.69935
12	8800	0.083833	84.48	14.25653
13	10,933.33	0.109833	81.63556	15.70569
14	12,533.33	0.0925	60.16	18.79167
15	5066.667	0.101167	56.74667	13.45084

Table 5 ANOVA table of latin hypercube design

Model	SS	df	MS	F
Regression	55.863	3	18.621	135.782
Residual	1.509	11	0.137	
Total	57.372	14		

useful life of the tool. Performance evaluation and the validity of these models have been evaluated by ANOVA and other statistical techniques. The results indicate that the models are reliable. As a part of future work, we are working to input the crack size as an attributable variable in modeling RUL.

References

1. Dolinšek, S., Šuštaršič, B., Kopač, J.: Wear mechanisms of cutting tools in high-speed cutting processes. Wear **250**(1–12), 349–356 (2001)
2. Sikorska, J.Z., Hodkiewicz, M., Ma, L.: Prognostic modelling options for remaining useful life estimation by industry. Mech. Syst. Signal Process. **25**(5), 1803–1836 (2011)
3. Si, X.S., Wang, W., Hu, C.H., Zhou, D.H.: Remaining useful life estimation–a review on the statistical data driven approaches. Eur. J. Oper. Res. **213**(1), 1–14 (2011)
4. Jardine, A.K., Lin, D., Banjevic, D.: A review on machinery diagnostics and prognostics implementing condition-based maintenance. Mech. Syst. Signal Process. **20**(7), 1483–1510 (2006)
5. Amaitik, S.M., Taşgin, T.T., Kilic, S.E.: Tool-life modelling of carbide and ceramic cutting tools using multi-linear regression analysis. Proc. Inst. Mech. Eng. Part B: J. Eng. Manuf. **220**(2), 129–136 (2006)
6. Meidert, M., Walter, C., Poehlandt, K.: Prediction of fatigue life of cold forging tools by FE simulation and comparison of applicability of different damage models. In: Tool Steels in the Next Century, 5th International Conference on Tooling (Leoben) (1999)
7. Li, L., Wang, P.: Numerical simulation and technology research on the precision forging for speed-reducer shaft of auto starter. In: 2010 International Conference on System Science, Engineering Design and Manufacturing Informatization (ICSEM), vol. 2. IEEE (2010)
8. Horita, A., et al.: Fatigue analysis of forging die. J. Therm. Stresses **35**(1–3), 157–168 (2012)
9. Gohil, D.: The simulation and analysis of the closed die hot forging process by a computer simulation method. J. Systemics **10**(3), 88–93 (2012)
10. Desai, S., et al.: Failure analysis of HSS punch tool: a case study. In: Proceedings of Fatigue, Durability and Fracture Mechanics, pp. 93–101. Springer, Singapore (2018)
11. Zhang, B., Bagchi, A.: Finite element simulation of chip formation and comparison with machining experiment. J. Eng. Ind. **16**, 289–297 (1974)
12. Shih, A.J.: Finite element simulation of orthogonal metal cutting. J. Eng. Ind. **117**, 85–93 (1995)
13. Cakir, M.C., Yahya, I.S.: Finite element analysis of cutting tools prior to fracture in hard turning operations. Mater. Des. **26**(2), 105–112 (2005)
14. Telford, J.K.: A Brief Introduction to Design of Experiments, vol. 27, Issue No. 3. Johns Hopkins APL Technical Digest (2007)
15. Chand, A.G., Bunyan, J.V.: Application of Taguchi technique for friction stir welding of aluminum alloy AA6061. Int. J. Eng. Res. Technol. **2**(6) (2013)
16. ANSYS® DesignXplorer, Release 15.0, Design Exploration User's Guide, Using Design of Experiments, ANSYS, Inc

Colon Cancer Stage Classification Using Decision Trees

M. Vidya Bhargavi, Venkateswara Rao Mudunuru
and Sireesha Veeramachaneni

Abstract Decision tree methodology is one of the most commonly used data mining techniques and is used for both classification and regression problems. The objective of this research is to perform a comparative study on stage classification of colon cancer data based on multiple covariates using various decision tree algorithms and to identify the best classifier for stage classification of colon cancer data. Decision tree algorithms are non-parametric and can efficiently deal with large datasets. We performed the analysis by splitting the data under study into training and testing datasets. This paper utilizes frequently used decision tree algorithms including CART, QUEST, and CHAID and identify on the appropriate tree size needed to achieve the optimal final model.

Keywords Decision tree · Data mining · Colon cancer · Classification · Prediction

1 Introduction

Most people do not know exactly what cancer is, they just know that it is never good thing when someone has it. Cancer can be life-threatening, if not detected and treated within a short amount of time. Most cancer tumors form by the rapid and uncontrolled reproduction and division of abnormal cells in a particular part of the human body [1]. This process normally results in a malignant growth, also known as a tumor. This disease also destroys the body's tissue in which the cells are reproducing. Since this disease can occur in any part of the body that only means

M. Vidya Bhargavi (✉)
Stanley College of Engineering and Technology for Women, Hyderabad, India
e-mail: bvidya@stanley.edu.in

V. R. Mudunuru
University of South Florida, Tampa, FL, USA

S. Veeramachaneni
GITAM University, Visakhapatnam, India

© Springer Nature Singapore Pte Ltd. 2020   599
K. S. Raju et al. (eds.), *Data Engineering and Communication Technology*,
Advances in Intelligent Systems and Computing 1079,
https://doi.org/10.1007/978-981-15-1097-7_50

that, there are several different types of cancer. However, out of all the cancers that are infecting the lives and bodies of millions of people, colon cancer is the most commonly diagnosed.

Colon cancer is the third most commonly diagnosed cancer and the second leading cause of cancer death in men and women combined in the USA. Colon cancer can occur in any part of the colon [2]. Colon cancer is cancer of the large intestine, the lower part of your digestive system. Rectal cancer is cancer of the last several inches of the colon. Together, they are referred as colorectal cancers. Most cases of colon cancer begin as small, noncancerous clumps of cells called adenomatous polyps. Over time, some of these polyps become colon cancers. With regular screening, colon cancer can be found early, when treatment is most effective. In many cases, screening can prevent colon cancer by finding can removing polyps before they become cancerous. Moreover, if cancer is present, earlier detection means a chance at a longer life.

Research shows that one in every twenty people will be diagnosed with some type of colon cancer [3]. This is roughly five percent of the population, who will contract colon cancer. However, the proven statistics on colon cancer between men and women are slightly different. The number of men and women diagnosed with colon cancer is about 4.6% and 4.2%, respectively [4, 5]. However, in recent years, this trend has been reduced due to the effective diagnostic techniques, which can cure the cancer if it is diagnosed in an appropriate time. The recent advancements of data-driven techniques have introduced new and effective ways in the area of cancer diagnostics. Data-driven techniques including Artificial Neural Networks, fuzzy systems, decision trees, KNNs, Support Vector Machines (SVM), etc. [6–8] are being actively used in many areas like engineering, medical, and pharmaceutical areas of research. It is understood that data evaluation, which has been attained from patients, can be considered as an important factor to develop an efficient and accurate diagnostic method. To this end, classification algorithms have been utilized to minimize the error of human errors that may happen during the treatment. In this paper, we will implement decision trees techniques on colon cancer data, to predict the stage of cancer with from the knowledge of the patient's demographic and medical characteristics.

2   Literature Review

Liu et al. [9] implemented C5 algorithm on breast cancer data to predict breast cancer survivability. Li et al. [10] used C4.5 algorithm on ovarian tumor data to diagnose cancer. Tsirogiannis et al. [11] applied various classifiers including artificial neural networks, support vector machines and decision trees on medical databases. Kourou et al. [12] presented a detailed review of recent machine learning approaches including artificial neural networks, Bayesian networks, support vector machines, and decision trees employed in the modeling of cancer progression. Tseng et al. [13] employed various statistical techniques and machine learning techniques on oral cancer data to establish a model for predicting the five-year disease-specific survival rate and five-year disease-free survival rate. PubMed search on "Colon cancer and decision trees" resulted in three articles as a good match for our search. "Colon cancer stage classification and decision trees," "Colon cancer SEER data analysis and risk factors and decision trees," and "Colon cancer SEER data analysis and stage classification and decision trees" resulted in zero ("0") articles. Clearly, there is a great potential for research in this analyzing colon cancer data using decision trees and any machine learning techniques in general.

The purpose of this study is to investigate the use of algorithms in the decision tree tool for data mining and classifying the stage of a colon cancer patient with the given attributable variables. We will implement the CART, QUEST, and CHAID classification algorithms with training–testing splits as 70–30% and 80–20%, respectively. We will also apply pruning for CART and QUEST algorithms. Finally, we identify the best classification algorithm as the one that is trained with less data, uses less nodes, and with high accuracy when classifying.

3 Data Description

For this research, we utilized patient and populace information from the SEER Registries Database. The Surveillance, Epidemiology, and End Results (SEER)—Medicare database links information from the National Cancer Institute's SEER disease vault program with cases information from Medicare, the governmentally financed protection program for the US older. This information is accessible to agents and has been utilized broadly in research. This asset is significant for leading exploration on tumors [14]. SEER is a National Cancer Institute-subsidized program gathering information on malignant growth occurrence and survival from US disease libraries (https://www.seer.cancer.gov). The SEER database consolidates tolerant dimension data on disease site, tumor pathology, stage, and reason for death [14]. In this work, we preprocessed the SEER information (time of 2004–2015 with all records named in colon.txt) for colon tumor to expel redundancies and missing data. The resulting data set had 30,251 records, which is a combination of four races, Caucasians, African Americans, American Indians, and others. The decision tree model used seven input variables and an output variable. The input variables age, tumor size, survival time, and number of positive lymph nodes are the numeric values for decision tree classification and the input variables grade of the tumor, survival status, and extension of neoplasm are the categorical variables. The variable stage is the output variable with levels I, II, III and IV, respectively. For numerical variables minimum, maximum, mean, and standard deviation were calculated and tabulated in Table 1. Frequency values of categorical variables are given in Table 2.

Table 1 Numerical input variables

No.	Variable	Statistics			
		Min	Max	Mean	Standard deviation
1	Age	15	106	67.74	13.97
2	Tumor size (in mm)	0	989	48.14	35.27
3	Survival time (in months)	0	143	52.16	38.8
4	Positive Lymph nodes	0	90	2	4.05

Table 2 Categorical input variables

No.	Variable	Frequency
1	Grade	1: Well differentiated—9.4% 2: Moderately differentiated—69.1% 3: Poorly differentiated—20% 4: Undifferentiated—1.6%
2	Survival status	0: Survived—72.7% 1: Not survived—27.3%
3	Extent of neoplasm	0: In situ—0.4% 1: Localized—38.3% 2: Regional—45.8% 4: Distant—15.4%
4	Stage—output	Stage I: 20.6% Stage II: 32.4% Stage III: 21.4% Stage IV: 25.7%

4 Decision Trees

A Decision tree is a decision supportive algorithm that utilizes tree-like diagram choices and their conceivable eventual outcome. On providing the values of independent variables, a decision tree, also called as classification tree, determines the value of dependent variable by learning through a classification function. This technique is considered as supervised classification algorithm because the count of classes and dependent variable are provided [15]. Decision trees are the most dominant methodologies in learning revelation and data mining. Unlike black-box algorithms like artificial neural networks, support vector machines, etc., decision trees will let us check and interpret the logic and rules implemented while doing the process of classification. It incorporates the innovation of research expansive and complex part of information to find valuable patterns. This thought is imperative since it empowers displaying and learning extraction from the majority of information accessible. All theoreticians are constantly hunting down strategies to make the procedure increasingly productive, practical, and precise. Decision trees are profoundly viable techniques fields like data and text mining, data extraction, machine learning, and pattern recognition. Decision tree offers numerous advantages to information mining, some are as per the following:

- Straightforward and easy to understand.
- Can deal with heterogeneous input data like Textual, Nominal, and Numeric.
- Able to process fallacious datasets or missing records.
- High execution with modest number of endeavors [15].

Data mining instruments are turned out to be effective in field of medical diagnosis. The blend of the two data mining techniques alongside decision trees is well-known and successful approach that gives justifiable and clear classifications that are applicable by doctors and therapeutic authorities. Data mining strategies help to diminish the false positive and false negative choices [15]. This will effectively produce promising outcomes in the colon cancer diagnosis. A decision tree includes—decision node or root node, leaf or terminal nodes, internal nodes. Decision nodes and leaf nodes are used to represent any classes whereas internal nodes as shown in Fig. 1 represent applied test conditions on variables.

The number of nodes, depth of the tree, number of terminal nodes, and number of characteristics utilized influence decision tree precision. The intricacy is controlled expressly by the ceasing criteria utilized and pruning strategies utilized. The target of the decision trees is to locate the ideal decision tree by limiting the general miscalculations. To take care of heuristic issues with substantial informational indexes decision tree inducers with developing and pruning are as a rule effectively utilized. The calculations utilized pursue the idea of "divide and rule" in assessing for the last ideal decision tree.

In this technique, based on the values of attributable variables, partition of the training set is executed. Splitting measures are employed to select the appropriate function. Internal nodes are further divided into sub-nodes and splitting measures are applied if needed or the process is stopped if the required criteria are satisfied [16].

Fig. 1 Example of a decision tree

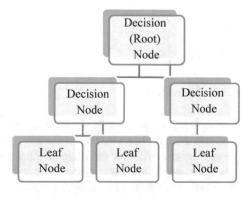

4.1 Growing Methods

4.1.1 CART

CART represents Classification and Regression Trees presented by Breiman. It is developed based on Hunt's algorithm. Both categorical and continuous variables can be handled by CART while constructing a decision tree. It handles missing records as well. CART utilizes Gini Index as an attribute selection measure to construct a decision tree. Gini Index measure does not utilize probabilistic presumptions like ID3, C4.5. CART utilizes cost complexity pruning to expel the inconsistent branches from the decision tree to enhance the precision [17].

4.1.2 QUEST

QUEST—Quick, Unbiased, Efficient Statistical Tree was proposed by Loh and Shih in 1997. QUEST is effective if the dependent variable is nominal. Comparing with other decision tree algorithms, QUEST is fast and avoids other methods bias. QUEST identifies the best contributing categorical predictors by method of chi-square and the numerical variables by F-statistic. Contributing variables is identified by comparing p-values of F and chi-square statistics of the variables.

4.1.3 CHAID

CHAID—Chi-squared Automatic Interaction Detection is considered as the best AID that reduces or nullifies the bias. Significant testing of independent variables using chi-square statistic carries out bias reduction. In each step, CHAID will identify the predictor variable that has a strong correlation with the dependent variable and will merge the categories if they are not significantly contributing to the response.

4.2 Validation

4.2.1 Splitting

The training segment, normally the biggest segment, contains the information used to construct the different models we are inspecting. The test segment is utilized to survey the execution of the selected model with new information [18]. This technique is preferred when we have a large datasets. In this work, we employed 70–30% and 80–20% for building the models and compared their performances.

4.2.2 Cross Validation

There are several forms of validation that can be used to assess how to go about statistically analyzing a data set, one of these types is k-fold cross validation. Cross validation is preferred when the dataset is small. Cross validation is a way of insuring that a researcher has chosen a proper statistics model to evaluate a set of data. K-fold cross validation is when the data set is split into "K" equal sized samples. Each of these samples is tested as either validation data or as training data ($K - 1$). This process is repeated "K" times and these results are averaged to obtain a single result [18].

5 Framework for the Modeling

The entire work in the paper is divided into two parts. The first step deals with the identifying the attributable variables for stage classification of colon cancer using Chi-squared, extra tree classifier, and logistic regression. We eliminated of the input variables, which have, less than 5% normalized importance in performance of cancer stage classification. In the second step, we used these attributable variables to develop decision tree classifier that can best classify the stages of the subjects. In our paper, we used CART, QUEST and CHAID techniques to develop decision tree classifiers. We will develop several models using CART, QUEST and CHAID techniques by first splitting the data into 70% for testing, 30% training sets (we write it 70–30%), and again splitting 80% for testing, 20% for training (we write it 70–30%). We will compare all the developed models and identify the best model for stage classification of colon cancer data. Figure 2 gives the workflow of this classification process.

Fig. 2 Workflow of colon cancer stage classification

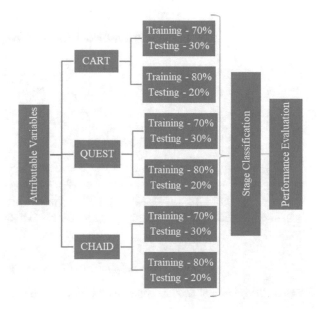

6 Results

After performing first step of the framework, we identified age, tumor size, survival time, number of positive lymph nodes, grade, survival status and extent of neo-plasm as contributing variables to construct decision tree classifier. We developed 10 models using the above-mentioned techniques. While developing the CART and QUEST models, we also considered and compared the models developed with pruning and models developed without pruning. We compared all the developed models with the time of processing, depth, number of terminal nodes, accuracy, and error rate. Figure 3 compares the number of terminal nodes and depth of the developed trees.

Comparing the trees with least number of nodes, terminal nodes, depth, and accuracy, for the CART model (70–30% split) with pruning, we obtained an overall accuracy of 84.9% for training and testing for the stage classification. This model also has less number of terminal and total nodes with a classification error of 0.151. When compared this CART model without pruning, it has an overall accuracy of 85% with 17 terminal nodes and a tree depth of 5. As anticipated, the QUEST model with 70–30% split and pruning has a very low execution time compared to CART and CHAID with same split. However, it has an accuracy of 80% with 11 terminal nodes and with a depth of 4. The overall accuracy of QUEST model without pruning is 82.5 with 20 terminal nodes and with a tree depth of 5. CHAID model with 70–30% split has an overall accuracy of 85.2% with tree depth as 3, but has 25 terminal nodes. For the split of 80–20% with pruning, the CART model has an overall accuracy of 84.7% with 15 nodes, 8 terminal nodes and a tree depth of 4, QUEST model has an overall accuracy of 80% with 11 terminal nodes and a tree

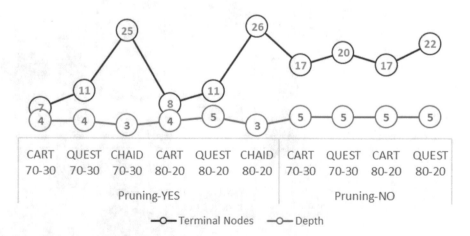

Fig. 3 Comparison of terminal nodes and the depth of the trees of the decision tree models

Table 3 Confusion matrix of training set of CART model

Observed	Predicted				
	I	II	III	IV	Percent correct (%)
I	2699	1725	0	1	61.0
II	832	5924	0	129	86.0
III	0	342	3978	138	89.2
IV	3	24	21	5441	99.1

Table 4 Comparison results of developed models

Pruning		Execution time	Number of nodes	Terminal nodes	Depth	Error
Pruning-YES	CART 70–30%	1.36	13	7	4	0.151
	QUEST 70–30%	0.84	21	11	4	0.197
	CHAID 70–30%	1.17	36	25	3	0.148
	CART 80–20%	1.3	15	8	4	0.151
	QUEST 80–20%	0.62	21	11	5	0.2
	CHAID 80–20%	0.91	38	26	3	0.148
Pruning-NO	CART 70–30%	1.28	33	17	5	0.15
	QUEST 70–30%	0.77	39	20	5	0.175
	CART 80–20%	1.17	33	17	5	0.152
	QUEST 80–20%	0.66	43	22	5	0.19

depth of 5. CHAID model achieved an overall accuracy of 85.2% with 26 terminal nodes and a tree depth of 3. Finally, for an 80–20% split without pruning, CART and QUEST has an overall accuracy of 84.8% and 81%, terminal nodes of 17 and 22, and a tree depth of 5, respectively. Few details of the modeled decision trees are in Table 4. We identify CART 70–30% with pruning as our best model which is well trained with less data and tested with best overall accuracy, less number of terminal nodes, and less depth as the preferred classification tree. The confusion matrix of the selected model is in Table 3. We further computed and compared the performance parameters including accuracy, precision, sensitivity, specificity, and F-Measure for all the models. The results of CART (70–30% split with pruning) alone are given in Table 5.

Table 5 Performance analysis of best performing CART decision tree

CART (70–30%)	Stage-1 (%)	Stage-2 (%)	Stage-3 (%)	Stage-4 (%)
Accuracy	87.57	85.53	97.30	98.28
Precision	76.37	73.91	99.47	95.31
Sensitivity	60.99	86.04	89.23	99.13
Specificity	94.84	85.28	99.85	97.92
F-score	67.82	79.52	94.08	97.18
ROC	93	92.2	98.5	99.4

References

1. What Is Cancer? National Cancer Institute. N.p., n.d. Web. 27 April 2017
2. Colon Cancer Treatment. Moffitt Cancer Center. N.p., n.d. Web. 27 April 2017
3. American Cancer Society: Colorectal Cancer Facts & Figures 2014–2016. American Cancer Society, Atlanta (2014)
4. American Cancer Society: Colorectal Cancer Facts & Figures 2017–2019. American Cancer Society, Atlanta (2017)
5. What Are the Survival Rates for Colorectal Cancer, by Stage? American Cancer Society. N.p., n.d. Web. 28 April 2017.
6. Mokhtar, M., Piltan, F., Mirshekari, M., Khalilian, A., Avatefipour, O.: Design minimum rule-base fuzzy inference nonlinear controller for second order nonlinear system. Int. J. Intell. Syst. Appl. **6**(7), 79 (2014)
7. Avatefipour, O., Piltan, F., Nasrabad, M.R.S., Sahamijoo, G., Khalilian, A.: Design new robust self tuning fuzzy backstopping methodology. Int. J. Inf. Eng. Electron. Bus. **6**(1), 49 (2014)
8. Shahcheraghi, A., Piltan, F., Mokhtar, M., Avatefipour, O., Khalilian, A.: Design a novel SISO offline tuning of modified PID fuzzy sliding mode controller. Int. J. Inf. Technol. Comput. Sci. (IJITCS) **6**(2), 72 (2014)
9. Liu, Y.-Q., Cheng, W., Lu, Z.: Decision tree based predictive models for breast cancer survivability on imbalanced data. In: 3rd International Conference on Bioinformatics and Biomedical Engineering (2009)
10. Li, J., et al.: Discovery of significant rules for classifying cancer diagnosis data. Bioinformatics **19**(suppl_2), ii93–ii102 (2003)
11. Tsirogiannis, G.L.: et al.: Classification of medical data with a robust multi-level combination scheme. In: 2004 IEEE International Joint Conference on Neural Networks, 2004. Proceedings, vol. 3. IEEE (2004)
12. Kourou, K., et al.: Machine learning applications in cancer prognosis and prediction. Comput. Struct. Biotechnol. J. **13**, 8–17 (2015)
13. Tseng, W.-T., et al.: The application of data mining techniques to oral cancer prognosis. J. Med. Syst. **39**(5), 59 (2015)
14. Venkateswara Rao, M.: Modeling and Survival Analysis of Breast Cancer: A Statistical, Artificial Neural Network, and Decision Tree Approach. University of South Florida, Scholar Commons, Tampa (2016)
15. Neeraj Bhargava, G.S.D.R.B.M.: Decision tree analysis on J48 algorithm for data mining. Int. J. Adv. Res. Comput. Sci. Softw. Eng. **3**(6), 1114–1119 (2013)

16. Mathan, K., Kumar, P.M., Panchatcharam, P., Gunasekaran, M., Varadharajan, R.: A novel Gini index decision tree data mining method with neural network classifiers for prediction of heart disease. Springer Link **22**(3), 225–242 (2018)
17. Lavanya, D., Rani, D.K.: Performance evaluation of decision tree classifiers on medical datasets. Int. J. Comput. Appl. **26**(4), 0975–8887 (2011)
18. Galit, S., Peter, C., Nitin, R.: Data Mining for Business Analytics. Wiley (2007)

Novel Performance Analysis of DCT, DWT and Fractal Coding in Image Compression

D. B. V. Jagannadham, G. V. S. Raju and D. V. S. Narayana

Abstract In digital image processing, image compression has prominent significance to improve the storage capacity, transmission bandwidth and transmission time. Basically, in image compression, the redundant information is reduced from the image by using proper compression techniques. The vital information in image is mined using different transformation techniques so that these transforms are used to restore image data without loss of information. In this paper, the compression of an image is done by three transform methods which are DCT, DWT and fractal coding using quadtree decomposition. Comparative analysis among these three transform methods is evaluated by finding performance evaluation parameters which are PSNR, MSE, CR and SSIM. Among the three different transform methods, DWT transform gives better PSNR, high compression ratio and low mean square error compared to other methods. Each transformation has its own merit and demerits. According to the application used for compression, the suitable method is used.

Keywords Compression ratio · Discrete cosine transform · Discrete wavelet transform · Inverse discrete cosine transform · Inverse discrete wavelet transform · Mean square error · Peak signal-to-noise ratio · Structural similarity index measurement

D. B. V. Jagannadham · D. V. S. Narayana
Department of ECE, GVP College of Engineering (A), Madhurawada,
Visakhapatnam, India
e-mail: dbvjagan@gmail.com

G. V. S. Raju (✉)
Department of CSE, Stanley College of Engineering and Technology for Women,
Abids, Hyderabad, India
e-mail: letter2raju@gmail.com

© Springer Nature Singapore Pte Ltd. 2020   611
K. S. Raju et al. (eds.), *Data Engineering and Communication Technology*,
Advances in Intelligent Systems and Computing 1079,
https://doi.org/10.1007/978-981-15-1097-7_51

1   Introduction

A two-dimensional function is defined as $f(u, v)$ where u, v are spatial coordinates. The intensity level of image is given by the amplitude of $f(u, v)$ at that point. Digital image processing [1] means processing a digital image on a computer. Image consists of some finite number of elements having a specific location and value for them. These finite elements are referred as pixels.

Digital image compression is used to reduce the amount of data which is required to represent a digital image. An increasing demand for digital images had led to excessive research in the study of techniques for compression. These compression techniques face the challenges in image quality. Several techniques for the improvement in high quality and less expensive image acquisition devices have come forward which steadily increases both image size and resolution. The design of efficient compression systems [2] had become very important.

In order to design a suitable compression system for transmission, processing, storage and providing acceptable computational complexity for practical implementation system, several techniques were used. For compression, aim is to decrease the number of bits needed for representing an image. Image compression algorithms are exploiting the idea of removing the unnecessary data in an image so that it can be represented using less numbers of bits which are able to maintain acceptable image quality. The factors needed for compression include the following:

Improvement in storage capacity for multimedia data;
Improvement in network transmission bandwidth;
Reducing the transmission time;
Improving the computational complexity of practical implementation.

The values of pixels are generally positive integers that range in between 0 and 255, which means that every pixel of a grayscale image occupies one byte on system. For pure black, the pixel value is 0. For a pure white, the pixel value is 255. For a gray color, the pixel value is around 128. Captured images [2] were rectangular in shape. Aspect ratio is the ratio between the width and the height of an image. It is written as a form of W:H format, where W is width and H is a height of the image. The aspect ratio is 4:3 among standard-definition television and 16:9 for high-definition television (HDTV). For movies, the aspect ratio is 21:9.

There are two aspect ratios given in Fig. 1, where one represents an aspect ratio of 4:3, while the other gives the aspect ratio of the same picture as 16:9. In 4:3 aspect ratio picture, some edges are not fitted to a television screen and picture is also viewed as zoomed picture. In 16:9 aspect ratio, the picture is fitted to the television screen.

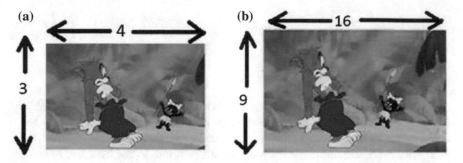

Fig. 1 **a** Aspect ratio of SDTV, **b** aspect ratio of HDTV

2 Objective

Image compression has prominent significance in improving the storing and transferring of images very efficiently. The evolution of electronic gadgets like smartphones, digital cameras, biometric applications and medical applications requires the storing of images with less memory, which is becoming necessary. For the requirement of less memory storage and efficient transmission, different compression technique is used accordingly and it should be considered that the compression technique must be able to rebuild the image with low loss or without loss of information and image quality as compared to the original image. "The objective of the Paper is to study such compression techniques and validate the results using MATLAB programming, and the objective of the Paper is to study of various image compression techniques and it also provides the motivations behind choosing different compression techniques."

3 Data Compression Model

Compression means removing the redundant data present in an image. Three major steps in the data compression system are as follows:

(a) Image transformation;
(b) Quantization;
(c) Entropy encoding.

In image transformation, an image is transformed from the spatial domain to frequency domain. The major data compression occurs through quantization. At high frequencies, an image contains low visual information so more quantization has to be done here to decrease the size in the transformed image representation. The transformed and quantized image data can further be reduced by using entropy coding. The data compression system can be labeled using block diagram which is shown in Fig. 2.

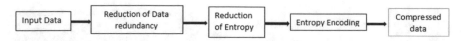

Fig. 2 General compression model

The principle usage of compressed data is to improve the data storage capability. It also improves both the transmission time and transmission bandwidth. When a high volume of data is transmitted over a channel, the data compression approach is used effectively to reduce the communication cost due to the data rate reduction. The usage of different compression techniques such as Internet usage and other digital-oriented things had become more graphic oriented instead of being text-centric phenomenon.

Higher compression techniques have enabled chances for different innovative ideas such as video teleconferencing, telemedicine, archiving and entertainment. In large database, these techniques obviously reduce the recovery cost and data backup cost in a system, as it stores in a compressed format and it enables more multimedia application with reduced cost. Data security is greatly improved by transmitting the encrypted decoded data parameters separately from the database containing compressed data file.

Depending on the area of particular application to be used and data sensitivity, this data compression has some drawbacks. During the process of encoding and decoding, extra overhead is incurred and this overhead is required for mostly identifying or interpreting the compressed data. So it is the most serious disadvantage of this compression, and thus, it depresses its applications in some areas. Data compression usually diminishes the reliability of the data archives [1].

Insensitive data transmissions using compressions such as medical information and satellite information across a noisy channel is risky because the burst errors which were introduced by the noisy channel can damage the data transmitted. Other major problem of this is the interruption in data properties as the original data is different from the compressed data.

In some applications such as very low-power VLSI implementation, the complexity introduced by these compression techniques will make cost of packing more and also decrease the system efficiency [3].

Image compression in digital domain is mainly based on theory of information. The concept of entropy in the information theory is used in an image compression. Entropy is defined as the amount of information produced by a source for every symbol, and the Entropy of a given symbol set is a negative sum of the products of all the symbols in a set.

Algorithm for compression is a procedure that reduces the number of data bits used to represent source information, consequently reducing the transmit time necessary for a given channel capacity. Grading to fewer target symbols from source symbols is referred to as "compression." It is also known as encoding process. The renovation from the target symbols back into the source symbols instead of a close approximation form of the original information is defined as "decompression." It is also known as decoding process. Generally, sampling and quantization of a signal are contained in compression system; meanwhile, an image is continuous not just in its coordinate axis (x-axis) but it also has an amplitude (y-axis). The digitizing of coordinate axis is known as sampling and the part that deals with digitizing the amplitude is known as quantization.

Compression in image can be categorized into two types. One is information preserving compression (lossless) and other is information loss compression (lossy). In lossless compression, the reconstructed information is nearly identical to original information. In lossy compression, the reconstructed information is not exactly similar to the input information. It gives high CR than information preserving compression.

4 Performance Evaluation Parameters

The significance of image assessment is given to know what amount of image is compressed and to measure the quality of decompressed image similar to the input image. For this, assessment was done on the basis of image distortion, which was measured using performance parameters such as MSE, PSNR, CR and SSIM. The image quality of reconstructed image can be measured in two types. One is subjective measure and other is objective measure. The quality of $\sum$images directly investigates by human viewers in subjective measure, whereas the numerical formulas are used to evaluate the image quality in objective measure. Generally, a better compression algorithm will have high PSNR, high CR, low MSE and better SSIM. The performance evaluation constraints are defined in the next sections.

5 Mean Square Error

It is defined as the mean of the square of the difference between the decompressed image and the input image. It is the measure of signal fidelity which means it compares the two signals by describing the measure of similarity. It incorporates both the variance and its bias of the estimator are measured. For standard deviation, taking the square root of MSE is known as root mean square error (RMSE) [4].

The equation for MSE can be written as:

$$\text{MSE} = \frac{1}{pq} \sum_{i=0}^{p-1} \sum_{j=0}^{q-1} [M(i,j) - N(i,j)]^2 \tag{1}$$

here $M(i, j)$ is the input image, $N(i, j)$ is decompressed image and p, q are the dimensions of the images. It is desired to have MSE value lower which means the error between the image is minimum and the quality of an image is better.

6 Peak Signal-to-Noise Ratio (PSNR)

PSNR is defined as the ratio between the maximum possible power of a signal and
the power of corrupting noise [4]. Here, an original image is represented as a signal
and the error in reconstruction as the noise. It is inversely proportional to error which
means the error is low and the quality of signal is high. It is articulated in terms of
decibel scale in (dBs). It is desired to have a higher PSNR value for better image
quality and it ranges between zero and infinity. Zero for dissimilarity between the
reconstructed and original images, whereas infinity for identical images.

$$\text{PSNR} = 20 \, \log_{10} \frac{255}{\sqrt{\text{MSE}}} \tag{2}$$

In Fig. 3, the different values of PSNR are measured using different transform
techniques, and the values are 29 dB, 36 dB and 33 dB, respectively.

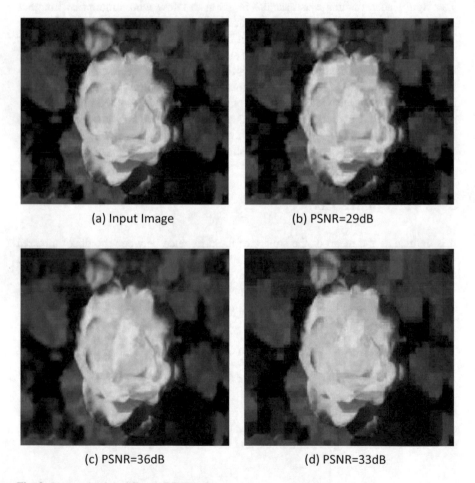

(a) Input Image   (b) PSNR=29dB

(c) PSNR=36dB   (d) PSNR=33dB

Fig. 3 Images having different PSNR value

7 Compression Ratio (CR)

CR is defined as the ratio between the original image size and reconstructed image size and it evaluates the decrease in detail coefficient information in images. The compression ratio is given by

$$\text{compression ratio} = \frac{\text{uncompressed image size}}{\text{compressed image size}} \tag{3}$$

To preserve the critical information present in the input data, we must know what amount of detailed coefficient it can remove from the original data. The compression ratio mainly depends on two controlling constraints; they are scaling parameter and the quantization matrix.

After quantization, every element of the transformed coefficient is divided by the corresponding coefficient and it is rounded to the adjacent integer. After rounding few coefficients resulting to be zero, these coefficients are discarded. So that, the more the detail coefficients are discarded, the higher the CR can be achieved. Different compression ratios have different image quality.

Images compressed higher will have lower quality reconstruction of the image. Comparisons of different CR are shown in Fig. 4. The higher an image is compressed, we get blurred image as compared to the image that has lesser compressed image.

(a) Input Image   (b) = 18%   (c) = 60%

Fig. 4 Images having different compression ratio

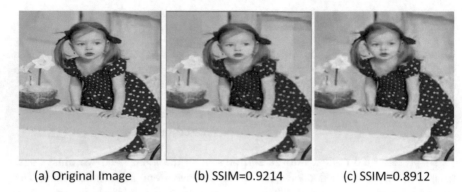

(a) Original Image   (b) SSIM=0.9214   (c) SSIM=0.8912

Fig. 5 Images having different SSIM

8 Structural Similarity Index Measurement (SSIM)

SSIM is used for measuring the similarity between original image and compressed. It is based on the comparison of three terms, namely the luminance term, the contract term and the structural term, and these indexes are considered on various spaces of an image [4]. The two windows u and v of common size $N \times N$ are measured as using Eq. (4)

$$\text{SSIM}\,(U, V) = \frac{(2\mu_U\mu_V + P1)(2\sigma UV + P2)}{(\mu_U^2 + \mu_V^2 + P1)(\sigma_U^2 + \sigma_V^2 + P2)} \tag{4}$$

where μ_u = Mean of u, μ_v = Mean of v, σ_u^2 = variance of u, σ_v^2 = variance of v, σ_{uv} = covariance of u and v. p_1, p_2 = two variables to stabilize the division with weak denominator (Fig. 5).

Generally, SSIM ranges between -1 and $+1$; if it is nearly $+1$, then the structure is nearly identical. If it is -1, then the structure is dissimilar. SSIM has been very widely used in the broadcast, cable and satellite television industries to measure the video quality.

9 Parameters Evaluated in Subjective Type

It is essential that the reconstructed image is visual observe the image quality. Objective measure does not give all the information about the reconstructed image quality. From this subjective type analysis the reconstructed image is analyzed by using subjective measure of the human perceptual system [1]. In subjective measure, the human observer focuses on the changes between original and reconstructed image and correlates the differences.

10 Results

Performance evaluation of various image compression algorithms and simulation results are presented here. Different types of images are applied to various algorithms which are natural and scale images. To evaluate the performance of image, algorithms can be tested for several applications. Scale images are the standard images which are Lena. Cameraman image is used in many image processing applications. The obtained results are presented in this section, and the results are compared with the different transform techniques.

The obtained results are presented in Tables 1, 2 and 3 for different images. The input and decompressed images are also shown in Figs. 6a–d, 7a–d and 8a–d. It can be observed from Tables 1, 2 and 3 that the PSNR is high for DWT Haar wavelet when related to different transforms. DWT Haar encompasses between CR and

Table 1 Comparative analysis college image

Image compression method	MSE	PSNR in dB	CR	SSIM
DCT	129.4680	27.0092	18.4262	0.9999
DWT (HAAR)	35.22	32.6965	22.1815	0.7593
DWT (DAUBECHIES)	37.63	32.4098	21.5249	0.7114
DWT (SYMLET)	37.95	32.3725	20.7524	0.7357
DWT (COIFLET)	37.01	32.4814	20.7677	0.7273
Fractal	37.06	32.4759	18.3130	0.7318

Table 2 Comparative analysis of baby image

Image compression method	MSE	PSNR in dB	CR	SSIM
DCT	64.3583	30.0448	17.7140	0.9999
DWT (HAAR)	27.06	33.8407	25.0207	0.8275
DWT (DAUBECHIES)	29.38	33.4848	24.0230	0.8103
DWT (SYMLET)	27.52	33.7680	23.8470	0.8356
DWT (COIFLET)	28.10	33.6783	23.5092	0.8318
Fractal	29	33.5413	21.6398	0.8081

Table 3 Comparative analysis of Lena image

Image compression method	MSE	PSNR in dB	CR	SSIM
DCT	108.02	27.7954	15.8287	0.9999
DWT (HAAR)	26.40	34.6311	23.5290	0.8275
DWT (DAUBECHIES)	30.17	33.3656	12.176	0.7932
DWT (SYMLET)	27.87	33.7180	12.2147	0.8130
DWT (COIFLET)	28.43	33.6248	12.3165	0.8096
Fractal	33.04	32.9748	3.9608	0.7378

(a) **(b)**

(c) **(d)**

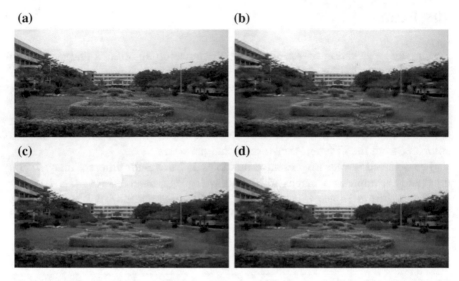

Fig. 6 **a** Original image, **b** DCT decompressed image, **c** DWT decompressed image, **d** fractal decompressed image

(a) **(b)**

(c) **(d)**

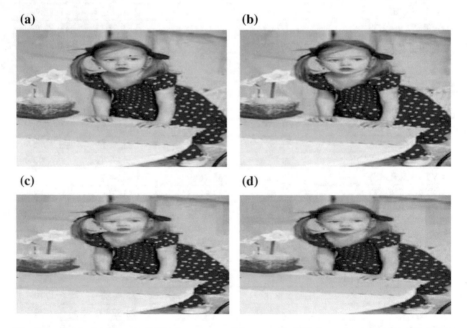

Fig. 7 **a** Original image, **b** DCT decompressed image, **c** DWT decompressed image, **d** fractal decompressed image

(a)   (b)

(c)   (d)

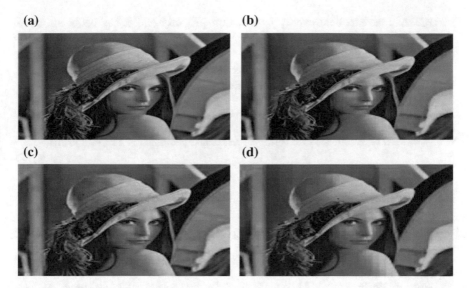

Fig. 8 **a** Original image, **b** DCT decompressed image, **c** DWT decompressed image, **d** fractal decompressed image

quality of reconstructed images. So it is used in satellite applications. DCT has an excellent energy compaction and gives lesser CR, and it has better structural similarities nearly equal to one. DCT is computationally efficient compared to other algorithms, and it requires less processing power. Fractal coding compressed images are resolution independent, so this technique is used in medical applications to find tumors, brain injuries, etc.

11   Conclusion

In this analysis, different compression techniques are used to various images to know the performance parameters such as CR, MSE, PSNR and SSIM.

The obtained results display that DWT Haar method attains higher CR without loss of quality image. The error between input image and decompressed image is low, and errors during transmission of information are minimum in wavelets due to multi-resolution nature. The drawback of these methods is that it needs more processing power.

Fractal compression gives resolution-independent image because pixels are not bound by a permanent scale. The disadvantage of fractal image compression requires more encoding execution time due to comparing each block in an image and it is image dependent, so this technique is used in medical applications to find tumors, brain injuries, etc.

DCT has an excellent energy compaction property and it has better structural similarities nearly equal to one, but it was blocking artifacts on the boundaries of the image. It requires low processing power and it generates low CR. While comparing these three methods, DWT gives better PSNR, CR and lesser MSE. Each transform technique has its own advantage and disadvantages.

Acknowledgements Authors are very much thankful to the secretary GVP College of Engineering (A), Madhurawada, Visakhapatnam, for given an opportunity to utilize the advanced version of MATLAB software to simulate the results and publishing in a reputed journal.

References

1. Veenadevi, S.V., Ananth, A.G.: Fractal image compression using quadtree decomposition and Huffman coding. SIPIJ **3**(2) (2012)
2. Cintra, R.J., Bayer, F.M.: A DCT approximation for image compression. IEEE Signal Process. Lett. **18**(10) (2011)
3. Wang, Z., Bovik, A.C., Sheikh, A.C., Simon Celli, E.P.: Image quality assessment: from error visibility to structural similarity. IEEE Trans. Image Process. **13**(4) (2004)
4. Gonzalez, R.C., Woods, R.E., Eddins, S.L.: Digital Image Processing Using MATLAB. PEARSON Education (2009)

CBIR using SIFT with LoG, DoG and PCA

Katta Sugamya, Pabboju Suresh, A. Vinaya Babu
and Rakshitha Akhila

Abstract Content based image retrieval using scale invariant feature remodel (SIFT) is employed to discover stable keypoint locations within the scale-space. The extraction of image options can be done by exploiting SIFT or K-means cluster. In the proposed work we can find feature extraction and locating scale-space extrema through SIFT-DoG & SIFT-LoG ways. Finally, planned ways, SIFT-DoG, SIFT-LoG, and PCA are compared.

Keywords Scale invariant feature transform (SIFT) · Difference of gaussians (DoG) · Laplacian of gaussians (LoG)

1 Introduction

The improvements in the content-based retrieval of images have been rapidly improving along with the enhancement of technology in the fast generation era.

The advancement in research is involved; augmentation in content-based photograph retrieval has been honestly rapid. In present-day years, there has been a significant exertion place in comprehension the $64000 world ramifications, applications, and confinements of the innovation. However, the use of the innovation is limited by certain parameters [1]. We dedicate this segment an attentive look on picture recovery inside the world and examine client desires, framework imperatives and necessities, and furthermore the push to make picture recovery region capacity inside the imminent future. In CBIR [2], visual contents [3] are extracted by scale invariant feature remodel [4] exploitation distinction of mathematician and Laplacian of mathematician with Feature remodel [5] exploitation and Laplacian distinction of mathematician and Laplacian of mathematician with

K. Sugamya (✉) · P. Suresh · R. Akhila
CBIT, Hyderabad, India
e-mail: sugamya.cbit@gmail.com

A. Vinaya Babu
SCET, Hyderabad, India

© Springer Nature Singapore Pte Ltd. 2020   623
K. S. Raju et al. (eds.), *Data Engineering and Communication Technology*,
Advances in Intelligent Systems and Computing 1079,
https://doi.org/10.1007/978-981-15-1097-7_52

Principal element Analysis. Next, we have to load the information of pictures [6] and for every image within the database, we have to section it and calculate its feature vectors and store them. Next, we' have to perform a similarity computation [7] of {the pictures |the pictures| the photographs} within the info with the question image and retrieve the photographs within the ascending order of the gap between the question image and also the images within the info [8].

1.1 Literature Survey

The published work entitled "Content based image retrieval system using clustered scale invariant feature transforms" by Montazer—2015 [9], address an efficient usage of memory and matching time, which are earlier drawbacks . But Cluttered and occuled background not recognised properly. "A comparison of SIFT, PCA and SURF' by Juan 2016 [10], the implementation of method depends on the requirement of different features. Improvement on blur and scale performance for PCA, and Illumination for SIFT and rotation for SURF need to improve." Principal Component Analysis to reduce dimension on digital image" by Ng—2017 [11] in this paper Sparing of capacity and transmission time for picture records while keeping up the integrity of the image is the demerit. It gives good dimension reduction for non-linear problem space. "A Comparative Study of Sift and PCA for Content Based Image Retrieval" [12] by Raghava Reddy, from expansive database of pictures, a key point to be chosen that effectively coordinates with the required picture. A Scale Invariant Transform recovers pictures with best coordinating, that have more corners and edges, shifting in size. Pictures that are transposed, turned, related and so forth. Be that as it may, it increasingly productive parameterization of highlight descriptors and Detection point is the disadvantage (Fig. 1).

Fig. 1 Block diagram content-based image retrieval

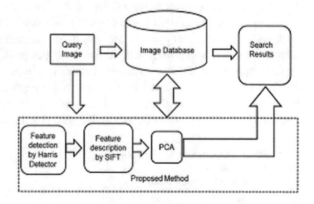

2 Scale Invariant Feature Work

SIFT was planned by David Lowe. The SIFT rule consists os four steps:

(i) Extremities Detection for Scale-space:

SIFT concentrates on the scale and site of the image and calculates all the come-at-able key points. This might be earned by implementing a distinction of mathematician (DoG). It does not amend to scale and orientation. The scale of an image $I(a, b)$ is printed as an operate $L(a, b, \sigma)$, that is made up of the convolution of $I(x, y)$ with a variable-scale mathematician

$$L(a, b, \sigma) = G(a, b, \sigma) * I(a, b) \tag{1}$$

where $*$ is that the convolution operation in a and b, and $G(a, b, \sigma)$ **can be** a variable-scale **mathematician** and $I(a, b)$ is that the input image.

$$G(A, B, \sigma) = \frac{1}{2TT\sigma^2} e^{-\frac{a^2 + b^2}{2\sigma}} \tag{2}$$

pair of $2TT\sigma^2$ gives stable keypoint [13] locations in scale house, it uses a scale house extreme a supported the difference-of-Gaussian operate, $D(a, b, \sigma)$, which can be computed from the excellence [13] of two shut scales separated by a seamless increasing issue k:

$$D(a, b, \sigma) = L(a, b, k\sigma) - L(a, b, \sigma) \tag{3}$$

This economical approach is construction of $D(x, y, \sigma)$. To watch the native maxima and minima of $D(a, b, \sigma)$, each purpose is compared with its eight neighbors at an identical scale and its 9 neighbors in one scale. If this assessment is the minimum or most of those points, then this point is associate extrema.

(ii) Locating a Keypoint

A keypoint candidate is detected from the previous step, which should standardize its correctness. For all interest points, elaborate method is found to envision location and scale. To support their stability, the keypoints are selected. A stable keypoint is a proof for image alteration. Invariant options [14] are forever detected among the foremost or least of the D matrix that's that the keypoints. Thus, on observing the extrema, we would wish to catch the aim that results in the second-order by-product of D to be zero. Higher results are obtained by interpolating instead of taking center of cell as location.

Taylor growth formula is used for this purpose:

$$D(a) = D + \frac{dD^t a}{da} + \frac{1}{2} at \frac{d^2 D a}{da^2} \tag{4}$$

where D is the excellence of mathematician. Then, the extrema a is set to eliminate the sting impact and acquire the native extrema; use Hassian matrix and notice the brink

$H = D\_aa\, D\_ab$

$"""Dab\, Dbb$

$Daa = D(a, a-1) - 2D(a, b) + D(a, b+1)$

$Dbb = D(b-1, b) - 2D(a, b) + D(a+1, b)$

$Dab = D(a, b-1) - 2D(a, b) + D(a-1, b) - D(a-1, b-1)$

$\text{Tr}(H) = D\_aa + D\_bb = c + B;$

$\text{Det}(H) = D_a a D_b b - (D_a b)^2 = c * B;$

$$\text{Tr}(H) = D\_aa + D\_bb = c + B; \tag{5}$$

$$\text{Det}(H) = D\_aa\, D\_bb - (D\_ab)^2 = c * B; \tag{6}$$

$$\text{Ratio} = (c + B)^2 / c * d; \tag{7}$$

Calculate the Hassian matrix for each key purpose selected by second-order by-product of D and remove the key purpose that is a smaller quantity than relation.

(iii) Assignment of Orientation

The direction of gradients is computed by SIFT for every of the familiar key points. One or extra orientations are assigned to each key properties of native image incline directions. By assigning a homogenous orientation to each key supported by properties of native image, its feature vector are typically delineated relative to the current orientation and therefore succeed invariability to image spin. The orientation is calculated from associate orientation chart of native gradients from the closest smoothed image $L(a, b, \sigma)$. For each image sample $L(a, b, \sigma)$ at this scale, the gradient magnitude $m(a, b)$ and orientation$^\theta(a, b)$ is computed for victimization component covariations.

The 360° vary of orientations covers the chart that has thirty six bins covering. Each purpose is additional to the chart weighted by the incline magnitude, $m(a, b)$, and by a circular mathematician with σ Variance that is five times the scale of the key purpose. Additional key purposes are generated for key point locations with multiple dominant peaks whose magnitude is at intervals eightieth of each completely different. For extra correct orientation assignment, the dominant peaks among the chart are interpolated with their neighbors.

(iv) Key point descriptor

At the chosen scale around each key region, the native image gradients are measured. These are transformed into a illustration that permits for vital levels of native type distortion and alter in illumination.

3 Principal Element Analysis

On the other hand, principal component analysis (PCA) is utilized for dimensionality reduction. PCA decide that to discover best picture classifier framework than other elective systems. The chief work of PCA is to remove main highlights of a picture. In predefined class or one module, these key decisions are incorporated. Specialists findings says that the PCA-based strategy offers higher arrangement and legitimate yield among the divisions of PC vision like meteorology, face recognizable proof, face acknowledgment, highlight basically based picture characterization, restorative medicine, remote detecting pictures, and information preparing.

The strategy PCA that utilizes fragile major scientific standards to adjust a spread of possibly associated factors into a littler sort of factors is named as vital segments. It is one in everything about principal important outcomes from logical unadulterated science. The advantage of PCA is disclosure the examples inside the information and pass information by linking the quantity of sizes without loss of data.

The planned thoughts that territory utilized for PCA are unit change, variance, co-difference, and eigenvectors [8]. The information film bliss to same class may differ in lighting conditions, clamor, and so forth, yet in specific zones where not completely irregular there might be a few examples despite their varieties. Such examples might be alluded as main segments. PCA is additionally a logical device familiar with concentrated prime segments of unique picture information. These prime parts may furthermore be alluded as eigen film. A vivacious factor of PCA is that a picture is remade by consolidating the Eigen film from the picture information. The standard to figure principal parts is as per the following.

Step 1: Prepare the data.

Let us assume we have X_i that contains N vectors of size M (rows of image columns of image) representing a set of images and P represents a element values.

$$X_i = (P_i \ldots P_m), 1 \ldots N \qquad (8)$$

Step 2: Get the mean.

Process the mean of the picture vector at that point set of pictures zone unit mean focused to keep with take off mean from each picture vector.

$$\text{Mean number sixty nine} = \frac{1}{M} \left(\sum K = 1 \text{ to } M, X_i \right) \qquad (9)$$

Step 3: Mean is subtracted from the initial image.

$$m(a,b) = \sqrt{(L((a+1,b) - L(a-1,b)^2 + (L(a,b+1) - (L(a,b-1))^2}} \quad (10)$$

$$\theta_{(a,b)} = \tan^{-1}((L(a,b+1) - L(a,b-1))/(L(a+1,b) - L(a-1,b))) \quad (11)$$

Let "A" be the matrix that is created newly by subtracting mean of image from the initial data.

Step 4: Determine variance matrix.

Let the eigenvectors and eigenvalues of the variance matrix C, and this variance matrix is calculated by multiplying matrix A with its transposing matrix of A.

$$C = A * A^T \quad (12)$$

Step 5: Eigenvectors and eigenvalues of variance matrix area unit are calculated, and principal components are selected.

The eigenvectors area sorted in descending order with their corresponding eigenvalues.

The {Eigen|Eigcn|ManfredEigen|Chemist} vector related to the foremost vital eigen worth is that the one that reflects the most effective variance among the image. The number of highest valued Eigenvectors is then picked to form an image house from the resultant variance matrix C.

4 Implementation and Results

4.1 Image Data

Corel-10 k dataset contains 100 categories, and there area unit 10, footage from varied contents like sunset, beach, flower, building, car, horses, mountains, fish, food, door, etc., each category contains 100 footage of sizc 192×128 or 128×192 among the JPEG format. Corel-5 K dataset consists of the first 5000 footage, and Corel-10 K dataset consists of the 10 thousand pictures (Figs. 2 and 3).

4.2 Guide

GUIDE is graphical interface development surroundings. We have a bent to use axes to indicate footage. We have a bent to use the push buttons for activity varied functions like loading the data, selecting a matter image, displaying the results, and clearing the screen.

Query Images:

Fig. 2 Corel dataset

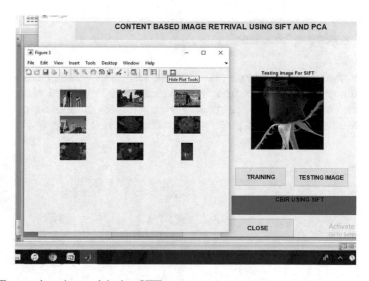

Fig. 3 Feature detection exploitation SIFT

4.3   Feature Detection

4.3.1. In SIFT Approach, Feature detection, using key descriptors are well known extraction technique. The rule for query image and every pictures in database can be calculated on an individual basis. As an example of binary exposure offers sixty-two key points, while a graph exposure offers nearly 13 key points if image has larger corners and edges.

4.3.2. In component analysis methodology, the foremost step is to pre-process the question image. Once done with pre-process, shape, and texture, type capabilities are going to extract the utilization of shape moment version in hue saturation and value coloration space, and then gray-level co-occurrence matrix and Fourier descriptor transformation are allotted. Once distinctive feature extraction is extracted, the PCA calculates essential parts from functions of every question image and every one info pics once that categories the question photograph to its several class. The methodology of principle component analysis classification makes our retrieval system more practical and strong. For testing question image exploitation in PCA, principal components are set for every image that is required to examine. All the steps of the PCA rule made public on top of are dead (Fig. 4).

4.4   Matching of Feature

In order to find identical feature classification, key point vectors from a question image and key point's vectors from info pictures are matched and verified for the nearest geometric distance between them. Per the geometric distance formula, the gap between two points within the plane with coordinates (x, y) and (a, b) is given as:

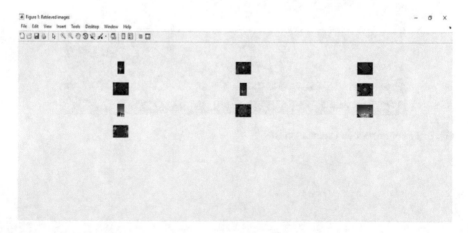

Fig. 4 Feature detection exploitation SIFT-PCA

$$\text{Dist}((x, y), (a, b) = \text{SQRT}(\llbracket x - a) \rrbracket^\uparrow 2 + \llbracket (x - b) \rrbracket^\uparrow 2 \tag{13}$$

The geometric distance formula for image matching is applicable for both SIFT and PCA.

4.5 Retrieving Results

The retrieval method after feature discovery the key points of feature descriptors of the question image are compared to those of the. During this prime 10, 15, 20, 25, 30 pictures are retrieved and results are analyzed.

4.6 Observation and Findings

The enactment of image retrieval utilizing SIFT or PCA strategy is checked by utilizing precision and recall charts. In this paper, we have tested our code on three diverse image databases. Filter code is kept running on parallel pictures, grayscale pictures, and shading pictures. The discoveries were caught independently for each databank classification and perceptions noted. PCA code is kept running on shading database pictures and results were recorded. A correlation of PCA and SIFT results is inspected for shading pictures and a specific significant appraisals are perceived. Precision versus recall is a standard assessment strategy of retrieval used in picture recovery. Mathematically describing the precision and Recall as

$$\text{Precision rate} = \frac{\text{No. of relevant image selected}}{\text{Total number of retrieved images}} \tag{14}$$

$$\text{Recall rate} = \frac{\text{No. of relevant images selected}}{\text{Total no. of similar images in the database}} \tag{15}$$

4.7 Comparison Between SIFT and PCA

The response graph given in Fig. 5 offers the results by exploitation SIFT with distinction of mathematician below the class of Roses. The analysis has done by computing preciseness and recall. Figure 6 depicts SIFT—DOG and PCA analysis (Tables 1, 2 and 3).

K. Sugamya et al.

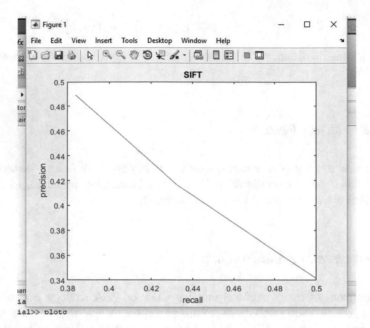

Fig. 5 Sift result analysis for Roses

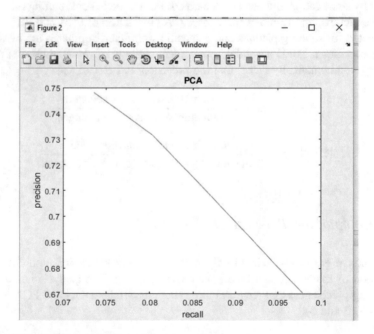

Fig. 6 SIFT-PCA result analysis for Roses

Table 1 Analysis of results for retrieved images = 10 of Roses

No. of images retrieved = 10	Scale invariant feature transform (SIFT)	Principal component analysis (PCA)	SIFT using LoC with PCA
Precision	0.4889	0.7401	0.3421
Recall	0.3840	0.0736	0.4519

Table 2 Analysis of results for retrieved images = 20 of Roses

No. of images retrieved = 20	Scale invariant feature transform (SIFT)	Principal component analysis (PCA)	SIFT using LoC with PCA
Precision	0.4164	0.7316	0.3241
Recall	0.4323	0.0803	0.4168

Table 3 Analysis of results for retrieved images = 30 of Roses

No. of images retrieved = 30	Scale invariant feature transform (SIFT)	Principal component analysis (PCA)	SIFT using LoC with PCA
Precision	0.3417	0.6703	0.2865
Recall	0.4998	0.0979	0.3743

4.8 Result Analysis for SIFT-PCA

The response graph given in Fig. 7 gives the results by using SIFT with Difference of Gaussian under the category of Roses. The analysis has done by computing precision and recall. Figure 8 depicts SIFT—DOG and PCA analysis (Table 4).

The graph given in Fig. 9 gives the results by using SIFT with PCA under the category of Mountains. The analysis has done by computing precision and recall. Figure 10 depicts SIFT-PCA analysis for Bus category.

5 Conclusion and Future Scope

PCA-SIFT is the best option, however, the application considers some sort of blur, and thus, it is required to relinquish PCA. PCA-SIFT some upgrading on blur performance. Otherwise, PCA could be a sensible methodology for color image retrieval in CBIR. Largely, the paper helps in lashing out a reasonable study of SIFT and PCA for CBIR in varied image situations and environments.

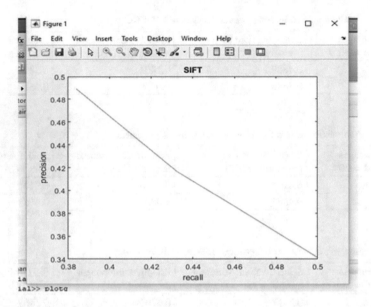

Fig. 7 SIFT result analysis for Roses

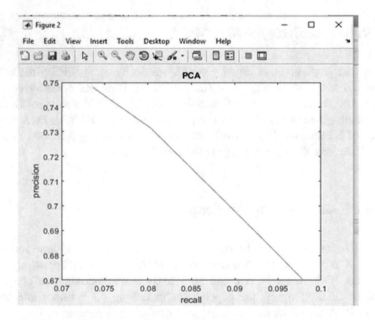

Fig. 8 SIFT-PCA results analysis for Roses

Table 4 SIFT-PCA

Category	No. of images retrieved = 10		No. of images retrieved = 15		No. of images retrieved = 20		No. of images retrieved = 25	
	Precision	Recall	Precision	Recall	Precision	Recall	Precision	Recall
Rose	0.6118	0.0607	0.5895	0.1049	0.5692	0.1106	0.4676	0.1190
Buildings	0.6660	0.0808	0.6042	0.1012	0.5573	0.1192	0.6549	0.1135
Horses	0.7576	0.0863	0.7377	0.0557	0.7130	0.0606	0.6555	0.0590
Buses	0.7337	0.0862	0.6987	0.0978	0.6543	0.1109	0.6231	0.1254
Beaches	0.7123	0.0745	0.7123	0.0897	0.6345	0.0923	0.5998	0.1005
Food	0.6134	0.0892	0.8765	0.0932	0.7234	0.1000	0.7001	0.1126
Mountains	0.7658	0.0834	0.6453	0.0995	0.5942	0.1132	0.5683	0.1286
Dinosaurs	0.7426	0.0820	0.7608	0.0981	0.6123	0.1123	0.5552	0.1132

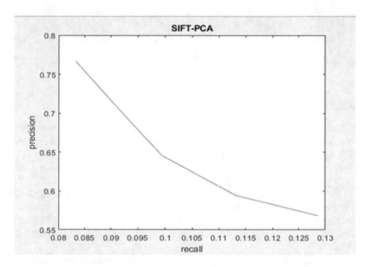

Fig. 9 SIFT-PCA for Mountains

The future work is to enhance the rule and apply these ways on single areas, like image sewing. SIFT retrieves pictures with higher performance that have a lot of corners and edges, varied in size, pictures that are converse, rotated, connected etc. compared to PCA. Future work can inspect a lot of inexpensive parameterization of feature descriptors and various ways and algorithms for higher retrieval performance.

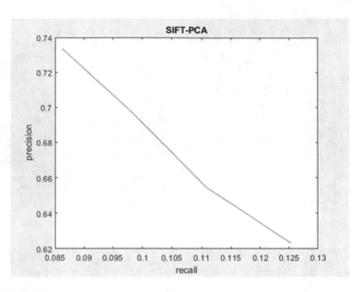

Fig. 10 SIFT-PCA results for Buses

References

1. Dutta, R., Joshi, D., et al.: Image retrieval: ideas, influences and trends of the new age. ACM Trans. Comput. Surv. 40(2), 1–77 (2008)
2. Bakar, S.A., Hitam, M.S., Yussof, W.N.J.H.W.: Content-based image retrieval using SIFT for binary and greyscale images. IEEE ICSIP (2013)
3. Lowe, D.G.: Object recognition. International Conference on Copp, pp. 1150–1157
4. Li, J.: Implement of scale-invariant feature transform algorithms
5. Li, J.: Implement of scale-invariant feature transform algorithms. ECE 734 Project
6. Velumurugan, K.: A survey of content based image retrieval systems using SIFT. ISSN: 2277-128X
7. Mandavi, R., Nagwanshi, K.K.: Content based image retrieval and classification using principal component analysis. ISSN (Online): 2319-7064
8. Joseph, D., Mathew, A.: Content based image retrieval of COREL images along with face recognition. IJCSMC 4(11). ISSN 2320-088X
9. Montazer, G.A.: Content based image retrieval system using clustered scale invariant feature transforms. Optik 126(18), 1695–1699 (2015)
10. Juan, L., Gwun, O.: A comparison of SIFT, PCA-SIFT and SURF. Int. J. Image Process. (IJIP) 3(4)
11. Ng, S.C.: Principal component analysis to reduce dimension on digital image. In: 8th International Conference on Advances in Information Technology, IAIT2016, 19–22 December 2016, Macau, China
12. Raghava Reddy, K., Narayana, M.: A comparative study of SIFT and PCA for content based image retrieval. Int. Refereed J. Eng. Sci. (IRJES) 5(11), 12–19 (2016). ISSN (Online) 2319-183X, (Print) 2319-1821
13. Lowe, D.G.: Distinctive image keypoints. Int. J. 2, 91–110
14. Wu, J., Cui, Z., Sheng, V.S., Zhao, P., Su, D., Gong, S.: A Comparative study of SIFT and its variants, vol 13, no. 3 (2013)

15. Lowe, D.: Object recognition from local scale-invariant features. In: Proceedings of the 7th International Conference on Computer Vision, vol. 2, p. 1150, September 20–25, 1999. IEEE (1999)
16. Lowe, D.: Distinctive image features from scale-invariant keypoints. January 5, 2004
17. Balaji, T., Sumathi, M.: PCA based classification of relational and identical features of remotesensing images. Int. J. Eng. Comput. Sci. **3**(7), July
18. Lowe, D.G.: Distinctive image features from scale invariant keypoints. Int. J. Comput. Vision **60**(2), 91–110 (2004)

RETRACTED CHAPTER: Design of Deep Learning Controller for Vector Controlled Induction Motor Drive

B. T. Venu Gopal, E. G. Shivakumar and H. R. Ramesh

Abstract This paper tends to the utilization of Deep Learning concept to structure the controller, to investigate achievability of using deep learning into control issues. Induction motors play an important role in industrial applications. If speed is not controlled properly, it is for all intents and purposes difficult to accomplish the preferred task for precise application. Induction motors are known for their simplicity and reliability. They are maintenance free and low cost electric drives. Due to the lack of ability of traditional control techniques such as PI and PID controllers to serve under wide scope of activity, AI controllers are broadly utilized in industries such as fuzzy logic controller, neuro-fuzzy controller (NFC), genetic algorithm. By learning PI controller, which is mostly utilized in the industries, the suggested Deep Learning-based controller is planned. The input and output of PI controller are utilized as learning informational index for Deep Learning (DL) system. To stay away from computational weight, just speed error is given as the input to the proposed deep learning controller, not at all like a conventional controller which utilizes speed error and its derivative. Deep belief network (DBN) control process is utilized to design the DL controller. DLC gives incredible speed exactness. By using MATLAB Simulink, simulation is carried out. Results of suggested DLC and NFC are compared; finally, it was led to show the effectiveness and usefulness of the suggested procedure.

The original version of this chapter was retracted. The retraction note to this chapter is available at https://doi.org/10.1007/978-981-15-1097-7_82

B. T. V. Gopal (✉) · E. G. Shivakumar · H. R. Ramesh
Department of Electrical Engineering, UVCE, Bangalore University,
Bengaluru, Karnataka 560001, India
e-mail: btvgopal@gmail.com

E. G. Shivakumar
e-mail: shivaettigi@gmail.com

H. R. Ramesh
e-mail: hrramesh74@gmail.com

© Springer Nature Singapore Pte Ltd. 2020, corrected publication 2024   639
K. S. Raju et al. (eds.), *Data Engineering and Communication Technology*,
Advances in Intelligent Systems and Computing 1079,
https://doi.org/10.1007/978-981-15-1097-7_53

Keywords Deep learning controller · Deep belief network · PI controller ·
Neuro-fuzzy controller · Induction motor · Vector control

1 Introduction

Because of their basic and solid development, induction motors (IM) possess 92%
of the modern electric drives among different ac motors. Be that as it may, due to
their nonlinear conduct and parameters of the IM, the control of IM is extremely
troublesome with various working conditions and chooses the best one for induction motors control of artificial intelligent controllers (AIC). This is on the grounds
that when compared with the traditional PI controllers, AIC demonstrates numerous
points of interest. Scientists have been working to apply AIC for IM drives more
than a couple of decades. Self-learning control systems utilizing fuzzy logic, neural
systems, and hybrid systems have recognized as the critical instrument for
enhancing the performance of electric drives in industrial applications [1].

During the 1970s, Haase and Blaschke proposed a vector control technique, which
depends on field orientation principle, separates the torque and flux control in an
induction motor drive. Thus, it impacts the control of IM drives like independently
energized dc motor while keeping up the common benefits of ac over dc motor, and
henceforth, appropriate for superior adjustable-speed drives. With the happening to late
power semiconductor developments and diverse keen control techniques, a fruitful
control system in view of vector control idea can be applied in real-time applications.
On account of the headways, today field-oriented control-based superior induction
motors have involved utmost of the situations that were in advance situated by dc
motors [2]. Starting late, vector control of IM drive is for the most part used as a piece of
predominant drive system. Flux sensors are absent in indirect vector control method.
In the course of the most recent couple of years, developments in neural networks and
upgradation in hardware have prompted progressively productive techniques for
preparing deep neural systems (DNNs) which have many layers of invisible and a vast
yield layer identified as DL algorithm. As of late deep learning has been attracting in a
huge consideration from extensive scope of applications. Compared with the traditional neural systems, the key highlights of DL algorithm are to have increasingly
invisible layers, neurons, and to enhance learning execution. Deep learning algorithms
can settle utilizing these highlights, huge, and difficult issues that could not be illu-
minated with conventional neural system. Subsequently, deep learning has connected
to different applications including human action recognition, pattern recognition,
speech recognition, etc. But, as we know, to the best information, not any outcome has
been published in AC drives field. The suggested research was intended to mirror the PI
controller utilizing a DBN algorithm [2–4].

Cheon et al. proposed deep learning controller technique to apply the deep
learning concept into dc motor control problems. The deep learning controller is
learnt by input and output data of PID controller [5]. Menghal et al. wrote a paper
on neuro-controller-based induction motor control. Results are compared with PI
and fuzzy controllers [6].

2 Deep Learning

Deep learning (DL) is the advanced version of neural networks, in which the amount of invisible layers and neurons are more than those of traditional neural networks. DL is a more superior algorithm than neural networks, particularly in big data. Deep learning concept uses deep architecture. There are many types of deep architecture design, among them deep belief architecture (DBN) is high performance yielding architecture.

DBN algorithm has two techniques, the pre-preparing system, and fine adjusting method. First, in restricted Boltzmann machine (RBM), observation vector v is pre-trained to obtain data vector v and in particular to create the starting weights of the second approach, known as fine tuning.

Throughout these techniques, the RBM which is essentially made out of three stages is center contrast of DBN control procedure compared with traditional neural network. But the RBM comes under unsupervised learning; hence, it has no objective information. RBM is in charge of creating the arrangement of initial weights that improve the learning. Figure 1 shows the working system of DBN concept. DBN algorithm has three stages:

Stage 1: The underlying information (that perceptive vector) drives within the observable stage of RBM, and after that by seeing first weight value, information will be exchanged to an invisible stage.

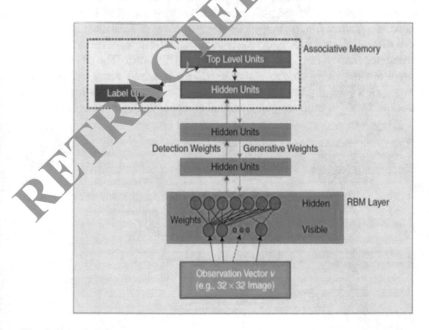

Fig. 1 Deep belief network framework

Stage 2: The first concealed stage winds up the second noticeable layer and exchanges information to the second invisible stage and second weights values also are taken into account. Similarly, the second concealed layer winds up the third hidden stage and exchanges the information to the third invisible stage. Output of the third invisible stage ends up beginning states of the training method. Stage one and stage two frames the pre-preparing technique that is going to create the beginning states of weights.

Stage 3: This progression signifies to fine-tuning method, where the training is done by varying weights of that input information pursues the output information, comparatively to many stages perception algorithm. Disclosed DBN is utilized to build up DLC [5] (Fig. 1).

2.1 Vector Control

The vector control calculation depends on two principal ideas. They are flux and torque producing currents. Induction motor can be controlled by separately excited dc motor by using two quadrature current components, instead of demonstrating it by three-phase currents fed to motor. Two currents I_d (direct) and I_q (quadrature) which we obtain from Clarke and Parke transformations are responsible for producing flux and torque differently in the motor. Id is in phase with stator flux and I_q is at right angles. The responsibility of transformation is to change a quantity that is sinusoidal in one reference frame to a constant value in rotating reference frame, which is rotating at same frequency. When a sinusoidal amount is changed to a steady by careful choice of reference frame, it winds up conceivable to control that amount with conventional PI controllers. In rotating reference frame, the output of electromagnetic torque in reference frame is equal as that of torque of DC motor. The advantages of vector control are improved torque response, dynamic speed accuracy, and low power consumption [4]. Figure 2 shows the block diagram of direct and indirect vector control method. In indirect vector control method, flux sensors are absent.

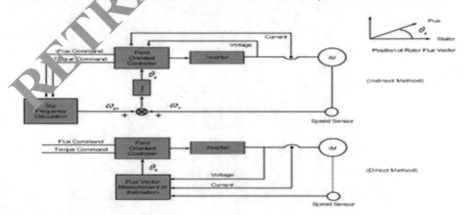

Fig. 2 Indirect and direct vector control methods

3 Design of Deep Learning Controller

To design the DLC, a PI controller was executed and verified, see Fig. 3a and after that by utilizing the executed PI's input/output information of the learning procedure separately. The DLC was adjusted to be fit for replacing all the other controllers, look at Fig. 3b; hence at long last DLC controls the induction motor drive, look at Fig. 3c. Earlier, deep neural system tool was used as a pattern recognition tool. Entire profound training calculation depends on deep neural system compartment; here, it is used as apparatus to structure DL controller, in which its invisible stage comprises two layers. There are diverse settling techniques for RBM, but the beta-Bernouli process RBM (BBPRBM) which yields the superior performance was viewed; here, the sigmoid function is utilized as actuation function [5].

3.1 Adaptive Neuro-Fuzzy Inference System

A neuro-fuzzy structure is a fuzzy structure that practices a derived learning algorithm or stimulated by the theory of neural networks to determine its factors (fuzzy set and fuzzy rules) by handling data models. NFC is the arrangement of fuzzy inference system (FIS) and NN. The fuzzy logic is activated permitting to the

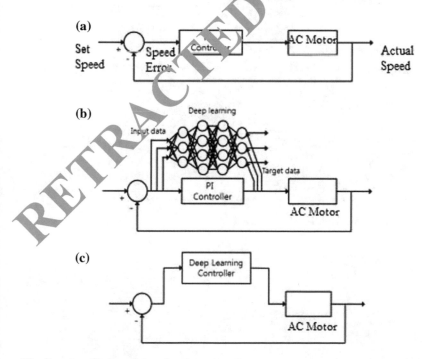

Fig. 3 a, b, c Design of deep learning controller

fuzzy rule and NN is accomplished based on the training data. The neural network formation dataset is generated by fuzzy rules [7, 8].

4   Development of Simulink Model

The recommended deep learning control strategy is implemented in MATLAB. At that point, the speed execution of deep learning method was tried with motor drive of 10 HP/400 V. Speed execution of deep learning method was compared with NFC. From created model, torque, current, speed, and voltages were examined. Simulink model of DLC is shown in Figs. 4 and 5, respectively.

5   Results

The comparative performance of deep learning controller and neuro-fuzzy controller is described. Speed responses of a drive with step change in speed reference from 100 to 150 rad/s and 150 to 60 rad/s are shown in Fig. 6.

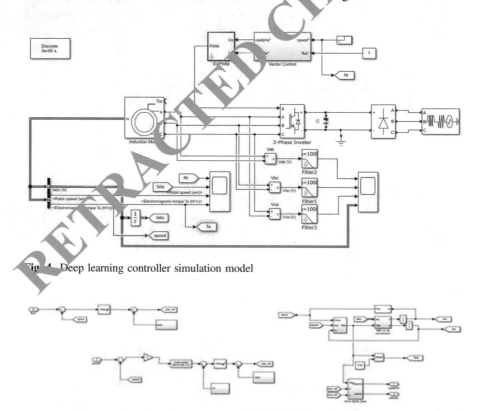

Fig. 4 Deep learning controller simulation model

Fig. 5 Simulink block of induction motor vector control system

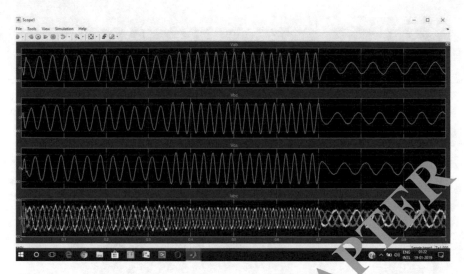

Fig. 6 Line-to-line voltages V_{ab}, V_{bc}, V_{ca}, and phase currents I_{abc} of deep learning controller

From the comparative analysis, the suggested deep learning controller offers superior performance when related to other controllers. In the case of DLC, overshoot is significantly less. Deep learning controller produces very less torque ripples compared to other controllers but takes more search space and CPU time for training. The performance of current, voltage, electromagnetic torque, and speed of deep learning controller is shown in Figs. 7 and 8, respectively. The settling time, rise time, peak time and steady state error of Deep Learning Controller (DLC) and Neuro Fuzzy Controller (NFC) are compared for better analysis (Table 1).

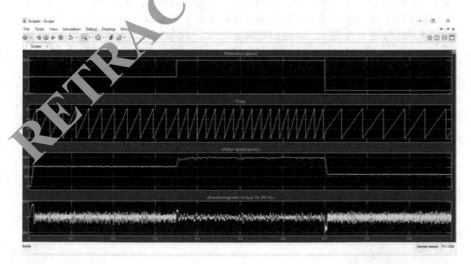

Fig. 7 Set speed, angle theta (in radians), rotor speed (wm), and torque (Nm) of DLC

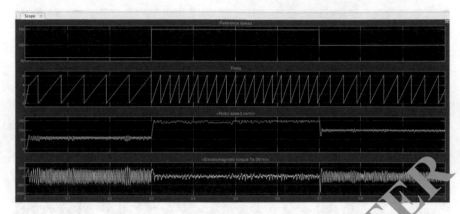

Fig. 8 Set speed, angle theta (in radians), rotor speed (wm), and torque (Nm) of NFC

Table 1 Result Analysis

Controller	Settling time (in s)	Rise time (in s)	Peak time (in s)	Steady state error (rad/s)
NFC	0.0200	0.015	0.0170	−1.80
DLC	0.0125	0.010	0.0115	+0.01

6 Conclusions

Another new advanced method proposed for the structure of deep learning controller for the induction motor drive is presented in this paper. From research findings, we can conclude that deep learning controller delivers superficially decreased torque and speed ripples compared with different controllers, in this way limiting losses in IM drives. Deep learning controller needs expansive pursuit space and more CPU time. Results demonstrated that DLC was able to do intently repeating the near execution. This method is quick in usage, powerful in identification, and fast in execution. Whenever the motor is loaded at any point, the speed of the motor falls down up to some extent, but in the case of DLC, this fall is very less. DLC must be preferred where speed exactness is particularly important. Torque and speed ripples are reduced significantly in the case of deep learning controller.

Future work consists of the following things

By using same optimization method, speed control system can be used to control the five-phase induction motor drive.

The hardware equipment might be implemented utilizing FPGA Spartan 7 to enhance the effectiveness of the induction motor drive.

Appendix

Parameters of induction motor: Nominal power 10 HP, line-to-line voltage 400 V, frequency 50 Hz, inertia 0.0226 kg m^2, friction 0.01 Nms, number of poles 2, stator resistance 0.19 Ω, stator inductance 0.21 mH, rotor resistance 0.39 Ω, rotor inductance 0.6 mH, mutual inductance 4 mH, parameters of PI speed controller: K_p 9.69, K_i 148.64.

References

1. Rushi Kumar, K., Sridhar, S.: A genetic algorithm based neuro fuzzy controller for the speed control of induction motor. Int. J. Adv. Res. Electr. Electron. Instrum. Eng. 4(9), 7837–7846 (2015)
2. Selvi, S., Gopinath, S.: Vector control of induction motor drive using optimized GA technique. In: IEEE Sponsored 9th International Conference on Intelligent System and Control (ISCO) (2015)
3. Keerti Rai, S, Seksena, B.L., Thakur, A.N.: A comparative performance analysis for loss minimization of induction motor drive based on soft computing techniques. Int. J. Appl. Eng. Res. 13(1), 210–225 (2018). ISSN 0973-4562
4. Douiri, M.R., Cherkaoui, M., Essadki, E.: Genetic algorithms based fuzzy speed controllers for indirect field oriented control of induction motor drive. Int. J. Circ. Syst. Signal Process. 6(1), 21–28 (2012)
5. Cheon, Kangbeom, Kim, Jaehoon, Hamadache, Moussa, Lee, Dongik: On replacing PID controller with deep learning controller for DC motor system. J. Autom. Control Eng. 3(6), 452–456 (2015)
6. Menghal, P.M., Jaya Laxmi, A.: Adaptive neuro fuzzy based dynamic simulation of induction motor drives
7. Srikanth, G., Madhusudhan Rao, G.: Adaptive neuro fuzzy based maximum torque control of three phase induction motor. Helix 8(2), 3067–3071
8. Venu Gopal, B.T., Shivakumar, E.G.: Design and simulation of neuro-fuzzy controller for indirect vector controlled induction motor drive. In: Nagabhushan, P., et al. (eds.) Data Analytics and Learning, Lecture Notes in Networks and Systems, vol. 43, pp. 155–167. Springer Nature, Singapore Pte Ltd (2019)

RETRACTED CHAPTER: A Survey on Face Recognition Using Convolutional Neural Network

M. Swapna, Yogesh Kumar Sharma and B. M. G. Prasad

Abstract Face recognition is an important concept, which has generally considered in the course of recent decades. Generally image location can be considered as an extraordinary sort of item recognition in PC vision. In this paper, we explore one of the vital and very effective systems for conventional item discovery using convolutional neural network (CNN) method, that is, a gentle classifier development for resolving the substance identification problem. That in recognizing the face of images as the problem is very difficult one, and so far no quality results are been obtained. Usually, this problem splits into distinctive sub-issues, to make it simpler to work predominantly identification of face or picture pursued by the face acknowledgment itself. There are several tasks to perform in between such as partial image face detection or extracting more features from them. Many years there are numerous calculations and systems have been utilized such as eigenfaces or active shape model, principal component analysis (PCA), K-nearest neighbour (KNN), and local binary pattern histograms (LBPH), but accurate results have not been identified. However because of the drawbacks of previously mentioned techniques in my study, I want to use CNN in deep learning to obtain best results.

Keywords Face recognition · Deep learning · Face detection · Convolutional neural network (CNN)

The original version of this chapter was retracted. The retraction note to this chapter is available at https://doi.org/10.1007/978-981-15-1097-7_82

M. Swapna (✉) · Y. K. Sharma
Department of CSE, JJT University, Jhunjhunu, India
e-mail: raviralaswapna@gmail.com

B. M. G. Prasad
Department of CSE, Holy Mary Institute, Hyderabad, India

© Springer Nature Singapore Pte Ltd. 2020, corrected publication 2024
K. S. Raju et al. (eds.), *Data Engineering and Communication Technology*,
Advances in Intelligent Systems and Computing 1079,
https://doi.org/10.1007/978-981-15-1097-7_54

1 Introduction

In the current computerized time, face recognition is an essential concept, which has broadly examined in the course of recent tenners. Face identification helped in developing many face-related applications, for example, picture authentication [1, 2], picture acknowledgment, picture clustering [3–5], and so on. Various procedures have been proposed for face recognition in the previous year, following the discovery work of Viola–Jones object discovery framework [6, 7]. Prior scrutiny examines in writing are for the most part centered around removing distinctive sorts of handmade highlights with area specialists in PC vision and preparing competing classifiers for location and acknowledgment with ordinary AI algorithms. These methods were restricted in regularly demand PC observation specialist of making successful highlights, and every individual segment is improved independently, making the entire identification pipeline frequently problematic. In conventional, picture identification can be considered as an exceptional sort of item discovery assignment in PC vision. Scientists have endeavored to handle face discovery by investigating some effective profound learning methods or conventional article location assignments. In this paper, we investigate one of the critical and very effective structures for conventional item identification utilizing convolutional neural network (CNN) strategy, which is a sort of classifier expansion for comprehending the article recognition assignments [8].

The improvements in PC innovation have made it conceivable to utilize face acknowledgment in applications where it is required to build up or affirm the personality of people. Applications for example, Apple, have acquired face ID as facial verification in iphones and a few banks are attempting to utilize facial validation for lockers. Some of the banks in Malaysia have introduced frameworks which use face acknowledgment to identify significant clients of the bank with the goal that the bank can give the customized administration. Along these lines, banks can induce more incomes by holding such clients and keeping them upbeat. Ample insurance organizations are utilizing face recognition to coordinate the essence of the individual with that gave in the personal ID confirmation. Along these lines, the secured composing process turns out to be much faster. Face recognition framework comprises of two levels [9].

1. **Face Detection**: Identify the people images within the digital images. It detects the human faces within the larger images.
2. **Face Recognition**: The detected images will check into the database and find out the correct faces.

The contrast between face recognition and face detection is that in location, we have to recognize faces in the picture; however in the face recognition, we should decide whose face it is.

2 Related Work

Face recognition is a technology efficient of recognizing or checking a human from a video's outline or an higher level picture. There are collective methods in which face recognition works, but in general, they work by correlate selected facial features from given image with faces within the databases. In olden days, many face recognition techniques implement which was not satisfied until Viola and Jones proposed work. The Viola and Jones are the first who are applying rectangular boxes for faces. It has lot of drawback. After getting so many problems using above technique, lot of people implemented new techniques like HOG, SIFT, and SURF. The histogram of arranged angles (HOG) is a character rubric employed in picture dealing and PC observation with item recognition being the end goal.

Eleyan et al. made and saw two face affirmation structures, one relied on the principal component analysis (PCA) sought after by a feed-forward neural framework (FFNN) called PCA-NN and the other subject to linear discriminant analysis (LDA) sought after by a FFNN called LDA-NN. The two structures contain two phases which are the PCA or LDA preprocessing stage and the neural framework gathering stage [10].

Li et al., K-set, an enormous vociferous neural system has been built which is situated in research on natural strong frameworks. This is not a merged neural system, and it reproduces the limits of natural intellect for flag preparing in example acknowledgment. Its exactness and proficiency are shown in this give an account of an implementation to the human face acknowledgment, with examinations of execution with traditional example acknowledgment calculations. Initially, a calculation for picture package is connected to separate the highlights of face pictures [11].

Ahonen et. al. proposed a book and effective facial delineation dependent upon isolating a facial image to compact districts and finding a depiction of every area usage street double design. These variants are then consolidated into a surfacially elevated histogram. The mix of every area representations encrypted the ubiquitous geometry of the face is depicted by the surface representation of an isolated area [12].

Gandhe et al. talk about and a realized unmistakable system for face affirmation, for instance, foremost component analysis, discrete wavelet transform cascaded with foremost component analysis, contour matching, and isodensity line maps cascaded with Hopfield neural network. All of these figurings are attempted on ORL database and BioID database [13].

Kute et al. examined in their overview paper 2D and 3D confront acknowledgment procedures, and numerous techniques have been dissected, demonstrating that all strategies guarantee acceptable acknowledgment rates, yet just when tried on standard databases or a few sections of them. On the opposite, it has been seen that straight/nonlinear methodologies defeat different strategies when enlightenment changes happen [14].

Bhuiyan et al. proposed a novel calculation for face acknowledgment utilizing neural systems prepared by Gabor highlights. The framework is started on convolving a face picture with a progression of Gabor channel coefficients at various scales and introductions. Two epic commitments of this paper are: scaling of rms complexity and presentation of fuzzily skewed channel. The neural system utilized for face acknowledgment depends on the multilayer perceptron (MLP) design with back-propagation calculation and joins the convolution channel reaction of Gabor fly [15].

Wright et al. proposed a technique for normally seeing human appearances from frontal points of view with change verbalization and illumination and also trouble and veil. They consider the affirmation issue as one of the collections among different direct relapse models, additionally, fight that new speculation from insufficient banner portrayal offers the best approach to keeping an eye on this problem [16].

Bashyal et al. proposed a programmed face demeanor acknowledgment system as outward appearance acknowledgment has potential applications in various parts of everyday life. In this paper, talks about the use of Gabor channel-based element extraction in blend with learning vector quantization (LVQ) for acknowledgment of seven distinctive outward appearances from still photographs of the human face. Firstly, it is seen that LVQ-based component grouping procedure proposed in this investigation performs preferred in perceiving dread articulations over multilayer perceptron (MLP) based characterization method [17].

Wright et al. examined the problem of appropriately observing personal expression from face aspect among differing articulation and enlightenment, moreover in extension drawback and concealment. The present original system commits original bits of knowledge within dual critical concern modern surface acceptance: feature expression moreover heartiness into drawback [18].

Latha ct al. have shown a visually established computation to recognize face points of view of faces. The capacity of the picture is diminished by the principal part examination (PCA) following which the affirmation is done by the back-inciting neural network (BPNN). Present 200 face pictures from Yale database are taken, and some execution estimations like aceptance extent and execution time are resolved. Neural-based face affirmation is generous and has better affirmation extent [19].

Anam et al. in his paper gave a face affirmation system for individual unmistakable verification and affirmation by making use of genetic estimation and back-inciting neural network. The structure contains three phases. At the particular starting some pre taking care of are associated on the data picture. Additionally, go up against natures are removed, which will be used as the commitment of the back spread neural system (BPN) and genetic calculation (GA) in the step three, and portrayal is finished by using GA and BPN [20].

Agarwal et al. introduced a structure for face recognition dependent on information of explanation approach about organize and deciphering the face picture. Expected system is cooperative of two stages—feature extraction applies guideline section investigation and compliance put to use the feedback well-developed engendering neural network. This conspire is self-determining of exorbitant calculation furthermore. The ratification of structure is executed dependent on eigenface, PCA, and ANN. Important segment of exploration for face recognition depends on the information of explanation approach in which the accruable data in a picture of a face are segregated as proficiently as it could be familiar covered by the precedence [21].

Gumus et al. Proposed an estimate of appropriate distinctive strategies for face acknowledgment. As highlight removing systems, they utilized wavelet deterioration and eigenfaces strategy which depends on principal component analysis (PCA). In the wake of creating highlight vectors, remove classifier and support vector machines (SVMs) are utilized for arrangement step. Then inspected the characterization precision as indicated by expanding measurement of preparing set, picked include extractor—classifier combines and picked part work for SVM classifier [22].

Paisitkriangkrai et al. proposed powerful and effective structure for taking in a versatile online ravenous scanty straight discriminant examination show for face acknowledgment. In this, the key commitments of this work are the paper put forth a productive gradual covetous meager. LDA classifier for building locator in a steady form. The online way of calculating incorporates the GSLDA-based component choice with a given adjustment plan for refreshing loads of the direct classifier edge and the discriminate capacities [23].

Hajati et al. proposed a productive posture invariant face acknowledgment strategy. This strategy is multimodal implies that it utilizes the 2D (shading) and 3D (profundity) data of a face for acknowledgment. In the initial step, the geodesic separations of all the face focus from a given reference point are processed. At that point, the face focuses are mapped through the 3D space to another 2D space. For highlighting the extraction, they utilize the Patch Pseudo Zernike Moments (PPZM) with another weighting approach to decrease the self-impediment brought about by inside and out revolutions [24].

Ahmad et al. proposed a strategy for programmed confront acknowledgment framework where ordinarily we have observation cameras at open spots for video catch and these cameras have the huge incentive due to security reason. The genuine preferences of face-based distinguishing proof over different biometrics are uniqueness and acknowledgment. In this paper, they tried the PCA, LDA, and LBP furthermore [25].

Kalaimagal et al. proposed a face acknowledgment framework dependent on picture space, scale, and introduction areas that can give profitable pieces of information not seen in either individual of the spaces. In this work, first they decayed the face picture into various introduction and scale by Gabor filter, and then consolidated nearby paired example investigation with Gabor which gives a decent face representation for acknowledgment. At that point for arrangement reason, they utilized middle histogram remove [26].

Kahou and Samira Ebrahimi in their paper gave the systems used for the University of Montréal's gathering sections to 2013. The test is to elucidate the sentiments imparted by the basic human subject in short video cuts isolated from full-length films [27].

Sharif Razavian in this paper add to the mounting proof that this is for sure the case. They write about a progression of examinations directed for various acknowledgment assignments utilizing the openly accessible code and model of the OverFeat organize which was prepared to perform object order on ILSVRC13. They use highlights removed from the OverFeat organize as a conventional picture representation to handle the differing scope of acknowledgment errands of item picture order, scene acknowledgment, fine-grained acknowledgment, quality location, and picture recovery connected to an assorted arrangement of datasets [28].

Schroff et al. in their work presented a framework, named FaceNet, that straightforwardly takes in a mapping from face pictures to a smaller Euclidean space which relate to a proportion of face similitude. The advantage of their methodology is a lot more prominent illustrative effectiveness [5].

Parkhi et al. presented the objective of this research is confront acknowledgment —from either a solitary photograph or from a set of countenances followed in a video. Late advancement around there has been because of two elements: (i) start to finish learning for the errand utilizing a CNN algorithm (convolutional neural system) and (ii) the accessibility of vast scale preparing datasets [4].

Mollahosseini et al. in this paper proposed automated facial expression recognition (FER) has remained a testing and fascinating issue with regard to PC vision. In spite of endeavors made in creating different techniques for FER, existing methodologies need generalizability when connected to inconspicuous pictures or those that are caught in wild setting [29].

Rothe et al. in this research proposed a profound learning arrangement to the age estimation from a solitary face picture by not using the utilization of facial tourist spots and present the IMDB-WIKI dataset, the biggest open dataset of face pictures with the gender and age label marks. On the off chance that the genuine age estimation inquire about ranges over decades the investigation of clear age estimation or the age as apparent by different people from a face picture is an ongoing undertaking they handle the two assignments with our convolutional neural systems (CNNs) of VGG-16 engineering which are pre-prepared on ImageNet for picture characterization. They represent the age estimation issue as a profound characterization issue pursued by a softmax expected esteem refinement [30].

Kamencay et al. proposed a new method for face recognition by making use of the convolutional neural network with three prominent image recognition methods such as principal component analysis (PCA), local binary patterns histograms (LBPH), and K-nearest neighbour (KNN). In their demonstration, they used accuracy of PCA, LBPH, KNN, and CNN. In this paper, they have succeeded 98% to reach accuracy by using CNN for the best results. [9].

Sharma and Rathore proposed in this paper can be very useful for removing noise from digital images using median filtering techniques of rectangular digital images. The corner defect detection from algorithm is very helpful for maintaining quality of ceramic tiles. This algorithm is very helpful for detection of corner defect from rectangle ceramic tile. The comparative analysis can be used four factors that are: accuracy level, consistency level, complexity of time, and production rate. Using these four factors can prove this algorithm is better than the previous method [31].

Liu et al. in this paper tend to profound face acknowledgment (FR) problem under open-set convention, where perfect face highlights are expected to have littler maximal intra-class separate than negligible between class separate under a reasonably picked metric space [32].

Sharma and Rathore in their research designed a model machine Upgraded Automated Quality Maintaining Machine (UAHQMM) that is fully automated by joining three phases: raw material, quality assurance, and packaging system [33].

Sung et al. in this paper presented another face identification plot utilizing profound learning and accomplish the best in class discovery execution on the outstanding FDDB confront location benchmark assessment. Specifically, they enhance the state-of-the-art faster RCNN system by joining various methodologies, counting highlight link, negative mining, multi-scale preparing, display pre-preparing, and legitimate alignment of key parameters [34].

Krishnamurthy et al. in this paper proposed a basic yet intense to beat multi-modular neural model for double-dealing identification. By consolidating highlights from various modalities, for example, video, sound, and content alongside micro-expression highlights, they demonstrate that recognizing double-dealing, in actuality, recordings can be more exact [35] (Table 1).

3 Need of Study

That in recognizing the face of images as the problem is very difficult one, and so far no quality results are been obtained. Usually, this problem splits into different subproblems, to make it easier to work mainly detection of face of an image followed by a face recognition. There are several tasks to perform in between such as partial image face detection or extracting more features from them. Many years there are many algorithms and methods have been used such as eigenfaces or active shape model, principal component analysis (PCA), K-nearest neighbour (KNN), and local binary pattern histograms (LBPH), but accurate results have not been identified. However because of the drawbacks of previously mentioned techniques in my study, I want to use CNN in deep learning to obtain best results (Fig. 1).

4 Dataset Collection

The following databases are used in this work:

4.1 Labeled Faces in the Wild Dataset (LFW)

This involves 13,233 images with 5,749 different identities, and it is the standard benchmark for automatic face verification.

Table 1 Comparison analysis of face recognition from 2006 to 2018

Author/year	Method	Image size	Dataset	Number of images	Recognition rate (%)
Alaa Eleyan (2006)	PCANN, LDANN	–	ORL	8, 8	75, 80
Ganguang Li (2006)	KIII SET, WPT-KIII, SVD-KIII, DCT-KIII	–	ORL	–	90, 91, 91.5
Timo Ahonen (2006)	LBP	–	FERET	–	97
Andrea F. Abate (2007)	2D-3D modal-based algorithm	–	FERET	–	
Tajmiluri Rahman (2007)	MLP	100 * 100	–	–	75
Jon Wright (2008)	SVM, NN, NS	192 * 168	–	–	97.7, 90.7, 94.1
Shishir Bashyal (2008)	LVQ	256 * 256	JAFFE	42	87.51
P. Latha (2009)	PCA, BPNN	–	Yale	200	90
SarawatAram (2009)	BPN, GA	–	–	30, 30	91.3, 82.61
Mayank Agarwal (2010)	PCA, FFBPN	112 * 92	ORL	400	97.018
Ergun Gumus (2010)	PCA, SVM	–	ORL	400	98.1
Farshid Hajati (2011)	2D + 3D multimodal, PPZM	–	BJUT	500 3D	98.8
Faizan Ahmad (2012)	LBP, SVM, HOG, Gabor	–	–	–	71.15, 77.90, 82.94, 92.35
S. Kalaimagal (2012)	LBP-Gabor	–	Yale	–	80
Samira Ebrahimi Kahou (2013)	CNN	–	Tornoto Face Dataset (TFD) AFEW2	–	96.73
Ali Sharif Razavian (2014)	CNN-SVM	221 * 221	Caltech-UCSD birds (CUB)	–	90.2
Omkar M. Parkhi (2015)	Deep-CNN	256, 384, 512 pixels	LFW, YTF	200	95.1
F. Schroff (2015)	Deep-CNN	96 * 96– 224 * 224	LFW, YTF	–	99.63, 95.12

(continued)

Table 1 (continued)

Author/year	Method	Image size	Dataset	Number of images	Recognition rate (%)
Mollahosseini (2016)	CNN	48 * 48	FERA, SFEW	–	95
Rasmus Rothe (2016)	CNN	–	IMDB-WIKI	–	–
Zhedong Zheng (2017)	GAN, LSRO	–	CUHK03	–	84.4
Xudong Sun (2018)	RCNN	64 * 64	FDDB	–	
Gangeshwar Krishnamurthy (2018)	Mutimodal deep neural network	–	–	–	6.4

Fig. 1 Comparison analysis of face recognition using CNN (2013–2018)

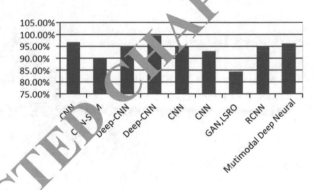

4.2 FERET or COLOR FERET

It involves around 1564 sets of images from a total of 14,126 images that further has 1199 individuals and 365 replacement sets of images.

4.3 Yale Face Database B

It has about 16,128 images out of 28 human subjects under 9 poses and 64 illuminating conditions.

4.4 ORL

It contains 400 images of 40 person and 10 images for each person.

5 Methodology

5.1 Image

This is the first module in this process. Here, we select the image that means training images. It is also called training AI dataset. The training data include both input and expected output. Based on this data, the algorithms can apply new technologies like neural networks and it will provide accurate results (Fig. 2).

5.2 Face Detection

The second module in this process is face detection. In this stage, we can detect faces via two approaches:

(1) image-based approach and (2) knowledge-based approach.

5.3 CNN

The convolutional neural network (CNN, ConvNet) is a class of deep learning. It is proposed to reduce the number of parameters and acquire the network architecture. It consists of multiple hidden layers. Each layer will take multi-dimensional array of number of inputs and multi-dimensional array of outputs. Each hidden layer consists of convolution layer, RELU layer that is activation function, pooling layer, fully connected (FC) layer, and normalization layers (Fig. 3).

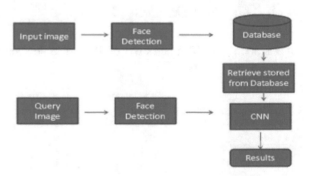

Fig. 2 Basic steps for face recognition

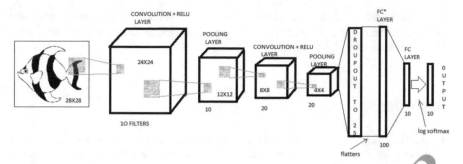

Fig. 3 CNN architecture

5.3.1 Convolution + Relu Layer

The dot production between weights and inputs is combined across channels. RELU stands for rectified linear unit for nonlinear operation.

5.3.2 Pooling Layer

Pooling layer applies nondirect down testing on actuation maps. Pooling layers segment would bring down the quantity of parameters when the pictures are excessively huge. Spatial pooling likewise called sub-sampling which downsize the dimensionality of each guide yet holds the vital data. Spatial pooling can be of various sorts:
   (1) max pooling, (2) average pooling, and (3) sum pooling.

5.3.3 Fully Connected

The FC layer, we flattened our matrix into vector and feed it into a fully connected layer like neural network.

6 Future Scope

That in this exploration, there is much extension to certain viewpoints that ought to be improved later on. The CNN is the procedure which is still failing to meet expectations contrasted and current condition of workmanship since this part is the reason for entire face acknowledgment framework. On the off chance that any improvement made on this strategy will profit the last execution. So as to get such outcomes might want to accumulate more information for execution to utilize diverse parameters and different arrangements and further to accomplish more tries different things with CNN procedure.

References

1. Sun, Y., Wang, X., Tang, X.: Deep learning face representation from predicting 10,000 classes. In: Proceedings of the IEEE Conference on Computer Vision and Pattern Recognition, pp. 1891–1898 (2014)
2. Taigman, Y, Yang, M., Ranzato, M., Wolf, L.: Deepface: closing the gap to human-level performance in face verification. In: Proceedings of the IEEE Conference on Computer Vision and Pattern Recognition, pp. 1701–1708 (2014)
3. Kumar, K.S., Semwal, V.B., Tripathi, R.: Real time face recognition using adaboost improved fast PCA algorithm, arXiv preprint arXiv:1108.1353
4. Parkhi,O.M., Vedaldi, A., Zisserman, A.: Deep face recognition. In: British Machine Vision Conference, vol. 1, p. 6 (2015)
5. Schroff, F., Kalenichenko, D., Philbin, J.: Facenet: A unified embedding for face recognition and clustering. In: Proceedings of the IEEE Conference on Computer Vision and Pattern Recognition, pp. 815–823 (2015)
6. Viola, P., Jones, M.: Rapid object detection using a boosted cascade of simple features. In: IEEE Conference on Computer Vision and Pattern Recognition (CVPR), vol. 1, p. 511. IEEE (2001)
7. Viola, P., Jones, M.J.: Robust real-time face detection. Int. J. Comput. Vision 57(2), 137–154 (2004)
8. Girshick, R., Donahue, J., Darrell, T., Malik, J.: Rich feature hierarchies for accurate object detection and semantic segmentation. In: Proceedings of the IEEE Conference on Computer Vision and Pattern Recognition, pp. 580–587 (2014)
9. Kamencay, P., et al.: A new method for face recognition using convolutional neural network (2017)
10. Eleyan, A., Demirel, H.: PCA and LDA based face recognition using feedforward neural network classifier. In: MRCS 2006, LNCS 4105, pp. 199—206. © Springer-Verlag Berlin Heidelberg (2006)
11. Li, G., Zhang, J., Wang, Y., Freeman, W.J.: Face recognition using a neural network simulating olfactory systems. UC Berkely Previously Published Works(2006)
12. Ahonen, T., Hadid, A., Pietikainen, M.: Face description with local binary patterns: application to face recognition. IEEE Trans. Pattern Anal. Mach. Intell. 28(12) (2006)
13. Gandhe, S.T., Talele, K.T., Keskar, A.G.: Intelligent face Recognition techniques: a comparative study. GVIP J. 7(2) (2007)
14. Abate, A.F., Nappi, M., Riccio, D., Sabatino, G.: 2D and 3D face recognition: a survey. Elsevier (2007)
15. Bhuiyan, A.-A., Liu, C.H.: On face recognition using gabor filters. World Academy of Science, Engineering and Technology (2007)
16. Wright, J., Yang, A.Y., Ganesh, A., Sastry, S.S.: Robust face recognition via sparse representation. IEEE Trans. (2008)
17. Bashyal, S., Venayagamoorthy, G.K.: Recognition of facial expressions using gabor wavelets and learning vector quantization. Elsevier (2008)
18. Wright, J., Yang, A.Y., Ganesh, A., Sastry, S.S, Ma, Y.: Robust face recognition via sparse representation. IEEE Trans. Pattern Anal. Mach. Intell. 31(2) (2009)
19. Latha. P., Ganesan, L., Annadurai, S.: Face recognition using neural networks. Signal Process. Int. J. (SPIJ) 3(5), 153–160 (2009)

20. Anam, S., Islam, M.S., Kashem, M.A.,Islam, M.N., Islam, M.R, Islam, M.S.: Face recognition using genetic algorithm and back propagation neural network. In: Proceedings of the International MultiConference of Engineers and Computer Scientists (2009)

21. Agarwal, M., Jain, N., Kumar, M., Agrawal, H.: Face recognition using Eigen faces and artificial neural network. IJCTE 2(4) (2010)

22. Gumus, E., Kilic, N., Sertbas, A., Ucan, O.N.: Evaluation of face recognition techniques using PCA wavelets and SVM. Elsevier (2010)

23. Paisitkriangkrai, S., Shen, C., Zhang, J.: Incremental training of a detector using online sparse eigendecomposition. IEEE Trans. Image Process. 20(1) (2011)

24. Hajati, F., Raie, A.A., Gao, Y.: Pose-invariant multimodal (2D + 3D) face recognition using geodesic distance map. J. Am. Sci. 7(10) (2011)

25. Ahmad, F., Najam, A., Ahmed, Z.: Image-based face detection and recognition: "state of the art". IJCSI 9(6, no. 1) (2012)

26. Kalaimagal, S., Singh, M.E.J.: Face recognition using Gabor volume based local binary pattern. IJARCSSE 2(6) (2012)

27. Kahou, S.E., et al.: Combining modality specific deep neural networks for emotion recognition in video. In: Proceedings of the 15th ACM on International Conference on Multimodal Interaction. ACM (2013)

28. Sharif Razavian, A., et al.: CNN features off-the-shelf: an astounding baseline for recognition. In: Proceedings of the IEEE Conference on Computer Vision and Pattern Recognition Workshops (2014)

29. Mollahosseini, A., Chan, D., Mahoor, M.H.: Going deeper in facial expression recognition using deep neural networks. In: 2016 IEEE Winter Conference on Applications of Computer Vision (WACV). IEEE (2016)

30. Rothe, R., Timofte, R., Van Gool, L.: Deep expectation of real and apparent age from a single image without facial landmarks. Int. J. Comput. Vision 126(2–4), 144–157 (2018)

31. Sharma, Y.K., Rathore, R.S.: Fabrication analysis for corner identification using algorithms increasing the production rate. Int. J. Comput. Sci. Eng. (IJCSE) 9(05), 248–256 (2017). ISSN: 0975-3397

32. Liu, W., et al.: Sphereface: deep hypersphere embedding for face recognition. In: The IEEE Conference on Computer Vision and Pattern Recognition (CVPR), vol. 1 (2017)

33. Sharma, Y.K., Rathore, R.S.: Advanced grading system for quality assurance of ceramic tiles based using digital image processing. Indian J. Sci. Technol. 10(14), 1–4 (2017). https://doi.org/10.17485/ijst/2017/v9i14/112248. ISSN (Online):0974-5645, (Print): 0974-6846

34. Sun, X., Wu, P., Hoi, S.C.H.: Face detection using deep learning: an improved faster RCNN approach. Neurocomputing 299, 42–50 (2018)

35. Krishnamurthy, G., et al.: A deep learning approach for multimodal deception detection. arXiv preprint arXiv:1803.00344 (2018)

A Hybrid Ensemble Feature Selection-Based Learning Model for COPD Prediction on High-Dimensional Feature Space

Srinivas Raja Banda Banda and Tummala Ranga Babu

Abstract Chronic obstructive pulmonary disease (COPD) can be identified by airflow limitation in lungs, and this condition is not at all reversible. Generally, the diagnosis process of COPD completely depends upon symptoms, medical history, clinical examination, and lung airflow obstruction. Hence, it is essential to develop an appropriate and effective automatic test in order to diagnose COPD for better disease management. Most of the traditional auto-classification models such as SVM, neural network, Bayesian model, expectation-maximization, and random tree are used to find the appropriate features and its decision patterns for COPD detection. However, as the number of COPD features increases, these models require high-computational memory and time for feature selection and pattern evaluation. Also these models generate high false-positive rate and error rate due to high feature space and data uncertainty. In order to overcome these issues, a hybrid ensemble feature selection-based classification model is proposed on high-dimensional dataset. Experimental results proved that the present model improves the true positivity and error rate compared to the traditional models.

Keywords Ensemble feature selection · COPD · Classification model

1 Introduction

Chronic obstructive pulmonary disease (COPD) is a special type of respiratory disease which is usually caused because of smoking or air pollution. Some important symptoms of COPD are: inflammation of the lung airways and degradation of lung tissue. This condition is called as emphysema, and this will restrict

S. R. B. Banda (✉)
Research scholar, Department of ECE, Acharya Nagarjuna University, Guntur, India
e-mail: srinivasarajacs@gmail.com

T. R. Babu
RVR & JC College of Engineering, Acharya Nagarjuna University, Guntur,
Andhra Pradesh, India

© Springer Nature Singapore Pte Ltd. 2020 663
K. S. Raju et al. (eds.), *Data Engineering and Communication Technology*,
Advances in Intelligent Systems and Computing 1079,
https://doi.org/10.1007/978-981-15-1097-7_55

the usual flow of air to and fro the lungs. This disease has different stages from non-harmful initial stage to worse terminal stage. If the disease will progress to the terminal stage, then it may lead to death of the patient. Identification of this disease is very much essential to increase the overall survival rate.

All asthma symptoms such as wheezing or coughing are considered for automatic prediction of the disease. Again, family history of disease also can affect the severity of asthma or bronchitis [1]. Logistic regression prediction approach is implemented in case of 1226 children. Among the above, 345 numbers of children have developed asthma after 5 years. The feature extraction strategy is capable of generating various fuzzy rules for the neural networks. Such pre-processing techniques have significant roles in order to achieve exact classification accuracy. There exist some other techniques that integrate machine learning approaches with feature significance ranking techniques. Both oscillometry and spirometry metrics are used for the assessment process of lung function. Oscillometry instruments are basically non-portable and costlier than that of spirometry metrics. The variables which are easier to obtain are generally used for the self-assessment of asthma and other respiratory diseases. Spirometer and oximeters are considered as most reliable tool for prediction.

Usually, these tools are integrated with smartphones and smartwatches. Some researchers have given emphasis on the risk of exacerbation in case of patients having different respiratory diseases. The above-presented approach implements classification as well as regression tree method. This approach is capable of classifying the whole data into two groups, those are: low risk and high risk. This approach results in 71.8% classification accuracy. Additionally, it is also capable to predict the condition of high risk much before the patient comes to know and observe great degradation in their condition.

In case of some other works, at first clinical factors are detected. These factors have the responsibility to modulate the risk of advancement of asthma patients toward COPD. BN technique is consists of age, sex, race, smoking history, and eight co-morbidity variables. This approach can predict all COPD diseases among set of patients. With the help of HRCT technique, patients having normal radiographs are studied thoroughly. It is very much efficient to identify symptoms as well as abnormalities of lung function. This technique is very much efficient to analyze the severity of abnormal modifications in case of visual examination of HRCT images.

All automatic diagnosis systems are required to apply robust pattern recognition approaches for non-linear and adaptive models. Artificial neural network (ANN) is the most common example of the above-mentioned group of classification schemes. Artificial neural network includes separate and non-linear processing units, those are known as neurons. All of these neurons can be classified into three distinct groups, those are: (a) input neurons, (b) processing neurons, and (c) output neurons. These three layers of neurons are arranged with the help of densely inter-connected space. These mentioned connections between neurons in case of a neural network contain some amount of weights.

In this research work, we have emphasized on the following features:

(1) The true-positive prediction rate of the proposed approach is optimized compared to conventional learning approaches.
(2) Since traditional approaches are sensitive to high-dimensional datasets with limited data size, proposed model is applicable to high-dimensional datasets with large size dataset.
(3) Proposed technique optimizes the feature selection mechanism from high-dimensional dataset to optimize training accuracy.

2 Related Works

Kuncheva et al. developed a new ensemble classification approach in order to diagnose COPD from volatile organic compounds in exhaled air [2]. The diagnosis process of COPD depends upon its symptoms, test results along with some other factors too. Large numbers of people are ignorant about the above disease so that they never pay attention to its effects. In the field of spirometry, different researchers have different views. In order to develop a successful diagnosis and treatment strategy, at first COPD must be detected at very early stage. Hence, accurate and reliable automated tests plan must be created.

Corlateanu et al. emphasized on multi-lateral assessment of COPD [3]. COPD can be included inside the category of multisystemic pathology. Airflow limitation problem is unable to diagnose and assess COPD. Hence, it is necessary to develop an advanced technique to resolve the diagnosis problems of COPD. In this paper, they have studied and analyzed various multidimensional classification schemes. They also considered different functional and physiological parameters. COPD is considered as multi-level disease from the perspective of clinical, cellular, and molecular view. In this paper, a thorough survey is performed from the basic model to high-level advanced model. According to the degree of airflow limitation, a new method is proposed that integrates the functional, physiologic, exacerbation, and health status domains. The above technique is known as new GOLD classification scheme. In this scheme, the effect of COPD on each and every patient is studied and analyzed.

van Rikxoorta et al. proposed a new ensemble classification which depends upon the basic concepts of novel classifier selection method [1]. COPD can be included under the category of chronic lung disease which severity is calculated by the airflow limitation. Pulmonary function testing (PFT) technique is generally implemented in order to recognize and observe COPD. It considers both global inspiration and expiration rate of patients. Again, it also considers the total amount of time required for the whole process. Generally, pairs of inspiration–expiration CT scans are generated.

Khan et al. developed an advanced approach for proficient lungs nodule detection and classification [4]. They basically used machine learning approaches for their research. In between various types of cancers, lung cancer is considered as one of the most dangerous cancers. Usually most of the time, this disease is not identified in the early stage. We can find huge numbers of image processing approaches in order to identify COPD at an early stage. But, till date there is no such optimized approach being developed that will overcome all the issues of these said traditional techniques.

Cheplygina et al. introduced an advanced classification scheme for COPD with the help of multiple instance learning [5]. COPD disease is considered as most dangerous kind of cancer. The chance of survival of patient increases, if the disease can be identified at an early stage. Abdalla et al. presented a new computer-aided diagnosis system for classification of lung tumors [6]. Lung cancer is well known for its dangerous effects worldwide. Nowadays with advancement of technology, many new research proposals are proposed for the detection of lung cancer. But there is no appropriate technique which can evaluate lung cancer more accurately. In this technique, artificial neural network concepts are used in order to classify among malignant or benign tumor.

Kuwahara et al. introduced classification of patterns for diffuse lung diseases in thoracic CT images through the implementation of AdaBoost algorithm [7]. M. Agarwal, et al., implemented local SIMPLE multi-atlas based segmentation technique to lung lobe identification on chest CT scan [8]. Md. R. Arbabshirani, et al., emphasized on accurate segmentation of lung fields on chest radiographs [9]. F. Chabat, et al., introduced a new texture classification scheme for differentiation at CT images [10]. L. Sørensen, et al., emphasized on transfer learning for multi-center classification scheme [11]. Garg, et al., introduced an automated segmentation of lungs out of high resolution CT images [12]. P. Ghosh, et al., developed a new and advanced unsupervised segmentation of lungs from chest radiographs [13]. G. Gill and R. R. Beichel developed a new strategy in order to decrease the overall error rate in case of automated lung segmentation technique [14]. D. Spathis et al., proposed an advanced machine learning approach for diagnosis of asthma and COPD [15]. R. J. Huijsmans, et al., focused clinical utility of the GOLD classification of COPD disease severity in pulmonary rehabilitation [16]. P. Kohlmann, et al., developed an automatic lung segmentation approach for MRI-based lung perfusion studies of patients with COPD [17]. CT scan images are found to be helpful and beneficial in case of differential diagnosis of diffuse lung diseases. The diagnosis process of lung diseases is considered as a complex issue. The complexity increases, because of the variance of patterns of CT scan images. The prime objective of the above-proposed method is to build a CAD system in order to classify patterns. This approach results in both quantitative and objective information in order to reduce the workload of radiologists.

3 Proposed Model

The biomarker should also be related to how the disease progresses and how the disease affects the life of the patient. The strength of consistency of the independent relationship between the COPD and biomarker is also analyzed. How the biomarker relates to the clinical outcomes such as the patient being hospitalized and death is also analyzed and the strength determined.

Determining if the biomarker can be modified using interventions should be evidently found out by carrying out controlled trials. Lastly, the facts from randomized and controlled tests should be able to indicate whether the changes in biomarker status can result in hard clinical outcomes such as death. Whether a biomarker is short term or long term should also be known. It should also be scientifically provable if changes in the biomarker have a significant effect.

Figure 1 illustrates the proposed ensemble COPD feature ranking model for multi-level COPD detection on high-dimensional dataset. In the proposed model, mixed types of attributes are used for COPD disease prediction with large number of feature set. In the first stage, COPD data is prepared using the index and class

Fig. 1 Proposed model

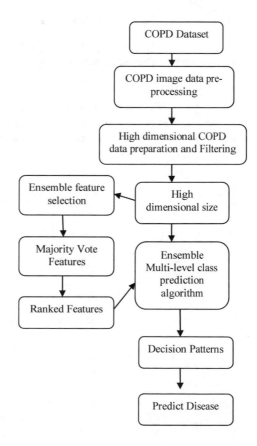

labels of the histogram intensities and kernel estimator. In the figure, initially dataset with large feature set is mapped to find the COPD class label using algorithm 2. After the preparation of the COPD data, data pre-processing algorithm is applied to fill the missing values to the numerical or nominal attributes. After the data pre-processing stage, ensemble feature selection process is performed to find the highest ranked attributes among the large COPD feature space.

These ensemble ranked features are used to predict the new COPD test instances using the proposed ensemble multi-level classification model.

Algorithm 1: COPD data preparation of index and class labels using histogram-based Gaussian kernel estimator

Step 1:   Input COPD images.
Step 2:   Filter the COPD images using Subject ID.
Step 3:   kernel Gaussian estimation using histogram H is computed as:

$$HKE = \sum_{i=0}^{N} \frac{1}{\sqrt{2\pi H}} \cdot e^{-(I[i]-H)^2/2H}$$

Step 4:   For each subject id in Image database
     Do
               For each block in image id
     Do
               Compute Block histogram intensities I.
               Compute average block histogram value H using eq.(1).
     Done
     Done
Step 5:   Each 256 histogram intensity feature values of ROI and kernel density estimation are store in a csv file. Each subject id and class label of the COPD either disease(1) or not disease (0) are stored in separate files as index and class.

Algorithm 2: Data Integration

Input: Input training dataset D_T, index dataset D_I, label dataset D_L.
Output Labelled COPD dataset.
Procedure:
Step 1: c=0;
   For i=1 to $|D_T|$
   Do
       c=c+1;
   done
Step 2: // Store index values in list
    For j=1 to $|D_I|$
   Do
       List I=D_I(j);
   Done
Step 3: Remove duplicate indices in List I as I'.
       {Cluster[],S[]}=CureCluster(D_T);
   For each pair of clusters i,j(where i!=j) in Cluster[]
   Do
       If(Similarity index S[i]==S[j])
       Then
            Remove cluster objects in D_T.
       End if
       Else
          List I'=D_T(i);
   Do
Step 5: Step 4: //Processing labels in D_L as L'.

For k=0 to $|D_L|$
do
   List L=D_L(k);
Done
Step 5: // Mapping clustered indices to labels
   For m=0 to $|D_L(k)|$
   Do
       D_{IL}(m)=(I'(m),L(m));
   Done
Step 6: // Label Training data using Mapped cluster indices.
   For r1=0 to $|D_T|$
      For r2=0 to $|I'|$

$$\textbf{If (I' (r2)} \in D_T)$$

      then

$$D_{TL}[r1] = getLabel(I'(r2))$$

      endif
      Done
Done

Input : Filtered COPD data D_{TL}

Output: Features with ranking weights.

Procedure:

Step 1: Read input data D_{TL} .

Step 2: Apply F1[]= PCA(D_{TL});

Step 3: Apply F2[]=HybridPSO(D_{TL})

$$Fitness_i = w_1 . \max \{ Correlated\, Probability(FS[], Classes) \}$$

$$+ w_2 \cdot \left(1 - \frac{\sum_{i=1}^{F} f_i}{N_f} \right)$$

f_i is the flag value 1 or 0 .

'1' represents selectd feature ,

'0' non-selected feature.

N_f represents number of features.

Step 4: Hybrid Ranking Measure

In the algorithm, labeled COPD data is filtered using the discretization and conditional probability measures. If the missing value of a feature in the feature space is numerical, then discretization operation is used to replace the missing value with the computed value. If the missing value of a feature in the feature set is nominal, then conditional probability measure is used to fill the missing value.

Partition the data into m classes as $D_1, D_2 .. D_m$.

For each partition D_i

Do

    Find the attribute ranking measure as

$$RFSM\left(D_i\left(A_j\right)\right) = \max\{\sqrt{\left(\sum_{p=1}^{|D_p|}\sum_{n=1}^{|D_n|}(\sqrt[3]{D_p\left(A_j\right)/\left|D_n\left(A_j\right)\right|} - \sqrt[3]{D_n\left(A_j\right)/\left|D_n\left(A_j\right)\right|})^2\right)},$$

$$RoughSet\left(D_p\left(A_j\right), D_n\left(A_j\right)\right), IG\left(D_p\left(A_j\right)/D, D_n\left(A_j\right)/D\right), \quad Where, is\, D_p\left(A_j\right) is$$

$$Chi-SquareFS\left(D_p\left(A_j\right), D_n\left(A_j\right)\right)\}$$

    where j=0...|A |;Featurespace

            the positive disease class data of attribute (A_j) .

    $D_n(A_j)$ is the normal class data of attribute (A_j)

    IG($D_p(A_j)$, $D_n(A_j)$)) is the information gain of (A_j)

    Chi − squareFS($D_p(A_j)$, $D_n(A_j)$) : Chi-square feature selection of attribute between two

classes (A_j)

Done

Candidate Features CF[]=Join(RFSM($D_i(A_j)$))

Find the features with highest ranking measure using threshold.

For each candidate feature CF[i] in CF

Do

For each feature (A_j) in CF[i]

if (RFSM($D_i(A_j)$) > λ)

then

    RankFS[(A_j)]=$D_i(A_j)$;

    else

    continue;

Done

Done

Proposed Model: Multi-layered Ensemble Decision Tree

Ensemble classification is defined as the training of multiple base classifiers to detect the disease severity in the test data.

Algorithm 3: Ensemble Steps

$D_0 \leftarrow <1/ D_1 ,1/ D_2 ,1/ D_3 ,.....1/ D_N >$ N uniform data partitions

$C_1 \leftarrow$ Parallel SVM +ELM

$C_2 \leftarrow$ Parallel Naive Bayes

$C_3 \leftarrow$ Neura l n etwork

$C_4 \leftarrow$ Improved RandomForest(Ensemble FS+RandomForest)

for each classifier k in C_i do

$\quad C_i = \text{classifier}(D, D_{k-1}) // \text{Training phase}$

$\quad P_n \leftarrow C_i(x_n), \forall n$ prediction on training instances

$\quad E_k \leftarrow \sum_n D_{k-1}[n] / P_n \neq P_n$

$\quad \phi_k = \log\left(\dfrac{1-E_k}{E_k}\right)$

end for

$$f(\hat{m}) = \text{sgn}\left[\sum_k \phi_k \cdot C_k(\hat{m})\right]$$

4 Experimental Results

Experimental results for COPD detection is taken from http://bigr.nl/research/projects/copd [5]. The dataset contains histogram intensity values and kernel density estimation values in separate files. A total of 256 features are taken in each COPD subject. In our experimental study, we have added two new fields, i.e., Subject ID and class attribute. Here, two classes are used COPD and normal for the classification problem.

Here, statistical classification measures are evaluated in the experimental results to find the true-positive rate and kappa error. In the proposed model, 99.16% of accuracy is achieved on the high-dimensional COPD dataset using the proposed model.

Correctly_Classified Instances	4958	99.16	%
Incorrectly_Classified Instances	42	0.84	%
Kappa_statistic	0.9829		
Mean absolute error	0.0139		

```
 a   b     <-- classified as
2792   8 |   a = Normal
  34 2166 |   b = COPD
```

Table 1 describes the performance analysis of the proposed model to the existing models in terms of statistical true-positive, false-positive, and ROC rates. From the table, it is clearly observed that the proposed model has high-computational accuracy compared to the existing models on high-dimensional dataset.

Figure 2 describes the performance analysis of the proposed model to the existing models in terms of statistical true-positive, false-positive, and ROC rates. From the figure, it is clearly observed that the proposed model has high-computational accuracy compared to the existing models on high-dimensional dataset.

Table 1 Performance evaluation of the proposed model to the existing models in terms of true-positive rate, ROC rate, and false-positive rate

Algorithm	TP rate	FP rate	ROC measure
Random forest	0.7843	0.245	0.7643
Neural network	0.845	0.1873	0.8194
AdaBoost	0.8425	0.1534	0.8314
PSO + SVM	0.9143	0.1943	0.9104
Proposed model	0.9893	0.00103	0.98943

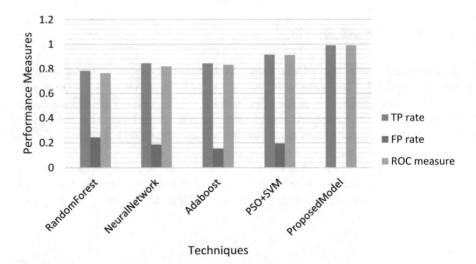

Fig. 2 Performance evaluation of the proposed model to the existing models in terms of true-positive rate, ROC rate, and false-positive rate

Table 2 describes the performance analysis of the proposed model to the existing models in terms of statistical true-positive, error, and ROC rates. From the table, it is clearly observed that the proposed model has high-computational accuracy compared to the existing models on high-dimensional dataset.

Table 3 describes the performance analysis of the proposed model to the existing models in terms of statistical true-positive, false-positive, and ROC rates. From the table, it is clearly observed that the proposed model has high-computational accuracy compared to the existing models on high-dimensional dataset.

Table 2 Performance evaluation of the proposed model to the existing models in terms of true-positive rate, ROC rate, and error rate when data size is 10 k

N = 10000			
Algorithm	TP rate	ROC measure	Error rate
Random forest	0.786	0.7852	0.247
Neural network	0.8343	0.8534	0.146
AdaBoost	0.854	0.8573	0.153
PSO + SVM	0.9143	0.9163	0.091
Proposed model	0.9824	0.9843	0.025

Table 3 Performance evaluation of the proposed model to the existing models in terms of true-positive rate, ROC rate, and error rate when data size is 30 k

N = 30000			
Algorithm	TP rate	FP rate	ROC measure
Random forest	0.753	0.353	0.763
Neural network	0.864	0.1463	0.8735
AdaBoost	0.8953	0.1153	0.8932
PSO + SVM	0.9435	0.0793	0.9463
Proposed model	0.9953	0.002	0.9894

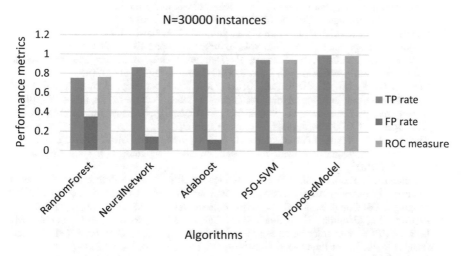

Fig. 3 Performance evaluation of the proposed model to the existing models in terms of true-positive rate, ROC rate, and error rate when data size is 30 k

Figure 3 describes the performance analysis of the proposed model to the existing models in terms of statistical true-positive, false-positive, and ROC rates. From the figure, it is clearly observed that the proposed model has high-computational accuracy compared to the existing models on high-dimensional dataset.

5   Conclusion

As the size of the COPD data increases, it becomes difficult to find the essential features for the disease prediction due to uncertainty and lack of weighted measures. Traditional standard feature selection models are applicable to limited number of feature space for efficient classification algorithms. Also, these model require static weighted parameters to filter the essential features from the list for disease prediction.

In order to overcome these issues, a hybrid ensemble feature selection-based classification model is proposed on high-dimensional dataset. Experimental results proved that the present model improves the true positivity and error rate compared to the traditional models. In the experimental results, 99.16% of classification rate is achieved on the high-dimensional COPD data using the proposed ensemble feature selection-based classification model compared to the existing models.

References

1. van Rikxoorta, E.M.. de Jongb, P.A., Metsb, O.M., van Ginnekena, B.: Automatic classification of pulmonary function in COPD patients using trachea analysis in chest CT scans. In: Proceedings of SPIE 8315, 83150P © 2012 SPIE 2012
2. Kuncheva, L.I., Rodriguez, J.J., Syed, Y.I., Phillips, C.O., Lewis, K.E.: Classifier ensemble methods for diagnosing COPD from volatile organic compounds in exhaled air
3. Corlateanu, A., Siafakas, N., Botnaru, V.: Defining COPD: from simplistic approach to multilateral assessment of COPD. Curr. Respir. Care Rep. 1, 177–182 (2012)
4. Khan, S.A., Kenza, K., Nazir, M., Usman, M.: Proficient lungs nodule detection and classification using machine learning techniques. J. Intell. Fuzzy Syst. 28, 905–917 (2015)
5. Cheplygina, V., Sørenseny, L., Tax, D.M.J., Pedersenz, J.H., Loog, M., de Bruijne, M.: Classification of COPD with multiple instance learning
6. Abdalla, A.S., Yusuf, I.A., Mohammed, S.H.A., Mahmoud, M.A., Mustafa, Z.A.: A computer-aided diagnosis system for classification of lung tumors. J. Clin. Eng. 40(3), 130–135 (2015)
7. Kuwahara, M., Kido, S., Shouno, H.: Classification of patterns for diffuse lung diseases in thoracic CT images by AdaBoost algorithm. In: Karssemeijer, N., Giger, M.L. (eds.), Medical Imaging 2009: Computer-Aided Diagnosis, Proceedings of SPIE, vol. 7260, p. 726037
8. Agarwal, M., Hendriks, E.A., Stoel, B.C., Bakker, M. E., Reiber, J.H.C., Staring, M.: Local SIMPLE multi Atlas-based segmentation. In: Haynor, D.R.. Ourselin, S. (ed.) Medical Imaging 2012: Image Processing, Proceedings of SPIE, Vol. 8314, p. 831410
9. Arbabshirani, M.R., Dallal, A.H., Agarwal, C., Patel, A., Moore, G.: Accurate segmentation of lung fields on chest radiographs using deep convolutional networks. In: Styner, M.A.,

Angelini, E.D. (eds.) Medical Imaging 2017: Image Processing, Proceedings of SPIE, vol. 10133, p. 1013305

10. Chabat, F., Yang, G., Hansell, D.M.: Obstructive lung diseases: texture classification for differentiation at CT, vol. 228, no. 3, p. 877

11. Cheplygina, V., Pino Peña, I., Pedersen, J.H., Lynch, D.A., Sørensen, L., de Bruijne, M.: Transfer learning for multi-center classification of chronic obstructive pulmonary disease. J. Biomed. Health Inform. 22(5), 1486–1496 (2018)

12. Garg, I., Karwoski, R.A., Camp, J.J., Bartholmai, B.J., Robb, R.A.: Automated segmentation of the lungs from high resolution CT images for quantitative study of chronic obstructive pulmonary diseases. In: Galloway, Jr., R.L., Cleary, K.R. (eds.), Medical Imaging 2005: Visualization, Image-Guided Procedures, and Display, Proceedings of SPIE, vol. 5744

13. Ghosh, P., Antani, S.K., Long, L.R., Thoma, G.R.: Unsupervised segmentation of lungs from chest radiographs. In: van Ginneken, B., Novak, C.L. (eds.) Medical Imaging 2012: Computer-Aided Diagnosis, Proceedings of SPIE, vol. 8315, p. 831532

14. Gill, G., Beichel, R.R.: An approach for reducing the error rate in automated lung segmentation. Comput. Biol. Med. 76, 143–153 (2016)

15. Spathis, D., Vlamos, P.: Diagnosing asthma and chronic obstructive pulmonary disease with machine learning. Health Inf. J., 1–17 (2017)

16. Huijsmans, R.J., de Haan, A., ten Hacken, N.N.H.T., Straver, R.V.M., van't Hul, A.J.: The clinical utility of the GOLD classification of COPD disease severity in pulmonary rehabilitation. Respir. Med., 102, 162–171 (2008)

17. Kohlmann, P., Strehlow, J., Jobst, B., Krass, S., Kuhnigk, J., Anjorin, A., Sedlaczek, O., Ley, S., Kauczor, H., Wielpütz, M.O.: Automatic lung segmentation method for MRI-based lung perfusion studies of patients with chronic obstructive pulmonary disease

RETRACTED CHAPTER: Comparative Study on Internet of Things: Enablers and Constraints

C. Kishor Kumar Reddy, P. R. Anisha, Rajashekar Shastry
and B. V. Ramana Murthy

Abstract As the part is changing over with the headways being made in computer technology, we are entering the new time of pervasiveness ruled by means of Internet of Things. IoT regards the sum as a shrewd thing and empowers correspondence among them be it substantial or virtual. With quick increment in wide assortment of hubs in IoT, the certainties this is gotten, spared, transmitted, and prepared to play out a little valuable training that could direct and oversee things make life less complex. Execution of IoT in different fields has realized the enhancement of numerous projects to give insurance and security comfortable, without a ton of human inclusion. This paper features how IoT as system being manufactured has gained. It shows a study made on various innovations how modernity is being completed, challenges, dangers being looked with the guide of IoT and its select projects like home automation, security system, smart irrigation system, environmental monitoring, and numerous others. The use of a successful and plausible gadget.

Keywords IoT architecture · RFID · WSN · Arduino · Raspberry Pi · IoT applications

1 Introduction

The overwhelming thought of IoT has been round for around 2 quite a while, drawing in various analysts, designers, and ventures as a result of its tremendous effect enhancing our general public. IoT has come in close by for execution of

The original version of this chapter was retracted. The retraction note to this chapter is available at https://doi.org/10.1007/978-981-15-1097-7_82

C. Kishor Kumar Reddy (✉) · P. R. Anisha · R. Shastry · B. V. Ramana Murthy
Stanley College of Engineering & Technology for Women, Hyderabad, India
e-mail: kishoar23@gmail.com

C. Kishor Kumar Reddy · P. R. Anisha · R. Shastry · B. V. Ramana Murthy
Department of Computer Science and Engineering, Hyderabad, India

© Springer Nature Singapore Pte Ltd. 2020, corrected publication 2024   677
K. S. Raju et al. (eds.), *Data Engineering and Communication Technology*,
Advances in Intelligent Systems and Computing 1079,
https://doi.org/10.1007/978-981-15-1097-7_56

some genuine universal applications, offering aid elite human games. The IoT contraptions oversee and screen various mechanical, computerized, and electric structures identified with a system. IoT lets in contraptions to feel or control remotely around the present network framework and makes a plausibility for direct incorporation of physical worldwide into PC-based gadget and impacts in improving the execution, exactness, and money-related gifts to decrease human mediations.

Working their switches by means of PCs and smartphones disposes of the issue of working the switches physically. Correspondence capacity and distant control finish in robotization of components, contingent upon settings. Execution of directions by means of people on gadgets through best in class handling is the last point of some IoT bundles. This makes IoT—an order that needs modern equipment and programming program. Research stays being developed and assorted innovation have emerged for detecting, systems administration, application, and preparing: RFIDs, WSN, Wi-Fi, Bluetooth, Arduino Uno, Raspberry Pi, Cloud Computing, and numerous others. With the unification of innovation, for example, tremendous insights, AI, distributed computing can put into impact IoT at canvases spots, houses, etc. by showing predominant contributions. At the point when things like family unit home gear are identified with systems, they can works of art together consolidation to give perfect contributions as a whole.

As a coin has two aspects, improvement in IoT finishes in numerous dangers and requesting circumstances comprising of insurance, control and control of frameworks, etc. Research has been executed and stays occurring to deal with those dangers, to embrace and put into impact the answers that make IoT secure, doable, and on the equivalent time green as well. The idea of IoT is essential and valuable for security, mechanization, help to handicapped, and numerous others. The consistent development in detecting and the modernity of the related preparing will make for an incredible subjective substitute by the way we depict and remain. This leads the focal point stream toward sharp gadgets, smart homes, shrewd towns, etc.

2   Relevant Work

Vandana Sharma et al. inside the paper "An audit paper on "IoT" and It's Smart Applications" portray the basic idea of the projects of IoT. It features that, in 2000s, we are going a fresh out of the box new period of pervasiveness, i.e., IoT. As the title recommends, this paper concentrates more on how mentally, IoT interfaces gadgets and frameworks remotely, containing smart machines intelligent and speaking with each extraordinary, to shape an overall system to make our lives significantly easier and more secure by utilizing bestowing prevalent administrations in various fields like security and crises, logical fields, household and home computerization, etc. [1]. John A. Stankovic inside the paper "Exploration Directions for the Internet of Things" features things that numerous networks, be it Internet of Things, mobile computing, cyber-physical system (CPS), and numerous others, are trying to get keen gadgets/houses/urban communities, by methods for relying on basic advances together with machine learning, big data, security, etc. This paper puts ahead exceptional investigations districts to be tended to for higher and propelled giving of administrations like structure and conditions, security, privateness, receptiveness, tremendous scaling, human in pool, and numerous others [2].

M. U. Farooq et al. in their paper "A Review on Internet of Things (IoT)" portray a six-layered Architecture of IoT. The paper records various innovations being used in IoT, applications of IoT in phenomenal fields like smart home, smart environment, smart hospital, and smart agriculture. Different security-related issues with IoT have also been tended to [3].

Ahmed Khalid in his paper "Web of Thing Architecture and Research Agenda" gives the three-layered design of IoT. These are named as perception, network, and application. He concentrates the need for research on different areas like identity management, communication and network innovation, software, services, and algorithms. Research in this control might be extremely crucial as these inconveniences will remain warm themes in future as there might be an expansion in assortment of gadgets to be connected to the Internet. Appraisals suggest that the assortment of gadgets with an end goal to be connected to the Internet will be 50 billion by method for the year 2020 [4].

Amine Rghioui et al. inside the paper "Web of Things: Visions, Technologies and Areas of Application" feature the advances being utilized in implementing IoT. Radio-frequency identification (RFID) is a key time in distinguishing things the utilization of radio cautions through staying a cement tag on things. Wireless sensor network (WSN) is whatever other period that is a network of sensors associated remotely used in identification of physical conditions, for example, temperature, contamination degrees, dampness, and so on. Other innovations being talked about are Middleware and Cloud Computing. Further challenges being gone up against by IoT and its applications in different fields likewise are referenced. A blackout idea of eventual fate of IoT is moreover portrayed [5].

Jekishan K. Parmar et al. of their paper "Systems administration Technologies and Research Challenges" center around the networking advances being used in IoT covering systems from Wi-Fi personal area networks like ZigBee to our Wi-Fi LAN Wi-Fi and cell dispatch period. It moreover talks about assorted network topologies. ZigBee, 6LoWPAN, and Bluetooth are utilized to expand vitality effective systems for IoT as those are low-power systems [6]. M. Ehsan Irshad et al. inside the paper "Extent of IoT: Performance and Hardware Analysis Between Raspberry Pi-three and Arduino Uno" bear the enhancement of IoT and speculations made through uncommon MNCs on IoT-Starting with Google shopping The Nest Labs for $three.2 billion making its most elevated venture ever inside the last decade. The fate commercial center of IoT and speculations which may be likely to be made inside the zone of IoT is additionally assessed. Microsoft then again is focusing on building up an Operating System for running equipment and sensors. This paper likewise offers the key highlights and contrasts between the two most indispensable and enamoring single-board gadgets Raspberry Pi-3 Model B and Arduino Uno. These single-board gadgets can make physical calculations and give a simple way of interaction between the individual and the unpredictable circuits. Raspberry Pi being a microchip requires an Operating System to include and is good with Operating Systems like Windows 10, LINUX, and UNIX. It additionally has its own one of a kind OS-Raspbian and NOOBS. Arduino Uno then again is a microcontroller and in this way does never again require an OS for its working. It is customized in its own IDE—Arduino IDE which can keep running on unique OS [7].

Nikola Petrol et al. in the paper "Instances of Raspberry Pi utilization in Internet of Things" feature the key elements of Raspberry Pi and its working. The formerly of Raspberry Pi is characterized with its Sense HAT-ready for this chip. It has five catch joysticks and six sensors. It features the ease of use of different stages for instructing IoT. It has isolated the IoT design into five layers: sensing, access, network, middleware, and application layers [8]. Wei Zhou et al. Inside the paper "The Effect of IoT New Features on Security and Privacy: New Threats, Existing Solutions and Challenges Yet to Be Solved": With IoT clients expanding and an assortment of gadgets being associated with the Internet, new Threats and Challenges had been analyzed. The creators talk about the dangers, challenges, and achievable answers for the key abilities of IoT that ar ability to dangers primarily cyber-threats. The highlights focus each on equipment and programming being utilized in IoT. With such huge numbers of dangers being perceived, managing and beating those issues has an absolutely basic position to play. Henceforth breaking down those abilities and taking a shot at it's far essential [9].

3   Analysis of Previous Work

The concept of IoT was first implemented in 1982 when for the first time a machine was connected to the Internet—a Coca Cola vending machine. This machine could report the number of drinks contained in it and keep a check on its temperature. In the year 1999, Kevin Ashton coined the term "Internet of Things." Currently, IoT is a field that is grasping every person's attention from consumers to researchers [10–20].

3.1   IoT Architecture

Coming to the Architecture of IoT, various authors have given different architecture. Here is analysis of different architectures given by

3.2 Technologies Being Used in IoT Implementation

IoT system is a vast system involving integration of hardware and software (communication between devices) with a user-friendly interface. Hence, it requires sophisticated and ubiquitous technologies to implement it effectively. Table 2 gives a list of existing technologies being employed in implementation of IoT (Table 1).

3.3 Single-Board Devices

3.3.1 Raspberry Pi

Raspberry Pi is a credit-size microprocessor. It was first built by a group of six students from University of Cambridge, UK, when these students realized that the number of applicants for Computer Science was decreasing due to lack of computer skills. This motivated them to build a physical computation device—the Raspberry Pi. Raspberry being a microprocessor requires an Operating System for its functioning. It is compatible with Operating Systems like Windows 10, LINUX, and UNIX. Raspberry has its own Operating System too—Raspbian and NOOBS.

3.3.2 Arduino Uno

Named after an Italian King—King Arduin—Arduino is a microcontroller unlike Raspberry Pi which is a microprocessor but quite similar to Raspberry Pi. Both the devices aim at providing ease for interaction between its user and complex circuits. It was developed by students of Interactive Design Institute, Ivrea, under the supervision of a Software Architect Banzi.

Arduino does not require an Operating System unlike Raspberry Pi which requires an OS its functioning. Arduino is programmed in its own IDE—Arduino IDE which can run on different Operating Systems.

4 Conclusions

IoT promises to improve the quality of life and productivity. It has the potential to provide the fundamental services in education, health care, security, transportation, etc. Safety and security can be provided without any kind of human intervention by the automation of devices using IoT. Development of security measures through IoT would prove to be advantageous and helpful for multipurpose. IoT and its services give us a wide range of applications, making it a part of our everyday lives and working. Research in this field is still going on and will continue to be an

Table 1 Different IoT Architecture given by research works considered

Number of tiers or levels	Level-1	Level-2	Level-3	Level-4	Level-5	Level-6
3		Perception layer: Gathers information, use of Sensors, RFID Tags, etc	Network layer: Called as "Central Nervous System," Transmission layer: Wi-Fi, CDMA, etc		Application layer: Concept of IoT being applied to real-time applications	
5	Sensing layer: Sensing, sharing, and gathering of information	Access layer: A mediator layer between the sensing and network layers	Network layer: Consists of wireless access point and gateways	Middleware layer: Processing and managing information	Application layer: Allows analyzing and monitoring data through applications	
6	Coding layer: Objects (Things) being considered are given unique IDs for identification and differentiation	Perception layer: Also known as recognition layer. Objects are identified; external parameters (physical) are sensed using WSN	Network layer: Signals are transmitted to the next layer for processing through this layer, e.g., ZigBee, Wi-Fi, and Bluetooth	Middleware layer: Information is processed, stores, and secures direct access to databases through cloud computing	Application layer: Helpful in large-scale development of IoT network	Business layer: Research on IoT, helps in effective business

Table 2 Technologies being used in IoT implementation

Technology	Used in layer (Architecture of IoT):	Description
Radio-frequency identification (RFID)	Perception layer	Used in identification of objects. An adhesive sticker like tag is used which is actually a transceiver microchip. Send location and other specifications of objects as data signals to devices called readers. These readers pass on the data for processing
Wireless sensor network (WSN)	Perception layer	Network of sensors that are connected wirelessly Connected bi-directionally which implies the sense and send environment-related information such as temperature, pollution levels and humidity and can also be used to control these parameters
Cloud computing	Middleware layer	With an estimate of 50 billion devices being connected to the Internet, the cloud seems to be the only technology that can be used to store and analyze large amounts of data Connecting to a cloud would facilitate a large scale of objects to be interconnected
Global sensor network (GSN)	Middleware layer	A middleware platform that allows development and deployment of sensor services with no programming effort
Wi-Fi	Network layer	A wireless local area network based on IEEE 802.11 standards The range of this wireless communication between Wi-Fi equipped devices and its access points are about 30 m indoors
Bluetooth	Network layer	Wireless personal area network protocol Low energy network and hence is energy efficient Aims at no cables being used for connecting devices
ZigBee	Network layer	Being an isolated network and implements different network topologies, it can be used to form an IoT Low Energy Network and hence is energy efficient
6LoWPAN	Network layer	It is an internet engineering task force (IETF) standard Provides direct connection to IPv6 internet which makes its use beneficial and efficient Low energy network and hence is energy efficient

ongoing process. Research in this field is very important as the significance of IoT is growing. In this paper, we discussed IoT Architecture, technologies being used, its threats, challenges, and possible solutions to it and ended up with its applications.

References

1. Sharma, V., Tiwari, R.: A review paper on IoT and its smart applications. J. Sci. Eng. Technol. Res. (IJSETR) **5** (2), 472–476 (2016)
2. Stankovic, J.A.: Research directions in IOT. IEEE IoT J. **1** (1) (2014)
3. Farooq, M. U., Waseem, M., Mazhar, S., Khairi, A., Kamal, T.: A review on internet of things (IoT). Int. J. Comput. Appl. **113** (1) (2015)
4. Ahmed K.: Internet of thing architecture and research agenda. Int. J. Comput. Sci. Mobile Comput. **5** (3), 351–356 (2016)
5. Rghioui, A., Oumnad, A.: Internet of things: visions, technologies, and areas of application. Autom. Control Intell. Syst. (2017)
6. Parmar, J.K., Desai, A.: IoT: networking technologies and research challenges. Int. J. Comput. Appl **154** (7) (2016)
7. Irshad, M.E., Feroz, M.M.: Scope of IoT: performance and hardware analysis between Raspberry Pi-3 and Arduino Uno. Int. J. Comput. Sci. Mobile Comput. **5** (5) 580–585 (2016)
8. Petrov, N., Dobrilovic, D., Kavalic, M., Stanisavljev, S.: Examples of Raspberry Pi usage in Internet of Things. Int. Conf. Appl. Internet Inf. Technol. (2016)
9. Zhou, W., Zhang, Y., Liu, P.: The effect of IoT new features on security and privacy: new threats, existing solutions, and challenges yet to be solved (2018)
10. Castillejo, P., Martinez, J.-F., Lopez, L., Rubio, G.: An internet of things approach for managing smart services provided by wearable devices. Int. J. Distrib. Sens. Netw (2013)
11. Harikiran, G.C., Menasinkai, K., Shirol, S.: Smart security solution for womwn based on internet of things (IOT). Int. Conf. Electr. Electron. Optim. Techn. (ICEEOT) (2016)
12. Li, B. Yu, J.: Research and application on the smart home based on component technologies and internet of things. Procedia Eng. **15** (2) (2011)
13. Mahalakshmi, G., Vigneshwaran, M. IOT based home automation using Arduino. Int. J. Eng. Adv. Res. Technol. (IJEART) **3** (8), 7–11 (2017)
14. Maheshwari, D.G., Umesh, I.M.: A study on internet of things based smart home. Int. J. Sci. Eng. Technol. Res. (IJSETR) **6** (8), 1286–1288 (2017)
15. Bhavna, Dr. Sharma, N.: Smart home automation using IoT. Int. J. Eng. Sci. Res. Technol. 435–437 (2018)
16. Mrs. Gbhane, J.P., Ms. Thakare, S., Ms. Craig, M.: Smart homes system using internet of things: issues, solutions and recent research directions. Int. Res. J. Eng. Technol. (IRJET) **4** (5), 1965–1969 (2017)
17. Iqbal, M.A., Olaleye, O.O., Bayoumi, M.A.: A review on internet of things (Iot): security and privacy requirements and the solution approaches. Glob. J. Comput. Sci. Technol. Netw. Web Secur. **16** (7) (2016)
18. Satapathy, L.M., Bastia, S.K., Mohanty, N.: Arduino based home automation using internet of things. Int. J. Pure Appl. Math. **118** (17), 769–777 (2018)
19. Sharma, M., Tripathi, D., Yadav, N.P., Rastogi, P.: Gas leakage detection and prevention kit provision with IoT. Int. Res. J. Eng. Technol. (IRJET), **5** (2), 2227–2230 (2018)
20. Yadav, G., Devi, M.S.: Arduino based security system-an application of IoT. Int. J. Eng. Trends Technol. (Special Issue), 209–212 (2017)

Telugu Movie Review Sentiment Analysis Using Natural Language Processing Approach

Srinivasu Badugu

Abstract In this modern age of digitization, everyone uses online services for various day-to-day activities like to know about product functionality or about movie-related information by numerous blogs. This short description about a movie is nothing but the review of a movie, which gives the opinion about the movie by several authors/authenticated persons. Opinion provides the sentiment of a movie, i.e., whether the movie is good or bad or normal/routine. Here in the sentiment, good resembles positive, and bad resembles negative. These reviews are very useful to improve and determine the pros and cons of movie. In India, there has been requirement for provincial dialects interfaces for better comprehension about the web substance and make it more easy to use to local clients. Here, we propose a framework to sort film surveys which are utilized transliteration plot, i.e., Telugu to English into positive, negative and nonpartisan for better basic leadership about the motion picture. The system is based on rule-based (natural language processing (NLP)) and machine learning approach. Our contributions in this approach are preprocessing (tokenization, sentence segmentation), tagging, feature extraction and opinion extraction. Feature extraction is based on tags mainly nouns. Opinion extraction is based on verb, adjacent and adverbs. It involves the concepts of natural language processing, machine learning for the development of the system. Our contributions in this paper are to create a seed list words based on category. We are finding the polarity of a sentence as positive, negative or neutral.

Keywords Opinion mining · Telugu · Natural language processing · POS tagging · Features · Opinion words · SentiWordNet

S. Badugu (✉)
Department of Computer Science and Engineering, Stanley College of Engineering and Technology for Women, Hyderabad, India
e-mail: drsrinivasu@stanley.edu.in

© Springer Nature Singapore Pte Ltd. 2020
K. S. Raju et al. (eds.), *Data Engineering and Communication Technology*,
Advances in Intelligent Systems and Computing 1079,
https://doi.org/10.1007/978-981-15-1097-7_57

1 Introduction

Presentation opinions are emotional explanations that mirror individuals' conclusions or recognitions about the elements and occasions [1]. Much of the existing research on text information processing has been focused on mining and retrieval of factual information [2]. The opinions differ from person to person and the statements they make are based on their experience, their feelings, and their perceptions. People can visualize real-world entities and the events which occur among them; a research is taking place in the field of text mining and the processing of information has been considered, and the retrieving of the actual information has been profound influence in text mining.

Conclusions are important to the point that at whatever point, one has to settle on a choice one needs to hear others' suppositions. This is not valid for people yet in addition valid for associations [3]. Opinions are hidden in long forum posts and blogs. People's influences and approach are so important that opinion becomes a key factor in decision-making. There were forums to find out one's opinion [4].

It is troublesome for a human peruser to discover important sources, separate relevant sentences, read them, abridge them and compose them into usable structures [5]. Basic components of an opinion are as follows: Individuals are finding it difficult to know the relevant sources and extracting the required sentences, understand, read and coming out with the summary and making it as a usable way.

Telugu language is one among the Dravidian language native to India. It stands alongside Hindi, English and Bengali collectively of the few languages with official status in more than one Indian state [6]. It is the first language within the states of Telangana, Andhra Pradesh and within the city of Yanam, Puducherry. It is additionally spoken by important minorities in Chhattisgarh, Maharashtra, Odisha, Karnataka, Tamil Nadu, the Andaman and Nicobar Islands and the Sri Lankan Gypsy individuals. It is one in every six languages selected a classical language of India by the Government of India [7]. Telugu language ranks third by the amount of native speakers in India (74 million, 2001 census) [8], fifteenth in the Ethnologue rundown of most-spoken dialects worldwide [9] and is the most broadly spoken Dravidian dialect. It is one of the twenty-two official dialects of the Republic of India [10].

2 Literature Survey

A wide range of inquires are conveyed for the expectation of motion picture audits assumptions by utilizing diverse methodologies, but a significant number have investigated through credits identified with a film. Presents prediction of a film's success through social media posts. The research shows that these articles can be used to get some future outcomes [11]. It likewise utilizes budgetary information of motion pictures from film industry magic by utilizing Pearson's connection coefficient and direct relapse. There is an exploration that predicts the opening end of the week income. It takes the film data like performing artists, executive, class and discharged date and so forth. From meta-critic and budgetary information like spending plan, opening week net income from the numbers. The pre-discharged articles about the films are gathered from seven unique articles sources.

"Joshi et al. [12] developed a fall-back strategy for sentiment analysis in Hindi; for a case study" in their paper, they follow three different approaches. Author's first step is to translate given document to English, and the second step, they develop a lexical resource called Hindi SentiWordNet (HSWN), Authors are used SentiWordNet 1.1. Wordnet linking is used to map synsets for different languages, authors also used POS tagging. Authors have tested 250 Hindi movie reviews. They are using RapidMinor 5.0 for document classification and LibSVM type-c learner for classification. TF-IDF features have given the highest accuracy of 78.19 and 60.31 for resource-based sentiment analysis."

Mittal et al. [13] planned a negation and discourse relation approaches for his or her paper Sentiment analysis of Hindi review supported negation and discourse relation. They have tested their rule victimization annotated dataset. anon they extracted rules for handling negation and discourse relation that extremely influence the feelings expressed within the review. Further, Hindi SentiWordNet (HSWN) is employed for polarity values of words. Technique for up the HSWN is additionally planned. Finally, overall linguistics orientation of the review document is decided by aggregating the polarity values of all the words within the document."

"Pandey et al. [14] developed "A framework for sentiment analysis in Hindi victimization Hindi SentiWordNet (HSWN). They have used HSWN to notice the general sentiment related to the document. Existing HSWN is improved with the assistance of English SentiWordNet, wherever sentimental words that are not gifted within the HSWN are translated to English so searched in SWN to retrieve their polarity. Sentiment is extracted for locating the general polarity of the document which may be positive, negative or neutral. Throughout preprocessing, tokens are extracted from sentence and spell check is performed. Rules are devised for handling negations and discourse relations that extremely influence the feelings expressed within the document. Finally, overall sentiment orientation of the document is decided by aggregating the polarity values of all the sentimental words within the document."

Pang et al. [15] proposed a novel machine learning strategy that applies content order methods to change in accordance with simply the emotional part of the record, which as in following procedure: (1) mark the sentences in the archive as either abstract or target, disposing of the last mentioned and after that (2) apply a standard machine learning classifier to the subsequent concentrate. The authors used an easy and intuitive graphic design based on identifying minimal cuts. Their tests include grouping motion picture surveys as either positive or negative. To accumulate emotional sentences or expressions, creators gathered 5000 motion picture audit pieces and for target information, they have taken from the web motion picture database. Both Naïve Bayes and SVMs can be prepared on subjectivity informational collection and afterward utilized as a fundamental abstract locator. Creators locate the better outcomes NB: 86.4% VS: 85.2% SVM: 86.15."

"Lina Zhou et al. [16] examined motion picture survey mining utilizing machine learning and semantic introduction. Supervised classification and text classification techniques are used in the proposed machine learning approach to classify the movie review. A corpus is framed to speak to the information in the archives and every one of the classifiers is prepared to utilizing this corpus. In this manner, the proposed procedure is progressively effective. However, the machine learning approach utilizes regulated learning. The proposed semantic introduction approach utilizes unsupervised learning, since it does not require earlier preparing so as to mine the information. Trial results demonstrated that the administered methodology accomplished 84.49% precision in the three-crease cross-approval and 66.27% exactness on hold-out examples. Along these lines, the examination reasons that the directed machine learning is increasingly productive; however, it requires a lot of time to prepare the model. Then again, the semantic introduction approach is somewhat less exact; however, it is increasingly proficient to use continuously applications."

3 Proposed System

Proposed approach for sentiment analysis of Telugu movie reviews using natural language processing [7, 8] approach works as follows. For testing the algorithm result, we have created annotated movie review dataset. First, we have pre-processed input data and later applied part of speech tagger. Then, extracted features and opinion words from the tagged corpus [17]. Further, Telugu SentiWordNet (TSWN) is used for polarity values of words. For this system, inputs are the movie reviews which are using transliteration from the Telugu script to roman (English) script [6, 18].

We have implemented our methods using core Python 2.7 and tested 60 Telugu movie reviews, which are collected from various online resources like 123telugu.com, grateandhra.com, etc. (Fig. 1).

For the rule-based approach after collecting the review text, the review text should be applied transliteration. Transliteration is not Translation; here, the language does not change, and only the script used to render the text is changed. After transliteration, we have applied sentence segmentation which is divided into individual sentences with the help of some set of rules. After sentence segmentation, the review text should be made into tokens. The tokens contain noise and irrelevant data like special characters and all those should be removed by using regular expressions patterns. It is called preprocessing. After preprocessing, we have applied part of speech (POS) tagging and feature extraction and opinion words

Fig. 1 Workflow of
rule-based Telugu movie
reviews

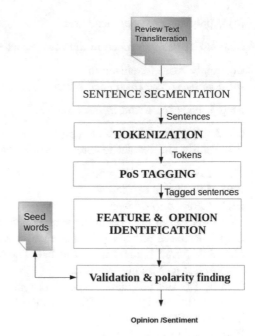

identification. Features extraction based on word categories like noun, verb,
adjective and adverb. For example:

చలనచిత్రంగా, జీనియస్ డైరెక్టర్, ఊహించడానికి (text)
CalanacitraMgaa, jiiniyas DairekTar uuhiMcaDaaniki (transliteration)
calanacitraMgaa||calanacitraM||N-COM-COU-N.SL-NOM-ADV.gaa
jiiniyas||jiiniyas||UNK   DairekTar||DairekTar||N-COM-COU-MF.SL-NOMuuhi
McaDaaniki||uuhiMcu||V-IN-gerund-.n-DAT (POS tagging).

CalanacitraMgaa is tagged as noun, jiiniyas is tagged as unknown (proper noun)
word, DairekTar is tagged as noun word again and uuhiMcaDaaniki is tagged as
verb. In our work, we are using verbs, unknown (proper noun), adjective, adverb
and nouns as considered opinion and feature words which carry the polarity.

We have created seed list from annotated corpus. Tagged data and seed list are
separately passed to the algorithm. The tags are compared against the seed list and if
the word is present in seed list, then that word is considered as opinion and feature.
Use Telugu SentiWordNet (TSWN) after considering opinion and feature terms and
finding the polarity from the extracted opinion words.

Algorithm for Telugu Movie Review Sentiment Analysis

Step 1: For each document in Telugu Movie Reviews corpus
Step 2: Apply Transliteration (change the script Telugu to Roman)
Step 3: Apply Preprocessing

(a) Apply sentence segmentation

Step 4: For each sentence in the movie review

(a) Apply word tokenization
(b) Remove noise from each token

Step 5: For each token in the sentence

(a) Apply Part of Speech tagging
(b) Identify and extract features and opinions

End of For Loop of Step 5;
End of For Loop of Step 4;
End of For Loop of Step 1;
Step 6: For each opinion in the sentence in each movie review
Step 7: Retrieve polarity (POL) from seed list
Step 8: If (word is present in seed list) Then go to Step 9
Else Add it to Missing Word List
Step 9: If (word is negated) Then word_plo = −1;
Step 10: If word is positive Then word_pol = 1
Else Word.POL = 0;
End of For Loop of Step 6
Step 11: Compute the aggregate polarity of the movie review by adding the polarities
values of all the token.
Step 12: If (movie review polarity > zero) Then label the movie review as positive
Else If (movie review polarity < zero) Then label the movie review as negative
Else Classify the movie review as neutral.
Step 13: Return the set of Labeled sentences.

Feature Extraction

We have observed that the feature carrying words is mainly nouns. In the process of feature selection, we have considered noun and unknown (proper noun) tags. All specified tags may not be useful, and to know the most specific feature words, we have taken more frequently occurred words in the review text. As a result, feature space does no longer include all the words, but instead, it only contains the feature words from the defined tagging. We have considered nouns as features in Telugu movie reviews text.

Example: CalanacitraMgaa||calanacitraM||N-COM-COU-N.SL-NOM-ADV.Gaa HaaliivuD||haaliivuD||N-COM-COU-N.SL-NOM Hiirooyin||hiirooyin||N-COM-UNC-F.SL-NOM

Opinion Extraction

For opinion extraction, we have considered verbs, adjectives, adverbs and few unknown (proper nouns) words from the tagged words.

Example: Naccakapoovaccu‖naccu‖V-IN-NPART-aux.Inceptive-INF-aux.permissive VisigistuMdi‖visugu‖V-IN-CAU-FUT.HAB-P3.FN.SL DuradRShTavashaattuu‖duradRShTavashattuu‖ADV-ABS KaShTamaina‖kaShTamaina‖ADJ-ABS.

Opinion Classification and Orientation

After extracting the feature and opinion words from tagged data, we have verified with feature and opinion seed words list. If the word contains in the feature list and opinion list, then we consider useful words. If the word contains in the feature list and not in the opinion list, then we consider it as neutral. The opinion word can be positive or negative or neutral words. Here, we have manually verified for the positive, negative and neutral words.

Opinion Orientation

Given a review which contains features and opinion words, here we identify the orientation of opinion using opinion word classification. Orientation specifies the deeper level of opinion with five different orientations after removing unusual opinion.

We calculate review which contains features and opinion words; here, we considered five different opinion categories (Table 1).

We have taken five different opinion categories (0) and this category specifies more negative opinion with opinion orientation ranging from 80 to 100.

1. This category specifies negative opinion with opinion orientation ranging from 60 to 80.
2. This category specifies normal opinion with opinion orientation ranging from 50 to 60.
3. This category specifies positive opinion with opinion orientation ranging from 60 to 80.
4. This category specifies more negative opinion with opinion orientation ranging from 80 to 100.
5. We are calculating the opinion orientation using.
6. Positive opinion = no. of positive sentences/total no. of sentences.
7. Negative opinion = no. of negative sentences/total no. of sentences.

Table 1 Opinion orientation analysis

Range	Positive	Negative
80–100	4 (more positive)	0 (more negative)
60–80	3 (positive)	1 (negative)
50–60	2 (neutral)	2 (neutral)

Table 2 Percentage of different opinion words

# of opinion words	+ve opinion words	−ve opinion words	Normal opinion words
783	401	187	195
Percentage (%)	51	24	25

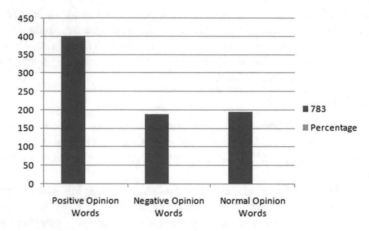

Fig. 2 Percentage of different opinion words

4 Result Analysis

All the reviews are taken from http://www.123telugu.com/, greatandhra.com. All of them are used as the training data to extract opinion and feature words. But we are using rule-based approach, POS tagging to extract opinion and feature words. These words are then used to extract movie review opinion and features from test reviews of these movies.

In this paper, we verify only thirty Telugu movie reviews as opinion and features, but we make sentiment orientation of opinion on that features as an ongoing process. The effectiveness of the proposed system has been verified with review set. All the results generated by our system are compared with the manually tagged results. We also assess the time saved by semi-automatic tagging over manual tagging.

Experiment is conducted on Telugu movie reviews from various online Web sites. Input is applied to the system, and output is given by the system by comparing the positive, negative and normal matches with negation handling.

Table 2 gives the percentage for different opinion words; we have considered 783 opinion words from which 51% are positive opinion words, 24% negative opinion words and 25% normal opinion words (Fig. 2).

From Table 3, we have calculated the opinion orientation of review using the opinion words (Figs. 3 and 4).

5   Conclusion and Future Work

Opinion mining plays a very important role in our daily life; life without opinion is like an empty vessel. In this paper, we have proposed an opinion extraction on Telugu movie review using natural language processing. With the proposed system, we are able to find opinion of the movie reviews of Telugu language. Telugu language is one of the highest spoken languages in India.

We have focused only on Telugu movie reviews, but movie review has many international Indian languages. It should be possible to use our approach in finding the opinion of other languages. In this approach, we apply preprocessing (sentence segmentation, tokenization,), part of speech tagging, feature and opinion extraction Feature extraction is based on POS tags mainly nouns. Opinion extraction is based on verb, adjacent and adverbs. After that, we are finding the opinion polarity (positive or negative). With the proposed system, we are able to find the orientation

Table 3 No. of sentences with polarity for a review

Review Id	No. of sentences	Positive sentences (+)	Negative sentences (−)	Performance (%)	Overall opinion
MR1	32	24	08	75	3
MR2	27	21	06	77	3
MR3	27	26	01	96	4
MR4	19	09	10	52	2
MR5	26	13	13	50	2
MR6	46	14	32	69	3
MR7	45	32	13	71	3
MR8	30	27	03	90	4
MR9	40	27	13	67	3
MR10	31	30	01	96	4

Fig. 3 No. of sentences with polarity for a review

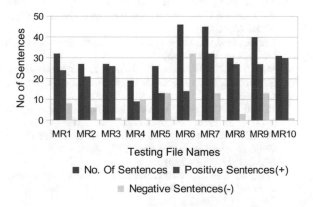

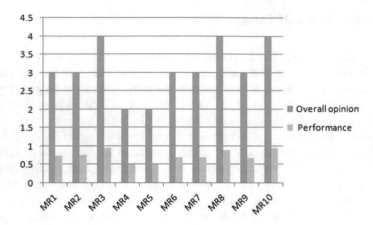

Fig. 4 Opinion orientation of reviews

of a movie review with good results. We also hope to understand the different domain data and draw the useful patterns. In future, we would like to create training-based text.

References

1. Ganesan, K., Zhai, C.: Opinion-based entity ranking. Inf. Retrieval **15**(2), 116–150 (2012)
2. Sohail, S.S., Siddiqui, J., Ali, R.: Book recommendation system using opinion mining technique. In: 2013 International Conference on Advances in Computing, Communications and Informatics (ICACCI), pp. 1609–1614. IEEE (2013)
3. Zhang, L., Liu, B.: Sentiment analysis and opinion mining. In: Encyclopedia of Machine Learning and Data Mining, pp. 1–10 (2016)
4. Indurkhya, N., Damerau, F.J. (eds.): Handbook of Natural Language Processing, vol. 2. CRC Press (2010)
5. Mishra, B.K., Sahoo, R.: Application of opinion mining in e-commerce. In: Improving E-Commerce Web Applications Through Business Intelligence Techniques, pp. 271–296. IGI Global (2018)
6. Krishnamurti, B., Gwynn, J.P.L.: A grammar of modern Telugu. Oxford University Press, Delhi, New York (1985)
7. Bird, S., Klein, E., Loper, E.: Natural language processing with Python. Copyright © 2009 Steven Bird, Ewan Klein, and Edward Loper. All rights reserved. Printed in the United States of America
8. Bharati, A., Chaitanya, V., Sangal, R.: Natural Language Processing: A Paninian Perspective, pp. 65–106. Prentice-Hall of India, New Delhi (1995)
9. Micheli-Tzanakou, E.: Supervised and Unsupervised Pattern Recognition: Feature Extraction and Computational Intelligence. CRC, Boca Raton, FL (2000), 371 pp. ISBN 0-8493-2278-2 (Reviewed by Ke Chen)
10. Pang, B., Lee, L., Vaithyanathan, S.: Thumbs up? Sentiment classification using machine learning techniques, In: Proceedings of Conference on Empirical Methods in Natural Language Processing (2002)

11. Manning, C.D., Schütze, H.: Foundations of Statistical Natural Language Processing (Stanford University and Xerox Palo Alto Research Center), xxxvii + 680 pp. The MIT Press, Cambridge, MA (1999), hardbound, ISBN 0-262-13360-1
12. Joshi, A., Balamurali, A.R., Bhattacharya, P.: A fall-back strategy for sentiment analysis in Hindi: a case study. In: Proceedings of ICON 2010: 8th International Conference on Natural Language Processing
13. Mittal, N., Agarwal, B., Chauhan, G., Bania, N., Pareek, P.: Sentiment analysis of Hindi review based on negation and discourse relation. In: International Joint Conference on Natural Language Processing, pp. 45–50, Nagoya, Japan, 14–18 Oct 2013
14. Pandey, P., Govilkar, S.: A framework for sentiment analysis in Hindi using HSWN. Int. J. Comput. Appl. (IJCA) **119**(19) (2015) (0975-8887)
15. Pang, B., Lee, L.: A sentimental education: sentiment analysis using subjectivity summarization based on minimum cuts (2004)
16. Zhou, L., Chaovalit, P.: Movie review mining: a comparison between supervised and unsupervised classification approaches. In: Proceedings of the 38th Hawaii International Conference on System Sciences (2005)
17. Singh, V.K., and Piryani, R.: Sentiment analysis of movie reviews: a new feature-based heuristic for aspect level sentiment classification. IEEE (2013)
18. Murthy, K.N., Srinivasu, B.: Roman transliteration of Indic scripts. In: 10th International Conference on Computer Applications, pp. 28–29, University of Computer Studies, Yangon, Myanmar, February 2012

RETRACTED CHAPTER: Short-Term Load Forecasting Using Wavelet De-noising Signal Processing Techniques

S. Anbazhagan and K. Vaidehi

Abstract Electrical load data are non-stationary and very uproarious on the grounds that an assortment of components influences power markets. The immediate prediction of electrical load with noisy information is typically subject to vast mistakes. This venture proposes a novel methodology for load forecasting by applying wavelet de-noising in a neural network models. The procedure of the proposed methodology initially deteriorates the chronicled information into an approximate part connected with low frequency and a detailed part connected with high frequencies through a wavelet transform (WT). A backpropagation neural network (BPNN) is built up by the low-frequency signal to estimate the future value. At last, the load is predicted by BPNN with and without utilizing WT. To assess the execution of the proposed methodology, the load data in New South Wales, Australia, are utilized as an illustrative model.

Keywords Backpropagation neural network · Combined model · Short-term load forecasting · Wavelet de-noising

The original version of this chapter was retracted. The retraction note to this chapter is available at https://doi.org/10.1007/978-981-15-1097-7_82

S. Anbazhagan
Department of Electrical Engineering, FEAT, Annamalai University,
Annamalai Nagar 608002, India
e-mail: s.anbazhagan@gmx.com

K. Vaidehi (✉)
Department of Computer Science and Engineering,
Stanley College of Engineering & Technology for Women, Hyderabad, India
e-mail: kvaidehi@stanley.edu.in; vainakrishna@gmail.com

© Springer Nature Singapore Pte Ltd. 2020, corrected publication 2024   697
K. S. Raju et al. (eds.), *Data Engineering and Communication Technology*,
Advances in Intelligent Systems and Computing 1079,
https://doi.org/10.1007/978-981-15-1097-7_58

1   Introduction

Load determining is an essential part of power system energy management system. Exact load prediction encourages the electric utility to settle on unit commitment choices, decrease turning reserve capacity and schedule device maintenance plan legitimately. Other than expecting a key employment in reducing the age cost, it is in like manner fundamental to the constancy of intensity frameworks. The framework executives use the heap expectation result as a reason of disconnected framework examination to choose whether the framework might be helpless. Provided that this is true, restorative activities ought to be readied, for example, load shedding, power buys and bringing peaking units on line.

Since exact load estimating remains an incredible test, the fundamental target of this venture work is to build up a mixture framework utilizing computational insight calculations. As of now, to the best of our knowledge there is no best prediction method available with regard to load forecasting. Along these lines, it is compulsory to pick a decent strategy or a blend of different techniques for various circumstances. This venture endeavors to deliver the upsides of the WT-based BPNN and subsequently to build up a half and half technique for exact load anticipating.

The research methodologies of load prediction can be for the most part partitioned into two classifications: statistical strategies and AI techniques. In statistical strategies, conditions can be acquired by demonstrating the connection among load and its relative factors in the wake of preparing the historical information, while artificial intelligence techniques attempt to impersonate the people's state of mind and thinking to get learning from the past experience and estimate the future load.

Engle et al. [1] displayed a few regression models for the short-term load forecasting (STLF). Their models join deterministic impacts, for example, holidays, stochastic impacts, for example, average loads and exogenous impacts, for example, climate. Ruzic et al. [2], Haida and Muto [3] depict different uses of regression models connected to load prediction. Fan and McDonald [4] and Cho et al. [5] depicted the usage of autoregressive integrated moving average with exogenous variables (ARMAX) models for load prediction. Yang et al. [6] utilized a evolutionary programming way to deal with recognize the ARMAX model parameters for one day to multi-week ahead hourly load prediction.

Papalexopoulos et al. [7] created and actualized a multilayered feed forward ANN for STLF. Khotanzad et al. [8] depicted a load prediction framework known as ANNSTLF. It depends on numerous ANN systems that catch different patterns in the information. Chen et al. [9] developed a three-layer totally related FFNN and a back proliferation calculation is adjusted for training technique.

Rahman and Hazim [10] depict utilizations of fuzzy logic to load prediction. The utilization of fuzzy logic takes out the requirement for a mathematical model. Li and Fang [11] built up an STLF model of WT systems to display the highly nonlinear dynamic behavior of the system loads and to enhance the execution of conventional ANN's. Transformative calculations like hereditary calculation, molecule swarm enhancement and artificial immune framework have been utilized for training neural network in STLF applications. These algorithms are superior to anything the backpropagation in convergence and search space capacity. These algorithms are past the extent of this undertaking work. Wang et al. [12] proposed a novel methodology for STLF by applying wavelet de-noising in a hybrid model that is a seasonal autoregressive integrated moving average (SARIMA) model and neural networks. Comparison of the outcomes with alternate models demonstrates that the SARIMA models can viably enhance the prediction precision. Tiwari et al. [13] considered the STLF issue and thought about the BPNN and RBFNN. Conejo et al. [14] proposed a novel system to predict day-ahead power prices dependent on the WT and ARIMA models. Anbazhagan [15] broke down top to bottom the different computational intelligence techniques for precise value prediction.

2 Hybrids of Individual Forecasting Models

2.1 Wavelet Transform

The WT strategy is finished reasoning about a predetermined number of positions and goals levels. Utilizing this procedure, the decomposition coefficients of the WT of the hourly load series are given.

A sufficient method to pick the wavelet capacities is the multi-goals framework subject to using a dad wavelet capacity and its complementary, a mother wavelet work. The dad work licenses deciding the low recurrence parts of the arrangement, while the mother one grants deriving fragments of high recurrence.

In the first place, Daubechies wavelets are the most regularly used in applications. For these gatherings of wavelets, the smoothness augments as the request of the capacities does; before long, the assistance interims additionally expands, which may disintegrate the figure. As needs be, low request wavelet capacities are regularly prudent. Second, a long estimation period may begin with botches since monetary circumstances advance with time. Then again, a short estimation period may start unpredictable estimations. In this way, a fitting estimation period ought to be chosen dependent on experimentation.

2.2 Backpropagation Neural Network (BPNN)

BPNN is made out of neurons organized in layers, as appeared in Fig. 1. The information is brought into the framework through an input layer. The output information rises up out of the network's last layer. It describes the mathematics behind the NN with Levenberg-Marquardt (LM) training algorithm.

When an input vector is presented to the NN, the output error can be computed by a squared error. The squared error is calculated as the sum of the squared differences between the target values and the output values. Conditions utilized in the BPNN structure with just a single concealed layer are appeared in (1) and (2).

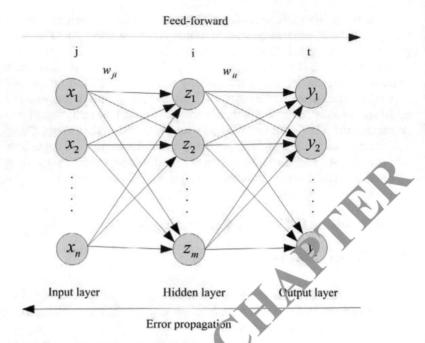

Fig. 1 The architecture of one hidden layer BPNN

$$f_h(x) = 1/(1 + \exp(-x)). \tag{1}$$

The yields of all neurons in the output layer are given as pursues.

$$y_t = f_t\left(\sum_{i=0}^{m} W_{ii}Z_i\right), \; t = 1, 2, \ldots, l \tag{2}$$

Here, $f_t (t = 1, 2, \ldots, l)$ is the transfer function, usually a linear function. In this venture, the three layer BPNN, which is a $24 \times 49 \times 24$ network design, is connected to make this forecast model.

2.5 Hybrids of WT and BPNN Models

The blend of conjecture is a completely researched issue in the quantifiable field. Different methods for ensemble desire have been shown. The hypothesis of the consolidated estimate technique depends upon a specific direct mix of different outcomes from various desire models.

In this research work, wavelet-based ensemble show that joins the WT and BPNN. The method of the approach at first separates the certain information into a harsh part connected with a low repeat and a low down part connected with high frequencies utilizing the WT. A BPNN show is set up by the low-repeat banner to predict the future regard. At long last, the future regard is figure by weighting the foreseen estimations of BPNN.

Stage 1: Wavelet: The unrefined power load information course of action is deteriorated into a low-repeat partition and a high-repeat part by the wavelet rot. The low-repeat segment tends to the critical highlights of the unrefined information course of action, and the high-repeat part is reliably called the tumult banner. The probability of this development is to evacuate the focal qualities and expel the optional disturbing effect from the unrefined information course of action.

Stage 2: Capture the non-straight point of reference. The BPNN was an exceptionally fundamental dimension used to get the non-straight instance power load information game plan.

Stage 3: The foreseeing estimations of the rough power load information course of action were constrained by the BPNN show dependent on input features (low-repeat segments).

2.4 Evaluation Criteria

Besides, three assessment criteria are additionally considered to assess the determining execution with respect to the real esteem. These are mean absolute error (MAE), root mean square error (RMSE) and mean absolute percentage error (MAPE) which were well-established statistical assessment indices, and the detailed discussion on MAPE is not presented here. Readers can refer [15] for more details.

3 Simulation Results and Discussion

Electric load prediction is a major subject in electric power framework tasks and arranging. Gauges are required for an age booking, the planning of fuel buys, support planning and security examination. Sadly, the power load is fundamentally novel, nonparametric and disarranged in nature. Frankly, various factors impact the power load in a region, for instance, the populace, budgetary improvement, social change, atmosphere conditions and mechanical creations in the area, control cost and occasion periods. This effect recommends that exact load assessing is not simply of remarkable interest yet furthermore hard to control framework officials.

3.1 Data Sets

The hybrids of model was utilizing the New South Wales (Australia) control load information for days of the week given by the National Electricity Market Management Company (NEMMCO). The power load information was amassed on an hourly reason (24 data focuses every day) for about a month.

To think about the choosing execution of various models, the perception estimations of force load in the basic 20 days were utilized to survey show parameters,
and the rest of the information was utilized for endorsement purposes.

3.2   Wavelet de-Noising Based BPNN

Power load time course of action is uproarious in light of the manner in which that an
accumulation of segments sway electrical markets. The prompt deciding of power
load with uproarious information is regularly subject to tremendous missions.
Around there, the racket is taken from the watched information utilizing DWT.
   A gathering of wavelets to do DWT has been applied in the composition. DWT
has charming targets point of confinement, capacity, and computational cost, and so
forth. In this examination, the db1 was gotten a handle on in the wavelet de-noising
process. Exactly when all is said in done, a formal method to pick the rot level does not
exist, and hence the choice depends upon commitment. Contemplating the qualities of
the exploratory information, after pre-testing specific estimations with db1, level 1
appears to work best with the information surveyed in the examination. Likewise, the
essential measurement was picked to fabricate figure shown in this examination. In this
examination, the low-repeat banner and is utilized to produce the desired show.
   To confirm the proposed technique, a prediction test is led with the same power
load. In this test, the BPNN and WT-BPNN demonstrate, which were set up utilizing the first information with and without de-noising preparing is thought about.
The comparisons of electric load esteems conjecture utilizing the BPNN and
WT-BPNN with the actual models are appeared in Figs. 2 and 3, separately.
   Also, the correlations among the BPNN and WT-BPNN models are displayed in
Table 1. Table 2 displays the correctnesses of these approaches dependent on

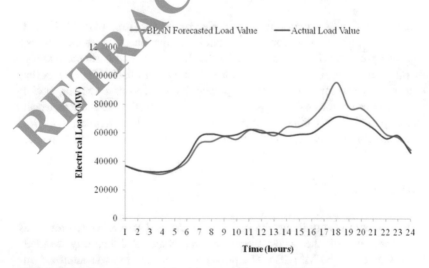

Fig. 2 Prediction of short-term load forecast values using BPNN

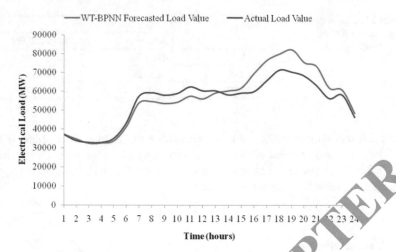

Fig. 3 Prediction of short-term load forecast values using WT-BPNN

Table 1 Comparison of the predicted values of BPNN and WT-BPNN models			
Time (h)	**Actual (MW)**	**Short term load forecasting**	
		BPNN (MW)	**WT-BPNN (MW)**
1	36956.23	36936.48	37236.41
2	33698.74	34095.78	34697.66
3	32753.12	31975.5	32474.56
4	32772.81	31221.26	32642.74
5	34689.08	34356.71	33759.78
6	43035.05	39891.37	41221.38
7	57209.7	52540.01	53721.49
8	59236.82	54097.49	54566.28
9	57910.82	57591.7	53510.45
10	58896.89	55602.68	54243.83
11	62409.25	61975.43	57394.73
12	60261.16	61677.85	55973.94
13	60242.86	58172.21	59182.97
14	58050.66	64193.46	60244.41
15	59056.87	64559.68	61777.85
16	59922.44	70242.43	69562.44
17	65838.98	80169.74	76510.5
18	71385.73	95273.53	79852.78
19	70266.51	77551.23	82067.14
20	68245.08	77206.04	75330.14
21	63108.8	69477.54	73243.51
22	56194.72	59196.91	61798.55
23	58069.88	56742.92	60716.38
24	46141.25	48127.64	48327.91

Table 2 Comparison of errors of the BPNN and WT-BPNN models

Errors	Model	
	BPNN	WT-BPNN
MAE	4701.9015	4388.3845
RMSE	7124.0455	5595.5516
MAPE	0.0768	0.0724

statistical assessment indices, and it shows that the slipups were smallest for the WT-BPNN among the rest of the models. An examination of the outcomes between the exclusive and essential ensemble models shows that the combined approach can possibly improve the desired accuracy.

4 Conclusion

Load gauging is especially fundamental for the power system planning and operation. The fundamental commitment of this venture is an approach for STLF that utilizes a consolidated model dependent on wavelet de-noising signal processing procedures. The absence of good estimating models propelled us to propose a half and half framework to enhance the prediction utilizing wavelet-based BPNN models. The rough power load information course of action is rotted into a low-repeat divide and a high-repeat part by the wavelet breaking down. The probability of this development is to disconnect the focal qualities and expel the capricious disturbing effect from the unrefined information game plan. The anticipating estimations of the crude power load data arrangement were dictated by the BPNN show reliant on WT input highlights. Execution of the single and hybrid models is evaluated reliant on the MAE, RMSE and MAPE which demonstrate that the errors were said to be the littlest for WT-based BPNN, and in this way, WT-BPNN model can be used for effectively upgrading the STLF precision.

References

1. Engle, R.F., Mustafa, C., Rice, J.: Modelling peak electricity demand. J. Forecast. **11**, 241–251 (1992)
2. Ruzic, S., Vuckovic, A., Nikolic, N.: Weather sensitive method for short term load forecasting in electric power utility of Serbia. IEEE Trans. Power Syst. **18**, 1581–1586 (2003)
3. Haida, T., Muto, S.: Regression based peak load forecasting using a transformation technique. IEEE Trans. Power Syst. **9**, 1788–1794 (1994)
4. Fan, J.Y., McDonald, J.D.: A real-time implementation of short-term load forecasting for distribution power systems. IEEE Trans. Power Syst. **9**, 988–994 (1994)
5. Cho, M.Y., Hwang, J.C., Chen, C.S.: Customer short term load forecasting by using ARIMA transfer function model. In: Energy Management and Power Delivery, pp. 317–322. IEEE Press (1995)

6. Yang, H.T., Huang, C.M., Huang, C.L.: Identification of ARMAX model for short-term load forecasting, An evolutionary programming approach. IEEE Trans. Power Syst. **11**, 403–408 (1996)
7. Papalexopoulos, A.D., Hao, S., Peng, T.M.: An implementation of a neural network based load forecasting model for the EMS. IEEE Trans. Power Syst. **9**, 1956–1962 (1994)
8. Khotanzad, A., Afkhami-Rohani, R., Lu, T.L., Abaye, A., Davis, M., Maratukulam, D.J.: ANNSTLF-a neural-network-based electric load forecasting system. IEEE Trans. Neural Networks **8**, 835–846 (1997)
9. Chen, B.J., Chang, M.W.: Load forecasting using support vector machines: a study on EUNITE competition 2001. IEEE Trans. Power Syst. **19**, 1821–1830 (2004)
10. Rahman, S., Hazim, O.: Load forecasting for multiple sites: development of an expert system-based technique. Electr. Power Syst. Res. **39**, 161–169 (1996)
11. Li, Y., Fang, T.: Wavelet and support vector machines for short-term electrical load forecasting. In: Wavelet Analysis and Its Applications, pp. 399–404 (2003)
12. Wang, J., Wang, J., Li, Y., Zhu, S., Zhao, J.: Techniques of applying wavelet denoising into a combined model for short-term load forecasting. Int. J. Electr. Power Energy Sys. **62**, 816–824 (2014)
13. Tiwari, A., Dubey, A.D., Patel, D.: Comparative study of short term load forecasting using multilayer feed forward neural network with back propagation learning and radial basis functional neural network. SAMRIDDHI J. Phys. Sci. Eng. Technol **7**, 14–27 (2015)
14. Conejo, A.J., Plazas, M.A., Espinola, R., Molina, A.B.: Day-ahead electricity price forecasting using the wavelet transform and ARIMA models. IEEE Trans. Power Syst. **20**, 1035–1042 (2005)
15. Anbazhagan, S., Kumarappan, N.: Binary classification of day-ahead deregulated electricity market prices using neural networks. In: IEEE Fifth Power India Conference, pp. 1–5. IEEE Press, India (2012)

Sentiment Extraction from Bilingual Code Mixed Social Media Text

S. Padmaja, Sameen Fatima, Sasidhar Bandu, M. Nikitha
and K. Prathyusha

Abstract In social media, code mixed data is getting exceptionally mainstream due to which there is an enormous development in noisy and inadequate multilingual content. Automation of noisy social media text is one of the existing research areas. This work focuses on extracting sentiments at word level for movie-related code mixed English–Telugu bilingual Roman script data extracted from Twitter. Initially, data was cleaned and slang words were replaced. Next, named entities and language of each word were identified. Further Telugu words were back transliterated to the native script. Finally, the text was classified either positive, negative or neutral sentiments based on the count of corresponding positive and negative lexicons present in each sentence and achieved an average accuracy of 79.9%.

Keywords Natural language processing · Sentiment extraction · Language identification

S. Padmaja (✉) · M. Nikitha
Keshav Memorial Institute of Technology, Hyderabad, India
e-mail: bandupadmaja@gmail.com

M. Nikitha
e-mail: mnikitha24@gmail.com

S. Bandu · K. Prathyusha
Prince Sattam Bin Abdul Aziz University, Al- Kharj, Saudi Arabia
e-mail: sasibandu@gmail.com

K. Prathyusha
e-mail: prathyushakairamkonda@gmail.com

S. Fatima
Osmania University, HYDERABAD, India
e-mail: sameenf@gmail.com

© Springer Nature Singapore Pte Ltd. 2020
K. S. Raju et al. (eds.), *Data Engineering and Communication Technology*,
Advances in Intelligent Systems and Computing 1079,
https://doi.org/10.1007/978-981-15-1097-7_59

1 Introduction

Social media allows users to facilitate the creation and sharing of information. These days, a substantial number of reviews posted by the users on the web provides significant information to different users. Social media texts such as Facebook posts, Twitter comments, and chats are interested in sentiment analysis where users simultaneously publish opinions about the same topic in different languages. In social networking sites, the Indian languages are composed in transliteration format and mixed with English which is known as code mixing. Code mixing is predominantly increasing in Indian social media text.

For example, '#Bahubali2 movie lo anni great songs, manchi love feel vuna songs, Sooooo nice'.

This statement describes the flavor of English–Telugu code mixed text, where user was expressing his/her opinion about Bahubali movie songs.

Natural language processing is helpful in understanding the usage of language and in detecting the sentiment behind the text. These days people consider sentiment as the most expressive value of user views shared in social media.

The remaining paper is discussed as follows. In Sect. 2, we discuss previous works of sentiment analysis on code mixed text. In Sect. 3, a methodology to identify the language of the word in English–Telugu code mixed text is showcased and further performed back transliteration of transliterated Telugu words to derive the sentiment of the text. In Sect. 4, result analysis of our proposed methodology is discussed. Section 5 concludes this work and a brief thought about future work.

2 Related Work

Sharma et al. in [10] introduced various strategies to analyze sentiment of text after normalizing the text. They utilized FIRE 2013 and 2014 Hindi–English data sets. They partitioned their work into two stages: In the principal stage, they identified the language of the words present in the code mixed sentences utilizing lexicons. In the second stage, they extracted the sentiments of these sentences utilizing SentiWordNet. They handled finding abbreviations, spelling rectifications, slang words, wordplay, phonetic typing and transliterations of Hindi words into Devanagari script. They accomplished a precision of 85%. In [9], they also handled named entity recognition, ambiguous words and got exactness of 80%.

Sitaram et al. in [11] proposed a procedure to determine the sentiment of sentences at phrase level and sub-phrase level written in a mixed dialect including Hindi and English dictionaries. They utilized Recursive Neural Tensor Network (RNTN) for sentiment classification as it deals with setting. They accumulated train and test information of Virat Kohli fan pages from Facebook. This classifier acquired a precision of 91.01% for trained information and got an exactness of 73.19% for test information.

Joshi et al. in [5] formally presented sub-word level representations utilizing long-short term memory network (LSTM) architecture rather than character level or word level representations in sentiment extraction of Hindi–English code mixed content. They gathered comments from Facebook public fan pages of Salman Khan and Narendra Modi. They developed a code mixed dataset with opinion polarity annotations. Their proposed approach figure outs how to recognize sentiment Lexicons. This sub-word LSTM framework gave an F-score of 65.8%.

Pravalika et al. in [7] have proposed a framework to analyze the sentiment of English–Hindi code mixed text which is extricated utilizing Facebook graph API. This framework incorporates both dictionary-based approach and machine learning-based approach. In dictionary-based approach, they made semi-automatic lexicons of words annotated with a semantic orientation polarity. They utilized trie data structure to keep up these lexicons with their polarity. They characterized text based on the count of positive and negative words, and they accomplished an exactness of 86%. In machine learning-based approach, they implemented SVM, Naive Bayes, decision tree, random tree, multilayer perceptron models with unigram words, list of negative words, list of positive words and list of negation words as features on WEKA tool. They accomplished an accuracy of 72% which is not as much as dictionary-based approach.

Malgaonkar et al. in [6] had extracted sentiment from Hindi–English live Twitter information utilizing lexicon-based approach. At first, they recognized the language of each word with the assistance of n-grams and tagged parts of speech of English–Hindi mixed sentences. They made two lexicons comprising of important words from the tweets and categorized the tweets as overall positive, negative and neutral. For looking through the words in lexicons, a linear search algorithm and dictionary search algorithm were tried among which dictionary-based search has better execution. While classifying data, they joined Breen's algorithm and Cholesky decomposition for deciding sentiment. They accomplished an accuracy of 92.68% for the positive case and 91.72% for negative case.

3 Proposed Methodology

The proposed work has been focused on a bilingual English–Telugu code mixed movie-related data which has been scraped from Twitter using Twitter API. The scraped data was then cleaned by removing punctuations, hashtags and further replaced short forms and slang words. Extraction of sentiment from bilingual code-mixed text has been tried in lexicon-based approach. The architecture of the same is shown in Fig. 1. The proposed methodology consists of following steps:

- Step 1: Language identification
- Step 2: Back transliteration
- Step 3: Sentiment extraction.

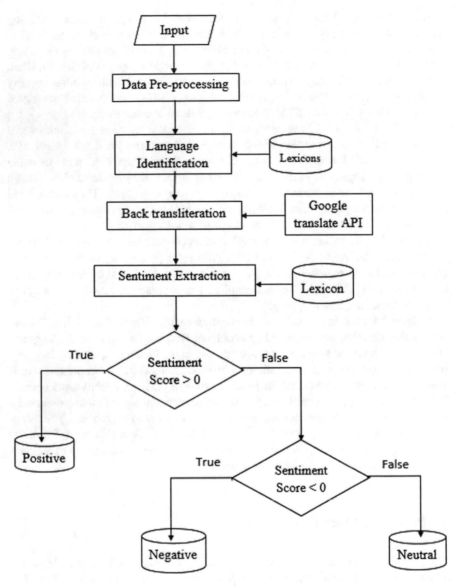

Fig. 1 Architecture of proposed methodology

4 Implementation

4.1 *Language Identification*

Identifying language of each word is the main and primary task for sentiment extraction of code mixed text. In this phase, language is identified through lexicon-based approach. Firstly, noise such as hashtags, punctuations and URLs is removed from the text. Slang words used by the users such as 'hru' for 'how are you?' are identified and replaced with the original.

Each word in the text is then looked up in language dictionaries and tagged with the corresponding language tag. For English, 'en' was tagged using British National Corpus [1], and for Telugu, 'te' was tagged using ITRANS format of Crubadan Corpus [8]. Further named entities in the text were tagged as 'ne' using lexicon which was created for movies, and remaining words are tagged as 'un' (unknown).

For example, 'just ippude Chalo movie chusanu, super movie annayya. pakka blockbuster' After language labeling:

'just||en ippude||te Chalo||ne movie||en chusanu||te super||en movie||en annayya||te pakka||un blockbuster||en'

In the above sentence, each word is tagged with its corresponding language.

4.2 Back Transliteration

After identifying language of words in text, 'te' tagged words are back transliterated to their native script using Google transliteration API [3]. The main reason for back transliteration is to find the sentiment of Telugu words which are in transliterated format.

After back transliteration:

just| |en ఇప్పుడే| |te Chalo| |ne movie| |en చూసాను| |te super| |en movie| |en అన్నయ్య| |te pakka| |un

blockbuster| |en

In the above sentence, all 'te' tagged words are back transliterated to Telugu script.

4.3 Sentiment Extraction

Sentiment extraction here is to classify each English–Telugu code mixed text either positive or negative or neutral through lexicon-based approach. Based on the language of text, the sentiment of code mixed text is determined. Each word is looked up into sentiment lexicons to find the corresponding positive and negative lexicons present in each sentence. We used two sentiment lexicons to extract sentiment of the text:

- Opinion lexicon [4] which consists of 2007 English positive words and 4783 English negative words.

– Telugu SentiWordNet [2] which consists of 2136 positive words, 4076 negative words, 359 neutral words and 1093 ambiguous words.

We determine overall sentiment based on statistical approach. If the count of positive words in code mixed text is more, then the text is classified as positive sentiment and inversely. If the count of positive and negative words is same, then the statement is classified as neutral sentiment.

After sentiment extraction:

just ఇప్పుడే Chalo movie చూసాను super movie అన్న య్య pakka blockbuster   pos

The above sentence is classified as positive (pos) sentiment by our methodology.

5 Results

The results have been drawn on 352 English–Telugu code mixed tweets and analyzed the proposed methodology. In the language identification phase, the language of the text has been identified through lexicon-based approach and back transliterated Romanized Telugu words into original script. Different evaluation metrics were used to analyze the performance of language labeling, which are tabulated in Table 1. In sentiment extraction phase, each text was classified either positive, negative or neutral. The performance of sentiment extraction is evaluated using precision, recall, F-measure and accuracy which are tabulated in Table 2 (Fig. 2).

Table 1 Result analysis of language identification

Language class	Precision	Recall	F-measure	Accuracy (%)
English	0.894	0.988	0.939	93.22
Telugu	0.891	0.803	0.887	92.60
Named entity	0.996	0.943	0.968	99.45
Unknown	0.29	0.640	0.399	97.18

Table 2 Result analysis of sentiment extraction

Sentiment class	Precision	Recall	F-measure	Accuracy (%)
Positive	0.846	0.779	0.811	78.12
Negative	0.626	0.573	0.598	82.1
Neutral	0.407	0.578	0.477	79.5

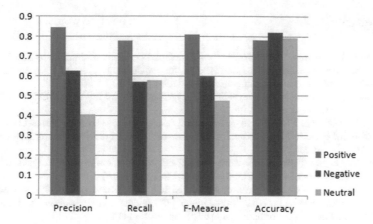

Fig. 2 Result analysis of sentiment extraction

5.1 Error Analysis

The results were analyzed to understand the flaws of the proposed methodology. It was identified that in language labeling, some of English and Telugu words were classified as unknown due to spelling variations. In sentiment extraction, there are some wrong classifications due to sarcasm and due to ambiguous of some words used in movie-related texts. For example, 'keka' is a negative word in Telugu and 'hit' is also negative in English, but such words express positive sentiment for movie-related data.

6 Conclusion and Future Work

The proposed methodology is able to classify a code mixed text into positive, negative or neutral. This approach involves identifying the languages of each word in text, back transliterating Telugu transliterated words into native script and then classifying sentiment of the text. Out of 352 English–Telugu code mixed tweets, the average accuracy achieved was 79.9%.

This work can be extended by handling spelling variation to improve the performance of language identification. Increasing sentiment lexicons for both languages and identifying more emotions will help in improving the performance of sentiment extraction.

References

1. Burnard, L.: Reference guide for the british national corpus (world edition) (2000)
2. Das, A., Bandyopadhyay, S.: Sentiwordnet for indian languages. In: Proceedings of the Eighth Workshop on Asian Language Resources, pp. 56–63 (2010)
3. Garcia, I., Stevenson, V.: Reviews-Google translator toolkit. Multilingual Comput. Technol. **20**(6), 16 (2009)
4. Hu, M., Liu, B.: Mining and summarizing customer reviews. In: Proceedings of the Tenth ACM SIGKDD International Conference on Knowledge Discovery and Data Mining, pp. 168–177. ACM (2004)
5. Joshi, A., et al.: Towards sub-word level compositions for sentiment analysis of Hindi–English code mixed text. In: COLING, pp. 2482–2491 (2016)
6. Malgaonkar, S., Khan, A., Vichare, A.: Mixed bilingual social media analytics: case study: Live Twitter data. In: 2017 International Conference on Advances in Computing, Communications and Informatics (ICACCI), pp. 1407–1412. IEEE (2017)
7. Pravalika, A., et al.: Domain-specific sentiment analysis approaches for code-mixed social network data. In: 2017 8th International Conference on Computing, Communication and Networking Technologies (ICCCNT), pp. 1–6. IEEE (2017)
8. Scannell, K.P.: The Crúbadán project: Corpus building for under-resourced languages. In: Building and Exploring Web Corpora: Proceedings of the 3rd Web as Corpus Workshop, vol. 4, pp. 5–15 (2007)
9. Sharma, S., Srinivas, P.Y.K.L., Balabantaray, R.K.: Sentiment analysis of code-mix script. In: 2015 International Conference on Computing and Network Communications (CoCoNet), pp. 530–534. IEEE (2015)
10. Sharma, S., Srinivas, P.Y.K.L., Balabantaray, R.K.: Text normalization of code mix and sentiment analysis. In: 2015 International Conference on Advances in Computing, Communications and Informatics (ICACCI), pp. 1468–1473. IEEE (2015)
11. Sitaram, D., et al.: Sentiment analysis of mixed language employing Hindi-English code switching. In: 2015 International Conference on Machine Learning and Cybernetics (ICMLC), vol. 1, pp. 271–276. IEEE (2015)

Incidence of Cancer in Breastfed Grownups-a Study

K. L. Vasundhara, Srinivasu Badugu and Y. Sai Krishna Vaideek

Abstract Cancer prevalence in India is 70–90/lakhs of population and patients around 2.5 million with over 8 lakhs new cases and 5.5 lakhs deaths occurring every year. More than 70% of the cases are in critical stage accounting to low chances of survival. There were many studies on rampant spread of cancer in the recent past, and one main reason is the lack of breastfeeding at the infant stage. Studies observed that Human Alpha-lactalbumin (HAMLET) controls tumor cells which are found in breast milk, can destroy cancer cells and safeguard against cancer. U.S. national library study on medicine found that HAMLET destroys cancer cells just like Programmed Cell Death. It was observed that HAMLET has very good anti-tumor activity. In this paper, we tried to find the impact of breast-feeding in the grownups in resisting the cancer and the relation between incidences of cancer in breastfed grownups and their resistant levels to cancer with particular reference to India through data (a survey of cancer patients). Chi-square test has been implemented on the survey data to check dependence of breastfed and non-breastfed cancer patients. We applied linear regression machine learning model for finding correlation between attributes.

Keywords HAMLET · Breast fed · Non breast fed · Age and family history

K. L. Vasundhara (✉)
Department of Mathematics, Stanley College of Engineering and Technology for Women,
Hyderabad, India
e-mail: vasundhara.yerasuri@gmail.com

S. Badugu
Department of Computer Science, Stanley College of Engineering and Technology
for Women, Hyderabad, India

Y. S. K. Vaideek
ECE-CBIT, Hyderabad, India

© Springer Nature Singapore Pte Ltd. 2020
K. S. Raju et al. (eds.), *Data Engineering and Communication Technology*,
Advances in Intelligent Systems and Computing 1079,
https://doi.org/10.1007/978-981-15-1097-7_60

1 Introduction

Among all diseases, cancer has become a big threat to human beings worldwide. As per Indian census data, rate of survival due to cancer in India was low and alarming with about 8 lahks existing cases by the end of the nineteenth century.

The cancer instigates that carcinogens can be present in air, water and food. It is yet to find a comprehensive, sure and safe treatment, in spite of many researches [1, 2]. This is the second most common disease after cardiovascular disorders for maximum deaths in the world [3]. Cancer stands next to cardiovascular diseases as far as global death rate is concerned. By 2020, 1.5 crores of cancer patients will be identified and 1.2 crores deaths [4]. Besides, studies have been done to describe the main causes of cancer along with their preventive measures. Various scientific methods can relate the breastfeed in the childhood and their relation to incidence of cancer after growing. Incidence of cancer in the breastfed grownup has been investigated. There is no substitute to the nutritional source of breast milk either organic or inorganic. In spite of little pollutants, breast milk remains superior to infant formula from the perspective of mother and child health.

Resistance of breastfed children is more when compared with the formula fed child. Mothers who breastfeed have a lower risk of breast, uterine and ovarian cancer. Large number of the health problems children face today is because of decreased or prevented breastfeed.

Studies also indicate that cognitive development is increased among children whose mothers choose to breastfeed. Researchers have observed a decrease in the probability of Sudden Infant Death Syndrome (SIDS) in breastfed infants. It is proved scientifically that when HAMLET is administered, tumor volume shrinks significantly.

Researchers in this study found that HAMLET to a greater extent controls incidence of cancer in the people after growing up, an attempt is also made on the impact of HAMLET on hereditary nature of cancer. In this paper, a statistical feature like chi-square test and regression analysis is used for data extraction [5].

1.1 Cancer Scenario in the World

Global cancer cases and deaths have been predicted in accordance with past available data. There is a rapid increase in the number of cancer cases by the end of 2025 which will be around 19 million, and it is probable to go further in the near future. Though there seems to be huge gap between mortality rate and cancer cases, it is prone to upsurge at any given point of time in the future [6] (Fig. 1).

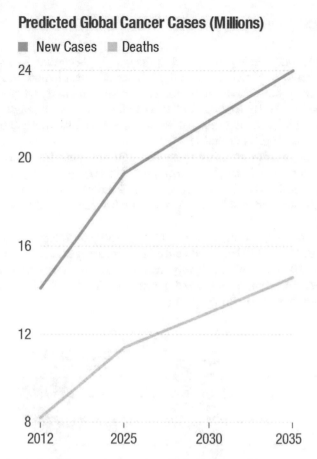

Fig. 1 Year-wise total cancer prevalence and prediction in the world

2   Data Set

Research survey was conducted on the patients who are already suffering with cancer, for the purpose of gathering required information. No clinical trials conducted or caused any pain to the patients during course of survey. We gathered the information from the patients like age, sex, cancer type, cancer since, family history and whether breast fed or not. No personal data of the patients was taken as they are not ready to give because of various reasons. Henceforth, ethics committee certificate is not needed and also cannot provide personal data of respondents as it was not given.

3 Analysis

A total of 300 individuals participated in the survey. According to a survey, incidence of cancer in breastfed is 150, and incidence of cancer in non-breastfed is 150.
A random survey was conducted on 300 cancer patients through a questionnaire at
various hospitals in Hyderabad, India. The data obtained through survey was
analyzed by using various statistical methods, and results were summarized for
appropriate steps (Fig. 2) (Table 1).

Out of 300 samples of cancer patients, 150 are breastfed, and 150 are not
breastfed. Out of 150 breastfed cancer groups surveyed, < 40 years of age are
twenty people, and > 40 years of age are 130. Same way out of 150 non-breastfed
cancer groups surveyed, < 40 years of age are thirty people, and > 40 years of age
are 120.

Out of 300 samples of cancer patients, 150 are breastfed, and 150 are
non-breastfed. Out of 150 breastfed cancer groups surveyed, patients having family
history are 110 and without family history are 40. Same way out of 150
non-breastfed cancer groups surveyed, patients having family history are 120 and
without family history are 30 (Fig. 3) (Table 2).

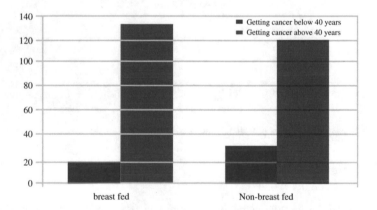

Fig. 2 Relation between age-specific rate and cancer patients

Table 1 Relation between age-specific rate and cancer patients

Different types	<40	>40	Total
Breastfed	20	130	150
Non-breastfed	30	120	150
Total			300

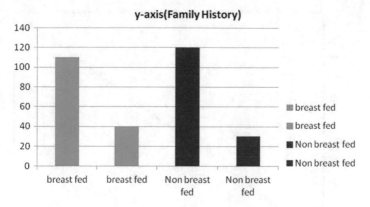

Fig. 3 Relation between family history rate and cancer patients

Table 2 Relation between family history rate and cancer patients

Different types	Family history	No family history	Total
Breastfed	110	40	150
Non-breastfed	120	30	150
Total			300

4 Linear Regression

Linear regression is a statistical approach for modeling relationship between a dependent variable with a given set of independent variables. Regression analysis is a form of predictive modeling technique which investigates the relationship between a **dependent** (target) and **independent variable (s)** (predictor). Simple linear regression has only one independent variable. Regression analysis is an important tool for modeling and analyzing data. It indicates the **significant relationships** between dependent variable and independent variable [7, 8]. In regression the dependent variable is binary (0/1, True/False, Yes/No) in nature. It indicates the **strength of impact** of multiple independent variables on a dependent variable. It calculates the best-fit line for the observed data by minimizing the sum of the squares of the vertical deviations from each data point to the line. Because the deviations are first squared, when added, there is no canceling out between positive and negative values [9, 10].

$$\min_{w}\|Xw - y\|_2^2 \tag{1}$$

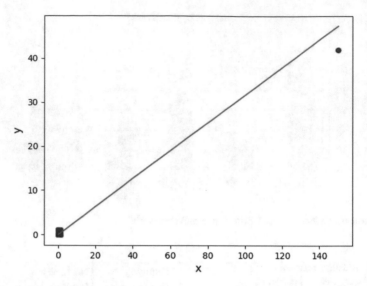

Fig. 4 Regression line for the relation between age-specific rate and cancer patients

We now use the above formula to calculate *a* and *b* as follows

$$a = \left(n\sum xy - \sum x\sum y\right) \Big/ \left(n\sum x2 - \left(\sum b\right)\& b = (1/n) - a\sum x\right)$$

Estimated coefficients:
b_0 = −0.00896597655382797
b_1 = 0.31596139569413495

   Please enter your new *X* value: 1
   Estimated *Y* value: 0

Estimated coefficients (Fig. 4):
b_0 = −0.002706935123042542
b_1 = 1.5454138702460851

   Please enter your new *X* value: 1
   Estimated *Y* value: 0

5 Chi-Square Test

Firstly, we discuss age and cancer patients [11].

H_0: Null hypothesis—both breastfed and non-breastfed are independent.
H_1: Alternative hypothesis—both breastfed and non-breastfed are not independent (Tables 3 and 4).

$$\text{Expected frequency}(E) = \frac{\text{Row total} * \text{column total}}{\text{grand total}} \tag{2}$$

Secondly, we discuss family history and cancer patients (Tables 5 and 6).

The calculated value of χ^2 (1.8630) is less than the table value (3.841). Thus, the null hypothesis is accepted in both the cases (age and family history). Hence, the breastfed cancer patients and non-breastfed cancer patients are independent (Fig. 5).

Table 3 Relation between age-specific rate and cancer patients

Patients	<40 age	>40 age
Breastfed	(50*150)/ 300 = 25	(250*150)/ 300 = 125
Non-breastfed	(50*150)/ 300 = 25	(250*150)/ 300 = 125

Table 4 Calculation of chi-square

Observed value (O) (age)	Expected value (E)	(O-E)²/E
20	25	1
130	125	0.2
30	25	1
120	125	0.2
		Total = 2.4

The critical value of the table value of x^2 at $\alpha = 0.05$ for one degree of freedom is 3.841 Since the calculated value of x^2 (2.4) is less than the table value (3.841)

Table 5 Relation between family history rate and cancer patients

Patients	Family history	No family history
Breastfed	(230*150)/ 300 = 115	(70*150)/ 300 = 35
Non-breastfed	(230*150)/ 300 = 115	(70*150)/ 300 = 35

Table 6 Calculation of chi-square

Observed (Family history)	Value (O)	Expected value(E)	$(O-E)^2/E$
110		115	0.2173
40		35	0.7142
120		115	0.2173
30		35	0.7142
			Total = 1.8630

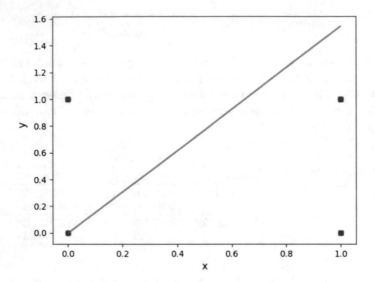

Fig. 5 Regression line for the relation between family history rate and cancer patients

6 Conclusion

Therefore, the incidence of cancer in the breastfed grownups is low which means mother milk will immunize people against cancer even after years of growth this is as well seen in the patients having family history. India is a fast-growing and fifth largest economy in the world and soon may become global address for health research.

It is imperative on our behalf conduct a research on all aspects controlling this epidemic in the near future. In view of these facts, attempts have been made to study the status of cancers in India including its causes, preventive measures, effect on Indian economy and comparison with global scenario. Our study is also an attempt in this direction as it is rightly said "prevention is better than cure".

This is only a simple study on impact of breast milk even after growing up and its resistance to cancer, and there is need for extensive study of cancer patients with reference to male–female, age groups with ten years span, impact on breastfeed on various cancers. etc. This will be taken up as project in near future with support of various National and International agencies.

Questionnaire Format

Age
_____

Sex
_____

Cancer Type
_____

Cancer since
_____

Family History – Cancer
_____

Habits
_____

Whether Breast Fed   O yes O No
If Yes How Many Months
_____

References

1. Ali I, Wani WA, Saleem K.: Cancer scenario in India with future perspective. Cancer Therapy **8**, 56–70 (2011)
2. Kotnis, A., Sarin, R., Mulherkar, R.: Genotype, phenotype and cancer: role of low penetrance genes and environment in tumor susceptibility. J. Biosci. **30**, 93–102 (2005)
3. Jemal, A., Siegel, R., Ward, E., Murray, T., Xu, J., Thun, M.J.: Cancer statistics. CA Cancer J. Clin. **57**, 43–66 (2007)
4. Brayand, F., Moller, B.: Predicting the future burden of cancer. Nat. Rev. Cancer **6**, 63–74 (2006)
5. Himabindu, G., Murty, R.M., et al.: Extraction of texture features and classification of renal masses from kidney images. Int. J. Eng.Technol. **7**(2.33), 1057–1063 (2018)
6. Fenley, J., Bray, F., Pisani D, Me.: World Health Organization. GLOBOCAN. Cancer Incidence, Mortality and Prevalence Worldwide. IARC, Lyon, France (2000)
7. Chatterjee, S., Hadi, A.S.: Regression analysis by example. Wiley (2015)
8. Murty, M.R.,et al.: A comparative analysis of breast cancer data set using different classification methods. In: International Conference and Published the Proceedings in AISC. Springer, SCI (2018)
9. Kaytez, F., Taplamacioglu, M. C., Cam, E., & Hardalac, F.: Forecasting electricity consumption: a comparison of regression analysis, neural networks and least squares support vector machines. Int. J. Electri. Power Energy Syst. **67**, 431–438 (2015)

10. Veronese, N., Cereda, E., Stubbs, B., Solmi, M., Luchini, C., Manzato, E., Sergi, G., et al. Risk of cardiovascular disease morbidity and mortality in frail and pre-frail older adults: results from a meta-analysis and exploratory meta-regression analysis. Ageing Res. Rev. **35**, 63–73 (2017)
11. Grewal, B.S.: Higher Engineering Mathematics. Khanna Publications

RETRACTED CHAPTER: A Study of Physiological Homeostasis and Its Analysis Related to Cancer Disease Based on Regulation of pH Values Using Computer-Aided Techniques

Rakesh Kumar Godi, G. N. Balaji and K. Vaidehi

Abstract In this modern life, many people are prone to cancer disease. According to the survey made by WHO (World Health Organization), the percentage of cancer found to increase up to 45 in middle-aged people and even found in children within age limit of 14; hence, death rate because of cancer is increasing day by day. This disaster is big challenge to medical world and has a massive amount of data to be analyzed within short time. It is not feasible for the doctors to predict the cancer at early stages based on the medical examination tests and reports. Hence in order to process the collected medical data, time reduction, data optimization, minimize service costs, quality of advanced treatment and efficiency of the clinical data, the health care industry relies on biomedical information technology. This paper aims to provide feasible solution for human health analysis related to cancer disease based on the electrolyte values extracted from the collected samples of urine and blood from the patient, and based on the output analysis, the patient condition is predicted. The electrolyte values from urine and blood samples are integrated with help of data fusion techniques to generate single-dimensional data. The desired pH homeostasis information is extracted using feature extraction technique from the data set. This optimized data is then classified using ANN (artificial neural network) and analyzed using SVM (support vector machine).

The original version of this chapter was retracted. The retraction note to this chapter is available at https://doi.org/10.1007/978-981-15-1097-7_82

R. K. Godi (✉) · G. N. Balaji
Department of Information Technology, CVR College of Engineering, Hyderabad, India
e-mail: drrakeshkumargodi@gmail.com

G. N. Balaji
e-mail: balaji.gnb@gmail.com

K. Vaidehi
Department of CSE, Stanley College of Engineering and Technology for Women,
Hyderabad, India
e-mail: vainakrishna@gmail.com

Keywords pH · Homeostasis · PCA reduction · Urine · Blood samples

1 Introduction

The cell is constituent component of human body, and its working is totally de-swinging on working of discrete cells. An ordinary cell works legitimately while encompassing outside conditions are steady. Every cell is encompassed by extracellular liquid, and its composition components are hydrogen, oxygen, calcium and phosphorus. The waterway is again separated into extracellular liquid and interstitial liquid. The extracellular is 20 L including blood and plasma about 3.4 L and interstitial liquid about 8.4 L of the all out TBW and around 23 L of liquid inside cells [1]. Minor change in the piece component will influence the typical working of the cell. The primary electrolyte in body water outside of cell is sodium and chloride. Though inside a cell potassium and other phosph... assume a vital job, the clinician chiefly focuses on ionic distribution, electrolyte piece and pH estimation of blood and pee. On the off chance that any minor distinction in electrolyte organization in unique individual from the typical range, it shows a strange wellbeing state of human body. An anomalous condition in the cells might demonstrate a malignancy sickness [2]. To inspect a sickness, they have to gather an individual electrolyte arrangement esteem from the patient for wellbeing examination.

For wellbeing examination, electrolyte substances gathered in tremendous volume in medicinal sector and this makes social insurance industry progressively complex to process the information. Henceforth, information handling strategy used to extricate a required information choice action to tolerant dependent on their restorative report [3]. To improve productivity and increment the basic leadership process there is necessity of new PC device to support the specialists and concentrate attractive therapeutic data. Information preparing strategies is progressively structured and executed to answer the necessities of the clinicians in their everyday basic leadership activity. Information mining or KDD (Knowledge Discovery in Databases) one of the information innovations which helps in basic leadership dependent on gathered data. It is a procedure of unwinding the valuable data from gigantic information [4].

The job of data mining methods is imperative in restorative segment regarding estimating and making a judgment in managing diverse illnesses like malignant growth, diabetes, cardiovascular variations from the norm. KDD is PC-helped procedure in which data mining is critical for information investigation. This information mining incorporates the induce ring calculations, with these calculations the information is examined, and demonstrate is created to discover beforehand obscure examples. This model is utilized for understanding marvels from the information, examination and forecast.

They are the Selection organize incorporates making an objective informational index and mostly focuses on various electrolytes (e.g., pH estimation of blood and pee) conveyance of the cell in the body for further wellbeing examination reason.

To catch homogenous information a KDD course through a pre-handling procedure, this part assumes a noteworthiness job. This stage incorporates target information cleaning procedure to eliminate the unwanted data present in the informational collection.

Information change is finished utilizing information decrease strategies or change methods. The data mining stage is helpful for isolating a pertinent information from the gathered information. The interpretation/evaluation arrange is useful in choosing [5, 6].

The fundamental point of information mining in therapeutic field is to improve the treatment quality and development of treatment assets. The utilization of information mining in medicinal services expands the work proficiency and decreases human mistake which manages undertakings for foreseeing a specific sickness where precision is critical. It lessens the time and cost related to illness expectation and therapeutic administration. Information mining strategy is useful in multi-process mechanization, for example, forecast models and master framework, and it is fundamentally useful in extraction of new information [7].

2 Methodology

The proposed framework of the system flow is shown in Fig. 1. The working of the proposed framework is separated into two sections: training and testing. In training part, unique electrolyte esteems are gathered from arbitrary patient's urine and blood tests. Here, patients might be a heavy drinker individual, diabetic individual or pregnant woman. These urine and blood tests esteems are gathered by the separate medicinal reports from therapeutic trial of individual patient. This procedure is known as information gathering where distinctive incentive from different patient therapeutic reports is gathered and put away in electronic records for computation [8]. This assembled different example esteems are coordinated to make single information base which lessens the computational multifaceted nature of the framework. This information coordination is performed by utilizing information combination procedure which speaks to most exact and steady data of gathered medicinal information. Extracting the concealed actualities from the information is finished by utilizing highlights extraction method, this calculation changes over the high dimensional information tests into low dimensional examples serious data contained in the information is secured. This reduced measurement of the example information will build the classifier execution that diminishes the computational many-sided quality [9]. This reduced dimensional information is given as input to ANN preparing to make learning base for the better database correlation.

Correspondingly in testing stage, information may be a pee or blood electrolyte esteems from therapeutic report of separate patient which is considered for well-being investigation. The pee and blood test esteem are gathered by thinking about a different number of patients. This accumulated information is coordinated to shape a solitary information base utilizing information combination system [10]. Again, this incorporated information measurement is lessened isolating alluring

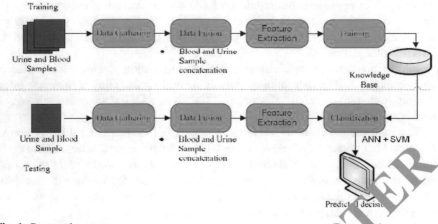

Fig. 1 Proposed system

information from the gathered; this diminishes the measurement of gathered information which is useful for better computational speed.

In testing, the SVM and ANN [11] calculations are utilized to break down the therapeutic and measure individual computational exactness in choice forecast. The two classifiers assume a criticalness part; the yield from every classifier is analyzed for characterizing framework exactness. The classifier yield result is contrasted and information based and introduces the yield. Figure 2 represents the general stream of chart of the venture.

2.1 Data Fusion

To decide the condition of the framework, the information combination relates informational index from different sources. The framework state and its working are consistently researched and the general structure of the information combination changes when present input informational index changes which totally rely upon the framework assets and ebb and flow condition of the framework. With a specific end goal to accomplish a superior execution and less memory use, an information combination strategy is utilized, to undercover high dimensional restorative into altering low-measurement information in medicinal part for information mining. Information combination is a procedure to consolidating the distinctive wellspring of information together. There are three primary sorts of information combination: information situated, undertaking focused (variable) and a blend of information and variable combination. The fundamental point of the information combination is to speak to incorporated outcome and create more definite, solid and precise data. In the proposed framework, blood and pee test esteems are gathered from medicinal

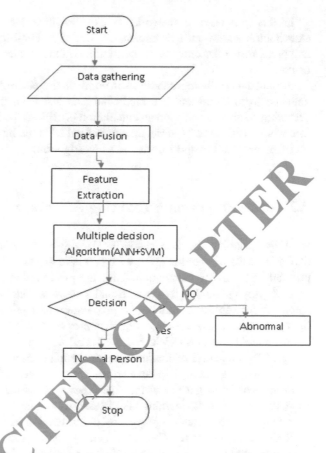

Fig. 2 Overall project flow diagram

report of the various patie... and this colossal restorative information is contribution to the information c...bination which speaks to therapeutic data in exact shape. This informatio... combination strategy expands the whole framework execution.

2.2 Ann Classifier

...rmation investigation, ANN grouping picks up a more importance. It is viewed as a most powerful research and application territory. ANN is a branch of man-made reasoning (AI). There are different procedures that are accessible in this characterization procedure. ANN shows the framework that as opposed to programming, it performs and executes unequivocal undertakings. In ANN, parallel and staggered arrangement strategies are considered as most essential methods. For preparing and testing of information order, ANN is considered as the most adaptable calculation and it is pertinent for better and exact grouping result.

In this proposed framework, ANN classifiers are utilized for information examination and to settle on choice, i.e., sound and unfortunate state of the patients and break down the unfortunate condition whether it identified with tumor ailment or not.

In unusual wellbeing condition in light of the electrolyte test esteem which is gathered from blood and pee medicinal Lab test from the specific individual. In preparing section, a back spread calculation is utilized for information examination. The whole utilization of arrangement method kills the human blunder related with maladies predication and information investigation.

2.3 Back Propagation Learning Algorithm

In ANN grouping, back engendering is a typical technique or educating and depicts how to play out a given errand. There are two operational stages in back proliferation learning calculation. They are proliferation and weight refresh.

Each propagation has two operational advances. They are forward and in reverse spread. A forward proliferation of a preparation layout creates the spread's yield in light of the neural system inputs. In back propagation keeping in mind the end goal to create the deltas of all yield and concealed neurons, proliferation yield actuation is done through neural system utilizing preparing designs. To get slope of weight, it is essential to duplicate yield delta and info. And afterward subtracting a proportion from the weight to get the weight the other way of the inclination; the resultant proportion expands the learning quality, and it is characterized as learning rate. The expanding of the mistake is shown by the indication of the weight slope.

Both the stages are rehashed until getting great system execution. To figure out change in angle regarding the modifiable weights of the system a back propagation calculation is primarily utilized. Basic execution and its speedy capacity to begin the system the back engendering is exceptionally well known. The whole utilization of characterization method disposes of the human blunder related with sicknesses prediction and information investigation.

From the prepared informational index, i.e., set of information is chosen and chosen informational collection design is exchanged to the system. In ANN classifier, the yield layer takes the concealed neurons esteem from shrouded layer and figures the flag yield. After pointing the yield neurons esteem the examination activity occur where display yield neuron esteems are contrasted and as of now produced yield neurons and the mistake manage that is ascertained. To lessen the mistake related with yield, neurons are conceivable by changing over the information neurons into shrouded layer neurons.

2.4   SVM Classifier

SVM is for the most part intended for characterization and relapse errand. The operational standard of SVM depends on VM measurement from factual learning and structural risk minimization. SVM is corresponded with the general classification of piece strategy. Direct, quadratic, polynomial and outspread premise are the basic piece work. A kernal function is a calculation that primarily in view of the information through spot item; these speck items can be changed by bit work which figures a dab item in some perhaps high dimensional element space. This has two points of interest.

To start with, utilizing techniques intended for liner classifier and piece calculation can produce non-direct choice limits. Second, piece capacity can likewise apply for a classifier to the settled dimensional vector space. SVM is groundbreaking apparatus for parallel order and creates quick classifier capacities amid preparing period. A case of a direct two-class classifier is support vector machines. SVM incorporates a two-separate protest names for information to the individual classes as shown in Fig. 3.

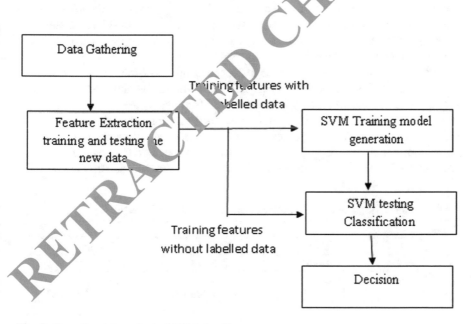

Fig. 3   Operational flow chart of SVM classifier

3 Experimental End Results

Framework is outlined and actualized for various number of individual's blood and pee tests for investigation of the growth ailments. The electrolyte organization of individual people is gathered and utilized as informational index. The gathered blood and pee test esteems are broken down by considering couple of imperative compound, e.g., potassium pH esteem. For better framework execution and to build the precision, demonstration is planned by utilizing SVM and ANN classifier. The yield from SVM and ANN classifier is looked at amid the execution investigation, when the consequence of the two classifiers is equivalent and meets the predefined condition than the individual regarded as typical generally individual named as anomalous. In view of strange condition, the therapeutic information broken down to decide if quiet identified with tumor infection. Medical results are considered when both the classifier gives distinctive outcome. In which yield of the SVM and ANN are contrasted and clinical report. On the off chance that clinical report matches with ANN yield, then choice is anticipated considering ANN classifier (Tables 1 and 2).

Essentially delicate registering is an investigative model utilized with SVM and ANN classifiers to examine the information and aides in information expectation. At the point when the classifier outcome is covered than is hard to settle on a choice in that circumstance delicate processing assume an essential part. Consequently, by utilizing delicate registering with bunching strategy will recognize classifier yield and make a compelling yield.

For better framework execution and expand the precision, display is planned by utilizing SVM and ANN classifier. The yield from SVM and ANN classifier is looked at amid the execution examination, when the consequence of the two classifiers is equivalent and meets the predefined condition than the separate individual regarded as typical generally individual named as unusual. In light of anomalous condition, the medicinal information investigated to decide if persistent identified with malignancy infection. Clinical reports are considered when both the

Table 1 Abnormal and normal electrolyte composition in fluid of a person

Blood pH value			
	Calcium	Potassium	Sodium
Normal person	4.4–5.4	133–145	3.5–4.5
No of healthy persons	<4.5 or > 5.5 mEq1	<133 or > 145 mEq 1	<3.5 or > 4.5 mEq1

Table 2 Abnormal and normal electrolyte composition in urine of a person

Fluid pH value			
	Calcium (mEq1)	Potassium (mEq1)	Sodium (mEq1)
Normal person	96–106	23–100	25–100
No of healthy persons	<98 or > 106	<23 or > 100	<25 or >

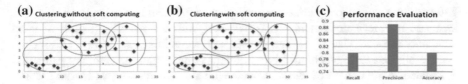

Fig. 4 **a** Soft computing for clustering, **b** Without soft computing, **c** Graph of performance evaluation

Table 3 Confusion matrix

N = 40	Forecast: No	Forecast: Yes	
Number of unhealthy people	TN = 12	FP = 4	16
Number of healthy people	FN = 6	TP = 18	24
Total	18	22	40

classifiers give diverse outcome. In which yield of the SVM and ANN are contrasted and clinical report. If clinical report matches with ANN yield, then choice is anticipated in light of ANN classifier yield or else SVM classifier yield is considered for wellbeing investigation.

A delicate registering strategy spoke to with grouping techniques in underneath Fig. 4a, b). Essentially delicate processing a logical model utilized with SVM and ANN [12] classifiers to break down the information and aides in information forecast. At the point when the classifier yield is covered than it is hard to settle on a choice in that circumstance delicate figuring assume a vital part. Consequently, by utilizing delicate registering with grouping technique will recognize classifier yield and make a powerful yield (Table 3).

Absolutely N number of individuals' therapeutic information is gathered. The input is examined, and result is spoken to in perplexity framework as appeared in Fig. 4c. Perplexity lattice is utilized to ascertain the execution exactness, accuracy and review estimation of the general undertaking.

The utilization of different calculation in wellbeing investigation will build the work proficiency and diminishes the unpredictability related to it. The trial result with characterization precision is appeared in Fig. 4b. The 78% framework precision observes the general powerful proposed framework in wellbeing examination.

4 Conclusion

The framework is planned, and a far-reaching model is actualized for the wellbeing investigation of ordinary and strange state of individual. Unusual condition may lead a malignancy, diabetic and pregnancy, and here, investigation is done in view of blood and pee test esteem. The database has been separated into two informational collections with the records in each of them haphazardly chosen. The database has been divided into two data sets with the records in each of them randomly selected. One of the data set was used for training and other for testing the data, using data fusing technique two different sample values of different patients is integrated. ANN and SVM classifier technique are used training and testing phase. There individual result accuracy is compared, based on the comparison result the overall system accuracy is defined. Finally based on the efficient knowledge base data, physician can make decision, which increase services accuracy in the medical sector.

Acknowledgements The authors would like to thank Dr. N. Chidambaram, Annamalai University for providing the data set required for this research.

References

1. Milovic, B.,Milovic, M.: Prediction and decison making in health care using data mining. Int. J. Public Health Sci. (IJPHS) **1**(2) 2012
2. Ahmad, P., Qamar, S., Rizvi, S.Q.A.: Techniques of data mining in healthcare: a review. Int. J. Comput. Appl. **120**(15) 2015
3. Chhikara, S., Sharma, P.: Data mining technique on medical data for finding locally. Int. J. Res. Appl. Sci. Eng. Technol. **1**(5) 2014
4. Hu, R.: Medical data mining based on decision tree algorithm. Comput. Info. Sci. **4**(5) (2011)
5. Lavanya, D.,Usha Rani, K.: Ensemble decision tree classifier for breast cancer data. Int. J. Info. Technol. Convergence Serv. (IJITCS) **2**(1) (2012)
6. Huang, Y.P., Lai, S.L., Sandnes, F., Lu, S.I.: Improving classification of medical data based on fuzzy ART2 decision trees, Int. J. Fuzzy Syst. **14**(3) (2012). Khaleel, M.A., Pradham, S.K., Dash, G.N.: A survey of data mining techniques on medical data for finding locally frequent dieseases. Int. J. Adv. Res. Comput. Sci. Softw. Eng. .**3**(8) (2013)
7. Durairaj, M., Ranjani, V.: Data mining application in health care sector: a study. Int. J. Sci. Technol. Res. **2**(10) (2013)
8. Chauhan, H., Chauhan, A.: Evaluating performance of decision tree algorithms. Int. J. Sci. Res. Publ. **4**(4) 2014.
9. Parvathi, I., Rautray, S.: Survey on data mining techniques for the diagnosis of diseases in medical domain. Int. J. Comput. Sci. Info. Technol. **5**(1), 838–846 (2014)
10. Surya, K.: Data mining in the field of agriculture banking and medical. Int. Res. J. Eng. Technol. (IRJET) **2**(9) (2015)
11. Yang, S., Atmosukarto, I., Franklin, J., Brinkley, J. F., Suciu, D., Shapiro, L.G.: A model of multimodal fusion for medical application
12. Balaji, G.N., Subashini, T.S., Chidambaram, N.: Detection and diagnosis of dilated cardiomyopathy and hypertrophic cardiomyopathy using image processing techniques. Eng. Sci. Technol. Int. J. **19**(4), 1871–1880 (2016)

Automatic Detection and Classification of Chronic Kidney Diseases Using CNN Architecture

R. Vasanthselvakumar, M. Balasubramanian and S. Sathiya

Abstract In medical imaging field, ultrasound imaging technique provides a tremendous service on disease recognition. It assists as non-catheterization, hazardous emission-free diagnosis at less expenses of cost. The main aim of this investigation is to detect and recognize the chronic kidney diseases CKD makes initiation and home for dysfunction of several organs of the homo sapiens. It may induce the heart disease, ischemic attack, cardiomyopathy, and cardiac disease through the hypertension. The early prediction of the kidney deformation would save the life from dreadful diseases. Renal sonography is the basic imaging technology used for kidney disease diagnosis. In this work, automatic detection and classification of kidney diseases using deep convolutional architecture have been proposed. For localizing kidney diseases, histogram-based feature along with AdaBoost algorithm is used. Deep convolutional neural network is used to recognize of kidney diseases, and TensorFlow batch prediction method is computed for recognition of diseases categories. The performance accuracy for detection of kidney disease is given as 89.79%. The performance of classification of chronic kidney diseases using CNN achieves an accuracy rate of 86.67%.

Keywords Adaboost · Convolutional neural network · Histogram of oriented gradient · Chronic kidney diseases · Ultrasonography

1 Introduction

The projection of acoustic shadow [1] in ultrasound findings would confirm the presence of kidney calculi. A thin-walled region with black acoustics over fluidic region on ultrasound image would represent the presence of cyst [2] in kidney region. Third major diseases are termed as suspected renal cancer. Ultrasound findings for RCC [3] is interpreted using hypoechoic shadow is present over the

R. Vasanthselvakumar (✉) · M. Balasubramanian · S. Sathiya
Annamalai University, Annamalainagar, Chidambaram 608002, India
e-mail: vasandth@gmail.com

© Springer Nature Singapore Pte Ltd. 2020
K. S. Raju et al. (eds.), *Data Engineering and Communication Technology*,
Advances in Intelligent Systems and Computing 1079,
https://doi.org/10.1007/978-981-15-1097-7_62

concentric porting of the kidney. Texture feature [4] provides the pattern recognition for measuring image inconsistency. Rectangle set of features floated over the input data would provide significant detection of object detection. [5] The deep convolution neural network [6] is the contemporaneous procedure to yield effective outcomes on different classification problem. In this work, deep convolutional network architecture is used to categorize the different types of kidney diseases.

1.1 Related Works

In ultrasound [7], traumatic portion of the body have been identified by moving transducer over the affected area, and acoustic shadows on the ultrasound image have been formed due to pitfalls present in the human anatomy. Each and every deformation have been represented with variety of image markers, and speckle noises [8] have created due to scattering of illumination that makes difficult to analyze the edge information in image, so segmentation would be strengthened by minimizing the speckle noise present in the input samples. In this work, special kind of diffusion filter has been used for noise reduction. Detection is the important process in medical image processing that localizes the anonymous object using some sort of algorithms. Boosting algorithm is mainly used for automatic detection and recognition based on the feature learning. Haar-like feature also learned through the AdaBoost classifier for object detection. Ripened stages of fruits [9] have detected using Haar features learning. Deep learning is the current study that processes the feature through the deep self-learning. The multiple number of convolutional operations are used to fabricate the feature for image operations. Generative adversarial network [10] with the convolution operations has computed for classification of various conditions liver medical images. Various kinds of CNN architecture [11] is analyzed for classifying the breast tissue images.

1.2 Contribution to the Work

This article investigates the detection and classification of various kidney diseases categories. The following sections describe the contributions made for this work. Section 1 illustrated the introduction and the literature review to this work. Section 2 represents the preprocessing and detection of kidney diseases using machine learning algorithm. Section 3 describes classification of diseases using deep learning methods. Section 4 evaluates the results and performance metrics to this work.

2 Detection of Kidney Diseases

2.1 Preprocessing

The database contains ultrasound b mode image with size of 640×480 dimension for processing. In this work, the speckle noise is treated with a significant anisotropic diffusion filter [12] which reduces noise without degrade the edge information. Given Fig. 1a shows the input ultrasound image to the anisotropic diffusion filter and Fig. 1b shows diffused output ultrasound image. The formulation for diffusion filter is given in Eq. (1)

$$I_t = \operatorname{div}(c(x, y, t)\, \nabla I) = c(x, y, t)\, \Delta I + \nabla_c \cdot \nabla I \qquad (1)$$

where div describes divergence operator, ∇ denotes gradient operator, Δ represents Laplacian operator concerning with space variables, $c(x,y,t)$ represents the diffusion coefficient, and I denotes input image.

2.2 Detection of Kidney Diseases

2.2.1 Histogram of Oriented Gradients

Histogram of Oriented Gradient (HOG) [13] feature is used to detect the anonymous portion of the kidney region. It is mainly constructed for vision detection and recognition problems. The essence of HOG featuring technique is orientation and its gradient magnitude of the images. It takes domestic information from the input image for segmentation processing. Initially, gradients of the image on both vertical and horizontal direction have been calculated using the single vectored kernel. For horizontal $[-1,0,1]$ and for vertical $[-1,0,1]^{\mathrm{T}}$.

(a)   **(b)**

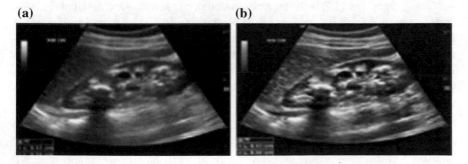

Fig. 1 **a** Input ultrasound image Fig. 2. **b** Preprocessed output image

(a)   (b)

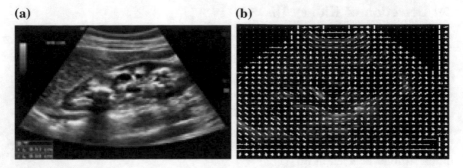

Fig. 2 HOG Feature extraction, **a** Input ultrasound image, **b** HOG feature extracted image

After convolution of kernels, gradient magnitude of the input image has to be calculated with the following Eqs. (2) and (3)

$$g = \sqrt{g_x^2 + g_y^2} \tag{2}$$

$$\theta = \arctan\left(\frac{g_y}{g_x}\right) \tag{3}$$

where g is the gradient component and θ is orientation of the corresponding element of an image. Each element of the calculated gradient array is mapped with orientation of the corresponding image based on the voting method. Divide the values of orientation if the median value occurs while mapping. Histogram containing nine number of bins is designed with positive angles from 0° to 160°. Below Fig. 2 shows the feature extraction using HOG features.

2.2.2 AdaBoost Classifier

AdaBoost algorithm [13] is the supervised composite learning method that finds the strong classifier from the weaker set of classifiers. The two categories of positive diseased image and negative non-diseased images are composed for learning process. Initially, the input training samples $(x_1, y_1, x_2, y_2 \ldots x_n, y_n)$ have taken for learning, set the weight for each samples w_t for each sample (x_1, y_1). In Eq. (4), train weak learner h_t using weight w_t. Identify the weak learner

$$\epsilon_t = \sum_{i=0}^{n} w_{t,i} \|h_t(x) - y_i\|^2 \tag{4}$$

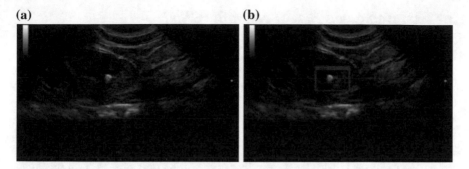

Fig. 3 Detection of kidney diseases using AdaBoost classifier, **a** input ultrasound image, **b** diseases detected image

Set threshold 0.5 if error rate is greater than threshold then update weights
Update weights by Eq. (5)

$$w_{t,i} = w_{t,i} \times \begin{cases} e^{-\alpha_t}, & \text{if } h_t(x_i) = y_i \\ e^{-\alpha_t}, & \text{if } h_t(x_i) \neq y_i \end{cases} \qquad (5)$$

where $w_{t,i}$ represents weight updates of weak learner for next iteration. Repeat the Eq. (5) until reaching the minimum error rate for the trained samples. Figure 3a denotes input ultrasound image to AdaBoost classifier and Fig. 3b denotes diseases detected image from the AdaBoost classifier. The final output of the strong classifier is given by Eq. (6)

$$H(x) = \begin{cases} 1, & \text{if } \sum_{t=1}^{\tau} \alpha_t h_t(x) \geq \frac{1}{2} \sum_{t=1}^{\tau} \alpha_t \\ -1, & \text{otherwise} \end{cases} \qquad (6)$$

3 Classification of Kidney Diseases

Deep learning technology [14, 15] is originated from conventional machine learning which have a significant role on medical imaging technology. It is a deep architecture possessing numerous of hidden layers, when compared to conventional layered architecture. Without any independent extraction method, it conceives features by processing large set of labeled data. In this work, CNN architecture is computed for classifying and recognizing different types of detected kidney diseases.

3.1 Convolutional Neural Network (CNN)

CNN [16] is a deep learning architecture rooted on feed forward network structure mainly used to investigate computer vision-oriented dilemmas. Architecture is designed with multi-layered model that has shift-invariant property. Relativity of multiple layers structured is adopted using human visual cortex. Three major tensors namely input layer, hidden layer, and output layer are composed for constructing CNN architecture. Input layer describes nature of image to be inputted, and it preprocesses input image into uniform height, width, and number of channels present in it. Second major layer describes convolution operation [17]. In this layer, a multitude uniform filter is convoluted to the input sample with user striding and padding. For this work, convolution layer is defined by Eq. (7)

$$D_{i,j,m} = \sum_{k=0}^{K-1}\sum_{p=0}^{H-1}\sum_{q=0}^{H-1} z_{i+p,j+q,k}^{l-1} h_{pq\,km} + b_{ijm} \tag{7}$$

where $D_{i,i,j}$ is the output of the convolution, I represents input image $i = 1\ldots m$ and $j = 1\ldots n$ represents row and column indexes of input image, $h_{pq\,km}$ represents weight applied to the convolution, and b_{ijm} represents bias applied to the convolution. Striding and padding are derived by (8)

$$\begin{array}{c} (W - 2[H/2]) \times (W - 2[H/2]) \\ [(W - 1/s) + 1] \times [(W - 1/s) + 1] \end{array} \tag{8}$$

where the W is the number of pixel and s is the stride

The above Fig. 4a represents values from the convolution layer, and Fig. 4b represents max-pooled values from given convolution layer. After extracting features from convolutional layer, the output maps are down sampled through pooling layer both (maximum and average values) and it is changed as one-dimensional vector using flattened layer. Flatten filter maps $2 \times 2 \times 64$ convoluted 2D vector to 256 flat vector dense layer is the final layer which connects output from the flattened layer with activation function. Each output node is connected by each input node Let w_k is weight function present in the kth node of flattened layer, and ReLU activation function is used to normalize the data in dense layer. In this work,

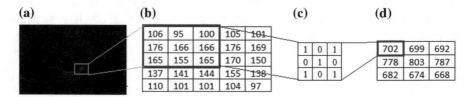

(a)   (b)   (c)   (d)

Fig. 4 Convolution, **a** input image, **b** input intensities for R component, **c** Filter, **d** Convolution output

multiclass diseases have classified using CNN. The trained model for the classification of CKD has done using prediction class of tensor flow model, and it takes trained data as input as NumPy arrays. In this work, multiple classes are used, so multiple inputs have mapped to NumPy arrays as dictionary. The class weights are assigned to indices of each sample during training processes. Probability distribution for each label is categorized using SoftMax activation function [18]. Classification results return prediction of NumPy arrays that indicating dependence class of samples SoftMax activation function is used for multiclass classification.

4 Experimental Results and Discussions

4.1 Detection of CKD Using AdaBoost Classifier

For training phases, ultrasound images have taken with dimension of 640 × 480 for detection, 60 images with different classes of kidney diseases are inputted to the feature extraction model. At first, 144-dimensional HOG features were extracted from the inputted samples. The extracted HOG features have used as input to the AdaBoost classifier for learning. For testing, 30 images with 640 × 480 dimension of hog features is given to the classifier. Based on the minimum error rated samples predicted by AdaBoost classifier, disease portion is drawn using bounding box.

4.2 Classification of CKD Using CNN

For training, detected portion of ultrasound is converted into grayscale and resized to 150 × 150 height and width, respectively, for classification. First convolution for input image has 32 number of filters with the size of 3 × 3, and it produces 896 param for activation function for this convolution is ReLU. Resultant two-dimensional vector from the convolution layer is max-pooled with 2 × 2 array matrix. Second convolution from the resultant data takes 32 number of filters with kernel size of 3 × 3 array, and it uses ReLU activation and produces 9248 params. Average pooling filter size for this convolution is 2 × 2. Third convolution onwards, the resultant data takes 64 number of filters with size of 3 × 3 array, and it also uses ReLU activation and produces 18,496 2D vector. The output from the third convolution 18,496 parameters is again convoluted, and resultant parameter is produced as 36,928 parameters. These 36,928 values are converted as single-dimensional vector using flattened layer. The final dense layer is the output layer for the network architecture. First dense layer processes nonlinear operation on flattened input vector and produces 16,448 parameters as the output. In this work, SoftMax activation function is used for multilevel diseases classification, it

Table 1 Performance analysis for classification of chronic kidney diseases (CKDs)

Classification Methods	Accuracy (%)	Precision (%)	Recall (%)	F-measure (%)
SIFT with SVM	82.60	73.80	73.80	73.80
SURF with SVM	84.30	83.25	83.25	79.63
CNN	85.21	93.70	93.70	93.70
DCNN	86.67	84.30	84.30	84.30

converts one-dimensional data as input from flattened layer. This CNN architecture combines feature learning (1, 19, 139 features from the kidney ultrasound images) and classification. Train the model using fit complied data from architecture. On training, 91% of accuracy is yielded. For testing, three categories of CKD images with the sizes of 150×150 have stacked to the batch. The prediction method with corresponding categories have identified through the TensorFlow prediction matrix and prediction matrix is showed by

$$\text{Prediction Matrix} = \begin{bmatrix} 1 & 0 & 0 \\ 0 & 1 & 0 \\ 0 & 0 & 1 \end{bmatrix} \begin{array}{l} \Rightarrow \quad \text{For kidney stones} \\ \Rightarrow \quad \text{For Cystic Kidney} \\ \Rightarrow \quad \text{For kidney Cancer} \end{array}$$

4.3 Performance Analysis

The performance analysis for the predicted matrix has calculated based on confusion matrix. For confusion matrix, true positive, true negative, false positive, and false negative, values are calculated from the prediction matrix on the validation of CKD. The performance measures have been calculated using following methods.

The above Table 1 depicts confusion matrix and performance metrics comparison between conventional and deep learning techniques for classification of CKDs.

5 Conclusion and Future Enhancement

In this work, automatic detection and classification of CKDs have done using both conventional and deep learning methods, for detection of kidney diseases, AdaBoost classifier is used and the portion of diseases is located by drawn a bounding box over it. The performance accuracy for detection of kidney diseases has predicted as 89.79%. For classification, convolutional neural network is used. Convolution layer directly observes the features form input images and also it has shift invariant output. Three convolutions with multiple number of filters have processed to the input images. The training stages are fine-tuned with different epochs. The final results for the classification of CKD are achieved and showed

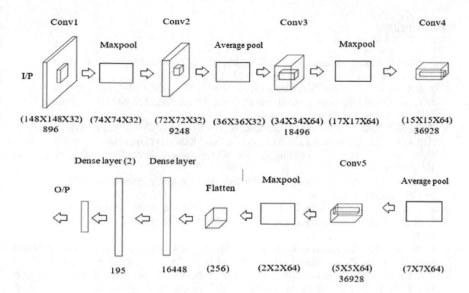

Fig. 5 Deep convolution neural network architecture

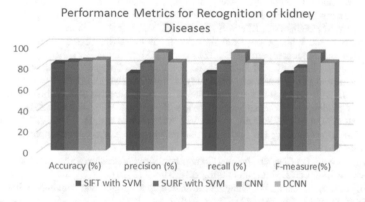

Fig. 6 Performance chart for detection and classification of kidney diseases

through the decision matrix. The performance accuracy for predicted diseases has given as 86.67% In future, conventional detection scheme may replaces current featuring techniques in order to yield better detection rates for CNN, improved network architecture with different hyperparameters will be worked out for better classification results. Figure 5 represents the architecture of proposed CNN model and Fig. 6 explores the comparative results between the conventional model and the deep learning model.

References

1. Gulati, M., Cheng, J., Loo, J.T., Skalski, M., Malhi, H., Duddalwar, V.: Pictorial review: Renal ultrasound. Clin. Imaging **51**, 133–154 (2018)
2. Karmazyn, B., Tawadros, A., Delaney, L.R., Marine, M.B., Cain, M.P., Rink, R.C., Jennings, S.G., Kaefer, M.: Ultrasound classification of solitary renal cysts in children. J. Pediatr. Urol. **11**(3), 149.e1–149.e6 (2015)
3. Calio, B.P., Lyshchik, A., Li, J., Stanczak, M., Shaw, C.M., Adamo, R., Liu, J.-B., Forsberg, F., Lallas, C.D., Trabulsi, E.J., Eisenbrey, J.R.: Long term surveillance of renal cell carcinoma recurrence following ablation using 2D and 3D contrast-enhanced ultrasound. Urology **121**, 189–196 (2018)
4. Nanni, L., Brahnam, S., Lumini, A.: Texture descriptors for representing feature vectors. Expert Syst. Appl. **122** (2019)
5. Lee, Y.-B., Choi, Y.-J., Kim, M.-H.: Boundary detection in carotid ultrasound images using dynamic programming and a directional Haar-like filter. Comput. Biol. Med. **40**(8), 687–697 (2010)
6. Diamantis, D.E., Iakovidis, D.K., Koulaouzidis, A.: Look-behind fully convolutional neural network for computer-aided endoscopy. Biomed. Signal Process. Control **49**, 192–201 (2019)
7. Arif-Tiwari, H., Kalb, B.T., Bisla, J.K., Martin, D.R.: Classification and diagnosis of cystic renal tumors: role of MR imaging versus contrast-enhanced ultrasound. Magn. Reson. Imaging Clin. N. Am. **27**(1), 33–44 (2019)
8. Singh, K., Ranade, S.K., Singh, C.: A hybrid algorithm for speckle noise reduction of ultrasound images. Comput. Methods Programs Biomed. **148**, 55–69 (2017)
9. Zhao, Y., Gong, L., Zhou, B., Huang, Y., Liu, C.: Detecting tomatoes in greenhouse scenes by combining AdaBoost classifier and colour analysis. Biosyst. Eng. **148**, 127–137 (2016)
10. Frid-Adar, M., Diamant, I., Klang, E., Amitai, M., Goldberger, J., Greenspan, H.: GAN-based synthetic medical image augmentation for increased CNN performance in liver lesion classification. Neurocomputing **321**, 321–331 (2018)
11. Saikia, A.R., Bora, K., Mahanta, L.B., Das, A.K.: Comparative assessment of CNN architectures for classification of breast FNAC images. Tissue Cell **57**, 8–14 (2019)
12. Joseph, J., Periyasamy, R.: A polynomial model for the adaptive computation of threshold of gradient modulus in 2D anisotropic diffusion filter. Optik **157**, 841–853 (2018)
13. Iwahori, Y., Hattori, A., Adachi, Y., Bhuyan, M.K., Woodham, R.J., Kasugai, K.: Automatic detection of polyp using Hessian Filter and HOG features. Procedia Comput. Sci. **60**, 730–739 (2015)
14. Yuan, Y., Xun, G., Suo, Q., Jia, K., Zhang, A.: Wave2Vec: deep representation learning for clinical temporal data. Neurocomputing **324**, 31–42 (2019)
15. Zhao, H., Liu, F., Zhang, H., Liang, Z.: Research on a learning rate with energy index in deep learning. Neural Netw. **110**, 225–231 (2019)
16. Li, Y., Pang, Y., Wang, J., Li, X.: Patient-specific ECG classification by deeper CNN from generic to dedicated. Neurocomputing **314**, 336–346 (2018)
17. Traore, B.B., Kamsu-Foguem, B., Tangara, F.: Deep convolution neural network for image recognition. Ecolog. Info. **48**, 257–268 (2018)
18. Vamplew, P., Dazeley, R., Foale, C.: Softmax exploration strategies for multiobjective reinforcement learning. Neurocomputing **263**, 74–86 (2017)

We Bring Your Identity: A Secure Online Passenger Identity Protocol (SOPIP) for Indian Railways Using Aadhaar Number

Sivakumar Thangavel, S. Gayathri and T. Anusha

Abstract Indian Railways offer instant ticket booking service to the passengers via online reservation portals (ORP). The major constraint to the passenger is to bring an original document as personal identity proof (PIP) at the time of travel to prove their identity. However, many passengers bring copy of the documents in order to avoid loss or theft of their original identity card during travel. But, it is not accepted as per the regulations of railway authority. Also, passengers may forget to bring the original ID card due to last-minute hurry. This leads to an unnecessary dispute between the passenger and the travelling ticket examiner (TTE). To address this issue, a secure online passenger identity protocol (SOPIP) is proposed in this paper. The SOPIP provides secure online identity service to authenticate the passenger to the TTE during travel. The proposed protocol also provides confidentiality, integrity and digital signature services to the personal details of the passenger during the transmission personnel details via the Internet. The protocol assumes that the ticket details are available in the ORP and the passenger's personal details are available in the Aadhaar database.

Keywords Indian railway · Original ID card · Aadhaar · Online authentication · Security protocol

List of Abbreviations

ORP   Online reservation portal
TTE   Travelling ticket examiner
RO   Regional office

S. Thangavel
Department of CSE, Dr. Mahalingam College of Engineering & Technology, Pollachi,
Tamilnadu 642003, India
e-mail: msg2sk2010@gmail.com

S. Gayathri (✉)
Oracle India Pvt. Ltd, Hyderabad 500084, India
e-mail: gayathri.selvaraaj@gmail.com

T. Anusha
Department of CSE, PSG College of Technology, Coimbatore, Tamilnadu 641004, India
e-mail: anu@cse.psgtech.ac.in

© Springer Nature Singapore Pte Ltd. 2020   745
K. S. Raju et al. (eds.), *Data Engineering and Communication Technology*,
Advances in Intelligent Systems and Computing 1079,
https://doi.org/10.1007/978-981-15-1097-7_63

CIDR   Central ID repository
PAN   Permanent account number
RSA   Rivest–Shamir–Adleman
DSA   Digital Signature Algorithm
ECDSA   Elliptic Curve Digital Signature Algorithm
PIP   Personal identity proof
PII   Personally identifiable information
CA   Certificate authority
RSP   Regional service provider
SSP   State service provider
NSP   National service provider
SDEx   Secure Data Exchange
C   Ciphertext
S   Signature

List of Notations

$C = E(K, M)$ Encryption of the message M using the key K
$M = D(K, C)$ Decryption of the ciphertext C using the key K
$H(M)$ Hash code of the message M using the hash function H()
PRx Private key of the user x
PUx Public key of the user x
$\|$ Concatenation

1   Introduction

Today, many online interactions necessitate disclosing personal identity information. This poses substantial challenges for both the people who disclose the
information and for those who receive it. Until now, this problem has been
addressed by a simplistic model in which a trusted third party or identity service
provider discloses identity information to a recipient, known as the relying party, on
behalf of a user. The identity service provider reduces risk and promotes efficiency
by confirming aspects of the exchange.

   The range of cybercrimes with the development and use of Internet is broad and
these can be generally classified to three types namely: (a) consumer threats:
identity theft, financial fraud, child endangerment, compromised computers being
used for unauthorised activities, (b) enterprise threats: stealing of financial data,
spying economic information, loss of personally identifiable information, extortion
via denial of service attacks and (c) government threats: information warfare and
cyber war. The four attributes that makes the Internet attractive to criminals are
global connectivity, anonymity, lack of traceability and valuable targets. A digital

identity is defined as attributes that are accessible by technical means. These attributes can be stored and retrieved by computer-based systems. E-mail address or a user name could be considered as a digital identity. Digital identity represents a set of unique digital characteristics that could establish the identity of the subject. It ensures that the subject is who or what it claims to be [1].

AADHAAR is a 12-digit unique number which the Unique Identification Authority of India (UIDAI) is issuing for all residents of India. The number will be stored in a centralised database known as central ID repository (CIDR) and linked to the basic demographics, such as name, date of birth, address, family details and biometric information such as photograph/face, fingerprints, palm prints and iris. Each Aadhaar number will be unique to an individual and will remain valid for life. It is a random number generated, devoid of any classification based on caste, creed, religion and geography. Once a person is on the database, the person will be able to establish his identity in online easily [2]. About 93% of adults in India have Aadhaar card [3]. The central ID repository (CIDR) is accessed by many government and private organisation due to the widespread applications of Aadhaar number. In order to implement the proposed protocol, it is necessary to link the ORP, which contains the ticket details of the passenger, with the Aadhaar database. Once the ticket reservation is confirmed, the personal information such as photograph and date of birth with ticket details of the passenger is transferred to the regional office. The TTE of that particular train will be receiving a list of passengers travelling along with their photograph and personal information coach wise in a handheld device available with him. TTE can check whether the photograph in the device matches with passengers face. If a match observed, then the passenger is authenticated.

The rest of the paper is organised as follows: Sect. 2 presents about UIDAI, identity systems, Aadhaar and the existing authentication protocols. The proposed protocol is discussed with illustrations in Sect. 3. The paper is concluded in Sect. 4.

2 Literature Survey and Existing Authentication Methods

In real world, identity is created based on social custom and followed by the creation of identity documents. A child's first identity document is his/her name at the time of birth. Further, the birth certificate will be created as the second identify. This document is later used to create additional public and private sector identity documents such as driving licence, bank account, passport, PAN card and Aadhaar number [4]. A national ID card can be any digital signature card which combines the digital identity of the user. It is essential to protect the information stored on the card, which enables an application to identify the card holder [5]. Digital signature cards emerge as a fundamental infrastructure element for e-voting due the growth of electronic voting system in real time. Digital signature card-based protocol for remote Internet voting and its requirements in terms of security are discussed in [5].

The impact of identity theft and fraud as a result of vulnerabilities in online security system is staggering. The biggest challenge to organisations is the high cost to implement and manage authentication. Personal privacy and the protection of personally identifiable information become more critical as almost everything about a user becomes accessible online [6]. Scope for developing digital signature signing servers is needed due to risk in integrity of digital assets by maliciously or intentionally. Hence, a common server for signing can be included in identity management system [7]. The two main social issues due to creation of an identity meta system are anonymity and privacy. Online identity systems could provide greater protections for security and privacy [4].

In electronic government where there is no complete trust between documents' sender and receiver, something more than authentication is needed. The most attractive solution to this problem is the digital signature which is analogous to the handwritten signature. The signature is formed by taking the hash of the message and encrypting the message with the creator's private key. It guarantees the source and integrity of the message [8]. In [8], the authors compared and analysed three main digital signature algorithms such as Rivest–Shamir–Adleman (RSA), Digital Signature Algorithm (DSA) and Elliptic Curve Digital Signature Algorithm (ECDSA) and concluded that RSA was found to be slower for signing and much faster for signature verification.

Digital documents are easy to generate, modify and manage. The easy modifiable property of digital document makes it more vulnerable to forgery. So, the challenge is to produce digital documents that are highly resistant to forgery and reliably confirms the real owner of the document. A biometric watermarking technique with secret key is introduced in [9] to digitise the authoritative documents issued by government/other organisations as a part of UID/Aadhaar card project of India using biometric watermarking.

Identity management has been a serious problem since the establishment of the Internet. Early identity management systems (IdMS) were designed to control access to resources and match capabilities with people in well-defined situations. With the advent of inter-organisational systems, social networks, e-commerce, m-commerce, service-oriented computing and automated agents, the characteristics of IdMS face numerous technical and social challenges [1]. The major challenges are (a) Security: computers and networks are susceptible to hacking. Once hackers obtain a digital identity, they can use it in the physical world, resulting in identity theft and (b) Privacy: The ability to map one's digital identity to a physical identity creates major privacy concerns. This is especially true due to the large number of databases containing private information such as health care, financial, and marketing [1]. Identity theft is one of the fastest growing crimes in America [10].

Online identity theft is being the primary reason for most of the cybercrime crimes that are happening today. Hence, there is a need for an efficient online identity system that provides unique identities to overcome online threats. These systems must provide suitable solution to speed, volume, security and accuracy issues [11]. Because of limited server and network capacities for streaming applications, multimedia proxies are commonly used to cache multimedia objects such that, by accessing nearby proxies, clients can enjoy a smaller start-up latency and receive a better quality-of-service guarantee [12]. Online identity management systems can be developed using Aadhaar unique identification number. Proxy server architecture could be utilised for providing solutions to security and speed [13].

2.1 Uidai

Due to economic globalisation, people's movement across the countries increased a lot, which increased security threat for the nation. So, there is a dire need for fool-proof mechanism to identify individuals correctly and securely. This led the Indian government to start the project Unique Identification Authority of India (UIDAI) to provide a unique identification number to each and every citizen of India.

In this paper, the verification of rail passenger's identity is done with the help of unique identification number in online. In the proposed protocol, the Aadhaar number is utilised to develop an online identity service system for authenticating the rail passengers. The major challenge is that the passenger identification details from the Aadhaar database (CIDR) must be communicated in a secure way. Hence, it is proposed to provide the security services like authentication, confidentiality, digital signature and integrity to the personal information of the passengers.

2.2 Existing Authentication Methods and Protocols

Authentication methods and protocols are capable of simply authenticating the connecting party or authenticating the connecting party as well as authenticating itself to the connecting party. Many of the existing protocols are specific to the domain or application and not generic.

Passwords are the widely used form of authentication and typically encrypted or hashed. To avoid the problems associated with password reuse, one-time passwords were developed. There are two types of one-time passwords, a challenge–response password and a password list. The challenge–response password responds with a challenge value after receiving a user identifier. The response is then calculated from either the response value or selected from a table based on the challenge [14, 15].

Confidentiality, integrity, authentication and digital signature services could be provided by public key cryptosystem. Both encryption and verification of signature are accomplished with the public key. Certification authority (CA) assures the integrity of a public key. SSL/TLS provides a secure method of communication for TCP connections [14–17].

Kerberos is used for user authentication and securing data exchange. The system connects to the Kerberos server while user logs into a system to retrieve a session key to be used between the user and the ticket granting service. The ticket expires after a set amount of time. The issue with Kerberos is its scalability [14, 15].

3 The Secure Online Passenger Identity Protocol (SOPIP)

In this section, the working principle of the proposed protocol is illustrated with sequence diagram, use case diagram and the network topology for real-time implementation.

3.1 Working Model of SOPIP

The major entities of the protocol include: passenger (P), online reservation portal (ORP), travelling ticket examiner (TTE), regional office (RO) and CIDR database. The sequence of communication between the entities is shown in Fig. 1 with the help of a sequence diagram.

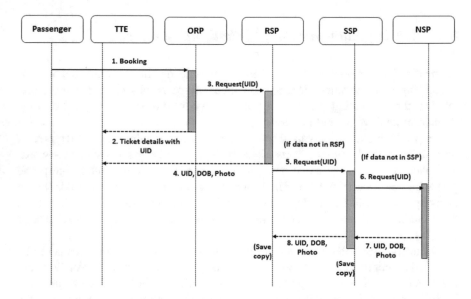

Fig. 1 Communication Sequence between the entities of SOPIP

1. Initially, the user logs into the ORP using the login credentials. Once the login is successful, user is permitted for booking the tickets. At the time of reservation, unique identification (UID) number must be provided by the user in addition to the details needed for booking tickets.
2. The ticket details of the passenger are sent from the online reservation portal (ORP) to the handheld device available with TTE.
3. The UID entered by the user is sent to the RSP in request of the user's date of birth (DOB) and photograph. If the detail corresponding to the UID is available in the RSP, then go to step 4 else go to step 5.
4. The RSP responds to the TTE of the particular coach/train with the requested passenger personal details (DOB and photograph).
5. The request is forwarded to the state service provider (SSP). If the details are available, then the response is sent to RSP as in step 8.
6. If the requested passenger details are not in the SSP database, then the request is forwarded to the national service provider (NSP), where the CIDR is located.
7. The CIDR response to the requested UID by sending the DOB and photograph corresponding to the UID. The copy of the details received from the NSP is stored in SSP for future reference and transferred to RSP.
8. The copy of the details received from SSP is stored in RSP for future reference and further the personal details are transferred to the TTE.
9. TTE can make use of this information to verify the passenger's identity and authenticate in online during travel.

In the above given communication sequence, step number 5–8 should be executed when a user reserve tickets first time under a particular regional office. For subsequent reservation, the personal details will be fetched from the RSP. When the same user reserves tickets in another region, the personal details will be fetched from the SSP and while booking tickets in another state, then the details will be fetched from the NSP.

3.2 Secure Message Exchanges Between Entities

The secure request message exchanges between the entities involved in the proposed protocol is shown in Fig. 2a. The portal requests for the passenger personal details (DOB, Photo) from the regional service provider (RSP). The UID sent to the RSP is encrypted with the public key of RSP to provide confidentiality service. In RSP, the encrypted UID is decrypted with the corresponding private key available with RSP. If the details are not available in RSP and SSP, then the request is forwarded to the NSP encrypted in the same way. Here, the asymmetric encryption algorithm RSA is utilised to provide confidentiality service to UID. For authentication, UID is encrypted with sender's private key and verified using the corresponding public key at the receiver end.

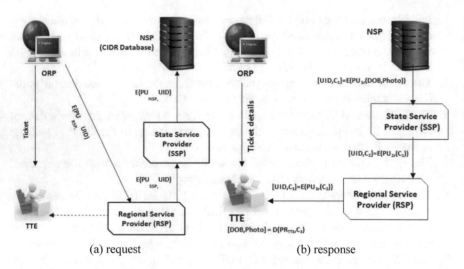

Fig. 2 Secure communications between the entities

Once the request reaches NSP, the details for the requested UID is fetched from the NSP (CIDR database) and the details are secured with multi-key RSA algorithm as shown in Fig. 2b. In multi-key RSA algorithm, the data are encrypted with multiple public keys and decrypted with only one corresponding private key.

The response from CIDR database, personal details of the passenger (DOB, Photo), is encrypted using the public key of NSP (PU_1). The copy of the encrypted details (C_1) with the corresponding UID is stored in the SSP database for future reference without decrypting. In SSP, the C_1 is again re-encrypted using public key of SSP (PU_2) and transferred to RSP. The encrypted data (C_2) received from SSP is stored in the RSP database for future reference. In RSP, C_2 is again re-encrypted using the public key of RSP (PU_2) and forwarded to the handheld device available with TTE. When TTE requests for the details of the passenger of the particular train, the encrypted data (C_3) received from RSP is decrypted using private key of TTE (PRTTE). Here, PRTTE is decryption key corresponding to the public keys PU_1, PU_2 and PU_3 available with NSP, SSP and RSP, respectively.

3.3 Secure Message Exchanges with All Possible Security Services

3.3.1 Confidentiality Service During Request Towards NSP

$$ORP \rightarrow RSP: \quad C = E(PU_{RSP}, \ UID); At RSP: UID = D(PR_{RSP}, C)$$
$$RSP \rightarrow SSP: \quad C = E(PU_{SSP}, UID); At SSP: UID = D(PR_{SSP}, C)$$
$$SSP \rightarrow NSP: \quad C = E(PU_{NSP}, UID); At NSP: UID = D(PR_{NSP}, C)$$

3.3.2 Confidentiality and Authentication Services During Request Towards NSP

$$\text{ORP} \rightarrow \text{RSP:} \quad C = E[\text{PR}_{\text{ORP}},[E(\text{PU}_{\text{RSP}}, \text{UID})]]$$
$$\text{AtRSP:} \text{UID} = D[\text{PR}_{\text{RSP}}, D(\text{PU}_{\text{ORP}}, C)]$$

$$\text{RSP} \rightarrow \text{SSP:} \quad C = E[\text{PR}_{\text{RSP}}, [E(\text{PU}_{\text{SSP}}, \text{UID})]]$$
$$\text{AtSSP:} \text{UID} = D[\text{PR}_{\text{SSP}}, D(\text{PU}_{\text{RSP}}, C)]$$

$$\text{SSP} \rightarrow \text{NSP:} \quad C = E[\text{PR}_{\text{SSP}}, [E(\text{PU}_{\text{NSP}}, \text{UID})]]$$
$$\text{AtNSP:} \text{UID} = D[\text{PR}_{\text{NSP}}, D(\text{PU}_{\text{SSP}}, C)]$$

3.3.3 Confidentiality, Authentication, Digital Signature and Integrity Services During Request Towards NSP

The services such as confidentiality and signature during request towards NSP are provided as shown in Fig. 3.

Where C is the encrypted UID to provide confidentiality and S is the signature to provide authentication and integrity services. At RSP, the signature verification is performed as shown in Fig. 4.

Similar operations are to be performed between RSP &SSP and between SSP & NSP with the appropriate private and public keys to provide confidentiality, authentication, signature and integrity services during request towards NSP.

3.3.4 Confidentiality Service During Response from NSP

$$\text{NSP} \rightarrow \text{SSP} \quad : \text{UID}\|C_1 \quad = E(\text{PU}_1, [\text{DOB, Photo}])$$
$$\text{SSP} \rightarrow \text{RSP} \quad : \text{UID}\|C_2 \quad = E(\text{PU}_2, C_1)$$
$$\text{RSP} \rightarrow \text{TTE} \quad : \text{UID}\|C_3 \quad = E(\text{PU}_3, C_2)$$

At TTE: (DOB, Photo) = $D(\text{PRTTE}, C_3)$, where PRTTE is the private key corresponding to the public keys PU_1, PU_2 and PU_3.

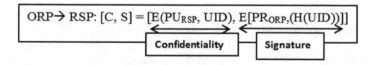

Fig. 3 Confidentiality and signature during request towards NSP

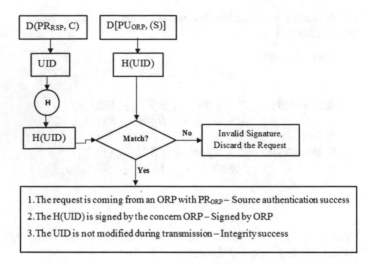

Fig. 4 Signature verification at RSP

3.3.5 Confidentiality, Authentication, Digital Signature and Integrity Services During Response from NSP to TTE

The confidentiality and digital signature services during response from NSP towards TTE are shown in Fig. 5.

At SSP, $H(UID)$ will be taken and this will be compared with the hash code obtained by decrypting the signature (S) using the public key (PUN_{SP}) of NSP. If a match found, then the source is authenticated and proceed further as shown in Fig. 6.

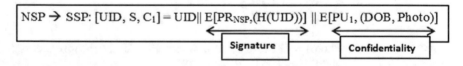

Fig. 5 Confidentiality and signature services during response from NSP

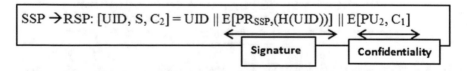

Fig. 6 Confidentiality and signature at SSP

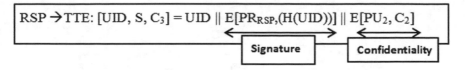

Fig. 7 Confidentiality and signature at RSP

At RSP, H(UID) will be taken and this will be compared with the hash code obtained by decrypting the signature S using the public key (PUSSP) of SSP. If a match found, then the source is authenticated and proceed further as shown in Fig. 7.

At TTE, H(UID) will be taken and this will be compared with the hash code obtained by decrypting the signature S using the public key (PU_{RSP}) of RSP. If a match found, then the source is authenticated and decrypt C_3 as given below.

$$At\ TTE: (DOB, Photo) = D[PR_{TTE}, C_3]$$

Thus, the Secure Online Passenger Identity Protocol (SOPIP) provides the essential security service between the various service providers such as NSP, SSP, RSP and TTE.

3.4 The Network Topology to Implement the Proposed Protocol

A suggested network topology to implement the proposed protocol in real time is presented in this section. The client/proxy/server architecture has a main server and many proxy servers. These proxy servers, in turn, have their appropriate clients or a proxy server again. A proxy server is a computer that sits between a client and server to handle requests and reply messages. The two most common needs for proxy server are to increase speed and reduce network traffic by caching data that are requested often. By doing so, the proxy server can deliver the request quickly, only polling the server when requested data are not available in the proxy. In this way, a proxy server not only speeds up data delivery, but also relieves server load [11–13].

Major Internet hubs and Internet service *providers* (ISPs) employ dozens of proxy servers. Thus, using proxy server, the overload on the main sever can be reduced and the speed of data transmission over network improves greatly. This serves the problem of scalability of the main server. In the proposed method, a network topology based on this architecture is suggested to implement the SOPIP protocol in real time. The suggested client-proxy-server network topology to implement the proposed protocol is shown in Fig. 8. It includes the entities such as:

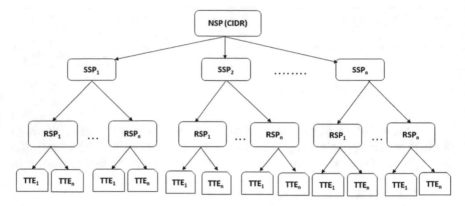

Fig. 8 Suggested network topology to implement SOPIP in real time

(a) National service provider (NSP): National (CIDR) server
(b) State service provider (SSP): Proxy server for each state
(c) Regional service provider (RSP): Proxy server for regional divisions
(d) Secure Data Exchange (SDEx): Provides the essential security services

In national service provider (NSP), the user details collected by the UIDAI project such as name, photograph, date of birth, family details, address and biometric attributes are stored and this is administrated by the Government authority. The state service provider (SSP) is the proxy server used to store, in encrypted format, the details necessary to identify and authenticate the passenger to the TTE. The RSP is level-2 proxy between the NSP and TTE to keep the data in encrypted format. Both SSP and RSP can be either administrated by the central government or railway authority. From the RSP, the details are transferred to the device available with the TTE and shown after decryption. The handheld device available with TTE should be administrated by the railway authority. The SDEx module is a part of NSP, SSP, RSP and TTE to provide the essential security services such as confidentiality, authentication, digital signature and integrity during transmission of data between the entities of SOPIP protocol.

4 Conclusion

The number cards maintained by a citizen are family card, voter ID, bank passbook, employee/school/college ID card, PAN card, Aadhaar card, passport, driving licence and many more. Avoiding the need of brining original identity card is a worthy initiative for digital India. Also, obtaining a duplicate card is time-consuming and hurdle process and many time it needs proof for loss/theft of original cards. Hence, this protocol is an initiative that would provide online identity service to authenticate the passenger to the TTE. It avoids the need of taking hard copy of the chart for the TTE and easy to track travelled/not-travelled passenger information. The proposed protocol also provides confidentiality, integrity and digital signature services to the personal details of the passenger during transmission of personnel details via the Internet. In real time, the process can be initiated before the scheduled departure time of the train to avoid delay in fetching the details. This protocol plays a significant role in real time across nation-wide.

Future Enhancements An application is being developed to experiment the proposed protocol. The notion of 'time' may be included as part of the messages exchanged during request and response to prevent reply and other possible attacks. The key management such as generation and distribution issues is to be addressed.

References

1. Hovav, A., Berger, R.: Tutorial: identity management systems and secured access control. Commun. Assoc. Info. Syst. **25** (2009)
2. Unique Identification authority of India website, www.uidai.gov.in
3. Firstpost Indian news and media website, www.Firstpost.com, 7th December 2017
4. Charney, S.: The evolution of online identity. IEEE Secur. Priv. **7**, 56–59 (2009)
5. Kofler, R., Krimmer, R., Prosser, A., Unger, M.K.: The role of digital signature cards in electronic voting. In: Proceedings of the IEEE 37th Hawaii International Conference on System Sciences, Big Island, USA, pp. 7 (2004)
6. Verizon, Online identity management needs a universal answer. White Paper (2012)
7. Romney, G. W., Parry, D. W.: A digital signature signing engine to protect the integrity of digital assets. In: International Conference on Information Technology Based Higher Education and Training (ITHET '06), Ultimo, NSW, Australia, IEEE, pp. 10–13 (2006)
8. Na, Z., Xi, X.G.: The application of a scheme of digital signature in electronic government. In: International Conference on Computer Science and Software Engineering, IEEE, Hubei, China, 3, pp. 618–621 (2008)
9. Anitha, V., Velusamy, R.L.: Authentication of digital documents using secret key biometric watermarking. Int. J. Commun. Netw. Sec. **1**(4), 5–11 (2012)
10. Social Security Administration, USA, SSA Publication No. 05-10064 (2016)
11. Narendra Kumar, M., Sivakumar, T.: Component based architecture for online identity management system. In: International Conference on Innovations in Engineering and Technology for Sustainable Development (IETSD-2012), Bannari Amman Institute of Technology, India. 3, pp. 28–32 (2012)
12. Yeung, S.F., Lui, J.C., Yau, D.K.: A multi-key secure multimedia proxy using asymmetric reversible parametric sequences: theory, design, and implementation. IEEE Trans. Multimedia **7**, 330–338 (2005)
13. Sivakumar,T., Anusha, T., Ummu Salma, A.: A novel approach for online identity management system using Aadhaar unique identification number. In: National Conference on Recent Trends in Information and Communication Technology (RTICT 2013), Bannari Amman Institute of Technology, India (2013)
14. Schneier, B.: Applied Cryptography: Protocols, Algorithms, and Source Code in C. Wiley, New Delhi (2008)

15. Stallings, W.: Cryptography and Network Security-Principles and Practice. Pearson Education, New Delhi (2013)
16. Zhao, X.M., Zhang, M.R.: Application of RSA digital signature technology in circulation of electronic official documents. Comput. Eng. Design **26**, 1214–1216 (2005)
17. Ylonen, T., and Lonvick, C.: Secure Shell (SSH) Protocol Architecture (2006)

Review of Decision Tree-Based Binary Classification Framework Using Robust 3D Image and Feature Selection for Malaria-Infected Erythrocyte Detection

Syed Azar Ali and S. Phani Kumar

Abstract We start with a famous proverb 'health is wealth.' Malaria is one of the most rapidly spreading and contagious diseases, mostly spread through microbes. Efficient treatment of the disease requires early and accurate estimation to ensure control from spreading and treatment in early phases. Accordingly, several studies have been put forward during the past decade. Analyzing the blood smear's images is one of the prominent works proposed in this context. This manuscript attempts to automate the process of diagnosis through machine learning techniques. The algorithm trains the model through different selected features of the input images and thereby uses the learning experience to classify the blood smears as disease prone or healthy. The cuckoo search algorithm is used for designing a heuristic scale, which is further assessed through multiple experiments to evaluate its accuracy. Different performance evaluation measures like precision, sensitivity, specificity, and accuracy are used to assess the robustness of the model toward early identification of malaria in the premature stage.

Keywords Malaria · Erythrocyte · Red blood cells (RBCs) · Machine learning (ML) approaches like SVM and bayesian · Blood smear · Principal component analysis (PCA) · K-means algorithm · Canonical correlation analysis (CCA)

S. Azar Ali (✉)
Asst. Prof., Department of information Technology, Muffakham Jah College of Engineering & Technology, Hyderabad, India
e-mail: syedazarali@gmail.com

S. Phani Kumar
Prof. & Head, Department of Computer Science & Engg., GITAM University, Hyderabad, India

© Springer Nature Singapore Pte Ltd. 2020   759
K. S. Raju et al. (eds.), *Data Engineering and Communication Technology*,
Advances in Intelligent Systems and Computing 1079,
https://doi.org/10.1007/978-981-15-1097-7_64

1 Overview

Diagnosing malaria at premature stage is a difficult task, which is a life-threatening disease mostly in Asia and Africa. Thus, detecting disease at premature stage improves chances of successful treatment. The researchers in [1] classified the malaria parasites that impact human beings into four categories-

Two protozoan parasite Species:

i. Plasmodium falciparum [2].
ii. Plasmodium vivax [2].

Two parasitic protozoa species:

i. Plasmodium oval [3].
ii. Plasmodium malaria [3].

Of the four types mentioned above, Plasmodium vivax [2] is observed to be the most influencing in hot and humid atmospheres [4]. Detecting the presence of this parasite in blood smears in premature stage is important for effective functioning of the prescribed drugs. The WHO guidelines [5, 6] advise practitioners to perform microscopic diagnosis of blood smears of those patients with likely symptoms to initiate early treatment of malaria. These tests enable identification of the parasite type and also estimate the count of their presence to determine the malaria severity. Alternatively, rapid tests are conducted in certain cases but these tests could not be used for premature detection and accordingly are not considered in this manuscript.

The diagnosis type adapted in this model is relatively easy and involves low costs. The process uses both thick and thin blood smears, which are used to determine the influence of parasite and the type of parasite respectively in the blood. In particular, to determine the scope of the disease in its premature stage requires analysis of the thick smears [6].

The diagnosis test results provide statistics of the parasite including its scope and type. Further, these outcomes are attested by a human expert with knowledge on the statistics. However, this exposes the malaria identification accuracy to vary with respect to the experience and knowledge of the individual attesting the results. In particular, the chances of accurate detection in the premature stages are often low. Accordingly, computer-aided efficient detection approaches are highly preferred. This manuscript proposes implementing machine learning techniques for detection purpose due to its significance among all computer-aided models. The image attributes generated from the microscopic images of erythrocytes (red blood cells) are vital for the training the model to differentiate infected and healthy cells in the premature stage.

Digital image-processing technique forms a crucial part of the model in optimal attribute selection stage. In specific, the edge-based segmentation is employed for obtaining benchmark attributes from the microscopic images [7–19]. Though the technique is highly efficient, it faces certain shortcomings including low contrast levels, unclear boundaries due to color similarity, irregular edges, and noise involved.

Accordingly, this paper put forward a machine learning approach developed on the principles of cuckoo search algorithm [20], which uses standard texture and morphological attributes. Further, the paper also presents solutions to handle certain challenges often faced in the process of implementing edge-based segmentation.

The next chapters in the paper present information on related literature (Sect. 2), suggested approach (Sect. 3), simulation results (Chap. 4) and conclusion and future research scope (Sect. 5).

2 Related Research

Computer-assisted diagnosis of malaria by assessing microscopic images of the blood smears has been an important research area over the past decade, attracting several researchers to develop new suggestions and improvements. Contemporary literature presents several related research studies, relying on supervised learning [21–23], decision support approaches [24], digital image analysis [24–26] and pattern detection [27, 28]. Further, the researchers in [29, 30] incorporated artificial NN algorithms for the diagnosis. Histogram equalizer was incorporated into [31], where segmentation differentiates overlapping cells.

Further, unsupervised approaches also are observed in the contemporary literature including the studies in [32–35]. ML techniques, image retrieval from content techniques [36, 37], and parasite prediction from the classified samples also are observed to be the prominent studies found in the literature.

Most of these standard approaches often face shortcomings of low contrast levels, same color intensity, irregular boundaries, noise levels, and normal regions of the processed images. Excluding the study in [9], rest of the approaches obtain signatures of affected images and incorporate them to detect the extent of parasites in the new image. This can result in high false alarm rates even if small changes occur in the signatures. Accordingly, machine learning approaches are expected to be superior in performance over the aforementioned approaches. However, efficient functioning of ML techniques requires a large number of images to be fed during the learning phase and also requires selecting the optimal attributes of the image.

The ML techniques presented in [9] utilize 94 attributes, and ML models SVM [38] and Bayesian approach [39] have been incorporated for learning and simulation stages. Optimal attributes have been chosen through the one-way-ANOVA [40] method. This approach showed 84% precision rates and was inconsistent for varying attribute count. Further, the attribute obtaining process in the approach involves image segmentation through marker-controlled watershed method [14], which is based on predicting gradients, which leads to under-segmentation. Accordingly, selection of optimal image attributes is not best.

To address these constraints, the model put forward in this manuscript suggests evolutionary computational ML techniques for selecting optimal features. The cuckoo search algorithm is employed for feature selection. Further, edge-based segmentation process is used for obtaining attributes from the considered blood smear images. This is due to the fact that the edge-based segmentation is regarded as the most efficient approach in contemporary studies [35].

3   Emerging Calculation Driven Scales for Predicting Malaria Span

The suggested scale is developed in the sequence of different phases. The phases included in our research work include obtaining existing images of blood smears for learning, preprocessing of input images, identifying optimal attributes based on existing studies, implementing decision tree techniques over these attributes to design scale for disease-free and infected blood samples.

A.   Representing Original Images as Grayscale Images

The techniques usable to represent a color (RGB) image to mono-color (grayscale) image include microscopic image to single channel (grayscale) image are perceptual luminance- preserving convertor [41], Luma convertor [42], green-channel convertor [42] and PCA convertor [43] (principal component analysis). Of these convertors, PCA convertor approach is strong and most desired fit for blood sample images [44–46]. Accordingly, this approach has been incorporated into our study to represent the color RGB images as grayscale images.

The highest contrast in post-conversion image is obtained through linear least-square technique through PCA model. This is executed by evaluating the primary axis of RBC color utilizing RGB coding system. The optimal suitable regression line is generated through PCA regression, which lowers the distance from the data point to axis on the image in the regression space. The pictorial depiction of this process is presented in Fig. 1 where the RGB image is fed to the model and required grayscale image is obtained as the outcome.

The angles between coordinates reflecting the red, green, and blue colors are identified initially and later the cosine of these angles are computed to convert the image into grayscale image data. The computation is made on the basis of the Eq. 1

$$\omega_{(x,y,z)} = \frac{\sum_{i=0}^{|P|}(r_i x + g_i y + b_i z)}{|x| + |y| + |z|} \cdots \tag{1}$$

From the Eq. (1), the output value $^\wedge(x, y, z)$ denotes the least the weights x, y, z applied to red, green, blue and $|P|$ denotes the volume of the image as the number of pixels and r_i, g_i, b_i denotes the R, G, B values of the corresponding ith pixel.

Fig. 1 Depiction of
converting colored image to
grayscale image on the basis
of PCA model

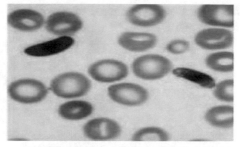

(a) Colored Image Fed to the Model

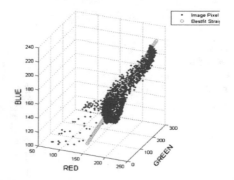

(b) PCA model based optimal fit coordinates graph

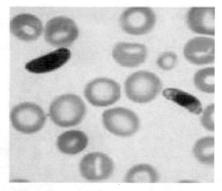

(c) Resulting Grayscale image generated by the model

B. Adjusting the Image Contrast Levels

Multiple diagnosis factors can impact the interpretation of images having low
contrast levels. Dull images pose an important challenge to categorize the blood
sample into malaria-prone or non-prone categories. Accordingly, enhancing illumination of the image is the first phase of the classification procedure. Driven by
the advantages of GE approach in enhancing image illumination levels [14, 47], this
manuscript opts for this contrast boosting technology.

Mathematically, the gamma equivalent to the given microscopic image $g_{(x,\ y)}$ is depicted through Eq. (2)

$$\left.\begin{array}{l} \text{diff}_{\min} = \left(g_{(x,y)} - g_{(x,y)}^{\min}\right)^{\gamma} \\ \text{diff}_{\min\leftrightarrow\min} = \left(g_{(x,y)}^{\min} - g_{(x,y)}^{\min}\right)^{\gamma} \\ f_{(x,y)} = g_{(x,y)}^{\max} * \left|\dfrac{\text{diff}_{\min}}{\text{diff}_{\max\leftrightarrow\min}}\right| \end{array}\right\} \tag{2}$$

The visualization of the fed low contrast image and the output image is achieved from GE ($\gamma = 0.5$) is presented in Fig. 2.

Fig. 2 Depiction of the functioning of the GE approach for contrast enhancement

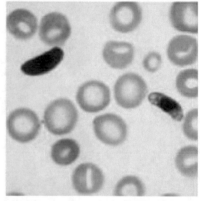

(a) Low contrast Image Fed to GE model

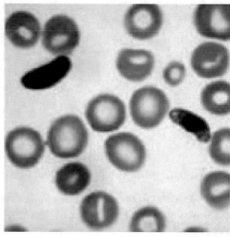

(b) Output image of the GE for ($\gamma = 0.5$)

C. Lowering Noise Levels

Resultant images from the above phases typically include salt & pepper noise. A typical median filter is applied for handling this noise type [48]. Another important noise observed in the process is super-imposed image noise [49]. The same median filter in combination with Gaussian filter [49] is considered as ideal for removing this type of noise. Accordingly, the manuscript used the combination of both these filters to reduce noise in the samples.

(1) *Identification of spectral peaks in pattern noise*

As discussed earlier, this research work uses spectral filtering to handle more pattern disturbances in the Fourier amplitude spectrum of the image including noises. The mathematical representation of the same is presented in Eq. (3)

$$A\left(c_{(k,l)}\right) = \sum_{i=0}^{L-1} \sum_{j=0}^{W-1} \left(a(i,j)^{*}e\right) \tag{3}$$

In the above equation, $a(i, j)$ denotes the pixel of the given image of size $L \times M$, whose Fourier spectrum is computed. The notation $A(c_{(k,l)})$ denotes the amplitude of the coefficient $c_{(k,l)}$. The illuminant dots depicted on amplitude spectrum projects as peaks. The notation e in Eq. (3) refers the Fourier error coefficient. As the median filter is important filter [50, 51] and is also able to identify impulses as noise [52], accordingly this peak identification approach utilizes median filter to identify peaks by presuming peaks as impulses. The same concept can be applied to our context in Fourier amplitude spectrum.

Further, the model adjusts both coefficients being concerncd along with the useful amplitude coefficients in the window; in regard to this, the model relies on Gaussian filter.

The Gaussian filter is implemented to identify Gaussian surface, which is coated through two connected peaks. The procedure overhead is observed to utilize this surface for filtering. In case the signal consists of different sets of noise peaks, the procedure overhead is much large.

Accordingly, both the median and Gaussian filters of different volumes are applied to best possible reduction of noise. The median filters limit the area surrounding the peaks and later the Gaussian filter functions over this area to generate optimal image (Fig. 3).

D. Edge Identification

This phase of edge identification attempts to reform the borders of the RBCs through Canny-based filters [53]. These filters are predominantly implemented to save the image continuous edges. Establishment of boundaries of the input image is smoothed first using the median filter with $q \times q$ pixels as mask. This image is noise-reduced image obtained from the previous stage. As malaria-prone section of the RBCs is depicted through the darkest part, the edges are the boundaries of the

Fig. 3 Blood sample images
fed into the proposed
combined model and resultant
images

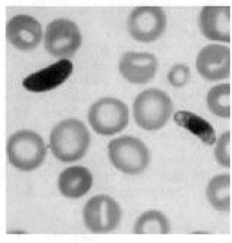

(a) Grayscale image fed to the model

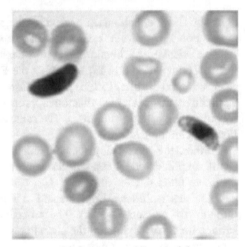

(b) Outcome of the model

darkest parts as can be observed in the figure below. Accordingly, the filters are able
to detect the thickness of the estimated RBC (Fig. 4).

E. RBCs Categorization Through K-Means Clustering Approach

The preprocessing phases as discussed in subchapters 4.1, 4.2, and 4.3 results in
grayscale images that are in the next levels fed as input to categorization phase
using the K-means clustering program with K value of 2. This model is believed to
optimal due to the fact that pixels detected in the input image are fall into the
category of malaria-prone RBCs and normal RBCs as depicted in Fig. 5.

The conventional and easiest clustering approach termed as K-means [54] has
been implemented to group the considered image database. Let database Z be

Fig. 4 Edge identification
over blood sample deploying
Canny edge model

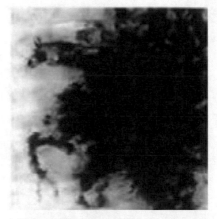

(a) Image Fed into the Edge Identification
model

(b) Output image of the RBC with appropriate
edge identification

grouped in d dimension area as k groups. In the current scenario, the number of groups is considered as 2 ($k = 2$). First, the healthy and affected RBCs are utilized to build prototypes so that healthy RBCs denote the corresponding group and affected ones denote the related group. Later, all the elements of database Z are grouped into the related group, on the basis of the closest prototype. In addition, the optimal prototype of every group is decided and if prototypes of either of the groups is found to be dissimilar from the previously decide prototype of the corresponding group, then the database Z will be regrouped based on the prototypes in the group. This process is iterated until no modification in prototypes of all groups is found and upon successful determination of no changes, groups with corresponding elements are concluded. The distance function that used by K-means detects the

Fig. 5 Images fed into
K-means algorithm and
outcome from the model

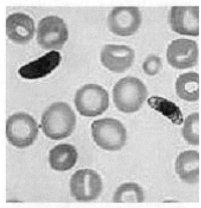

(a) Image fed into K-Means cluster model

(b) Outcome of K-cluster model which depicts
two affected groups-white (infected) and black
(healthy)

distance between each entry of the given dataset D against the cluster centroids as depicted below: Eq. (4)

$$\overset{|D|}{\underset{ji=1}{\forall}} \left\{ f(x) = \max\left(\overset{|c|}{\underset{j=1}{\forall}} \left\{ \frac{|e_i \cap c_j|}{|c_i|} \right\} \right) \right\} \tag{4}$$

Intersecting of the elements exists in each entry e_i of the dataset D, and each centroid c_j listed in set C should be found first, further, the ratio of these common elements against the count of elements exists in centroid is considered as the similarity $\frac{|e_i \cap c_j|}{|c_i|}$.

The max similarity of a dataset entry against all of the centroids is considered fittest and moves that entry to the corresponding cluster represented by the respective centroid. Sequences involved in execution of the program are depicted below:

I. Choose k-elements into the area denoted by points/objects, which are to be grouped. These elements denote first cluster prototypes.

II. Allocate each point o the corresponding cluster, which has nearest matching prototype

III. Upon successful completion of allocation of all points, recomputed the locations of k-prototypes

IV. Iterate sequence II and sequence III till such point that the values of both the prototypes are unchanged. The outcome is the optimal classification of points into clusters, from which the parameter for minimization is computed.

Theoretically, the iteration stops at some specific point of time but in practice, the K-means clustering does not guarantee detection of best configuration related to determination of universal minimum of the function. The program is also highly variable to first choose group centers. Accordingly, execution of K-means algorithm for grouping the selected images is as below:

(a) Groups to be selected must be 2, due to the infected RBCs are the darkest spaces.

(b) As the fed image is grayscale image, the pixels are separated based on the pixel intensity.

(c) Accordingly, both the groups are built by evaluating all the pixels if they possess the darkest space or not.

To execute this, the primary centroids of the group 1 and 2 are pixels chosen from the most shaded portions of the image and the rest of the portions of the image in the corresponding order.

Further process augments the malaria causing virus generated black and white image produced through FCM approach. Afterward, a morphological process, erosion was implemented to nullify certain disturbances in the image. Finally, hole-filling technique was engaged to fill the holes to enable the data ready for classification process.

Most of the times, the output grayscale images of the malaria evinced RBCs include extra spread areas around the infected RBCs. These extra spread areas must be removed and for accomplishing this, the morphological binary removal process [14] is employed. The structuring component $s_{(x, y)}$ employed to given input image $f_{(x, y)}$ yields resulting the image $g_{(x, y)}$

In the current research work, a 3×3 square STREL is chosen for the removal function. Numerous structuring components of different volumes, a $3 \times 3, 5 \times 5$ and 7×7 have been evaluated. The following series of such evaluations, we found that 3×3 square SE recorded the highest performance among the three sizes.

(2) *Filling Holes identified in grouped RBC regions*

Malaria influenced blood smears depicted through K-means clustering, often evinces the holes in the nucleus of the RBCs. These holes must be shaded to ensure that the model provides accurate classification of affected RBCs for moving on to next stages [55]. The filling image $h_{(x, y)}$ marks as dark in all regions of the given image $f_{(x, y)}$ that excludes the boundaries, where it will complement the image $f_{(x, y)}$ evincing the holes. The resultant image $g_{(x, y)}$ reflects as the original $f_{(x, y)}$ with no holes (see Fig. 6)

F. Obtaining Desired Attributes

The attributes in the scenario of classifying the texture along with morphological structure of the corresponding grayscale images are well researched and presented in the previous studies [7–18] are some of the prominent works focusing on the context.

(3) *Entropies*

Entropy is defined as the extent of ambiguity observed in any context. With respect to sensitiveness of the unpredictability toward distinguishing malaria-prone and healthy RBCs, the proposed approach takes into account all existing entropies published in existing literature [15, 16]. Further, to calculate the efficiency of these entropies, we constructed the normalized histograms. Five most prominent and suitable ones are used in this manuscript.

(4) *GLCM characteristics*

The structure data exploited through such attributes is significant [8]. The total 19 attributes associated with power, entropy, variation, and extent of correlation [27, 28] can be presented through GLCM characteristics [8]. The GLCMs are $n \times n$ matrices, which denote n different dark colors observed in the input image. Moreover, the GLCM is utilized to understand the design parameters.

(5) *Gray-level harmony matrix-driven textural characteristics* [8]

These attributes enable us to understand the coarse formation of the grayscale image. Consider the matrix $SM_{(i, j)}$ denotes the continuous occurrence of the grayscale value i of the given image $g_{(x, y)}$, which is in sequence of length j. Further, this matrix is in use to define all possible texture features [8, 10, 11].

(6) *Fractal dimension*

The texture formation of any image is often analyzed through the fractal dimension [18]. So as to detect the malaria RBC through the texture formation, the grayscale image is regarded as the extra- d of the 2D image. In addition, the fluctuations in this 3D confirm the formation of the malaria RBC. The model implemented to detect these fluctuations is termed as MDBC with sequential programming [9, 12].

Fig. 6 Images fed to the
CCA model and the resultant
outcomes

(a) Image fed to 'Connected-Component'
Phase

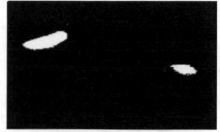

(b) Outcome image after deletion of misgrouped
RBCs

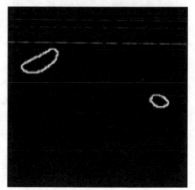

(c) Output image of 'Connected-Component'
Phase

(7) *LBP structure*

The surface characteristics which depict the likeness of local nearby areas of the
image are denoted as LBP [13, 17]. Consider P as the group of pixels present in the

circle having radius R of the RBC image I and gp_c denotes the gray value of the pixel p_c at the nucleus of R, and the gray value of each near around pixel p_i of p_c denotes as gp_i. In regard to this, each pixel p_i projects as 0 or 1, which is in accordance with the gp_c.

(8) *Morphological characteristics*

The characteristics put forward in [14] and characteristics of non-variant moments [7, 56] are considered as Morphological characteristics that are highly useful for demonstrating inconsistent RBCs detection. These characteristics are based on analyzing the variations in shape and volume of the healthy and malaria-prone RBCs.

G. Choosing Attributes

- Let the set N represents the features of normal erythrocytes depicted as a cluster during K-means algorithm, and the set M represents the features of malaria-prone erythrocytes that also depicted as a cluster in K-means process.
- Apply OFS-Z [57] to select optimal features in regard to infected and healthy blood smears as follows:

    - A feature said to be optimal if that feature coverage for both infected and normal erythrocytes are highly covariant. If feature challenges to this property such that coverage for both infected and normal erythrocytes is not diverse, then that feature is discarded. The feature optimization is done as follows:
    - The values observed for all considered features in corresponding normal and infected erythrocytes are represented in matrix format of the respective order.
    - The corresponding matrix of the normal erythrocytes are represented as follows:
    - Each row is the values observed for all of the considered texture and morphological features corresponding to a normal erythrocyte.
    - Similarly, matrix will be built for infected erythrocytes, such that each row represents the values observed for all the features considered in respective of infected erythrocyte.
    - Further, the significance of each feature is explored by estimating the Z-score between the values observed for corresponding feature in respective to normal and infected erythrocytes. If Z-score found to be significant at given degree of probability threshold, then the respective feature will be considered as optimal.

- Z-Score

The Z-score that denotes the given two vectors are distinct or not is assessed as follows:

- Let mo_f represents the frequency that observed for feature f in set M
- Let no_f represents the frequency that observed for feature f in set N
- Let mo_f represents the set of records, which does not contain feature f in set M
- Let no_f represents the set of records, which does not contain feature f in set N

Then, the Z-score Z of feature f can be assessed as follows: Eq. (5)

$$\left. \begin{array}{l} cr = mo - (mo + mo) * \frac{(mo_f + no_f)}{mo_f + no_f + mo_f + no_f} \\ dv = \sqrt{(mo_f + mo_f) * (mo_f + no_f) * (1 - mo_f + no)_f)} \\ Zscore = \frac{cr_f}{dv_{(f)}} \end{array} \right\} \tag{5}$$

Further, find the degree of probability (p-value) of the Z-score Z (f) from the Z-table. If degree of probability is less than the given degree of probability threshold, then the feature f is optimal.

4 Binary Classification by Decision Trees

A. Training Phase: Formatting Decision Tree

This section explores the training phase of the binary classification process that carried through decision tree. The training phase builds the branches and links appropriate entries as nodes. The depicted branches are in hierarchy, where the feature patterns with max size n that selected as optimal feature will be connected to the branch as nodes that reside at 1st level of the tree hierarchy. Similarly, the feature sets of size $n–1$ that selected as the optimal features will be connected to branch exists at second level of the hierarchy, such that the feature sets of size $n–i$ will be connected to branch that exists at $(i + 1)$th level of the hierarchy. The feature sets of size 2 (the minimum size) will be connected to branch that exists at the last level of the hierarchy.

The tree hierarchy is formed such that the first level of the hierarchy contains a branch that represents the nodes formed by the features of size n, second level of the hierarchy contains branches that represent the nodes formed by the features of size $n–1$, and mth level of the hierarchy contains branches that represent the nodes formed by the features of size $n–m–1$. Such that, last level of the hierarchy contains branches that represent the nodes formed from the feature sets of size 2.

Hence, the depicted tree hierarchy contains the branches, the mth level of the hierarchy contains branches such that each branch represents feature patterns of size $n–(m–1)$, here n is the max length of the feature pattern depicted, such that first branch of this hierarchy represents the feature patterns of size $n–(m–1)$. According to the description given about the branch formation, in regard to represent the feature patterns of the normal blood smears, and infected erythrocytes, the depicted model builds two branch hierarchies *IH*, *NH* corresponding to the malicious, and benevolent feature patterns.

Upon building the tree hierarchies with appropriate feature patterns connected as nodes in the respective branches, the classification will be initiated to assess the fitness of the given record to predict the malaria scope that explored in the following section.

B. Classifying

The fitness of the given record estimates based on the number of compatible branches noticed in respective hierarchies *IH*, *NH*. Concerning this, for each branch, any egg of the respective branch is identical to the values observed in given record for the feature patterns in corresponding branch representative set, then the fitness of the given record in related to corresponding hierarchy will increment by the 1.

This practice delivers the fitness related to malicious and benevolent state of the given record. Further, the fitness ratio of the given record about to both hierarchies will measure, which is the average of the fitness related to number of corresponding hierarchies. Then, the root-mean-square distance of the fitness values corresponding to both hierarchies should measure. Then, these fitness ratios and root-mean-square distances corresponding to both hierarchies will use to confirm the state of the given record is prone to coronary vascular disease or not that explored in the following section. The mathematical model to assess the fitness follows:

step 1: Let R be the blood smear to be labeled as infected or normal

step 2: Let eR be the set of possible feature patterns projected from the given blood smear R.

step 3: $mf = \sum_{h=1}^{|IH|} \sum_{j=1}^{|IH_H|} \sum_{i=1}^{|eR|} \left\{ \begin{matrix} 1 \exists e \in br & \wedge\, br & \in IH \\ i & \{h,j\} & \{h,j\} \end{matrix} \right\}$ //add 1 to fitness mf

of the given record R related to malicious scope, if feature pattern e_i is found as node to the branch $br_{\{h,\,j\}}$ in hierarchy *IH*.

step 4: ihnc $= \int_{h=1}^{H|} |IH_h|$ //number of branches in *IH*

step 5: $\left\langle mf = \frac{mf}{\text{ihnc}} \right\rangle$ //Finding the fitness ratio $\langle mf \rangle$ of the given record in related to malicious scope

step 6: $d_{mf} = \sum_{i=1}^{|mf|} \underbrace{\left\{ \sqrt{\left(1 - \langle mf \rangle^2\right)} \right\} + \langle mf \rangle^*(\text{ihnc} - mf)}_{\text{ihnc}}$ The sum of absolute dif-

ference of the fitness ratio depicted, and max fitness in regard to each branch (which is 1) for number of branches having nodes compatible to the feature patterns exist in eR and the fitness ratio multiplies by the number of incompatible branches, which is the difference between total number of branches and number of compatible branches that denoted as ihnc–*mf*

step 7: $nf = \sum_{h=1}^{|NH|} \sum_{j=1}^{|NH_h|} \sum_{i=1}^{|eR|} \{1 \exists e_i \in br_{\{h,j\}} \wedge br_{\{h,j\}} \in NH\}$ //add 1 to normal scope *nf* related to hierarchy *NH* if feature pattern e_i is compatible to any of the node connected to branch n_i in hierarchy *NH*.

step 8: nhnc $= \sum_{h=1}^{|NH|} |NH_h|$ // number of branches in *NH*

step 9: $\langle nf \rangle = \frac{nf}{\text{nhnc}}$ //Finding the fitness ratio $\langle nf \rangle$ of given record *R*, in related to normal scope

step 10: $d_{nf} = \dfrac{\sum_{i=1}^{nf} \left\{ \sqrt{(1-\langle nf \rangle)^2} \right\} + \langle nf \rangle^* (\text{nhnc} - nf)}{\text{nhnc}}$ //finding the root-mean-square distance of the fitness in regard to normal scope using the similar process defined in step 6

C. Discovering the record state

The fitness ratios $\langle mf \rangle$, $\langle nf \rangle$ and root-mean-square distances d_{mf}, d_{nf} obtained for given input record *R* should use to label the record *R* is prone to disease or normal. The label should define using the conditional flow that follows:

step 1: if $(\langle mf \rangle \cong \langle nf \rangle)$ Begin
step 2: if $(d_{mf} < d_{nf})$ Begin
step 3: Label the record as infected
step 4: End //of step 2
step 5: Else if $(d_{mf} < d_{nf})$ Begin
step 6: Label the record as normal
step 7: End //of step 5
step 8: Else //of condition in step 5
step 9: Record state is ambiguous//since the fitness ratios and root-mean-square distance obtained for both hierarchies is same
step 10: End //of step 1
step 11: Else Begin //of condition in step 1
step 12: if $(\langle mf \rangle \langle nf \rangle)$ Begin
step 13: Label the record as infected
step 14: End //of step 11
step 15: Else if $(\langle mf \rangle \langle nf \rangle)$ Begin
step 16: Label the record as normal
step 17: End //of step 14
step 18: Else Begin //of condition in step 15
step 19: Record state is ambiguous //since the fitness ratios and root mean square distance obtained for both hierarchies are not meeting the prescribed conditions
step 20: End //of step 18
step 21: End //step 11

5 Simulation Phase and Evaluation of Outcomes

A. The Corpus

In order to perform the experimental study, the labeled (malaria-prone, normal) microscopic images from Ma Mic [58], and bio-Sig data [59] databases were collected under statistical guidelines [60]. The total samples are 1600, and among these only 1127 samples were used. The rest of the samples were ignored based on microscopic visible factors.

The data statistics of the samples used in experiments are depicted in Table 1.

B. The Experimental Setup

The experiments conducted on i5-generation Intel processor with 4 GB ram and windows operating system. The implementation of the proposed model carried out using FIJI [61], which is a medical image-processing tool. The overall process includes preprocessing, segmenting, feature extraction, feature optimization, and binary classification implemented using relevant Java APIs provided in FIJI.

C. Performance Analysis

The experiments were conducted for proposed model DTBC and another contemporary model called SEMPS [62], which is using the dataset and experimental setup stated in Section A and Section B.

The results obtained for SEMPS and DTBC are explored in Table 2.

The experimental outcomes are presented in Table 2. The precision that denotes the infected erythrocyte detection value is significantly high as it evinced 97% (0.966), whereas the precision observed for other contemporary model SEMPS is 94% (0.943). The sensitivity that denotes the right detection rate of infected erythorcytes is observed for DTBC is 93% (0.931), whereas the counterpart model SEMPS evinced 88% (0.875). The value depicted for metric specificity that denotes the right detection rate of normal erythrocytes are 91% (0.912), and 86% (0.857) from corresponding proposed model DTBC, and counterpart SEMPS. The missing rate that denotes the failure rate of detecting the infected erythrocytes evinced for DTBC, and SEMPS in respective order are 7% (0.07), and 13% (0.126). According to the metric values depicted from experimental study that carried on both proposed model DTBC and SEMPS, it is obvious to confirm the significance of the DTBC toward infected erythrocyte detection from microscopic images of the blood smears. Process complexity of found to be low in DTBC that compared to SEMPS. This is since, the proposed DTBC is using 11 features to build the decision tree, also

Table 1 Dataset statistics

	Normal	Infected	Total
Training	213	576	789
Testing	91	247	338
Total	304	823	1127

Table 2 Comparison of performance metric values obtained from SEMPS and proposed model

Test statistics	DTBC features count: 11	SEMPS features count: 17
Positives	247	247
Negatives	91	91
TP	230	216
FP	8	13
TN	83	78
FN	17	31
Precision	0.966387	0.943231
Sensitivity	0.931174	0.874494
Specificity	0.912088	0.857143
Accuracy	0.926036	0.869822
Missing rate	0.068826	0.125506

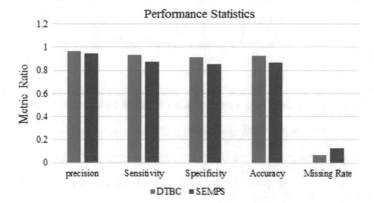

Fig. 7 Comparison of the values depicted for statistical assessment metrics from DTBC, and SEMPS

performs ordered search to assess the class label of the given erythrocyte. In contrast to this, the contemporary model is using 17 features to train the model and performs random search on nest hierarchy. The values depicted for statistical metrics from experiments carried on DTBC, and SEMPS are also visualized in a column chart the depicted below (see Fig. 7).

6 Conclusion

This research paper proposed the ECHS scale for assessing the extent of parasites in the input blood smear images. The complete procedure is executed in different stages, which include preprocessing of images to obtain attributes, segmentation, attribute generation, choosing optimal ones and ECHS definition.

Contrary to standard ML approaches, which use SVM and Bayesian approaches for training and testing, the suggested approach determined identification accuracy in proportion to optimal attribute count falling within different hamming distance thresholds. Simulation results also confirmed that the proposed approach involved linear process overheads and significant, stable estimation accuracy rate.

The research findings of the study encourage researchers to conduct in-depth research in different aspects like establishing the correlation among different attributes, possible impact on scale definition and incorporating emerging approaches like GA for optimal attribute selection.

References

1. Rougemont, M., et al.: Detection of four plasmodium species in blood from humans by 18S rRNA gene subunit-based and species-specific real-time PCR assays. J. Clin. Microbiol. 5636–5643 (2004)
2. Florens, L., et al.: A proteomic view of the plasmodium falciparum life cycle. Nature **419** (6906), 520–526 (2002)
3. Pain, A., et al.: The genome of the simian and human malaria parasite Plasmodium knowlesi. Nature **455**(7214), 799–803 (2008)
4. Snow, R.W., et al.: The global distribution of clinical episodes of Plasmodium falciparum malaria. Nature **434**, 214–217 (2005)
5. World Health Organization: Guidelines for the treatment of malaria, World Health Organization, pp. 9–12. Switzerland, Geneva (2010)
6. Reyburn, H.": New WHO guidelines for the treatment of malaria. c2637 (2010)
7. Hu, Ming-Kuei: Visual pattern recognition by moment invariants. Info. Theory IRE Trans. **8** (2), 179–187 (1962)
8. Galloway, M.M.: Texture classification using gray level run length. Comput. Graph. Image Process **4**(2), 172–179 (1975)
9. Mandelbrot, B.B.: The fractal geometry of nature/Revised and enlarged edition, 495 p. WH Freeman and Co., New York (1983)
10. Chu, A., Sehgal, C.M., Greenleaf, J.F.: Use of gray value distribution of run lengths for texture analysis. Pattern Recogn. Lett. **11**(6), 415–419 (1990)
11. Dasarathy, Belur V., Holder, Edwin B.: Image characterizations based on joint gray level-run length distributions. Pattern Recogn. Lett. **12**(8), 497–502 (1991)
12. Sarkar, N., Chaudhuri, B.B.: An efficient differential box-counting approach to compute fractal dimension of image. IEEE Trans. Syst. Man Cybernetics **24**(1), 115–120 (1994)
13. Ojala, Timo, Pietikaeinen, Matti, Maeenpaeae, Topi: Multiresolution gray-scale and rotation invariant texture classification with local binary patterns. IEEE Trans. Pattern Anal. Mach. Intell. **24**(7), 971–987 (2002)
14. Gonzalez, R.C., Richard E.W.: Processing (2002)

15. Pharwaha, A.P.S., Singh, B.: Shannon and non-shannon measures of entropy for statistical texture feature extraction in digitized mammograms. In: Proceedings of the World Congress on Engineering and Computer Science, vol. 2 (2009)

16. Ghosh, M., Das, D., Chakraborty, C.: Entropy based divergence for leukocyte image segmentation. In: 2010 International Conference on Systems in Medicine and Biology

17. Krishnan, M., Muthu, R., et al.: Textural characterization of histopathological images for oral sub-mucous fibrosis detection. Tissue Cell **5.43**, 318–330 (2011)

18. Krishnan, M., Muthu, R., et al.: Statistical analysis of textural features for improved classification of oral histopathological images. J. Med. Syst. **2.36**, 865–881 (2012)

19. Celebi, M.E., et al.: An improved objective evaluation measure for border detection in dermoscopy images. Skin Res. Technol. Offic. J. Int. Soc. Bioeng. Skin (ISBS); Int. Soc. Digital Imaging Skin (ISDIS); Int. Soc. Skin Imaging (ISSI) **15.4**, 444 (2009)

20. Yang, X.-S.: Trumpinton Street, and Suash Deb. Cuckoo Search via Lévy Flights. arXiv preprint arXiv:1003.1594 (2010)

21. Ross, N.E., et al.: Automated image processing method for the diagnosis and classification of malaria on thin blood smears. Med. Biol. Eng. Comput. **44.5**, 427–436 (2006)

22. Kaewkamnerd, S., et al.: An automatic device for detection and classification of malaria parasite species in thick blood film. BMC Bioinform. **13**, Supple 17 (2012)

23. Díaz, G., González, F.A., Romero, E.: A semi-automatic method for quantification and classification of erythrocytes infected with malaria parasites in microscopic images. J. Biomed. Inform. **42**(2), 296–307 (2009)

24. Lai, C.H., et al.: A protozoan parasite extraction scheme for digital microscopic images. Comput. Med. Imaging Graphics Official J. Comput. Med. Imaging Soc. **34**(2), 122 (2010)

25. Le, M.T.: A novel semi-automatic image processing approach to determine Plasmodium falciparum parasitemia in giemsa-stained thin blood smears. BMC Cell Biol. **9**(1), 15 (2008)

26. Díaz, G., Gonzalez, F., Romero, E.: Infected cell identification in thin blood images based on color pixel classification: comparison and analysis. In: Proceedings of the Congress on pattern recognition 12th Iberoamerican conference on Progress in pattern recognition, image analysis and applications. Springer-Verlag (2007)

27. Tek, F.B., Dempster, A.G., Kale, I.: Computer vision for microscopy diagnosis of malaria. Malaria J. **8**, 153–153 (2009)

28. Tek, F.B., Dempster, A.G., Kale, I.: Parasite detection and identification for automated thin blood film malaria diagnosis. Comput. Vision Image Underst. **1.114**, 21–32 (2010)

29. Memeu, D.M., et al.: Detection of Plasmodium Parasites from Images of Thin Blood Smears (2013)

30. Yunda, L., Alarcón, A., Millán, J.: Automated image analysis method for p-vivax malaria parasite detection in thick film blood images. Sistemas Telemática **10**(20), 9–25 (2012)

31. Sio, S.W., et al.: MalariaCount: an image analysis-based program for the accurate determination of parasitemia. J. Microbiol. Methods **68**(1), 11–18 (2007)

32. Tek, F.B., Andrew, G.D., Izzet, K.: Malaria parasite detection in peripheral blood images. In: Proceedings of British Machine Vision Conference (2006)

33. Makkapati, V.V., Rao, R.M.: Segmentation of malaria parasites in peripheral blood smear images. In: 2009 IEEE International Conference on Acoustics, Speech and Signal Processing

34. Purwar, Y., et al.: Automated and unsupervised detection of malarial parasites in microscopic images. Malaria J. **10** (2011)

35. Somasekar, J., Eswara Reddy, B.: Segmentation of erythrocytes infected with malaria parasites for the diagnosis using microscopy imaging. Comput. Electr. Eng. **45.C**, 336–351 (2015)

36. Das, D.K., et al.: Machine learning approach for automated screening of malaria parasite using light microscopic images. Micron **45**, 97–106 (2013). (Oxford, England: 1993)

37. Khan, M.I., et al.: Content based image retrieval approaches for detection of malarial parasite in blood images. Int. J. Biometr. Bioinform. (IJBB) **5.2**, 97 (2011)

38. Hearst, M.A., et al.: Support vector machines. IEEE Intell. Syst. Appl. **13.4**, 18–28 (1998)

39. Langley, P., Sage, S.: Induction of selective bayesian classifiers. Conf. Uncertainty Artificial Intel (1994)
40. Tabachnick, B.G., Fidell, L.S.: Using Multivariate Statistics. Ally and Bacon Pearson Education, Boston (2001)
41. Iwaki, Y.:. U.S Patent No. 8,861,878 (2014)
42. Kanan, C., Cottrell, G.W.: Color-to-grayscale: does the method matter in image recognition. PLoS ONE **7**(1), e29740 (2012)
43. Kovačević, J., Chebira, A.: An introduction to frames. Found. Trends Signal Process. **2**(1), 1–94 (2008)
44. Abdul-Nasir, A.S., Mashor, M.Y., Mohamed, Z.: Colour image segmentation approach for detection of malaria parasites. WSEAS Trans. Biol. Biomed. **10**, 41–55 (2013)
45. Yeon, J., et al.: Effective Grayscale Conversion Method for Malaria Parasite Detection. (2014)
46. Kim, J.-D., et al.: Comparison of grayscale conversion methods for malaria classification. Int. J. Bio-Sci. Bio-Technol. **7.1**, 141–150 (2015)
47. Lai, C.H., et al.: A protozoan parasite extraction scheme for digital microscopic images. Computer. Med. Imaging Graphics Official J. Comput. Med. Imaging Soc. **34**(2), 122 (2010)
48. Chokkalingam, S.P., Komathy, K., Sowmya, M.: Performance Analysis of Various Lymphocytes Images De-Noising Filters over a Microscopic Blood Smear Image.
49. Wei, Z., et al.: Median-Gaussian filtering framework for Moiré pattern noise removal from X-ray microscopy image. Micron (2012)
50. Astola, J., Kuosmanen, P.: Fundamentals of Nonlinear Digital Filtering, vol. 8. CRC press (1997)
51. MathWorks. (2011) medfilt2. Retrieved from mathworks.com: http://www.mathworks.com/help/toolbox/images/ref/me dfilt2.html
52. Aizenberg, I., Bregin, T., Paliy, D.: New method for impulsive noise filtering using its preliminary detection. In: SPIE Proceedings, vol. 4667 (2002)
53. Gonzalez, R.C., Woods, R.E., Eddins, S.L.: Digital image processing using MATLAB. Pearson Education India (2004)
54. Hartigan, J.A., Wong, M.A.: Algorithm AS 136: A K-Means Clustering Algorithm. Appl. Stat. **28**(1), 100–108 (1979)
55. Christ, M.J., Parvathi, R.M.S.: Segmentation of medical image using K-Means clustering and marker controlled watershed algorithm. European J. Sci. Res. **71.2**, 190–194 (2012)
56. Das, D., et al.: Invariant moment based feature analysis for abnormal erythrocyte recognition. In: 2010 International Conference on Systems in Medicine and Biology
57. Sadiq Jaffer M.D., Balaram, V.V.S.S.S.: OFS-Z: Optimal Features Selection by Z-Score for Malaria Infected Erythrocyte Detection using Supervised Learning. In: Proceedings of the First International Conference on Computational Intelligence and Informatics. Springer Singapore (2018)
58. http://fimm.webmicroscope.net/Research/Momic/mamic
59. http://www.biosigdata.com/?download=malaria-image
60. Altman, D.G., et al.: Statistical guidelines for contributors to medical journals. British Med. J. (Clin. Res. ed.) **287**(6385), 132–132 (1983)
61. Schindelin, J., et al.: Fiji: an open-source platform for biological-image analysis. Nature Methods **9.7** 676–682 (2012)
62. Jagtap, C.D., Usha Rani, N.: Heuristic scale to estimate premature malaria parasites: scope in microscopic blood smear images. Indian J. Sci. Technol **10.8** (2017)
63. Sadiq, M.J., Balaram, V.V.S.S.S.: DTBC: decision tree based binary classification using with feature selection and optimization for malaria infected erythrocyte detection. Int. J. Appl. Eng. Res. **12**(24), 15923–15934.

Prediction of Phone Prices Using Machine Learning Techniques

S. Subhiksha, Swathi Thota and J. Sangeetha

Abstract In this modern era, smartphones are an integral part of the lives of human beings. When a smartphone is purchased, many factors like the display, processor, memory, camera, thickness, battery, connectivity and others are taken into account. One factor that people do not consider is whether the product is worth the cost. As there are no resources to cross-validate the price, people fail in taking the correct decision. This paper looks to solve the problem by taking the historical data pertaining to the key features of smartphones along with its cost and develop a model that will predict the approximate price of the new smartphone with a reasonable accuracy. The data set [12] used for this purpose has taken into consideration 21 different parameters for predicting the price of the phone. Random forest classifier, support vector machine and logistic regression have been used primarily. Based on the accuracy, the appropriate algorithm has been used to predict the prices of the smartphone. This not only helps the customers decide the right phone to purchase, it also helps the owners decide what should be the appropriate pricing of the phone for the features that they offer. This idea of predicting the price will help the people make informed choice when they are purchasing a phone in the future. Among the three classifiers chosen, logistic regression and support vector machine had the highest accuracy of 81%. Further, logistic regression was used to predict the prices of the phone.

Keywords Support vector machine · Logistics regression · Smartphone prices · Random forest

S. Subhiksha (✉) · S. Thota · J. Sangeetha
School of Computing, SASTRA Deemed University, Tirumalaisamudram, Thanjavur, Tamil
Nadu 613401, India
e-mail: subhisru@gmail.com

© Springer Nature Singapore Pte Ltd. 2020
K. S. Raju et al. (eds.), *Data Engineering and Communication Technology*,
Advances in Intelligent Systems and Computing 1079,
https://doi.org/10.1007/978-981-15-1097-7_65

781

1   Introduction

On a daily basis, we encounter a lot of trade-off in life, for instance, what to opt tasty food versus healthy food, cost of product versus features of product, durability versus reliability and lot more. The complexity of such situations increases day by day and common man faces a lot of tough time to cope up with it. Moreover, taking correct decision in limited time has always been crucial in such a scenario. Nowadays, in this modern digital age, social media is substantially evolving at a very fast pace. The growth has always been tremendous by connecting everyone across the globe in a quick, secure and convenient manner. Smartphones are one of the most readily accessible devices by every individual in this platform. It is one of the most common devices which many individuals possess and it is quite impossible for anyone to survive without it. With various customizations, features along with add-on plugin have enhanced its position in market.

The increase in demand for smartphones has simultaneously led to increase in manufacturers all over the world. The manufactures have started increasing features of their product to securely speed up their position in market. This arises problem for people as to which smartphone to purchase with appropriate features. Overall, purchase of smartphone has always been an issue encountered by all in some instance of time. People spent a lot of time thinking and cross-checking with their peers about product. People are often in dilemma whether the features provided by the manufacturer of the phone are really worth the cost of buying. The attempts to purchase a phone by people in dilemma led to disappointing results of spent significant amount of time as well as their money. "These above-mentioned factors fascinate up the thought of 'Predicting the price of Mobile Phone'". This helps people in making correct decision as well as for manufacturer for validating cost of phone with features provided to their customers. At the end of day, both customer and manufactures get satisfied with product on the whole with valid statistical proofs.

In social media, many people tend to post rating of product without any hesitation. So, in this research, the historical sentiments of people are taken into consideration, i.e. their opinion of price whether it is high, medium or low. Apart from the important factors like display, processor and memory, other features like camera, thickness, battery and connectivity have also been taken into account to determine the approximate price for a phone.

For predicting the price, we are using data set [1] from the popular data set platform Kaggle. The train and test data set are given as two different CSV files. So, to train the model and for predicting which model gives the most accuracy, the train data set is used. The trained model is then used on the test data set to predict the price.

For this concept, due to unavailability of related specific papers, papers regarding recommender systems, sentiment analysis and other predicting system have been used for the purpose of literature survey.

In the paper [2], the authors have mined for historical data from many sources like IMDB and Rotten Tomatoes. After processing the data, the authors have implemented SVM and neural networks to find the accuracy of the prediction. For the data set chosen by the team, the neural network gave the most accurate predictions.

Mansouri et al. in the paper [3], have made prediction on the useful battery charge that can be used in a UAV. The authors have used a variation of SVM, an advanced tree-based algorithm, a linear sparse model and a multilayer perceptron to make the prediction. Using all these algorithms, a preliminary investigation was done.

In the paper [4], the authors look to predict the performance of Karachi Stock Exchange. Various machine learning techniques like single-layer perceptron, SVM, radial basis function and multilayer perceptron are used on the data set to make the prediction. The performance of the multilayer perceptron was the best. The final conclusion drawn from the research was that KSE performance can be predicted using machine learning techniques.

In the paper [5], neural network is implemented and an accuracy of 75% is achieved. This sentiment analysis is very useful for mining useful knowledge from all the data available.

Paper [6] aimed to recommend users the ingredients for different cuisines. The recipes were checked using classifiers like support vector machine and associative classification. The accuracy used in classifiers was used for comparison.

The paper [7] predicts factors that will be affecting future usage of tags. Tags are essentially used to easily search for the answer in that particular domain. Machine learning techniques are used to automate as well as classify them popularity of tags based on structural and non-structural features. The classifiers used were logistic regression, SVM, random forest and AdaBoost. Random forest is the best among the other classifiers based on accuracy.

The paper [8] helps in proper usage of correct algorithm in recommended system and identifying machine learning techniques, new research areas to invest upon. It also focuses on main and alternative performance metrics.

The paper [9] attempted to research efficient models and compared their performance in predicting the direction of movement of daily Istanbul stock Exchange (ISE) National 100 index. Artificial neural network and support vector machine were used for classification. Among them, artificial neural network model (ANN) is a better performance model compared to support vector machine (SVM).

The paper [10] helps in generating forecast of online content. Prediction methods of existing system are used in algorithm for real-time forecasting of popularity with minimal training information.

The paper [11] helps in predicting severity of crash that is expected to occur in future. It uses multinomial logit (MNL), nearest neighbor classification (NNC), support vector machines and random forest (RF) for predicting the traffic crash severity. The proposed approach showed NNC as the best prediction performance in overall and is in more accurate in severe crashes.

Based on the literature, an efficient and accurate way for predicting the price of a smartphone has been determined. These are discussed in the corresponding sections of the paper. Section 2 of the paper talks about the various classifiers used. Sections 3 and 4 discuss about the various classifiers used along with the proposed work. Section 5 performs accuracy analysis of the result obtained. Finally, the paper is concluded in Sect. 6.

2   Classifiers

2.1   Random Forest Classifier

Random Forest Classifier is one of the decision tree algorithms. As the name forest suggests, the name forest refers to the fact that the algorithm is a collection of many decision trees. When a number of cases are present in the training set, the equivalent number of samples is taken at random and is used for training the growing tree. Each of the decision tree is let to grow for the maximum extent. Pruning is not done [12, 13].

Random forest is an aggregate of decision trees, supervised classifier. Each decision tree is in turn a subset of forest which is trained on different training data set with some random selected subset of features. The final result is collection of results of different subset model.

2.2   Support Vector Machine

Among the various classification algorithms available, this is a very powerful algorithm. Here, the data are first plotted as a scatter plot. Then, multiple lines are drawn separating the various data points. Then, a line is modeled in such a way that it splits the data points into two distinct groups [12].

Support vector machine is primarily used for modeling binary classifiers, which classifies data into two categories by developing hyper plane in multidimensional space. Moreover, its decision of predicting is based on linear function [14].

$$y = f(x) = a + bx \tag{1}$$

It is supervised model wherein it learns from training data and predicts the output. The best hyper plane is known by name called margin hyper plane which maximizes sum of distance on either.

The Rbf parameter in support vector machine means radial basis function. It is widely used in the algorithms that use kernels for execution. One such algorithm that uses Rbf is the SVM algorithm. The equation representing the equation is given below [15].

$$\exp\left(-\frac{1}{2}|x - x'|^2\right) = \sum_{j=0}^{\infty} \frac{1}{j!}(x^T x')\exp\left|\frac{-1}{2}\right|x^2|\exp|\frac{-1}{2}|x^2| \qquad (2)$$

The above equation shows the equation of radial basis function kernel.

2.3 Logistic Regression

The name regression for this algorithm is a misnomer. This is a classification algorithm. By using a given set of independent variables, this algorithm predicts discrete values as output. Mathematically, the logarithm of probability of outcome is modeled as a combination of various predictor variables. This combination is linear in nature. Here, the algorithm chooses parameters in such a way that odds of observing the sample quantity is increased. Mathematical equation is as follows [16]:

Linear Equation

$$y = a + bx$$

$$\text{Outcome probability} = \frac{P}{(1 - P)}$$

$$\log e\left(\frac{p}{(1 - p)}\right) = a_0 + a_1 x_1 + a_2 x_2 + \cdots + a_n x_n$$

The function is a step function, and so, logarithmic value is taken to make replication easier [12].

3 Data Description and Methodology

The data set [1] was downloaded from Kaggle. As the preprocessed data were already available no preprocessing has been done on the data set. The data set has the following columns. Each of this is an attribute associated with phone. The test data set that has the same attributes as the train dataset except the price column is not available. This data set takes into account all the parameters associated with a smartphone as they are all a deciding factor on the price of the phone. Listed below is the various names of the column in the table.

- Battery_power: Total energy that a battery can store at a time (measured in mAh)
- Blue: if phone as Bluetooth or not

- Clock_speed: speed at which the microprocessor in the phone executes the instructions
- Dual_sim: If dual-sim support is offered or not
- Fc: Front camera focus in megapixels
- Four_g: If 4G is available or not
- Int_memory: Internal memory of phone in gigabytes
- M_dep: Mobile depth in cm
- Mobile_wt: Weight of the phone
- N_cores: Number of cores in the processor
- Pc: Primary camera focus in megapixels
- Px_height: Pixel resolution height
- Px_width: Pixel resolution width
- Ram: Random access memory (megabytes)
- Sc_h: Screen height of phone in cm
- Sc_w: Screen width of phone in cm
- Talk_time: maximum time that a single battery charge will last when used for talking
- Three_g: Has 3G or not
- Touch_screen: is the display touch screen or not
- Wi-Fi: Has Wi-Fi connectivity or not
- Price_range: The target variable of this data set. This has four discrete values of 0(low cost), 1(medium cost), 2(high cost) and 3(very high cost).

Here, the approximate price range is predicted. We use discrete value for the price in the place of exact price. The price of the column is predicted based on 21 different parameters. The data set contains data of close to 2000 odd phones.

4 Proposed Work

The technique proposed in this paper is to take different features of phone affecting the price in market as the input of problem instance. The output of model predicts whether the price of phone is high, medium or low. Prior to classification of model, data preprocessing task is to be performed on the data set [1]. But the data set considered here was already preprocessed and was ready for use. The supervised classifiers like logistic regression, random forest and support vector machine were used for training the data set. The various parameters associated with the classifiers were modified so as to make sure that the make sure that the model was not overstrained. The accuracy of the trained model was observed for all the supervised algorithms used. Logistic regression algorithm was found to be giving the most accurate result of 81% alongside SVM which also equivalently gave the same result.

After this training based on the accuracy, the price range of the phone was predicted using the logistic regression model. The price range was returned as an

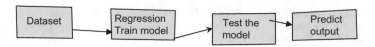

Fig. 1 A diagram representing the flow of events in the proposed model of execution of this project

array. All the coding for the implementation was done using Python language. Various libraries like pandas, sklearn were used to help visualize the data so that the correct model could be predicted for use. Implementation of the supervised algorithm on the data set [1] gave the following accuracies for the different algorithms implemented. Figure 1 represents the accuracy portrayed by the various classifiers along with the changes made in their parameter used during the implementation has been drawn.

- **Support Vector Machine**

  SVM Linear Accuracy: 0.81
  SVM Rbf Accuracy: 0.77

- **Logistic Regression**

  Linear model accuracy: 0.81

- **Random Forest**

  Accuracy: 0.77
  Best max depth: 13 (Fig. 2).

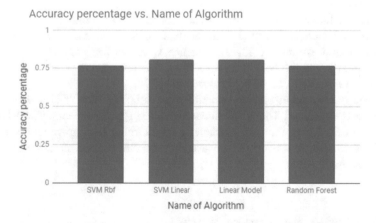

Fig. 2 A graph representing the accuracy percentage as a measure against the algorithm for which the accuracy was obtained. Even the accuracy that was obtained upon changing the parameter for the algorithms is represented here

Fig. 3 A graph representing the accuracy percentage as a measure against the algorithm for which the accuracy was obtained

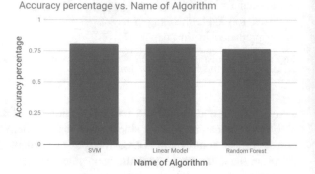

Accuracy percentage vs. Name of Algorithm

5 Performance Measures and Analysis

Accuracy is the measure of the correctness of result when a test is performed upon a set of values. In machine learning, accuracy refers to the correctness with which a trained model is able to predict the output value in the test data set. First accuracy is measured so that we can choose a data set appropriate algorithm and train the algorithm to our needs to give the most accurate prediction. For the same reasons, on the data set chosen here, three different algorithms were used.

Based on the literature reading performed on the various papers, the various algorithms chosen to predict the accuracy of the trained model were logistic regression, support vector machine and random forest.

For logistic regression, a variation in parameter was done. The accuracy obtained was the same when the inverse of regularization parameter was used. The accuracy obtained was 81%.

In random forest classifier, the primary accuracy was found for all the depths. From the result obtained, branch with the most accuracy was chosen along with its predecessor and successor. These branches were further expanded to find the branch which gives the best accuracy. When this classifier was used on the train data set, the maximum accuracy obtained was 77%.

For support vector machine, two different variations out of two were implemented on the data set. The SVM linear algorithm gave an accuracy of 81%. The algorithm of SVM with the Rbf parameter on the other hand gave an accuracy of only 77%. In Fig. 3, the maximum accuracy of the various classifiers has been represented graphically.

6 Conclusion and Future Work

The logistic regression algorithm and SVM were found to give the most accuracy of 81%. Then, the prediction of the price is done using logistic regression. This is just a preliminary paper. The future work to be done on this paper lies in predicting the approximate price of a second-hand phone sale. The same project can further be extended to predict the exact price of the phone instead of the discrete values that is being predicted in this paper. By further pursuing this project, we will be able to help people to spend money wisely and also mine other data like the how the price is affected by the brand, how long a phone can work, etc.

References

1. https://www.kaggle.com/iabhishekofficial/mobile-price-classification#train.csv
2. Quader, N., Ganim M.O., Chaki, D., Ali, M.H.: A machine learning approach to predict movie box-office success. In: 2017 20th International Conference of Computer and Information Technology (ICCIT), pp. 1–7. IEEE (2017)
3. Mansouri, S.S., Karvelis, P., Georgoulas, G., Nikolakopoulos, G.: Remaining useful battery life prediction for UAVs based on machine learning. IFAC-Papers OnLine 50(1), 4727–4732 (2017)
4. Usmani, M., Adil, S.H., Raza, K., Ali, S.S.: Stock market prediction using machine learning techniques. In: 2016 3rd International Conference on Computer and Information Sciences (ICCOINS), pp. 322 327. IEEE (2016)
5. Ramadhani, A.M., Goo, H.S.: Twitter sentiment analysis using deep learning methods. In: 7th International Annual Engineering Seminar (InAES) pp. 1–4. IEEE (2017)
6. Jayaraman, S., Choudhury, T., Kumar, P.: Analysis of classification models based on cuisine prediction using machine learning. In: International Conference on Smart Technologies For Smart Nation (SmartTechCon) pp. 1485–1490. IEEE (2017)
7. Fu, C., Zheng, Y., Li, S., Xuan, Q., Ruan, Z.: Predicting the popularity of tags in StackExchange QA communities. In: 2017 International Workshop on 2017 Complex Systems and Networks (IWCSN) pp. 90–95. IEEE (2017)
8. Portugal, I., Alencar, P., Cowan, D.: The use of machine learning algorithms in recommender systems: a systematic review. Expert Syst. Appl. 1(97), 205–227 (2018)
9. Kara, Y., Boyacioglu, M.A., Baykan, Ö.K.: Predicting direction of stock price index movement using artificial neural networks and support vector machines: the sample of the Istanbul Stock Exchange. Expert Syst. Appl. 38(5), 5311–5319 (2011)
10. Lymperopoulos, I.N.: Predicting the popularity growth of online content: model and algorithm. Inf. Sci. 10(369), 585–613 (2016)
11. Iranitalab, A., Khattak, A.: Comparison of four statistical and machine learning methods for crash severity prediction. Accid. Anal. Prev. 30(108), 27–36 (2017)
12. https://www.analyticsvidhya.com/blog/2017/09/common-machine-learning-algorithms/
13. https://thedatalass.com/2018/04/17/random-forest/
14. https://www.researchgate.net/figure/Figure-3-SVM-classification-scheme-H-is-the-classification-hyperplane-W-is-the-normal_fig3_286268965
15. https://en.wikipedia.org/wiki/Radial_basis_function_kernel
16. https://www.saedsayad.com/logistic_regression.htm

An Extended Polyalphabetic Cipher Using Directional Shift Patterns

Karishma Yadav and Muzzammil Hussain

Abstract Security is a crucial requirement of any communication system, and it becomes a predominant as most of the businesses are done online, and confidentiality of data is a prime security requirement and is being addressed through innumerous mechanisms. Secrecy of data in significant communication is being ensured from ancient times. There are various techniques to provide confidentiality services to a system. In this paper, we discussed various classical techniques of encryption and proposed an enhanced version of polyalphabetic cipher, which is more secure than the later one. Even though the proposed work may not be accepted in comparison with current encryption mechanisms, it will secure as an attempt to revisit the classical techniques and design new techniques inline to them.

Keywords Security · Cryptography · Polyalphabetic cipher · Vigenere cipher

1 Introduction

In today's world, computers are being used in every field or area, and this has reduced many complexities and cost and is prime contributor toward ease of doing business. Most of the businesses are done online, and this gave birth to a new challenge, i.e., security of transactions and data. Today most of the communications are online and hence attract malicious entities to attack, gain unauthorized access to crucial data and misuse for commercial and noncommercial activities. Every year, the business community is losing billions of dollars due to security attacks. Hence, security is a predominant challenge in any online system. Ever since the evaluations of communications, there have been innumerous contributions by the security

K. Yadav · M. Hussain (✉)
Department of Computer Science and Engineering, Central University of Rajasthan,
Ajmer, India
e-mail: mhussain@curaj.ac.in

© Springer Nature Singapore Pte Ltd. 2020   791
K. S. Raju et al. (eds.), *Data Engineering and Communication Technology*,
Advances in Intelligent Systems and Computing 1079,
https://doi.org/10.1007/978-981-15-1097-7_66

experts to secure the communications. As presumed secrecy of data is not the only security requirement rather the security system has to ensure authenticity of entities, integrity of data, availability of resources, nonrepudiation services, etc. The security experts are tirelessly designing mechanisms to ensure the security of communications; on the other hand, the attackers are investing inexhaustible affords to break the security systems, thereby ensuring security of an online system has become a challenging effort. The security experts are working tirelessly in designing new mechanisms to provide security services to the communication system; on the other hand, the attackers are applying new strategies to break the security, and there are numerous security attacks like release of message content, modification attack, replay attack, man in middle attack, and denial of service attack. Each of these security attacks can be either prevented or mitigated through a security service or mechanism.

Although there are multiple security requirements of a system, confidentiality always remains a predominant security requirement. The confidentiality service is provided through different mechanisms which encrypt and decrypt the data. Encryption is a technique of enciphering a plain message into an unreadable format known as cipher. The cipher deciphered into plain form known as plaintext through a technique known as decryption.

There are numerous mechanisms for encryption of data, and they are broadly classified into three categories:

1. Substitution techniques,
2. Transposition techniques,
3. Computational techniques.

Substitution techniques are one in which the letters in the plaintext are replaced by other letters or symbols in ciphertext.

Transposition techniques are one in which the ciphertext is constructed by changing the order or location of the plaintext characters in ciphertext.

Computational techniques are one in which the ciphertext is computed mathematically or logically or both from the plaintext. Further, computational techniques are classified into two that are symmetric-key and asymmetric-key algorithms.

In symmetric-key algorithm, a single key is used for encryption of plaintext and decryption of ciphertext.

In asymmetric-key algorithm, two different keys are used for encryption of plaintext and decryption of ciphertext.

In this paper, we discuss various substitution techniques for encryption of plaintext and shall propose an advanced version of polyalphabetic cipher with enhanced security. We shall also discuss the security of the proposed mechanism.

2   Literature Review

Security in communications between two and more parties is not a recent thought; rather, it has been in practice from ancient times that is before Christ. In fact, Julie's Caesar was the first person to think and design a mechanism for securing the messages during communication. However, the techniques that have been designed until the nineteenth century were designed for manual transmission of messages between the communicating persons and messengers. These techniques are widely called as classical techniques. The mechanism designed after the nineteenth century was designed for electronic transmission of data between computers. Here, we discuss few significant classical techniques of encryption.

2.1   Caesar Cipher

Caesar cipher [1–4] is a substitution technique for encryption of messages, and it is simple and most widely known. It is a type of substitution cipher in which each letter in the plaintext is "shifted" a certain number of places down the alphabet set. The alphabets are numbered as $a = 0, b = 1, c = 2 \ldots z = 25$. Let k be the number of shifts.

$$\text{Ciphertext}(C) = (P + k) \bmod 26$$
$$\text{Plaintext}\ (P) = (C - k) \bmod 26$$

2.2   Mono-Alphabetic Substitution Cipher

It works by replacing each letter of the plaintext with another letter. Due to this reason, a mono-alphabetic cipher is also called a simple substitution cipher [2, 4, 5]. It relies on a fixed replacement structure, which means the substitution is fixed for each letter of the alphabet. Thus, if the letter "a" is encoded as letter "Q," then every time the letter "a" appears in the plaintext, and it is replaced with the letter "Q." There are two implementations of mono-alphabetic cipher:

- Caesar cipher, where each letter is shifted based on a numeric key, as we discuss above in substitution cipher.
- Atbash cipher, where each letter is mapped to the letter symmetric to it about the center of the alphabet.

2.3 Play Fair Cipher

Play fair cipher [1, 2, 4, 6, 7] is a practical substitution cipher. This scheme was introduced by Charles Wheatstone in 1854, but was named after Lord Play fair who promoted the use of this cipher. It is harder to break as it has 25*25 = 625 possible digraphs. It was used for practical purposes by British forces in the Second Boer War and in World War I and for the same purpose by the Australians during World War II. In this, key is a 5*5 matrix and it is constructed using a long keyword. The characters of the key are filled in the matrix from left to right and top to bottom without repetitions. Characters I and J are recommended in the same cell. After filling the keyword in the remaining cells of the matrix, we filled by remaining alphabets from a to z that were not in the keyword (Table 1).

2.3.1 Encryption

Example:-plaintext "BALLOON"

Key "UNIVERSITY"

Step 1-Fill the key in key square.

Step 2-Identify any double letters in the plaintext and any character that is repeating need to be separated by a "Filler character which is X."

Step 3-Pair of character on same row of the square needs to be substituted by a character to its right example: *L* right is *M*, *O* circular right is *H* so, BAMXMHXHN.

Step 4-Pair of character on same column of the key square needs to be substituted by a character beneath it.

Example: BAMXMHXHN, output: HAXVXPVHN.

Step 5-Any pair neither on same row nor on same column of the key square needs to be substituted by a character on the same row or same column.

Example: HAXVXPVHN, output: HEXVXPVHS.

Table 1 Play fair cipher

U	N	I/J	V	E
R	S	T	Y	A
B	C	D	F	G
H	K	L	M	O
P	Q	W	X	Z

2.4 Vigenere Cipher

It is an encryption technique which is more secure than any other simple substitution cipher in this plaintext characters that are replaced by other character in Vigenere table [8] (the table consists of the alphabets written out 26 times in different rows, each alphabet shifted cyclically to the left compared to the previous alphabet, corresponding to the 26 possible Caesar cipher). The cipher uses a different alphabet from one of the rows. The alphabet used at each point depends on a repeating keyword [1–4, 6, 9].

2.4.1 Encryption

Let plaintext be (p_1, p_2, p_3, p_4) and key be (k_1, k_2, k_3, k_4) then ciphertext is found out with p_1 as a row, k_1 as a column in table; and the cross point of p_1 and k_1 is c_1 which is the first letter of ciphertext, so ciphertext shall be (c_1, c_2, c_3, c_4).

2.4.2 Decryption

ciphertext is (c_1, c_2, c_3, c_4) key is (k_1, k_2, k_3, k_4) so we find plaintext by table go for k_1 row and search for c_1 then c_1 appear in which column that is p1 and so on for second character so we get plaintext p_1, p_2, p_3, p_4.

2.4.3 Polyalphabetic Cipher

Polyalphabetic cipher is based on substation cipher, and it works as like Vigenere cipher, as discussed earlier.

3 Proposed Work

Polyalphabetic cipher

A polyalphabetic cipher [1, 4, 6, 9–11] is any cipher based on substitution using multiple substitution alphabets, which is more secure then mono-alphabetic cipher because a mono-alphabetic cipher is any cipher in which the letters of the plaintext are mapped to ciphertext letters based on a single alphabet key. Polyalphabetic cipher uses Vigenere table [8], but it uses different technics for encryption and decryption as described below.

3.1 Encryption

Let plaintext be p_1, p_2, p_3 ... pn.
Key be k_1, k_2, $k3$ ... km.

To generate ciphertext, we use Vigenere matrix [8]. For the first letter of plaintext p_1, we take first letter of key k_1 and map on Vigenere matrix [8] and the cross point of p_1 and k_1 is c_1, as it is the first letter of plaintext displacement which is zero and the cipher is c_1. For subsequent plaintext letters pi, we take corresponding key letter ki map on Vigenere matrix [8] and the cross point of pi and ki be ci then the cipher is cj which is $(i-1)$ rotation right to ci on the same row (Fig. 1).

Example

Input:
Plaintext: CAESAR
Keyword: CIPHER
Output: Ciphertext: EJVCIN

STEP:-1 For the first letter of the ciphertext first letter of the key is used, for first letter of plaintext which is C as row and first letter of key is C than go for column C and displacement to right is 0. So, first letter of ciphertext is C.
STEP:-2 select second letter of plaintext and key so select row A and column I we get I than displacement to right by 1 which is J. So, second letter of ciphertext is J.
STEP:-3 select third letter of plaintext and key so select row E and column P the letter is T then displacement to right by 2 which is V. So, third letter of ciphertext is V and so on $(i-1)$ displacements.

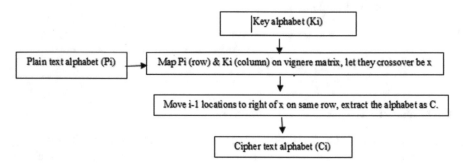

Fig. 1 Flow diagram of encryption process

3.2 Decryption

Decryption is performed by going to the row in the Vigenere table [8] corresponding to the key, finding the position of the ciphertext letter in this row, and then using the column's label as plaintext.

To generate the plaintext Vigenere matrix [8] is used again. For the first letter of ciphertext c_1, we take first letter of key k_1 and on the Vigenere matrix [8] we locate c_1 in the row of k_1, the corresponding column is identified and corresponding letter of the column s plaintext letter p_1 for subsequent letter ciphertext ci, we take corresponding letter ki locate ci in row ki of Vigenere matrix [8] and the plaintext pi is the column $(i-1)$ location left to the column of ci (Fig. 2).

Example

Ciphertext: EJVCIN
Key: CIPHER

STEP:-1 Select the first letter of key which is C, then select row C, and search for the letter E which is first letter of ciphertext which appears in column C and displacement is 0. So plaintext first letter is C.

STEP:-2 For second letter of plaintext, select second letter of key which is I as row and search for second letter of ciphertext which is J then displacement left by 1 which is I, appear in column A.

STEP:-3 For third letter of plaintext, select third letter of key which is P as row and search for third letter of ciphertext which is V then displacement left by 2 which is T appear in column E and so on.

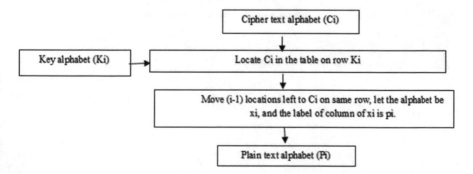

Fig. 2 Flow diagram of decryption process

The proposed mechanism is more complex and secure then both polyalphabetic cipher and vigenere cipher, as a shift has to be computed at each substitution. The attacker may not be able to easily identify the shift and the shift at each substitution can be determined by complex mathematical function $f(\cdot)$, which further secures the mechanism.

4   Implementation and Results

The Vigenere cipher and proposed mechanism were implemented in java, and it was found that the proposed cipher is more secure than Vigenere cipher but consumes more computational time due to shift operations (Figs. 3 and 4).

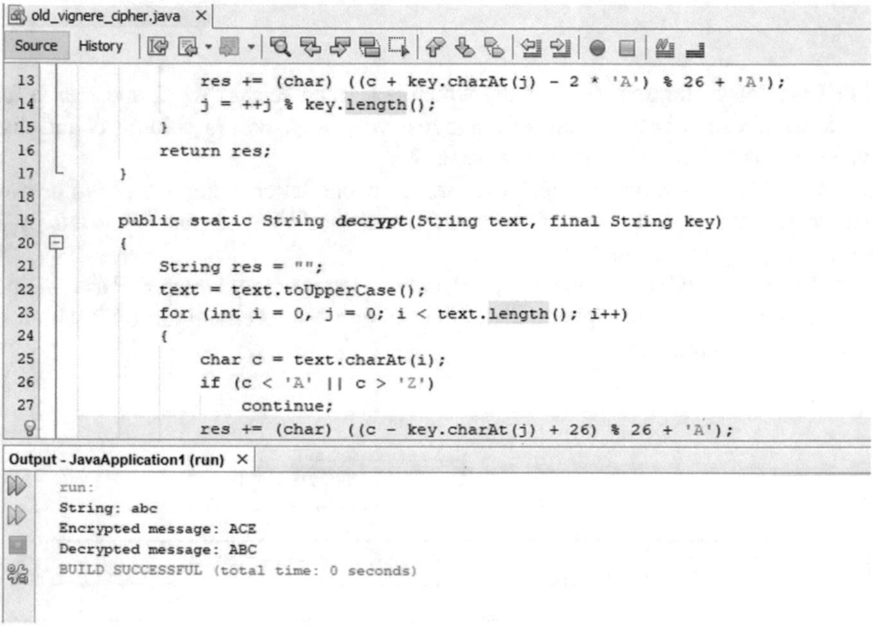

```
old_vignere_cipher.java  ×

Source   History

13                  res += (char) ((c + key.charAt(j) - 2 * 'A') % 26 + 'A');
14                  j = ++j % key.length();
15              }
16              return res;
17          }
18
19          public static String decrypt(String text, final String key)
20          {
21              String res = "";
22              text = text.toUpperCase();
23              for (int i = 0, j = 0; i < text.length(); i++)
24              {
25                  char c = text.charAt(i);
26                  if (c < 'A' || c > 'Z')
27                      continue;
                    res += (char) ((c - key.charAt(j) + 26) % 26 + 'A');

Output - JavaApplication1 (run)  ×

    run:
    String: abc
    Encrypted message: ACE
    Decrypted message: ABC
    BUILD SUCCESSFUL (total time: 0 seconds)
```

Fig. 3 Vigenere cipher output

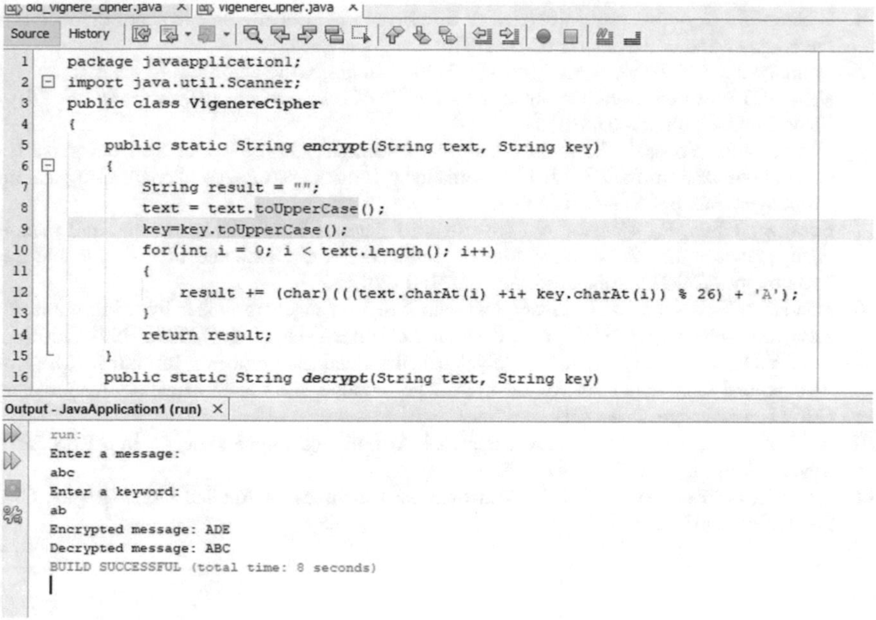
old_vignere_cipher.java
VigenereCipher.java
Source
History
package javaapplication1;
import java.util.Scanner;
public class VigenereCipher
{
public static String encrypt(String text, String key)
{
String result = "";
text = text.toUpperCase();
key=key.toUpperCase();
for(int i = 0; i < text.length(); i++)
{
result += (char)(((text.charAt(i) +i+ key.charAt(i)) % 26) + 'A');
}
return result;
}
public static String decrypt(String text, String key)
Output - JavaApplication1 (run)
run:
Enter a message:
abc
Enter a keyword:
ab
Encrypted message: ADE
Decrypted message: ABC
BUILD SUCCESSFUL (total time: 8 seconds)

Fig. 4 Proposed cipher output

5 Conclusion

Ensuring confidentiality of messages in communications is an ancient practice; over time, innumerous techniques and mechanisms have been designed to ensure security of data. Substitution techniques are one among the oldest ones to generate the cipher out of plain text, and they were the base and inspiration for the techniques and mechanism designed later. As an attempt to revisit the classical technique, an extended polyalphabetic cipher has been proposed which is more secure than the actual one. The proposed mechanism can be further secured through random mathematical function to compute the shift at each substitution.

References

1. Avinash, K.: Lecture and notes on computer and network security. https://engineering.purdue.edu/kak/compsec/. Accessed 21 Jan 2019
2. Stinson, D.R.: Cryptography, theory and practice, Third Edition. Chapman & Hall/CRC (2006)
3. Vidakovic, D.: Analysis and implementation of asymmetric algorithms for data secrecy and integrity protection. Master Thesis (mentor Jovan Golic). Faculty of Electrical Engineering, Belgrade, Serbia (1999)

4. William, S.: Cryptography and Network Security: Principles and practice 5ed, Prentice Hall Press Upper Saddle River, NJ, USA (2010)
5. Omran, S.S., Al-Khalid, A.S., Al-Saady, D.M.: Using genetic algorithm to break a mono—alphabetic substitution cipher. In: *2010 IEEE Conference on Open Systems (ICOS 2010)*, Kuala Lumpur, pp. 63–67 (2010)
6. Uddin, M.F., Youssef, A.M.: Cryptanalysis of simple substitution ciphers using particle swarm optimization. In: 2006 IEEE International Conference on Evolutionary Computation, Vancouver, BC, pp. 677–680 (2006)
7. Hans, S., Johari, R., Gautam, V.: An extended Playfair Cipher using rotation and random swap patterns. In: 2014 International Conference on Computer and Communication Technology (ICCCT), Allahabad, pp. 157–160 (2014)
8. Mandal, S.K., Deepti, A.R.: A cryptosystem based on vigenere cipher by using mulitlevel encryption scheme. (IJCSIT) Int. J. Comput. Sci. Info. Technol. **7**(4), 2096–2099 (2006)
9. Ten, V.D., Dyussenbina, A.B.: Polyalphabetic Euclidean ciphers. In: 2014 IEEE 8th International Conference on Application of Information and Communication Technologies (AICT), *Astana,* pp. 1–4 (2014)
10. Vidakovic, D., Simic D.: A novel approach to building secure systems. In ARES 2007, Vienna, Austria, pp 1074–1084 (2007)
11. Menezes, A., van Oorschot, P.C., Vanstone, S.: Handbook of Applied Cryptography. CRC Press, New York (1997)

Real-Time Aspect-Based Sentiment Analysis on Consumer Reviews

Jitendra Kalyan Prathi, Pranith Kumar Raparthi
and M. Venu Gopalachari

Abstract The rise of e-commerce websites, as new shopping channels, led to an upsurge of review sites for a wide range of services and products. This provides an opportunity to use aspect-based sentiment analysis and mine opinions expressed from text which can help consumers decide what to purchase and businesses to better monitor their reputation and understand the needs of the market. Aspect-based sentiment analysis (ABSA) is a technique aimed to foster research beyond sentence or text-level sentiment classification. The goal is to identify opinions expressed about specific entities (e.g., laptops) and their aspects (e.g., price, performance, build quality, etc.). There exist very few techniques which can generate such results based on customer ratings, however usually for a limited set of pre-defined aspects and not from free-text reviews. The other challenge in this process is cold start problem because of the lack of enough review data for a product. In this paper, a methodology is proposed to automatically compute sentiments of dynamic aspects from user-generated reviews from the web scraping from multiple sources to overcome the cold start problem. Therefore, this methodology is devising a better solution for understanding sentiments in e-commerce than existing methods.

Keywords Aspect-based sentiment analysis · Data integration · Cold start · Web scraping

J. K. Prathi (✉) · P. K. Raparthi · M. V. Gopalachari
Department of CSE, Chaitanya Bharathi Institute of Technology, Hyderabad, India
e-mail: kalyanjithendra27@gmail.com

P. K. Raparthi
e-mail: praneeth970@gmail.com

M. V. Gopalachari
e-mail: venugopal.m@cbit.ac.in

K. S. Raju et al. (eds.), *Data Engineering and Communication Technology*,
Advances in Intelligent Systems and Computing 1079,
https://doi.org/10.1007/978-981-15-1097-7_67

1 Introduction

In recent times, online reviews are very common to be found in lots of websites across the Internet, one particular kind is e-commerce reviews. People who shop online use reviews to get a brief explanation of the products or any information they need; however, preferences of the readers differ from one another. Some would like to find a product of specific brand, while others would be more interested in aspects such as quality, price, service, etc. On the other hand, the reviews written by different customers contain preferences of the respective reviewer. Therefore, readers will have to spend more time to go through the content and understand the opinions expressed on products. It will be of great use to customers if the websites provide the product with rating based on aspect categories rather than an overall rating. However, only few websites such as Amazon and Flipkart provide such rating system at present. To overcome this challenge, an automatic rating generator based on sentiment for each aspect is needed, where the technique to achieve this is aspect-based sentiment analysis [1]. Retrieving information from a user-generated content, particularly retrieving the sentiment in it requires the use of specialized NLP techniques [2]. Added the task of categorizing into aspects and retrieving sentiment for each aspect has proved to be quiet difficult. The user-generated reviews are of high value to the organizations as well as customers. Cold start is another challenge that degrades the quality of the information retrieval. Cold start refers to the new or less data in terms of size or users though there are several alternatives to remedy cold start problem such as using demographic information [9].

In this paper, an automated information retrieval system has been proposed that generates the sentiment analysis report on real-time reviews. The proposed system uses natural language processing techniques to mine the aspects and supervised learning techniques for sentiment classification. This system collects the name of a product as input and then uses spiders to crawl popular e-commerce websites and scrape the reviews from them. Then, the proposed methodology automatically extracts the potential aspects and corresponding entities and finally determines the sentiments expressed. These entities are matched to its aspect, and the system will measure the value of positive and negative sentiments for each aspect. The experiments conducted on the proposed system shown significant improvement in retrieving the information from multiple sources.

Rest of the paper is organized as follows: Sect. 2 contains the related work of proposed system, Sect. 3 contains the architecture and essential steps of automated ASBA, Sect. 4 describes results and discussions of proposed system, and Sect. 5 concludes research work with future directions.

2 Related Work

ABSA is a complex technique which is usually split into aspect extraction and sentiment analysis sub-tasks. Previous approaches to aspect extraction framed the task as a multiclass classification problem, and it relied on techniques with large corpora, e.g., named entity recognizer, parsing, semantic analysis, bag-of-words, as well as domain-dependent ones, such as word clusters [11]. The previous sentiment analysis approaches have used different classifiers with a wide range of features based on n-grams, POS, negation words, and a large array of sentiment lexica. We observed there is no technique which can automatically extract aspects from raw text without any previous information about the aspects. Sentiment analysis is classified into three categories: document level, sentence level, and aspect level [3, 4]. Sentiment analysis done at document and sentence level does not represent the specific view of the reviewers instead provides generalized to entire product. Otherwise, the user has to be prompted manually for each specific aspect to review makes the process static. This work focuses on automation for aspect-level sentiment analysis.

There also exists another approach with supervised learning that focuses on opinion target extraction (OTE) and sentiment polarity towards a target or a category [8]. For OTE, a conditional random field model is proposed with several groups of features including syntactic and lexical features. For polarity detection, a logistic regression classifier is utilized along with the weighting schema for positive and negative labels, and several groups of features are extracted including lexical, syntactic, semantic, and sentiment lexicon features [5–7]. In the work conducted using bag-of-words [10, 13, 14], a semantic-based approach is used to list the word clusters, and supervised machine learning algorithms are used to generate the results.

3 Automated ASBA

Our goal is to generate sentiment rating for each aspect from the real-time reviews. The main challenges for the proposed system are limited resources, and the use of informal language in the reviews, that being the reason, proposed system using the available resources. The architecture of the proposed system is shown in Fig. 1, which contains five main tasks.

The architecture illustrates the general structure of the system and each component of the system which deals with each sub-task of the aspect-based sentiment analysis. This system is mainly developed using natural language processing and machine learning techniques in its core. The main aim of this work is to develop a dynamic approach to aspect-based sentiment analysis. The work is conducted exploring various techniques used in opinion mining and sentiment analysis. The

Fig. 1 Architecture of the
proposed system

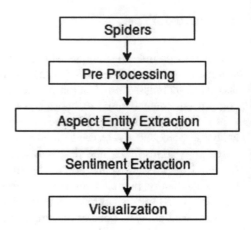

tasks in the architecture can be divided as four modules named data scrapping, pre-processing, aspect entity extraction, and sentiment extraction.

The data scraping module mainly concentrates on an automated process to crawl the web, to find the product details and reviews from multiple domains, and scrape them. The pre-processing module concentrates to prepare the data in an efficient manner to achieve high accuracy while extracting aspect entity pairs. Aspect entity extraction deals with mining features of the products as aspects and opinions expressed on them as entities. We used pattern mining techniques to identify and extract aspect entity pairs. Sentiment polarity is obtained by utilizing Naive Bayes classifier which is trained on a large movie review corpus.

3.1 Data Scraping

Web scraping is a technique used to retrieve data from the web using scripts called as crawlers or scrapers. In our work, we implemented scraping using Scrapy, a python package. The goal is to automate our system in order to scrape the data of same product in multiple domains (Amazon, Flipkart). We give real-time data from web sources as an input using this process. Web scraping begins with a target URL to the data source. The ids used in HTML tags will act as indices for the scraper. Those ids are used to extract the information required which can be text, links, images, etc. There are two formats, one uses CSS tags, and the other uses Xpath. The code snippet 'response.xpath('//title/text()').extract()' is used to scrape the data using Xpath.

3.2 Pre-processing Reviews

Pre-processing the scrapped reviews is of highest importance. The accuracy of the classifier majorly depends upon the preparation done in this stage. Real-time data is dirty, unstructured, and not properly formatted [12]. Given these challenges, we implemented the following techniques to get data cleaned.

(1) *Removing unnecessary characters*: Real-time reviews contain unnecessary characters such as exclamations (!), apostrophises ('), underscores (_), hyphens (-), and Unicode characters. We used regular expressions to retrieve only text from a review. [a-zA-Z0-9_\'] characters which do not match the above regular expression will be removed from the review. Conventional NLP techniques such as stop word removal can be used, but comparatively our proposed model yields better outputs.

(2) *Stop word removal*: This is a technique in natural language processing where unnecessary words are removed from the text if they are listed in stop words, e.g., that, them, a, an, etc. This step drastically improves the efficiency of the system. The goal of this step is to narrow down the choice of words to get the focus words of the sentence, here aspects and their corresponding entities. We use stanford-NLTK package in python, stands for natural language toolkit. It provides a set of stop words and also provides the option to add additional stop words which can be domain specific.

(3) *Stemming*: For grammatical reasons, we use same root word differently at different occasions product, products–product [root word]

Example usage: '*I love this product, We loved using this product*'

In each of the aforementioned cases, product, products are the targets and love, loved, loving express the opinion on it. Instead on having two noun forms of the same word 'product', stemming will give the root word for it. By following this process, we can accumulate different noun forms under the same word, thus reducing the number of aspects and increasing accuracy.

4) *Parts of speech tagging*: This is most important technique to be followed when processing textual data. We use Stanford POS-tagger, CoreNLPTagger from the NLTK package. The functionality of the above-mentioned taggers is to tag a word in a sentence with its respective parts of speech. The following Fig. 2 depicts an example of parts of speech tagging.

For a well-structured sentence, parts of speech will be tagged perfectly. So, it is highly important to clean the reviews before tagging them. Stop word removal can be done even after the reviews are tagged. This helps us to preserve the POS tag of important words and get high accuracies while further processing. In the next step, we use pattern matching techniques to mine aspect entity pairs.

Fig. 2 An instance of POS
tagging example

```
[ ( 'Each' ,    'DT' ) ,
  ( 'of' ,    'IN' ) ,
  ( 'us' ,    'PRP' ) ,
  ( 'is' ,    'VBZ' ) ,
  ( 'full' ,    'JJ' ) ,
  ( 'of' ,    'IN' ) ,
  ( 'stuff' ,    'NN' ) ,
  ( 'in' ,    'IN' ) ,
  ( 'our' ,    'PRP$' ) ,
  ( 'own' ,    'JJ' ) ,
  ( 'special' ,    'JJ' ) ,
  ( 'way' ,    'NN' ) ]
```

D. *Aspect Entity Extraction*

This section focuses on the core part of the system aspect entity pairs extraction. We used a pattern mining-based approach to achieve this. The POS tags 'NN', 'NNP' stand for noun and proper noun. We identified that aspects are nouns in the sentences and corresponding adjectives 'JJ', adverbs 'RBR/RB', verbs 'VBN' will act as entities to it. The following is an example:

Review: RAM is good, processor is amazing. Best laptop.

POS tagged: [('RAM', u'NNP'), ('is', u'VBZ'), ('good', u'JJ'), ('processor', u'JJ'), ('is', u'VBZ'), ('amazing', u'JJ'), ('Best', u'NNP'), ('laptop', u'NN')]

A-E pairs: ['RAM—good', ('NN','JJ')], ['processor—amazing', ('NNP','JJ')], ['laptop—best', ('NN','NNP')].

The system will first check if the pos tagged review contains any nouns, and then it checks for the corresponding entities associated with it. A word is considered as aspect if it has to be a noun, or it must not be a stop word, or it must appear more number of times. We observed that nouns which are repeated more number of times tend to become more important aspects. A counter is initialized, and appearance of aspects is counted. Similar aspects are grouped together along with their corresponding entities.

3.3 Sentiment Polarity Extraction

Sentiment of an aspect entity pair is predicted using TextBlob, a package developed on nltk and pattern libraries. This is achieved with a pre-trained sample on large dataset of movie reviews by Stanford. We used it to predict sentiments of smaller versions of reviews. Since the smaller versions of reviews only contain [Aspect + Entity], they are classified by textblob irrespective of the domain. For example, almost all the A-E pairs are in this form.

Example: ['camera','awesome'] ['battery','bad'] ['quality','great'].

The sentiment property returns a named tuple of the form sentiment (polarity, subjectivity). The polarity score is a float within the range $[-1.0, 1.0]$. The subjectivity is a float within the range $[0.0, 1.0]$ where 0.0 is very objective, and 1.0 is very subjective. Here, we consider the polarity only. We display the result based on the polarity values, 0.0 is termed as neutral, values in the range [0.0, 1.0] are termed positive, and values between [−1.0, 0.0] are termed negative.

4 Results and Discussions

Previous works contributed to ABSA were performed on tagged data which is used to train and test using the models. The performance is measured accordingly with testing data. However, the work done on real-time techniques is low and is mostly confined to restaurants domain. To put it simply, our proposed system will extract sentiments of aspect entity pairs automatically without any definite class or category of the sentence.

We deployed our application in cloud to use it in real time. We used Amazon web services platform to deploy our application. There were numerous challenges to achieve this task. Some of them are package permissions, access permissions, cache management, etc. All these challenges were met with different strategies. Timestamps were used to keep the data intact in the system for a limited period. After the expiry, data (reviews of a product) are scraped again so that the system is updated with new reviews.

Given a product name as input, our system scraped the reviews, processed them, and generated results. The total time taken is about 14–15 s. To reduce the overall runtime, we implemented caching technique discussed earlier. We chose Amazon as the standard and scraped it first, and the spider utilizes the search results of Amazon. They contain the product name and top ten links for the product. The product name which is returned by Amazon spider is used to search Flipkart for the same reviews. This is how the system is maintaining its integrity in scraping data. Following are outputs of the system for the product 'iPhone 8'.

Figure 3 depicts the plot of positive and negative sentiments of top 15 aspect (X-axis) for combined reviews (Amazon and Flipkart combined). After extracting the aspect entity pairs, they are stored as (key, value) pairs. Key being the aspect and values is corresponding entities. A counter will be initialized to list the top n aspects. Then, each aspect is concatenated with its corresponding entities and is fed to the classifier to get the sentiments. The outputs vary from a range [−1.0, 1.0], −1.0 being absolute negative and 1.0 being absolute positive. Means of positive and negative sentiments of aspects are taken separately and displayed in the graph. This will help the users to better understand the positive and negative sentiments on each aspect of a product. The following figures Fig 4 are word clouds generated from all the aspects and their sentiments by the system. As discussed earlier, the focus of the system is to mine the aspects dynamically. On viewing the results, we can say that our system has mined potential aspects successfully. Following section contains evaluation of the results.

We use different dataset to evaluate our model. It is used in SemEval-16 task-5 [4]. The dataset consists of 3048 laptop reviews out of which 2373 reviews are

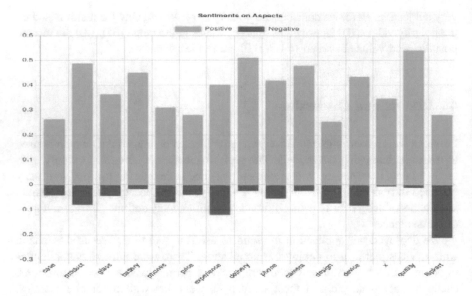

Fig. 3 Bar plot of aspects versus sentiments

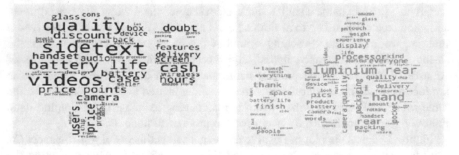

Fig. 4 Positive and negative word cloud

tagged with aspects of each review. We gave the data as input to our system and recorded the results. Interestingly, our system listed out new aspects which were not listed in the dataset earlier. Some of these new aspects include HP, Macbook, OS, laptop, etc. We processed the dataset and listed out top 50 aspects and compared the results of top 50 dynamic aspects of our system. We evaluated the results based on precision, recall, and F1-measure for aspect extraction.

Following table will list out the results. Our model is termed as ABSA-dy which is compared with the best results from the work done by Nipuna and Chamira [2]. The analysis [2] performed on laptop review dataset is done by using supervised learning approaches including unigrams, stemming, pos tagging, and Headword POS. The best results observed are mentioned in the above table with the

Model	Domain	Recall	Precision	F-Measure
ABSA-dy	laptop	0.862	0.735	0.793
ABSA-su	laptop	0.508	0.451	0.478
TABLE I: Flipkart and amazon data				

Fig. 5 Comparing static and dynamic aspects

label ABSA-su. We observed a great enhancement in the measures using the dynamic approach to mine aspects. Following Fig. 5 will show the comparison between static (mentioned in the dataset) and dynamic aspects (generated from our system).

We compared the results of our system for the product 'iPhone 8' in three modes: using only Amazon reviews, only Flipkart reviews, and using both the reviews. We observed that there are slight differences in positive and negative sentiments expressed. Dynamic aspects generation can be used by various vendors to support their marketing strategies and customers to better understand the products.

5 Conclusion

In this paper, we present a composite method based on pattern mining combined with natural language processing to solve the ABSA problem in real time. To improve the accuracy, we implemented a technique to append new stop words to the system. We were able to bring down the running time to mere 12 s including the tasks of scrapping and processing from two domains (web sites). Dynamic aspect classification can be viewed as a better technique than static and supervised techniques as it is efficient with an F1-measure of 0.793. The system obtained very satisfying results for dynamic aspect extraction from real-time reviews on multiple domains.

References

1. Gojali, S., Khodra, M.L.: Aspect based sentiment analysis for review rating prediction. In: Proceedings of International Conference on Advanced Informatics: Concepts, Theory And Application, IEEE, Malaysia (2016)
2. Pannala, N.U., Nawarathna, C.P.: Supervised learning based approach to aspect based sentiment analysis. In: Proceedings of IEEE International Conference on Computer and Information Technology (CIT), Fizi (2016)
3. Pontiki, M., Galanis, D., Papageorgiou, H., Manandhar, S., Androutsopoulos, I.: Aspect based sentiment analysis. In: Proceedings of Proceedings of the 9th International Workshop on Semantic Evaluation SemEval 2015, Denver, Colorado (2015)
4. Pontiki, M., Galanis, D., Papageorgiou, H., Androutsopoulos, I.: Aspect Based Sentiment Analysis. SemEval (2016)
5. García-Pablos, A., Cuadros, M., Rigau, G.: W2VLDA: almost unsupervised system for aspect based sentiment analysis. Expert Syst. Appl. **91**, 127–137 (2018)
6. Ruder, S., et al.: INSIGHT-1 at SemEval-2016 Task 5: deep learning for multilingual aspect-based sentiment analysis. SemEval@NAACL-HLT (2016)
7. Jebbara, S., Cimiano, P.: Aspect-based sentiment analysis using a two-step neural network architecture. In: Cognitive Interaction Technology Center of Excellence (2016)
8. Hamdan, H.: Opinion target extraction and sentiment polarity detection. In: Oroceedings of SentiSys at SemEval-2016 Task 5 (2016)
9. Xu, L., Lin, J., Wang, L., Yin, C., Wang, J.: Deep convolutional neural network based approach for aspect based sentiment analysis. Adv. Sci. Technol (2017)
10. Guha, S., Joshi, A., Varma, V.: SIEL: aspect based sentiment analysis in reviews. In: 4th Joint Conference on Lexical and Computational Semantics (2015)
11. Go, A., Bhayani, R., Huang, L.: Twitter sentiment classification using distant supervision. CS224N Project Report. Stanford **009**, 112 (2017)
12. Pradhan, V.M., Vala, J., Balani, P.: A survey on sentiment analysis algorithms for opinion mining. Int. J. Comput. Appl. (2016)
13. Zhang, W., Xu, H., Wan, W.: Weakness finder: find product weakness from chinese reviews by using aspect based sentiment analysis. Expert Syst. Appl. **39**(11), 10283–10291 (2012)
14. SoufianJebbara, P.C.: Aspect-based sentiment analysis using a two-step neural network architechture. Bielefeld University, Germany (2016)

RETRACTED CHAPTER: Performance Analysis on IARP, IERP, and ZRP in Hybrid Routing Protocols in MANETS Using Energy Efficient and Mobility Variation in Minimum Speed

Chatikam Raj Kumar, Uppe Nanaji, S. K. Sharma
and M. Ramakrishna Murthy

Abstract MANET represents its gadgets behavior in its network structure to relocate any movement of time without drawing near any topological approval in multilateral guidelines, which ensuing the runtime link status quo with different gadgets that belongs to the identical zone. The important problem with building the MANET is to keep runtime place facts of the participated gadgets for managing the routing facts to examine traffic. MANET has the possibility to preserve one or more than one nature of transceivers. Strength control in wi-fi networks deals with the technique of managing power resource through controlling the battery discharge, adjusting the transmission electricity, and scheduling of strength assets so that it will boom the life of the nodes of a advert hoc wi-fi network. Right here, in our recommend paintings area, Intra Sector Routing Protocol (IARP), Zone Routing

---

The original version of this chapter was retracted. The retraction note to this chapter is available at
https://doi.org/10.1007/978-981-15-1097-7_82

C. R. Kumar (✉)
Department of Computer Science and Engineering, Vignan's Institute of Information
Technology, Visakhapatnam, India
e-mail: chatikam@gmail.com

U. Nanaji
Department of Master of Computer Applications, Vignan's Institute of Information
Technology, Visakhapatnam, India
e-mail: nanajistiet@gmail.com

S. K. Sharma
Department of Computer Science and Engineering, Andhra University, Visakhapatnam, India
e-mail: sharma.santosh83@gmail.com

M. R. Murthy
Department of Computer Science and Engineering, Anil Neerukonda Institute of Technology
and Sciences, Visakhapatnam, India
e-mail: ramakrishna.malla@gmail.com

© Springer Nature Singapore Pte Ltd. 2020, corrected publication 2024   811
K. S. Raju et al. (eds.), *Data Engineering and Communication Technology*,
Advances in Intelligent Systems and Computing 1079,
https://doi.org/10.1007/978-981-15-1097-7_68

Protocol (ZRP), and Inter Quarter Routing Protocol (IERP) are simulated with dedicated small networks with 90 nodes the usage of EXATA emulator to examine QOS for application and electricity efficiency.

Keywords MANETS · ZRP · IARP · IERP · EXATA · Emulator

1 Introduction

Wireless network sensors supply an annex between actual domain and effective structures. They have played a vital function in message functions inside the latest existences. Wireless structures are collected into numerous classes created at the community and communication kind. Sensor devices use batteries to function and consume a lesser amount of strength. They are widely distribute into modules, one is arrangement-aided and other is substructure much less. Mobile avert ad hoc structures are identity categorization of adaptable nodes with no base. It is considerably a structure-less machine. Intermediate or transfer nodes are used to dimension up a communication among nodes. The routing rules are compulsory allowing for the fact, so that the particular nodes are essential to transport within the typical technique and the routing method plays a vital quantity in advert ad hoc structures. The nodes can interconnect inside of the device at of any kind time. Alongside these traces, the relationship can be established among every node to exceptional nodes in the machine. MANET varies from the visitors necessities. In MANET, the conversation is M2M networks, that has a reduced flexibility and its structure much less.

MANETS is an arrangement of sensors and actuator which is accustomed to remove the records as a rule from the predefined area for particular thought process. Sensors might be utilize to get various sort of information with various targets esteem, for example, interloper base sensors, sound sensors, temperature sensor, weight sensor, and so on yet it demonstrate the static structure of sensor usefulness. Here, MANET works in unique condition in which gadgets may change their area at any development of time which is fundamentally altogether different to deal with non-uniform structure of hubs. MANETs have a place with space of isolated unprepared system that for the maximum part has a routable arrangements management condition and course notice, course disclosure data best of a link layer specially appointed system. Self-organizing behavior of wireless ad hoc networks.

The improvement of remote correspondence of Tabs, shrewd gadgets, Wi-Fi PCs and 802.11 or Wi-Fi isolated systems management requires completed MANETs, a most requesting research zone since 1990s. An immense no of academicians have anticipated there inquire about view in remote systems and versatile ad hoc systems. Numerous scholarly papers measure agreements and their volumes, supposing differing degrees of portability confidential a restricted space, normally with altogether hubs classified a join of jumps of each other. Diverse agreements are then restrained in light of procedures, for example, the parcel drop rate, the above accessible by the leading convention, end-to-end bundle interruptions, arrange quantity, volume to scale, and so on. We utilize a TTL (time to live) in ad hoc systems to course ask for, each took an interest hub lessens an opportunity to live by one after the each hit of course ask for and finishing up the course ask for by dropping parcels.

Mobile ad hoc networks can be utilized as a part of numerous applications, reaching out from sensors for environment, street wellbeing, vehicular impromptu open administrations, home, distributed informing, rockets, calamity save activities, robots, air/arrive/naval force barriers. One arrangement is the utilization of tools like OMNeT++, NetSim, OPNET, NS2, and NS3. A study of different test systems for VANETs such as compelled road topology, multi-path declining and roadside hindrances, traffic flow prototypes, trip models, variable vehicular speed and flexibility, traffic lights and jamming, drivers' behavior, etc.

2   The Zone Routing Protocol

Prior a zone is established as holding the objective hub, the proactive convention, or put away directing table, is utilized to appropriate the packet. In case a packet's goal node is in the same zone as per the inchoation, the applied protocol IARP is used for routing the packet. This moderates the handling overhead for individuals routes. If the route is external, the zone the packet's initiating, a sensitive protocol IERP is recycled for routing.

The packets in this way with destinations inside the same zone as the initiating zone remain distributed to deposit or particular routing tables. A hub device that takes an input (i.e., a frame's bits) and retransmits the input on the hub's outgoing ports, the sending zone dodge the overhead of examination steering tables adjacent the strategy intertwining the touchy convention to confirm whether separately zone experienced holds the objective hub. The ZRP eliminates the kill time for routing classified zone through route recognition methods with sensitive and active routing protocol which diminishes the mechanism overhead for extended routes.

In Fig. 1, the IARP is used as exclusive for direct regions and it is operated between routing zones by using routing table.

In Fig. 2, some route to a goal that is surrounded by the identical resident region is expeditiously predictable starting the bases proactively stored in a routing table by IARP. If the initial place and goal packet of a data remain in the identical region,

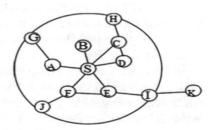

Fig. 1 Example routing zone with $r = 2$

Fig. 2 ZRP architecture

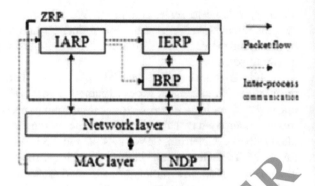

the packets can attain instantaneously. Maximum surviving the active direct algorithms can be recycled as the IARP for ZRP distributed openly.

3 Overview of Routing Protocols

3.1 Zone Routing Protocol Tables

ZRP is a cross protocol and it is a sensitive arrangement and active routing protocols. Proactive routing norms additional bandwidth to keep direction-finding information, whereas responsive routing includes extended route demand suspensions and also ineffectually abundances the whole network for route purpose.

In ZRP, for every node is expected to retain routing data [1]. For the reason that informs are individual circulated locally, the quantity of modernize traffic is essential to keep a routing region does not rest on all the system nodes.

3.2 Intra Zone Routing Protocol (IARP)

The ZRP is defined in three separate Internet drafts: IARP, IERP and BRP. ZRP is one of the protocols that are currently under evaluation and standardization by the IETF MANET working group [2, 3] using QUALNET simulator for the performance for evaluation [4].

This protocol is used to link among nodes inside the routing zones. It might be some link-state or distance-vector routing protocol. A path to a target contained by the resident region can be recognized beginning the routing table of the initial node. Consequently, the packet will be directly distributed if the source and destination are in identical region.

In this context, the transmission method would be concluded in a transitory time setting to deliver adequate interval to a motor automobile to latch up on the data, which comprises security guided communication. The technique of a routing is an authoritative stage for the nodes grouping in the classification. There are numerous directing conventions in MANET, these are demanded into two categories on an actual simple level active and reactive, which is table and on request determined. Several steering convention traditions are proposed for ad hoc frameworks (a) Dynamic Source Routing(DSR) and (b) Number Ad hoc On-Demand Distance Vector(AODV) need stood utilized on the MANETs are used as a piece of Vehicular ad hoc Network in the past centuries [5, 6].

ZRP is a relationship of active and on-request conventions. A hubs nearby region is called directing zone. In a hub established, steering region for the hub is the minimum jump separation and it ought not to be more noteworthy than the region range. A hub retains courses to every one of the goals in the steering region in its directing table. To set up a directing area, first the hub must distinguish all its close-by individuals which are far from an achieving separation. The two conventions inside ZRP are IARP which utilizes directing table the other is IZRP. The IARP is responsible for protection hubs privileged the directing region and IERP is accountable for definition and keeping up the courses to the hubs external the steering region. A noteworthy purpose of inclination of this convention is that a solitary course demand can bring about numerous course answers.

The speed bounds is from 40 to 90 MPH. Automobile treated as nodes, are positioned arbitrarily on the path at a 2500 feet distance. If we consider a public road, which has double routes, and 150 vehicles go in it. In this, 50 automobiles travel in one path and 50 are in opposite path. High wayside components are recycled to create transmission among the automobiles. Completely the parts will be linked to the network of WiMAX, also predictable as the base station. The transmission will be through Web and the vigorous storage is the cloud. The position of the automobiles is tracked by GPS, which drive supportive for vehicle-to-vehicle transmission. There are various constraints in this situation; the message impairment and the load of routing are completely threadbare in this examination. Interruption is the total period occupied by packets flanked by the delivery node on or after the transfer node. Load routing resources the quantity of switch packets which has been communicated to supply the documents, facts, and data packet at the destination from Fig. 3.

The ZRP is in use for distributing the data competently from source node to the destination node through various intermediate nodes. The routing region for node 1 and 11 is shown in Fig. 4. Take up if the data nodes from 1 to 10 crave to interconnect, i.e. in its routing region it uses IZRP in which it usages the routing table to discover the target node. In additional instance, if N1 requestor wants to send a data packet external the routing region, the IZRP is consumed which is on request.

Let a state somewhere node 1 wants to send a data packet to node 20. Now, the source node is N1 and the terminal node is N20. To propel the data packet to the destination, it has to determine the route first. Node drives done its routing table for 20 developing IZRP. Given that the node 20 is not in its distinct routing region it starts the path needs utilizing the IZRP. The requirements are margin casted to the peripheral nodes. Here, the tangential nodes of N1 are N9 and N10. Now, N9 and N10 go during its particular routing table for the node N20.

In Fig. 5, the routing zone aimed at N9 and N10 are visible. The destination of N19 cannot be identified in routing tables, it wants to propel to tangential nodes utilizing border casting. The tangential node aimed at N10 is N16 and aimed at N9

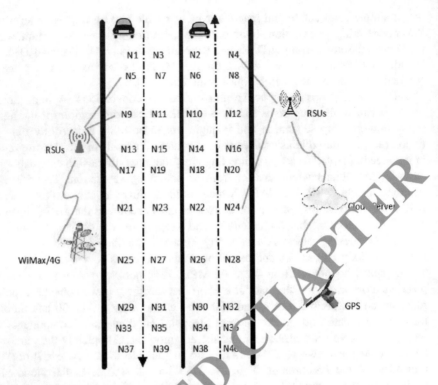

Fig. 3 Zone Environment

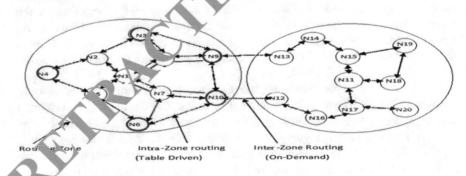

Fig. 4 Detection the path

its N14. Now, routing table for N19 was identified by the nodes. While N19 is in the zone of N14 the N19 exists and N14 adds the path from N14 to N1 by the route request the path. At last, it sends the created response route opposite to the N1. An extra path is also created by the node N16 and directed to N1. The node N1 catches some route responses. Among the responses, it uses the direct path to the N19 and N19 sends the data packet. The shortest path is N19-N15-N14-N-13-N9-N8-N1. In this process, the route-finding technique is completed and it decreases the obstacle

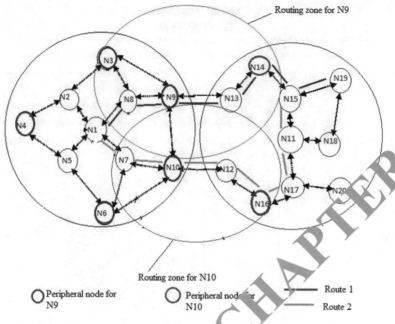

Fig. 5 Flow of data packet

Table 1 Parameters for simulation setup scenarios	Channel frequency	2.4 GHz
	Simulation Time	900 s
	Physical layer radio type	802.1 1b
	MAC protocol	802.11
	Application	CBR
	Packet size	512 bytes
	Hybrid routing protocol	ZRP, IARP, IERP
	Transport layer protocol	UDP
	No. of nodes	90
	Minimum speed	1–6 mps
	Pause time	0 s
	Simulation range	1000 m × 1000 m
	No. of channel	One
	Mobility	Random way point
	Simulator	EXATA
	Data rate	2 Mbps
	Antenna model	Omni-directional
	Nodes placement	Random
	Pathloss model	Two-ray

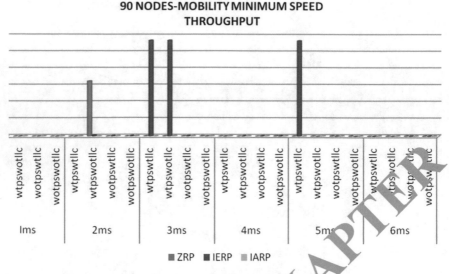

Fig. 6 Throughput (bits/s)

for N19 and N1 to interchange limited words [7, 8]. Also a smaller amount control packets are sent to AODV to discover the path to the destination node, when it decreases the Routing Load.

4 Simulation Setup

In this text, we need to estimate the performance variance of mesh networks hybrid routing protocols ZRP, IARP, and IERP by varying the minimum speed of nodes with stir or move in the network, in excess of an area of 1000×1000 m^2 from Table 1.

5 Simulation Results

5.1 Throughput

It is clearly observed from Fig. 6, which depicts the throughput of the network with variation in nodes. The network throughput for IERP is better as compared to other hybrid routing protocols.

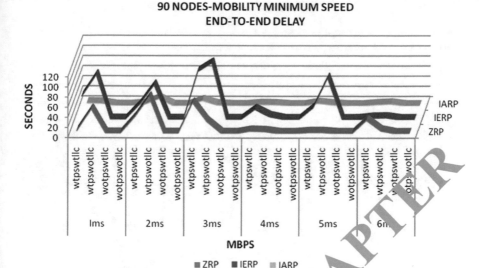

Fig. 7 Source to destination delay

5.2 *Source Node to Destination Node Delay*

It is clearly observed from below Fig. 7 which depicts the source to destination node delay of the network with variation of nodes. The network source to destination delay for IERP is enhanced as compared to other hybrid routing protocols.

5.3 *Average Jitter*

It is clearly observed from Fig. 8, which depicts the average jitter of the network with variation of nodes. The average jitter for IERP is better as compared to other hybrid routing protocols.

5.4 *Energy Consumed*

It is clearly observed from Fig. 9, which depicts the Energy consumed of the network with variation of nodes. The Energy Consumed for IARP is better as compared to other hybrid routing protocols.

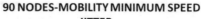

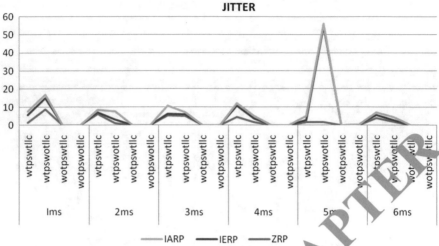

Fig. 8 Average jitter

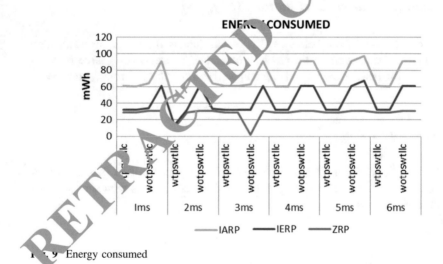

Fig. 9 Energy consumed

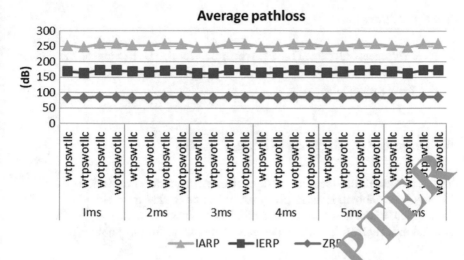

Fig. 10 Average pathloss

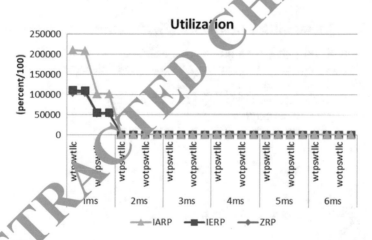

Fig. 11 Utilization

5.5 *Average Path Loss*

It is clearly observed from Fig. 10, which depicts the average pathloss of the network with variation of nodes. The average pathloss for IARP is better as compared to other hybrid routing protocols.

5.6 Utilization

It is clearly observed from Fig. 11, which depicts the Utilization of the network with variation of nodes. The Utilization for IARP is better as compared to other hybrid routing protocols.

6 Design of Experiment Using Taguchi

Taguchi is used for factorial design of experiment for full factorial design to identify all combination from given set of factors to find most significant factor from given a small set from possibility is selected to reduce large number of simulations. Taguchi is used in design of experiment for our application.

6.1 Designing an Experiment

The design of an experiment involves the following steps

1. Selection of independent variables
2. Selection of number of level settings for each independent variable
3. Selection of orthogonal array
4. Assigning the independent variables to each column
5. Conducting the experiment
6. Analyzing the data
7. Inference

6.2 A Typical Orthogonal Array

While there are many standard orthogonal arrays available, each of the arrays is meant for a specific number of independent design variables and levels. From Table 2, if one wants to conduct an experiment to understand the influence of four different independent variables with each variable having three set values (level values), then an L9 orthogonal array might be the right choice. The L9 orthogonal array is meant for understanding the effect of four independent factors each having three factor level values. This array assumes that there is no interaction between any two factors.

Based on Table 2 and Fig. 12, it is found that the strongest factor was the Mobility 10 m/s since its estimated effect the biggest value and degree of the scope was the highest, followed by energy consummation and node size. The optimal

Table 2 Simulation result using Taguchi

	C1	C2	C3	C4	C5	C6
↓	NW	MOB	TTL	THROUGHPUT	DELAY	JITTER
1	20	10	3	164.389	0.003949	0.000612
2	20	15	4	109.592	0.003864	0.000938
3	20	20	5	77.628	0.004301	0.000860
4	40	10	4	191.787	0.004929	0.001___
5	40	15	5	159.822	0.004386	0.001_20
6	40	20	3	118.725	0.009206	_02_37
7	60	10	5	210.052	0.006_._	0.0_2490
8	60	15	3	132.420	0.0__099	0.000599
9	60	20	4	95.890	0.0__739	0.001646

```
Larger is better

Level      NW      MO_      TTL
1        40.97    _5.4_    42.75
2        43.__    42.44    42.03
3        4_.__    39.64    42.77
Delta    _._7      5.83     0.74
Rank        2         1        3
```

factor level com_ nati_n was A2B1C3. Where Ai notion represents ith level of the factor.

7 C_nclusion and Future Work

Th_s paper displays an execution contrast of ZRP, IERP, and IARP steering convention for versatile ad hoc systems with variable least portability speed and energy effectiveness. We compute source to destination delay (s), average jitter (s), throughput, source to destination node delay, average pathloss, vitality devoured, and utilization as execution measurements. Our reproduction comes about shows ZRP execution is best under all execution measurements for Minimum speed arrange condition and vitality effectiveness.

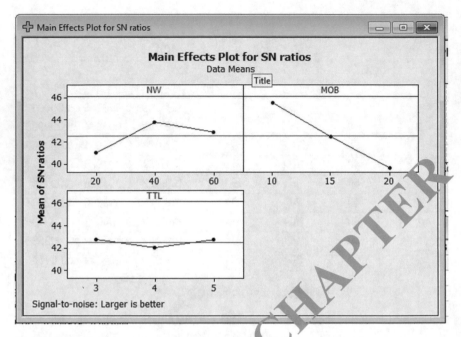

Fig. 12 Main effect plot for SN ratios

In upcoming, this paper can be improved by break down the other MANET directing conventions under genuine situations.

References

1. Schaumann, J.: analysis of zone routing protocol (2002)
2. Haas, Z.J., Pearlman, M.R.: The Zone routing protocol (ZRP) for Ad Hoc networks. IETF Internet Draft, draft-ietf-manet- zone-zrp-04.txt (2002)
3. Internet draft-draft-ietf-manet-aodv-08.txt and draft-ietf-manet- zrp-04.txt and draft-ietf-manet-tbrpf-01.txt
4. QualNet Documentation.: QualNet S.O Model library: Advanced wireless. http://www.scalablenetworks.comiproducts/Oualnetidownload.php#doc
5. Suresh, K., Jogendra, K.: Comparative performance study of zone routing protocol over AODV and DSR routing protocols on constant bit rate (CBR). Volume 4S-NoA, IJCA (097S-8887) (2012)
6. Subramanya Bhat, M., Shwetha, D., & Devaraju, J.T.: A performance study of proactive, reactive and hybrid routing protocols using qualnetsimulator. Volume 28-no.S, IJCA (2011)
7. Hiertz, G.R., Denteneer, D.: IEEE 802.11s: The WLAN mesh standard. IEEE Wireless Commun (2010)
8. Cornils, M., Bahr, M., Gamer, T.: Simulative analysis of the hybrid wireless mesh protocol (HWMP). European Wireless Conf. (2010)

Analysis of Big Data in Healthcare and Life Sciences Using Hive and Spark

A. Sai Hanuman, R. Soujanya and P. M. Madhuri

Abstract Big data is a declaration used to recognize the database whose area is afar the potential of typical database software tools to store, organize and examine. Big data has shown a new path toward the mankind. With several theoretical and technological obstacles in health huge processing, it is onerous to transfer knowledge into fortunate and valuable applications. Meeting the challenge of handling big data in healthcare information construction procedure, this paper proposes a referential architecture on the Hive and Spark platform to overcome the problems in healthcare big data process. Hive is a noteworthy project as a result of it permits exposing the simplest components of Hadoop, specifically map reduce and knowledge storage. Spark may be a memory-based computing framework that features a higher ability of computing and fault tolerance, supports batch, interactive, iterative and flow calculations. Experiment results of data upload, data query and data analysis show that the performance of the proposed framework is greatly improved, and a brief summary of the performance and the differences between two methods of Hive and Spark is also discussed.

Keywords Big data · Hadoop · Healthcare · Hive · Spark and Scala

A. Sai Hanuman · R. Soujanya · P. M. Madhuri (✉)
Department of Computer Science and Engineering, GRIET, Bachupally, Hyderabad,
Telangana, India
e-mail: pmmadhuri18@gmail.com

A. Sai Hanuman
e-mail: a_saihanuman@hotmail.com

R. Soujanya
e-mail: soujanya96@gmail.com

1 Introduction

1.1 Big Data and Hadoop

Big data: Big data is "a data that is in huge size, high speed and variable data that wants innovative techniques and new technologies to enable the capture, distribution, management, memory and analysis of the information."

Some characteristics of big data are:

- **Volume**: It is the amount of data produced by organizations or individuals. All organizations are searching for ways to handle the ever-increasing data volume that is being created every day [2].
- **Velocity**: It is the frequency and speed at which data is captured, produced and shared. Consumers as well as businesses now producing lots of data and in shorter cycles, from hours, minutes, seconds down to milliseconds.
- **Variety**: It is the creation of new data types together with social, mobile and machine resources. New types include metrics, content, physical data points, mobile, process, location or geo-spatial, machine data, radio-frequency identification (RFID), hardware data points, search and web. It also includes unstructured data.
- **Veracity**: It is outlined because of the exactness of information. Incorrect knowledge will cause tons of issues for organizations. Hence, organizations want to ensure that the data is correct and analyses performed on the data are precise. In robotic decision-making, where no human is involved we need to be sure that both the data and the analyses are correct [7, 12] (Fig. 1).

Hadoop 2.0. An open-source software framework is provided by Apache software foundation, which is used store and processing the large-scale data sets with clusters of commodity hardware. It is designed for scaling up from a single server to

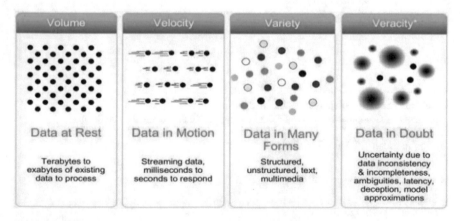

Fig. 1 Big data

Fig. 2 Hadoop framework

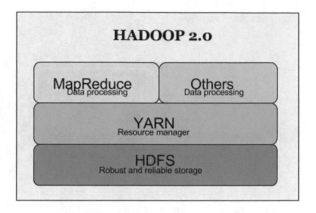

thousands of servers in a cluster with local computation and storage on every individual server. There are three main core components inside this run-time environment (from bottom to top) [1] (Fig. 2).

1.2 Big Data Analytics

Big data analytics is a strategy, for analyzing giant volumes of information. This huge knowledge is gathered from a good style of sources, including sensors, social networks, videos, sales transaction records and digital images. The aim of analyzing all this data is to find out new patterns and connections that may somewhat be imperceptible and that might offer valuable insights about the users who created it.

1.3 Types of Analytics in Big Data Analytics

There are four types of analytics. Here, we tend to begin with the only one and go all the way down to a lot of subtle. As it happens, the more complex an analysis is, the more value it brings (Fig. 3).

- **Descriptive** analytics, which uses data aggregation and data mining to provide insight into the past and answer: "What has happened?"
- **Diagnostic** analytics, which uses historical data, can be measured against other data to answer the question of "why something happened?"
- **Predictive** analytics, which uses statistical models and forecasts techniques to understand the future and answer: "What could happen?"
- **Prescriptive** analytics, which uses optimization and simulation algorithms to advice on possible outcomes and answer: "What should we do?" [13]

Fig. 3 Types of analytics

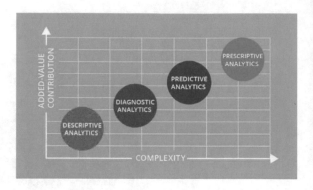

2   Big Data Analytics in Healthcare

Regular methods can be changed in any industry by using big data. By applying data analytics in healthcare industry, we can reduce the cost for treatments, can prevent avoidable diseases and quality of life can be improved. Based on treatment methods using nowadays, challenges are also increasing as there is continuous increment in the population count. Just like business people, healthcare professionals are also accumulating the information of patients and are using those data for the research.

By gathering large volume of information in healthcare, how it is going to help? We can show huge knowledge examples in attention that exist already which we tend to take pleasure in [14].

2.1   How to Analyze Large Data in Healthcare

- There are many positive and lifesaving outcomes when we apply big data analytics in healthcare each and every information creates huge volume of data that is used by specific technologies to get analyzed. Application toward healthcare data gets facilitated by avoiding cost issues and early detection.
- Now that we tend to live more, treatment models have altered and a lot of those progressions are explicitly determined by data. Specialists need to get a handle on the greatest sum as they will a couple of patients and as right off the bat in their life as potential, to pick up notice indications of critical sick—treating any ailment at a beginning period is way a ton of clear and less pricy. With medicinal service data investigation, the obstacle is best than fix and figuring out how to draw a far-reaching picture of a patient can give protections, a chance to offer a custom-fitted bundle. This is regularly the business's imagine to handle the storehouse's issues a patient's data has: Everyplace are gathered bits and

bytes of it and filed in emergency clinics, centers, medical procedures and so on with the difficulty to talk appropriately.

- A considerable length of time gathering tremendous measures of information for therapeutic use has been expensive and long. With the present continually enhancing advances, it winds up less demanding not exclusively to accumulate such data anyway conjointly to change over it into important pivotal bits of knowledge that may then be acclimated offer higher consideration. This is regularly the point of medicinal service data investigation: abuse information-driven discoveries to foresee and understand a pull before it is past the point of no return, anyway conjointly evaluate procedures and covers speedier, monitor stock, include patients a great deal of in their very own well-being and engage them with the instruments to attempt [14].

2.2 Analyzing the Data Flow in Healthcare

The business is growing fast and it has several categories of information which are setting out to remodel it—in the other hand, there is still availability of labor and the field slowly implements the different technologies that may be helpful in the long run and for effective business operations. The below are some of the examples laid out to analyzing the data in healthcare.

- Analyzing the patients' footprint will helpful to recruitment staff accordingly.
- Usage of Electronic Health Records (EHRs).
- Periodic alerts for health checkup.
- Improvise research and development to cure cancer.
- Usage of predictive analytics.
- Help in averting narcotic maltreatment in the USA.
- Improve patient awareness in their own health.
- Use health information for a better-informed strategic coming up with.
- Improve the data security.
- Exercise telemedicine.
- Fit in medical imaging for broader diagnosis.
- Avoid needless ER visits.

These samples of big data in medicinal services demonstrate that the event of medical applications of information ought to be the apple inside the eye of knowledge science as they require the possibility to spare heaps of money and most essentially, individuals' lives. As of now these days, it permits for early identification of sicknesses of individual patients and financial groups and accepting preventive activities as a result, we have a tendency to all understand, hindrance is best than cure [14].

2.3   The Most Effective Method to Use Big Data
in Healthcare

All things considered, these applications of big data in healthcare 3 principle patterns: the patient's ability may enhance drastically, together with best quality of medication and satisfaction; the overall healthcare of the residents should improved after sometime; and thus, the general cost should be reduced. We have a look for best approach to utilize enormous information in medicinal services, in a hospital for case [14] (Fig. 4):

This healthcare dashboard furnishes you with the outline required as a hospital director or as a facility manager. Assembling in one focal reason all the data on each division of the hospital, the gathering activity, its temperament, the costs acquired, etc., which is a good facilitate to run it swimmingly.

Here, the foremost necessary principal regarding varied aspects: The quantity of patients existed in your facility, on the other hand they stayed long time, how much cost to check patients, and the waiting time in crisis rooms. Such a holistic read helps top administration to determine potential bottleneck and patterns over time. This key will improve the common operations performance, improving the patient treatment and having the exact requirements for good staffing.

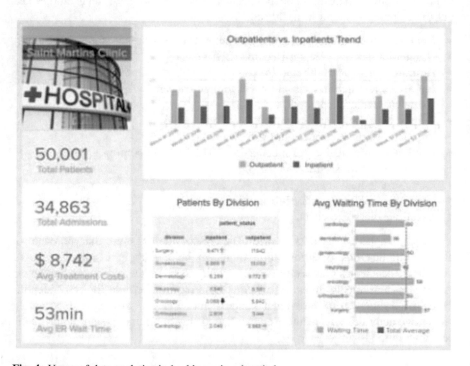

Fig. 4 Usage of data analytics in healthcare in a hospital

3 Diverse Tools to Analyze Big Data in Healthcare

Hadoop Ecosystem is an open-source framework by Apache software foundation written in Java for processing and storing large volumes of data with clusters of commodity nodes which solve big data problems. To analyze the huge data in less time and choose the right tool, here some of top big data tools in the areas of data mining, storing, integrating, cleaning, visualizing, extracting and analyzing [15].

- **HDFS (Hadoop Distributed File System)**: This is a core component of Hadoop framework and used for distributing data across a cluster of commodity hardwares.
- **Apache Sqoop**: A tool designed to transferring data between Hadoop and other relational database servers.
- **Apache Flume**: It is a framework which will efficiently collecting, aggregating, moving large amount of streaming data into HDFS.
- **Apache HBase**: A column-oriented database model which runs on top of Hadoop.
- **Apache Pig**: A platform/tool for analyzing massive amount of data in Hadoop in parallel.
- **Zookeeper**: A admin tool for managing the jobs in Hadoop clusters.
- **Apache Hive**: It is an open-source data warehouse, and it uses language called HiveQL which is similar to SQL.
- **Apache Ambari**: A tool which is responsible for keeping track of applications and their status.
- **Apache Spark**: A new way of running algorithms even faster than Hadoop.
- **Apache Mahout**: It is open-source framework, primarily used for creating machine learning algorithms.

4 A Conceptual Architecture of Big Data Analytics

The abstract structure for a big data analytics project is a standard health analytics project. In this health analytics project, a business intelligence platform will perform the analysis of data and put data into a complete system, like a portable computer, or a desktop. By definition, big data is data it is large in volume, variety and velocity and executes this massive data across multiple nodes. The distributed computing process has occurred for decades. Use terribly large data sets for analyzing new patterns and achieve big insight for creating better health-related decisions. Moreover, some of the open-source platforms like Hadoop or map reduce, available on the cloud, for developing any new application for analyzing massive data in healthcare [3, 4] (Fig. 5).

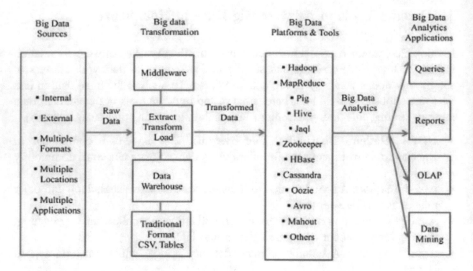

Fig. 5 Conceptual architecture of big data analytics

5 Methodology and Implementation

This paper described healthcare data set at first, later about the analysis of data provided using various tools like Hive and Spark. Then, discussed how efficiently the data will be processed.

Hive is an open-source data warehouse system designed on top of Hadoop. It uses language called HiveQL (HQL) which is similar to SQL. HQL is used for querying, analyzing table data which are put in storage as flat files on Hadoop Distributed File System. HQL automatically translates SQL like queries into a group of map reduce jobs which will execute on a Hadoop Cluster. Hive supports data partitioning and schemas to keeps the metadata in a relational database. Basically, Hive supports partitioning the table on a precise dimension. For instance, we could store patient particulars and partition the table on disease info. This allows for creating queries in an organized data model in the future [11].

Spark is a memory-based computing framework model, and it supports fault tolerance iterative, batch processing, interactive, and flow calculations. It absorbs the benefits of Hadoop Map Reduce model, but unlike Map Reduce, the data is kept in random access memory (RAM) instead of some slow disk drives and is processed in parallel, it is called as Memory Computing. This will improve the efficiency of data computing [8, 9].

5.1 *Implementation*

Healthcare data sets:

The data consists of different patient details which are used to analyze data in healthcare [6, 16]. The data set attributes are:

1. **Patient_Pid**: Unique identification of patients
2. **Patient_Pname**: Name of the patient
3. **Age**: Patient's age at diagnosis
4. **Gender**: Patient gender
5. **Disease_Info**: Type of the disease to analyze
6. **Hospital_Name**: Name of the hospital for treatment
7. **Admited_Date**: Patient admitted in hospital
8. **Address:** For communication (Fig. 6)

Below are some implementation steps for analyzing healthcare data set [10].

Step 1: Collect different disease information of the patients from various hospitals.

Step 2: Convert the data set into .csv format and copy the file to cloud X lab.

Step 3: Create a table using Structured Query Language, i.e., SQL.

`Query: create table patient_details (Patient_Idstring, Patient_Pname string,AGE int, GENDERvarchar (6), DISEASE_INFO string, HOSPITAL_NAME string, ADMITED_DATE varchar(15), ADDRESS string);`

Step 4: Load the data into MYSQL.

Fig. 6 Different patient details

Query: Load local file 'project/data/dataset.csv' into table patient_details FIELDS TERMINATED BY ',' ENCLOSED BY "" LINES TERMINATED BY '\n' ignore 1 lines;

Step 5: Import data from MYSQL to HDFS using SQOOP.

Query: sqoop import –connect jdbc:mysql://ip-172-31-20-247/sqoopex –username sqoopuser –password NHkkP876rp –table patient_details –target-dir project/healthcare/data/input/stage -m 1 –direct

Step 6: Execute commands in HUE Environment.

6 Results and Analysis

6.1 Using Hive

As we discussed earlier, Hive environment is used for the purpose of analyzing the data. In Hive, the patients' data set should be first loaded into it. The uploaded file is simply a comma separated file. Below figures show how the raw data is uploaded into Hive.

Step 1: Create table in Hive and load the patient details from HDFS using Query:

```
create table patient_details(Patient_Id string, Patient_
Pname string, AGE int, GENDER varchar(6), DISEASE_INFOstring,
H OSPITAL_NAME string, ADMITEDDATE varchar(15), ADDRESS
string);
```

Once the file is uploaded into a Hive environment, analysis can perform on the given data set. The given data set has 2999 records and eight attributes (Fig. 7).

Step2: Analysis of data in Hive without partitioning

Without partition in Hive, the query will give the result within 6.60 s for analyzing the cancer disease info. Analyze the patient details by using Query:

```
select HOSPITAL_NAME, count(DISEASE_INFO) as cancer_-
count from patient_details where DISEASE_INFO = 'Cancer'
group by HOSPITAL_NAME order by cancer_count desc;(Fig. 8)
```

Step 3: Create one more table in Hive with partition method using below Query:

```
create table patient_data (Patient_Id string,Patient_
Pname
string, AGE int, GENDER varchar(6), HOSPITAL_NAME
string, ADMITED_DATE varchar(15), ADDRESS string) parti-
tioned by (DISEASE_INFO string)location 'hdfs://ip-172-31-35-
141.ec2.internal:8020/user/mbabu60499425/project/healthcare/data/output/
disease';
```

Step 4: Load patient details into patient_data table using Query:

set hive.exec.dynamic.partition.mode = nonstrict;

```
insert into table patient_data partition (disease_info)
```

Fig. 7 Loading of raw data to Hive

Fig. 8 Query without partition in Hive: number of cancer patients visited to hospital

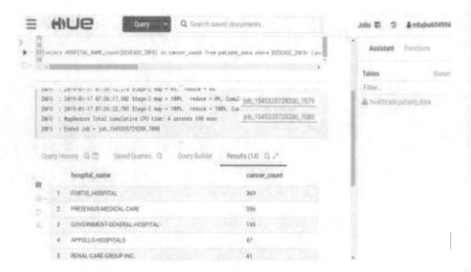

Fig. 9 Query with partition in Hive: number of cancer patients visited to hospital

```
select patient_id, patient_name, age, gender,
hospital_name, admited_date, address, disease_info from
patient_details;
```

Step 5: Analyze the partitioned table using below Query:

```
select  HOSPITAL_NAME,count(DISEASE_INFO)  as  cancer_-
count from patient_data where DISEASE_INFO = 'Cancer' group
by HOSPITAL_NAME order by cancer_count desc;
```
(Fig. 9)

The time taken in Hive environment to perform this analysis was recorded as 4.690 s.

6.2   Using SPARK

In Spark, the results are usually held on in memory rising potency of data computing. It also supports seamless data sharing between applications hence best suited for batch processing, ad hoc querying and streaming processing applications.

6.2.1   Spark SQL

Spark SQL may be a portion on top of Spark Core that presents a new data abstraction denoted as Schema RDD, which is responsible for structured and semi-structured data.

6.2.2 Spark Streaming

Spark Streaming influences Spark Core's for fast planning to perform streaming analytics. It ingests data in small batches and completes RDD (Resilient Distributed Datasets) transformations on those small batches of data.

6.2.3 Machine Learning Library

MLlib may be a distributed machine learning framework higher than Spark. Spark MLlib is nine times faster than Hadoop disk-based version of Apache mahout.

6.2.4 GraphX

GraphX may be a distributed graph processing framework on top of Spark. It provides an API for expressing graph computation that can model the user defined graphs by using Pregel abstraction Python.

6.2.5 Scala

Many of the high-performance data science frameworks will build on top of Hadoop usually which are written and use Scala or Java. Scala has amazing concurrency support. It also runs on the JVM, which makes it almost a no-brainer when paired with Hadoop. Similar to Java, Scala is an object oriented, and uses curly brace syntax like C programming language. Scala has many features of functional programming languages like Scheme, including currying, Standard ML and Haskell, immutability, lazy evaluation, type inference and pattern matching [5].

Here, we discussed some of the steps to analyze the data in spark.

Step1: Import Hive context to Spark and Load data into patient_input using Query:

```
import org.apache.spark.sql.hive.HiveContext
val    hiveContext = new    org.apache.spark.sql.hive.Hive
Context(sc) import hiveContext._
valpatient_input = sc.textFile
("/user/mbabu60499425/project/healthcare/data/input/
stage
")
```

Step 2: Analyze the data using Query:

```
val   cancer_hospitals = hiveContext.sql("select   hospi-
tal_name,count(patient_id) as patient_count from health-
care.patient_details where disease_info = 'Cancer' group by
hospital_name order by patient_count desc"); spark.time(-
cancer_hospitals.show) (Fig. 10).
```

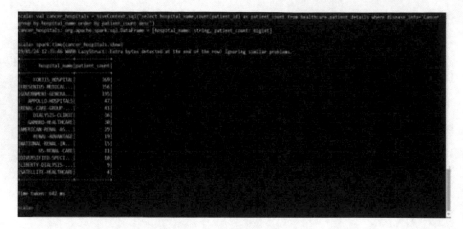

Fig. 10 Query in Spark: number of cancer patients visited to hospital

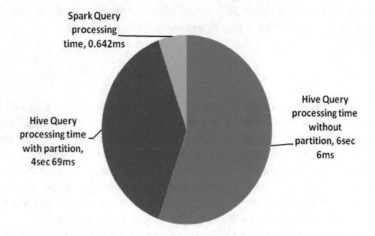

Fig. 11 Pie chart for Query processing time using Hive and Spark

Table 1 Time taken to process the healthcare data in Hive and Spark	S. No	Time (in s)	
	Hive	6.06 (without partition)	4.69 (with partition)
	Spark	0.642 (in memory)	

Spark will give the result within 0.642 s for analyzing the cancer disease info.
Results: In this exploration, it is observed that for processing the different patients' data with Spark tool taking less time than Hive (Fig. 11 and Table 1).

7 Conclusion

Big data analytics in healthcare will become a capable field for providing insight from very big data sets and improving outcomes on reducing costs. Its potential is so great; however, there also remain challenges to overcome.

Big data is a collection of data elements whose size, speed, complexity required to adopt software and hardware mechanisms to successfully store, analyze and visualize data. Healthcare is an important example to show how four Vs of data velocity, veracity, variety and volume are vital aspects of the data it produces. The data spread among multiple healthcare centers should provide a platform for global data transparency. Researchers are reviewing the complexity in healthcare data in terms of both characteristics of the data itself and the taxonomy of analytics that can be expressively performed on them. The goal of using Hive and Spark in healthcare is to collect and analyze data from public health trends in a region of people to identify better hospital options for each patient.

References

1. Zaveri, C.: Use of big-data in healthcare and lifescience using Hadoop technologies. Copyright. 978-1-5090-3239-6/17/$31.00©2017IEEE
2. Bhosale, H.S., Gadekar, D.P.: A Review paper on big data and hadoop. Int. J. Sci. Res. Publ. 4(10),1 (2014). ISSN 2250-3153
3. Raghupathi, W., Raghupathi, V.: Big data analytics in healthcare: promise and potential. Health Info. Sci. Syst. 2, 3 (2014)
4. Parimala, S.: A survey on security and privacy issues of big data in healthcare industry and implication of predictive analytics. Int. J. Inn. Res. Comput. Commun. Eng. 0504098. https://doi.org/10.15680/ijircce.2017
5. JayaLakshmi, G., Srisaila, A., MadhaviLatha, P.: Enhancment of healthcare outcomes using big data analytics. IJLTET 7(3) (2016). ISSN:2278-621X
6. Panda, R.P., Barik, P.P., Prusty, P.A.K.: A review paper on big data in lung cancer big data analytics in lung cancer. Int. J. Trend Res. Develop. 3(5) (2016). ISSN: 2394-9333 IJTRD
7. Nandhini, S.G., Nandhini, V., Lavanya, K., Kokilam, V.: Big data analytics in health care. IJIRT 2(4), (2015). ISSN: 2349-6002
8. Shrutika Dhoka, R., Kudale, A.: Use of big data in healthcare with spark. Int. J. Sci. Res. (IJSR). ISSN (Online): 2319-7064 Index Copernicus Value (2013): 6.14| Impact Factor (2015): 6.391
9. Liu, W., Li, Q., Li, X.: A prototype of healthcare big data processing system based on spark. In: 2015 8th International Conference on Biomedical Engineering and Informatics (BMEI). https://doi.org/10.1109/bmei.2015.7401559
10. Durga Sri, B., Nirosha, K., Padmaja, M.: Healthcare Analysis Using Hadoop. 4(6), 2017. ISSN (PRINT): 2393-8374, (ONLINE): 2394-0697
11. Sadhana, S., Shetty, S.: Analysis of diabetic data set using Hive and R. Int. J. Emer. Technol. Adv. Eng. 4(7) (2014). ISSN 2250-2459, ISO 9001:2008 Certified Journal. Website: www.ijetae.com

12. https://www.scribd.com/document/107279699/Big-Data-in-Healthcare-Hype-and-Hope
13. https://www.scnsoft.com/blog/4-types-of-data-analytics
14. https://www.datapine.com/blog/big-data-examples-in-healthcare/
15. https://www.kdnuggets.com/2014/08/18-essential-hadoop-tools.html
16. https://www.kaggle.com/rajr16/diseaseinfo

Long Short-Term Memory with Cellular Automata (LSTMCA) for Stock Value Prediction

N. S. S. S. N. Usha Devi and R. Mohan

Abstract Time series analysis is a difficult task as the data changes dynamically with time. Stock value prediction is one of the examples of highly variable data which cannot be averaged for computation purpose. Even though many approaches are available for predicting stock values, still there is plenty of room for thinking of a better classifier with more accuracy and adaptability. Many statistical analysis methods are available for predicting the stock where the predictability low when tested with different data sets. We have studied the existing literature of both statistic and dynamic classifiers and arrived at a classifier developed with a deep learning technique augmented with cellular automata. We propose a long short-term memory cellular automata (LSTMCA) classifier for news data set in order to predict stock level either increase or decrease in next ten days. We have tested our classifier LSTMCA with existing literature and found an improvement of 2.67% when tested with news data sets.

Keywords Stock prediction · Deep learning · Cellular automata · LSTM · Time series

1 Introduction

The prediction of the stock market is usually carried through algorithms, which are automated, and most of the investors have an option to use statistical methods like average, relative strength indexing and other indexes. These indicators may give some future intuition about the stock value variations. Assets share or price is decided by the demand and supply, likewise, stock market variations also change accordingly [1]. Political, social, economic and cultural events have particular

N. S. S. S. N. Usha Devi (✉) · R. Mohan
Department of CSE, NIT, Tiruchirappalli, India
e-mail: usha.jntuk@gmail.com

R. Mohan
e-mail: rmohan.nitt@gmail.com

© Springer Nature Singapore Pte Ltd. 2020
K. S. Raju et al. (eds.), *Data Engineering and Communication Technology*,
Advances in Intelligent Systems and Computing 1079,
https://doi.org/10.1007/978-981-15-1097-7_70

Fig. 1 Feed-forward neural
network

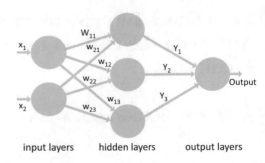

input layers   hidden layers   output layers

effects on public mood in all the dimensions. The moods can be predicted using Twitter tags, Facebook discussions and many other people involvement platforms.

Stock market prediction is dependent on various mechanism like political announcements, political implications, global financial condition, holidays, rumors, organization status, news [2], Twitter analysis, corporate communication, communication between CEO'S and employees, etc. Stock market variations can also be found from Yahoo finance, Facebook, Google trends, close price, etc.

Deep neural network (DNN) [3] is a wide and deep neural network which has proven power to deal with such dynamic problems like stock market prediction.

In Feed-forward neural network (FFNN), the flow of information will be in one direction, i.e., from the input to hidden to outputs layers as shown in Fig. 1. The entire information will travel directly from beginning to end of the network. The information never taps a node more than once because of this architecture. Feed-forward neural networks have no memory of the input they received previously and are therefore if fares bad in predicting what is coming next. Because a feed-forward network only considers the current input, it has no notion of order in time. They simply cannot remember anything about what happened in the past, except their training.

Long short-term memory (LSTM) networks [4] are extensions of recurrent neural networks where basically memory is extended. Therefore, it is well suited to learn from important experiences that have very long time lags in between. The units of an LSTM are used as building units for the layers of a RNN, which is then often called an LSTM network.

A cellular automation consists of a number of cells structured in the form of a grid with a finite number of states, which operates on a computational grid. It is a computing model that provides a good platform for performing complex computations with the available local information.

$$CA \text{ is defined as four tuple} <G, Z, N, F>$$

where

G Grid (Set of cells)
Z Set of possible cell states

N Set that describes cells neighborhoods
F Transition function (rules of automata)

We have used this CA with rules to find the nearest neighbors (semantic relevance) which will have the similar meaning during sentimental analysis. Different types of CA are used in pattern classification, fuzzification, NLP, bioinformatics, speech recognition, image processing, big data analytics, cloud computing, etc., for identifying similarity features using fuzzy rules.

2 Literature Review

Some of the authors have used inducing emotional stabilizer mechanism to predict stock movement [5]. Few of them have tested sentiment analysis in which unstructured textual documents are converted into daily sentiment indexes [6]. Some of the machine learning model addresses stock movement prediction [6–8]. Support vector regression (SVR) algorithm addresses non-normality, high-noise in high-frequency stock data [7]. SVR with particle swarm optimization [7] can predict the stock values. Mining email communications and historical records [9] can assess the values of stock. The summary of the pros and cons of the literature survey is listed in Table 1.

Sheng et al. [10] has studied the impact of media attention, prominence and valence in case of stock market. Authors have summarized the influence web media on potential financial markets with some quantification with some constraints. Sushree et al. [11] have applied sentimental analysis on Twitter data for predicting

Table 1 Pros and cons of various existing methods

References	Problem addressed	Method	Limitations
[10]	Web media impact on stock movement	Linear and regression model	Term vectors do not work as this social media contain slag data. Rumor news may lead to inaccuracy
[5]	Emotion stabilizer mechanism	Investor agent algorithm	Done statistical analysis using closing price restricted to a emotional bandwidth
[7]	Analyzing stock data at three different time scales	Particle swarm optimization and SVR	Adaptive SVM is required to meet the required efficiency
[6]	Investor sentiment analysis	SVM model	Static due to statistical analysis
[9]	Corporate communication	Pattern recognition using DM	Sample space is limited

the variations in the stock. The study says that sentimental analysis is inefficient due to domain-independent dictionary [12], and the media prioritizes huge markets only [13]. The reliability of predictions is more for these works.

3 Design of LSTMCA

The abstract design of LSTMCA is shown in Fig. 2. As we are in the process of predicting stock values from new data sets, we will initially take the news data sets and process them, and these data sets are used to form a financial dictionary to measure the sentimental analysis. Then, we assign values to these words by embedding vector method [14]. Each word is converted into a word by mapping with 100 spaces. We have used CA to find the similarities between two words faster and efficiently. We then train the proposed classifier LSTMCA initially with these data sets. Finally, the trained classifier is tested with the remaining data sets.

Word embedding is one of the most popular representations of document vocabulary. It is capable of capturing context of a word in a document, semantic and syntactic similarity, relation with other words, etc. It is vector representations of a particular word. Word2Vec is one of the most popular techniques to learn word embeddings using shallow neural network. The Word2vec representation supports in identifying the analogies of word set W with distance 10 to get score to each $w_i \epsilon W$. It was developed by Tomas Mikolov in 2013 at Google. CA is used for semantic relevance based on score of each word.

John Von Neumann has proposed cellular automata [15] which consist of 29 states for each cell. After his proposal, so many researchers have proposed different structural variations to CA [16]. A self-reproducing CA is introduced by Arbib [17]. Eight states per cell are introduced by Codd [18]. The five-neighborhood is proposed by Von Neuman [15], and nine neighborhood is proposed by Moore [1].

Fig. 2 Design of LSTMCA

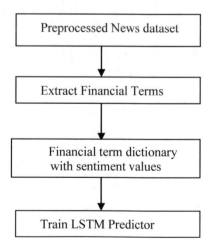

Here, LSTMCA is used as a predictor on training with statements with their scores obtained in previous step. While testing LSTMCA, this sentiment score factor is added to actual price improves the prediction accuracy.

4 Results and Discussion

We have collected 75,000 of data sets from news data repository and used half of the data sets for training and other half for testing. We have used a cost function framed with CA which measures the efficiency of similarity in between two words as shown in Fig. 3 (Figs. 4 and 5).

Visualizing word2vector embedding is done by dimensionality reduction using principal component analysis. PCA is used for dimensionality reduction in order to visualize the word embeddings. To understand the semantic relevance between w_i and w_j belong to W for any i, j in dictionary as shown in Fig. 6.

We have tested the developed classifier with sample AXIS bank data from NSE to represent the actual fluctuations in stock and obtained the following predictions as shown in Fig. 4 by our proposed framework. We have compared our approach with the standard approaches ANN and DNN and found that we have a better accuracy compared with these with respect to accuracy and stock value prediction as shown in Fig. 5. The average accuracies comparison was depicted in Fig. 7.

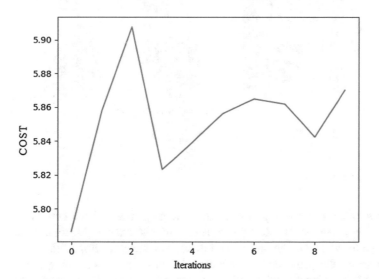

Fig. 3 Cost function variations with iterations

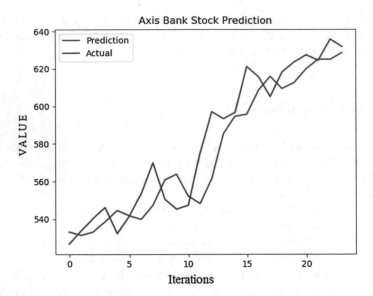

Fig. 4 Axis bank stock value prediction

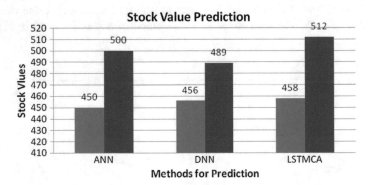

Fig. 5 Stock value prediction comparisons

5 Conclusion

We have examined the volatility of the emerging financial markets that affect the global market and developed a versatile classifier which can predict the stock values dynamically with the changes in the news. We have developed the classifier and implemented the same and obtained an increase in the classifier accuracy over 2.67%. We are trying to adapt this classifier for more real-time data sets and obtain the accuracy so that this can be in future can be made as an benchmark.

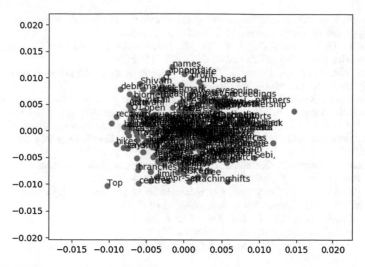

Fig. 6 Visualizing word semantic similarity

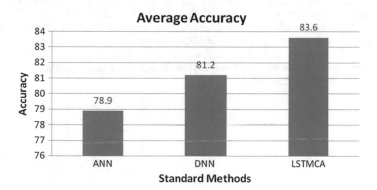

Fig. 7 Average accuracy comparisons

References

1. David, G., Moore, C. (eds.): New constructions in cellular automata. Oxford University Press on demand (2003)
2. Sheng, J., Lan, H.: Business failure and mass media: an analysis of media exposure in the context of delisting event. J. Bus. Res. (2018)
3. LeCun, Y., Bengio, Y., Hinton, G.: Deep learning. Nature **521**(7553), 436 (2015)
4. Martin, S., Schlüter, R., Ney, H.: LSTM neural networks for language modeling. In: Thirteenth Annual Conference of the International Speech Communication Association (2012)
5. Cabrera, D., Cubillos, C., Cubillos, A., Urra, E., Mellado, R.: Affective algorithm for controlling emotional fluctuation of artificial investors in stock markets. IEEE Access **6**, 7610–7624 (2018)

6. Ren, R., Wu, DD., Liu, T.: Forecasting stock market movement direction using sentiment analysis and support vector machine. IEEE Sys. J. (2018)
7. Guo, Y., Han, S., Shen, C., Li, Y., Yin, X., Bai, Y.: An adaptive SVR for high-frequency stock price forecasting. IEEE Access **6**, 11397–11404 (2018)
8. Li, X., Xie, H., Chen, L., Wang, J., Deng, X.: News impact on stock price return via sentiment analysis. Knowl.-Based Sys. **69**, 14–23 (2014)
9. Zhou, P.-Y., Keith CCC., Ou, CX.: Corporate communication network and stock price movements: insights from data mining. IEEE Trans. Comput. Soc. Sys **5**(2), 391–402 (2018)
10. Li, Q., Chen, Y., Wang, J., Chen, Y., Chen, H.: Web media and stock markets: a survey and future directions from a big data perspective. IEEE Trans. Knowl. Data Eng. **30**(2), 381–399 (2018)
11. Sushree, D., Behera, R.K., Rath, S.K.: Real-time sentiment analysis of twitter streaming data for stock prediction. Procedia Comput. Sci. **132**, 956–964 (2018)
12. Katayama, D., Tsuda, K.: A method of measurement of the impact of Japanese news on stock market. Procedia Comput. Sci. **126**, 1336–1343 (2018)
13. Wu, G.G.-R., Hou, T.C.-T., Lin, J.-L.: Can economic news predict Taiwan stock market returns? Asia Pacific management review (2018)
14. Mikolov, T., Chen, K., Corrado, G., Dean, J.: Efficient estimation of word representations in vector space. arXiv:1301.3781 (2013)
15. Aspray, W.: John Von Neumann and the origins of modern computing, vol. 191. Mit Press, Cambridge, MA (1990)
16. Sree, P.K, Babu, I.R.: Identification of protein coding regions in genomic DNA using unsupervised FMACA based pattern classifier. arXiv:1401.6484 (2014)
17. Arbib, M.A.: Simple self-reproducing universal automata. Inf. Control **9**(2), 177–189 (1966)
18. Codd, E.F.: Cellular automata. Academic Press (2014)
19. Kimoto, T., Asakawa, K., Yoda, M., Takeoka, M.: Stock market prediction system with modular neural networks. In: 1990 IJCNN International Joint Conference on Neural Networks, pp. 1–6. IEEE (1990)

Detection of Lightening Storms in Satellite Imagery Using Adaptive Fuzzy Clustering

Gurram Sunitha and J. Avanija

Abstract Natural hazards are the environmental managers that balance nature and environment. The earth's environmental system has become a cardinal area of research and development. As such, the meticulous and methodical studies of natural disasters are gaining its prominence in all environmental science institutions and industries. Despite the small size, most of the thunderstorms are usually dangerous. This makes the automated detection of lightning storms in satellite imagery not only important but also indispensable. Many different techniques have been proposed and evaluated to detect lightning storms in satellite imagery. Nevertheless, the accuracy of detection is still a challenging concern. Here, a method has been proposed for efficient detection of the lightning storms in order to colossally improve the accuracy of the automated lightning storm detection rate, thus supporting the study, observation and characterization of the lightning storms in the satellite imagery.

Keywords Image processing · Lightning storm detection · Adaptive fuzzy k-means clustering · Image segmentation · Haar wavelet transform

1 Introduction

Natural hazards are the prime processes of nature which cause highest levels of natural disasters. A natural hazard either a geographical hazard or biological hazard leaves its long-term negative traces on humans as well as environment. There are various factors of concern for the occurrence of natural hazards like geological, hydrological, and climatic factors. The enormous revolution in technology, industry

G. Sunitha · J. Avanija (✉)
Department of CSE, Sree Vidyanikethan Engineering College,
Tirupati, Andhra Pradesh, India
e-mail: avanija03@gmail.com

G. Sunitha
e-mail: gurramsunitha@gmail.com

© Springer Nature Singapore Pte Ltd. 2020
K. S. Raju et al. (eds.), *Data Engineering and Communication Technology*,
Advances in Intelligent Systems and Computing 1079,
https://doi.org/10.1007/978-981-15-1097-7_71

and lifestyle of humans also has severe effect on the occurrence and intensity of natural hazards. Each natural disaster has its own factors and complications. Understanding the basic principles of ecology can provide keys to lessening their effects. The natural hazards like famines, drought, epidemics, etc., have slow onset on their contribution to death tolls. Natural hazards like floods, tsunami, storms, hurricanes, earthquakes, avalanches, etc., have rapid onset on their contribution to death tolls. Even though the slow onset type of hazards cause 50–60% [1] more deaths than rapid-onset type of hazards, the indirect and high-order losses caused by rapid onset hazards are tremendous.

Many disaster losses are the predictable result of interactions among three major systems—the physical environment, social characteristics and demographic characteristics of the ground zero communities [2]. Also, most of the geographical hazards stand as the base cause for biological hazards like diseases, infections, etc. Natural hazards that overpower the human capacity to respond and manage, lead to long-term damage to humans and environment. The effects to human society include social, political, financial, economical, mental and epidemical. The natural disaster management process comprises of a multi-layered approach with the primary target to minimize the loss of human life and prevent spread of infectious diseases. Not all natural disasters can be prevented. But, efficient and timely prediction of natural hazards can give time to the government to pioneer the attitude of natural disaster management departments. It allows to reduce the exposure and vulnerability of human society and to improve resilience and emergency responses to a range of disasters. In this scenario the researchers, government and private industry are striving with cumulative effort toward the development of knowledge, forecasting and decision tools for disaster prevention and management. The disaster management departments are initiating efficient alert systems in the event of natural hazards [3].

Disasters are influence of natural hazards and human-created environment [1]. The first step is for researchers and disaster management departments to recognize that climate change impacts do not occur in isolation, but are strongly coupled. Considering compound disasters and cascading disasters as two different events and studying them is no more apposite. Approaching the disasters as a chain of events can give prominent insights into the occurrence of natural hazards. This allows us to take up preventive measures for a short term and the countries to design policies to reduce the communal influence on nature. Global systems need to be developed to assess compounding hazards and cascading hazards. Researchers should go beyond calculating the statistics of hazards in isolation and delve into their interactions with natural and built environments.

Natural hazards have been studied and researched all over the world, for understanding their basic nature, their intensity, aggression, timeline, influence, extent of damage, etc. But in many cases there are also deeper causes, which offer a more powerful explanation. The rapidly changing nature of hazards and climate shifts due to industrial revolution, urbanization and population growth is challenging the researchers to develop new techniques for assessing coupled risks. The efforts to minimize the extent of damage caused by disasters have become more multi-disciplinary and in line. Thus, the proximate cause of research in this paper is on understanding the causes of natural disasters with rapid onset to identify clues to their prevention. The work pursued in this paper aimed at efficient detection of lightning storms in the given satellite imagery dataset.

The contents in the paper are organized as follows: Sect. 2 presents the related work. Section 3 presents the proposed methodology. Section 4 presents the experimental evaluation of the proposed methodology with Sect. 5 presenting the conclusions and the future directions.

2 Related Work

Prevention of natural hazards is not possible. But the efficacy of disaster management can be rigorously improving by timely prediction of hazards and by proficient preparation of plans to control the intensity of their onsets, their effects on humans and environment. The disaster management departments must have a vast and sustainable alerting system for prior intimation of emergencies to the people. This enables the local disaster management teams for faster evacuation of communities to safer places. The proposed work in [3] has assessed the major parameters for a best alert and warning system. Also, usage of various communication channels like mass media, social networks, cell broadcast, etc., has been evaluated for the alert and warning system.

In [4], techniques have been proposed for information sharing, maintenance and collaboration thus proposing a digital platform for disaster planning, management and recovery. A prototype of Web application and mobile application has been developed to support the processes of data collection from victims, authorities, social services, hospitals, etc. Then, using appropriate data mining techniques, summarized report generation at varying granularities and perspectives can be generated to provide insights into recovery process over a period of time. This will help disaster management departments to collaborate recovery activities between victim communities, social services, various local authorities and utility agents. Effective information management and flow between various entities are critical issues in disaster management.

In [5], it has been prioritized that one of the long-term impacts of natural disasters is on economy. They discussed about the assessing the long-term consequences of disasters on economy. Even though researchers claim that they have models for the problem, they are not sufficient considering the scale and complexity of the problem at hand.

In [6], a Web service WEB-IS has been conceptualized. NaradaBrokering system acts as middleware between WEB-IS client and various servers like computation server, visualization server, etc. They have also developed a high-performance visualization package, data analysis tools with Web integration system toolkit. The system supports the clustering schemes, feature generation and extraction techniques, visualization algorithms, and faster access of earthquakes data distributed among data acquisition servers located at different geographical regions. A grid framework has been considered, assessed and evaluated for the purpose of developing an integrated system for faster and up-to-date collection of seismic data from various centers in the world. The WEB-IS provides three-dimensional interactive visualization and exploration of earthquake events.

In [7], the authors have considered six parameters of thunderstorms and have designed a thunderstorm prediction model. The models are based on adaptive neuro-fuzzy inference system with fuzzy c-means clustering and fuzzy subtractive clustering. The models push interactive analysis deeper into the process. The domain expert can push his domain knowledge deep into the prediction process. Experimental evaluation of the models has been done using real-world data from Tawau meteorology station

Numerical weather prediction models have always been the challenging problem for the researchers. Even though the prediction models have been extensively explored and designed, because of the epic change in the environment and the natural hazards, the existing models have become insufficient and in appropriate. Also, the models have to be customized as per the needs of the local regions. This problem has been of major concern for the government and weather forecasting departments to efficiently minimize the effects the disasters. In this regard, Australian Bureau of Meteorology has launched a thunderstorm prediction system Calibrated Thunder [8]. The Calibrated Thunder system produces coarse predictions of thunderstorms for upcoming 48 h and fine predictions for upcoming 24 h.

Under Severe Thunderstorm Observations and Regional Modeling (NASA STORM) project, the ground-based lightening detection capabilities have been extended with portable lightning mapping array (LMA) package [9]. Five LMA sensors have been deployed for data collection. The lightning strikes data from five LMA sensors are integrated to enhance the lightning strikes prediction and regional modeling.

The dynamics of a mesoscale convective cloud system are much more complex and quickly time-varying, hence making the process of characterizing such system a tough task. Also, they are widely affected by the weather characteristics of the local regions. The mesoscale convective cloud system quickly changes its volume and intensity by the process of cloud splitting and cloud merging causing heavy hailfall and rainfall events. Models for region-based and real-time forecasts are needed for prospective risk assessment. A storm-scale model is designed for short-range prediction of heavy rainfall and hailfall to generate sufficiently prior flood warnings [10].

Sawaitul et al. [11] used back propagation algorithm in order to predict the future weather. It is focused on three parts; different models used in weather forecasting, introducing a new wireless kit which is used for weather forecasting and the back propagation algorithm which can be applied on different parameters of weather forecast. Image retrieval and content retrieval are two methods used for retrieving the data in case of weather forecasting. Back propagation algorithm is used to find the variations in weather forecasting dataset with respect to the change in one parameter. Wireless sensors are also described which are used for weather forecasting. Wireless sensors are used for recording the parameters, such as wind, rain fall, temperature and humidity, respectively.

3 Detection of Lightning Storms in Satellite Imagery

Thunderstorms are highly unpredictable in nature. Their occurrence may cause severe weather complications in local regions bringing flash floods, hailfalls, strong winds and tornadoes. Thundershowers are weaker thunderstorms which are more difficult to detect than thunderstorms. The proposed methodology supports efficient detection of lightning storms in the given satellite imagery.

The satellite images are initially placed through an image segmentation process for identifying the presence of the lightning storm in the imagery. For the purpose of image segmentation, the adaptive fuzzy k-means algorithm [12] has been used. In this process, segmentation is performed by considering various color factors, because color possesses wavelength values of an image. Similar wavelength values of an image are grouped into different clusters. Figure 1 shows an example of a given image and the clusters formed after applying fuzzy k-means clustering technique.

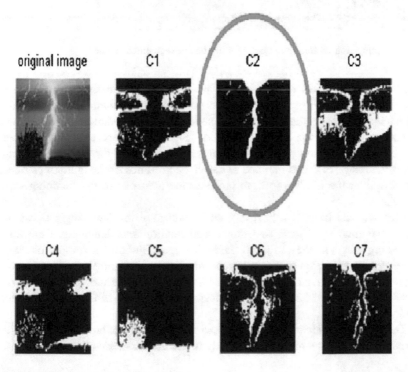

Fig. 1 Original image and segmented images

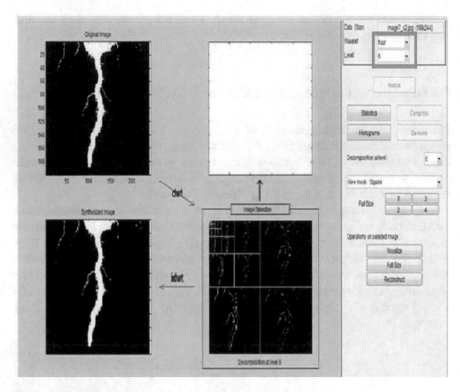

Fig. 2 Application of Haar wavelet transform on the segmented image

Among the clusters formed, one of the clusters is used for further processing. Haar wavelet transform [13] is used to further perform denoising process on the segmented imagery. These two steps, segmentation and denoising, form the pre-processing stage in the proposed method. Figure 2 shows a sample of clustered image after application of Haar wavelet transform. Then, the wavelength range of each such image is then computed to detect the presence of the lightning strike. The wavelength range of 320–450 nm indicates the presence of the lightning strike in the image.

Haar wavelet transform [13] is used to then perform denoising process on the segmented imagery. These two steps, segmentation and denoising, form the pre-processing stage in the proposed method. Figure 2 shows a sample of clustered image after application of Haar wavelet transform. Then, the wavelength range of each such image is then computed to detect the presence of the lightning strike. The wavelength range of 320–450 nm indicates the presence of the lightning strike in the image.

The eventual objective of the proposed method is to perform an investigation on the satellite images in order to accurately detect the existence of lightning storms in

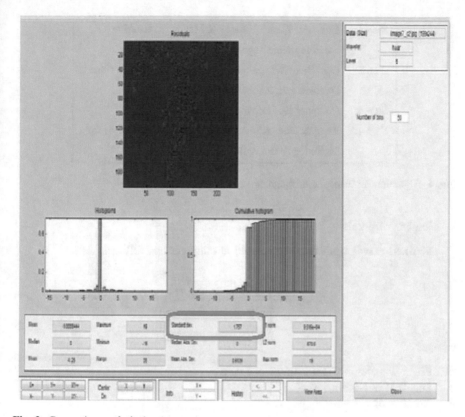

Fig. 3 Computing analytical values

the given satellite imagery. Figure 3 shows a sample of computing analytical variable values from the image. The algorithm for the proposed work is shown in Fig. 4.

3.1 Computing Analytical Values

In Haar wavelet transform technique [13], the threshold value is computed as follows.

$$\tau = \sqrt{\frac{2\sigma^2 \log(n)}{n}} \tag{1}$$

where

τ Threshold value
σ Standard deviation

Step 1:	Use Adaptive Fuzzy K-Means Clustering for Satellite Image Segmentation.
Step 2:	Use Haar Wavelet Transform Method to Perform Denoising on Segmented Image.
Step 3:	Compute the Variable Values from the Image.
Step 4:	Analyze the Results to Detect the Existence of Lightning Storm in the Image.

Fig. 4 Algorithm for detection of lightning storms in satellite imagery

n Number of pixels

The fixed wavelength factor threshold is computed as follows.

$$\tau_k = \int\limits_0^{255} \frac{d\Delta}{d\lambda} \sqrt{\frac{2\sigma^2 \log(n)}{n}} dn \qquad (2)$$

where

τk Fixed wavelength factor threshold value
σ Standard deviation
n Number of pixels

3.2   Determining Existence of Lightning Storm

On computing the wavelength, the wavelength range between 320 and 450 nm indicates the presence of lightning storm in the considered image.

4   Experimental Evaluation

To prove the efficiency of the proposed approach, experimental evaluation is performed using MATLAB 2013a. Consider the image shown in Fig. 3. The image in hand has been segmented using adaptive fuzzy k-means clustering technique and one of the segments (clusters) is selectively chosen that specifically contains cluster of the lightning storm (Fig. 1). It is further processed by applying Haar wavelet transform with Level 6 to convert the data from spatial domain to frequency domain (Fig. 2). Further, the analytical values are computed as discussed in Sect. 3 to determine the existence of lightning storm in the considered image (Fig. 3). For the image dataset in hand, the above process is performed on each image.

The following accuracy measures have been considered to prove the accuracy of the proposed approach for detection of lightning storms.

$$\text{Sensitivity} = \frac{TP}{TP + FN} \tag{3}$$

$$\text{Specificity} = \frac{TN}{TN + FP} \tag{4}$$

$$\text{Accuracy} = \frac{TP + TN}{TP + TN + FP + FN} \tag{5}$$

$$\text{Precision} = \frac{TP}{TP + FP} \tag{6}$$

To prove the efficiency of the proposed approach, experimentation has been performed on the dataset containing of eighteen images. Table 1 shows the computational results. From the results of Table 1, the above-specified performance measures have been computed and tabulated as shown in Table 2.

Table 1 Detection of lightening storms

Image ID	σ	τ	τ_k	λ	Detection of thunderstorm by proposed approach	Detection of thunderstorm [3]
Image 1	1.4424	10	61.66	375.33	Yes	Yes
Image 2	0.976	8.125	53.22	549.23	No	Yes
Image 3	1.41	10.25	36	371.52	Yes	Yes
Image 4	1.027	10.87	50.58	521.98	No	No
Image 5	0.7942	10.75	65.41	675.03	No	No
Image 6	1.5	11	29.56	360.05	Yes	Yes
Image 7	1.358	11.25	38.25	394.74	Yes	Yes
Image 8	1.243	10.25	41.79	431.27	Yes	Yes
Image 9	1.43	12.38	36.32	374.822	Yes	Yes
Image 10	1.48	11.38	35.1	362.23	Yes	Yes
Image 11	1.412	11.38	36.79	379.67	Yes	Yes
Image 12	1.199	11.25	43.32	447.06	Yes	Yes
Image 13	1.17	10.75	44.4	458.2	Yes	Yes
Image 14	0.8698	7.813	59.72	616.31	No	No
Image 15	1.779	10.25	29.2	301.344	Yes	Yes
Image 16	1.465	10.88	35.45	365.844	Yes	Yes
Image 17	1.115	10.75	46.59	480.8	Yes	Yes
Image 18	1.632	10.56	31.83	328.48	Yes	No

Table 2 Performance of proposed approach

Performance measure	Accuracy (%)
Sensitivity	95
Specificity	90
Accuracy	93
Precision	92

The proposed approach is compared by using fuzzy c-means [14] for image segmentation. The experimental results of the proposed approach have been compared with the results of the approach proposed in [14]. The in-depth evaluation proved that the proposed approach is more accurate than the compared approach for the dataset considered. The experimental results clearly show that adaptive k-means clustering gives better detection accuracy than fuzzy c-means.

5 Conclusions and Future Scope

The work proposed here, uses adaptive fuzzy k-means clustering and Haar wavelet transform techniques for the detection of lightning storms in the satellite imagery. A statistical analysis has been performed for efficient detection of lightning storms in the satellite imagery. The experimental results show that the proposed method outperforms the compared method in the detection of lightning storms in satellite imagery. An average accuracy of 93% is observed while experimenting on the considered dataset using the proposed method.

Additional experiments in the same direction of research work can be attempted, however, with different wavelet transform applications, such as DB, SYM, COIF, BIOR, RBIO and DMEY to further improve the efficiency of the approach with regard to varying performance measures. Further, the results can be analyzed and better applications could be achieved.

References

1. Blaikie, P., Terry, C., Davis, I., Wisner, B.: At risk: natural hazards, people's vulnerability and disasters. Routledge (2004)
2. Haque, C.E. (ed.): Mitigation of natural hazards and disasters: international perspectives. Springer (2005)
3. Hirst, P., Ramírez J., Mendes, M., Ferrer-Julià, M.: Alert systems in natural hazards: best practices. In: Exploring Natural Hazards, pp. 111–140. Chapman and Hall/CRC (2018)
4. Zheng, L., Shen, C., Tang, L., Zeng, C., Li, T., Luis, S., Chen, S.C.: Data mining meets the needs of disaster information management. IEEE Trans Hum.-Mach. Sys. **43**(5), 451–464 (2013)
5. Noy, I., DuPont IV, W.: The long-term consequences of natural disasters—a summary of the literature. (2016)

6. Yuen, D.A., Kadlec, B.J., Bollig, E.F., Dzwinel, W., Garbow, Z.A., da Silva, C.R.: Clustering and visualization of earthquake data in a grid environment. Vis. Geosci. **10**(1), 1–12 (2005)
7. Suparta, W., Putro, W.S.: Parametric studies of ANFIS family capability for thunderstorm prediction. In: Space Science and Communication for Sustainability, pp. 11–21. Springer, Singapore (2018)
8. Loo, Y.Y., Billa, L., Singh, A.: Effect of climate change on seasonal monsoon in Asia and its impact on the variability of monsoon rainfall in Southeast Asia. Geosci. Front. **6**(6), 817–823 (2015)
9. Schultz, C.J., Gatlin, P.N., Lang, T.J., Srikishen, J., Case, J.L., Molthan, A.L., Jedlovec, G.J.: The NASA severe thunderstorm observations and regional modeling (NASA STORM) project (2016)
10. Spiridonov, V., Curic, M.: A storm modeling system as an advanced tool in prediction of well organized slowly moving convective cloud system and early warning of severe weather risk. Asia Pac. J. Atmos. Sci. **51**, 61–75 (2015)
11. Sawaitul, S.D., Wagh, K.P., Chatur, P.N.: Classification and prediction of future weather by using back propagation algorithm—an approach. Int. J. Emerg. Technol. Adv. Eng. **2**(1), 110–113 (2012)
12. Sulaiman, S.N., Isa, N.A.M.: Adaptive fuzzy-K means clustering algorithm for image segmentation. IEEE Trans. Consum. Electron. **56**(4) (2010)
13. Qiang, Y.: Image denoising based on Haar wavelet transform. In: 2011 International Conference on Electronics and Optoelectronics (ICEOE), vol. 3, pp. V3–129. IEEE (2011)
14. Kannan, S.R., Ramathilagam, S., Chung, P.C.: Effective fuzzy c-means clustering algorithms for data clustering problems. Expert Sys. Appl. **39**(7), 6292–6300 (2012)

Detection of Concept-Drift for Clustering Time-Changing Categorical Data: An Optimal Method for Large Datasets

K. Reddy Madhavi, A. Vinaya Babu, G. Sunitha and J. Avanija

Abstract Finding optimal method for clustering of time-changing categorical data in large datasets is always lagging and has attracted significant interest in clustering domain. As the data in the World Wide Web has been growing exponentially, change in the concept occurs between every cluster called concept-drift. Firstly, the given dataset is divided into subsets, called sliding windows. Since the new data enters always into the database, effectiveness of the clustering method needs to be evaluated continuously and dynamically. This paper focuses on an optimal method for clustering time-changing categorical data. This method evaluates the effectiveness of the clustering process for each of new input data subset and considers the stability with respect to the previous clustering method. The proposed algorithm follows iterative procedure to obtain optimal solution of the clustering method with few constraints. Further, it detects drifting concepts from the optimal method. This paper proposes a detection of drift and the optimal method for catching the progression trend of cluster structures in time-changing categorical domain. Finally, the effectiveness of proposed algorithm is illustrated using the experimental study on KDD-CUP'99 dataset for various sizes of sliding windows.

Keywords Concept-drift detection · Sliding window · Optimal method and categorical data

K. Reddy Madhavi (✉) · G. Sunitha · J. Avanija
Sree Vidyanikethan Engineering College, A. Rangampet, Tirupati, India
e-mail: kreddymadhavi@gmail.com

A. Vinaya Babu
Stanley College of Engineering and Technology for Women,
Abids, Hyderabad, India

© Springer Nature Singapore Pte Ltd. 2020   861
K. S. Raju et al. (eds.), *Data Engineering and Communication Technology*,
Advances in Intelligent Systems and Computing 1079,
https://doi.org/10.1007/978-981-15-1097-7_72

1 Introduction

Data mining is an interdisciplinary subfield of computer science and statistics and defined as the process of analyzing hidden patterns from large data sets with intelligent methods according to different perspectives involves methods at the intersection of statistics, machine learning, and database systems, then categorizing into comprehensible structure facilitating decision-making and other information requirements for ultimate cut costs and increase revenue [1]. Data mining algorithms deal with hundreds or even thousands of various types of attributes like ordinal, numerical, categorical/nominal, and ratio-scaled.

Data clustering is an important technique and is a part of unsupervised machine learning and statistical multivariate analysis has been the center of attention of generous research in numerous domains for decades. The process of generating clusters is based on the property that: Given a set of objects are partitioned into groups taking objects in the same cluster are with high similarity while objects in other cluster are dissimilar. Hence, clustering analysis gains insight into the data distribution [2, 3]. Clustering numerical data is an easy task in normal conditions than categorical data [4]. With the advances in data storage technology, sensor systems and continuous-query applications is based on Internet, the data becoming huge. Performing clustering on this time-changing categorical data is of great need in most of real applications like network-traffic monitoring, forecasting, and credit card fraud detection. The concepts that are tried to learn from these applications drift with time [5], designations of an employee in an organization, searching in social networks, Web documents, weather predictions like HudHud occurred in 2014, buying and seasonal preferences of customer, clustering entire time-changing data reduces the quality of clusters [5]. Storing data and performing clustering on the entire data set is expensive. Therefore, the techniques for efficiently cluster time-changing data are required.

The rest of the paper is organized as: Sect. 2 discusses relevant study, Sect. 3 provides an optimal model for clustering time-changing categorical data and detecting concept-drift. Section 4 presents implementation results and finally Sect. 5 presents conclusion with further work.

2 Relevant Study

This section provides brief discussion of clustering algorithms that motivated toward this research work. First, numerical algorithms for time-changing data were reviewed and next categorical algorithms for time-changing data which has still a room to enhance.

The process of clustering numerical time-changing data which was explored in previous works as follows: This concept was first introduced by Aggarwal et al. [6], considering continuous time-changing data set. It first stores the microclustering results over the window and clusters are got over user-defined time slots with flexibility. A density-based technique [7] discovers arbitrary shape clusters and considers core-microclusters in every sliding window. An evolutionary clustering does clustering over time and challenges to optimize two conflicting conditions [8, 9]. An approach to study the drifting concepts that using mean and standard deviation of centers to form clusters and changes can be identified in the data stream was presented in [10]. A density-based approach forgetting old models, taking new models and counts the presence of outliers [11]. It concentrates on robust clustering having noisy and evolving data streams leading to high quality.

Various clustering frameworks for time-changing categorical data have been [5, 12–14]. COOLCAT algorithm measures information entropy to portray cluster structures and assigns new incoming objects without considering drifting concepts. Node Importance Representative is used to detect the drifting concepts in the changing categorical data [5]. Rough membership function can be used to detect drifting concepts using distance-based measures [12]. How to perform clustering on different types of data like categorical and mixed data using partitioning procedure is discussed in [15]. The performance of partition-based clustering algorithms using genetic algorithms was analyzed in [16]. A new validity measure for clustering using homogeneity is presented in [17]. If the clusters generated for a new incoming data are poor, then detection of drifting concept maybe not true. Thus, this paper provides detection method based on an optimal model for clustering time-changing categorical data to augment the reliability of the detection results.

3 Optimal Model for Clustering Time-Changing Categorical Data

An optimal model for clustering time-changing categorical data presents objective function. On each incoming data subset with few constraints, this method uses iterative optimization algorithm for drifting concepts, aims at finding new clusters which has good certainty, continuity with the latest clustering results.

3.1 Framework for Clustering Time-Changing Categorical Data

The process of clustering concept-drift time-changing categorical data is initiated as: Let the given categorical dataset with each data item Xi having number of tuples divided into number of sliding windows. After forming equal size sliding windows (or subsets) and selecting the first S^1(equal to the size of the window) random data items and perform clustering to generate initial clusters. Select the next incoming subset of data items as other sliding window for clustering. Continue the process for

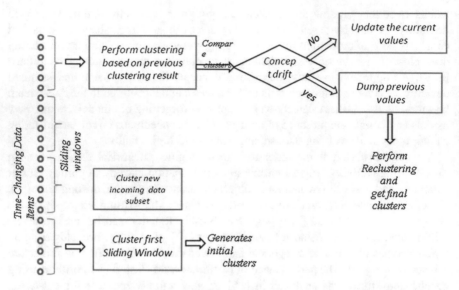

Fig. 1 Framework for clustering concept-drift time-changing categorical data

remaining sliding windows of the dataset. The framework of performing clustering on concept-drift time-changing categorical data is shown in Fig. 1. Sliding window most popularly used in clustering time-changing numerical data to implement drifting concepts. If a change is detected between present clustering result of present sliding window and previous clustering result, then drift is said to occur and then it is subjected to re-clustering process.

3.2 Steps of an Optimal Method

Let S^d be the dth subset taken as input from time-changing data, $X_i^d \in S^d$ be a data item for $1 \leq i \leq |S^d|$, L^{d-1} be the previous cluster method, M^d is membership function of S^d, L^d be the cluster method of incoming subset S^d, k^d = number of clusters of subset S^d. When clustering new incoming subset, assume that the concepts do not drift. Then, the number of clusters is same as that of the previous window.

The objective function is defined as:

$$F(M^d, L^d) = P_n(M^d, L^{d-1}) + P_n(M^d, L^d) + D(L^d, L^{d-1}) \tag{1}$$

where

$$P_n\left(M^d, L^{d-1}\right) = \sum_{m=1}^{k^d} \sum_{i=1}^{|s^d|} m_{mi}^d d_n\left(X_i^d, C_m^{d-1}\right) + \alpha \sum_{m=1}^{k^d} \sum_{i=1}^{|s^d|} m_{mi}^d \sum_{p=1}^{m} \sum_{q=1}^{n_p} \left(v_{mp_q}^{d-1}\right)^2 \quad (1a)$$

$$P_n\left(M^d, L^d\right) = \alpha \sum_{m=1}^{k^d} \sum_{i=1}^{|s^d|} m_{mi}^d d_n\left(X_i^d, C_m^d\right) + \alpha \sum_{m=1}^{k^d} \sum_{i=1}^{|s^d|} m_{mi}^d \sum_{p=1}^{m} \sum_{q=1}^{n_p} \left(v_{mp_q}^d\right)^2 \quad (1b)$$

$$D\left(L^d, L^{d-1}\right) = \beta \sum_{m=1}^{k^d} \sum_{i=1}^{|s^d|} m_{mi}^d \sum_{p=1}^{m} \sum_{q=1}^{n_p} \left(v_{mp_q}^d - v_{mp_q}^{d-1}\right)^2 \quad (1c)$$

d_n Dissimilarity measure of an object
C_m^d The mth cluster (on S^d)
$v_{mp_q}^d$ The representability of $a^{(q)}{}_m$ to c_m (after Sp is input)
α, β The parameters used
$F(..)$ The objective function for clustering subset of sliding window
m_{mi}^d The membership degree of $\mathbf{X}i$ to c_m (after S^d is input)

$P_n\left(M^d, L^{d-1}\right)$ = Computes an effectiveness of the clustering result when previous clustering is used to represent the new incoming subset, shown in Eq. (1a).
$P_n\left(M^d, L^d\right)$ = Computes effectiveness of cluster results when new clustering method is used for representing new incoming data subset, shown in Eq. (1b). The new clustering method develops from the previous clustering method.
$D\left(L^d, L^{d-1}\right)$ = Computes the difference between the new and previous clustering methods, shown in Eq. (1c).

The above-mentioned Eq. (1) can be transformed into an optimal method for the new incoming subset as follows:

$$\min_{M^a, L^d} F\left(M^d, L^d\right)$$

subject to

$$\qquad (2)$$

$$v_{mp_q}^d \in [0, 1], \sum_{q=1}^{n_p} v_{mp_q}^d = 1 \quad m_{mi}^d \in \{0, 1\}, \sum_{m=1}^{k^d} m_{mi}^d = 1, 1 < \sum_{i=1}^{n} m_{mi}^d < |S^d|$$

This is used iteratively for solving this optimal problem. For this, it is divided into two minimization subproblems: fixing M^d and solving for L^d and then fixing L^d and solving for M^d.

3.3 Detection of Concept-Drift

To detect the concept-drift between the new and previous clustering results, two measures like variation in distribution (VD) and variation in certainty (VC) are used and as follows:

$$\text{VD}\left(L^d, L^{d-1}\right) = \sum_{m=1}^{k^d} \frac{C_m^d}{S^d} \sum_{p=1}^{m} \sum_{q=1}^{n_p} \left(v_{mp_q}^d - v_{mp_q}^{d-1}\right)^2 \tag{3}$$

The higher the VD value, the new and previous clustering results are different. If it reaches threshold value then they are said to drift.

$$\text{VC}\left(L^d, L^{d-1}\right) = \sum_{m=1}^{k^d} \frac{C_m^d}{S^d} \sum_{p=1}^{m} \sum_{q=1}^{n_p} \left(\left(v_{mp_q}^{d-1}\right)^2 - \left(v_{mp_q}^d\right)^2\right) \tag{4}$$

The higher the VC value, the more be the uncertainty of the clustering results is enhanced. Now, these two measures VD and VC are integrated to define a detection parameter for the drifting concepts. Formula for detection of concept-drift is as follows:

$$\Theta\left(L^d, L^{d-1}\right) = \frac{2}{\eta_1 + \eta_2} \left(\eta_1 \text{VD}\left(L^d, L^{d-1}\right) + \eta_2 \text{VC}\left(L^d, L^{d-1}\right)\right)$$

where

$\Theta(..)$ Detection parameter
η_1 and η_2 Are the weights

The higher the Θ value, more chance for occurring concept-drift. If $L^d = L^{d-1}$, then $\Theta = 0$, hence, we need to set threshold value γ.

If $\Theta\left(L^d, L^{d-1}\right) > \gamma$, then concept-drift said to occur.

The value of Θ depends on effectiveness of new clustering results L^d and setting of parameters η_1 and η_2.

4 Results and Discussions

Now, objective function and drifting equation are combined while implementation. Here, the parameters η_1, η_2 and γ need to be set, because these may have impact on the performance of optimal method used for clustering time-changing categorical data. Here, η_1, $\eta_2 = 0.5$, value of γ is set based on the domain knowledge of datasets. For experimental analysis, KDD-CUP'99 dataset is considered which is available in UCI machine learning repository, as this dataset is used earlier by number of researchers for time-changing categorical data and fit for assessing proposed algorithms. It is a network dataset having time-changing data and consists of 494,020 records with their time stamp. These records are classified into 23 classes, where 22 classes have types of network attacks and one class for normal connection. Each record has 41 attributes, where 34 are continuous and 7 are categorical. To convert the continuous values into discrete values a method of uniform quantization is applied and 10% of the dataset is considered as a subset for testing. We used three measures for validating the proposed algorithm, accuracy, precision, and recall. For the given dataset D, with the clusters $C = \{C_1, C_2, C_3,...C_k\}$ and the divisions of the original classes of D as $\{E_1, E_2, E_3......E_k^1\}$. n_{ij} be the common number of objects in C_i and E_j where $i = 1$ to k, $j = 1$ to k^1. Ui is the number of objects in C_i and vj is the number of objects in the E_j.

$$\text{Accuracy} = \frac{1}{n}\sum_{i=1}^{k}\max_{j=1}^{k^1}nij \quad \text{precision} = \frac{1}{k}\sum_{i=1}^{k}\frac{\max_{j=1}^{k^1}nij}{ui}$$

$$\text{Recall} = \frac{1}{k}\sum_{i=1}^{k}\frac{\max_{i=1}^{k}nij}{vj}$$

KDD-CUP'99 dataset is tested with respect to various sizes of sliding windows 1000, 2000, 3000, and 5000 to compare the three measures using Chen's algorithm and proposed optimal algorithm. Accuracy, precision, and recall values between two methods are shown in Tables 1, 2, and 3, respectively. The results drawn in Figs. 2, 3, and 4 show that proposed optimal method is effective and stable than Chen method for time-changing categorical data. The computational time of both the methods with varying number of clusters is tested and rounded to nearest integer, shown in Table 4 and Fig. 5. Obviously, the proposed optimal method will take more time because it is run more than one time for getting an optimal solution. And Chen method uses 'data labeling' for new subset, whereas optimal method follows 'iterative process'. But the speed of convergence is high, that is, between 5 to 10 iterations, an optimal solution is achieved. And also the proposed optimal method is scalable because it has linear time complexity with respect to number of attributes, objects, and clusters.

Table 1 Accuracy of clustering results

S. No.	Size of sliding window	Chen method	Optimal method
1	1000	0.843	0.918
2	2000	0.851	0.920
3	3000	0.864	0.924
4	5000	0.846	0.913

Table 2 Precision of clustering results

S. No.	Size of sliding window	Chen method	Optimal method
1	1000	0.823	0.885
2	2000	0.815	0.875
3	3000	0.832	0.892
4	5000	0.829	0.887

Table 3 Recall of clustering results

S. No.	Size of sliding window	Chen method	Optimal method
1	1000	0.641	0.818
2	2000	0.652	0.821
3	3000	0.675	0.814
4	5000	0.649	0.822

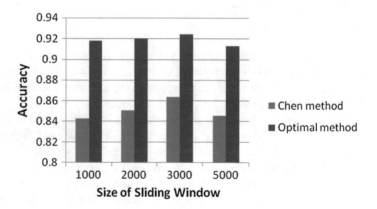

Fig. 2 Accuracy of clustering results

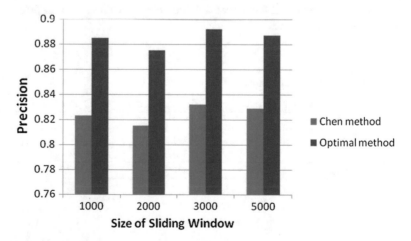

Fig. 3 Precision of clustering results

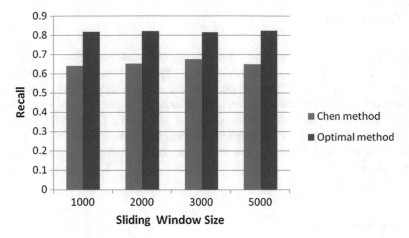

Fig. 4 Recall of clustering results

Table 4 Computation time in seconds for both clustering methods

S. No.	Number of clusters	Chen method	Optimal method
1	5	48	171
2	10	50	262
3	15	51	383
4	20	53	470
5	25	54	560

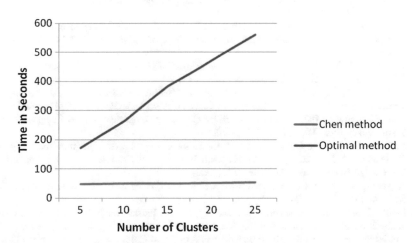

Fig. 5 Computational time of clustering results

5 Conclusion

In this paper, an optimal method for clustering time-changing categorical data has been presented. This method combines the objective function and validity and detection of concept-drift functions on a new incoming sliding window/subset. This process is followed iteratively to attain an optimal solution. The proposed method attains an optimal solution in less number of iterations with fast convergence speed. An experimental results of the proposed optimal method on KDD-CUP'99 dataset shown that our method is effective and robust for clustering time-changing categorical data.

References

1. Han, J., Kamber, M.: Data mining: concepts and techniques. Morgan Kaufmann (2001)
2. Kaufman, L., Rousseeuw, P.J.: Finding groups in data: an introduction to cluster analysis, Wiley, pp. 344 (2009)
3. Jain, A.k., Dubes, R.: Algorithms for clustering data. Prentice Hall (1988)
4. Gibson, D., Kleinberg, J.M., Raghavan, P.: Clustering categorical data: an approach based on dynamical systems. VLDB J. 8(3–4), 222–236 (2000)
5. Chen, H., Chen, M., Lin, S.: Catching the trend: a framework for clustering concept-drifting categorical data. IEEE Trans. Knowl. Data Eng. 21(5), 652–665 (2009)
6. Aggarwal, C., Han, J., Wang, J., Yu, P.: A framework for clustering evolving data streams. In: Proceedings of 29th International Conference on Very Large Data Bases (2003)
7. Cao, F., Ester, M., Qian, Q., Zhou, A.: Density-based clustering over an evolving data streams with noise. In: Proceedings of SIAM Conference, pp. 328–339 (2006)
8. Chakrabarti, D., Kumar, R., Tomkins, A.: Evolutionary clustering. In: Proceedings of ACM SIGKDD '06, pp. 554–560 (2006)
9. Chi, Y., Song, X.-D., Zhou, D.-Y., Hino, K., Tseng, B.L.: Evolutionary spectral clustering by incorporating temporal smoothness. In: Proceedings of ACM SIGKDD '07, pp. 153–162 (2007)
10. Gaber, M.M., Yu, P.S.: Detection and classification of changes in evolving data streams. Int. J. Inf. Technol. Decis. Mak. 5(4), 659–670 (2006)
11. Nasraoui, O., Rojas, C.: Robust clustering for tracking noisy evolving data streams. In: Proceedings of Sixth SIAM International Conference on Data Mining (SDM) (2006)
12. Cao, F., Liang, J., Bai, L.: A framework for clustering categorical time-evolving data. IEEE Trans. Fuzzy Syst. 18(5), 872–885 (2010)
13. Barbara, D., Li, Y., Couto, J.: COOLCAT: an entropy-based algorithm for categorical clustering. In: Proceedings of the Eleventh International Conference on Information and Knowledge Management. McLean, VA (2002)
14. Aggarwal, C., Yu, P.: On clustering massive text and categorical data streams. Knowl. Inf. Sys. 24, 171–196 (2010)
15. Madhuri, R., RamakrishnaMurty, M., Murthy, J.V.R., Prasad Reddy P.V.G.D, et al.: Cluster analysis on different data sets using K-modes and K-prototype algorithms. In: International Conference and Published the Proceeding in AISC and Computing, vol. 249, pp. 137–144. Springer (indexed by SCOPUS, ISI Proceeding DBLP etc). ISBN: 978-3-319-03094-4 (2014)

16. Lahari, K., RamakrishnaMurty, M., et al: Partition based clustering using genetic algorithms and teaching learning based optimization: performance analysis. In: International Conference and Published the Proceedings in AISC, vol. 2, pp. 191–200. Springer. https://doi.org/10.1007/978-3-319-13731-5_22
17. RamakrishnaMurty, M., Murthy, J.V.R., Prasad Reddy, P.V.G.D, et al.: Homogeneity separateness: a new validity measure for clustering problems. In: International Conference and Published the Proceedings in AISC and Computing, vol. 248, pp. 1–10. Springer (indexed by SCOPUS, ISI Proceeding DBLP etc). ISBN: 978-3-319-03106 (2014)

A Model for Securing Institutional Data Using Blockchain Technology

D. Durga Bhavani and D. Chaithanya

Abstract Educational institutions store their valuable data in registers where it can be exposed for tampering. We can store those transactions such as attendance, marks and certificates and secure them using a blockchain and can be viewed by anyone in the institution. Traditional methods of record keeping like storing marks in a register or a physical ledger are not secured. An instructor or a student may allegedly change the data for their own selfish needs which are unethical in educational field. Blockchain has a decentralized architecture which stores and secures transactions transparently, and there will be no central authority over those occurring transactions. Therefore, a novel model is proposed in smart contracts which secures and stores each transaction in a blockchain with hashing algorithm making them tamperproof.

Keywords Blockchain · Consensus algorithm · Hashing techniques · Smart contracts · Institutional data

1 Introduction

A blockchain is a ledger without a central authority having control. It is an enabling technology for individual companies and institutions to collaborate with trust and transparency. It normally contains a growing list of publicly accessible records which are cryptographically secured from tampering.

Normally in any organization or institution records of all their transactions are stored in books, ledgers in offices and certificates in educational institutions. So they can be accessed easily and is prone to data manipulation [1]. This is the main

D. Durga Bhavani (✉) · D. Chaithanya
Department of Computer Science and Engineering (Autonomous),
CVR College of Engineering, Hyderabad, India
e-mail: drddurgabhavani@gmail.com

D. Chaithanya
e-mail: dchaithu94@gmail.com

© Springer Nature Singapore Pte Ltd. 2020 873
K. S. Raju et al. (eds.), *Data Engineering and Communication Technology*,
Advances in Intelligent Systems and Computing 1079,
https://doi.org/10.1007/978-981-15-1097-7_73

disadvantage of current record-keeping formats. Not only losing the originality of the data, there also exists a problem in storing those records. Storage of large sets of data and accessing them becomes a tedious task both in paper and digital format because the incoming data needs to be organized continuously and maintained securely [2]. At present, we are facing three problems for record keeping. They are:

1. Data manipulation
2. Storage of data
3. Data accessing.

Therefore, to overcome those problems, we will implement a novel idea with the help of blockchain technology which is a peer-to-peer record-keeping system with the addition of cryptographic methods like digital signatures to ensure transparency, data security and easy storage of the records for educational institutions that will make sure no manipulation of data at any given point of time. Since digital signatures are created by hashing technique with the use of public and private keys, the transactions which are created by owner cannot be modified by third parties [1]. Thus, blockchain uses a consensus algorithm which is a fault-tolerant mechanism and is used to achieve the necessary agreement on a single data value over a system where all the drawbacks of traditional methods of record keeping are overcome.

2   Model Design

To secure the transmission of data from source to destination, a process called as hashing needs to be done using an algorithm. MD5 algorithm was previously used for this purpose, but sometimes it generates the same hash value for different data. In our proposed model, we will implement a new type of record-keeping system which uses SHA-256 algorithm to secure and store transactions in a series of blocks which are connected to each other with the help of hashes. The first block is called the genesis block which contains initial data like simple timestamp of creating that block [3]. Next block onwards, transactions are added while storing the hashes of previous block, therefore propagating the chain as shown in Fig. 1.

- B1 is called the genesis block and contains timestamp of its creation.
- B2 contains the hash of B1 and has a transaction.
- B3 contains the hash of B2 and a transaction.

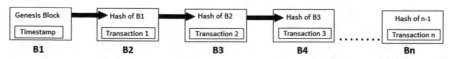

Fig. 1 Block propagation

The blocks are connected to each other in a chain model as long as the transactions are being made. This is how transactions are stored securely. The block, which has been altered in case of tampering, can be easily detected by checking its hash in the next block.

Currently, educational institutions are storing their valuable data in paper and digital format where it can lose its integrity. Therefore, a novel idea is proposed to implement an institution management system that stores and secures transactions like attendance, marks of students, personal details of students and faculty and securing library books as shown in Fig. 2. Examination cell will issue digital marks memo to the eligible students which can be accessed with their private key from anywhere in the world without coming to the institution and going through all the formal procedures for obtaining them.

3 Transaction Mechanism

The blockchain is a type of distributed network that follows a consensus approach of single data value over a system. In a distributed network, the transactions cannot be simply stored. They have to go through a step-by-step procedure of transaction mechanism [3]. It can be defined in five stages such as Transaction Encryption, Transaction Decryption, Block Creation, Block Authentication and Block Propagation as shown in Fig. 3.

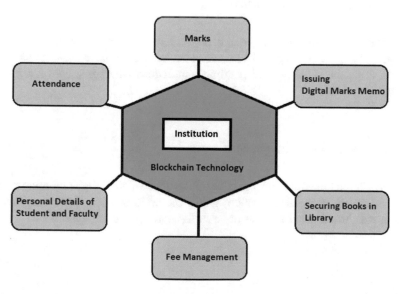

Fig. 2 Types of transactions being implemented

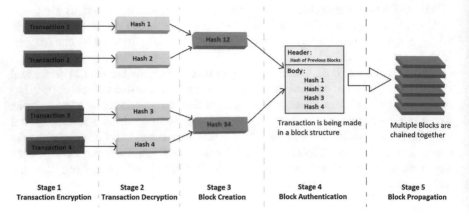

Fig. 3 Transaction mechanism in a blockchain

3.1 Transaction Encryption

The data owner will initiate the transaction in which the details of the receiver's public key, receiver's address and the transactional value are specified. After initiating this transaction, it will be encrypted with the sender's cryptographic digital signature, which authenticates the stored transactions.

3.2 Transaction Decryption

The signed transaction is sent to the network and received by all nodes of the network. Then, the message will be validated by decrypting the digital signature. Further, it is sent to the pool of pending transactions where it waits till a block is created.

3.3 Block Acknowledgement

Any node in the network takes the initiation to create a block by combining with all other decrypted transactions which are in the waiting state. Once the block is created, it is broadcasted to each node in the network for authentication.

3.4 Block Authentication

The nodes, after receiving the block copies, start an interactive process to validate it and communicate with each other to check for single data value over a system. However, there might be a difference among blockchain's branches when they do not share the same data value due to network issues. Therefore, it is necessary to reach a consensus on the block authenticity among all the nodes.

3.5 Block Propagation

Once a block is authenticated, it will be registered in the network as a verified block. Next block will be linked to the recently verified block. Further, these two blocks are formed as a chain and broadcasted over the network, where the future blocks will be propagated as a verified version of blockchain [4].

4 Participants Involved

All the nodes in the network act as participants of the blockchain. In our model, there will be four participants who communicate with each other as shown in Fig. 4.

4.1 Blockchain Manager (BM)

Blockchain Manager is responsible for managing all the transactions, database services and blockchain services.

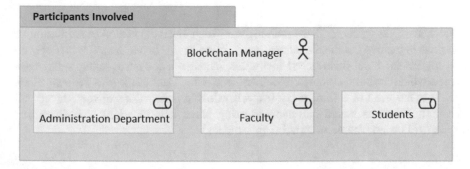

Fig. 4 Participants involved in proposed model

4.2 Administration Department

It deals with the current activities of the educational institute. It records the relevant information of all the current employees and students and guarantees the smooth execution of all the processes. In this model, we considered the processes like enrolling the student for a specific course, storing the personal information and crediting the marks. The educational institution develops student records that are indestructible and universal. This facilitates developing of a global standard of accreditation that will not need any verification, thus eliminating the time to do such activities. It will expand the opportunities for the students on a global scale. If, in any case, a student loses his academic records, then he can get it from his home without going through the formal procedures at any given point of time.

4.3 Faculty

The teaching staff verify and stores their student particulars like attendance, assessment, behavior, leadership qualities and results obtained, etc.

4.4 Student

The student is the pivoting participant in this model. He can always access his own records using a private key, but cannot modify the data. He can approach any organization for internship, scholarship and employment without the barrier of country, language or time by providing his block's hash to the required person for his verification purposes [5].

5 Conclusion

Blockchain has a decentralized architecture which has the potential to revolutionize various industries besides the educational sector. In this paper, a model is proposed to implement blockchain-based institution management system using smart contracts which hashes a complete block and integrates that hash key in the next block without the need of a third party [6]. All valuable transactions of the educational institutions, when stored and secured using blockchain technology, ensures that there will be no chance of data being tampered.

References

1. Zheng, Z., Xie, S., Dai, H., Chen, X., Wang, H.: An overview of blockchain technology: architecture, consensus, and future trends. In: Proceedings of IEEE International Congress on Big Data (Big Data Congress), pp. 557–564. [Online]. Available: http://ieeexplore.ieee.org/document/8029379/ (2017)
2. Realworld Uses for Blockchain. https://www.fool.com/investing/2018/04/11/20-real-world-uses-for-blockchaintechnology.aspx
3. Nakamoto, S.: Bitcoin: a peer-to-peer electronic cash system. [Online]. Available: https://bitcoin.org/bitcoin.pdf (2008)
4. Consensus Algorithms: A Brief overview. Available: https://medium.com/coinmonks/blockchain-consensus-algorithms-an-early-days-overview2973f0cf49c6
5. Schools are Using Blockchain: Blockchain Goes to School. [Online]. Available: https://www.cognizant.com/whitepapers/blockchain-goes-to-school-codex3775.pdf
6. Bahga, A.: Blockchain applications: a hands-on approach. Available: https://books.google.co.in/books?id=M7sanQAACAAJ&redir_esc=y&hl=en (2017)

Cloud-Based Dempster-Shafer Theory (CDST) for Precision-Centric Activity Recognition in Smarter Environments

Veeramuthu Venkatesh, Pethuru Raj, T. Suriya Praba
and R. Anushiadevi

Abstract Smart environments such as smart homes, offices, hotels, and cities are increasingly being furnished with a variety of sensors and actuators to constantly monitor all kinds of accidents and incidents in order to make correct inferences and right decisions. However, the recurring challenge here is that disparate and distributed sensors and their ad hoc networks collectively could produce some wrong sensor data. There are several mathematical techniques to deal with the impurity and the uncertainty issues of heterogeneous sensor data. Bayesian, rough set theory, hidden Markov model, and Dempster–Shafer theory (DST) take the lead. DST is being positioned as one of the most powerful methods to deal with the perpetual problem of inaccuracy and uncertainty. In this paper, a novel framework is proposed for activity recognition such as fire/flame/fall detection for smart environments. The framework comprises a cloud-based execution of Dempster–Shafer theory to increase the overall accuracy of incident monitoring and activity recognition. A fire detection application is implemented using several upcoming technologies like RESTful services, Jess-based rule engine, cloud, and mobility.

Keywords Wireless sensor network · Cloud Dempster–Shafer theory (CDST) · Sensor fusion · Context-awareness · Event detection

1 Introduction

The noteworthy advancements in the IT space help us in equipping our everyday environments with differently capable sensors and actuators to extract value-adding data for enabling decision-making. The quality, timeliness, and trustworthiness of

V. Venkatesh · T. Suriya Praba · R. Anushiadevi (✉)
School of Computing SASTRA Deemed University, Thirumalaisamudram,
Thanjavur 613401, Tamilnadu, India
e-mail: anushiavenkat@gmail.com

P. Raj
Reliance Jio Cloud Services (JCS), Bangalore 560025, India

© Springer Nature Singapore Pte Ltd. 2020   881
K. S. Raju et al. (eds.), *Data Engineering and Communication Technology*,
Advances in Intelligent Systems and Computing 1079,
https://doi.org/10.1007/978-981-15-1097-7_74

sensor data are also ensured through the fast-evolving mathematical concepts. With the ready availability of smart sensors [1, 2] for different purposes and with faster maturity and stability of technologies for forming ad hoc networks of sensors, the capturing, transmission, storage, processing of environment, asset, and machine data get speeded up. On the other side, the cloud paradigm has gone through a number of delectable transformations. That is, clouds are being articulated as the one-stop IT infrastructure solution for all kinds of business and IT applications and platforms. The data collection, cleansing, and analytics are being accomplished through cloud-based platforms [3, 4]. These days, multi-sensor environments are now being much elevated thanks to the highly matured sensor data fusion algorithms [5–7] and implementations. These are capable of producing accurate results as we are combining the values from heterogeneous sensors to arrive at the result. Hence, in this paper, we are using five different types of sensors to monitor for a particular activity. The sensors chosen are smoke, temperature, gas, fire, and humidity sensors. The sensor fusion is carried out using Dempster–Shafer theory (DST) [8–10]. The underlying concept of DST is to merge crucial probability assignments into a common basic probability assignment (BPA) with DST grouping conventions in an appropriate scaffold. From the assigned BPA values, we can estimate the influence of a particular sensor in making the decision and thus improving the overall accuracy of the system.

Despite having many advantages, multi-sensor environments are still weighed down due to their computing, storage, and communication capabilities. So we propose a cloud-based DST application. Typically any cloud environment presents an illusion of infinite compute and storage resources. We can compensate on the computing, storage, and scaling capabilities of the cloud environment to better serve the user and help in timely recognition of the activity. We can essentially scale the performance of the system by assigning more resources automatically in the cloud. The improvement in speed and throughput is evident as we are not limited to processing capabilities of the hardware anymore.

We are making use of the upcoming REST [11, 12] technology for connecting with the cloud environment. The decision of detecting fire, flame, or fall is taken by a rule engine which is running in the cloud environment. We have utilized java-based rule engine (JESS) for this purpose which takes the processed data from DST and decides if the fire/flame/fall event has occurred.

The prototype is implemented using Raspberry pi board which is connected with all the needed sensors and it monitors the values regularly. The value is periodically sent to the Amazon Web Services (AWS) cloud. The server running in the cloud uses the DST implementation to calculate value and return the result to the android application from which we can constantly monitor the environment for any kind of event or activity such as fall or fire detection. If the event is identified, then we can inform the fire station about the stage of fire in the environment to take quick preventive actions. If the user is away when the accident takes place, he can even use the android application to send commands to the hardware which can enable prevention method for containing the fire.

2 Related Work

One among the main arenas of the wireless sensor networks for initial fire detection is based on sensor fusion [13]. This has used smoke mass CO mass and temperature beliefs and technique called neural networks and fuzzy interpretation to define if any fire has occurred are not. Another approach is proposed in [14] as a model for forest fire detection by using wireless sensor networks (WSN). It includes a set of motes like wind speed sensors, temperature, humidity, and smoke. These motes used in WSN are systematized in cluster format and also assign a single node as cluster head. The data sensed from each cluster head and each node makes weather information using a neural network. The information referred by each cluster head to coordinator will decide if there is a hazard of fire is not. The strategy and calculation of a WSN [15] for forest fires are posturing and primary revealing. The idea and assessment of a WSN for forest fires [16] sculpting and starting identification through image processing by applying the Gaussian mixture model (GMM), which can be implemented to discover smoke at the earliest time possible in a forest.

2.1 Dempster–Shafer Theory (DST)

Monitoring produces a lot of data and then data systematically get transitioned into information and knowledge towards decision-making. However, there are some critical challenges associated with sensor data due to several reasons such as sensors become inactive, network outage, sensor multiplicity, and heterogeneity. Therefore, some data sometimes turn out to be unreliable, inaccurate, and incorrect and correspondingly the decisions being made out of wronged data could be risky.

Context-awareness in any multi-sensor environments is being achieved by modelling the data acquired prior to the event. But in real world, it is impossible to have information of all contexts in advance. So there must be a method to determine BPA for achieving context-awareness without any prior information.

For performing multi-sensor fusion, the Dempster–Shafer Theory (DST) [9] uses basic probability assignment (BPA). In DST, a frame of discernment is nothing but a universal set θ. It consists of exclusive proportions with m as BPA

$$m{:}2^{\theta} \rightarrow [0, 1],$$
$$0 \leq m\{A\} \leq 1,$$
$$m(\emptyset) = 0, \text{and}$$
$$\sum_{A \subseteq \theta} m(A) = 1$$

The BPA can be calculated by using each sensor's S_i data that contributed its observation by assigning beliefs over θ. Belief$_i(A)$ Gives the probability that A is definitely true and Plausibility$_i(A)$ gives the probability that A is definitely not false.

$$\text{Belief}_i(A) = \sum_{E_k \subseteq A} m_i(E_k) \tag{1}$$

$$\text{Plausibility}_i(A) = \sum_{E_k \cap A \neq \phi} m_i(E_k) \tag{2}$$

The probability is designated by a "sureness interval" $[\text{Belief}_i(A), \text{Plausibility}_i(A)]$.

2.2 BPA Determination for Fire Detection

In order to calculate BPA, first the variation rates of sensor data have to be found out. Let us say the data from the gas sensor is $G(t)$, then the variation rate is found out by using the below formula:

$$\Delta G(t) = \frac{G(t_n) - G(t_{n-1})}{t_n - t_{n-1}} \tag{3}$$

The time from 1 to 30 s was measured per second for each sensor as follows:

$$t_n = 1, 2, 3, \ldots, 30.$$

To have BPA, the variation rate, α, is to be calculated which is given by the formula as follows:

$$\alpha = \frac{\sum_{t=0}^{n} \arctan(\Delta G(t_t))}{\sum_{t=0}^{n} \arctan(|\Delta G(t_t)|)} \cdot \left(\frac{t_n}{T}\right)^2, \quad \text{where } T = \text{total time.} \tag{4}$$

The same method has to be followed in order to determine BPA for all the sensors [7, 8]. Different sensors have different ranges of value and different threshold values and so we must normalize the sensor values [9]. Once BPA is calculated for all the sensors, they have to be normalized. While normalizing, all the combinations of the sensor values should also be taken into consideration. Finally, the normalized sensor values of different sensors are taken and normalized again with each other to find a percentage of influence by each sensor. The decision is found out by sending these normalized values to the rule engine. The result from the rule engine determines the different stages of fire. According to the stage of fire, the fire station is alerted. If the fire has a high chance of spreading, then the nearby buildings are also alerted to save them as much as possible.

3 The Proposed Architecture

We can monitor the event of fire using the following proposed architecture Diagram 1. It consists of collection of sensors [5–7], cloud services [3, 4] for storage, and processing of data, a rule engine to make decisions based on the processed data, an android application to access and control the sensor remotely [11].

The sensor transmits the data to the cloud storage using the ZigBee protocol [2, 3]. We are using the Amazon Web Service to store and process the data. The Amazon EC2 service provides virtual machine which retrieves the data from the database and process the data.

REST protocol [11] is used for the communication between the database and the cloud machine. The RESTful architecture has some of the HTTP methods which are associated with database. Android application is used to monitor the data. The Amazon Web Service (AWS) software development kit (SDK) for android is used in the android application to communicate with the Amazon Web Service. It used REST protocol to communicate and the information is retrieved and displayed on

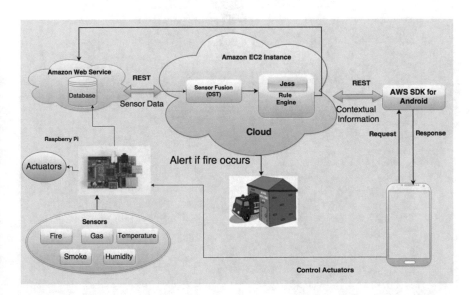

Fig. 1 Cloud-based Dempster–Shafer theory (CDST) architecture

the application. If there is a fire, alert is sent to the fire station which has a similar application with the location of the building and the stage which the fire is in.

We have defined the different stages of fire below and so different alerts are provided in the android application. Then the application can also be used to control the sensor and other devices which are connected to the system. For example, we can connect a sprinkler system and control the system from the android application.

Rules for finding the stage of fire based on [13–15, and 16]:

- If the highest influencing sensor is smoke or temperature sensor, then fire is at the initial stages. In that case only, the fire station is alerted.
- If the highest influencing sensor is light (fire) sensor then fire is in developed stage. Then fire station and people residing in same building are alerted.
- If the highest influencing sensor is humidity sensor or gas sensor then chance of fire spreading is high. So, the nearby building is also alerted. Finally, normalized sensor values are sent to the rule engine which makes the decision and notifies the fire station.

Algorithm: It shows how CDST is useful to make the right decision in smarter environment

Def Fire_Detection_Impl (F):

Total_time = 30;
for i in F:

Get values of sensors and store in Sensor $[i]$ $[t]$

for i in Sensor:

Calculate Alpha value for each sensor where

$$\alpha = \frac{\sum\limits_{t=0}^{n} \arctan(\Delta F(t_t))}{\sum\limits_{t=0}^{n} \arctan(|\Delta F(t_t)|)} \cdot \left(\frac{t_n}{T}\right)^2, \quad \text{where } T = \text{Total\_time}$$

for i in Sensor:

Calculate Belief for each sensor where

$$\text{Belief}_i(A) = \sum_{E_k \subseteq A} m_i(E_k)$$

for i in Sensor_Belief:

Normalize the values

```
for i in Sensor:
out [i] = Sensor_Belief [i] [0]
return out
```

Rule defined for level of fire along with CDST

```
Rules:
if Belief(Smoke) or Belief(Temp) is maximum:
Fire in initials Stage
if Belief(Fire) is maximum:
Fire in developed Stage
if Belief(Humidity) or Belief(Gas) is maximum:
Chances of fire spreading to nearby area is high
```

4 Experiment and Evaluation

A variety of sensors such as gas, fire, temperature, humidity, and smoke sensors are deployed and used inside a smart environment for fire detection shown in Fig. 2.

Multiple sensors like temperature, smoke, humidity, fire, gas. are connected to Raspberry pi which collects the data periodically and these data are sent to the cloud using JSON format to calculate the belief value for the sensors using DST. The result of the belief calculation for all possible combination of sensors at a particular time interval (T8 ∼ T9) is shown in Table 1.

After performing the sensor data fusion by the cloud-hosted DST application, the blended data is sent to the rule engine (JESS); in order to precisely detect what is going on in the environment and the extracted information gets updated to the android application as shown in Fig. 3; in order to notify all the right stakeholders about the fire level and the intensity to take necessary counteraction to avoid any kind of major loss. The CDST in association with real-time knowledge discovery and dissemination comes handy in taking correct and quick action.

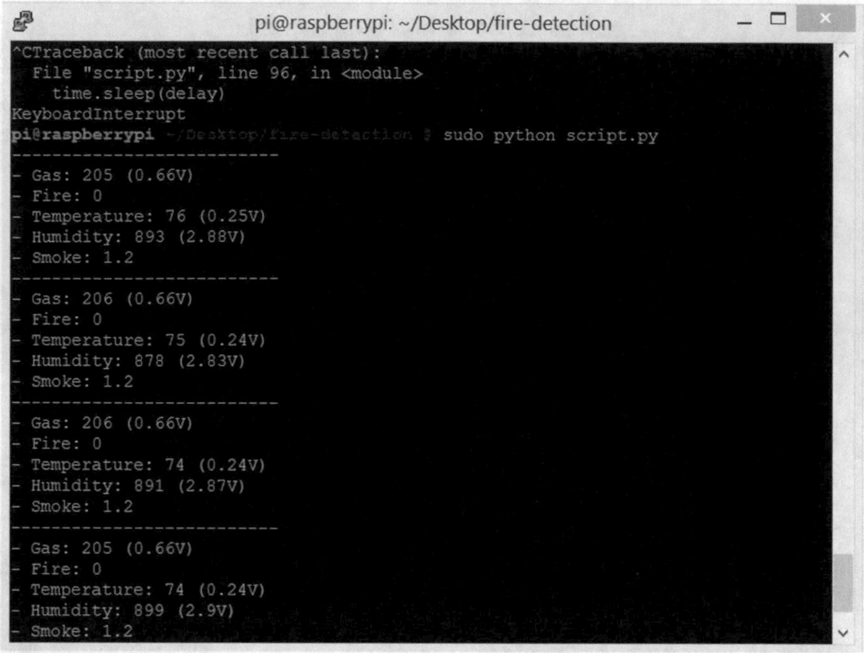

```
⌨                    pi@raspberrypi: ~/Desktop/fire-detection          –  ☐  ✕
^CTraceback (most recent call last):
  File "script.py", line 96, in <module>
    time.sleep(delay)
KeyboardInterrupt
pi@raspberrypi ~/Desktop/fire-detection $ sudo python script.py
-------------------------
- Gas: 205 (0.66V)
- Fire: 0
- Temperature: 76 (0.25V)
- Humidity: 893 (2.88V)
- Smoke: 1.2
-------------------------
- Gas: 206 (0.66V)
- Fire: 0
- Temperature: 75 (0.24V)
- Humidity: 878 (2.83V)
- Smoke: 1.2
-------------------------
- Gas: 206 (0.66V)
- Fire: 0
- Temperature: 74 (0.24V)
- Humidity: 891 (2.87V)
- Smoke: 1.2
-------------------------
- Gas: 205 (0.66V)
- Fire: 0
- Temperature: 74 (0.24V)
- Humidity: 899 (2.9V)
- Smoke: 1.2
```

Fig. 2 Sensing values through Raspberry pi

5 Conclusion and Future Enhancements

As our everyday environments are getting digitized with the leverage of smart sensors and actuators, the amount of data getting generated is significantly growing. It is evident that data is very important for extracting actionable insights. However, for perfect and precise decision-making, the source data needs to be accurate and dependable. Any kind of data impurity, ambiguity, and errors need to be removed at the source level. It is clear that processing of bad data leads to risky decisions and hence there are pragmatic mechanisms and solutions in order to ensure the trust-worthiness and reliability of data. In this paper, we have leveraged the proven CDST approach to arrive at data that is precise and the cloud environment which offers scalable solution to the problem of increase in the number of sensor nodes which indirectly increases the data generation rate. The framework inherently involves and invokes the DST technique to significantly enhance the correctness of decision-making and timely processing of data using cloud environment. The data generation, capture and usage grow at exponential level. We have used the DST mechanism to come out with a precision-centric framework for effective

Table 1 Belief values for all sensor at interval T8–T9

T8 ~ T9	Variation rate	BPA	BPA (Normalized)	Belief
G	5	0.004262	0.001589332	0.001589332
F	0	0	0	0
T	2	0.026396	0.009843267	0.009843267
H	3	0.034788	0.012972707	0.012972707
S	0.3	0.09	0.033561677	0.033561677
$G + F$		0.00426	0.001588586	0.001589332
$G + T$		0.030658	0.011432599	0.011432599
$G + H$		0.03905	0.014562039	0.014562039
$G + S$		0.094262	0.035151009	0.035151009
$F + T$		0.026396	0.009843267	0.009843267
$F + H$		0.034788	0.012972707	0.012972707
$F + S$		0.09	0.033561677	0.033561677
$T + H$		0.061184	0.022815974	0.022815974
$T + S$		0.116396	0.043404944	0.043404944
$H + S$		0.124788	0.046534384	0.046534384
$G + F + T$		0.030658	0.011432599	0.011432599
$G + T + H$		0.065446	0.024405306	0.024405306
$G + H + S$		0.12905	0.048123716	0.048123716
$G + F + H$		0.233546	0.08709106	0.08709106
$G + F + S$		0.094262	0.035151009	0.035151009
$G + T + S$		0.120658	0.044994276	0.044994276
$F + T + H$		0.061184	0.022815974	0.022815974
$F + T + S$		0.116396	0.043404944	0.043404944
$F + H + S$		0.124788	0.046534384	0.046534384
$T + H + S$		0.151184	0.056377651	0.056377651
$G + F + T + H$		0.065446	0.024405306	0.024405306
$G + F + T + S$		0.120658	0.044994276	0.044994276
$G + T + H + S$		0.155446	0.057966983	0.057966983
$G + H + S + F$		0.12905	0.048123716	0.048123716
$F + T + H + S$		0.151184	0.056377651	0.056377651
$G + T + F + H + S$		0.155446	0.057966983	0.057966983

decision-enablement. In future, we have planned to leverage a number of proven mathematical concepts and advanced data analytics technologies collaboratively to substantially improve the fault-tolerance, availability, and dependability of the sensor-centric sophisticated systems.

Fig. 3 Result updated to android application from cloud

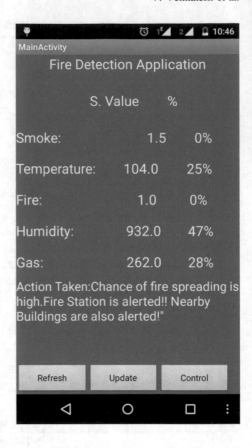

References

1. Rashidi, P., Cook, D., Holder, L., Schmitter-Edgecombe, M.: Discovering activities to recognize and track in a smart environment. IEEE Trans. Knowl. Data Eng. **23**(4), 527–539 (2011)
2. Sixsmith, A., Johnson, N.: A smart sensor to detect the falls of the elderly. IEEE Pervasive Comput. **3**(2), 42–47 (2004)
3. Kurschl, W., Beer, W.: Combining cloud computing and wireless sensor networks.In: Proceedings of 11th International Conference on Information Integration and Web-based Applications and Services, pp. 512–518 (2009)
4. Alamri, A., Ansari, W.S., Hassan M.M., Shamim Hossain M., Alelaiwi, A., Hossain M.A.: A survey on sensor-cloud: architecture, applications, and approaches. Int. J. Distrib. Sensor Netw. 1–18 (2013)
5. Chen, S., Deng, Y., Wu, J.: Fuzzy sensor fusion based on evidence theory and its application. Appl. Artif. Intell. **27**(3), 235–248 (2013)
6. Khaleghi, B., Khamis, A., Karray, F.O., Razavi, S.N.: Multisensor data fusion: a review of the state-of-the-art. Inf. Fus. **14**(1), 28–44 (2013)
7. Dallil, A., Oussalah, M., Ouldali, A.: Sensor fusion and target tracking using evidential data association. IEEE Sensors J. **13**(1), 285–293 (2013)

8. Hong, X., Nugent, C.D., Mulvenna, M.D., McClean, S.I., Scotney, B.W., Devlin, S.: Evidential fusion of sensor data for activity recognition in smart homes. Pervasive Mob. Comput. **5**, 236–252 (2009)
9. Chen, S., Du, Y., Deng, Y.: A decision-making method based on Dempster–Shafer theory and prospect theory. J. Inf. Comput. Sci. **11**(4), 1263–1270 (2014)
10. Ma, J.: A novel wireless sensor networks event detection method based on evidence theory. J. Converg. Inf. Technol. **7**(16), 305–314 (2012)
11. Lee, S., Jo, J., Kim, Y., Stephen, H.: A framework for environmental monitoring with arduino-based sensors using restful web service. Paper presented at the proceedings—2014 IEEE international conference on services computing, SCC 2014, pp. 275–282 (2014)
12. Datta, S.K., Bonnet, C., Nikaein, N.: An IoT gateway centric architecture to provide novel M2M services. IEEE World Forum on Internet of Things (WF-IoT), pp. 514–519, 6–8 Mar 2014
13. Chen, S., Bao, H., Zeng, X., Yang, Y.: A fire detecting method based on multi-sensor data fusion. In: Proceedings of the IEEE International Conference on Systems, Man and Cybernetics, pp. 3775–3780. https://doi.org/10.1109/icsmc.2003.1244476 (2003)
14. Bernardo, L., Oliveira, R., Tiago, R., Pinto, P.: A fire monitoring application for scattered wireless sensor networks: a peer-to-peer cross-layering approach. In: Proceedings of the International Conference on Wireless Networks and Systems, pp. 28–31. Barcelona, Spain (2007)
15. Hefeeda, M., Bagheri, M.: Forest fire modeling and early detection using wireless sensor networks. Ad Hoc Sensor Wirel. Netw. **7**, 69–224 (2009)
16. Yoon, S., Min, J.: An intelligent automatic early detection system of forest fire smoke signatures using Gaussian mixture model. J. Inf. Process. Sys. **9**(4), 621–632 (2013)

Brilliant Corp Yield Prediction Utilizing Internet of Things

R. Vijaya Saraswathi, Sravani Nalluri, Somula Ramasubbareddy, K. Govinda and E. Swetha

Abstract A decent yield from horticulture relies upon various parameters or integral elements. Water assumes an essential part for the best possible yield of product since water prerequisite of harvest shifts as it develops, i.e., water prerequisite changes with developing periods of a harvest. Different parameters can be considered like supply of synthetic substances and the amount in which it is utilized, as a rule because of absence of learning ranchers utilize synthetic substances in either gigantic or low sum and once in a while utilize it in a period when it is minimum required, and this absence of information prompts colossal misfortune in the yearly yield of a harvest. This part proposes a shrewd method to deal with and handle this issue, by the utilization of IOT. In this framework, different sensors are sent to the rural field and the motivation behind these sensors is to consistently screen the readings of different parameters for which they are utilized. These readings are then sent to a microcontroller which thusly advances the information to cloud, where it broke down and the move is made in view of information. At the cloud, the readings are contrasted the use of supplemental with the limit readings of every sensor information that is chosen by the agriculturist in the wake of counseling with a specialist, if some anomaly is distinguished, a message is sent to the rancher specifying about the issue and furthermore the actuators associated with the microcontroller gets initiated, the area where the supplements are required is passed on to the actuators utilizing the GPS module associated with the microcontroller. This in this manner expands the odds of having a decent yearly yield of yield. In this manner, a framework naturally recognizes and makes a move as indicated by the regularly changing parameters to at last help in expanding the yield cleverly to handle the issue in the field.

R. Vijaya Saraswathi · S. Nalluri · S. Ramasubbareddy (✉)
VNR Vignana Jyothi Institute of Engineering and Technology, Hyderabad, India
e-mail: svramasubbareddy1219@gmail.com

K. Govinda
VIT University, Vellore, India

E. Swetha
SV College of Engineering, Tirupati, India

K. S. Raju et al. (eds.), *Data Engineering and Communication Technology*,
Advances in Intelligent Systems and Computing 1079,
https://doi.org/10.1007/978-981-15-1097-7_75

893

Keywords Agricultural yield · Internet of Things · Sensors · Cloud · Microcontroller · Smart information system

1 Introduction

India is a nation which is known for its agribusiness. Horticulture is a noteworthy business area in our province. A huge number of harvests are developed in our nation like wheat, cereals, and so forth which assume a critical part in our everyday life by giving us a feast to eat. India is the greatest producer, buyer, and exporter of flavors and get-up-and-go things. India's common item creation has turned out to be speedier than vegetables, making it the second greatest natural product item generation on the planet. India's farming yield is assessed to be 287.3 million tons (MT) in 2016–17 after the primary push survey. It positions third in property and agriculture yields. The farming based business in India is confined into a couple of subareas, for instance, canned, dairy, and so forth.

1.1 Chemicals in Agriculture

The use of supplemental supplements to grow trim yield started as experimentation as wood red hot remains, ground bones, salt lessen, and gypsum. Justus von Liebig (1803–1873), a German physicist, built up the framework for the usage of engineered excrements as a wellspring of plant supplements starting in 1840. He saw the essentialness of various mineral segments got from the earth in plant sustenance and the need of supplanting those parts in order to keep up soil readiness. Two British scientists, J. B. Lawes and J. H. Gilbert, hence settled the agrarian test station at Rothamsted, in the UK. They in view of crafted by Liebig and likely demonstrated the importance of compound fertilizers in upgrading and keeping up soil wealth. Without a doubt, the utilization of made excrements was the commence of the overall addition in rustic creation after World War II.

1.2 Significance of Irrigation in Air Business

The water framework encourages the agriculturists to have less dependence on rainwater with the ultimate objective of cultivating. The need and hugeness of water framework are featured underneath in centers.

1.3 Grouping Climate

Indian environment and atmosphere conditions experience a moved extent of air. There is uncommon warmth at a couple of spots, while the climate stays to an awesome degree cold at various spots. While there is over the best precipitation at a couple of spots, distinctive spots experience unprecedented dryness. Thusly, water framework is required in India.

1.4 Capricious and Flawed Tempest

India is a place that is known for rainstorm. Regardless, storm is sporadic and erratic in nature. Now and again, it comes successfully and brings overpowering precipitation, anyway sometimes it comes late and brings insufficient precipitation. Encourage, there is irregularity in the allocation of precipitation reliably. The water framework system encourages the farmers to have less dependence on rainwater. In the midst of the seasons of lacking precipitation, the items are given water through water framework systems.

1.5 Cultivation-Based Economy

Indian economy relies upon cultivating. A generous piece of Indian people depends on upon cultivation. Without water framework, cultivating is implausible in dry zones or in the midst of the seasons of inadequate precipitation. Regularly, for the cultivating practices, transversely finished different regions, there is a necessity for fitting water framework structure.

1.6 Soil Character

In many spots, the water holding capacity of the soil is very low thereby requiring a need of regular irrigation. Proper irrigation facilities play a huge role in the annual yield of the crops in our country. Scarcity of the same during critical time and overwatering due to improper knowledge can ultimately affect the annual yield of the crop. Shortage in irrigation of water and overwatering at basic stages are influencing the crop's quality and yield. Manures and chemicals additionally assume an essential part to build its profitability. As ranchers ought to know in their product fields where they are developing the harvest what are the supplements display in the dirt and in what amount, with the goal that they can include supplements remotely in a legitimate sum which will bring about great yield and great quality. Be that as it may, the genuine issue at present is that the ranchers do not have the learning that will influence their harvest's yield. Because of their absence of learning about the correct time and how much water ought to be included, distinctive basic stages like Pre-blooming period, the yield is diminished. Due to over watering or because of absence of water, once in a while supplements are absent in the right extent that result in low quality of harvest yield. On account of the absence of information, ranchers end up plainly in charge of their misfortune themselves. Accordingly, they require somebody who can inform them regarding this [1, 2].

2   Literature Survey

India's rank in the field of horticulture, development, backwoods, and so on is widely acclaimed adding to around 8% of the world's organic product creation and numerous more probable. As known India has an immense scope of soil condition prompting extensive variety of harvests that can be developed yet it implies that there will be territories with substantial precipitation and in the meantime there are zones with just about zero precipitation. The development of products relies upon the methods for procedures utilized by the cultivators to handle these issues and consequently guaranteeing non-stop development of the yield. Presently, for this, dirt examination is important, and so, the normal or known perusing that is accessible about the dirt condition in different territories is given The open N, P, K, C, and Mg obsessions in wheat of Punjab reached out from 80 to 300, 1.67 to 7.10, and 51.0 to 160.4 kg/ha in different wheat plants. The most vital mean open N (2.01 t/ha) was recorded in high yielding plant than that of low yielding (0.8 t/ha). The P moved from 1.79–1.98 t/ha. Similarly, Mn changed from 0.66–4.20. Fe and Zn changed from 0.96–4.21. The data also exhibited that all examples contained low N, P, and K as differentiated and the characteristics uncovered by various wheat cultivators and analysts. Duan in his paper discussed how IoT can be used as a piece of agriculture store arrange organization. They have in like manner inspected how IoT can manage the farmland to grow the effectiveness of the harvest [3]. They moreover exhibited that, IoT can in like manner help farmers for agribusiness developed supply. Mechanical improvement and contention have provoked changes in supply arrange organization for rural things. Nonetheless, overall market checks are stringent. Purchasers ask for secured and restorative sustenance, staggering quality and at the last possible second conveyance. Thusly, joint exertion between trade accessories has ended up being continuously basic for the achievement of cross-edge exchange in the market. Cultivating supply arrange administration is a competent gadget to achieve this participation. In particular, for the ASCM, the investigation of composed information arrange is the key to upgrading level of ASCM. The making of information anchoring and process development for the fundamentally relies on the change of present day Information Technology, for instance, PC advancement, devices, satellite course development, RS, sensor advancement and framework development et cetera. In the monetarily stable nations, horticulture robotization has been utilized for the most part, and the level of agribusiness informationization and frameworks organization has achieved high states. China, as a created country, the investigation of cultivating information-ization still has a place within the beginning level. Regardless, after the presentation of IoT, it will push the making of automated horticulture, and the examination of brilliant agribusiness in perspective of IoT will make gigantic sharp rural ages and supply sort out and relate the entire farm, develop towns, and exchange market of significant city, for instance, Beijing together. It cannot simply improve the idea of horticultural thing, the profitability of cultivating creation, and the level of farming thing supply and moreover gainfully handle the rising dispatch of nourishment in excellent situation. Thus, Patil et al. talked about Internet of Things and cloud registering in the field of horticulture [4]. As indicated by them, the single rancher can deal with their products and can convey the yields specifically to the shoppers not in a little locale but rather additionally in business sectors. Panigrahi et al. in their paper "Shortfall water system scheduling and yield and its forecast in a semiarid area" [5] have given the most extreme worry to the way that a plant necessity for water is distinctive in its diverse developing stages subsequently giving insights about different developing periods of plants.

They additionally talked about the water rare regions and how the creation of plants can be balanced in order to get the greatest yield even on account of a water shortage, and for this, they included different parameters like evapotranspiration, leaf vanishing, and so on in this manner giving a proper method to alter the development procedure of the plants. Cheng-Jun in his paper "Exploration and Implementation of Agricultural Environment Monitoring Based on Internet of Things" [6] said that as demonstrated by the examination of field condition watching framework in application scene, EPC's engineering structure is alluded, ponder framework structure is arranged, and perception layer of ZigBee distin-guishing center is in like manner laid out. The identifying center point should accumulate distinctive basic genuine opportune condition information which is related to cultivating generation and perform controllable picture getting on gather's condition. By then, different data are sent to Web server by strategies for remote module to design Web server at application layer.

Sharma and Mittra in their paper "Impact of manuring and mineral compost on development and yield of harvests in rice-construct trimming in light of corrosive lateritic soil" [7] have talked about the correct planning of manuring have altogether enhanced the yearly development of yields. Facilitate they likewise examined that accessibility of required centralization of supplements at certain season of devel-opment of plants for e.g. 15 kg N/ha or 15 kg N and 30 kg P2O5/ha additionally expanded yield along these lines enhancing the yearly yield of the yield. N fertilizers are usually better than any other because it causes almost no impact on the fertility build up. Kharrufa et al. in their published work "Irrigation and Agriculture Development" experimentally proved that water stress, i.e., water is not supplied to the crops at the proper time plays a major part in yield [8]. Relationship between yield and relative evapotranspiration deficit is given by yield response factor:

$$(1 - Sa/Sb) = ky(1Ea/Em)$$

Sa = Actual output received, Sm = Maximum output received
Ky = Total growth time constant, Ea = real evapotranspiration
Em = Max evapotranspiration

Thereby proposing the idea how IoT can help to improve or help in increasing the annual growth and quality of crops sowed. Firstly, different sensors like moisture sensor, temperature sensor, pH sensor, and chemical sensor are set in the crop fields which will collect the data in a regular interval of time, and these data are processed by the device which will give signal to alert device to alert the farmers. It will alert about various scenarios like water is needed, proportion of fertilizers to be added, etc.

Various Sensors used: Moisture sensor, temperature sensor, pH sensor, Chemical sensor.

For transmission of notification: GSM module/Wi-Fi module.

Mechanical devices: Sprinklers, circuit breakers, chemical restoring machines servo motors.

Microcontroller: Arduino UNO, i.e., Fig. 1.

3   Proposed System

As stated before, IOT along with the use of various sensors like moisture, pH, etc., can provide a perfect solution for the issue faced, i.e., the lack of knowledge and resources to resolve the issue, thereby not only reduce the loss but also help in increasing the annual yield of a crop. The connection and interaction among various sensors are explained with the help of a diagram Fig. 2.

Algorithm

1. Various Store House is created namely a, b, c each with specific storage of water, chemicals increasing pH and organic materials respectively.
2. Input the threshold values according to the crop cultivated
3. Input the Threshold Moisture, Threshold pH, Threshold organic material
4. If S_moisturec < Sm_moisturec Alert mechanism initiated
5. Storage a is selected, containing water
6. GPS Module is activated sending the precise location where the moisture is less
7. Motors of storage a is activated and the water container is moved to the deficient location.
8. If c < 3
9. Send sms
10. Addition of water_addition
11. Water_addition = Sm_moisturec - S_moisturec

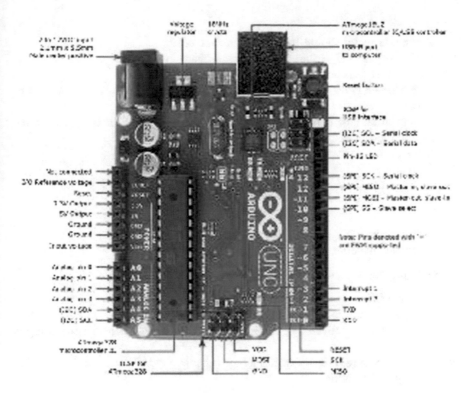

Fig. 1 Arduino UNO

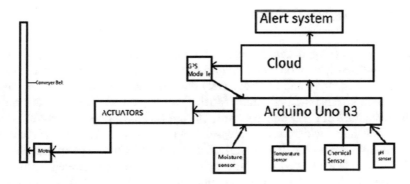

Fig. 2 Proposed architecture

Else

12. Precaution major should be taken to reduce effect
13. If pH_present- pH_required > 0 Alert mechanism initiated
14. Storage b is selected, containing pH improving chemicals
15. GPS Module is activated sending the precise location where the pH is less or more
16. Motors of storage b is activated and the chemical container is moved to the required location.
17. If c < 3
18. Send sms
19. Addition of chemical or water to reduce pH

Else

20. Add chemical to increase pH.
21. If organic_material_present < organic_material_required

   &&P_present < P_required&&Na_present < Na-required

22. Alert mechanism initiated
23. Storage c is selected, containing organic materials.
24. GPS Module is activated sending the precise location where the manuring is required
25. Motors of storage c is activated and the organic material container is moved to the required location.
26. If c < 3

Send sms

27. Then add organic compound.
28. Generate notification

In this engineering, different sensors like the temperature, moisture, pH, and so on are associated with a microcontroller, i.e., an Arduino UNO. The sensors gather information frequently and inconclusively and are put away for future references. The Arduino is transferred with the code that will request that the client determines the limit readings of every single sensor. The information gathered by the sensors is sent to the Arduino which additionally sends it to the server. Presently, at the server side, principle part of the design works, i.e., the examination. The information got by the sensors is contrasted and the edge esteems which are indicated by the code composed. On the off chance that the information got or the information got from different sensors is over the limit readings determined by a specialist, and an alarm message is transmitted to the agriculturist indicating the zone that needs the consideration. This part is finished with the assistance of a GSM module which can be associated with the microcontroller. At the point when they got esteem surpasses the limit readings, the flag is sent to the Arduino to send a SMS to the client in this way GSM module is utilized to do as such. The lacking supplement or watering time is advised to the client with the assistance of the SMS.

For the prompt reaction by the keen framework, different transport lines will be introduced in the field which is associated with storage facilities to be specific a, b, and c containing capacity of water, synthetics, and natural materials. These compartments are associated with the transport line for development which in turn is associated with the engine or the actuator. At the point when the edge is crossed, the GPS module is enacted indicating, the area of the sensor from which the readings are gotten. Presently, as indicated by the need, the transport of particular storage facility is enacted utilizing actuators, i.e., assume dampness content is not as much as the transport line of storage facility an is enacted making the compartment containing water to the area determined by the GPS module accordingly recharging the water level by discharging water at that area. Different mechanical gadgets like engines that will start gadgets like sprinklers are associated with the Arduino's simple or computerized pins as indicated by their sort. These gadgets consequently react when the limit readings are crossed or surpassed. Since these gadgets are associated with and fueled by the Arduino's 5V yield in this manner when the flag is gotten by the microcontroller, control is provided to the gadgets accordingly turning them on. Until the point that the information or readings got by the Arduino from the sensors or end hubs are standardized as for the coded esteems, the gadgets will stay fueled and work, and when the qualities are standardized, a flag is sent to the microcontroller slicing the ability to the gadgets subsequently ceasing them.

The information with respect to the edge readings and the no. of time it is crossed and the typical period of the year when it happens is recorded on the server. These perusing are additionally handled and fed at the server to create basic groupings and tenets which will in turn help maintain a strategic distance from basic issue and give ranchers better information about their dirt and how to handle basic horticulture-related issues that as a rule happens in that kind of soil. Microcontroller as we probably aware contains different information pins. Different sensors like temperature, moisture, humidity, pH, and so forth are associated with these pins, and information or readings got from the sensors are sent to the microcontroller which in turn sends it to the proposed calculation for, a great many processing's flag is sent to the Arduino indicating the fitting move to be made, i.e., cautioning the agriculturists and taking certain prompt activities.

4 Result Analysis

In the wake of actualizing the accompanying strategies, we watched an extraordinary increment in the agrarian items creation. Different parameters were viewed as like pH level, moisture focus in the dirt, and so on with the end goal of examination. The prerequisite of water by the yields or legitimate planning of applying the synthetic concoctions and the measure of synthetic compounds to be connected to the harvests were educated to the cultivators consequently, and prompt activity of enacting the actuators associated with the transport lines of the storage space is initiated if required in this manner guaranteeing the better and appropriate development of the products.

5 Conclusion

With the powerful use to end hubs, i.e., the sensors and the rising and consistently developing idea of IoT glancing over to the state of harvests, soil, and so on can be kept under a customary check, detailing any inconsistency on the off chance that it happens or exists and up to some degree attempt to determine the issue consequently. Educating the ranchers about the proper time and amount of water required by the harvests and furthermore cautioning them about changing conditions can help in avoiding crop disappointments as well as in expanding the general yield of the product.

References

1. Rahman, H.-Ur., Ghulam, N., et al.: Growth and yield of fruits and soil physical properties as affected by orchard floor management practices in Punjab, Pakistan (2012)
2. Khokhar, Y., et al.: Soil fertility and nutritional status of orchards grown in aridisol of Punjab, India. Afr. J. Agric. Res., 4692–4697 (2012)
3. Duan, Yan E.: Research on integrated information platform of agricultural supply chain management based on Internet of Things. JSW 6(5), 944–950 (2011)
4. Patil, V.C., et al.: Internet of things (IoT) and cloud computing for agriculture: an overview. In: Proceedings of Agro-Informatics and Precision Agriculture (AIPA 2012), pp. 292–296. India (2012)
5. Panigrahi, P., Sharma, R.K., Hasan, M., Parihar, S.S.: Deficit irrigation scheduling and yield and its prediction in a semiarid region. Agric. Water Manag. J. (Elsevier) (2014)
6. Cheng-Jun, Z.: Research and implementation of agricultural environment monitoring based on Internet of Things. In: Proceedings of the 2014 Fifth International Conference on Intelligent Systems Design and Engineering Applications, IEEE Computer Society (2014)
7. Sharma, A., Mittra, B.: Effect of green manuring and mineral fertilizer on growth and yield of crops in rice-based cropping on acid lateritic soil. J. Agric. Sci. 110(3), 605–608 (1988). https://doi.org/10.1017/S0021859600082198
8. Schleiff, U., Kharrufa, N.S., Al-Kawaz, G.M., Ismail, H.N., Doorenbos, J., Kassam, A.H., Abdel-Al, Z.E., et al.: Crop water requirements and yield response in Iraq and Saudi Arabia. In: International Expert Consultation on Irrigation and Agricultural Development. Pergamon Press, Baghdad, 24 Feb 1979 (1980)

Chronic Heart Disease Prediction Using Data Mining Techniques

Sravani Nalluri, R. Vijaya Saraswathi,
Somula Ramasubbareddy, K. Govinda and E. Swetha

Abstract Cardiovascular disease has become one of the most widespread diseases in the world at present. It is estimated to have caused around 17.9 million deaths in 2017 which constitutes about 15% of all natural deaths. One major type of cardiovascular disease is chronic heart disease. CHD can be detected at the initial stages by measuring the levels of various health parameters like blood pressure, cholesterol level, heart rate and glucose level. Other characteristics of a person like number of cigarettes smoked per day and BMI level also help to diagnose CHD. This paper focuses on utilizing data mining techniques to predict whether a person is suffering from CHD based on data about various symptoms of CHD. This paper proposes two prediction models namely XGBoost algorithm and logistic regression approach to predict the CHD values. This prediction of CHD at its early stage will help in reducing the risk of CHD on a person.

Keywords Chronic heart disease · Logistic regression

1 Introduction

Prediction of CHD has been one of the major areas of research for health personnel since heart diseases became prevalent during the 1940's. Heart disease is the most common natural cause of death in the world at present. The sudden increase in heart disease was mainly due to rapid changes in the day-to-day activities of people like

---

S. Nalluri · R. Vijaya Saraswathi · S. Ramasubbareddy (✉)
VNR Vignana Jyothi Institute of Engineering and Technology, Hyderabad, India
e-mail: svramasubbareddy1219@gmail.com

K. Govinda
VIT University, Vellore, India

E. Swetha
SV College of Engineering, Tirupati, India

© Springer Nature Singapore Pte Ltd. 2020   903
K. S. Raju et al. (eds.), *Data Engineering and Communication Technology*,
Advances in Intelligent Systems and Computing 1079,
https://doi.org/10.1007/978-981-15-1097-7_76

consumption of junk foods or foods with high-fat content, high-stressful lifestyle and smoking habits. Chronic heart disease is also known as atherosclerotic heart disease, ischaemic heart disease and coronary artery disease. It occurs when cholesterol deposits on the walls of major arteries which results in the formation of plaques. These plaques narrow or block the coronary arteries, and this process is known as atherosclerosis. This leads to a reduction in the amount of oxygen transferred to the heart chamber. Low level of oxygen causes the heart muscles to stress and leads to heart failure or cardiac arrest which can be fatal. Data mining techniques differ from standard statistical methods over the fact that the former demands the researcher to introduce a hypothesis whereas the latter can determine the existing pattern without any prior hypothesis. On the whole, data mining operations are of two categories: descriptive data mining operations are the ones in which conventional characteristics of the current data and predictive data mining operations function by predicting and arriving at some conclusion on the basis of inference derived from the current data. The statistical techniques are less efficient, less influential and less adaptable than these techniques for exploratory analysis. One of the essential epidemiological studies ever conducted was the Framingham study, and the analytics of its data was very important for our current understanding of cardiovascular disease. The major motivation for the study to be conducted was the health condition of former US President Franklin Delano Roosevelt, FDR, who was the President of the United States from 1933 to 1945. He died while President on April 12, 1945. Today, healthy blood pressure is considered to be less than 120/80. Two months before his death, his blood pressure was 260/150, and the day of his death was 300/190. This motivated the US Government to conduct more research on cardiovascular diseases. Hence, in the late 1940s, the US Government started to conduct more research on cardiovascular disease. The plan was to monitor a large cohort of patients who are healthy initially over the course of their lifetimes. The city of Framingham in Massachusetts was selected to conduct the study. Framingham is of ideal size with a moderate level of population which is also stable as its people do not move too much. The doctors and residents of Framingham were also quite cooperative. So, the Framingham Heart Study was started in 1948. The study consisted of 4240 patients, aged 32–70. Patients were monitored with the help of a questionnaire and an examination every two years. This examination recorded their physical characteristics, their behavioural characteristics, as well as test results. Over the time, exams and questions expanded, but the major objective of the study was to monitor the trajectory of the health of the patients over their entire lifespan.

2 Literature Survey

S. No.	Title	Author	Remarks
1	XGBoost: a scalable tree boosting system [1]	Tianqi Chen, Carlos Guestrin	This paper describes the working of XGBoost algorithm. It also focuses on cache access patterns, data compression and sharing to build a scalable tree boosting system
2	Ordinal logistic regression [2]	Frank E. Harrell Jr.	This paper explains the functioning of logistic regression model in medical or epidemiologic studies
3	The epidemiology of heart failure: the Framingham study [3]	Kalon K. L. Ho, Joan L. Pinsky, William B. Kannel, Daniel Levy	This paper explains the various parameters of Framingham study and their impact on heart disease study. An analysis of the data from the study is also provided

3 Proposed Method

The steps involved in implementing this project are listed below as follows.

Preprocessing

Missing Values

Since the study was conducted in the 1940s, due to lack of advanced measuring equipment and loss of data over time, the dataset consists of many values missing from the attributes. As the presence of missing values prevents accurate prediction, preprocessing was performed to fill these missing values. Each missing value for the attributes was replaced by the mean of the other values in the attribute.

Splitting of Dataset

For a prediction to be successfully carried out, the dataset has to be split into the training set and test set. The training set is a subset of the original dataset which is used to train the prediction model. The test set is also a subset of the original dataset and is used to evaluate the performance of the trained prediction model. The model initially undergoes supervised learning with the help of the training set and later uses its gained knowledge to predict the values of the dependent variable in the test set. In order for this to be accomplished, the original dataset is split into training set and test set in a particular ratio. The accuracy of the prediction is usually higher if the original dataset is split with a higher value for the training set than the test set.

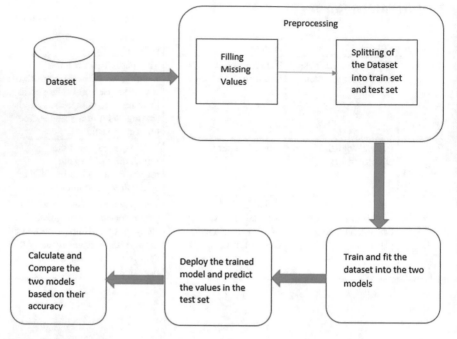

Fig. 1 Major processes of the project

Generally, the dataset is split in the ratio of 50–80% for the training set. In this project, the original dataset has been split in the ratio of 8:10 for the training set and the test set, respectively (Fig. 1).

4 Implementation of Prediction Model

Two prediction models namely XGBoost algorithm and logistic regression are implemented to predict the dependent variable ten year CHD in this project.

XGBoost Algorithm

XGBoost stands for Extreme Gradient Boosting. It is a supervised learning algorithm. It is a library for developing fast and high-performance gradient boosting tree models. The major advantage of XgBoost is parallel computation. It has risen to prominence recently due to its versatility, scalability and efficiency. It supports all three forms of gradient boosting namely: gradient boosting, Stochastic Gradient Boosting and Regularized Gradient Boosting. XGBoost performs these gradient boosting techniques parallel and determines the best model.

Gradient boosting is explained as follows:

1. Let us consider a binary classification problem. Small decision trees are considered as weak learners. Each data point is assigned a weight (all equal initially), and the error is the sum of weights of misclassified examples.
2. Here, we will have a weak predictor which classifies the data points.
3. Now, we will increase the weights of the points misclassified by the previous predictor and learn another predictor on the data points. As a result, the residual is minimized.
4. This process is repeated for the specified number of iterations till the residual is minimum.

In order to increase the accuracy of the model, K-fold cross-validation method is applied during training. In this method, the training set is divided into k subsets. For each iteration, one subset is considered as testing fold, and remaining $K - 1$ subsets are considered as training folds. The training folds are trained using the model, and the model is evaluated using the testing fold. All K subsets are considered as testing folds during different iterations. This increases the accuracy of the model as a whole before being tested on the testing set.

Logistic Regression

Logistic regression is a type of classification algorithm involving a linear discriminant. Unlike actual regression, logistic regression does not try to predict the value of a numeric variable given a set of inputs. Instead, the output is a probability that the given input point belongs to a certain class.

In the case of the normal linear model, there is a chance for the probability of the dependent variable being negative or greater than 1. If this probability is limited between 0 and 1, still there are chances of breaks which lead to inaccurate outputs. Logistic regression overcomes this drawback. Logistic regression converts the probability to be always positive as follows:

The linear discriminant is

$$P = a_1 + a_2 x_1 + a_3 x_2 + \ldots + a_{n+1} x_n \tag{1}$$

where P is probability, a_n is co-efficient and x_n is the independent variable.

The (Eq. 1) is modified as

$$P = \exp(a_1 + a_2 x_1 + a_3 x_2 + \ldots + a_{n+1} x_n) \tag{2}$$

where exp is the exponential function which converts the probability into its exponential form, which is always positive.

The Eq. 2 is further modified as

$$P = \exp(a_1 + a_2 x_1 + a_3 x_2 + \ldots + a_{n+1} x_n) \\ / \exp(a_1 + a_2 x_1 + a_3 x_2 + \ldots + a_{n+1} x_n) + 1 \tag{3}$$

This can be rewritten as

$$P = P/P + 1 \tag{4}$$

Equation 3 ensures that the probability (P) is always less than 1 by performing division with P as the dividend and $P + 1$ as the divisor.

On applying the logarithmic function on Eq. 3, the final logistic regression function is derived.

$$(P/1 - P) = a_1 + a_2 x_1 + a_3 x_2 + \ldots + a_{n+1} x_n \tag{5}$$

where ln is the logarithmic function.

The probability of a data point belonging to a particular class is calculated by substituting the independent attribute values of the data point in Eq. 5. This probability is used to determine the classification.

5 Result Analysis

Early prediction and diagnosis of CHD are highly crucial for protecting a person from coronary diseases. This project utilizes the Framingham dataset in order to predict the chances of CHD in a person. Prominent prediction models namely XGBoost algorithm and logistic regression have been implemented to predict the CHD value of each person. The project has been implemented using RStudio. The dataset consists of 16 attributes which comprise the risk factors and the prediction or dependent attribute. Risk factors are the variables that increase the chances of developing a disease. They are also known as independent variables. This dataset consists of about 15 risk factors which are described below.

This dataset includes several demographic risk factors:

a. The sex of the patient, male or female.
b. The age of the patient in years.
c. The education level coded as

    I. for some high school
   II. For a high school diploma or GED
  III. For some college or vocational school
  IV. For a college degree.

The dataset also includes behavioural risk factors associated with smoking—whether or not the patient is a current smoker and the number of cigarettes that the person smoked on average in one day.

Fig. 2 Accuracy model

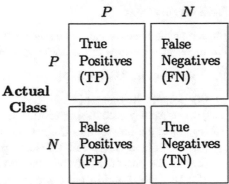

Medical history risk factors were also included. These were

a. Whether or not the patient was on blood pressure medication
b. Whether or not the patient had previously had a stroke
c. Whether or not the patient was hypertensive
d. Whether or not the patient had diabetes.

Lastly, the dataset includes risk factors from the first physical examination of the patient. They are

a. The total cholesterol level
b. Systolic blood pressure
c. Diastolic blood pressure
d. Body mass index or BMI
e. Heart rate
f. Blood glucose level of the patient.

I. **Calculation of Accuracy**

The accuracy of each model is calculated with the help of confusion matrix. The confusion matrix is a table which depicts the number of correct and incorrect predictions in the test set by comparing the predicted values with the values in test set which are already known.

Figure 2 shows the accuracy of each model calculated by:

$$\text{Accuracy} = (\text{TP} + \text{TN})/(\text{TP} + \text{FP} + \text{FN} + \text{TN})$$

II. **Comparative Analysis**

The two models implemented are compared based on their accuracy to determine the most efficient model among them.

Result

A. Preprocessing

Implement an if-else condition stating that if the value 'NA' is found replace it with the mean of all values in the corresponding variable.

B. Execution of XGBoost [5]

(a) Import the dataset [4].
(b) Perform preprocessing to fill the missing values.
(c) Split the dataset into test set and training set.
(d) Download and install **XGBoost** package.
(e) Fit the classifier and train the model using **XGBoost** function.
(f) Download and install caret package to implement K-fold cross-validation method.
(g) Use K-fold cross-validation method to calculate the accuracy of the model.

Fold01	double [1]	0.9102564
Fold02	double [1]	0.8255814
Fold03	double [1]	0.7792208
Fold04	double [1]	0.8958333
Fold05	double [1]	0.8846154
Fold06	double [1]	0.7951807
Fold07	double [1]	0.8111111
Fold08	double [1]	0.7831325
Fold09	double [1]	0.8674699
Fold10	double [1]	0.893617

Final   Accuracy = (0.91 + 0.82 + 0.77 + 0.89 + 0.88 + 0.79 + 0.81 + 0.78 + 0.86 + 0.89)/10

Final Accuracy = 0.8446

This indicates that the **XGBoost** model is able to predict with an accuracy of 84.46% with respect to the Framingham dataset.

C. Execution of Logistic Regression [6]

(a) Import the dataset.
(b) Perform preprocessing to fill the missing values.
(c) Split the dataset into test set and training set.
(d) Use the Generalized Linear Model (GLM) function to train the model.
(e) Deploy the trained model for prediction in the test set.
(f) Compute the confusion matrix (CM) and calculate the accuracy (Table 1).

Table 1 Confusion matrix

CM	0	1
0	1071	8
1	174	18

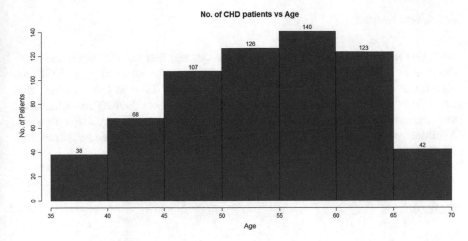

Fig. 3 Histogram analysis

- The final accuracy is calculated by the formula:

$$\text{Accuracy} = (\text{cm}[1,1] + \text{cm}[2,2])/(\text{cm}[1,1] + \text{cm}[2,2] + \text{cm}[1,2] + \text{cm}[2,1])$$
$$= (1071 + 18)/(1071 + 8 + 174 + 18)$$
$$= 0.8568$$

This indicates that logistic regression has an accuracy of 85.68% with respect to the Framingham dataset.

D. Histogram Analysis

A histogram based on the age of the patients suffering from CHD was developed. It depicted the number of patients belonging to each age category with an interval of five years. As shown in Fig. 3, there are a total of 644 patients suffering from CHD in the Framingham dataset.

This histogram indicates that the age category of 55–60 consists of most patients suffering from CHD while the patients belonging to the age category of 35–40 are the least likely to be a CHD patient. We can also observe from the histogram that the majority of people suffering from CHD are in the age interval 45–65 while people belonging to the age category of 65–70 having CHD is significantly low. This indicates that apart from ageing, other factors like lifestyle activities of a person also greatly influence a person's chances of suffering from CHD.

6   Conclusion

With respect to the Framingham dataset, we can say that logistic regression algorithm with an accuracy of 85.86% is marginally more accurate than **XGBoost** algorithm which has an accuracy of about 84.46%. This project indicates that Logistic Regression is slightly more accurate and suitable than Xgboost for small and medium-sized datasets. But in the case of analysis of very large datasets, **XGBoost** is preferred over logistic regression due to its shorter computation time which leads to significant saving of time.

References

1. Chen, T., Guestrin, C.: Xgboost: A scalable tree boosting system. In: Proceedings of the 22nd acm sigkdd international conference on knowledge discovery and data mining, ACM, pp. 785–794. Aug 2016
2. Harrell, F.E.: Ordinal logistic regression. In: Regression modeling strategies, pp. 331–343. Springer, New York, NY (2001)
3. Ho, K.K., Pinsky, J.L., Kannel, W.B., Levy, D.: The epidemiology of heart failure: the Framingham Study. J. Am. Coll. Cardiol. **22**(4), A6–A13 (1993)
4. https://www.kaggle.com/amanajmera1/framingham-heart-study-dataset/data
5. https://dimensionless.in/introduction-to-xgboost/
6. https://codesachin.wordpress.com/2015/08/16/logistic-regression-for-dummies/

Server Security in Cloud Computing Using Block-Chaining Technique

Illa Pavan Kumar, Swathi Sambangi, Ramasubbareddy Somukoa, Sravani Nalluri and K. Govinda

Abstract Data security is a key concern in cloud computing. With the increase in attempts to hack a virtual cloud server in order to access the data stored there privately, more and more measures are being taken to ensure a reliable and secure cloud. Enhancements are made on traditional cryptography algorithms so that they are more suited to the distributed, virtual and multi-tenant environment of the cloud. Block-chaining is the newest technique in the fields of computing. It is basically a growing list of records, called as "blocks", which are linked and secured using cryptography. In this project, we are aiming to implement an encryption algorithm which works on the principles of block-chains, to ensure the security and integrity of a cloud-based system. The infallible nature of block-chains is what makes it perfect for cloud server security.

Keywords Cloud computing · Security · Block-chain · Encryption

1 Introduction

Cloud computing is one of the most recent advances in the technology which provides a common, distributed environment to its users. It is a very flexible paradigm of computing which allows the users to efficiently access their resources from a third-party service provider. The cloud users can use the cloud for a variety of purposes, but their common goal is same—minimization of infrastructure cost and the utmost security of their data. As the cloud users outsource the data, the load and hence the cost of storing and maintaining the data are reduced.

Being a distributed storage framework, a cloud stores a huge amount of information in its servers. As it being done over the Internet, long-term storage of data

I. Pavan Kumar · S. Sambangi · R. Somukoa (✉) · S. Nalluri
VNR Vignana Jyothi Institute of Engineering and Technology, Hyderabad, India
e-mail: svramasubbareddy1219@gmail.com

K. Govinda
VIT University, Vellore, India

© Springer Nature Singapore Pte Ltd. 2020
K. S. Raju et al. (eds.), *Data Engineering and Communication Technology*,
Advances in Intelligent Systems and Computing 1079,
https://doi.org/10.1007/978-981-15-1097-7_77

makes it vulnerable and exposes it to the hackers to steal or modify. As data confidentiality is one of the key concerns for a cloud user, maintaining the security is one of the key aspects of a cloud.

The method described in this paper uses block-chaining method (used in the popular bitcoin system of e-currency) to ensure server security in a cloud. It uses a hash function (which can be changed to incorporate any encryption algorithm) which is block-chained across each server. Any event which can hack a server results in a change of its block-chain. At the end of each transaction, the analysis of block-chains of all the servers can help identify the compromised server and appropriate measures can be taken to restore the system. This particular method can be used when "prevention of data loss" is preferred over "prevention of data theft"; that is, a hacker can at most steal the file from the cloud but can not change the contents of the file in the actual cloud.

2 Literature Survey

Information technology is advancing day by day; both processes and information are now moving to cloud. By this, not only the way of computing is changing but fundamental changes are also seen. This new advancement in the world of technology has brought on various challenges as well, like information security, ownership and storage. This paper discusses about how to deal with security issues in cloud computing. The algorithm discussed here is RSA which is basically used for encryption and decryption consisting of a public and private key, used for encryption and decryption, respectively. Another algorithm is the Blowfish algorithm, encrypting a 64-bit block with a key of 128–448 bits (variable). DES algorithm uses a 64-bit encryption method too but here the key is 56-bit long. All these algorithms aim on solving the most important concern of the hour which is data security. After comparing all the algorithms, AES is found out to be the one that executes the fastest [1, 2]. A set of IT services like software system, software, hardware and storage all come under cloud computing, and these services, over a network, are provided to a number of customers all around the world. This makes accessing knowledge anywhere anytime very easy, also cost efficiency, scalability and reliability of the information. All these advancements are now convincing every organization to move to cloud to store all sorts of data and information. This means we need all this data to be kept secure which this paper talks about, and the method discussed here is cryptography. Data cryptography is mainly the scrambling of data which maybe text, audio, image, etc., thus making it unrecognizable and in a way, useless! This, as we know, is called encryption and its opposite is decryption, used to restore our meaningful data. There are a variety of symmetric and asymmetric algorithms to perform this task but the latter is slower than the former. Multilevel encryption and decryption algorithm proposed here is a combination of DES and RSA which encrypts the text files uploaded by the user in the cloud itself, and inverse of this combination of algorithms is used to decrypt when the user downloads the file [3, 1]. Cloud computing being flexible in nature allows easy access of data and resources anywhere and anytime; hence, we know there is a lot of data that has to be stored for a very long time so we can't afford internet to lose its data and information confidentiality; therefore, here too, we discuss cryptography as a method to keep our cloud and its data secure to ensure efficiency. Cryptographic techniques talked about here are searchable encryption and holomorphic encryption. The former technique is applied at a high level, encrypting the data in search index to hide it from others except the ones that provide tokens while the latter scheme allows computations to be executed on the data which is encrypted but is a little slow. Such techniques retain properties like confidentiality, reliability and integrity. Data can be secured and shared with the authorized customers hence keeping away the notorious hackers. Cryptography is found to be one of the most reliable ways to secure information as a lot of our big organizations have already adapted it [2, 4].

3   Methodology

The project deals with enabling cloud security across multiple servers hosting a cloud. The algorithm makes the following assumptions. The cloud is hosted by multiple servers which are all connected to each other dedicatedly so as to enable communication between them. Each server contains all the files in encrypted form and their corresponding keys. Each server has a block-chain as a record of the cloud transactions. The servers start with the default block-chain, and with every valid transaction, they add a block to the block-chain. Each server, file and user has a unique identifier. For this reason, the block of the chain can be uniquely specified by the server identifier, user identifier, file identifier and a random number based on which server sent which file to which user [5, 6].

The architecture of the cloud for a five server cluster is shown in Fig. 1, the following diagram.

The algorithm is inspired by block-chaining method that is used in the near infallible security of the bitcoin currency. Its operations are shown in a console-based application where a certain situation can be constructed by run-time choices. The application allows a user to set the number of servers, number of files and their keys in those servers, the choice of hacking those servers for the sake of the situation, etc. It then shows how the algorithm works to validate the user and protect data from destruction and theft.

3.1   The Algorithm Works as Follows

Each server has the same block-chain initially where each block is a combination of the file, server, user ids and a random number. The default chain when the system

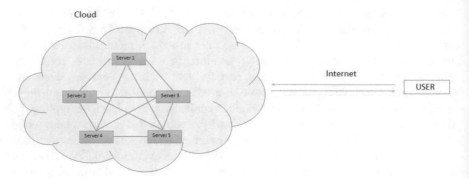

Fig. 1 Cloud architecture diagram

first starts is "start;" where ";" is used as the block separator in the chain. Each server contains the encrypted file and its key for redundancy; so that even if the other servers were hacked, the entire system cannot be compromised and no data is ever lost. The cloud asks the user for login details to validate the user and perform the first level security check so as to thwart direct attacks. If user is authentic, a random server selected from the group sends the encrypted file to the user. To determine if the transaction can be compromised, the servers validate the cloud security before sending the decryption key for the file. To validate, it checks the block-chain of every other server against itself. This is done by every server and, thereby, checks if all the servers have the same transaction record. If any server has a different block-chain, it has probably been hacked and that server is removed from the cloud temporarily for repair. The validation process starts again with the remaining servers to reduce the chances of unauthorized access to data. It uses a new random number each time so that a hacker which has already hacked a server cannot hack another server again with the same information. This validation process repeats till either all remaining servers are determined safe or more than half the servers are hacked. In the latter case, the transaction is cancelled to protect data in other servers so that sensitive data is not lost and transaction can be repeated later. Once all servers are determined safe and secure, a random secure server sends the key to the user and updates its block-chain for that transaction. The other servers copy its block-chain. The servers that were hacked are repaired, and their block-chains are also updated to the same as every other server. This completes one transaction and then the next transaction can start.

This algorithm relies on the fact that hacking a server will change the block-chain of the target server but not the other servers. So, when the other servers find that the target server has a different block-chain, the target server is immediately taken offline and repaired. Each server helps validate every other server.

4 Run-Time Analysis

The five major operations involved in server security are:

The condition check runs in $O(n)$ linear complexity. The determination of valid and invalid servers that runs in $O(n)$ linear complexity by checking each server once. Segmentation into valid and invalid servers is again an $O(n)$ linear complexity operation. Sending the decryption key to the user is an $O(1)$ complexity operation. Updating chains of all the servers is another $O(n)$ linear complexity operation. Thus, the average run-time complexity of these operations is $O(n)$ and is linear in the number of servers in the cloud.

The algorithm here uses a linear system of data encryption which contributes to $O(1)$ to the whole program. If any other encryption algorithm is used, the overall run-time complexity will change accordingly but will make the data more secure against theft.

5 Results and Discussions

This section shows some sample test cases that were run and the screenshots of the corresponding program run.

The **first test case** will describe the application of the system when no server is hacked. The inputs specify three servers with the files—dpple, zall and zat (encrypted). The user enters authentication details and is logged in. User then enters the file he/she needs, that is file 2. So user with id 3 has file 2. Since no servers are hacked, all servers are valid servers. They started with a primary block-chain of "start;". After the transaction, the block-chains become "start;1-3-2-63;" where 1 is the server chosen to send the key to the user, 3 is the id of the user, 2 is the file id required by the user and 63 is a random number to show the uniqueness of the transaction as shown in Fig. 2. The program shows the current block-chain and contents of each server after this transaction. So user got the decrypted file.

The second test case is where one of the three servers is hacked during the transaction with the user. We assume the id of the user to be 3 and the id of the hacker to be 4. Let the servers have a primary block-chain of "start;1-3-2-63;". The user asks for file 1. Here, the hacker hacks the server with id 1. Thus, the hacked server chain becomes "start;1-3-2-63;1-4-1-21", whereas the other servers have unchanged chains. The program correctly classifies that the valid servers are the ones with id 0 and 2, whereas the server with id 1 is the invalid server. After this, the invalid server is taken off the network and the cloud only consists of the two valid servers for the rest of the transaction. Then a random server out of the valid ones is chosen to send the key to the user. Thus, the user gets the decrypted file without any problems even when there was an attack. Moreover, after the transaction, the server 1 was repaired and put online again. The current block-chain is "start;1-3-2-63;2-3-1-15;". Here, the server 2 sent the decryption key. Thus, data was not lost or stolen, and the user got the data required as shown in Fig. 3.

```
Enter Username:
manyu
Enter Password:
manyu
Access granted!
Enter id of required file:
2
User 3 has file zat
Enter 1 to allow hacking a server in cluster:
0
The valid servers are: 0, 1, 2,
The invalid servers are:
Selected server is 1
User 3 has file vat
Server - 0 contains:
        1 - dpple & 3
        2 - zall & 4
        3 - zat & 4
Current Blockchain = start;1-3-2-63;
End of server content
Server - 1 contains:
        1 - dpple & 3
        2 - zall & 4
        3 - zat & 4
Current Blockchain = start;1-3-2-63;
End of server content
Server - 2 contains:
        1 - dpple & 3
        2 - zall & 4
        3 - zat & 4
Current Blockchain = start;1-3-2-63;
End of server content
Enter 0 to end session:
```

Fig. 2 Block-chain transaction details

6 Conclusion

The algorithm was successfully implemented and tested for numerous cases. It was found to be near infallible. The only way to compromise the system is to hack every server hosting the cloud in the time it takes to complete one transaction. Since this is computationally very hard, it ensures security. Hence, the system can be made more secure if the number of servers is increased. Moreover, the files on the servers follow an encryption function which may use public key encryptions like the RSA algorithm to ensure that even if the encrypted file was stolen, no data will be leaked.

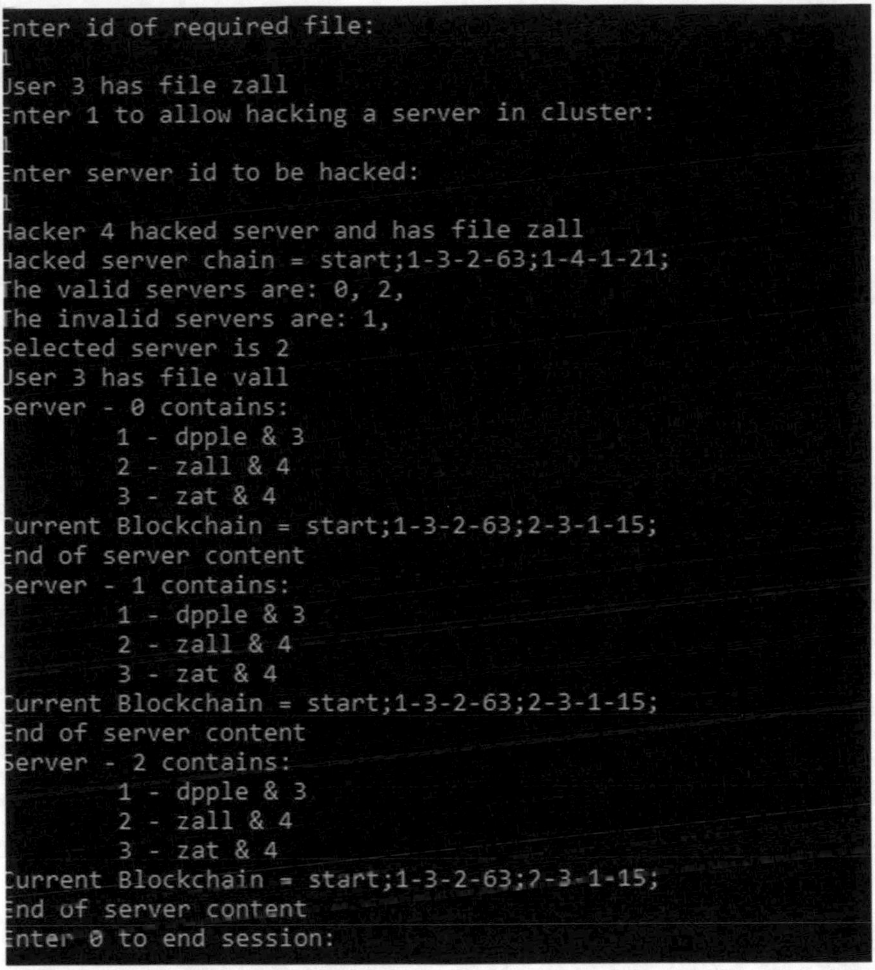

```
Enter id of required file:
1
User 3 has file zall
Enter 1 to allow hacking a server in cluster:
1
Enter server id to be hacked:
1
Hacker 4 hacked server and has file zall
Hacked server chain = start;1-3-2-63;1-4-1-21;
The valid servers are: 0, 2,
The invalid servers are: 1,
Selected server is 2
User 3 has file vall
Server - 0 contains:
        1 - dpple & 3
        2 - zall & 4
        3 - zat & 4
Current Blockchain = start;1-3-2-63;2-3-1-15;
End of server content
Server - 1 contains:
        1 - dpple & 3
        2 - zall & 4
        3 - zat & 4
Current Blockchain = start;1-3-2-63;2-3-1-15;
End of server content
Server - 2 contains:
        1 - dpple & 3
        2 - zall & 4
        3 - zat & 4
Current Blockchain = start;1-3-2-63;2-3-1-15;
End of server content
Enter 0 to end session:
```

Fig. 3 Decryption file

References

1. Arora, R., Parashar, A., Transforming, C.C.I.: Secure user data in cloud computing using encryption algorithms. Int. J. Eng. Res. Appl. 3(4), 1922–1926 (2013)
2. Kirubakaramoorthi, R., Arivazhagan, D., Helen, D.: Survey on encryption techniques used to secure cloud storage system. Indian J. Sci. Technol. 8(36) (2015)
3. Tuteja, R.R., Khan, S.S.: Security in Cloud Computing using cryptographic algorithms. Int. J. Innov. Res. Comput. Commun. Eng. 3, 2320–9798 (2015)
4. Sanka, S., Hota, C., Rajarajan, M.: Secure data access in cloud computing. In: 2010 IEEE 4th international conference on internet multimedia services architecture and application, IEEE, pp. 1–6. Dec 2010

5. Hwang, K., Li, D.: Trusted cloud computing with secure resources and data coloring. IEEE Internet Comput. **5**, 14–22 (2010)
6. Wang, C., Wang, Q., Ren, K., Cao, N., Lou, W.: Toward secure and dependable storage services in cloud computing. IEEE Trans. Serv. Comput. **5**(2), 220–232 (2012)

RETRACTED CHAPTER: Dominant Color Palette Extraction by K-Means Clustering Algorithm and Reconstruction of Image

Illa Pavan Kumar, V. P. Hara Gopal, Somula Ramasubbareddy, Sravani Nalluri and K. Govinda

Abstract In our present age, there are numerous applications and devices for converting a standard picture into a picture, which is built from its predominant shading palettes. 'predominant colors' are the hues in the picture, in which the pixels of the picture contain the particular shading. These hues are then removed and another picture is then framed from this picture. However, this is just a speculation of the issue. In this undertaking, we dive into much more proficient techniques with a specific end goal to remove the prevailing shading palettes from the picture. In this manner, utilizing a grouping calculation and a significantly more ideal instatement technique, effective shading palette is extracted. The grouping calculation being utilized here is K-implies bunching calculation, which is much more effective calculation than the other bunching calculations out there. The bunching calculation groups the pixels in bunches in view of their shading. The palettes separated from the picture are then utilized and another picture is reproduced.

Keywords RGB space · HSV space · Clustering · Clusters · Centroids · K-implies · Kaufman introduction · Shading containers

1 Introduction

In spite of the fact that there are many spots online where we can remove the 'prevailing' hues from the picture, we adopt an alternate strategy with a specific end goal to play out the extraction of the predominant hues. A grouping calculation is

The original version of this chapter was retracted. The retraction note to this chapter is available at https://doi.org/10.1007/978-981-15-1097-7_82

I. Pavan Kumar · V. P. Hara Gopal · S. Ramasubbareddy · S. Nalluri
VNR Vignana Jyothi Institute of Engineering and Technology, Hyderabad, India

K. Govinda (✉)
VIT University, Vellore, India
e-mail: kgovinda@vit.ac.in

utilized for uncovering the structure in the information, i.e., prevailing hues. In any case, first, let us investigate what are overwhelming hues and the criteria to locate the prevailing hues [1].

1.1 Dominant Colors

Predominant hues are the hues which can be characterized in two methodologies

1. The strength of shading can be controlled by what a number of pixels have that particular shading.
2. Utilizing the separation work and end result is determined accordingly.

There are two methodologies for removing the predominant hues from the picture they are

1. HSV space.
2. K-implies grouping calculation.

The over two strategies accomplish a similar exact result. Be that as it may, HSV actually implies a space, for example, RGB and k-implies is a grouping calculation [2].

1.2 HSV Space

HSV stands for hue, saturation, and value. HSV being a color area allows for the saturation of the pixels values in awesome entities. The two entities namely hue and saturation are connected to color, while the other one is related to illumination or brightness.

1.3 K-Means Clustering

All the clustering algorithms do no longer lead to the same result, the result varies in particular in case of the k-means algorithms on the initialization method being used and additionally on the way the centers are positioned in a one-of-a-kind location.

With centers defined, we take every point belonging to given records set and partner it with the nearest center. But originally, clustering and its underlying causes need to be noted.

1.4   K-Means Clustering Procedure

The simple notion of the k-means is to pick out and outline the centers, one for every of the clusters. Clusters in the case of photographs are just an aggregation of the facts which in our case is pixels. Thus, the end result relies upon the way the centers are placed in a specific location. If the facilities are placed in a line, then it is regarded as linear clustering [3, 4].

2   Literature Survey

In our current technology, there are many functions and equipment for changing an ordinary image into a photo which is constructed from its dominant color palettes.

2.1   K-Means Clustering Algorithm

Here, we will give the cluster number and we acquire the new centroids; a new binding has to be completed between the equal dataset and nearest center. This loop keeps rotating.

As an end result of this loop, we may additionally note that the ok facilities trade their region step by step until no extra changes or the centers of each of the clusters do now not trade their function anymore.

Thus, we, in the end, obtain a minimized objective function known as the squared error function given with the aid of

$$J(V) = \sum_{i=1}^{C} \sum_{j=1}^{C_i} \left(||x_i - v_j|| \right)^2$$

'$||x_i - v_j||$' is the Euclidean distance between xi and vj.
'c_i' is the number of data points in ith cluster.
'c' is the number of cluster centers.

Peaks in the color histograms can be associated with the major colors.

Techniques

1. Decides number of pixels having unique color.
2. Two colorations are distinct if the distance between its RGB components is high.

The first option might also seem logical, but it leads to a problem: In a real photo, it is uncommon to discover two pixels of equal colors. If suppose any color is 'red,' then it will have hundreds of pixels of unique shades of red. Thus, it is increasingly challenging to divide into a precise color.

The 2D option is wanted to determine which hues are 'close' to red [5].

We accordingly use k capacity clustering in order to divide the picture into range of discrete areas such that the pixels have excessive similarity in every area. In an RGB space, the pixels can be diagnosed and consequently clustered however, this is no longer as top quality as using it in the HSV area and consequently, this is the procedure we follow [6].

The two error functions are discussed as follows

Root Mean Square Error: it is one of the standard performance measures of the output image; it gives how much the output image is deviated from the input image given by the formula.

$$\mathrm{RMSE} = \sqrt{\frac{1}{n_x n_y} \frac{\sum_0^{n_x-1} \sum_0^{n_y-1} [(r(x,y))]^2}{\sum_0^{n_x-1} \sum_0^{n_y-1} [r(x,y) - t(x,y)]^2}}$$

Peak to Signal Noise Ratio: it is the ratio between the maximum powers that can be obtained and the corrupting noise that influences likeness to image.

$$\mathrm{PSNR} = 10 \cdot \log_1 0 \left[\frac{\max(r(x,y))^2}{\sqrt{\frac{1}{n_x n_y} \frac{\sum_0^{n_x-1} \sum_0^{n_y-1} [(r(x,y))]^2}{\sum_0^{n_x-1} \sum_0^{n_y-1} [r(x,y) - t(x,y)]^2}}} \right]$$

2.2   *Flowchart of the Steps Involved in the Dominant Color Palette Extraction*

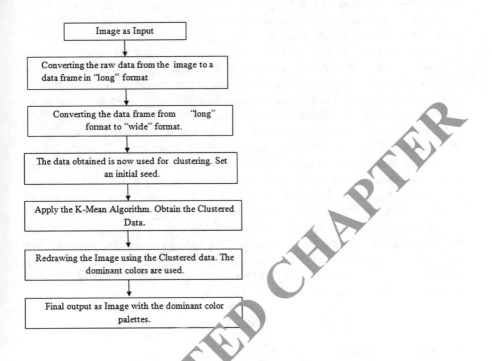

3   Collecting Image and Reading the Image

Firstly, any Jpeg image which we are involved in is taken and then read. The simple idea is to read the picture and convert it into an information frame layout so that the K-means clustering algorithm can be used in order to extract the dominant color palettes. The photo is then transformed into an information frame, and then the statistics which is in the large layout is then 'melted.' This melting is performed because the uncooked records that we get hold of from the inputted picture may additionally no longer be in a format that is no longer prepared to dive straight to facts analysis or any statistical methods.

4 Initialization and Applying the Algorithm

Now that we have bought the data, we need to cluster it. Since the initial cluster assignments are random, we provide the most reliable preliminary seed. Thus, relying upon the magnitude of the starting assignments, the algorithm will attempt the number of one-of-a-kind random starting assignments and then choose the one with rare cluster representation. A more top-quality method of initialization regarded as the Kaufman Initialization technique can be used. In this technique, first, the centrally most located point is taken then for every information factor, we discover most of the distinction between two points and the distance between factor and the centroid. If the value is positive, then we take it into consideration else the price is zero. The point with the maximum price is selected. Thus the cluster centroids, the clusters that each statistics point was once assigned to and the inside cluster variant are obtained. We can also specify the variety of colors to which we cluster to. The gold standard case may be three. But seemingly as there are many variations of the identical color, we may trade the number of colorings which we favor to cluster to greater than that of three. Thus, our K-means algorithm clusters the data obtained from the statistics set to these numbers of colors. Thus, through the clustering algorithm, we attain a histogram depicting the dominant colorations from the image. Thus, we finally obtain output as a redrawn image with the extracted dominant colors.

5 Implementation

The dominant color palette extraction from snapshots is achieved by the usage of a K-means clustering algorithm. We use R Studio in order to function our operation of extraction. The implementation is pretty easy and as a consequence can be explained easier. First, the packages required for performing our operations are loaded; next the pictures from which we want to extract our dominant colorings are loaded and read. The photograph is then converted to a records frame, and then the records which are in the large layout are 'melted.'

We now follow the K-means in order to perform the process. The first step of clustering is initialization. Since the preliminary cluster assignments are random, we apply the most reliable initial seed. Thus, depending upon the magnitude of the starting assignments, the algorithm will try the wide variety of distinct random beginning assignments and then pick out the one with the low-level cluster. First, the centrally most located point is taken then for every records point, we locate the most of the distinction between two factors and the distance between point and the centroid. If the price is fantastic, then we take it into consideration else the value is zero. The point with the most prices is selected. We specify the numbers to which we prefer to cluster to. Finally, we achieve a histogram plot which depicts the dominant hues obtained (Figs. 1, 2 and 3).

The Input Image

Fig. 1 Input image

The Histogram of the dominant colors

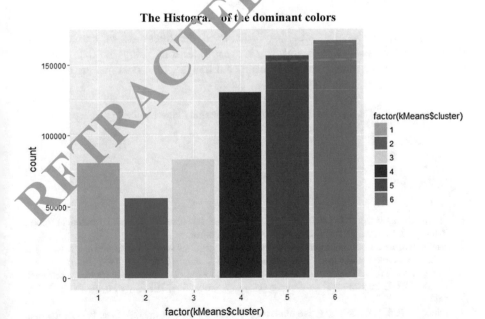

Fig. 2 The dominant colors in the from of graph

Fig. 3 Output image

6 Conclusion

In this paper, we have successfully completed the operation of extracting the dominant colors and then obtain the histogram and thus using the data obtained from the image, we construct the reconstructed image which contains the dominant colors that we extracted.

Declaration We declare that we have taken due permission to use images in this paper and take all responsibility for same in future.

References

1. Bellot, P., El-Beze, M.: A clustering method for information retrieval (technical report IR-0199). Laboratoired 'Informatiqued' Avignon, France
2. MacQueen, J.B.: On convergence of k-means and partitions with minimum average variance. Annals of Mathematical Statistics, Abstract only E. Vidal, vol. 36, pp. 1084 (1986)
3. Wong, K.M., Po, L.M., Cheung, K.W.: Dominant color structure descriptor for image retrieval. In: 2007 IEEE international conference on image processing, IEEE, vol. 6, pp. VI-365. Sep 2007
4. Hsieh, S., Fan, K.C.: An adaptive clustering algorithm for color quantization. Pattern Recogn. Lett. **21**(4), 337–346 (2000)

5. Semmo, A., Limberger, D., Kyprianidis, J.E., Döllner, J.: Image stylization by oil paint filtering using color palettes. In: Proceedings of the workshop on computational aesthetics, Eurographics Association, pp 149–158. June 2015
6. Smith, J.R., Chang, S.F.: Single color extraction and image query. In: Proceedings of international conference on image processing, IEEE, vol. 3, pp. 528–531. Oct 1995

Spot a Spot—Efficient Parking System Using Single-Shot MultiBox Detector

Yashvi Thakkar, Anushka Sutreja, Ashutosh Kumar, Sanya Taneja
and RajeshKannan Regunathan

Abstract In this paper, a method has been proposed to allot a parking spot after detecting it using single-shot multibox detector (SSD). The CCTV cameras will capture n number of frames and pass it through single neural network that is SSD. This will help in fetching details of the vacant parking spot. Subsequent convolution and non-maximum suppression will make the most accurate box around the detected object. Further, the coordinates will be fetched and will be allotted to the incoming cars. This will save time and fuel especially in leveled parking areas. The industrialization of the world, increment in population, moderate-paced city improvement, and bungle of the accessible parking spot have brought about parking-related issues. There is a critical requirement of clever, efficient, fast, protected, dependable, and secured framework for this paper which will be used for getting the vacant parking spots. The method proposed in this paper is better than the already existing one which is ultrasonic radiations and sensors as it is faster, and the drawback of an object/vehicle not being recognized due to its presence in the blind spot is overcome. SSD captures more number of frames compared with YOLO and CNN. It is best for large object recognition and real-time processing which will provide a secured framework for faster and efficient parking allotting.

Keywords Single-shot multibox detector · Car detection · Artificial intelligence · Efficient parking · Deep neural network · Computer vision

1 Introduction

Nowadays, deep learning has become a more chosen and prioritized method for image classification, surpassing traditional computer vision methods. In computer vision, convolution neural networks excel at classification of images, which usually

Y. Thakkar · A. Sutreja · A. Kumar · S. Taneja · R. Regunathan (✉)
School of Computer Science and Engineering, Vellore Institute of Technology,
Vellore, Tamil Nadu, India
e-mail: rajeshkannan.r@vit.ac.in

© Springer Nature Singapore Pte Ltd. 2020
K. S. Raju et al. (eds.), *Data Engineering and Communication Technology*,
Advances in Intelligent Systems and Computing 1079,
https://doi.org/10.1007/978-981-15-1097-7_79

consist of images categorized on the basis of a set of classes, and then, the network tends to determine the strongest class in the image [1]. In this paper, a novel method has been proposed to detect objects in an image or a video using single deep neural network. The proposed algorithm, single-shot multibox detector, scales the output bounding box according to the feature map and uses default boxes for various aspect ratios. Single-shot multibox detector (SSD) is designed for object detection in the real time. SSD is a popular algorithm for object detection [2]. It is one of the best object detection algorithms with both fast speed and high accuracy [3]. SSD broadly consists of two parts: extracting feature maps and applying convolution filter to detect objects. After detecting the vehicles using the proposed algorithm, single-shot multibox detector, the number of empty parking spots in the lot will be sent to the driver. The algorithm will continuously check through the parking lot using the cameras for empty parking spots. Therefore, the method proposed in this paper will work in the real-world environment.

2   Literature Survey

One of the authors had proposed smart parking reservation system using short message services (SMS), for that global system for mobile (GSM) with micro-controller was used to enhance security. The Zigbee technique was used along with the GSM module for parking management and reservation [4]. Intelligent transport system (ITS) and electronic toll collection (ETC) using optical character recognition (OCR) create a record for all entering vehicle. This creates tag-less entry for all vehicles in the parking lot, but it does not assign a slot to the user. A universal OCR algorithm is not available, making it difficult to create said records [5]. Robotic garage (RG) using bluetooth was proposed which would be used to fully automate the placement of a car in the slot without the aid of the driver. The system auto-matically checks the unique registration number stored in the bluetooth chip to check if the new vehicle needs to be parked. This system is a vertical parking arrangement for the vehicles with sensors that confirm placement of the car. Various other sensors are used to confirm that there are no passengers left in the vehicles, and then, the system moves the vehicle to storage area employing rack and pinion (RaP) mechanism [6]. An upgraded system has been proposed, which is deployed with radio-frequency identification (RFID) and light detection and ranging (LIDAR) to authenticate at the gate management service (GMS) to assign a definitive slot [7]. In this paper, a method has been presented to detect objects in an image or a video using single deep neural network. Single-shot multibox detector (SSD) broadly consists of two parts: extracting feature maps and applying con-volution filter to detect objects. SSD300 provides 74.3% mean average precision (mAP), with 59 frames per second [8]. It is one of the best object detection algo-rithms with both fast speed and high accuracy [9]. After detecting the vehicles using the proposed algorithm, single-shot multibox detector, the number of empty parking spots in the lot will be sent to the driver or the security guard can intimate.

The algorithm will continuously check through the parking lot using the cameras for empty parking spots. There are many object recognition algorithms that can be used such as SSD, YOLO, faster R-CNN, R-FCN, etc., but we have chosen SSD over the rest. Single-shot detectors have impressive frames per second (FPS) as compared with others and are relatively faster [10]. Also, SSDs are comparatively best for large object recognition and real-time processing which in our case is parking spot detection.

3 Methodology

The currently existing models for efficient parking use sensors to detect the presence of any object. Ultrasonic waves, Zigbee, optical character resolution, and LIDAR techniques are used which are not as efficient as compared with single-shot multibox detector (SSD) that uses the already-defined space of default boxes considering the aspect ratio, feature maps, and a set of bounding boxes are produced. During prediction, the network will generate scores for objects of each category which are present. It is depicted using default boxes, and certain adjustments are made further to better match the object shape.

The methodology proposed is as follows:

- Firstly, the driver enters the campus and requests for the parking spot. The cameras will get initialized, and photos will be taken of different parking spots, and the photos taken will be stored in the central server. We have used Google's TensorFlow on top of the SSD network for image processing and detecting the cars present in the image.
- Secondly, the image processing will take place on the SSD network, and the data will be acquired which will be updated in the database. After detecting the number of cars in the image and where it is located, it will then detect if the detected car is present inside the parking spot or not. Depending on the output of detection in the parking spot, the database will be updated automatically.
- Thirdly, the nearest empty parking spot will be allocated to the driver entering the campus directing the driver to the allocated parking spot. The obtained parking slot details can be transferred to a central server through the internet using a Wi-Fi module or ethernet. We follow this process for each parking slot.

The following is the procedure through which the whole system works:

- **Continuous Monitoring**: When the program is set to run, the program is directly connected to the database. So, it continuously monitors weather each and every parking spot is full or not. Once, all the parking spot is full, the program is initialized.
- **User Side**: When the user enters the campus spot for parking, data is collected from the database and driver is provided with the nearest available parking spot. If all the spots are scene full in the database, the program is initialized.

- **Program**: When the program is initialized, firstly, all the photos are taken from all the parking spots and stored in a folder. Then, the photos are parsed through the SSD TensorFlow network, and the spots are detected. Once that is done, the database is updated accordingly. So, the new database will be ready for use.

Single-shot multibox detector (SSD) uses single convolutional neural networks to detect an object in one shot. It uses top-k filtering and non-maximum suppression technique [11]. We have used PASCAL VOC dataset which consists of twenty objects and one background and therefore 21 classes. This predefined dataset helps in identifying any object though here we have restricted it to a car. SSD comprises certain steps like training the dataset by generating priors, matching priors with the ground truth boxes, selecting the ground truth boxes, data augmentation, and prediction of classes. Here, the image is divided into a grid, and further, there is a possibility that each box considers itself to be center of the object and thus depicts the object accurately. In order to be sure of the process, non-maximum suppression technique is used which will consider boxes with certain threshold value and then match it with the ground truth boxes. The one having the highest intersection over union value (IoU) is then taken to produce reliable output. The input image undergoes convolution n number of times. It is simultaneously matched with the fully complex (FC) classifier. With every extraction of features, it is compared with the trained dataset in fully complex classifier. These make the ground truth boxes.

Fig. 1 Single-shot multibox detector architecture

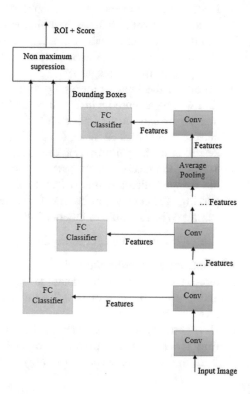

Right after, the boxes having precision value greater than the threshold are considered, and non-maximum suppression filter is applied. Non-maximum suppression returns the region of interest (RoI), and the score gives the precision value. Figure 1 shows the architecture of SSD. In this paper, through SSD, we identify the cars parked in the parking area.

3.1 Pseudocode (Single-Shot Multibox Detector)

Training:

Generate Priors
Match priors with ground truth boxes
      For every gtb:
            match the gtb with prior_value having the biggest IoU
      For every prior_value:
            ious = IoU(prior , g_t_f)
            maxim_iou = maxim(ious)
            if maxim_iou > threshold:
                 i = argmax(ious)
                 match the prior_value with g_t_f[i]
Scale ground truth boxes

$$center_{predicted} * variance0 = \frac{center_{real} - center_{prior}}{hw_{prior}}$$

$$\exp\left(hw_{predicted} * variance1\right) = \frac{hw_{real}}{hw_{prior}}$$

Hard Negative Mining
Loss function
Data augmentation

Prediction:

For every class:
      while predictions are not empty:
            select the bounding box with the greatest prob
            insert it along with the class to the result.
            eliminate rest of the bounding boxes having IoU greater
      than the threshold value from the predictions.

4 Implementation

The images shown below in Figs. 2 and 3 represent the output. Output consists of class 7 as the class of cars, which is being represented with the boxes around them. Next to the class is the precision value. The precision value depicts how accurately SSD has identified the object after carrying repeated convolution and non-maximum suppression.

Fig. 2 Output for parking lot snapshot

Fig. 3 Output for parking lot with vacant parking spots

5 Graph

The graphs of the output of the above depicted are shown below. It is the number of iterations Vs precision of the image. Number of iterations depicts the number of time it scans the images and passes it through SSD network. The precision value depicts with what correctness it identifies the object class. So, basically, for whatever amount one shows the picture in front of the webcam or in real time let us say the frames which are captured, the graph with its precision value will be plotted. Let us consider an example, say, in a parking area, there are six parking spots which are being viewed by one CCTV camera. Once it scans through and captures frames, these are passed to single-shot multibox detector. This output box consists of the class say 7 in this case as shown in Figs. 2 and 3. Right next to it is the precision value which ranges from 0.5 to 1.0 (since the threshold value in most cases is greater than 0.5).

The number of blue lines in the graph depicts the number of the times the image is scanned by one's webcam (in this paper); otherwise, in real-time application, it will plot the graph of frames captured by live CCTV cameras. X-axis depicts the number of iterations per scan. Each pair of lines represents the number of times the image has been scanned. Here, as shown in Fig. 3, it has been scanned only once and three blue lines represent the three cars which have been identified and plotted for its precision value.

Here, as shown in Fig. 4, it has been scanned four times and we see four pairs of line. Hence, 16 lines are plotted. Y-axis depicts the accuracy of the number of cars depicted. Class 7 is the class for cars in our trained dataset (Fig. 5).

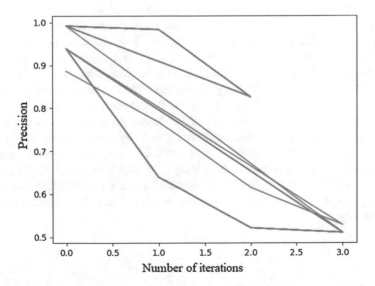

Fig. 4 Number of iteration versus precision for outputs of parking lot

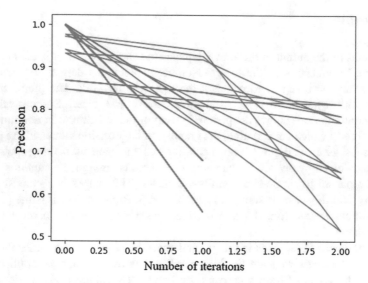

Fig. 5 Number of iterations versus precision for outputs of vacant parking lot

6 Conclusion

After training and carrying out SSD using TensorFlow, we could successfully identify all the basic 20 classes say, for example, trees, animals, and cars. Since this is an automated parking spot allocation system, we restricted the plotting of other classes except cars. Further, we will fetch the coordinates of the vacant spot and allot it to the incoming cars. This will save time and fuel especially in leveled parking areas. Single-shot detectors have impressive frames per second (FPS) as compared with others and are relatively faster. SSDs are comparatively best for large object recognition and real-time processing which in our case is parking spot detection. The challenge that was faced was fluctuations with speed of internet and low graphic processing unit making the system a tad slower. The major advantage is that it does not require the use of expensive equipment such as big sensors and can be effectively incorporated in the daily activities of the common user. Further, a mobile application can be made which can guide the driver directly to the vacant spot using Google/Apple maps. This will be very user-friendly and less time-consuming. Parking reservation concept can also be used using this model in further development stages. Spot A Spot work focuses on artificial intelligence and uses single-shot multibox detector to solve our day-to-day life problems with ease and accuracy.

References

1. He, K., Zhang, X., Ren, S., Sun, J.: Deep residual learning for image recognition. In: Proceedings of the IEEE conference on computer vision and pattern recognition, pp. 770–778 (2016)
2. Liu, W., Anguelov, D., Erhan, D., Szegedy, C., Reed, S., Fu, C.Y., Berg, A.C.: Ssd: Single shot multibox detector. In: European conference on computer vision, pp. 21–37. Springer, Cham. Oct 2016
3. Ning, C., Zhou, H., Song, Y., Tang, J.: Inception single shot multibox detector for object detection. In: 2017 IEEE international conference on multimedia & expo workshops (ICMEW), IEEE, pp. 549–554. July 2017
4. Lalitha, K., Sowjanya, M. N., Pavithra, S.: A Secured vehicle parking management and reservation system using Zigbee and GSM Technology (2018)
5. Rao, N.V., Sastry, A.S.C.S., Chakravarthy, A.S.N., Kalyanchakravarthi, P.: Optical character recognition technique algorithms. J. Theor. Appl. Inf. Technol. **83**(2) (2016)
6. Singh, H., Anand, C., Kumar, V., Sharma, A.: Automated parking system with bluetooth access. Int. J. Eng. Comput. Sci. (2014). ISSN: 2319-7242
7. Karbab, E., Djenouri, D., Boulkaboul, S., Bagula, A.: Car park management with networked wireless sensors and active RFID. In: 2015 IEEE international conference on electro/information technology (EIT), IEEE, pp. 373–378. May 2015
8. Jeong, J., Park, H., Kwak, N.: Enhancement of SSD by concatenating feature maps for object detection. arXiv preprint: arXiv:1705.09587 (2017)
9. Huang, J., Rathod, V., Sun, C., Zhu, M., Korattikara, A., Fathi, A., Fischer, I., Wojna, Z., Song, Y., Guadarrama, S., Murphy, K.: Speed/accuracy trade-offs for modern convolutional object detectors. In: Proceedings of the IEEE conference on computer vision and pattern recognition, pp. 7310–7311 (2017)
10. Qiong, W.U., LIAO, S.B.: Single shot MultiBox detector for vehicles and pedestrians detection and classification. In: DEStech transactions on engineering and technology research, (APOP) (2017)
11. Liu, W., Anguelov, D., Erhan, D., Szegedy, C., Reed, S., Fu, C.-Y., Berg, A.C.: SSD: single shot multibox detector. Cornell University, Computer Science, Computer Vision and Pattern Recognition (2016)

Optimization of Railway Bogie Snubber Spring with Grasshopper Algorithm

Abhishek G. Neve, Ganesh M. Kakandikar, Omkar Kulkarni and
V. M. Nandedkar

Abstract Swarm intelligence is a branch which deals in research that models the population of interacting agents or swarms that are self-organizing in nature. Grasshopper optimization algorithm is a modern algorithm for optimization which is inspired from the swarm-based nature. This algorithm simulates the behaviour of the grasshopper in nature and models that mathematically for solving optimization problems. Grasshopper optimization algorithm is used for the optimization of mechanical components and systems. Snubber spring is a kind of helical spring which is a part of suspension system in railway bogie. In this work, the design of snubber spring is optimized by using grasshopper optimization algorithm. The suspension system of railway bogie consists of inner spring, outer spring, and snubber spring. Optimization is done for the weight minimization of snubber spring. Wire diameter, number of active turns and mean coil diameter are the design parameters for the optimization. These parameters are optimized by using grasshopper optimization algorithm according to bounds, loading, and boundary conditions. The optimized parameters are validated experimentally and also by using a software. The spring is modelled in CATIA V5 and analyzed in ANSYS 17.0. The comparison of results is done and is validated with results experimentally in which the spring is tested on universal testing machine for compression test.

Keywords Swarm-based optimization · Grasshopper optimization algorithm · Weight minimization · Snubber spring

A. G. Neve (✉) · G. M. Kakandikar · O. Kulkarni
Department of Mechanical Engineering, MAEER'S MIT, Pune 411038, India
e-mail: neveabhi@gmail.com

G. M. Kakandikar
e-mail: kakandikar@gmail.com

O. Kulkarni
e-mail: omkarkul9@gmail.com

V. M. Nandedkar
Department of Production Engineering, Shri Guru Gobind Singhji Institute of Engineering & Technology, Nanded, India
e-mail: vilas.nandedkar@gmail.com

© Springer Nature Singapore Pte Ltd. 2020
K. S. Raju et al. (eds.), *Data Engineering and Communication Technology*,
Advances in Intelligent Systems and Computing 1079,
https://doi.org/10.1007/978-981-15-1097-7_80

1   Introduction

"The process in which finding the most effective or favorable value or condition" is defined as optimization by Lockhart and Johnson [1]. To achieve the best for the given constrained conditions, optimization is used. Several metaheuristics have been proposed for optimization. Nature/bio-inspired optimization techniques are included in metaheuristics such as swarm intelligence and evolutionary algorithms [2]. There are most standard and single solution-based algorithms which are simulated annealing [3–5] and hill climbing [6]. Other recently developed single solution algorithms are (TS) tabu search [8, 9] and (ILS) iterated local search [7]. Genetic algorithms (GA) [10–13], (PSO) particle swarm optimization [14, 15], cohort intelligence algorithm (CI) [16], (ACO) ant colony optimization [17], and (DE) differential evolution [18] are various multi solutions-based algorithms. The most common thing in such types of algorithm inspired by nature is that the actual solution obtained does keep on improving until the end criterion is satisfied or obtained [19].

A mechanical spring is a mechanical device which has property of elasticity which performs the distortion of deflection action under the applied load and also retains its original shape on removal of the load [20]. Helical spring is divided into tension, compression, and torsion springs on the basis of ways of loading on spring. Springs usually are square or round in section [21, 22]. Snubber spring is a helical spring which is a part of secondary suspension system of railway bogie. This suspension system is very important in the ride index of bogie [23]. The optimization for the weight minimization of snubber spring is considered in this paper. The literature indicates numerous algorithms for optimization which are framed on the base of natural occurrence [24]. (GOA) grasshopper algorithm is constructed on the behavior of insect grasshopper, while its search for food in this similar phenomenon is utilized for the algorithm for obtaining solution which is optimum for any real-life problem. The rest of the research paper is lined as follows:

The problem for weight minimization of snubber spring is formulated in Sect. 2. Section 3 presents the grasshopper optimization algorithm which is used as optimization tool. The finite element analysis of the optimized spring is given in Sect. 4, while the experimentation is explained in Sect. 5. The results of GOA, FEA, and experimentation are discussed in Sect. 6. Conclusions are mentioned in next section, and references are given in last section of the paper.

2   The Problem Formulation

2.1   Secondary Suspension Spring System

Snubber spring is a part of secondary suspension spring assembly in a railway bogie. This kind of secondary suspension actually contains mainly forms three

Fig. 1 Secondary suspension springs arrangement

Wedge Block Snubber Spring

Inner Spring

Outer Spring

Table 1 Spring details [23]

Description	Snubber spring
Free height (mm)	293
Wire diameter (mm)	18
Mean coil diameter (mm)	86
Number of effective coils	8.5
Working height (mm)	211
Stiffness (kg/mm)	19.79

springs. Seven sets of outer and inner spring arrangement in such a way that the inner springs are covered by outer springs as shown in Fig. 1. The material of spring is taken as spring steel. The total load of wagon is supported by 4 snubber springs. The maximum allowable deflection with shear stress is given in Table 2. As the spring is closed and squared, the number of inactive coils is 2.

The spring details are given in Table 1.

The information required to design a snubber spring is given in Table 2.

Table 2 Information to design a snubber spring [23]

Notation	Data
Deflection along the axis of spring	δ (mm)
Mean coil diameter	D (mm)
Wire diameter	d (mm)
Number of active turns	N
Gravitational constant	$g = 9.8$ m/s^2
Shear modulus	$G = 8007$ kg/mm^2
Mass density of material	$\rho = 7.8 * 10^{-6}$ kg/mm^3
Allowable shear stress	$\tau_a = 385$ Mpa
Number of inactive coils	$Q = 2$
Applied load	$P = 225$ kg
Maximum spring deflection	$\Delta = 33$ mm
Limit on outer diameter of coil	$D_o = 114$ mm

2.2 Objective Function

The main objective of the optimization problem is minimization of weight. The mass of the spring is given as volume*mass density, so the objective function is:

$$\text{Weight}(f) = \left(\frac{\pi}{4}d^2\right)[(N+Q)\pi D]\rho = \frac{1}{4}(N+Q)\pi^2 Dd^2\rho \qquad (1)$$

Let us consider the design variables d, D, and N as x_1, x_2, and x_3, respectively. Putting the values of constant terms from Table 2 in Eq. 1, the objective function and problem formulation are obtained as

$$\min f(x) = 1.9246 \times 10^{-5} x_1^2 x_2 (x_3 + 2) \qquad (2)$$

Subjected to

$$g_1(x) = 0.0668 x_2^3 x_3 - x_1^4 \leq 0 \qquad (3)$$

$$g_2(x) = \frac{0.37205(4x_2^2 - x_1 x_2)}{x_1^3(x_2 - x_1)} + \frac{0.91524}{x_1^2} - 1 \leq 0 \qquad (4)$$

$$g_3(x) = x_2 + x_1 - 114 \leq 0 \qquad (5)$$

The upper and lower bounds for design variables are as follows:

$$16 \leq x_1 \leq 24$$
$$80 \leq x_2 \leq 90 \qquad (6)$$
$$6 \leq x_3 \leq 12$$

2.3 Constraint Handling Technique

Mostly, the nature-inspired algorithms are designed to solve the unconstrained optimization problems. But, most of the real engineering problems are constrained to optimization problems [16].

Penalty functions

Static penalty

A most simple methodology is to penalize the solution which does not fulfill the feasibility criterion by a constant penalty. The penalty function for a problem with equality and inequality constraints is added to form the pseudo-objective function $f_p(x)$ as follows [16]

$$f_p(x) = f(x) + \sum_{i=1}^{n} q_i \times S \times (g_i(x))^2 + \sum_{j=1}^{m} B_j \times S \times h_j(x) \qquad (7)$$

where $f_p(x)$ is the expanded penalized objective function.
   S is a penalty applied for constraint violation.

$q_i = 1$, if constraint i is violated.

$q_i = 0$, if constraint i is satisfied.

$B_j = 1$, if constraint i is violated.

$B_j = 0$, if constraint i is satisfied.
   m and n are the number of equality and inequality constraints, respectively.
   The penalized objective function is utilized in algorithm to gain the best solution for the defined problem.

3   Grasshopper Optimization Algorithm (GOA)

GOA signifies a metaheuristic based on population [25] which is designed for solving optimization problems, i.e., finding answer x^*, actually reduces objective function $f: S \rightarrow R$. Formulation of it can be written as:

$$x* = \arg \min_{x \in S} f(x), \qquad (8)$$

with $S \subset R^D$. The heuristic algorithms which are based on population try to solve (8) using a group or swarm of P independent grasshoppers, in per iteration of k of algorithm, which is denoted by a set $\{xp\}_{P=1}^{P}$ with $xp = [xp1, xp2, \ldots xpD]$. The most important perception for the creation of measures is also a degree of nearness between the two members in swarm $p1$ and $p2$, represented here by Euclidean distance $dist(xp1, xp2)$. The solution opted is best which is gained by swarm inside the limit of k-iterations is reserved as $x * (k)$. The expectation here that the space S utilized for search is constrained, and such type of constraints is signified by the (LB) lower bound $LB1, LB2, \ldots, LBD$ and (UB) upper bound $UB1, UB2, \ldots, UBD$. Effectively, it means that:

$$LBd \leq xpd(k) \leq UBd \qquad (9)$$

for all the $k = 1, 2, \ldots, p = 1, 2, \ldots, P$ and $d = 1, 2, \ldots, D$.
   (GOA) grasshopper optimization algorithm is actually derived from the socialized behavior or nature of insects known as Orthoptera order (Caelifera). Every

member from swarm or group indicates a sole insect grasshopper which is located in the search space S which is moving and exploring within its bounds. This algorithm implies two different properties of grasshopper's movement approaches for the food search. First of all, the communication of grasshoppers is in which it validates itself through very gentle actions (when in young insect stage) and a drastic active gesture (when in the adult stage). The other second most corresponds to its behavior to approach the food source. The insect forthcoming toward food and ultimately eating it is also considered as the important concept behind it.

The movement in which separate p in generation k (index k is obsoleted for the reason of perception can be expressed via following equation:

$$X_i^d = c \left(\sum_{\substack{j=1 \\ j \neq 1}}^{N} c \frac{ub_d - lb_d}{2} s\left(\left| x_j^d - x_i^d \right| \right) \frac{x_j - x_i}{d_{ij}} \right) + \overline{T_d} \tag{10}$$

with $d = 1, 2, \ldots, D$. Factor c is reducing referring to the formula:

$$c = c_{max} - l \frac{c_{max} - c_{min}}{L} \tag{11}$$

The pseudo code of GOA algorithm is shown below.

Initialize the swarm Xi (i = 1, 2, 3, ..., n)
Initialize cmax, cmin, and maximum number of iterations
Calculate the fitness of each search agent using static penalty by using eq. (2.7)
T = the best search agent
While (l < Max number of iterations)
   Update c using eq. (3.4)
   for each search agent
     Normalizes the distance between grasshoppers
     Update the position of current search agent
bring the current search agent back if it goes outside the boundaries
   end for
   update T if there is a better solution
   l = l+1
End while
Return

---

where c_{max} indicates the maximum value, c_{min} denotes the minimum value, l specifies the iteration (current), and L denotes the maximum number of iterations. In this work, we use l and 0.0001 for c_{max} and c_{min}, respectively.

4 Finite Element Analysis (FEA)

Finite element analysis gives the predicted results for actual system or a component. Optimized parameters were obtained from grasshopper optimization algorithm with search agents as 100 and maximum iterations as 500. The coding of the GOA algorithm is performed in MATLAB (R2016a) on the platform of Windows 7 having processor I% of 3.2 GHZ speed and RAM of 4 GB. The problem was solved, and the results are recorded 30 times to generate statistical results. CATIA V5 R19 and ANSYS 17.0 were used for modeling and analysis of optimized spring.

4.1 Modeling

The CAD model of spring is created using the software CATIA V5. The model has 6 active turns with 80 mm mean coil diameter and 21.5 mm wire diameter. The CAD model from CATIA is then imported to ANSYS for the analysis purpose.

4.2 Meshing

After importing the CAD model, meshing is done in ANSYS. The spring is a solid part, so 3D meshing is required for spring. The sizing of mesh is kept fine and adaptive. In meshing, the nodes used are 20,546 and the elements are 10,447.

4.3 Boundary Condition for Static Calculations

In the boundary condition, the fixed supports are applied to one end of the spring. Vertical load of 2250 N is applied on the other end of the spring. The load is applied in components where other than vertical component is kept as 0 N. The displacement of spring is restricted in other than vertical direction.

4.4 Post Processing

In post processing, the maximum shear stress, maximum principal stress, von Mises stress, and total deformation and directional deformation are calculated.

5 Experimentation

Snubber spring is manufactured using optimized parameters obtained from GOA. (UTM) universal testing machine is used for compression test of spring. Spring is mounted on UTM with the help of mountings. Weight applied is in kg, and the deformation is given in mm. The measuring software used in UTM testing is developed by STAR TESTING SYSTEMS. After the test, the load is released and spring is taken out of the UTM. From this test, complete deformation on load is obtained. From this test, reading of the load on every unit time and every unit displacement is obtained.

6 Result and Discussion

GOA is used for optimization of snubber spring for weight minimization. The deflection, allowable shear stress, and outer diameter limits are the constraints taken into consideration. The parameters settings used in grasshopper optimization algorithm (GOA) are search agents as 100 and maximum iterations as 500. The coding of the GOA algorithm is performed in MATLAB (R2016a) on the platform of Windows 7 having processor I% of 3.2 GHZ speed and RAM of 4 GB.

Figure 2 shows that how the trajectory of the grasshopper and how the grasshopper reach the global optimum solution. From Fig. 2, the following things are observed.

- Parameter space: It shows the space in which the minimum function value lies. The search space is dependent on design variables and changes accordingly to the change in bounds.
- Search history: The figure shows the location history for the artificial grasshoppers during optimization.
- Trajectory obtained for the first grasshopper in the very first dimension: this diagram shows the value obtained for the first grasshopper for the each iteration.

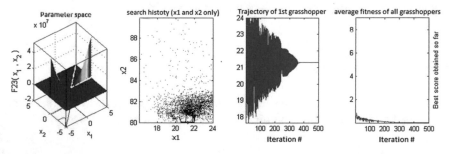

Fig. 2 Parameter space, search history, trajectory of first grasshopper, average fitness of all agents, and convergence curve

Table 3 Comparison of GOA and standard parameters of snubber spring

Technique	Wire diameter (x_1) (mm)	Mean coil diameter (x_2) (mm)	Number of active turns (x_3)	Weight of spring (kg)
Standard [21]	18	86	8.5	5.63084
GOA	21.5	80	6	5.56937

- Average fitness: This figure shows the average objective function values of all insects (grasshoppers) for per iteration.

GOA results are compared with standard results [21] as given in Table 3. Optimized parameters obtained by grasshopper optimization algorithm are $x1 = 21.5$, $x2 = 80$, and $x3 = 6$ mm.

The weight of the spring with new values is reduced, so it can be definitely said that the GOA algorithm gives better and optimized results than the existing results. The snubber spring with GOA obtained parameters (modified) is analyzed in ANSYS 17.0 for the given loading and constrained conditions. Maximum shear stress, maximum principal stress, and total deformation in the spring are calculated. Results from FEA clearly show that the stress and deformation in the spring are in given limit. So, the constraints maximum allowable stresses condition, deformation condition, and outer diameter limit are fully satisfied.

The finite element analysis is shown in Fig. 3. The spring is tested on (UTM) universal testing machine for compression test. The deformation for increasing compressive load is measured, and from that the deformation on the given load is measured. Time required for the complete test is 53.9 s where the load is increased up to 524.6 kg on the spring, till the displacement is reached to 10.3 mm. Required load on spring as per the problem is 225 kg. At 225 kg, the displacement observed is 5.6 mm.

The curve represents the nature of spring deformation on increasing load. The slope of the curve indicates stiffness of the spring. The results of the experimentation and CAE analysis comparison are given in Table 4.

In Table 4, experimental results are compared with CAE analysis. In other words, CAE results are validated using experimentation. From the comparison, it is observed that the variation between the results of CAE and experimentation is very less and the results are validated and are in acceptable range.

7 Conclusions

Suspension system has very much impact on the ride index of the vehicle. Snubber spring is a part of secondary suspension spring in railway bogie. In this project, the optimization problem for snubber spring with weight minimization as an objective function is formulated. Maximum allowable deformation, maximum allowable

(a)

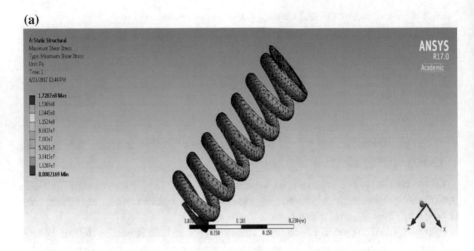

(b)

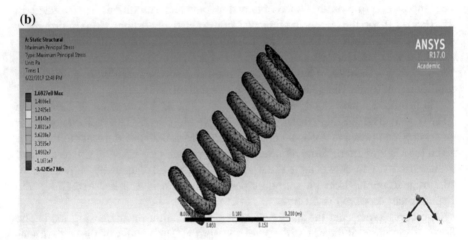

Fig. 3 Snubber spring showing **a** maximum shear stress **b** maximum principal stress

Table 4 Comparison of CAE and experimental results

Load applied (kg)	Spring parameters			Weight of spring (kg)	Deformation (mm)		% error
	d (mm)	D (mm)	N		Analysis	Experiment	
225	21.5	80	6	5.56937	5.825	5.6	3.86

shear stress, and outer diameter limit are the constraints taken into consideration for the problem. Mean coil diameter, wire diameter, and number of active turns of spring are chosen as design variables. Static penalty function approach is used as constraint-handling technique to convert the optimization problems from constrained into unconstrained ones. The GOA is used to solve the optimization problem, and optimized parameters are obtained for the snubber spring hence minimizing weight. Optimized parameters are used to design a spring in CATIA and then are analyzed in ANSYS for given loading conditions. Results from analysis are obtained and found to be in the given limit. Snubber spring with optimized parameters is manufactured and tested on universal testing machine (UTM) for compression test. The deformation of the spring for the given load is noted, and it is found to be in the limit. The results from experiment and analysis are compared. From these results, it can be concluded as the GOA can be applied to the real engineering design problems efficiently.

Disclosure Statement No potential conflict of interest was reported by the author(s).

References

1. Todd, R., Kelley.: Optimization, an important stage of engineering design. In: The Tech Teacher (2005)
2. Dasgupta, D., Michalewicz, Z.: Evolutionary algorithms in engineering applications. Springer (1997)
3. Yang, X.-S.: Nature-inspired metaheuristic algorithms. Luniver Press (2010)
4. Mirjalili, S., Lewis, A.: S-shaped versus V-shaped transfer functions for binary particle swarm optimization. Swarm Evol. Comput **9**, 1–14 (2013)
5. Davis, L.: Bit-climbing, representational bias, and test suite design. ICGA, 18–23 (1991)
6. Kirkpatrick, S., Gelatt, C.D., Vecchi, M.P.: Optimization by simmulated annealing. Science (1983)
7. Lourenço, H.R., Martin, O.C., Stutzle, T.: Iterated local search. In: arXiv preprint (2001)
8. Fogel, L.J., Owens, A.J., Walsh, M.J.: Artificial intelligence through simulated evolution (1966)
9. Glover, F.: Tabu search-part I. ORSA J. Comput. **1**, 190–206 (1989)
10. Holland, J.H.: Genetic algorithms. Sci. Am. **267**, 66–72 (1992)
11. Kakandikar, G.M., Nandedkar, V.M.: Spring-seat forming optimisation with genetic algorithm. Int. J. Comput. Aided Eng. Technol. **5**(4) (2013)
12. Bhoskar, T., Kulkarni, O.K., Kulkarni, N.K., Patekar, S.L., Kakandikar, G.M., Nandedkar, V. M.: Genetic algorithm and its applications to mechanical engineering: a review. Mater. Today Proc. **2** (2015)
13. Kakandikar, G.M., Nandedkar, V.M.: Prediction and optimization of thinning in automotive sealing cover using genetic algorithm. J Comput. Des. Eng. **3**(1), 63–70 (2016)
14. Eberhart, R.C., Kennedy, J.: A new optimizer using particle swarm theory. In: Proceedings of the sixth international symposium on micro machine and human science, pp. 39–43 (1995)
15. Kulkarni, N.K., Patekar, S., Bhoskar, T., Kulkarni, O.K.: Particle swarm optimization application to mechanical engineering—a review. Mater. Today Proc. **2**(4), 2631–2639 (2015)
16. Kulkarni, O., Kulkarni, N., Kulkarni, A.J., Kakandikar, G.: Constrained cohort intelligence using static and dynamic penalty function approach for mechanical components design. Int. J. Parallel Emerg. Distrib. Syst. (2016). https://doi.org/10.1080/17445760.2016.1242728
17. Colorni, A., Dorigo, M., Maniezzo, V.: Distributed optimization by ant colonies. In: Proceedings of the first European conference on artificial life, pp. 134–42 (1991)

18. Storn, R., Price, K.: Differential evolution—a simple and efficient heuristic for global optimization over continuous spaces. J. Glob. Optim. **11**, 341–359 (1997)
19. Eiben, A.E., Schippers, C.: An evolutionary exploration and exploitation. Fundamental Info (1998)
20. Prawoto, Y., Ikeda, M., Manville, S.K., Nishikawa, A.: Design and failure modes of automotive suspension springs. Eng. Fail. Anal. **15**, 1155–1174 (2008)
21. Chiu, C.-H., Hwan, C.-L., Tsai, H.-S., Lee, W.-P.: An experimental investigation into the mechanical behaviors of helical composite springs. Compos. Struct. **77** (2007)
22. Taktak, M., Omheni, K., Aloui, A., Dammakb, F., Haddar, M.: Dynamic optimization design of a cylindrical helical spring. Appl. Acoust. **77** (2014)
23. Bajpai, N.: Suspension spring parameter's optimization of an indian railway freight vehicle for better ride quality. IEEE Int. Conf. Adv. Eng. Tech. Res. (2014)
24. Boussaï, D.I., Lepagnot, J., Siarry, P.: A survey on optimization metaheuristics. Inf. Sci. **237**, 82 (2013)
25. Sarema, S., Mirjalili, S., Lewis, A.: grasshopper optimisation algorithm: theory and application. Adv. Eng. Softw. **105**, 30–47 (2017)

QoS Aware Group-Based Workload Scheduling in Cloud Environment

Suneeta Mohanty, Suresh Chandra Moharana, Himansu Das
and Suresh Chandra Satpathy

Abstract Cloud computing paradigm provides dynamically scalable resources as a service over the Internet. It promises reduction in capital expenditure as well as operational expenditure. In cloud computing, there are still some challenges need to be solved. Among them, job scheduling is one of the important issues, since the objective of service provider and the cloud user is to utilize the computing resources efficiently. Job scheduling is the process of assigning user jobs to appropriate resources so that it can utilize the cloud resources efficiently. In cloud computing environment, the communication time between the users and resources can have a great impact on job scheduling. The primary objective in cloud environment is to process the tasks provided by user with higher responsiveness. So there is a need of an efficient group-based job scheduling algorithm that assemble the independent fine-grained jobs into groups by taking the communication time into consideration. In this work, we have proposed a group-based scheduling strategy for scheduling fine-grained jobs according to the processing capability and bandwidth of available resources in cloud computing environment.

Keywords Cloud computing · Task scheduling · QoS

S. Mohanty (✉) · S. C. Moharana · H. Das · S. C. Satpathy
School of Computer Engineering, KIIT Deemed to be University, Bhubaneswar, Odisha,
India
e-mail: suneetamohanty@gmail.com

S. C. Moharana
e-mail: sureshmoharana@gmail.com

H. Das
e-mail: himanshufcs@kiit.ac.in

S. C. Satpathy
e-mail: suresh.satpathyfcs@kiit.ac.in

© Springer Nature Singapore Pte Ltd. 2020 953
K. S. Raju et al. (eds.), *Data Engineering and Communication Technology*,
Advances in Intelligent Systems and Computing 1079,
https://doi.org/10.1007/978-981-15-1097-7_81

1   Introduction

Cloud computing is a new computing paradigm where the cloud provider provides resources on demand to the cloud users. In cloud environment, users get its services from the virtual machines and virtual machines are created from the physical resources. So to assign the user jobs to the virtual machines and virtual machines to the physical resources, we need a scheduling strategy. So scheduling in cloud environment is done at two levels: One is job-level scheduling, and another one is resource-level scheduling. Job-level scheduling is the deployment of user jobs on the virtual machines, and resource-level scheduling is deployment of virtual machines on physical resources. As per users point of view, user pays according to the consumption, so it always tries to use the resources at lower cost. As per the provider's point of view, it always wants to utilize resources efficiently and get maximum profit. So to satisfy both user and provider requirement, we need an efficient scheduling algorithm. In scheduling strategy, user submits its job to the data center broker, where broker is the mediator between the user and the provider. It assigns the user jobs to the virtual machines by collecting the information about resources from the cloud information service, and then virtual machines are allocated to appropriate physical resources.

The rest of the paper is organized as follows: Sect. 2 consists of literature review that provides an overview of the existing scheduling approaches in cloud as well as grid computing, Sect. 3 discusses the system architecture, Sect. 4 presents the algorithm of the proposed approach, and finally, Sect. 5 concludes the work.

2   Literature Review

In cloud environment, users get its services from the virtual machines and virtual machines are created from the physical resources. In order to assign the user jobs to the virtual machines and virtual machines to the physical resources, we need a scheduling strategy. So scheduling in cloud environment is done at two levels: One is job-level scheduling, and another one is resource-level scheduling. Job-level scheduling is the deployment of user jobs on the virtual machines, and resource-level scheduling is deployment of virtual machines on physical resources. The main purpose of an efficient scheduling algorithm is to use the resources efficiently and gain maximum profit. So for this reason, job scheduling problem is a very important and challenging work in the field of research. Many scheduling algorithms take some common factor like system utilization, QoS, SLA, security, fault tolerance, reliability, and deadline into consideration to measure the performance of their algorithm. In this paper, we have made survey of many immediate and batch mode type scheduling algorithms of independent and dependent jobs both in cloud and in grid.

Hemamalini [1] considered different factors like execution time, makespan, load balancing, and completion time to analyze the performance of different scheduling algorithms such as Min–Min, Max–Min, Minimum Completion Time, and Minimum Execution Time. Through this analysis, they found that Max–Min grid task scheduling algorithm effectively utilizes the resources and minimizes the makespan over other scheduling algorithms.

Xiaoshan et al. [2] discussed the importance of quality of service (QoS) constraints for the scheduler in the grid environment. They proposed a QoS guided Min–Min heuristic algorithm to provide an efficient match among different level of QoS request/supply by embedding QoS information into the scheduling algorithm for grid system.

A balanced ant colony optimization (BACO) algorithm to schedule the dependent jobs in grid environment is proposed by Chang et al. [3]. This algorithm performs job scheduling in grid computing by applying ant colony optimization (ACO) algorithm. ACO uses the behavior of the ant colony and finds the optimal path to schedule the jobs by using the pheromone update rule. Here, it assumes each job is an ant and pheromone means the weight of resource in grid.

Cloud service provider should provide satisfactory service to different users on demand. So to measure the satisfaction of users using the cloud computing services, quality of service (QoS) is the standard. Different users have different QoS requirements. Qi-Yi et al. [2] proposed a job scheduling strategy and algorithm based on QoS, which could meet user requirements on time and cost considering the given deadline and budget. This is a dependent task scheduling strategy. This paper proposed a scheduling strategy which considers two factors: One is completion time, i.e., the job should be completed in the time as soon as possible, and another one is cost, i.e., user wants the services at low cost.

Zhang et al. [4] proposed a multi-stage task allocation mechanism in cloud environment. Initially, jobs are classified using Bayes classifier model. In next stage, jobs are dynamically mapped to appropriate virtual machine. It gives an job mapping technique to limit the completion time of jobs subject to deadline imperatives while keeping up the most elevated balanced workload among all the cloud assets in a dynamic cloud model. The proposed technique can be conveyed to the cloud environment in order to lessen energy utilization and achieve guaranteed service quality as well.

Sahni et al. [5] addressed the challenges of workflow scheduling problem in cloud environments. The authors have proposed a deadline aware heuristics for dynamically scheduling workflows in public cloud. The objective of the proposed approach is to exploit the favorable circumstances offered by cloud computing models while taking the performance variations of the virtual machines and instance obtaining delay into consideration. This enables us to acquire a just-in-time scheduling scheme over deadline aware workflows in limited cost. The proposed scheme outperforms the existing scheduling models in terms of performance.

There are many applications which required small amount computation; when these types of applications are sent to the cloud, it gives a poor computation/communication ratio. So for the better result, these types of applications are grouped together and sent to the resources for computation. So a scheduling strategy in grid that performs dynamic job grouping activity at runtime is proposed by Muthuvelu et al. [6] and gives out the detailed analysis of algorithm by running simulations.

In cloud environment, users get its services from the virtual machines and virtual machines are created from the physical resources. So to assign the user jobs to the virtual machines and virtual machines to the physical resources, we need a scheduling strategy. So scheduling in cloud environment is done at two levels: One is job-level scheduling, and another one is resource-level scheduling [7]. Job-level scheduling is the deployment of user jobs on the virtual machines, and resource-level scheduling is deployment of virtual machines on physical resources. In this work, we have considered only job-level scheduling.

3 Proposed System Architecture

In cloud environment, mapping of tasks toward virtual machines plays an important role for effective use of available conventional resources. The efficient use of the resources leads to the achievement of QoS requirements of the application. The proposed task scheduling scheme can be visualized as a model as shown in Fig. 1.

The different components of the proposed model are:

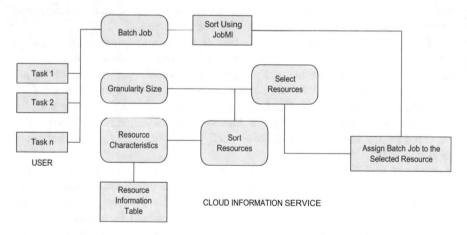

Fig. 1 Proposed system model

3.1 Task

User applications are made out of numerous jobs. These jobs are isolated into different autonomous tasks. So a task is a nuclear unit to be selected by the scheduler and allotted to a resource [8].

3.2 Cloud Information Service (CIS)

A CIS is an entity that gives resource enlistment, ordering, and discovering abilities. CIS underpins two essential activities: firstly, it permits resources to enlist themselves with CIS, and secondly, it permits entities like cloud coordinator and brokers in finding status and endpoint positioning of computing resources.

3.3 Data Center Broker

Data center broker is the arbiter between the user and the cloud service provider [9]. User presents its tasks to the cloud managers for execution on cloud resources. The broker plans the mapping of task to resources with the goal of maximizing profit.

3.4 Resource Information Table

It is available inside the cloud information service. This table contains the attributes of resources, for example, resource id, MIPS, bandwidth, etc. Data center broker gathers the data about the resources and afterward plans the mapping of task to resources.

The workflow of the proposed model can be described formally as below. All user tasks are submitted to the data center broker of cloud information service. When tasks are arrived at data center broker, it batches all the tasks and then sorts them as per processing requirement of each job (MI) in decreasing order. In order to group all the tasks, data center broker first collects the information about available resources of cloud from information table and arranges the resources according to the bandwidth of resources in decreasing order.

From that point forward, it chooses a specific accessible resource and computes the complete MI the asset can process by finding the product of the MIPS of the resource and the predetermined granularity estimate. Granularity estimate is the time inside which a task is handled at the resources or a specific time for which a resource is accessible. It is utilized to quantify the aggregate sum of tasks that can be finished inside a predetermined time in a specific resource. The scheduler begins grouping of the user tasks by aggregating the MI of each activity while looking at the subsequent activities total MI (Group MI) with the asset all out MI. In case, the total MI of user workload is more than the resource absolute MI; at that point, it scans for a suitable task whose MI fulfills the resource complete MI appropriately. Subsequently, another job group with collected absolute MI will be made with a unique ID and put away inside a list with its relating resource ID to be executed in the chosen resource. This procedure proceeds until all the user tasks are booked into some group and allotted to the accessible resources. The scheduler at that point sends the task gatherings to their comparing resources for further calculation. The resources process the received task group and then send the processed task to the user.

4   Proposed Algorithm

The objective of the proposed work is to schedule tasks with computing processing requirements in order to efficiently make use of available resources resulting in achievement of QoS requirements. The proposed algorithm is an extension to the existing grouping-based dynamic job scheduling strategies as provided below (Table 1).

Table 1 Proposed scheduling algorithm

```
1.    Create a Rlist
2.    Create a Jlist
3.    Sort the Rlist as per the Bandwidth in Descending Order
4.    Sort the Jlist as per the Job Size in Descending Order
5.    Repeat
6.    for i = 1 to size(Rlist) do
7.     TotalRMIᵢ = MIᵢ * Granularity
8.     TotalGMI = 0
9.      for j = 1 to size(Jlist)
10.       while TotalGMI<=TotalRMIᵢ && MIⱼ<=(TotalRMIᵢ-TotalGMI) do
11.         TotalGMI = TotalGMI + MIⱼ
12.         remove jth task from the Jlist
13.       end while
14.     end for
15.     select a new task whose MI equals to TotalGMI
16.     assign a unique ID to it and place it in the Glist
17.     place Resᵢ into target resource list
18.   end for
19.   until the resources are available
20.   for i = 1 to size(Glist)
21.     send ith task to Resᵢ from the target resource list for
                                          the computation
22.   end for
```

The general clarification of our proposed calculation is as per the following. At the point when the user tasks are submitted to the broker or scheduler, the scheduler assembles the attributes of the accessible resources. First scheduler sorts, all the user jobs as per its processing requirement (job MI) and available resources as per the bandwidth of the network in descending order. Further, it chooses one of the resources from the list and discovers its handling capacity by multiplying the total MI of the resource with a predefined granularity measure.

Further, the scheduler begins gathering the user workloads by aggregating the MI of each task while looking at the subsequent tasks total MI (Group MI) with the resource absolute MI. In the event that the absolute MI of user tasks is more than the resource total MI, it discovers for an appropriate job whose MI satisfies the resource total MI properly. At that point, a new activity (job group) of aggregated complete MI will be chosen with an independent ID and put away inside a list with its comparing resource ID for further execution. This process continues until all the user jobs are scheduled into few groups and assigned to the available resources. The scheduler then sends the job groups to their corresponding resources for further computation. The resources process the received job and then send the processed job to the user.

5 Conclusion and Future Directions

The primary objective in cloud environment is to process the tasks provided by user with higher responsiveness. In this direction, we have proposed a grouping-based job scheduling algorithm that takes the independent tasks into consideration. The proposed approach groups the jobs as per required MIPS and schedules to the appropriate computing resources. Our objective is not only efficient use of available resources but also achievement of user QoS requirements. In future, our plan is to analyze the efficiency of the proposed approach. Further, we will simulate the proposed model in Cloudsim toolkit, a popular environment for testing cloud applications. After that, we will converge to test the efficacy of the proposed approach by comparing it with the existing approaches based on different parameters.

References

1. Hemamalini, M.: Rreview on grid task scheduling in distributed heterogeneous environment. In: Int. J. Comput. (2012)
2. Qi-yi, H., Ting-lei, H.: An optimistic job scheduling strategy based on QoS for cloud computing. In: IEEE Comput. (2010)
3. Chang, R., Chang, J., Lin. P.: An ant algorithm for balanced job scheduling in grids. In: IEEE Comput. (2010)

4. Zhang, P., Zhou, M.: Dynamic cloud task scheduling based on a two-stage strategy. In: IEEE Trans. Autom. Sci. Eng., 772–83 (2018)
5. Sahni, J., Vidyarthi, D.P.: A cost-effective deadline-constrained dynamic scheduling algorithm for scientific workflows in a cloud environment. IEEE Trans. Cloud Comput., 2–18 (2018)
6. Muthuvelu, N., Liu, J., Soe, N., Venugopal, S.: A dynamic job grouping based scheduling for deploying applications with fine-grained tasks on global grids. In: IEEE Comput. (2004)
7. Moharana, S.C., SD, M.K.: An efficient approach for storage migration of virtual machines using bitmap. In: International conference on information processing, pp. 438–447. Springer (2011)
8. Dong, F., Akl, S. G.:Scheduling algorithms for grid computing. In: Technical report (2006)
9. Mohanty, S., Pattnaik, P.K., Mund, G.B.: Framework for Auditing in Cloud Computing Environment. J. Theor. Appl. Inf. Technol., 261–267 (2014)

Retraction Note to: Data Engineering and Communication Technology

K. Srujan Raju, Roman Senkerik, Satya Prasad Lanka
and V. Rajagopal

Retraction Note to:
K. S. Raju et al. (eds.), *Data Engineering*
and Communication Technology, **Advances in Intelligent**
Systems and Computing 1079,
https://doi.org/10.1007/978-981-15-1097-7

The Series Editor and the Publisher have retracted this chapter. An investigation by the Publisher found a number of chapters, including this one, presenting various concerns, including but not limited to compromised editorial handling, incoherent text or tortured phrases, inappropriate or irrelevant references, or a mismatch with the scope of the series and/or book volume. Based on the findings of the investigation, the Series Editor therefore no longer has confidence in the results and conclusions of this chapter.

The authors have not responded to correspondence regarding this retraction.

The retracted version of these chapters can be found at
https://doi.org/10.1007/978-981-15-1097-7_28
https://doi.org/10.1007/978-981-15-1097-7_44
https://doi.org/10.1007/978-981-15-1097-7_53
https://doi.org/10.1007/978-981-15-1097-7_54
https://doi.org/10.1007/978-981-15-1097-7_56
https://doi.org/10.1007/978-981-15-1097-7_58
https://doi.org/10.1007/978-981-15-1097-7_61
https://doi.org/10.1007/978-981-15-1097-7_68
https://doi.org/10.1007/978-981-15-1097-7_78

Author Index

© Springer Nature Singapore Pte Ltd. 2020
K. S. Raju et al. (eds.), *Data Engineering and Communication Technology*,
Advances in Intelligent Systems and Computing 1079,
https://doi.org/10.1007/978-981-15-1097-7

Printed in the United States
by Baker & Taylor Publisher Services